Methods in Paleoecology

Vertebrate Paleobiology and Paleoanthropology Series

Edited by

Eric Delson
Vertebrate Paleontology, American Museum of Natural History
New York, NY 10024, USA
delson@amnh.org

Eric J. Sargis
Anthropology, Yale University
New Haven, CT 06520, USA
eric.sargis@yale.edu

Focal topics for volumes in the series will include systematic paleontology of all vertebrates (from agnathans to humans), phylogeny reconstruction, functional morphology, Paleolithic archaeology, taphonomy, geochronology, historical biogeography, and biostratigraphy. Other fields (e.g., paleoclimatology, paleoecology, ancient DNA, total organismal community structure) may be considered if the volume theme emphasizes paleobiology (or archaeology). Fields such as modeling of physical processes, genetic methodology, nonvertebrates or neontology are out of our scope.

Volumes in the series may either be monographic treatments (including unpublished but fully revised dissertations) or edited collections, especially those focusing on problem-oriented issues, with multidisciplinary coverage where possible.

More information about this series at http://www.springer.com/series/6978

Methods in Paleoecology

Reconstructing Cenozoic Terrestrial Environments and Ecological Communities

Edited by

Darin A. Croft

Department of Anatomy, Case Western Reserve University, Cleveland, OH, USA

Denise F. Su

Department of Paleobotany and Paleoecology, Cleveland Museum of Natural History, Cleveland, OH, USA

Scott W. Simpson

Department of Anatomy, Case Western Reserve University, Cleveland, OH, USA

Springer

Editors
Darin A. Croft
Department of Anatomy
Case Western Reserve University
Cleveland, USA

Scott W. Simpson
Department of Anatomy
Case Western Reserve University
Cleveland, OH, USA

Denise F. Su
Department of Paleobotany and Paleoecology
Cleveland Museum of Natural History
Cleveland, OH, USA

ISSN 1877-9077 ISSN 1877-9085 (electronic)
Vertebrate Paleobiology and Paleoanthropology Series
ISBN 978-3-319-94264-3 ISBN 978-3-319-94265-0 (eBook)
https://doi.org/10.1007/978-3-319-94265-0

Library of Congress Control Number: 2018945445

Cover illustration: Montage of modern terrestrial habitats and Cenozoic terrestrial fossil localities, illustrating a small portion of the diversity seen in both. A primary goal of many paleoecological studies is reconstructing ancient habitats: using data from modern environments and ecological communities to envision how a landscape would have appeared thousands to millions of years ago. From left to right: Contamana, Peru (middle Eocene); western Nebraska, USA; Los Queñes, central Chile (late Eocene); Cordillera de Sama, southern Bolivia; Cerdas, Bolivia (middle Miocene); Uvita, Costa Rica. Photo credits: D. Croft.

This Springer imprint is published by the registered company Springer Nature Switzerland AG
The registered company address is: Gewerbestrasse 11, 6330 Cham, Switzerland

Foreword

The past has always fascinated people once we became self-aware. In early days, it was the study of written history that engaged the best minds, but the ideas of prehistory, that humans had a history going beyond this, did not really take shape until after the work of T. H. Huxley. Darwin speculated on the subject, but at the time when he was working there was no fossil record for apes or humans other than the few dryopithecine specimens in France and the Neanderthal skull from Gibraltar. Huxley could also draw on paleontological studies in the 19th century and the work of Charles Lyell, who wrote extensively on the burial and 'imbedding' of plant and animal remains as well as the geological processes producing them. Similarly, the descriptions by William Buckland and Boyd Dawkins of British cave sites generally included comments on the preservation of animal remains and the processes by which they entered the caves. None of these 19th century authors commented directly on paleoecology, but Darwin's (1881) book *The Formation of Vegetable Mould Through the Action of Worms: With Observations on Their Habits* is a classic (and delightful) description of what happens to bones after burial, a taphonomic consideration essential to interpreting paleoecology.

As fossils accumulated in the 19th and 20th centuries, the prehistory of life began to fill out. It was mostly concerned with what kinds of animals and plants were present in the past, and how they were related to each other and to modern biota. Early works on the Carboniferous floras or the London Clay mainly consisted of taxonomic descriptions, but they also included interpretations of the type of environment the plants were living in, and although the word 'paleoecology' was not used, they were in fact paleoecological descriptions of past environments. In the case of plants, it is reasonably straightforward inferring the vegetation type from the species present, even as far back as the Eocene, and the plants growing in the London Basin at this time can easily be interpreted as tropical forest. However, most sites with fossil animals do not have fossil floras preserved, and many interpretations of paleoecology are based on species identifications of the preserved animals.

Once people started thinking about paleoecology, it was recognized that it has two aspects, the physical world and the biological world. The former consists of properties of the Earth and climate, and the latter the properties of the plants and animals. The paleoecology of a fossil site is therefore based on the geological attributes of the site, its inferred climate, and the animals and plants occupying the space and time of the site. As paleontology developed as a science, questions began to be asked about how fossils came to be preserved and what kind of environment they represented. Pei (1938) in *Le rôle des animaux et des causes naturelles dans la cassure des os* described marks on bones that he related, sometimes mistakenly in the light of present knowledge, to the agents that produced them and the ways in which the agents altered the composition of bone assemblages. This was the starting point for numerous taphonomic studies in the 20th and 21st centuries that have laid the groundwork for paleoecology.

At this stage, the interpretations of paleoecology were based on identifications of the animal species (and plants where available) present at a site. This was, and continues to be, a major source of evidence for paleoecology for general paleontologists, for it is the most straightforward approach to reconstructing past environments. It depends on James Hutton's principle of uniformitarianism that changes in the past came about by the same processes that operate today. If a large species of deer is present at a fossil site, this 'indicates' that woodland was present, as most deer today live in woodland; if an alcelaphine antelope is present, this 'indicates' grassland, as alcelaphines today feed on grass; and so on. Sometimes paleontologists take account of relative abundances of these indicator species, and sometimes the numbers of species indicating the same environment, but that is usually as far as it goes. Only in the past 50 years or so has paleoecology come of age and used established ecological principles in trying to reconstruct past communities of animals, but whereas ecology can be studied by direct observation, paleoecology is based on inferences about the past, assuming that processes in the past were the same as they are today.

Ecological communities consist of groups of animals and plants that coexist at one place and time and form an interacting unit. The place may range from a few meters, such as an area of forest floor, to several kilometers, for example the ecology of a forest unit. The time may be days, years, or even centuries. Paleoecology attempts to reconstruct these past ecologies by means of the fossil record, and in this case, the place means an excavation area, a stratigraphic level, or a fossil locality, taking into account the possibilities of intrusive or transported species. The time unit is the best estimate of the period over which the fossils were assembled, which may be measured in thousands of years. Stable isotopes, phytoliths, and fossil soils may provide direct insights into the paleoecology, but the evidence may be spatially limited to the immediate extent of the site. Similarly, fossil assemblages are often biased by taphonomic processes, which may exclude rare species or may bring about the mixing of faunas from two or more sources.

Based on these principles, Shotwell (1955) set out several criteria by which fossil mammal 'communities' differ from the communities from which they were derived. Selection may be indicated by biases in bone density, by which some bones are more likely to be preserved than others, biases in body size (modern faunas are approximately 80% small mammals and 20% large mammals), biases in state of preservation (mammals living close to the fossil site may be better preserved than those from far away), and biases from predator selection. These potential sources of error in reconstructing past communities provided the stimulus for taphonomic work in the second half of the 20th century, but at the same time it steered paleontologists toward considering community paleoecology.

Building on many discussions, Judy Van Couvering (now Judith Harris) and I tried out several different approaches to using entire faunas to reconstruct paleoecology. We devised two methods based on weighted averages ordination of entire fossil faunas, but these had two drawbacks. Firstly, they assumed that the faunas provided good representations of past communities, i.e., that they were not biased. I tried to estimate potential bias by experimentally altering living faunas either by rarefaction, for example eliminating species below a certain size limit from faunas of known habitat, or by mixing faunas from known habitats in different proportions. Secondly, both methods are based on interpreting past ecological preferences of fossil species from known preferences of the same or related species. Reliance on species equivalence between present and past is always going to be questionable for mammals. For invertebrates and plants, which had established their evolutionary niches by the beginning of the Tertiary, it is possible to reconstruct past communities in this way, setting out the interrelationships between all members of the community, but mammals lack a long evolutionary history, and modern families of mammal have rather fluid habitat preferences. Mammals have radiated quickly and are still diversifying, making it difficult to reconstruct their paleoecology on the basis of uniformitarianism. I then came across a new method of ecological analysis called species diversity, which I thought could be applied, not just to a few species, but to an entire mammal community. Instead of the taxonomic identifications of the species, the method depended on analyzing key features in the anatomy of the species in a community: their body sizes, using regressions of body parts against size in modern mammals; their types of locomotion, based on the morphology of their limbs; and their diet, based on their tooth morphology. In theory, it should be possible to extend this method to all animals, particularly including amphibians and birds, but here the fossil record lets us down, for it is unusual to find other vertebrates, to say nothing of invertebrates, preserved alongside mammals.

The principle of ecological diversity is based on the observations made by many naturalists, including Darwin and Wallace. Species of mammals and birds living in similar environments but on different continents evolve similar adaptations because they occupy similar ecological niches. Thus, the distributions of mammalian body sizes in tropical forests are similar on different continents even though the species, genera and even families are all different; there are anteaters in South America that have evolved feeding adaptations similar to those of anteaters in Africa even though they are not related phylogenetically; there are species of hedgehog, flying squirrel, and carnivore which are unrelated but which have evolved similar limb structure in parallel from different ancestral stocks. (Although Darwin observed this phenomenon, he was more concerned with showing the phylogenetic connectivity of faunas and floras *within* continents as opposed to similarities *between* continents.) These similarities come about because mammals adapt in similar ways to similar environments by convergent evolution, independently of phylogeny, so that by analyzing by multivariate analysis the range of adaptations present in all members of a fossil mammal fauna and comparing it with analyses of modern faunas from known habitats, an indication of the environment to which they are adapted can be obtained free of taxonomic bias.

This taxon-free approach is based on the summation of all adaptations present in entire floras and faunas, but ecological diversity relied on subjective interpretations of mammalian skeletal anatomy. Quantitative methods based on ecomorphology have been devised to relate measurements of anatomy directly with bodily function. Ecomorphology is a time-consuming method, however, and it can realistically be applied only to segments of faunas. In addition, while some morphologies are directly related to equivalent behaviors in one group, the same behavior may have different anatomical traits in a different group of mammals. For example, running bovids have different adaptations of the femur from leaping bovids, and these differences can readily be identified in fossil bovids, but the same anatomical features are different in equids. In other words, ecomorphology is not always a taxon-free approach to paleoecology, for it may depend on identifying the phylogenetic position of the groups before they can be analyzed. On the other hand, some dental ecomorphologies such as microwear and mesowear on teeth may cross phylogenetic boundaries.

Ecomorphological criteria applied to large databases of mammals in the northern hemisphere may indicate environmental and climate change. For example, molar hypsodonty distinguishing grazing from browsing herbivores is used as a proxy for changing vegetation patterns in Europe. The change from browser-dominated faunas in the early Miocene to grazer-dominated faunas today indicates loss of forest browse and increase in grasslands, and this has been linked with decreases in temperature and precipitation. This is a good indication of how the herbivore communities have evolved through time, but it does not provide any insights into how the interrelationships within the herbivore/carnivore communities changed.

Another method of paleoecology is now based on physiological constraints of mammals in relation to their ecological place in communities. This grew out of a seminal paper by Olson (1966), who distinguished pelycosaur-therapsid communities on the basis of physiological adaptations, for example between terrestrial and water-based food chains. These principles were applied to mammals by Damuth (1982), who investigated community structure by analysis of body size in relation to species abundances. He found that there are energetic and metabolic limits to present-day mammalian communities; for example, very large mammals have limited numbers of species, and insectivores could not survive above a certain weight limit. These ideas were developed by Chistophe Soligo and me (Soligo and Andrews 2005) by linking the physiological constraints with ecological diversity variants. We were able to show that there are limits to species numbers based on metabolic and size constraints and that all modern communities conform to these limits. Assuming the same constraints apply to fossil mammals, fossil faunas could be analyzed by the same criteria, and those that were outside those limits were not true communities.

The present volume describes and extends many of these historical topics in a timely fashion, describing techniques applied to individual mammal species, species richness, analyses of sediments and plants (in the broadest sense of the word), and community analysis. The significance of body size, functional morphology of limb structure, and tooth structure all have chapters. Dietary inferences are made from microwear and mesowear, and growth patterns of teeth record ecological pressures on individuals and communities. Trace fossils contribute both to biological aspects of paleoecology, by identifying the agents that produced them, and the physical aspects by their effects on the sediment in which they are preserved. Physical evidence of paleoecology is based on analyses of fossil soils, biomarkers, and minerals preserved in sediment, isotopes, phytoliths, pollen, and trace fossils, for they all contribute to understanding paleoecology in addition to plant macrofossils. Geometric morphometrics is a form of shape analysis that is here applied to ecomorphological data, and ecometrics is applied to the distribution of functional traits. Finally, there are chapters on ecometrics and community structure analysis, which shows the striking similarities in communities sharing similar habitats. These contributions bring paleoecology up to date and show that the subject is alive and kicking.

London, UK

Peter Andrews
Natural History Museum

References

Damuth, J. (1982). Analysis of the preservation of community structure in assemblages of fossil mammals. *Paleobiology*, *8*, 434–446.

Darwin, C. (1881). *The formation of vegetable mould through the action of worms: With observations on their habits*. London: John Murray.

Olson, E. C. (1966). Community evolution and the origin of mammals. *Ecology*, *47*, 291–302.

Pei, W.-C. (1938). *Le rôle des animaux et des causes naturelles dans la cassure des os*. Beijing: Geological Survey of China.

Shotwell, J. A. (1955). An approach to the paleoecology of mammals. *Ecology, 36,* 327–337.

Soligo, C., & Andrews, P. (2005). Taphonomic bias, taxonomic bias and historical non-equivalence of faunal structure in early hominin localities. *Journal of Human Evolution, 49*, 206–299.

Preface

The field of paleontology has changed dramatically and rapidly over the past several decades. Although discovering, naming, and determining the evolutionary relationships of extinct species are still a priority for paleontologists, understanding the biology of extinct species (paleobiology) and reconstructing the environments in which they lived have become increasingly prominent goals of such investigations. This is partly due to new technologies (e.g., microtomographic imaging, laser surface scanning), greater computational abilities, and innovative analytical methods (e.g., texture analysis in tooth wear, morphometrics, novel isotopic analyses), but it also results from an increased emphasis on the interconnectedness of organisms and their environment: the field of paleoecology. This greater emphasis on paleoecological studies, in turn, has resulted in a proliferation of techniques for understanding how extinct species lived, moved, and what they ate, the habitats in which they lived, and the climatic regime that affected them and their environment. These advances have revolutionized our understanding of the paleobiology and paleoecology of past organisms. They have also enabled us to better document the timing and context of adaptations and to more confidently reconstruct the evolutionary history of life.

Keeping up with the rapidly changing field of paleoecology is a formidable task for professionals and students alike. Equally challenging is integrating the many techniques developed by specialists in different disciplines to create a coherent picture of an ancient environment and the animals that inhabited it. To this end, we convened a three-day conference in Cleveland, Ohio, in September 2015 to bring together a diversity of early- and mid-career researchers with strong interests in paleobiology and paleoecology. (Conference participates are designated by '#' in the contributors list.) The principal goals of this conference were to: (1) summarize and synthesize currently available tools for paleoecological investigations; (2) facilitate new research collaborations; and (3) create a plan for publishing a volume that would be a 'how to' guide on paleoenvironmental reconstruction for researchers and students. We focused on terrestrial ecosystems and limited the scope to the Cenozoic Era, since communities and habitats from this interval can generally be reconstructed with greater confidence than those from more ancient ones. Methods in Paleoecology: Reconstructing Cenozoic Terrestrial Environments and Ecological Communities is the product of that conference.

Our Cleveland conference was generously supported by the Cleveland Museum of Natural History (CMNH), an anonymous donor to the CMNH, Case Western Reserve University's Institute for the Science of Origins, and Bob Jackson. All chapters were reviewed by at least three external referees in addition to the editors, and we thank all of the reviewers for taking time to provide valuable feedback. We are grateful to Peter Andrews for crafting an insightful and historically illuminating foreword to this volume. Thanks to Caitlin Schwarz for editorial assistance. We are also indebted to the editors of Springer's Vertebrate Paleobiology and Paleoanthropology Book Series, Eric Delson and Eric Sargis, for their help and guidance in bringing this volume to fruition.

Contents

Contributors

Note * indicates preferred address for correspondence

W. Andrew Barr #

Center for the Advanced Study of Human Paleobiology, Department of Anthropology, The George Washington University, Washington, DC, USA

Aly Baumgartner

Terrestrial Paleoclimatology Research Group, Department of Geosciences, Baylor University, Waco, TX, USA

Melissa A. Berke #

Department of Civil and Environmental Engineering and Earth Sciences, University of Notre Dame, Notre Dame, IN, USA

Emily J. Beverly

Department of Geosciences, Georgia State University, Atlanta, GA, USA; Terrestrial Paleoclimatology Group, Department of Geology, Baylor University, Waco, TX, USA; *Earth and Environmental Sciences, University of Michigan, Ann Arbor, MI, USA

Benjamin Blonder

School of Life Sciences, Arizona State University, Tempe, AZ, USA

Camilla Crifò

*Department of Biology, University of Washington, Seattle, WA 98195, USA; Burke Museum of Natural History & Culture, University of Washington, Seattle, WA, USA

Darin A. Croft #

Department of Anatomy, Case Western Reserve University, Cleveland, OH, USA

Sabrina C. Curran #

Department of Sociology and Anthropology, Ohio University, Athens, OH, USA

Rachel H. Dunn #

Department of Anatomy, Des Moines University, Des Moines, IA, USA

Regan E. Dunn #

Department of Botany, University of Wyoming, Laramie, WY, USA; *Integrative Research Center, Field Museum of Natural History, Chicago, IL, USA

Jussi T. Eronen

*Ecosystems and Environment Research Programme and Helsinki Institute of Sustainability Science (HELSUS), University of Helsinki, Helsinki, Finland

Alistair R. Evans #

*School of Biological Sciences, Monash University, Melbourne, Australia; Geosciences, Museums Victoria, Melbourne, Australia

Andrew Flynn

Terrestrial Paleoclimatology Research Group, Department of Geosciences, Baylor University, Waco, TX, USA

Jeremy L. Green #

Department of Geology, Kent State University at Tuscarawas, New Philadelphia, OH, USA

Elisha B. Harris

*Department of Biology, University of Washington, Seattle, WA 98195, USA; Burke Museum of Natural History & Culture, University of Washington, Seattle, WA, USA

Jason J. Head

Department of Zoology and University Museum of Zoology, University of Cambridge, Cambridge, UK

Daniel Hembree #

Department of Geological Sciences, 316 Clippinger Laboratories, Ohio University, Athens, OH, USA

Pennilyn Higgins #

Department of Earth and Environmental Sciences, University of Rochester, Rochester, NY, USA

Russell Hogg #

Department of Rehabilitation Sciences, Florida Gulf Coast University, Ft. Myers, FL, USA

Samantha S. B. Hopkins #

University of Oregon Clark Honors College and Department of Geological Sciences, Eugene, OR, USA

Kris Kovarovic #

Department of Anthropology, Durham University, South Road, Durham, UK

A. Michelle Lawing #

Department of Ecosystem Science and Management, Texas A&M University, College Station, TX, USA

Kari Lintulaakso

Department of Geosciences and Geography, University of Helsinki, Helsinki, Finland

William E. Lukens

Department of Geosciences, Georgia State University, Atlanta, GA, USA; *Terrestrial Paleoclimatology Group, Department of Geology, Baylor University, Waco, TX, USA

Luke Mander

Department of Environment, Earth and Ecosystems, The Open University, Milton Keynes, UK

Silvia Pineda-Munoz

Evolution of Terrestrial Ecosystems (ETE), Smithsonian Institution, Washington, DC, USA

Daniel J. Peppe #

Terrestrial Paleoclimatology Research Group, Department of Geosciences, Baylor University, Waco, TX, USA

P. David Polly

Departments of Earth & Atmospheric Sciences, Biology, and Anthropology, Indiana University, Bloomington, IN, USA

Surangi W. Punyasena #

Department of Plant Biology, University of Illinois, Urbana, IL, USA

Scott W. Simpson #

Department of Anatomy, Case Western Reserve University, Cleveland, OH, USA

Gary E. Stinchcomb #

Department of Geosciences and Watershed Studies Institute, Murray State University, Murray, KY, USA

Caroline A. E. Strömberg

*Department of Biology, University of Washington, Seattle, WA 98195, USA; Burke Museum of Natural History & Culture, University of Washington, Seattle, WA, USA

Denise F. Su #

Department of Paleobotany and Paleoecology, Cleveland Museum of Natural History, Cleveland, OH, USA

Wesley A. Vermillion

Department of Ecosystem and Management, Texas A&M University, College Station, TX, USA

# indicates participant in the 2015 workshop (see Preface above)

Chapter 1
Introduction to Paleoecological Reconstruction

Darin A. Croft, Denise F. Su, and Scott W. Simpson

Abstract Ancient terrestrial ecosystems cannot be observed directly, but a wide variety of approaches and techniques have been developed that provide indirect evidence of many aspects of ecosystem functioning. By integrating multiple lines of evidence about the climate, vegetational structure, and fauna of a given location at a particular time, a relatively complete paleoecological picture of a fossil locality can be generated. This volume reviews some of the most commonly used techniques for paleobiological and paleoecological analysis; in this chapter, we briefly introduce these approaches and the insights they can provide. They include techniques for: inferring attributes of particular mammal species and/or individuals (body mass, locomotor adaptations, diet, life history variables); interpreting ancient soils (paleosols), trace fossils (ichnofossils), organic biomolecules, and plant remains of various types (pollen, phytoliths, macrofossils); analyzing isotopic and geometric morphometric data to answer a range of ecological and environmental questions; and seeking patterns in data across broad taxonomic, geographic, and/or temporal scales (ecomorphology, ecometrics, and ecological diversity analysis).

Keywords Ecosystems • Ecometrics • Paleobiology • Paleoecology • Paleoenvironment

Although many aspects of modern ecological communities are not yet understood, describing the animals, plants, and climate of a particular point on the globe – at least in broad strokes – is relatively straightforward. For example, we have a pretty good idea of the number and types of birds, mammals, amphibians, and other animals that inhabit Great Smoky Mountains National Park in the eastern United States. We know that the area is covered by temperate deciduous forest, has a mean annual temperature of about 13°C, and generally receives 140–220 cm of precipitation per year. Similar data are readily available for such sites as the Iberá Wetlands in northeast Argentina, Krau Wildlife Reserve in Malaysia, and Kruger National Park in South Africa. As a result, much of the research that goes on in these areas focuses on clarifying details of these ecosystems, such as documenting inconspicuous and/or unrecognized species, unraveling the complex relationships among organisms and between them and their environment, and predicting how small and large-scale human activities many affect these areas in the future.

Understanding ecosystems of the ancient past requires a different strategy. Since it isn't possible to directly observe extinct organisms or measure rainfall or ambient temperature in the past, such ecosystems must be reconstructed from the physical evidence and chemical signatures left in the geological record by their biotic (plants and animals) and abiotic (rain, temperature) components. Instead of starting from the big picture and drilling down to the details, a paleoecologist must start with the details – such as fossil teeth and bones, phytoliths, or soil carbonates – and work up. This bottom-up approach is also necessitated by the nature of the fossil record itself, which preserves only a subset of the plants and animals that comprised an ancient ecosystem. The many methods that are available for reconstructing particular aspects of ancient ecosystems are all based on certain assumptions and generate some degree of error. Therefore, maximal confidence comes from combining analyses of entire faunal and floral assemblages with geological, geochemical, and other types of contextual data. Integrated paleoecological studies like these provide a historical

D. A. Croft (✉) · S. W. Simpson
Department of Anatomy, Case Western Reserve University, 10900 Euclid Ave., Cleveland, OH 44106, USA
e-mail: dcroft@case.edu

S. W. Simpson
e-mail: sws3@case.edu

D. F. Su
Department of Paleobotany and Paleoecology, Cleveland Museum of Natural History, Cleveland, OH 44106, USA
e-mail: dsu@cmnh.org

D. A. Croft et al. (eds.), *Methods in Paleoecology: Reconstructing Cenozoic Terrestrial Environments and Ecological Communities*, Vertebrate Paleobiology and Paleoanthropology, https://doi.org/10.1007/978-3-319-94265-0_1

perspective on the evolution of Earth's ecosystems that permits macroecological analyses of patterns of climate change, the evolution of lineage and guild-level adaptations, and large scale biogeographic patterns.

The goal of this volume is to review some of the most common techniques that are used to reconstruct ancient terrestrial environments and ecosystems in order to illustrate the myriad sources of data and types of analyses that can inform such an effort. The chapter authors are experts in fields that span chemistry, geology, and biology. They describe approaches that provide varying levels of resolution about the environmental conditions and the plants and animals that existed thousands to millions of years ago. They also provide glimpses into the future of paleoenvironmental reconstruction by discussing potential ways that these approaches may be refined.

While many of the techniques discussed in this volume can be applied to any part of the fossil record, we focus here on examples from Cenozoic terrestrial localities. Mammals are among the most conspicuous elements of modern ecological communities, and extinct mammals have long been a prime source of information about the ancient habitats in which they lived. This is primarily due to the fact that their teeth and skeletons withstand fossilization far better than those of most other organisms. Thus, most terrestrial paleontological sites from the past 66 million years yield mammal fossils, often in abundance. The first five chapters of this volume discuss techniques used to elucidate the paleoecology of particular mammal species or individuals by studying the structure or attributes of their remains. Although understanding the ecologies and adaptations of the mammals themselves is often the ultimate goal of such studies, these data can also be integrated into studies of the broader mammal community in order to characterize the paleoenvironment in which they lived, a topic addressed near the end of this volume.

Three attributes describe much of an animal's ecological niche: body mass, locomotor style, and diet. The next four chapters address these topics in turn. Chapter 2, by Samantha Hopkins (2018), discusses how one goes about determining the size (body mass) of an extinct species through analyses of their skeleton. Body size is probably the most immediately recognizable aspect of any mammal, and it can easily be quantified in a living individual by measuring its body mass. Measuring body mass is much less straightforward when one is dealing with fossil remains – which are often fragmentary – particularly for mammals with a body form quite unlike any species alive today. Body mass estimates have traditionally relied on regression equations, and although the mathematics involved is relatively simple, many important considerations must be taken into account to obtain reasonably accurate estimates. More complex methods have been developed in recent years, but whether these are widely applicable remains to be seen.

The limbs are the load-bearing structures of most mammals, and thus there is a reasonably tight correlation between their structure and body mass. However, limb bone morphology is also tightly linked to function. Chapter 3, by Rachel Dunn (2018), describes the process whereby a functional morphologist deduces limb function in an extinct species by reconstructing the sizes and positions of muscles based on their bony attachments. This chapter provides a solid theoretical grounding for comparative anatomical studies, whether one uses traditional descriptive approaches or more computationally-intensive investigative techniques such as three-dimensional morphometrics. As noted in this chapter, a species' evolutionary (phylogenetic) relationships can strongly influence many aspects of its anatomy and physiology, including its limb structure. For this reason, the influence of phylogeny in paleoecological studies is a recurring theme throughout this volume. It is likely that correcting for underlying phylogenetic biases in data will become increasingly important and feasible as the structure of the tree of life is ascertained with greater confidence and precision.

Chapters 4 and 5 both tackle the subject of determining what an extinct species ate, but they do so from different perspectives. Chapter 4, by Alistair Evans and Silvia Pineda-Munoz (2018), describes a mechanical and material properties approach to that question that strives to determine the types of foods a tooth (or dentition) is optimally structured to process. In other words, for what type of diet have the teeth been shaped by natural selection? This is a "long-view" strategy for determining the average diet to which a species is adapted that parallels methods used to reconstruct locomotor style, and certain aspects of tooth form, most notably hypsodonty, can be (and have been) applied quite broadly across space and time to deduce large-scale environmental and climatic patterns. But not all aspects of tooth form are dictated at birth. Teeth undergo wear through interactions with one another as well as with consumed foods. How one reads that wear record is described in Chap. 5 by Jeremy Green and Darin Croft (2018), who focus on mesowear and microwear analyses. Mesowear and microwear can be thought of *a posteriori* strategies, as they strive to correlate tooth wear (macroscopic and microscopic, respectively) with the foods an individual animal consumed prior to its death. Thus, they represent direct evidence of behavior, though they do so at different time scales. (The analogous approach for limb function might be analyzing the distribution of internal trabeculae in limb bones, but a standardized, broadly-applicable technique for interpreting such data has not yet been developed.) The techniques for inferring diet described in Chaps. 4 and 5 are complementary and, when combined, can assess diet variation among individuals and populations in addition to species.

Teeth and bones are affected by internal physiological processes as well as external ecological ones, and their microscopic growth proceeds like a biological metronome, evenly marking time in variable durations ranging from days to weeks to years. Chapter 6, by Russell Hogg (2018), details how such changes can be read in the teeth and bones of extant and extinct species and applied to a variety of paleobiological and paleoecological topics such as life history traits of particular species and seasonal environmental variation. The fields of odontochronology and skeletochronology are still in their relative infancy, at least in terms of their application to non-primates, and have the potential to provide fine-scale growth and development information about individuals and populations analogous to that provided by microwear and mesowear for diet.

Teeth and bones can also be analyzed to reconstruct diet on an even finer scale by documenting the relative compositions of the stable isotopes they contain. Such analyses most often focus on oxygen and carbon isotopes, and this topic is reviewed by Penny Higgins (2018) in Chap. 7. The theoretical basis behind stable isotope analysis is that the proportions of different isotopes vary depending on the environment in which an animal lived and the plants that it ate. These isotopes are incorporated into the animal's tissues in similar proportions – though the values are altered to varying degrees by physiological processes – and can be preserved in the fossil record. Measuring these isotopes in fossil teeth and bones requires destructive sampling, but much less material is required now (typically 10+ mg for tooth or bone) than was necessary only a couple of decades ago. Moreover, newer microsampling techniques have enabled researchers to study changes in isotopic compositions within a single specimen, a strategy equivalent to sampling the composition of a mammal's tissues at different points during its life. Isotopic sampling is usually performed on fragmentary specimens that have relatively little value as comparative specimens and, as a result, this technique has been applied to a wide variety of species and sites across the globe. This, in turn, has resulted in many new insights into past climate and habitat and the development of new hypotheses to be tested by other means.

Other micro-scale sources of information are organic molecules preserved in fossil-bearing sediments, which can provide clues about past moisture, temperature, and vegetation. These "chemical fossils" represent records of past life (biomarkers), often from organisms that are not themselves preserved in the fossil record of a particular area, such as bacteria and plants. Their study lies at the intersection of biology, geology, and chemistry and comprises a relatively young (less than a century old) field of scientific research known as organic geochemistry. In Chap. 8, Melissa Berke (2018) describes a portion of the wide diversity of organic biomarkers that have been discovered thus far, how they are collected and analyzed, and the types of information they can provide about ancient terrestrial environments.

Organic biomarkers, like fossils themselves, are commonly preserved in ancient ("fossilized") soils, which are termed paleosols. As detailed by Emily Beverly and colleagues (2018) in Chap. 9, studying the macroscopic, microscopic, and chemical characteristics of these paleosols – the purview of paleopedology – can provide a wide range of complementary information about past landscapes and climatic conditions. Paleosols are the product of local conditions and thus act as archives of the paleoenvironment of a specific site over a relatively constrained period of time. They provide information about a site's climate (mean annual temperature and precipitation), vegetation, sedimentary regime, and landscape stability and can be used to assess regional variation in these variables when studied throughout a basin.

Among the more commonly encountered fossils in paleosols are ichnofossils: traces of organisms and records of their behavior, often organisms not directly preserved in the fossil record. In Chap. 10, Daniel Hembree (2018) reviews various kinds of animal-soil interactions that can result in ichnofossils, such as nests, burrows, and trails, and describes the types of interpretations that can be gleaned from them regarding habitat and environmental conditions. One of the key challenges in continental (terrestrial) ichnology is correlating ichnofossils with the organisms that made them. In recent decades, researchers have begun to build an ichnological library based on laboratory studies of traces made by living organisms, a pursuit that will undoubtedly provide new information about otherwise unrecognized inhabitants of ancient ecosystems.

Plants are closely adapted to abiotic factors such as climate and soil composition, and any ecological interpretation of a terrestrial paleontological site is incomplete without considering its predominant plant types and vegetational structure. Unfortunately, the conditions that permit the preservation of plant macrofossils such as leaves are seldom the same ones that preserve vertebrate bones and teeth. However, many sites that preserve fossil vertebrates also preserve microscopic plant remains. Such remains include pollen and spores, reviewed by Luke Mander and Surangi Punyasena (2018) in Chap. 11, as well as phytoliths, which are reviewed by Caroline Strömberg and colleagues (2018) in Chap. 12. Pollen and spores have the advantage of being relatively widespread, but this can also be a liability when they record plants living far away from the site in question. Moreover, they are usually only found in relatively fine-grained sedimentary layers. Phytoliths, which are silica bodies that form in plant cell walls, are durable and can often be found alongside vertebrate fossils, but unlike pollen and spores, a single type of plant can contain many forms of phytoliths. Thus, a primary challenge of phytolith investigations is taxonomically identifying the types of plants these phytoliths represent. Studying phytoliths in deep time is a recent phenomenon, and much work remains to be done in documenting their diversity in modern taxa. Despite the challenges posed by both paleopalynology and phytolith-based paleoecology, these approaches are extremely useful for providing primary evidence about the plants that were living at a particular site in the absence of (or as a complement to) macrofossil evidence. It should come as no surprise that both techniques have effectively developed and tested a diversity of geohistorical and macroecological hypotheses.

In sites that preserve plant macrofossils, these specimens can be a source of detailed information about climate and habitat in addition to species diversity, as summarized by Daniel Peppe and colleagues (2018) in Chap. 13. The close linkages between plants and their environment, combined with the taxonomic information contained in fossil leaves, flowers, and other plant parts, means that associations of plant macrofossils are among the best methods of documenting environments of the past. Paleobotany is a comparatively old field, and much recent progress in this area has focused on quantifying fossil assemblages and creating more accurate algorithms for correlating assemblage characteristics with abiotic variables such as temperature and precipitation. The quest to refine paleoecological interpretations by quantifying form is a second conspicuous cross-cutting theme in this volume and a major focus of Chaps. 14–17.

Thanks to increases in computational power that continue to accelerate, it is now relatively easy to quantify form in ways that were scarcely dreamt about even 10 years ago. Although linear measurements remain more than satisfactory for describing many aspects of a fossil specimen, complex surfaces or morphologies can only be adequately described using more sophisticated methods. Chapter 14 by Sabrina Curran (2018) reviews techniques and applications of what is known as three-dimensional geometric morphometrics: describing morphology in three axes (or sometimes two) using devices such as 3D digitizers, laser surface scanners, and computed tomography (CT) scanners. There is no limit to the types of paleoecological studies that can incorporate such techniques, and the diversity of proven applications continues to expand as the devices used to acquire three-dimensional data become less expensive and more accessible.

The final chapters of this volume focus on broader themes of paleoecological reconstruction that build on the results of the types of studies reviewed in Chaps. 2–7. In Chap. 15, Andrew Barr (2018) describes how one aspect of form – such as a particular aspect of tooth or limb morphology – can be quantified across a broad range of extant species, correlated with an ecological variable such as diet or locomotor style, and used to interpret a paleoecological attribute. The shorthand name now used for such studies is ecomorphology. In our usage of the term, ecomorphological studies parallel paleobiological studies of diet and locomotion such as those described in Chaps. 3–5 but differ in their goal; rather than inferring the habits of a particular species, ecomorphological studies typically use form-function relationships to characterize an entire fauna or to compare associations within a particular group across broad spans of space or time.

Chapter 16, by Kris Kovarovic and colleagues (2018), describes a process whereby ecomorphological data from modern and fossil mammal communities are compared to one another with the goal of reconstructing past habitats and ecological associations, a process known as community structure analysis or ecological diversity analysis (EDA). A typical ecological diversity analysis examines the three primary determinants (or reflections) of an animal's ecological niche mentioned earlier in this chapter – body mass, diet, and locomotor strategy – and uses the collective ecological niches of all members of a mammal community to infer the habitat in which they were living. Like many of the approaches described in this volume, this strategy for paleoecological interpretation relies on observed relationships between living mammals and their environment. Thus, a key consideration in EDA is selecting an appropriate dataset of modern comparative faunas, a task that is much simpler in principle than in practice. Some recent EDAs have also begun to suggest that certain mammalian communities in the past may have been structured differently from any that exist today, a finding that has important implications for understanding niche partitioning, ecosystem functions, and the diversity of mammalian ecological communities.

The final methods chapter in the volume, Chap. 17, by Wesley Vermillion and colleagues (2018), describes an approach to interpreting paleoclimate and paleovegetation known as ecometrics. Like ecomorphology, ecometrics strives to correlate particular traits with environmental characteristics, but it differs in its broader taxonomic scope (typically entire communities), more extensive geographic and temporal focus, and its use of mathematical modeling to understand long-term evolutionary and ecological processes.

Which of the many methods detailed in this volume can and should be applied to reconstructing ancient terrestrial ecosystems depends in large extent on the data that are available. But assuming the best-case scenario – a paleontological site with abundant and well-preserved fossils, a diversity of willing and able-bodied researchers, and generous financial and institutional support – how can results from an array of analyses be combined to create an integrated understanding of a particular place at a specific time in the past? What additional considerations should be taken into account, and what should one do in the case of conflicting results? The final chapter in this volume, by Denise Su and Darin Croft (2018), addresses these and other questions and attempts to outline a path forward so that knowledge of past ecosystems may one day begin to approach those of the present.

Acknowledgements We thank R. Engelman and J. Green for reading and providing comments on this chapter.

References

Barr, W. A. (2018). Ecomorphology. In D. A. Croft, D. F. Su & S. W. Simpson (Eds.), *Methods in paleoecology: Reconstructing Cenozoic terrestrial environments and ecological communities* (pp. 337–347). Cham: Springer.

Berke, M. A. (2018). Reconstructing terrestrial paleoenvironments using sedimentary organic biomarkers. In D. A. Croft, D. F. Su & S. W. Simpson (Eds.), *Methods in paleoecology: Reconstructing Cenozoic terrestrial environments and ecological communities* (pp. 121–149). Cham: Springer.

Beverly, J. E., Lukens, W. E., & Stinchcomb, G. E. (2018). Paleopedology as a tool for reconstructing paleoenvironments and paleoecology. In D. A. Croft, D. F. Su & S. W. Simpson (Eds.), *Methods in paleoecology: Reconstructing Cenozoic terrestrial environments and ecological communities* (pp. 151–182). Cham: Springer.

Curran, S. C. (2018). Three-dimensional geometric morphometrics in paleoecology. In D. A. Croft, D. F. Su & S. W. Simpson (Eds.), *Methods in paleoecology: Reconstructing Cenozoic terrestrial environments and ecological communities* (pp. 317–335). Cham: Springer.

Dunn, R. H. (2018). Functional morphology of the postcranial skeleton. In D. A. Croft, D. F. Su & S. W. Simpson (Eds.), *Methods in paleoecology: Reconstructing Cenozoic terrestrial environments and ecological communities* (pp. 23–36). Cham: Springer.

Evans, A. R., & Pineda-Munoz, S. (2018). Inferring mammal dietary ecology from dental morphology. In D. A. Croft, D. F. Su & S. W. Simpson (Eds.), *Methods in paleoecology: Reconstructing Cenozoic terrestrial environments and ecological communities* (pp. 37–51). Cham: Springer.

Green, J. L., & Croft, D. A. (2018). Using dental mesowear and microwear for dietary inference: a review of current techniques and applications. In D. A. Croft, D. F. Su & S. W. Simpson (Eds.), *Methods in paleoecology: Reconstructing Cenozoic terrestrial environments and ecological communities* (pp. 53–73). Cham: Springer.

Hembree, D. (2018). The role of continental trace fossils in cenozoic paleoenvironmental and paleoecological reconstructions. In D. A. Croft, D. F. Su & S. W. Simpson (Eds.), *Methods in paleoecology: Reconstructing Cenozoic terrestrial environments and ecological communities* (pp. 183–212). Cham: Springer.

Higgins, P. (2018). Isotope ecology from biominerals. In D. A. Croft, D. F. Su & S. W. Simpson (Eds.), *Methods in paleoecology: Reconstructing Cenozoic terrestrial environments and ecological communities* (pp. 99–120). Cham: Springer.

Hogg, R. (2018). Permanent record: the use of dental and bone microstructure to assess life history evolution and ecology. In D. A. Croft, D. F. Su & S. W. Simpson (Eds.), *Methods in paleoecology: Reconstructing Cenozoic terrestrial environments and ecological communities* (pp. 75–98). Cham: Springer.

Hopkins, S. S. B. (2018). Estimation of body size in fossil mammals. In D. A. Croft, D. F. Su & S. W. Simpson (Eds.), *Methods in paleoecology: Reconstructing Cenozoic terrestrial environments and ecological communities* (pp. 7–22). Cham: Springer.

Kovarovic, K., Su, D. F., & Lintulaakso, K. (2018). Mammal community structure analysis. In D. A. Croft, D. F. Su & S. W. Simpson (Eds.), *Methods in paleoecology: Reconstructing Cenozoic terrestrial environments and ecological communities* (pp. 349–370). Cham: Springer.

Mander, L., & Punyasena, S. W. (2018). Fossil pollen and spores in paleoecology. In D. A. Croft, D. F. Su & S. W. Simpson (Eds.), *Methods in paleoecology: Reconstructing Cenozoic terrestrial environments and ecological communities* (pp. 213–232). Cham: Springer.

Peppe, D. J., Baumgartner, A., Flynn, A., & Blonder, B. (2018). Reconstructing Paleoclimate and Paleoecology using Fossil Leaves. In D. A. Croft, D. F. Su & S. W. Simpson (Eds.), *Methods in paleoecology: Reconstructing Cenozoic terrestrial environments and ecological communities* (pp. 213–232). Cham: Springer.

Strömberg, C. A. E., Dunn, R. E., Crifò, C., & Harris, E. B. (2018). Phytoliths in paleoecology: analytical considerations, current use, and future directions. In D. A. Croft, D. F. Su & S. W. Simpson (Eds.), *Methods in paleoecology: Reconstructing Cenozoic terrestrial environments and ecological communities* (pp. 233–285). Cham: Springer.

Su, D. F., & Croft, D. A. (2018). Making sense of the evidence: synthesizing paleoecological data. In D. A. Croft, D. F. Su & S. W. Simpson (Eds.), *Methods in paleoecology: Reconstructing Cenozoic terrestrial environments and ecological communities* (pp. 393–402). Cham: Springer.

Vermillion, W. A., Polly, P. D., Head, J. J., Eronen, J. T., & Lawing, A. M. (2018). Ecometrics: a trait-based approach to paleoclimate and paleoenvironmental reconstruction. In D. A. Croft, D. F. Su & S. W. Simpson (Eds.), *Methods in paleoecology: Reconstructing Cenozoic terrestrial environments and ecological communities* (pp. 371–392). Cham: Springer.

Chapter 2
Estimation of Body Size in Fossil Mammals

Samantha S. B. Hopkins

Abstract Body mass is a fundamental ecological parameter of mammals with implications for a variety of other ecological characteristics. While it cannot be directly measured in fossil taxa, it can be inferred using allometric relationships between skeletal dimensions and mass derived from extant species. Many such relationships have been described, primarily for dental and limb dimensions. Methods of statistical analysis vary widely, however, in ways with substantial implications for the inferred masses of fossil species. The subset of extant species from which the relationship is derived must be representative of the evolutionary and ecological scope of the fossil taxa for which mass is to be estimated. Increasing computing power and an explosion of phylogenetic comparative methods offer the opportunity to gain an understanding of the processes driving these important empirical relationships.

Keywords Body mass • Proxy • Mammalia • Regression • Allometry • Paleoecology

Introduction

Body size is an essential characteristic of animals, with substantial implications for a number of other ecological and evolutionary parameters (Eisenberg 1981; McMahon and Bonner 1983; Peters 1983; Calder 1984; Schmidt-Nielsen 1984; LaBarbera 1989; Blackburn and Gaston 1994). Size is measured in a variety of ways; among animals it is commonly described in terms of either body length or mass, as these parameters offer the greatest predictive value for ecology. Both of these characteristics translate into the constraints on ecology and evolutionary history, from life history to trophic strategy, evolutionary rates to preferred habitats. As a result, reconstructing body size from extinct organisms is an essential step in understanding their ecology.

Among mammals, body size is commonly described in terms of mass; body length is primarily used for mammals too large to be easily weighed, such as whales (Lockyer 1976), although it has been used for other taxa as well (Iskjaer et al. 1989). Body mass does vary significantly through the lifetime of an organism in response to health, lactation, pregnancy, age, food availability, and other factors (Lindstedt and Boyce 1985; Schulte-Hostedde et al. 2005; Toigo et al. 2006), but paleontologists are generally concerned primarily with estimating average adult size for a species. While mass is easy to measure in living mammal species, requiring only a spring scale for the vast majority of mammals, it cannot be directly measured in extinct organisms. There are, however, a variety of skeletal and dental proxies available for the estimation of body mass. All of these proxies depend on isometric or allometric relationships between various skeletal dimensions and body mass.

The most commonly-applied proxies for mass make use of either dental dimensions (e.g., Legendre 1986; Hopkins 2008; Freudenthal and Martín-Suárez 2013) or the diameters of limb bones (e.g., Scott 1983; Gingerich 1990; Rafferty et al. 1995). Ruff (1988, 2003; Ruff et al. 1991) has also had success estimating mass with the articular dimensions of limb bones. Dental proxies, most commonly used for mammals, take advantage of the relationship between food-processing capacity and metabolism (which is driven primarily by body mass; McNab 1988; Whittaker 1999) and benefit from using the most identifiable and preservable elements in the mammalian skeleton, the teeth. Limb proxies, on the other hand, have the advantage of relying on the more direct relationship between body mass and the load borne by the limbs as they support the body on land. In the case of aquatic mammals, for which neither of these proxies may be effective, other proxies such as cranial dimensions

S. S. B. Hopkins (✉)
Department of Geological Sciences, Clark Honors College, 1293 University of Oregon, Eugene, OR 97403-1293, USA
e-mail: shopkins@uoregon.edu

D. A. Croft et al. (eds.), *Methods in Paleoecology: Reconstructing Cenozoic Terrestrial Environments and Ecological Communities*, Vertebrate Paleobiology and Paleoanthropology, https://doi.org/10.1007/978-3-319-94265-0_2

and estimated body volume are necessary. Multiple proxies are often necessary for a robust estimate of mass, and a variety of approaches have been applied to determining which of the various mass estimates are most reliable.

While the allometric relationships between mass and skeletal dimensions have been studied for more than 40 years, there remain substantial problems still to be solved in body mass estimation. In particular, there is a need to confront the difficulties created by using regressions on log-transformed data for estimation. Another factor that complicates mass estimation is the lack of a consistent physical constraint on morphology that can be generalized across all mammals; it is nearly impossible to use the same proxy when studying a diverse assemblage of mammals, as the relationships between mass and skeletal proxies change across mammalian phylogeny. Reconstruction of the body mass of extinct mammals requires careful consideration of phylogenetic constraints, locomotion, functional morphology, and the statistics of regression analysis, prediction intervals, and error propagation. Nonetheless, despite the challenges in the approaches available to us today, there is reason for hope that new computational approaches will solve or at least control for existing problems, and even the strategies currently available can generate ecologically meaningful information about fossil ecosystems and organisms.

Terms

Accuracy: The degree to which an estimated value approximates the actual value. In the estimation of mass, this is probably the primary basis with which to judge the value of a method for paleoecological reconstruction. This is in contrast with Precision; see below.

Allometric scaling: A description of the relationship of overall size to morphology. As size changes, some aspects of morphology change faster or slower than others, leading to changes in shape with changing size.

Body size: A measure of the total size of an organism. Common measures of size in vertebrates include mass, total length, and snout-vent length.

Body mass: The total mass of a live or recently dead organism, conventionally measured as weight with a spring scale.

Geometric scaling: A scaling relationship in which two characters are related mathematically as predicted by their geometry. That is, a volumetric measurement (such as mass) should be proportional to the cube of length (with or without coefficients arithmetically modifying the slope or intercept of the relationship).

Heteroscedastic: Non-normally distributed and skewed. Commonly used with reference to the shape of a distribution of variables.

Isometric scaling: A scaling relationship in which two characters of interest retain the same proportions with size change, resulting in retention of consistent shape over a range of sizes.

Jackknifing: The statistical practice of establishing the predictive value of a regression by removing the data points one by one, analyzing the relationship between the x and y variables, and then using that regression to predict the y-value of the excluded datum. Iterated over the entire dataset, this approach generates an estimate of the capacity of the regression to predict values for new data.

Metabolic scaling: A scaling relationship predicted by the mathematical relationship between size and metabolic rate. The metabolic theory of ecology (Brown et al. 2004) predicts that metabolism scales with $\text{mass}^{3/4}$. Hence, other features, such as tooth area, used in fueling metabolism are expected to scale with $\text{mass}^{3/4}$. This is in opposition to volumetric scaling, which would predict area scaling with $\text{mass}^{2/3}$.

Multiple regression: Linear regression that simultaneously considers multiple independent variables to account for the correlations among those variables in estimating the strength of their linear relationships with the dependent variable. The regression equation takes the form $y = m_1x_1 + m_2x_2 + m_3x_3 + \cdots + b$, rather than $y = mx + b$ as in an ordinary linear regression.

Overfitting: A statistical error in which a regression attributes more of the variation in the dependent variable to the independent variable(s) than is actually causally driven as a result of violations of the distribution assumptions of the statistical method.

Power law regression: A method of regression analysis that, rather than fitting a linear relationship between the independent and dependent variables, looks for the exponential relationship that best explains the distribution of values; the regression equation takes the form $y = ax^m$ rather than $y = mx + b$ as in an ordinary linear regression.

Precision: The uncertainty in an estimated value; the exactitude to which an estimated value is articulated. In its simplest terms, this could be the number of decimal places to which the value is estimated. Note that precision is not the same as accuracy, defined above. A mass estimate value can be highly precise, in terms of being highly replicable and not particularly variable in magnitude, but also quite inaccurate, in being quite distant from the actual mass of the animal.

Rectangular area: The estimated surface area of a roughly quadrangular feature (for the purposes of this chapter, tooth surface area) estimated by multiplying length by width.

Standard error of the estimate: A measurement of the uncertainty in a dependent variable value estimated using a regression. It is commonly calculated as $\sigma_{est} = \sqrt{\frac{\sum (Y-Y')^2}{N}}$

Training data: The data used to generate a predictive statistical relationship. In the case of body mass estimation, these are commonly derived from skeletal measurements of museum specimens of extant species associated with individual body masses. The allometric equations produced from the training data can then be used to estimate body mass where only skeletal dimensions are known.

Volumetric estimation: An approach to body mass estimation wherein the 3-dimensional outline of an organism is reconstructed from its skeleton, and that volume is used to infer mass given either an assumed density or an allometric relationship between volume and mass.

Theoretical/Historical Background

Current approaches to the estimation of body size in extinct mammals trace their roots to a paper by Gould (1975) in which he pointed out that the area of mammalian postcanine teeth scales against mass with positive allometry, likely as a result of metabolism and/or the changes in dietary strategies with increased habitat grain. This work provided a basis for straightforward reconstruction of body mass in any fossil mammal known from teeth. While Gould's descriptions of these allometric relationships for a few key mammalian clades were foundational, subsequent work discussed the drivers of variation in the relationship between size and dental dimensions (Gingerich et al. 1982; Fortelius 1990; Smith 2002). Certainly, the assumption underlying the use of allometric relationships derived from extant mammals to infer mass in extinct species is that the processes shaping those relationships in living species were the same as those in place when the fossil of interest evolved. Whether or not this assumption is true depends on whether our explanations for this allometric relationship are correct (Fortelius 1990).

Many authors have suggested that the relationship between tooth size and body mass is constrained by the energetic demands of size and their relationship to food processing needs (e.g., Gingerich et al. 1982; Fortelius 1985). If this is true, then the fossil taxa for which we reconstruct mass must be similar in physiology (so that energetic demands scale similarly with size), the types of food consumed, and digestive strategy to the range of extant mammals used to produce the allometric curve from which fossil mass is inferred. Tooth size-body mass allometry was proposed by Gould (1975) to represent ¾ power metabolic scaling and hence presumably be similar across all crown group mammals. For a discussion of the complexities of metabolic scaling in mammal teeth, see Fortelius (1990). It has also been proposed that the relationship between tooth area (which is a function of length squared) and body mass (which is directly related to volume, a function of length cubed) should show ⅔ power geometric scaling (Gingerich et al. 1982). There is enough noise in the relationship between mass and tooth area (as a result of measurement error, individual variation in tooth size and mass, and interspecific variation in the relationship between the two), and enough different ways of measuring tooth dimensions that the nature of the scaling relationship remains controversial (a problem that plagues the geometric/metabolic scaling problem in general; White et al. 2009). As a consequence, most approaches to mass reconstruction are empirically-based rather than having a strong theoretical foundation.

While early efforts used the area of the entire postcanine tooth row to infer allometric relationships with mass (Gould 1975), Gingerich et al. (1982) and several subsequent authors (Legendre 1986; Van Valkenburgh 1990; Gordon 2003) demonstrated a strong relationship between body mass and the areas of individual teeth, in particular the first molar. The first molar is particularly useful because it tends to have relatively low levels of intraspecific variation and lower levels of sexual dimorphism, likely as a result of its eruption early in ontogeny, prior to puberty (Gingerich 1974). Inferring mass from individual teeth adds to the metabolic scaling assumption by assuming a consistent relationship between the size of that tooth and that of the rest of the tooth row. This assumption can often be assumed to hold true where the dental formula and gross dental morphology remain similar but becomes more questionable with changes in the functional role of the individual tooth within the tooth row. Changes in diet that emphasize the function of particular teeth cause increases or decreases in the relative sizes of teeth, changing these allometric relationships. A widely-used set of regression equations of lower first molar (m1) rectangular area (that is, length × width, rather than the actual area measured within the tooth outline) against body mass for a variety of mammalian clades was developed by Legendre (1986), which led to the adoption of m1 area as the most common proxy for mammalian body size. This proxy has the additional advantage that the lower first molar of the vast majority of mammalian species is among the most taxonomically diagnostic of skeletal remains.

The allometric scaling of limb bones with mass has seen a great deal of work (e.g., Gingerich 1990; Biknevicius 1999; Christiansen 2002), but one key realization is that the load-bearing strength of a columnar limb is proportional to its cross-sectional area (Ruff 1990), so one can use the cross-section of limb bones to infer mass in extinct mammals. This method was shown to predict mass much more reliably than dental dimensions in an array of studies that arose out of a workshop on body mass held at the University of Florida (Damuth and MacFadden 1990), as well as some subsequent work (e.g., Egi 2001). However, this approach has a different set of assumptions that can be violated by extinct mammals. For example, using humeral cross-sectional area as a predictor of body mass assumes similar posture to the modern dataset from which the fossil species' mass was inferred. This has been problematic in some past studies (Sánchez-Villagra et al. 2003; Millien and Bovy 2010; Basu et al. 2016), which have concluded that weight-bearing posture differed in the extinct species under study. Other studies have shown that this scaling relationship may change over the range of extant mammals (Biewener 1990; Christiansen 2002) as the physical constraints change in importance allometrically, so inferences should be made with reference to species within the size range of extant comparators. Ruff (2003) showed that the distribution of bone within the shaft (measured as the section modulus) is more effective at predicting mass than the breadth of the bone shaft, so where precision is important and fossil material is well-preserved, this more labor-intensive approach may yield greater accuracy.

Breadths and areas of limb bone articular surfaces provide another functionally constrained parameter; while they may be affected by locomotor behavior (Ruff 1990, 2002), there is some evidence (Ruff 1988, 1990, 2003) to suggest that they may be more reliable than many other proxies for mammalian body mass. The articular surface area is assumed to be constrained by mass in that excessively large forces exerted on articular surfaces in the joints increase the chances of damage to cartilage and soft tissue. These methods require preservation of identifiable limb elements with complete articular surfaces, but have proven to predict mass well in primates (Ruff 1990, 2003). They have not yet been widely applied among other mammalian groups.

Recently, there has been a dramatic increase in the application of phylogenetic comparative methods to paleoecology. Given the limitations mentioned above to body mass estimation, it is worth asking whether evolutionary history has a substantial influence on the outcome of body mass reconstructions, given that these methods are often applied to fossil taxa with no close living relatives. Campione and Evans (2012) determined that limb bone circumferences do show some component of phylogenetic signal in the fit between limb dimensions and mass but concluded that it was not significant for reconstructions of ecology. No study has made a comprehensive effort to examine the influence of phylogeny in dental reconstructions of body mass, but studies at the ordinal level within mammals (Legendre 1986), between families of primates (Kay and Ungar 1997; Scott 2011), and within families and superfamilies of rodents (Lindsay 1988; Morgan et al. 1995; Hopkins 2008) have shown that regression equations can differ substantially between clades, suggesting the importance of phylogeny in constraining these relationships.

Approaches

Given the variety of proxies available, any study of fossil mammal body mass has a number of methodological options. Choice of an appropriate proxy for body mass reconstruction should be dictated by the biology of the group under study. Where possible, multiple proxies should be used and compared to determine which estimate is most likely to be correct and the range of possible values for body mass. While one can obtain a very precise (but not necessarily very accurate) estimate of mass from previously-published allometric regressions, even a very good mass proxy generally has notable (though often poorly reported) uncertainty, and an honest reconstruction of mass for a fossil mammal includes an assessment of this uncertainty.

The methodology is similar for almost all proxies commonly used. Data are collected on the proxy measurements and body mass from a training dataset of extant species, commonly from mammalogy collections, which often have live mass data for many of their museum specimens. There is disagreement about whether or not it is necessary to collect data from a sample of individuals for each species or not; the argument has been made (Legendre 1986; Hopkins 2008) that the goal is a species average, so individual variation isn't essential to the outcome. However, if the single individual measured is unusual, it can be misleading, especially if the training dataset is small. Even a small sample of 3–5 individuals can increase the robustness of estimates. Values for the sample are often averaged to generate a data point for the species. Many studies use individual masses from the specimens measured (e.g., Reynolds 2002; Hopkins 2008), but others simply use published average masses for the species (e.g., Gingerich et al. 1982; Janis 1990; Mendoza et al. 2006) or a mix of published and individual masses (e.g., Ruff 1990; Freudenthal and Martín-Suárez 2013). Published values are adequate when necessary, but the correspondence is likely to be less accurate; the individuals sampled may or may not be representative of the species average for dental measurements, and a mismatch could lead to an inaccurate placement in the allometric equation. The species data, both mass and proxy, are generally log transformed (either natural log or base 10), and then a linear regression is applied to establish the allometric relationship between the proxy and mass in log-linear space (Fig. 2.1). For some fossil taxon of interest, the proxy is then measured, and the regression equation can be used to infer mass.

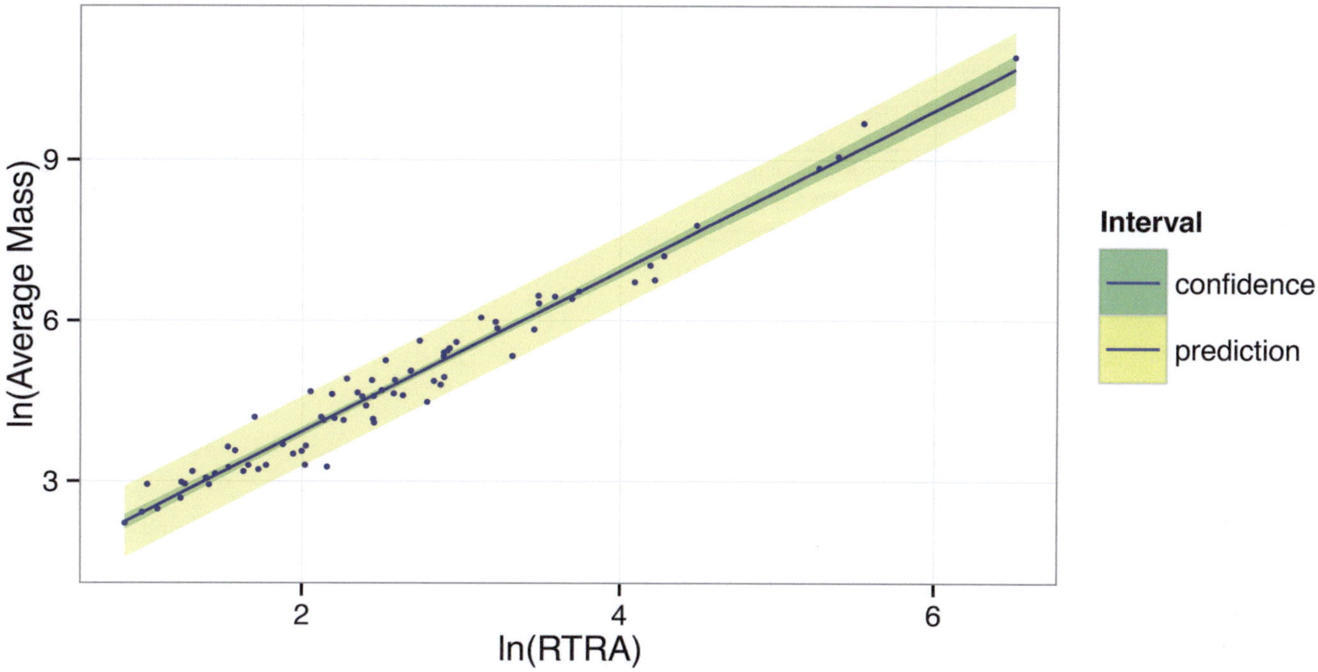

Fig. 2.1 Example regression in ln-ln space of a mass proxy (in this case RTRA, or lower cheek tooth row area) against body mass for a training dataset of 76 extant rodents. Data from Hopkins (2008). Points represent species average values. The green shading shows the 95% confidence interval for the regression line, while the yellow shading represents the 95% prediction interval

Examples of widely-used studies generating mass estimation equations are referenced in Table 2.1 with information about the taxonomic group and morphological proxy or proxies they use as well as the methods applied. The regression equations can be found in the papers, along with information about the training data used to generate them.

Dental Proxies

The most common proxy used to estimate the body mass of fossil mammals is the size of the first lower molar, m1 (Alroy 1998; Gingerich et al. 1982; Legendre 1986; Van Valkenburgh et al. 2004). This proxy is often measured in terms of area, approximated by multiplying the maximum mesio-distal length by maximum bucco-lingual width, assuming a roughly rectangular occlusal shape (Fig. 2.2). Other authors have found length alone to be adequate and, in fact, to be less sensitive to dietary differences when estimating mass of large mammals (e.g., Damuth 1990). Regressions using this proxy are available for a variety of mammalian groups (Legendre 1986) including ungulates (Janis 1990; Mendoza et al. 2006), marsupials (Gordon 2003), carnivorans (Van Valkenburgh 1990), and especially for primates (Gingerich et al. 1982; Conroy 1987; Dagosto and Terranova 1992; Vinyard and Hanna 2005; Copes and Schwartz 2010; Scott 2011). The m1 has the least variation in its fit to body mass of all the cheek teeth (Gingerich et al. 1982; Janis 1990; Gordon 2003) and therefore should be the most precise single tooth dental proxy for reconstructing body mass, though regressions are available for most of the other teeth in the dentition as well (see Table 2.1 for examples).

Because dental formula varies among mammals, another proxy uses the area of the entire postcanine tooth row for body mass reconstruction, taking advantage of the partitioning of function among the mammalian teeth (see also Evans and Pineda-Munoz 2018). As for a single tooth proxy, this value is most commonly estimated by multiplying the maximum anteroposterior length of the postcanine tooth row by the maximum width of the cheek teeth (Gould 1975; Hopkins 2008; Fig. 2.2). Note that this proxy would not be expected to work as well as single-tooth proxies in taxa with substantial curvature to the postcanine dental arcade, as occurs in some primates. Tooth row length can be used similarly (Hopkins 2008) but only in a clade for which the proportions and numbers of teeth are relatively constant. Both of these proxies are useful for taxa such as muroid or mylagaulid rodents with unusual first molars but require a fossil record that preserves undistorted tooth rows. Tooth row area and length can also be approximated from the sum of isolated cheek teeth (Freudenthal and Martín-Suárez 2013), although given mesio-distal overlap of some cheek teeth, this method will necessarily introduce some noise into the estimates of mass.

Table 2.1 Example body mass proxy studies. This is not intended to be a comprehensive list; rather, it is intended to show some of the diversity of available proxies, particularly those that are widely used or methodologically novel. Abbreviations in methods: MLR = multiple linear regression; OLS = ordinary least squares; PIC = phylogenetically independent contrasts; RMA = reduced major axis

Reference	Type of proxy	Proxy measurement(s)	Regression method	Taxonomic scope
Gingerich et al. (1982)	Dental	Area of each tooth of the tooth row; Cheek tooth row area	Ln transform, OLS	Primates
Anderson et al. (1985)	Postcranial	Total stylopodial circumference	Log_{10} transform, OLS	Ungulates
Legendre (1986)	Dental	m1 area	Ln transform, OLS	Mammals, Chiroptera, Eulipotyphla, Carnivora, Primates, Rodentia, Ungulates
Damuth (1990)	Dental, Whole body	Head-body length, Length, width, and area of individual teeth, molar row, and cheek tooth row	Log_{10} transform, OLS, MLR	Ungulates
Gingerich (1990)	Postcranial, Multivariate	Length and diameter of femur, humerus, tibia, metacarpal, and metatarsal; Ulna length	Log_{10} transform, OLS, MLR	Mammalia
Janis (1990)	Dental, Cranial	Lengths, widths, and areas of each individual postcanine tooth; Premolar row length; Molar row length; Muzzle width; Masseteric fossa length; Occipital height; Posterior skull length; Basicranial length; Anterior jaw length; Posterior jaw length; Depth of mandibular angle; Width of mandibular angle; Jaw length	Ln transform, OLS	Artiodactyla, Perissodactyla, Procaviidae (Hyracoidea), Macropodidae (Diprotodontia)
Martin (1990)	Dental	m1 area	Ln transform, OLS	Cricetinae
Ruff (1990)	Postcranial	Length and breadth of femur and tibia; Cross-sectional shape and area of femur and tibia; Proximal and distal articular surface areas of tibia and femur	Log_{10} transform, OLS	Anthropoidea (Primates)
Van Valkenburgh (1990)	Dental, Cranial, Whole body	Head-body length; Skull length; Occiput-to-orbit length; m1 area	Log_{10} transform, OLS	Carnivora, Dasyuridae (Dasyuromorphia)
Reynolds (2002)	Cranial, Postcranial	Condylobasal length; Skull length; Femur length	Log_{10} transform, OLS and RMA	Beavers and giant beaver analogs
Mendoza et al. (2006)	Cranial, Dental, Multivariate	Lower tooth row length; Premolar tooth row length; Lengths and widths of individual postcanine teeth; Anterior jaw length; Posterior jaw length; Depth of mandibular angle; Width of mandibular angle; Length of coronoid process; Length of masseteric ridge; Occipital height; Posterior length of skull;	Log transform, MLR, phylogenetic weighting	Artiodactyla, Perissodactyla, Procaviidae (Hyracoidea)

(continued)

Table 2.1 (continued)

Reference	Type of proxy	Proxy measurement(s)	Regression method	Taxonomic scope
		Depth of face under orbit; Length of paraoccipital process; Muzzle width; Palatal width; Basicranial width		
Hopkins (2008)	Dental	Tooth row area; Tooth row length	Ln transform, OLS	Rodentia, and several families therein
Millien and Bovy (2010)	Dental, Postcranial, Multivariate	Skull length; Upper cheek tooth row; Upper and lower incisor length and width; m1 length; m1 width; Humerus length; Femur length; Humerus diameter; Femur diameter	Log_{10} transform, OLS and MLR	Caviomorpha (Rodentia)
Campione and Evans (2012)	Postcranial	Total stylopodial circumference	Log_{10} transform, OLS, PIC	Mammalia, Ungulates, Carnivora, Marsupialia, Euarchonta, Glires
Freudenthal and Martín-Suárez (2013)	Dental	Tooth row area	Ln transform, OLS	Rodentia, and several families therein

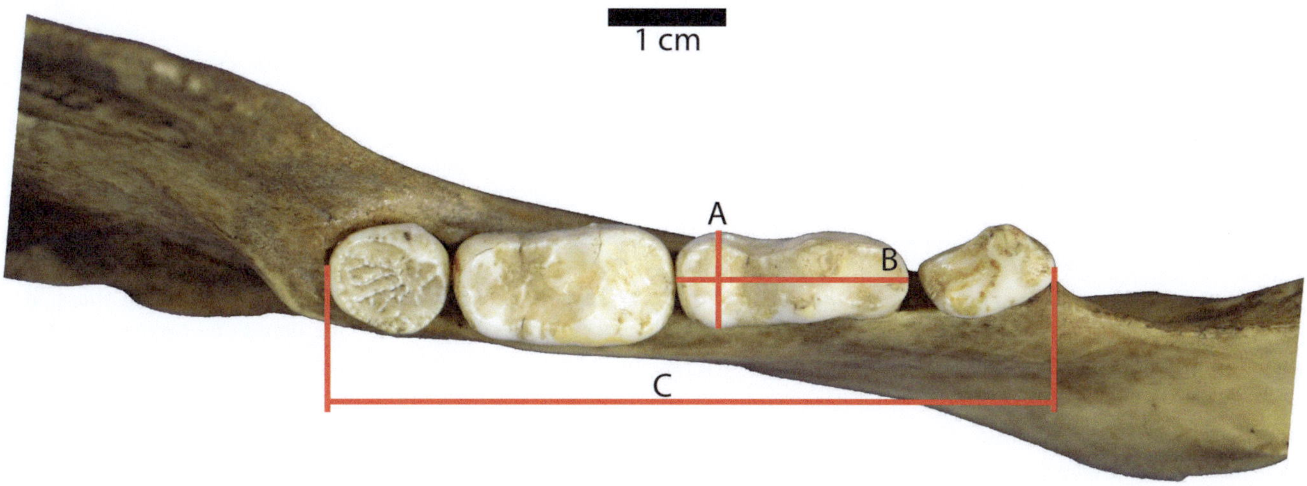

Fig. 2.2 Examples of three dental dimensions commonly used for mass regressions, illustrated on the lower tooth row of a black bear (*Ursus americanus*), University of California Museum of Vertebrate Zoology mammal specimen #8816. **A** = m1 width; **B** = m1 length; **C** = lower tooth row length

Limb Bone Proxies

Limb-based mass estimation of body mass uses a greater diversity of proxies than dental estimation. The most commonly-employed proxies use either lengths of long bones (Gingerich 1990; Reynolds 2002) or the midshaft cross-sectional dimensions of the humerus (Gingerich 1990; Millien and Bovy 2010), femur (Ruff 1990; Reynolds 2002), or both (Anderson 1985). Bone lengths are generally measured with calipers or measuring tapes (Gingerich 1990; Reynolds 2002); this proxy is useful because it is also commonly published in taxonomic descriptions and can be measured from figures in publications with reasonable accuracy. In contrast, cross-sectional dimensions (including area, perimeter, and diameter) usually require direct measurement of museum specimens; these values are seldom published except in studies of body mass estimation, and only rarely are published images sufficiently accurate to allow measurement of even diameter. In some cases, these proxies are measured digitally from bone scans (Ruff 1990), but they are also commonly measured directly with a measuring tape (Campione and Evans 2012). Limb bone cross-section as a mass proxy depends on the physical relationship between the load-bearing capacity of a cylinder and its cross-sectional area (Campione and Evans 2012), and as long as the work adequately addresses postural differences among species in the training data (Sánchez-Villagra et al. 2003; Basu et al. 2016), these proxies can be among the most universally applicable (Campione and Evans 2012). Long bone lengths require a tight morphological, functional, and phylogenetic match to provide reliable mass estimates; good analogs for locomotor habits of fossil taxa must be used and, as pointed out by Biewener (1990) and Christiansen (2002), large mammals (above 50 kg) have relatively shorter limb bones for their mass than smaller species due to a shift in the primary constraint on limb shape from compressional stresses to bending strain. Because limb-based estimates have been used most often for larger mammals (in part because identifying postcrania of large mammals is often easier), many of which are above this threshold, this change is less problematic than it would be if the proxy were applied more widely. This limitation should be an important consideration in any study reconstructing a wide range of masses.

While limb shaft dimensions are the most widely used proxy for mass estimation, the ends of the limb bones can also be used. Articular surface dimensions, often as linear or approximate area measures, can offer an alternative approach (Ruff 1988, 1990, 2003). However, their reliability as a mass proxy is somewhat contingent on posture (Ruff 1990, 2002), and at least one study (Ruff 2003) found that cortical bone distribution (obtained from CT scans of limb bones) is a better predictor of body mass that external dimensions. In a case where higher precision is required, these more labor-intensive approaches can increase the robustness of the physical relationship on which the inference depends.

Volumetric Estimation

In taxa with no good living analogs, several authors (Bates et al. 2015; Brassey et al. 2015; Basu et al. 2016) have argued that the best method of body mass estimation is to make a complete skeletal reconstruction, overlay soft tissue on that reconstruction, and estimate body mass based on volume and estimated average density. Such an approach enables reconstruction of mass in some taxa for which other proxies fail but requires exceptionally complete fossil preservation and substantial modeling effort. Hence, it is not efficient or even possible for many applications. It has, however, allowed reconstruction of mass in animals such as dinosaurs (Bates et al. 2015; Brassey et al. 2015) and the extinct giraffe relative *Sivatherium* (Basu et al. 2016) for which simpler mass proxies are suspect.

Considerations

Available fossil record: A decision about mass estimation methods begins with the skeletal elements preserved and well-identified for the taxon of interest. If postcrania cannot be confidently identified to species, dental estimates must be preferred in spite of their decreased accuracy. If the study requires reconstructing mass of a large number of distantly-related species, it may be necessary either to use multiple proxies or to choose a proxy that does not necessarily reconstruct mass with the greatest accuracy or precision but that does allow all the species of interest to be included or minimizes the violations of regression assumptions for taxa in the analysis.

Physical constraints: Regression equations only apply where the training data are representative of the species for which mass is being estimated. For example, it is challenging to reconstruct mass for taxa well outside the size range of extant relatives such as the extinct giant rodents *Phoberomys* and *Josephoartigasia* (Hopkins 2008; Rinderknecht and Blanco 2008; Millien and Bovy 2010), as the physical constraints that shape the relationship may change between extant and extinct species (Millien and Bovy 2010; Basu et al. 2016). In these cases, it can be more accurate to use more distantly related but more physically similar species as training data.

Dentition: In using dental proxies for body mass, it is necessary to consider how the dentition compares with that of the taxa from which the regression equation was derived. Taxa with dramatically different diets tend to have somewhat different constraints on the shape and scaling of the cheek teeth (Gingerich et al. 1982; Legendre 1986; Janis 1990; Scott 2011; Evans and Pineda-Munoz 2018). Regressions based on training data including taxa with similar dietary ecology are most likely to reconstruct mass accurately. There are even a few ecologies explored by mammals (for example, colonial insect specialists) that release the constraint on dentition size because teeth are not used in food processing. In such a case, dental proxies are not useful or cannot even be used for mass reconstruction.

Locomotion: In using proxies based on limb bone dimensions, we make the assumption that the limbs are loaded in the same way as those in the training data. If the distribution of weight or the limb loading in locomotion is dramatically different in study species relative to training data, mass reconstructions will be inaccurate. If the problem is simply distribution of weight (for example, a species that bears more weight on the forelimbs than its extant relatives), combining the limb cross-sectional areas can still yield accurate estimates of mass. However, if there is a difference in locomotor modes, such as a ricochetal (hopping) species whose extant relatives are all terrestrial, reconstructing mass from limbs is much more challenging; the magnitude of forces experienced during locomotion relates to mass quite differently, with the possible result that dental proxies are a better option. Even complex methods that reconstruct 3D body geometry may not be any more accurate in these cases, as reconstructing musculature and body shape requires reference to modern analogs (Carrano and Hutchinson 2002; Hutchinson and Garcia 2002; Basu et al. 2016). Limb posture also dictates which physical constraints are important for limiting limb dimensions, so mammals with unusual limb postures (for example, Mesozoic mammals) should be compared to species with analogous posture where possible.

Evolutionary relatedness: The shape of mammalian teeth, limb posture, locomotor habits, and even body density and muscle mass vary over evolutionary time and show substantial phylogenetic constraints, so it is important to consider evolutionary relatedness in choosing which of the available regression equations to use in reconstructing body mass. Studies of dental regressions, in particular, have demonstrated substantial and significant differences in regression coefficients between different taxa (Legendre 1986; Damuth and MacFadden 1990; Hopkins 2008). There are also some taxa for which these regressions are simply more variable and hence less precise in reconstructing mass. Limb bone dimensions offer a less phylogenetically-constrained proxy that has been argued to be consistent across wide swaths of terrestrial vertebrates (Campione and Evans 2012).

Statistical Methods: For those developing their own mass estimation regressions, a few statistical best practices are worth considering. Choice of regression methods, data transformation, and realistic examination of statistical errors are all very important for generating robust mass estimates for fossil mammals. Existing regressions differ widely in the degree to which these recommendations have been followed; it is worthwhile to consider what effects these methodological differences could have on the resulting mass estimates.

Regression methods: The most commonly used method for linear regression is Ordinary Least Squares (OLS) Regression. While some authors have argued (Ricker 1973, 1984; Reynolds 2002) that a different method, such as Major Axis or Reduced Major Axis Regression, is more appropriate given that they allow for equal levels of uncertainty in the x and y axes, this argument applies primarily to determining the strength and direction of allometric relationships and not to mass prediction. OLS regression assumes that the independent variable (in this case the proxy measurement) is known absolutely, while the dependent variable contains all the error in the system (Zar 2010); this is true for body mass estimation, because the measurement of the proxy is generally accurate to within a very small error, while the mass is generally much less certain, given the issues discussed previously.

Multiple regression methods have sometimes been applied in cases where multiple skeletal elements are available for a given taxon (Gingerich 1990; Mendoza et al. 2006; Millien and Bovy 2010). While it seems intuitive that adding information from multiple proxies would improve the accuracy of mass estimates, this is not always true (Kaufman and Smith 2002); unless the number of species sampled is large, there are problems in overfitting when the ratio of sampled species to predictors (that is proxy measurements) is less than 10 (Smith 2002; Kaufman and Smith 2002). Multiple regression is also not appropriate to selecting the best proxy from among a variety of candidate mass proxies, because the multicollinearity of the proxies makes such a process unreliable (Kaufman and Smith 2002). The practice of generating multiple bivariate regressions and averaging the resulting mass estimates (statistically an unjustifiable approach) does not consistently escape the problems of multiple regression and involves an additional assumption about the independence of measurements that rarely holds true (Smith 2002). In some cases, averaging proxies improves the answer where the errors of the different estimates average out (Scott 1990), but because this relies on the included proxies erring roughly equally in either direction of the true mass, it cannot be relied upon consistently. Instead, it is generally better to choose the most reliable univariate proxy based on the biology of the taxon of interest or to choose a small number of independent proxies for multivariate analysis (Kaufman and Smith 2002).

Data transformation: Size data in living systems often have a heteroscedastic distribution of variance because in larger organisms, the magnitude of variation in size is greater. Hence, the residuals from a regression line suggest that the variance is not evenly distributed, as assumed by the vast majority of linear regression methods. In the past, the solution has commonly been to log-transform the values both for the proxy measurements and the masses (see Table 2.1). This does solve the problem of heteroscedastic variance in many cases, although it is necessary to examine regression residuals to determine whether $\log_{10}$ or natural log adequately account for differences in variance (Zar 2010). In some cases, they do not perform equally well, although both methods have been applied to body mass proxy data. Unfortunately, log-transformation of the mass proxy data creates some analytical problems, a topic discussed below (see "Challenges"); there is currently no clear answer about whether log-transformation leads to values that would be more or less reliable than untransformed results (Smith 2002; Kerkhoff and Enquist 2009; Packard 2009, 2013).

Statistical uncertainty: Because these relationships between body mass and skeletal dimensions are empirically determined, they have error. In order to use these regressions meaningfully, it is necessary not only to generate a mean estimate for mass but also to make an honest statement of the uncertainty in that estimate. Sources of error that must be included in that uncertainty include measurement error in the mass proxy, uncertainty in the regression line that relates the proxy to the inferred mass, variance between individuals in a population, and uncertainty in which proxy best predicts mass. The first two points are statistical; the last two are biological. Many studies do a good job of exploring the uncertainty in mass depending on the proxy chosen (e.g., Gingerich et al. 1982; Sánchez-Villagra et al. 2003; Freudenthal and Martín-Suárez 2013). It is also reasonably common to measure a range of specimens and take a mean of that sample before inserting that value into a mass regression (e.g., Janis 1990; Ruff 1990; Delson et al. 2000), although it is somewhat less common for those regressions to include measurements of multiple individuals of the species in the training dataset. Few studies discuss the measurement error in the mass proxy, likely because the error of measurement is assumed to be proportionally much smaller than the error in the prediction equation, although this may be less true in very small taxa. While a regression analysis commonly generates a standard error of the estimate that summarizes the error across all these sources of uncertainty, the error in a mass regression is often not evenly distributed across the range of values; it is commonly greater at the extremes of the distribution. An accurate assessment of uncertainty in the mass estimate for a given taxon depends on where in the distribution its size falls, and hence requires a predictive equation of its own generated through the use of statistical error propagation. It is very rare, though, to properly propagate error through the regression equation (although see Hopkins 2008), even though the error in the x and y dimensions of the original training data are compounded by the error in the x measurement for the fossil species of interest (Zar 2010). It is worth mentioning that biological sources of error may also generate statistically non-normal distributions of variation, which causes many conventional statistical estimates of error, such as standard error, to skew uncertainty inappropriately. There is little research on the role this problem may play in generating inaccurate mass estimates, although it may be important to the persistent skew in prediction errors found by Freudenthal and Martín-Suárez (2013).

One critical dimension of statistical uncertainty that is desperately underreported is the uncertainty in an estimate produced by the equation; this uncertainty varies across the range of sizes encompassed by the training data (with lower uncertainties in the middle of the data range), but reported equations rarely give users the information needed to estimate the uncertainty in a mass prediction generated by a given equation. This is a critical parameter to mass estimates that cannot be replaced by a simple average prediction error value over the whole training dataset.

Empirical uncertainty: Given the challenges unique to the different systems for which these regressions have been developed, it is vital for a regression analysis to generate meaningful error estimates. One of the most valuable strategies is jackknifing (Wu 1986), in which each of the data points in the training dataset is removed one at a time and then predicted using the remaining data. Comparing these predictions with the observed data will make it possible to assess, within the data used for training the regression, how much error is likely in mass estimates and whether there are any systematic biases in the regression result. This helps with both the log-averaging problem and the problem of error estimation presented above, though it is an estimate relevant only to the breadth of the training data, and hence still limited by the degree to which those data are representative of the fossil taxa being reconstructed.

Strengths

Mass is correlated to many other aspects of a mammal's biology, possibly more so than any other biological variable, making it a vitally important aspect of the biology of an extinct mammal. Mass has demonstrated correlations with diet (Price and Hopkins 2015), habitat use (Peters 1983), rates of evolution (Martin and Palumbi 1993; Gillooly et al. 2005; Liow et al. 2008), metabolic rates (McNab 1988; West et al. 1997; Whittaker 1999), life history strategies (Sibly and Brown 2007; Liow et al. 2008, 2009), home range size (McNab 1963; LaBarbera 1989), and a variety of other biological variables that are much more difficult to infer from the fossil record than is mass. Hence, a robust estimate of body size has the capacity to predict other aspects of biology that may not otherwise be available.

Many commonly-used skeletal proxies are very highly correlated with mass, with correlation coefficients generally between 0.95 and 1, yielding surprisingly precise estimates, at least of species average mass values. Mass can be reconstructed from a variety of proxies, depending on the available material, because overall body size is one of the most important predictors of the size of individual skeletal elements. This variety of potential proxies makes it possible to estimate mass for almost any fossil mammal, even those for which preserved remains are extremely rare or fragmentary or those that have evolved unusual limb or tooth morphologies (Biknevicius 1999; Hopkins 2008; Pyenson and Sponberg 2011). Because these methods are non-destructive, they allow reconstruction of a critical ecological dimension even for a species known from a single isolated tooth (although estimates are obviously more robust for a larger sample).

Mass data are also frequently available for extant mammals in museum collections, in ecological publications, and in a variety of databases, making it easy to compare fossil ecological assemblages with modern ones. One of the classic papers in mammalian body mass reconstruction (Legendre 1986) made use of this phenomenon to reconstruct habitat types from body mass distributions among fossil mammal communities, although this inference remains somewhat controversial (Rodriguez 1999). There are certainly some significant ecological patterns in body mass distributions that can be valuable for understanding large-scale ecological processes.

Biases and Shortcomings

While these methods of body mass estimation are powerful, there are also some noteworthy problems with reconstructing body mass in fossil mammals that must be acknowledged. First, we are generally limited to proxies commonly available from the fossil record. While there are essentially limitless possibilities for mass proxies, many of those that may be some of the best predictors of mass may not apply usefully to many taxa of fossil mammals because the elements required are rarely preserved. For example, the breadth of the occipital condyles offers an excellent mass proxy that may be consistent across a wide swath of mammalian taxa with minimal variation even in very different ecologies, sizes, or phylogenetic lineages (Martin 1980; Pyenson and Sponberg 2011). However, because the majority of mammalian taxa are identified using dental features, reconstructing mass with this proxy requires exceptional preservation of an undistorted basicranium with associated dentition, so this proxy is less useful for many mammalian species. Alroy (2012) showed that the size of the glenoid fossa of the scapula is one of the best predictors of mass in mammals; sadly, scapulae are rarely identified to species and almost never preserved intact, so these proxies are of very limited utility in estimating fossil mammal masses in spite of their accuracy and precision. While they may provide a useful check on other proxies in cases of exceptional preservation, they are not applicable to the majority of fossil mammals.

A second shortcoming, as previously mentioned, arises when proxies disagree in their estimate of body mass (Hopkins 2008; Millien and Bovy 2010; Basu et al. 2016). Determining which, if any, of the estimates are correct is a challenge that requires additional study of the paleobiology of the species of interest. The discussion in this chapter of the processes underlying different proxies should provide some basis for such a study, but in the end, this problem makes the simple estimation of this fundamental ecological parameter much more complex in some species.

Phylogeny can bias results in a variety of ways. First, the allometric relationships between various skeletal dimensions and overall size are generally strongest within narrow clades and much less consistent as species diverge evolutionarily, although Freudenthal and Martín-Suárez (2013) argue that these differences are not significant. Within families of mammals and even within orders, regressions are often robust and quite accurate (Legendre 1986; Hopkins 2008; Freudenthal and Martín-Suárez 2013), but morphological constraints (such as dental formulae, locomotor modes, etc.) can change enough between families and orders to render mass estimates from these proxies less reliable (Legendre 1986; Hopkins 2008; Millien and Bovy 2010). In addition, there are some narrow clades for which particular proxies or even all proxies may be less reliable. For example, muroid rodents differ from most other rodents in having lost the fourth premolar in the upper and lower dentition. As a result, the first molar, to which many mass estimates are tied, might be expected to be more variable, as it has become the anteriormost cheek tooth; this position is generally expected to be released from stabilizing selection by the absence of an articulating tooth on the mesial side (Gingerich 1974). Indeed, muroid mass is much less reliably predicted by m1 dimensions than other rodents (Lindsay 1988; Morgan et al. 1995). However, this expected release of constraint on m1 is not consistent with the findings of Freudenthal and Martín-Suárez (2013) wherein variation in muroid teeth was lower in m1 than m2, nor does it explain why the tooth row length and area were also more variable in their relationship to mass in muroids than in any other rodent clade (Hopkins 2008). It appears that the relationship between dental dimensions and mass is simply more variable in this clade, suggesting that this correlation may also be a property that varies phylogenetically. Glirids also present challenges to mass prediction via dental dimensions. Freudenthal and Martín-Suárez (2013) present the hypothesis that differences in digestive efficiency may explain confusing results of fossil and extant glirid body mass estimates. The dentitions and limbs of whales are also poor predictors of mass (Pyenson and Sponberg 2011), likely the result of a similar release of constraint in these aquatic creatures. These issues become particularly important in reconstructing mass for mammals well outside of extant orders and families. A better understanding of the reasons for variation in the relationship between mass and various size proxies will make estimates of mass more robust.

One of the more controversial problems in mass estimation is the log-averaging problem. In short, the use of log transformation to reduce heteroscedasticity in the variance of training data leads to consistent under-prediction of mass by proxy data (Smith 1993, 2002). However, the distribution of error in dental and skeletal mass proxy data is multiplicative (Huxley 1932; McMahon and Bonner 1983; Gingerich 2000; Kerkhoff and Enquist 2009; Xiao et al. 2011; Glazier 2013), and hence requires some kind of transformation to correct for the heteroscedasticity of this error. There are some mathematical solutions to the problem (Smith 1993) that approximate a correction and work effectively in some cases (Myers 2001), but the bias cannot be completely and consistently removed as long as regressions are performed on log-transformed data. The bias is relatively small in magnitude, perhaps 5–15% for most regressions, but gets larger as the range of sizes included in the regression equation increases or as the prediction error increases. Freudenthal and Martín-Suárez (2015) considered the problem when estimating rodent head-body length and found that predictions from log-transformed data were low at the high and low extremes and high in the center of the distribution, from which they concluded that log transformation should be avoided when possible. This problem implies that statistics that are robust to non-normal distributions may be necessary in some cases. Packard (2009, 2013) has argued that regression algorithms are now robust enough that the data transformation may not be necessary and that a power law regression can be run on untransformed data to study such allometric relationships. It remains to be seen whether this solution is applicable to mass regressions; in general, the non-normality of the data in most mass regressions is large enough that standard statistical practice (Zar 2010) would demand data transformation prior to using parametric regression methods. Kerkhoff and Enquist (2009) argue that the geometric averaging inherent to regressions on log data actually yields a more biologically realistic description of variation and that log-transformed data are actually more appropriate for allometric regressions. The question of whether arithmetic regression can be applied to allometric relationships such as this one has not yet been adequately tested in body mass proxy regressions, so most workers continue to apply log transformation to body mass regressions (White et al. 2009; Millien and Bovy 2010; Field et al. 2013; Freudenthal and Martín-Suárez 2013; Brassey et al. 2015).

Examples of Applications

Size Evolution over Deep Time

Our understanding of mass evolution across the mammalian lineage relies heavily on skeletal proxies for mass reconstruction. Smith et al. (2010) used mass estimates for fossil terrestrial mammal species to find the largest species in each mammalian order through the Cenozoic. A study so broad-reaching required application of a variety of dental and limb-based mass proxies (depending on the identified material available for each of the described species) although the effort admittedly was not comprehensive, focusing only on the largest (in size, not diversity) clades in each order. While the use of varied proxy data, including some for which the consistency of the proxies was not assessed, is less than optimal, this study aimed for a broad-brush picture of maximal size by order and for mammals as a whole through time. As a result, the differences in precision and accuracy between the varied proxies were not problematic for the purpose to which the size estimates were applied. Roth (1990, 1992) solved the problem of intercomparability of proxies and of increasing error with size by simply comparing the raw proxy values in her studies of dwarfing in fossil elephants. Gould and MacFadden (2004) took advantage of the similar constraints on mass proxies in fossil horses to use a single proxy, m1 length, to reconstruct all the fossil genera in their study of Cope's Rule, dwarfism, and nanism in the family Equidae. Finarelli and Flynn's (2006) study of ancestral body mass in carnivorans used m1 area to reconstruct masses in fossil carnivores, demonstrating the important contribution of fossil data to accurately reconstructing the ecological history of a lineage.

Reconciling Multiple Mass Proxies

The largest extant rodent, the capybara, is dwarfed by the extinct rodents *Phoberomys* and *Josephoartigasia*. The magnitude of this size difference, however, has been a bit difficult to tie down, in part because different proxies yield different answers. *Josephoartigasia* is known primarily from craniodental remains, but the partial skeleton known for *Phoberomys pattersoni* reveal the complexity of reconstructing body sizes well outside the range encompassed by living analogs. Sánchez-Villagra and colleagues (2003) used diameters of the humerus and femur to infer mass of a specimen of *P. pattersoni*. Regressions yielded two very different answers, 436 kg for the humerus and 741 kg for the femur. The authors chose the larger size reconstructed by the femur, arguing that posture in *Phoberomys* was likely unusual relative to other caviomorphs and the femur was expected to be weight-bearing. Hopkins (2008) obtained a much smaller mass estimate (between 164 and 334 kg) from dental dimensions in a study developing tooth row dimensions as a mass proxy for rodents. Millien and Bovy (2010) reconciled the issue by comparing multiple proxies and the allometries of the proxy values relative to other skeletal measurements and by considering other information about the biology of this species. They came to the conclusion that

Phoberomys had very robust limb bones relative to the rest of its skeleton, perhaps as a result of the important postural and locomotor differences necessitated by growing rodent skeletal morphology to this extraordinary size; they concluded that the dental estimates were likely more accurate, yielding a mass estimate around 200–300 kg rather than the 741 kg proposed based on femur diameter.

Finding the Right Body Mass Regression

Freudenthal and Martín-Suárez (2013) explored some of the challenges in reconstructing body mass in extinct rodents from teeth. Using a large database of dental measurements and masses from fossil and extant rodents of a variety of lineages, the authors sought to understand which dental proxies most effectively predict mass in rodents, given that the rodent fossil record is almost exclusively dental and cranial, with few identified postcrania. The authors explored the issues around changing dental formula, comparing m1 and tooth row length as predictors of mass in rodents with and without a fourth premolar. They found that tooth row length is a strong predictor of body mass, even when calculated from the summed lengths of isolated teeth, but that there is a persistent problem with minor discrepancies in the mass reconstructed from upper and lower tooth rows. The authors explored various methods, including phylogenetic differences in the performance of the different proxies and the role of log averaging, and generated a set of regression equations that are useful for reconstructing body mass in a variety of rodent species.

Community Paleoecology

Mass proxies are commonly used in ecological diversity analyses (see Kovarovic et al. 2018) and sometimes used to reconstruct size distributions in fossil mammal communities. One technique commonly used in fossil assemblages is the cenogram (Valverde 1964), a rank-order plot of body sizes within a community sometimes inferred to reflect habitat (Legendre 1986; Travouillon and Legendre 2009) on the basis of the presence and size of breaks in slope or a gap in the size distribution. For such an application, masses are generally reconstructed using a single skeletal proxy, although the regressions (e.g., Legendre 1986) may be clade-specific. In most cases, in order to maximize the number of taxa from the fossil assemblage included, most workers have used dental proxies for mass. The cenogram method has been widely applied (Legendre 1989; Morgan et al. 1995; Croft 2001; Costeur 2004; Travouillon et al. 2009), although interpretation often requires awareness of the phylogenetic limitations of the regional species pool (e.g., Croft 2001) and may be more useful as a descriptive technique than an interpretive one (Maas and Krause 1994; Rodriguez 1999).

Future Prospects

Increasing computational power in recent years has led to an increase in the sophistication of body mass reconstructions in fossil animals. Certainly, accurate volumetric estimation would not be practical without high-powered computing (Bates et al. 2015; Brassey et al. 2015; Basu et al. 2016), although early versions of these approaches used much simpler methods with physical models (Alexander 1985). It is likely that the increasingly widespread availability of such computing power as well as the much wider access to 3D scanning will lead to more such reconstructions for species for which there are no good living analogs, either phylogenetically, morphologically, or ecologically. For the more divergent species from the fossil record such as *Sivatherium* (Basu et al. 2016), this is likely the best way to achieve a meaningful estimate of size. For most species, however, allometric equations for single proxies will remain most practical. Even these estimates can be improved by current computing capabilities, with greater access to 3D and CT scanning making it increasingly inexpensive and practical to add proxies such as articular dimensions and limb bone cortical volume to the efforts to predict mass in a greater range of species. Nonlinear regression algorithms are more sophisticated now than they were in the past, making it possible to test suggested unlogged allometric regressions (Packard 2013) to address the controversy in the log-averaging problem. It remains to be seen whether that problem will prove tractable given the substantial problems presented by heteroscedastic variance over a wide range of sizes, and whether the problem will actually translate to inaccurately-estimated masses (Kerkhoff and Enquist 2009). Finally, the growth of phylogenetic comparative methods in the last decade or two (Pennell and Harmon 2013) demands a more sophisticated analysis of the interaction between physical and physiological constraints and evolutionary history in holding these correlations together; a recent analysis of cetacean body size proxies (Pyenson and Sponberg 2011) suggests that the inclusion of phylogenetic information enhances the precision of body size estimates significantly. Given the abundance of data available and the increasingly accurate and detailed phylogenies of living species, the problem is overdue for a phylogenetic comparative treatment. A recent analysis suggests physical constraints are the primary driver of limb bone proxies (Campione and Evans 2012), but dental proxies will likely prove more complex, given the genetic constraints on dental morphology (Kangas et al. 2004; Evans et al. 2007; Kavanagh et al. 2007). A more nuanced understanding of the underlying processes can only benefit our ability to reconstruct accurate masses for extinct species.

Acknowledgments Thanks are due to Darin Croft, Scott Simpson and Denise Su for organizing the mammal paleoecology symposium that led to this volume, inspiring numerous exciting and productive conversations. I also owe thanks to Samantha Price for endless patience in teaching me phylogenetic comparative methods. Thanks to undergraduate and graduate students past and present for questioning assumptions, inspiring research, and actually getting science done. In particular, I will be ever grateful to John Orcutt for deep discussions of body size evolution in the course of his dissertation research. Finally, thanks to Edward Davis for years of discussion of body mass reconstruction, statistics of fossil samples, and regression analysis.

References

Alexander, R. M. (1985). Mechanics of posture and gait of some large dinosaurs. *Zoological Journal of the Linnean Society, 83,* 1–25.

Alroy, J. (1998). Cope's rule and the dynamics of body mass evolution in North American fossil mammals. *Science, 280,* 731–734.

Alroy, J. (2012). Simple equations for estimating body mass in mammals (and dinosaurs). *Journal of Vertebrate Paleontology, SVP Program and Abstracts Book, 2012,* 55–56.

Anderson, J. F., Hall-Martin, A., & Russell, D. A. (1985). Long-bone circumference and weight in mammals, birds and dinosaurs. *Journal of Zoology, 207,* 53–61.

Basu, C., Falkingham, P. L., & Hutchinson, J. R. (2016). The extinct, giant giraffid *Sivatherium giganteum*: skeletal reconstruction and body mass estimation. *Biology Letters, 12,* 20150940.

Bates, K. T., Falkingham, P. L., Macaulay, S., Brassey, C., & Maidment, S. C. R. (2015). Downsizing a giant: re-evaluating *Dreadnoughtus* body mass. *Biology Letters, 11,* 20150215.

Biewener, A. A. (1990). Biomechanics of mammalian terrestrial locomotion. *Science, 250,* 1097–1103.

Biknevicius, A. R. (1999). Body mass estimation in armoured mammals: cautions and encouragements for the use of parameters from the appendicular skeleton. *Journal of Zoology, 248,* 179–187.

Blackburn, T. M., & Gaston, K. J. (1994). Animal body size distributions: patterns, mechanisms and implications. *Trends in Ecology & Evolution, 9,* 471–474.

Brassey, C. A., Maidment, S. C. R., & Barrett, P. M. (2015). Body mass estimates of an exceptionally complete *Stegosaurus* (Ornithischia: Thyreophora): comparing volumetric and linear bivariate mass estimation methods. *Biology Letters, 11,* 20140984.

Brown, J. H., Gillooly, J. F., Allen, A. P., Savage, V. M., & West, G. B. (2004). Toward a metabolic theory of ecology. *Ecology, 85,* 1771–1789.

Calder, W. A. I. (1984). *Size, function, and life history*. Cambridge, MA: Harvard University Press.

Campione, N. E., & Evans, D. C. (2012). A universal scaling relationship between body mass and proximal limb bone dimensions in quadrupedal terrestrial tetrapods. *BMC Biology, 10,* 60.

Carrano, M. T., & Hutchinson, J. R. (2002). Pelvic and hindlimb musculature of *Tyrannosaurus rex* (Dinosauria: Theropoda). *Journal of Morphology, 253,* 207–228.

Christiansen, P. (2002). Mass allometry of the appendicular skeleton in terrestrial mammals. *Journal of Morphology, 251,* 195–209.

Conroy, G. C. (1987). Problems of body-weight estimation in fossil primates. *International Journal of Primatology, 8,* 115–137.

Copes, L. E., & Schwartz, G. T. (2010). The scale of it all: postcanine tooth size, the taxon-level effect, and the universality of Gould's scaling law. *Paleobiology, 36,* 188–203.

Costeur, L. (2004). Cenogram analysis of the Rudabánya mammalian community: palaeoenvironmental interpretations. *Palaeontographia Itallica, 90,* 303–307.

Croft, D. A. (2001). Cenozoic environmental change in South America as indicated by mammalian body size distributions (cenograms). *Diversity and Distributions, 7,* 271–287.

Dagosto, M., & Terranova, C. J. (1992). Estimating the body size of Eocene primates: a comparison of results from dental and postcranial variables. *International Journal of Primatology, 13,* 307–344.

Damuth, J. (1990). Problems in estimating body masses of archaic ungulates using dental measurements. In J. Damuth & B. J. MacFadden (Eds.), *Body size in mammalian paleobiology: Estimation and biological implications* (pp. 229–253). Cambridge: Cambridge University Press.

Damuth, J., & MacFadden, B. J. (Eds.). (1990). *Body size in mammalian paleobiology: Estimation and biological implications*. Cambridge: Cambridge University Press.

Delson, E., Terranova, C. J., Jungers, W. L., Sargis, E. J., Jablonski, N. G., & Dechow, P. C. (2000). Body mass in Cercopithecidae (Primates, Mammalia): estimation and scaling in extinct and extant taxa. *Anthropological Papers of the American Museum of Natural History, 83,* 1–159.

Egi, N. (2001). Body mass estimates in extinct mammals from limb bone dimensions: the case of North American hyaenodontids. *Palaeontology, 44,* 497–528.

Eisenberg, J. F. (1981). *The mammalian radiations*. Chicago, IL: University of Chicago Press.

Evans, A. R., & Pineda-Munoz, S. (2018). Inferring mammal dietary ecology from dental morphology. In D. A. Croft, D. F. Su & S. W. Simpson (Eds.), *Methods in paleoecology: Reconstructing Cenozoic terrestrial environments and ecological communities* (pp. 37–51). Cham: Springer.

Evans, A. R., Wilson, G. P., Fortelius, M., & Jernvall, J. (2007). High-level similarity of dentitions in carnivorans and rodents. *Nature, 445,* 78–81.

Field, D. J., Lynner, C., Brown, C., & Darroch, S. A. F. (2013). Skeletal correlates for body mass estimation in modern and fossil flying birds. *PLoS ONE, 8,* e82000.

Finarelli, J. A., & Flynn, J. J. (2006). Ancestral state reconstruction of body size in the Caniformia (Carnivora, Mammalia): the effects of incorporating data from the fossil record. *Systematic Biology, 55,* 301–313.

Fortelius, M. (1985). Ungulate cheek teeth: developmental, functional, and evolutionary interrelations. *Acta Zoologica Fennica, 180,* 1–76.

Fortelius, M. (1990). Problems with using fossil teeth to estimate body sizes of extinct mammals. In J. Damuth & B. J. Macfadden (Eds.), *Body size in mammalian paleobiology: Estimation and biological implications* (pp. 207–228). Cambridge: Cambridge University Press.

Freudenthal, M., & Martín-Suárez, E. (2013). Estimating body mass of fossil rodents. *Scripta Geologica, 14,* 1–130.

Freudenthal, M., & Martín-Suárez, E. (2015). Estimating head and body length in fossil rodents. *Scripta Geologica, 149,* 1–158.

Gillooly, J. F., Allen, A. P., West, G. B., & Brown, J. H. (2005). The rate of DNA evolution: effects of body size and temperature on the molecular clock. *Proceedings of the National Academy of Sciences, USA, 102,* 140–145.

Gingerich, P. D. (1974). Size variability of the teeth in living mammals and the diagnosis of closely related sympatric fossil species. *Journal of Paleontology, 48,* 895–903.

Gingerich, P. D. (1990). Prediction of body mass in mammalian species from long bone lengths and diameters. *Contributions from the Museum of Paleontology, the University of Michigan, 28,* 79–92.

Gingerich, P. D. (2000). Arithmetic or geometric normality of biological variation: an empirical test of theory. *Journal of Theoretical Biology, 204,* 201–221.

Gingerich, P. D., Smith, B. H., & Rosenberg, K. (1982). Allometric scaling in the dentition of primates and prediction of body weight from tooth size in fossils. *American Journal of Physical Anthropology, 58,* 81–100.

Glazier, D. S. (2013). Log-transformation is useful for examining proportional relationships in allometric scaling. *Journal of Theoretical Biology, 334,* 200–203.

Gordon, C. L. (2003). A first look at estimating body size in dentally conservative marsupials. *Journal of Mammalian Evolution, 10,* 1–21.

Gould, G. C., & MacFadden, B. J. (2004). Gigantism, dwarfism, and Cope's Rule: "Nothing in evolution makes sense without a phylogeny". *Bulletin of the American Museum of Natural History, 285,* 219–237.

Gould, S. J. (1975). On the scaling of tooth size in mammals. *American Zoologist, 15,* 351–362.

Hopkins, S. S. B. (2008). Reassessing the mass of exceptionally large rodents using toothrow length and area as proxies for body mass. *Journal of Mammalogy, 89,* 232–243.

Hutchinson, J. R., & Garcia, M. (2002). *Tyrannosaurus* was not a fast runner. *Nature, 415,* 1018–1021.

Huxley, J. S. (1932). *Problems of relative growth*. London: Methuen & Co., Ltd.

Iskjaer, C., Slade, N. A., Childs, J. E., Glass, G. E., & Korch, G. W. (1989). Body mass as a measure of body size in small mammals. *Journal of Mammalogy, 70,* 662–667.

Janis, C. M. (1990). Correlation of cranial and dental variables with body size in ungulates and macropodoids. In J. Damuth & B. J. MacFadden (Eds.), *Body size in mammalian paleobiology: Estimation and biological implications* (pp. 255–300). Cambridge: Cambridge University Press.

Kangas, A. T., Evans, A. R., Thesleff, I., & Jernvall, J. (2004). Nonindependence of mammalian dental characters. *Nature, 432,* 211–214.

Kaufman, J. A., & Smith, R. J. (2002). Statistical issues in the prediction of body mass for Pleistocene canids. *Lethaia, 35,* 32–34.

Kavanagh, K. D., Evans, A. R., & Jernvall, J. (2007). Predicting evolutionary patterns of mammalian teeth from development. *Nature, 449,* 427–432.

Kay, R. F., & Ungar, P. S. (1997). Dental evidence for diet in some Miocene catarrhines with comments on the effects of phylogeny on the interpretation of adaptation. In D. R. Begun, C. V. Ward & M. D. Rose (Eds.), *Function, phylogeny, and fossils: Miocene hominoid evolution and adaptations* (pp. 131–151). Dordrecht: Springer.

Kerkhoff, A. J., & Enquist, B. J. (2009). Multiplicative by nature: why logarithmic transformation is necessary in allometry. *Journal of Theoretical Biology, 257,* 519–521.

Kovarovic, K., Su, D. F., & Lintulaakso, K. (2018). Mammal community structure analysis. In D. A. Croft, D. F. Su & S. W. Simpson (Eds.), *Methods in paleoecology: Reconstructing Cenozoic terrestrial environments and ecological communities* (pp. 349–370). Cham: Springer.

LaBarbera, M. (1989). Analyzing body size as a factor in ecology and evolution. *Annual Review of Ecology and Systematics, 20,* 97–117.

Legendre, S. (1986). Analysis of mammalian communities from the late Eocene and Oligocene of southern France. *Palaeovertebrata, 16,* 191–212.

Legendre, S. (1989). Les communautés de mammifères du Paléogène (Eocène supérieur et Oligocène) d'Europe occidentale : Structures, milieux et évolution. *Münchner Geowissenschaftliche Abhandlungen A, 16,* 1–110.

Lindsay, E. H. (1988). Cricetid rodents from Siwalik deposits near Chinji Village. Part 1: Megacricetodontinae. Myocricetodontinae and Dendromurinae. *Palaeovertebrata, 18,* 95–154.

Lindstedt, S. L., & Boyce, M. S. (1985). Seasonality, fasting endurance, and body size in mammals. *The American Naturalist, 125,* 873–878.

Liow, L. H., Fortelius, M., Bingham, E., Lintulaakso, K., Mannila, H., Flynn, L., et al. (2008). Higher origination and extinction rates in larger mammals. *Proceedings of the National Academy of Sciences, USA, 105,* 6097–6102.

Liow, L. H., Fortelius, M., Lintulaakso, K., Mannila, H., & Stenseth, N. C. (2009). Lower extinction risk in sleep-or-hide mammals. *The American Naturalist, 173,* 264–272.

Lockyer, C. (1976). Body weights of some species of large whales. *Journal du Conseil / Conseil Permanent International pour l'Exploration de la Mer, 36,* 259–273.

Maas, M. C., & Krause, D. W. (1994). Mammalian turnover and community structure in the Paleocene of North America. *Historical Biology, 8,* 91–128.

Martin, A. P., & Palumbi, S. R. (1993). Body size, metabolic rate, generation time, and the molecular clock. *Proceedings of the National Academy of Sciences, USA, 90,* 4087–4091.

Martin, R. A. (1980). Body mass and basal metabolism of extinct mammals. *Comparative Biochemistry and Physiology – Part A. Physiology, 66,* 307–314.

Martin, R. A. (1990). Estimating body mass and correlated variables in extinct mammals: travels in the fourth dimension. In J. Damuth & B. J. MacFadden (Eds.), *Body size in mammalian paleobiology: Estimation and biological implications* (pp. 49–68). Cambridge: Cambridge University Press.

McMahon, T., & Bonner, J. T. (1983). *On size and life*. New York: Scientific American Books – W. H. Freeman and Co.

McNab, B. K. (1963). Bioenergetics and the determination of home range size. *The American Naturalist, 97,* 133–140.

McNab, B. K. (1988). Complications inherent in scaling the basal rate of metabolism in mammals. *The Quarterly Review of Biology, 63,* 25–54.

Mendoza, M., Janis, C. M., & Palmqvist, P. (2006). Estimating the body mass of extinct ungulates: a study on the use of multiple regression. *Journal of Zoology, 270,* 90–101.

Millien, V., & Bovy, H. (2010). When teeth and bones disagree: body mass estimation of a giant extinct rodent. *Journal of Mammalogy, 91,* 11–18.

Morgan, M. E., Badgley, C., Gunnell, G. F., Gingerich, P. D., Kappelman, J. W., & Maas, M. C. (1995). Comparative paleoecology of Paleogene and Neogene mammalian faunas: body-size structure. *Palaeogeography, Palaeoclimatology, Palaeoecology, 115,* 287–315.

Myers, T. J. (2001). Prediction of marsupial body mass. *Australian Journal of Zoology, 49,* 99–118.

Packard, G. C. (2009). On the use of logarithmic transformations in allometric analyses. *Journal of Theoretical Biology, 257,* 515–518.

Packard, G. C. (2013). Is logarithmic transformation necessary in allometry? *Biological Journal of the Linnean Society, 109,* 476–486.

Pennell, M. W., & Harmon, L. J. (2013). An integrative view of phylogenetic comparative methods: connections to population genetics, community ecology, and paleobiology. *Annals of the New York Academy of Sciences, 1289,* 90–105.

Peters, R. H. (1983). *The ecological implications of body size*. Cambridge: Cambridge University Press.

Price, S. A., & Hopkins, S. S. B. (2015). The macroevolutionary relationship between diet and body mass across mammals. *Biological Journal of the Linnean Society, 115,* 173–184.

Pyenson, N. D., & Sponberg, S. N. (2011). Reconstructing body size in extinct crown Cetacea (Neoceti) using allometry, phylogenetic methods and tests from the fossil record. *Journal of Mammalian Evolution, 18,* 269–288.

Rafferty, K. L., Walker, A., Ruff, C. B., Rose, M. D., & Andrews, P. J. (1995). Postcranial estimates of body weight in *Proconsul*, with a note on a distal tibia of *P. major* from Napak, Uganda. *American Journal of Physical Anthropology, 97,* 391–402.

Reynolds, P. S. (2002). How big is a giant? The importance of method in estimating body size of extinct mammals. *Journal of Mammalogy, 83,* 321–332.

Ricker, W. E. (1973). Linear regression in fishery research. *Journal of the Fisheries Research Board of Canada, 30,* 409–434.

Ricker, W. E. (1984). Computation and uses of central trend lines. *Canadian Journal of Zoology, 62,* 1897–1905.

Rinderknecht, A., & Blanco, R. E. (2008). The largest fossil rodent. *Proceedings of the Royal Society B: Biological Sciences, 275,* 923–928.

Rodriguez, J. (1999). Use of cenograms in mammalian palaeoecology. A critical review. *Lethaia, 32,* 331–347.

Roth, V. L. (1990). Insular dwarf elephants: a case study in body mass estimation and ecological inference. In J. Damuth & B. J. Macfadden (Eds.), *Body size in mammalian paleobiology: Estimation and biological implications* (pp. 151–180). Cambridge: Cambridge University Press.

Roth, V. L. (1992). Inferences from allometry and fossils: dwarfing of elephants on islands. *Oxford Surveys in Evolutionary Biology, 8,* 259–288.

Ruff, C. (1988). Hindlimb articular surface allometry in Hominoidea and *Macaca*, with comparisons to diaphyseal scaling. *Journal of Human Evolution, 17,* 687–714.

Ruff, C. B. (1990). Body mass and hindlimb bone cross-sectional and articular dimensions in anthropoid primates. In J. Damuth & B. J. Macfadden (Eds.), *Body size in mammalian paleobiology: Estimation and biological implications* (pp. 119–150). Cambridge: Cambridge University Press.

Ruff, C. B. (2002). Long bone articular and diaphyseal structure in old world monkeys and apes. I: locomotor effects. *American Journal of Physical Anthropology, 119,* 305–342.

Ruff, C. B. (2003). Long bone articular and diaphyseal structure in Old World monkeys and apes. II: estimation of body mass. *American Journal of Physical Anthropology, 120,* 16–37.

Ruff, C. B., Scott, W. W., & Liu, A. Y.-C. (1991). Articular and diaphyseal remodeling of the proximal femur with changes in body mass in adults. *American Journal of Physical Anthropology, 86,* 397–413.

Sánchez-Villagra, M. R., Aguilera, O., & Horovitz, I. (2003). The anatomy of the world's largest extinct rodent. *Science, 301,* 1708–1710.

Schmidt-Nielsen, K. (1984). *Scaling: Why is animal size so important?* Cambridge: Cambridge University Press.

Schulte-Hostedde, A. I., Zinner, B., Millar, J. S., & Hickling, G. J. (2005). Restitution of mass-size residuals: validating body condition indices. *Ecology, 86,* 155–163.

Scott, J. E. (2011). Folivory, frugivory, and postcanine size in the cercopithecoidea revisited. *American Journal of Physical Anthropology, 146,* 20–27.

Scott, K. M. (1983). Prediction of body weight of fossil Artiodactyla. *Zoological Journal of the Linnean Society, 77,* 199–215.

Scott, K. M. (1990). Postcranial dimensions of ungulates as predictors of body mass. In J. Damuth & B. J. Macfadden (Eds.), *Body size in mammalian paleobiology: Estimation and biological implications* (pp. 301–336). Cambridge: Cambridge University Press.

Sibly, R. M., & Brown, J. H. (2007). Effects of body size and lifestyle on evolution of mammal life histories. *Proceedings of the National Academy of Sciences, USA, 104,* 17707–17712.

Smith, F. A., Boyer, A. G., Brown, J. H., Costa, D. P., Dayan, T., Ernest, S. K. M., et al. (2010). The evolution of maximum body size of terrestrial mammals. *Science, 330,* 1216–1219.

Smith, R. J. (1993). Logarithmic transformation bias in allometry. *American Journal of Physical Anthropology, 90,* 215–228.

Smith, R. J. (2002). Estimation of body mass in paleontology. *Journal of Human Evolution, 43,* 271–287.

Toigo, C., Gaillard, J. M., Van Laere, G., Hewison, M., & Morellet, N. (2006). How does environmental variation influence body mass, body size, and body condition? Roe deer as a case study. *Ecography, 29,* 301–308.

Travouillon, K. J., & Legendre, S. (2009). Using cenograms to investigate gaps in mammalian body mass distributions in Australian mammals. *Palaeogeography, Palaeoclimatology, Palaeoecology, 272,* 69–84.

Travouillon, K. J., Legendre, S., Archer, M., & Hand, S. J. (2009). Palaeoecological analyses of Riversleigh's Oligo-Miocene sites: implications for Oligo-Miocene climate change in Australia. *Palaeogeography, Palaeoclimatology, Palaeoecology, 276,* 24–37.

Valverde, J. A. (1964). Remarques sur la structure et l'évolution des communautés de vertébrés terrestres. *La Terre et La Vie, 111,* 121–154.

Van Valkenburgh, B. (1990). Skeletal and dental predictors of body mass in predators. In J. Damuth & B. J. Macfadden (Eds.), *Body size in mammalian paleobiology: Estimation and biological implications* (pp. 181–206). Cambridge: Cambridge University Press.

Van Valkenburgh, B., Wang, X., & Damuth, J. (2004). Cope's Rule, hypercarnivory, and extinction in North American canids. *Science, 306,* 101–104.

Vinyard, C. J., & Hanna, J. (2005). Molar scaling in strepsirrhine primates. *Journal of Human Evolution, 49,* 241–269.

West, G. B., Brown, J. H., & Enquist, B. J. (1997). A general model for the origin of allometric scaling laws in biology. *Science, 276,* 122–126.

White, C. R., Blackburn, T. M., & Seymour, R. S. (2009). Phylogenetically informed analysis of the allometry of mammalian basal metabolic rate supports neither geometric nor quarter-power scaling. *Evolution, 63,* 2658–2667.

Whittaker, R. J. (1999). Scaling, energetics, and diversity. *Nature, 401,* 865–866.

Wu, C. F. J. (1986). Jackknife, bootstrap and other resampling methods in regression analysis. *The Annals of Statistics, 14,* 1261–1295.

Xiao, X., White, E., Hooten, M., & Durham, S. (2011). On the use of log-transformation vs. nonlinear regression for analyzing biological power-laws. *Ecology, 92,* 1887–1894.

Zar, J. H. (2010). *Biostatistical analysis (5th ed).* Englewood Cliffs: Prentice-Hall.

Chapter 3
Functional Morphology of the Postcranial Skeleton

Rachel H. Dunn

Abstract The field of vertebrate functional morphology grew out of traditional comparative vertebrate morphology. By the 17th century, scientists were modeling the actions of muscles on the skeleton as simple levers, but this remained uncommon until the latter half of the 20th century with the introduction of the concepts of mechanical advantage and speed vs. power systems in limb morphology. Current studies in this field are largely unchanged from those of the late 1900s, the largest steps forward being in the ability to analyze large, multivariate datasets, and quantify complex shapes due to leaps in computing power, and in the development of methods to account for phylogenetic signal in morphological data. This chapter will describe the process of a functional morphological analysis of lever mechanics including: identification of dominant muscles and assessing the action of those muscles based on comparative anatomy, identification of in-levers and out-levers, measurement of lever arms, and statistical analysis.

Keywords Functional anatomy • Comparative anatomy • Biomechanics

Introduction

The goal of functional morphology is to define the relationship between form and function. Postcranial functional morphology is most commonly applied to understanding the relationship between the shape of the limbs and their function but also applies to analysis of the vertebral column and ribs.

The field of vertebrate functional morphology grew out of traditional comparative anatomy, Aristotle (384–322 BC) and Galen (129–201 AD) being some of the best known early comparative anatomists. In the 17th century, Giovanni Borelli (1608–1679) was among the first to model locomotion mechanically as lever systems made up of muscle actions and long bones (Ashley-Ross and Gillis 2002), but this remained uncommon until the 1900s. Vertebrate paleontology has relied heavily on comparative anatomy to interpret morphologies and predict behaviors of extinct organisms; the most well-known early vertebrate paleontologists such as Georges Cuvier and Sir Richard Owen were first and foremost comparative anatomists. Functional morphology offers vertebrate paleontologists a means of adding quantifiable support to their comparative observations.

W. K. Gregory was one of the first researchers to examine biomechanical principles of limbs from an evolutionary perspective, examining limb proportions in an array of extant and fossil ungulates (Gregory 1912). Following directly from Gregory (1912), Smith and Savage (1956) introduced the concept of mechanical advantage and discussed the trade-offs of speed vs. power systems in limb morphology. These two fundamental concepts continue to dominate how we approach functional analyses of limbs to this day. The practice of functional morphology as we know it was codified in the wake of this paper by researchers such as Milton Hildebrand, R. McNeill Alexander, and Farrish Jenkins, to name but a few. These researchers were instrumental in the measurement of the biomechanics of locomotion in living mammals in an experimental setting (see Hildebrand et al. 1985; Alexander 2003).

Although early pioneers in functional morphology were largely neontologists (with the exception of Farrish Jenkins, whose research extended into paleontology), their contributions to the field of paleontological functional morphology cannot be overstated; much of what paleontologists can infer about the behavior of past organisms is by analogy with living ones and depends on the modern biomechanical

R. H. Dunn (✉)
Department of Anatomy, Des Moines University,
3200 Grand Ave., Des Moines, IA 50312-4198, USA
e-mail: rdunn@dmu.edu

D. A. Croft et al. (eds.), *Methods in Paleoecology: Reconstructing Cenozoic Terrestrial Environments and Ecological Communities*, Vertebrate Paleobiology and Paleoanthropology, https://doi.org/10.1007/978-3-319-94265-0_3

literature. Paleontologists rely on biomechanical, behavioral, and performance data of modern mammals to interpret fossil morphology (e.g., Biewener 1990; Hildebrand and Goslow 1998; Alexander 2003). However, functional interpretations of fossil morphology can also highlight gaps in the biomechanical literature that lead to new neontological studies (e.g., Boyer et al. 2007; Sockol et al. 2007; Barak et al. 2013; Warrener et al. 2015). Thus, the fields of neontological biomechanics and paleontological functional morphology complement and reinforce each other.

In the absence of detailed modern behavioral data, paleontologists often use the comparative method to statistically test associations between form and behavior in extant taxa and extrapolate to fossil organisms. This method relies in large part on assuming uniformitarianism of morpho-functional relationships, a relatively safe strategy when dealing with more recent fossil groups that are closely related to extant organisms but one that is more problematic in older and more distantly-related fossil groups (Van Valkenburgh 1994). Moreover, statistical associations between function and morphology can be misleading if they are not supported by biomechanical principles and, ideally, by performance testing (Koehl 1996; Warrener et al. 2015). Without explicit testing of the mechanistic link between morphology and function or performance, statistical analysis can fail to correctly identify a strong association under a number of conditions (e.g., non-linearity). Conversely, a purely statistical analysis may incorrectly identify a morpho-functional link if function is associated with an unmeasured but highly correlated variable.

There is considerable overlap between the fields of functional morphology and ecomorphology, and this warrants a short discussion of what differentiates the two. Functional morphology seeks to correlate form with "function" (Ashley-Ross and Gillis 2002), and ecomorphology seeks to correlate form with "ecology" (Gould 2014). However, what is meant by "function" and "ecology" can differ from one researcher to the next. When studying the limbs of an extinct organism, "function" usually signifies "locomotor pattern." Functional morphological studies often investigate gross differences in limb function—how the limb of a cursorial (i.e., running) organism differs from that of an ambulatory (walking) one, for example—although many studies address finer divisions within broad locomotor categories. The word "ecology" also has a broad range of definitions in this context, from denoting a general "locomotor pattern" to a more specific "habitat structure." Walter Bock, a champion of ecomorphology, asserted that the difference between functional morphology and ecomorphology is that ecomorphology requires knowledge of how an organism actually behaves in its natural environment whereas functional morphology primarily seeks to determine the functional capabilities of a structure rather than how an organism functioned during its lifetime (Bock and Wahlert 1965; Bock 1990). More recently, Kappelman et al. (1997) defined ecomorphology ("ecological functional morphology" or "eco-funk") as: "the linkage between certain morphological patterns and their functional expression in specific habitats (p. 230)." In current practice (at least within the context of paleoecology), ecomorphology largely uses functional traits from fossil faunas to predict ecological characteristics (Barr 2018), whereas functional morphology focuses on predicting behavior of extinct animals based on functional principles. However, it should be noted that locomotion is inherently ecological, as the locomotor pattern of an organism is dictated in part by the environment in which it lives or in which it evolved.

Approaches

The primary focus of this chapter is the analysis of the musculoskeletal system as a series of levers. This approach assumes that the postcranial skeleton can be modeled as a series of beams (either individual bones, entire limbs, or something in between) on which muscles act as the in-force to produce an out-force that results in movement against a resistant substrate. Most of the literature cited herein relates to the functional morphology of mammals since this is my research focus and this volume as a whole deals with reconstructing Cenozoic terrestrial ecosystems (of which mammals are a conspicuous part). The vast literature of non-mammalian functional morphology will not be discussed here. It should also be noted that methods other than lever mechanics can be used either alone or in concert with this technique to provide complementary functional interpretations.

Skeletal lever mechanics focuses on external bone morphology (such as length, articular facet shape, positions of major landmarks, etc.), which reflects function in deep time: a combination of phylogeny (ancestral function) and ecology that, in theory, corresponds to the "adaptation" of an organism (Ruff and Runestad 1992; Ruff et al. 1994; Lieberman 1997; Lieberman et al. 2001; Kivell 2016). Thus, analyses of lever mechanics and joint shape will give insights into the conditions under which the animal evolved but will also reflect its phylogeny. The external structure of a bone changes relatively little during the life of an organism and therefore provides little information about the performance of that particular individual. In contrast, the internal structure of bone (the trabecular architecture) undergoes frequent remodeling and reflects actual stresses incurred by the bone during the life of an organism (Lieberman 1997; Lieberman et al. 2001; Kivell 2016). Analyses of both internal and external structure can answer questions of function, but they differ in scale: external structure reflects evolutionary functional "adaptation" over thousands to millions of years, whereas internal structure reflects function during an individual's lifetime (Hogg 2018). As such, results from these analyses are sometimes contradictory (Ryan and Ketcham 2002b). Both approaches are addressed further below.

Internal Structure of Bone

Two methods of measuring internal structure are relatively common: cross-sectional geometry and analysis of trabecular structure. Cross-sectional geometry is the biomechanical analysis of the cross-sectional properties of bone. It is primarily concerned with the amount of diaphyseal cortical bone (cortical area) and the distribution of cortical bone relative to a specified axis, both of which vary as a consequence of different loading regimes (Ruff 1989; Demes et al. 1991; Swartz 1993; Lieberman et al. 2004; Marelli and Simons 2014). Cortical area (CA), the second moment of area (I), and the polar second moment of area (J) are the most commonly analyzed variables but by no means the only ones available (see Ruff 2009). Cortical area is a relative measure of the amount of cortical bone in a given cross-section. An increase in cortical area indicates increased resistance to axial loading and often occurs as a consequence of increased body mass (Demes et al. 1991; Marelli and Simons 2014). As such, cortical area can be used to estimate body mass (Ruff 2003a). The second moment of area (I) and polar second moment of area (or polar moment of intertia; J) relate to the distribution of bone in a cross section. In general terms, I represents the maximum (I_{max}) and minimum (I_{min}) displacement of cortical bone from the neutral bending axis (Swartz 1993; Marelli and Simons 2014). The more oval the bone (i.e., the greater the difference between I_{max} and I_{min}), the greater the resistance to bending with respect to the axis along I_{max} (Ruff 1989; Swartz 1993; Connour et al. 2000; Marelli and Simons 2014). J is calculated as the sum of I_{max} and I_{min} and represents resistance to torsional forces or overall bending stress (Lieberman 1997; Connour et al. 2000; Lieberman et al. 2004; Marelli and Simons 2014). Cross-sectional geometry has been used in primates to differentiate bipeds from quadrupeds (Ruff 2009) and hind limb-driven leapers from quadrupedal climbers (Ruff 1989; Runestad 1997; Connour et al. 2000); it has also been used to trace ontogenetic changes in locomotor mode or exercise (Ruff 2003b; Ruff et al. 2006) and to examine differences in flight pattern and prey capture in raptors (Marelli and Simons 2014).

Trabecular (or cancellous) bone remodels faster than cortical bone, making it more responsive to external loading than either cortical bone or external morphology (Kivell 2016). It responds to mechanical loading by increasing in density and aligning trabeculae to the direction of principal stress and thus reflects the magnitude and direction of external loads incurred throughout the life of an organism (Swartz 1993; Scherf 2008; Ryan and Walker 2010; Kivell 2016). Numerous in-vivo studies of humans and other mammals have shown that trabecular morphology consistently responds to loading in a predictable way (see Kivell 2016 for an overview). Within primates, the trabecular structure of the proximal femur has a strong functional signal in strepsirrhines but a more ambiguous signal in anthropoids (MacLatchy and Müller 2002; Ryan and Ketcham 2002a; Shaw and Ryan 2011). An excellent summary of trabecular bone analysis is given by Kivell (2016).

One potential limitation of the analysis of internal bone structure is that it must be visualized either through physical sectioning of specimens, biplanar radiographs, or, more recently, through CT and microCT scans (which may not be accessible to all researchers). Additionally, cancellous bone is often missing, distorted, or otherwise obscured in fossil specimens, precluding analysis of trabecular morphology (Ryan and Ketcham 2002b; Kivell 2016).

Joint Shape

Joint shape is an important functional component of the postcranial skeleton. However, joint shape can be difficult to quantify, with the result that there have historically been relatively few standalone analyses of joint surface shape (e.g., Yalden 1970, 1971, 1972; Hamrick 1996; but see Curran 2018). Nevertheless, joint shape is often qualitatively considered along with lever mechanics for inferring joint mobility. For example, strongly grooved "female" joint surfaces (i.e., those that are concave) are often indicative of a stable, hinge-like interlocking joint that constrains motion to a single plane (Schaeffer 1947; Hildebrand and Goslow 1998; Barr 2014). A female joint surface with a more open curve is often considered indicative of mobility (e.g., Gebo and Rose 1993; Heinrich and Rose 1997), although the shape and degree of congruence of the corresponding male surface impacts the mobility as well (Yalden 1972; Hamrick 1996).

Traditional functional indices of joint surface shape include radial head index (max. diameter × 100/min. diameter) and humeral head index (head width × 100/head length). In both of these cases, an index close to 100 indicates an essentially spherical surface and a very mobile joint; higher values indicate that the joint is more constrained to movement in a single plane (Rose 1990). In the case of the radius, a wider radial head correlates with decreased ability for the radial head to pivot on the humeral capitulum during supination, and thus a higher radial index indicates an elbow joint that is restricted to flexion-extension (MacLeod and

Rose 1993). A higher humeral head index generally indicates a relatively wider head, as is often seen in climbing mammals which have a more mobile glenohumeral joint (Larson 1993; Schmitt 1996; Hildebrand and Goslow 1998).

Although joint shape has been (and still is) difficult to quantify, there have been several scholarly attempts. Yalden (1970, 1971, 1972) published a series of primarily descriptive papers dealing with joint shape and carpal mobility in a variety of mammals including primates, xenarthrans (armadillos, sloths, anteaters), carnivorans (cats, dogs, bears, and relatives), and ungulates (hoofed mammals). Several studies (e.g., Kappelman 1988; Schmitt 1996) have used a series of linear measurements taken along an articular surface to estimate overall shape. Hamrick (1996) used digitized landmarks to measure articular length and curvature of carpal bones, and MacLeod and Rose (1993) used eigenshape analysis to capture outlines of radial heads and correlate them with locomotor pattern in mammals. Increased access to 3D imaging techniques has led to the ability to quantify more complex features and a corresponding increase in analyses of joint shape (see Curran 2018). For example, Tocheri and colleagues (Tocheri et al. 2011; Dunn et al. 2014) used 3D models of foot bones to analyze the curvature of joint surfaces in gorillas, and Barr and Dunn (2015) used digital elevation models to analyze slope of the proximal trochlea in artiodactyl astragali.

Procedure

The following is an outline of a traditional functional analysis consisting of linear measurements and ratios. However, there are other methods of making functional inferences that require different types of data. Geometric morphometric methods (discussed by Curran 2018) can be used to make functional inferences from landmark and other dimensional data. Additionally, finite element analysis has been used in postcranial functional morphology, especially in the analysis of internal bone structure (Huynh Nguyen et al. 2014; Kivell 2016).

Collecting Data

The first step in an analysis is to identify a region of interest (e.g., forelimb, hind limb, shoulder, ankle, etc.). In paleontology, this is often dictated by available fossil material. Second, identify measurements that will be informative about the function of the area of interest. Most often, this second step is done through surveying the literature for measurements (usually ratios of linear measures) that have been used previously to differentiate locomotor or behavioral groups (Table 3.1) and for comparative anatomical data on what muscles are present in the region, their attachments, and primary actions. While most traditional functional morphological analyses use ratios or indices as an approximation of the mechanical advantage of muscles and/or muscle groups (Hildebrand 1985, 1988; Hildebrand and Goslow 1998; Elissamburu and Vizcaíno 2004; Shockey et al. 2007; Croft and Anderson 2008), this is not necessarily explicitly stated, so a thorough understanding of the anatomy in the region of interest will help in choosing which measurements are the most appropriate to answer the research question.

There are many good overviews of lever systems and their application to biomechanics (e.g., Swartz 1993; Hildebrand and Goslow 1998) so I will not provide a comprehensive review here. However, the concept of mechanical advantage is worth mentioning because many of the ratios that are calculated in traditional functional analysis are meant to model the mechanical advantage of individual muscles or muscle groups. Mechanical advantage as applied to the limbs of mammals was defined by Smith and Savage (1956) as the ratio of the moment arm (or in-lever) to the load arm (out-lever). All else being equal, a relatively larger mechanical advantage in one animal compared to another signifies a relatively longer moment arm, and thus more powerful movement (Hildebrand and Goslow 1998). In contrast, a relatively smaller mechanical advantage would signify a relatively longer load arm and less powerful, but faster movement (Fig. 3.1). This approach represents a simplification relative to previous treatments of limb mechanics (e.g., Gregory 1912) but has the advantage that detailed information about the origins and insertion angles of the surrounding muscles is not necessary for calculations and comparisons.

Sample Selection

The construction of an appropriate comparative sample is a crucial part of functional morphological analysis. What constitutes an "appropriate" sample can depend on many factors including phylogenetic affinity and body size of the animal(s) under study and the research question itself. In general, a comparative sample from the same phylogenetic group as the study fossil provides the best comparative material. Often, at least some extant members of a clade share broadly similar body sizes and/or ecologies with fossil members of the same group, providing close living analogues and easily interpretable results. In cases in which the phylogenetic affinity of the study fossil is unknown or in which a lineage has left no descendants, other criteria should be used to determine the most appropriate extant sample. For example, the sample may consist of its closest living

Table 3.1 Description of some common functional ratios and their functional and behavioral significance. * ratios are usually multiplied by 100 to produce an index

Name	Index*	Description	Functional Significance	Behavioral Significance
Forelimb				
Brachial Index	length of ulna / length of humerus	measure of the relative length of the distal forelimb	higher values indicate higher velocities	runners and jumpers have higher values
Deltoid Index	length of deltopectoral crest / length of humerus	measure of the mechanical advantage of the deltoid muscle	higher values indicate more powerful shoulder protraction	diggers have higher values
Epicondyle Index	width of distal humerus / length of humerus	measure of the relative area of origin of wrist and digital flexors/extensors	higher values indicate larger muscle volume	diggers have higher values
Humeral Head Index	mediolateral width of humeral head / proximodistal length	measure of the shape of the humeral head	higher values indicate a more spherical humeral head and a more mobile joint	climbers have higher values
Olecranon Process Index	length of olecranon process / length of ulnar shaft	measure of the mechanical advantage of the triceps brachii muscle	higher values indicate more powerful elbow extension	diggers have higher values
Radial Index	maximum diameter of radial head / minimum diameter of radial head	measure of the shape of the radial head	higher values indicate a more oval radial head and more restricted humero-radial joint	climbers have lower values
Hind Limb				
Crural Index	length of tibia / length of femur	measure of the relative length of distal hind limb	higher values indicate higher velocities	runners and jumpers have higher values
Distal Femoral Index	distal femoral width / distal femoral depth	measure of the mechanical advantage of the quadriceps muscles	higher values indicate more powerful knee extension	runners and jumpers have higher values
Gluteal Index	length from femoral head to distal end of / length of femur	measure of the mechanical advantage of the gluteus medius muscle	higher values indicate more powerful hip extension	runners and jumpers have higher values
Hip Flexor Index	length from femoral head to lesser trochanter / length of femur	measure of the mechanical advantage of the iliopsoas muscle	higher values indicate more powerful hip flexion	runners and jumpers have higher values
Metatarsal Femur Index	length of third metatarsal / length of femur	measure of the relative length of pes	higher values indicate higher velocities	runners and jumpers have higher values
Tibial Crest Index	length of tibial crest / length of tibia	measure of the mechanical advantage of the hamstring muscles	higher values indicate more powerful knee flexion	diggers have higher values

relatives (sister-group), distantly related species of similar body size, or species presumably occupying the same general ecological niche (e.g., other ungulates or carnivores). It is important to emphasize that close phylogenetic relatedness should not trump other considerations, particularly body size; since smaller animals are subject to different physical constraints than larger ones, problems in interpretation can arise if there is a large body size difference between the study object and the comparative sample. For example, modern tree sloths (<10 kg) would make a poor comparative sample for understanding the functional morphology of extinct ground sloths (>100 kg to >1,000 kg), even though they are their closest modern relatives. A detailed treatment of body size is presented elsewhere (Hopkins 2018).

The research question should also play a role in the construction of the comparative sample. If relatively little is known about the fossil of interest, the research question will be broad (e.g., what was its locomotor behavior?) and will require a comparative sample with a broad range of locomotor behaviors to adequately address it. However, if a good deal is known about the object of study, it may be possible to ask a more precise question (e.g., was it more likely an ambush or pursuit predator?) that requires a comparative sample with a narrower range of locomotor behaviors. In most cases, each taxon in the dataset should be categorized by either behavior, such as locomotor mode (e.g., cursorial, ambulatory, etc.) or the type of habitat it occupies (e.g., arboreal, terrestrial, aquatic, etc.). In some cases, it may be

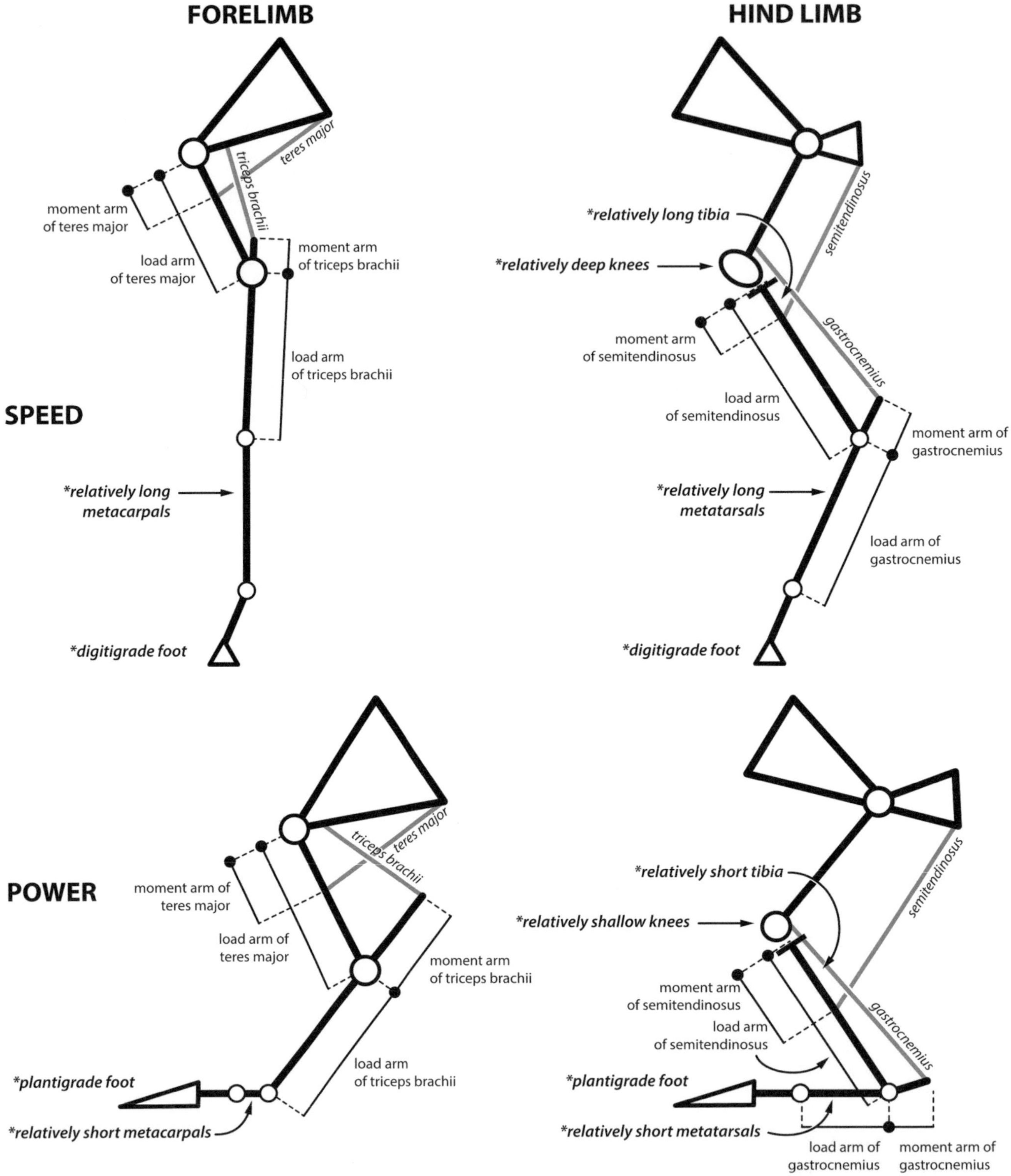

Fig. 3.1 Schematic illustration of a forelimb (left column) and hind limb (right column) adapted for speed (top row) and power (bottom row). Bones are illustrated by thick lines and triangles. Joints are represented by circles. Forelimb bones depicted from proximal to distal are: scapula (top triangle), humerus, ulna, metacarpals, and phalanges (distal phalanx depicted by bottom triangle). Forelimb joints depicted from proximal to distal are: shoulder (glenohumeral), elbow (humeroulnar), wrist (carpometacarpal), and proximal finger (metacarpophalangeal) joints. Hind limb bones depicted from proximal to distal are: innominate (hip; top two triangles), femur, tibia, metatarsals + calcaneus (foot and heel), and phalanges (distal phalanx depicted by bottom triangle). Hind limb joints depicted from proximal to distal are: hip (acetabular), knee (tibiofemoral), ankle (astragalotibial), and proximal toe (metatarsophalangeal) joints. Gray lines connect the attachment sites of the two muscles (noted in italics along gray line) for the highlighted lever systems. Note that moment arms are relatively greater relative to load arms for power-adapted systems compared to those for speed-adapted systems. Other features considered indicative of power or speed-adapted limbs are marked with an asterisk

possible to use a continuous numerical variable such as leaping frequency or top running speed rather than a categorical one.

Taking Measurements

One of the advantages of an analysis that primarily relies on linear measurements is that they are relatively easy and inexpensive to collect. It is generally best to take linear measurements on the bones themselves using digital calipers, but this sometimes is not possible either due to specimen size (very large or very small) or inaccessibility. In such cases, measurements can be taken using other tools (e.g., an optical micrometer), from photographs with a scale bar or, increasingly commonly, from virtual (computer) models of a specimen constructed from surface or CT scanning. Measurements other than straight-line distances are sometimes relevant. Joint surface areas can be informative about the loading patterns of specific joints, and certain joint surface areas can be used as a proxy for body size (Dagosto and Terranova 1992; Tsubamoto 2014; Yapuncich et al. 2015). Angular data are relevant when dealing with questions such as joint range of motion (Jacobs et al. 2009; Tocheri et al. 2011; Hammond et al. 2016). Free software such as ImageJ (Abramoff et al. 2004) and R (R Core Team 2015) can be used to measure linear distances and angles from digital files.

Analyses

There are many ways to analyze functional morphological data. The following is a brief description of some of the more common analyses but is by no means comprehensive. As for assembling a comparative sample and selecting measurements, the most appropriate type of analysis will follow largely from the specific research question being addressed.

Data exploration: The most basic approach, and often the first step in exploring the data, is to simply compare functional ratios between behavioral categories within the comparative sample and between the sample and the object of study. This can be done with a table of measurements, boxplots, or some other method of visualizing distributions (Fig. 3.2). Bivariate plots of calculated ratios are also useful for comparing the fossil in question to more than one variable at a time and visualizing areas of morphospace occupied by the modern comparative sample.

Multivariate data may be visualized using principal components analysis (PCA; Fig. 3.3). PCA is an excellent exploratory technique that facilitates understanding the underlying structure of the data but does not recognize groups or classify individuals a priori. PCA reorganizes the individual, correlated variables into new ones (principal components; PCs) that are independent from each other such that the first principal component explains the greatest proportion of variation within the sample, the second principal component explains the second greatest proportion, etc. (Zelditch et al. 2012). Variables that influence the same component can be interpreted as highly correlated with each other, and those that influence the first component most strongly are those that tend to explain the most variation within the data set (Fig. 3.3). Thus, PCA can help identify variables of interest. However, PC1 tends to be highly correlated with body size unless variables have been size corrected, so testing for significant correlations with body size before interpreting variables as functionally relevant is advisable. PCA can load a single variable onto multiple components, but when this happens, the variable is typically most strongly correlated with one component and influences the others more weakly. PCA is designed to investigate variation among individual data points rather than among groups of points. This means that variables that are highly correlated with the first few PCs (and thus explain the most variation within the sample) may not contribute significantly to differences between groups. To identify variables that best differentiate groups, a canonical variates analysis (CVA) is more appropriate (discussed below).

Univariate analysis: After identifying variables of interest, one should test for significant differences among groups. There are many methods for doing this depending on the sample size, the distribution of the data, and the number of groups in the sample (i.e., the number of comparisons being made). For normally-distributed data with relatively large samples (>30 specimens per group), parametric tests such as the t-test (for two groups) and ANOVA (Analysis of Variance; for more than two groups) are most commonly used. For data that deviate from a normal distribution or for small samples, nonparametric tests such as a Mann-Whitney U test or Kruskal-Wallis test are often more appropriate. Nevertheless, ANOVA is also commonly used with smaller samples, as data can often be transformed to conform to the necessary assumptions.

The t-test and Mann-Whitney U test are used to evaluate significant differences between two samples. ANOVA and Kruskal-Wallis simultaneously test for significant differences among all of the groups in the sample but do not indicate which groups differ from one another. Thus, a positive result from either requires further testing to identify, specifically, which groups differ from one another. However, because multiple comparisons are being made (e.g., with four groups, seven pairwise comparisons are made) the significance level must be adjusted to avoid falsely rejecting the null hypothesis of equivalence (type I error). There are many methods of adjusting the significance level, one of the simplest and most popular being Bonferroni correction, which adjusts the

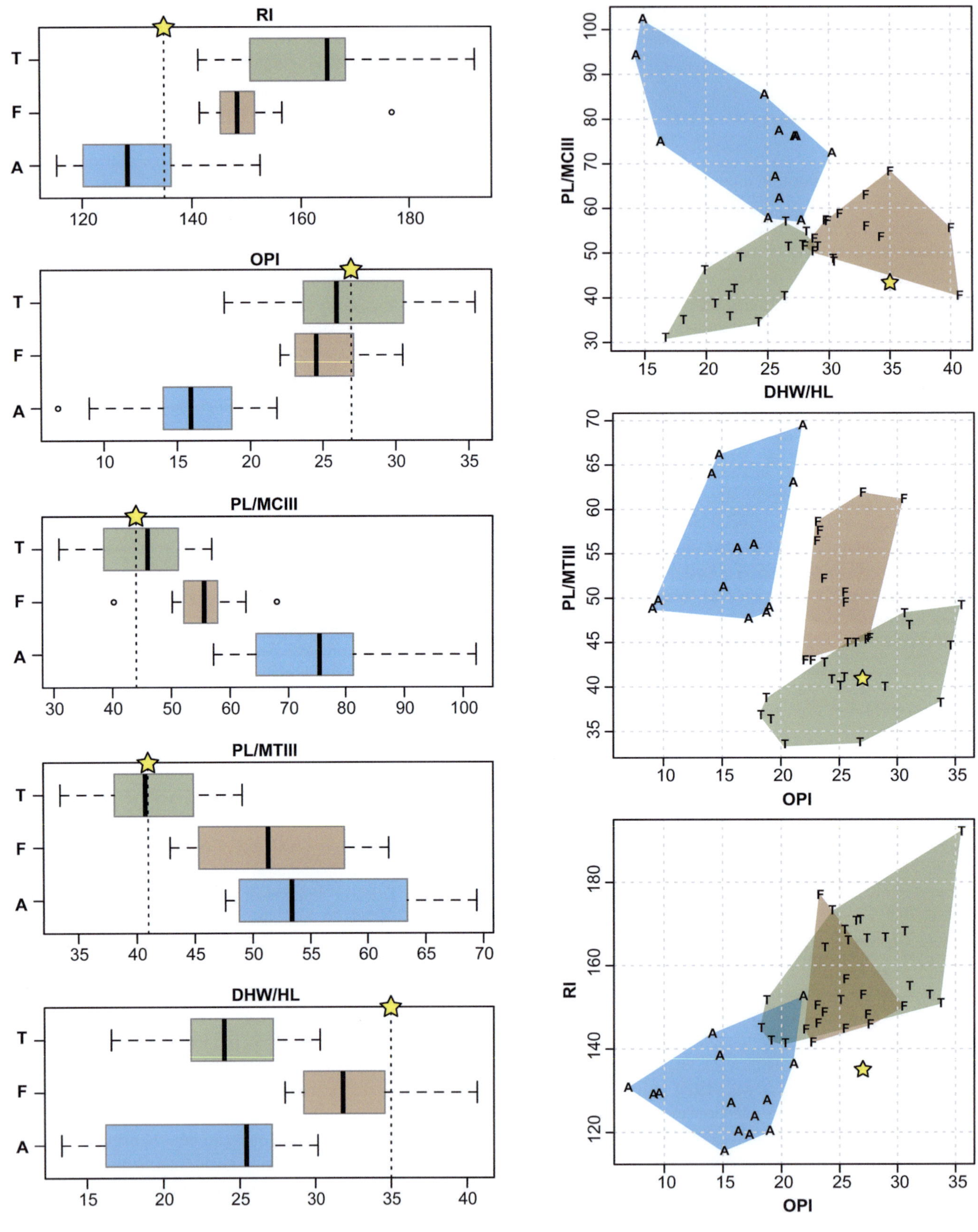

Fig. 3.2 Plots of five functional indices in rodents (adapted from Dunn and Rasmussen 2007). Left: box plots of functional indices by behavioral category. **A** = arboreal (blue), **F** = fossorial (brown; displays some morphological specialization for digging), **T** = terrestrial (green). Right: scatter plots of the same functional indices with convex hulls for each behavioral category. The yellow star indicates the fossil rodent *Pseudotomus eugenei*. Note that the fossil species shares some functional indices with each locomotor category and that it mostly does not overlap ranges seen in extant rodents in the scatter plots. The authors concluded that the fossil rodent was likely terrestrial with some digging ability but was unlikely to be arboreal despite some overlap with this category. **RI** = radial head index; **OPI** = olecranon process index; **DHW/HL** = epicondyle index; **PL/MCIII** = length of manual proximal third phalanx × 100/length of third metacarpal; **PL/MTIII** = length of pedal proximal third phalanx × 100/length of third metatarsal. Indices not defined here can be found in Table 3.1

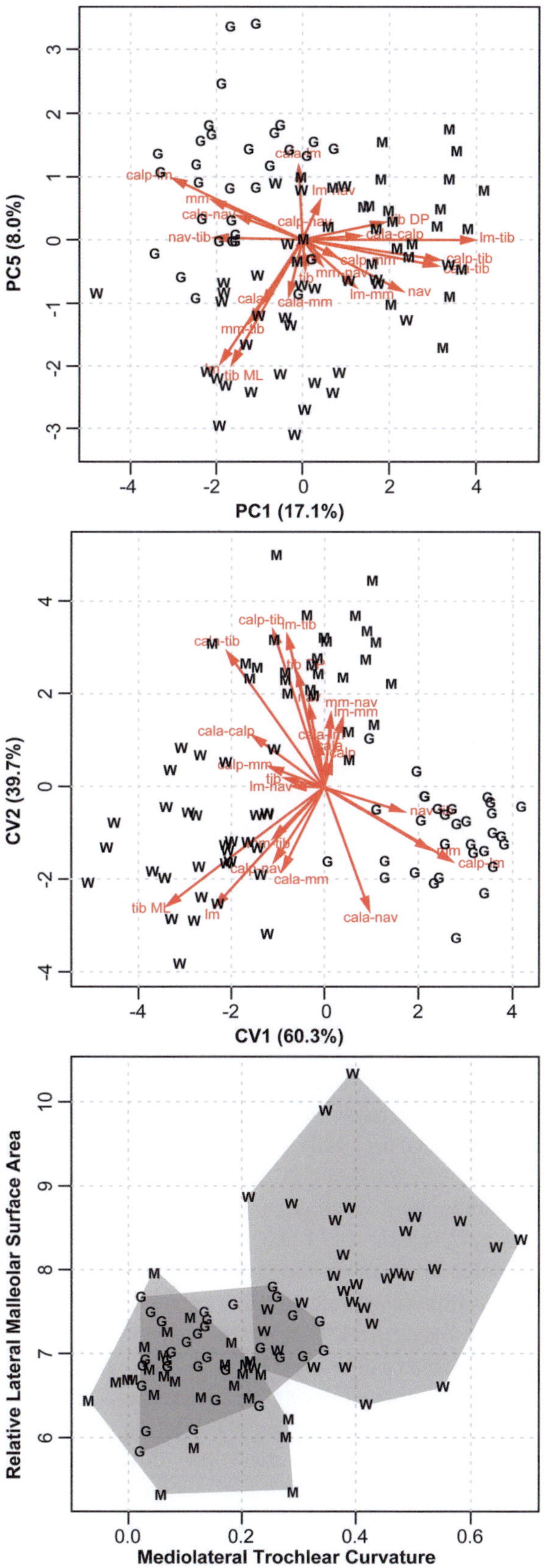

◀ **Fig. 3.3** Plots of relative articular facet areas and angles in three gorilla populations with different locomotor tendencies based on a PCA (top), CVA (middle), and two high-loading variables (bottom). **G** = *Gorilla beringei graueri* (grauer gorilla), **M** = *Gorilla beringei beringei* (mountain gorilla), **W** = *Gorilla gorilla* (western lowland gorilla). Western gorillas spend more time in the trees than mountain gorillas, which are predominantly terrestrial. Grauer gorillas are more closely related to mountain gorillas but are intermediate relative to the other two taxa in frequency of arboreal behaviors. Red arrows indicate loadings of individual variables on the plotted PCs or CVs. Note that, while some separation of groups can be seen in all methods, maximal separation among groups occurs with CVs, a method that maximizes the differences among pre-defined groups. Also note that grauer gorillas overlap both western and mountain groups, reflecting their intermediate locomotor behavior. An explanation of variables and specific loadings, as well as a more thorough discussion of gorilla locomotion and morphology, can be found in Dunn et al. (2014)

significance level by dividing the critical value by the number of comparisons. The simplicity of Bonferroni correction is attractive, but it is also more conservative than some other methods, which may not be desirable (Sokal and Rohlf 1995). Many statistical packages that perform ANOVA offer post-hoc tests (commonly Tukey's HSD) that automatically adjust the significance level. There is no standard post-hoc test for nonparametric multiple-group analyses, though several methods of evaluating unplanned pairwise comparisons have been proposed (Sokal and Rohlf 1995; Hammer and Harper 2006). The R package PMCMR includes several non-parametric post-hoc tests (Pohlert 2014).

Multivariate analysis: Discriminant analyses such as discriminant function analysis (DFA) and canonical variates analysis (CVA) are some of the most commonly used multivariate techniques in postcranial functional morphology (in addition to PCA, discussed above). These methods are also very commonly used in ecomorphology and geometric morphometrics (see: Barr 2018; Curran 2018). Like PCA, DFA and CVA recombine original variables to form new, uncorrelated variables (canonical variates [CVs] or discriminant functions [DFs]). However, unlike PCA, these methods emphasize differences between groups that are defined a priori by creating mathematical functions that can be used to classify unknown individuals (Hammer and Harper 2006; Zelditch et al. 2012). Modern statistical texts differentiate DFA from CVA by the number of groups in the analysis, with DFA simply being a special case of CVA that deals with only two groups (Hammer and Harper 2006; Zelditch et al. 2012). Nevertheless, the scientific literature tends to use the term CVA for any descriptive discriminant analyses and DFA for analyses that predict group membership of unknown specimens (e.g., Gunz et al. 2005; Tocheri et al. 2011). In the paper that first formally described DFA, Fisher (1936) provided examples using both two and three groups, suggesting that, at least initially, DFA was used in cases involving more than two groups. In current practice, the terms CVA and DFA are essentially identical.

CVA and DFA are not simply used to separate groups and classify individuals; examining the loadings of the original variables on the discriminant variables (CVs or DFs) can help identify features that best differentiate groups (Fig. 3.3). For example, Dunn et al. (2014) examined size-scaled surface areas, angles, and curvatures of talar (ankle) joint surfaces in three populations of gorillas with different climbing behaviors to determine if differences in talar morphology among the populations might correlate with differences in behavior. A CVA significantly separated all three groups, and two variables in particular (mediolateral curvature of the trochlea and area of the fibular facet) had a disproportionately high influence on the CVs. Boxplots and a bivariate plot of those two variables also showed separation of all three groups, and non-parametric tests for differences in the mean confirmed that the differences were statistically significant. The researchers concluded that these two variables reflect differences between a more inverted ankle in gorillas that climb more frequently and a more everted ankle in those that spend more time locomoting terrestrially (Dunn et al. 2014).

Strengths and Shortcomings

One advantage of traditional functional analysis is that the data – usually a series of linear measurements combined into ratios – are relatively easy to collect. In addition, the equipment needed to collect these data (e.g., calipers) is relatively inexpensive, as is the software needed to analyze the data (e.g., R [R Core Team 2015], MorphoJ [Klingenberg 2011], and PAST [Hammer et al. 2001] are free). Using combinations of linear measurements to describe joint shape and the lever mechanics of limbs is also a relatively well-established method that continues to produce easily interpretable results comparable to those that use more complex coordinate data (Gould 2014). The statistical use of ratios has been controversial (e.g., Atchley et al. 1976; Hills 1978; Smith 2005). This partly stems from unattractive statistical properties of ratios, such as not being normally distributed, a basic assumption of most statistical tests. This problem can be avoided by transforming the data. This is most commonly done through logarithmic transformation, although many other strategies exist (Hills 1978; Sokal and Rohlf 1995; Carrano 1999; Smith 2005). A greater cause of concern has been the improper use of ratios as a method of statistical control: attempting to remove the effect of a confounding variable such as body size (see Smith 2005 for a review and discussion). While ratios do not control for the effects of the denominator on the numerator, they are necessary for describing the relationship between the two variables ("proportionality" or "relative size"; Smith 2005), such as two lever arms when quantifying mechanical advantage (Carrano 1999).

There are several confounding factors that should be considered when designing an analysis or interpreting results. The two main factors for which researchers often try to "control" are body size and phylogeny. As mentioned previously, ratios do not correct or control for body size in a statistical sense. In fact, most functional ratios are highly correlated with body size because animals of different body size have different structural constraints and must differ proportionally to some extent in order to move about in their environment (Fleagle 1985; Jungers et al. 1995). These differences in shape that co-occur with differences in size are known as allometric differences, and there is a vast literature dealing with the study of allometry (e.g., Gould 1966; Smith 1980; Hildebrand and Goslow 1998; Alexander 2003). There are many ways to account for allometry in a functional analysis. A simple approach is to use a comparative sample of individuals of similar body size. This can help to mitigate (but not remove entirely) the influence of body size on the analysis, as samples with a higher range of body sizes have a stronger allometric component (Smith 1980).

Even when a comparative sample is made up of similarly sized individuals, it is important to test for allometry. This is done by running correlations between the variables of interest and some measure of body size, usually body mass data taken from the literature or a well-established proxy for body size in the case of fossil species. If the variable of interest shows a significant allometric effect, further functional interpretation must be done in a way that takes body size into account. An excellent example of such as study was conducted by Boyer et al. (2013) on the allometry of calcaneal elongation in primates. Among primates, tarsiers and galagos have extremely elongated distal (anterior) calcanei, which has been interpreted as a modification to increase the load arm of the triceps surae to increase the speed of the foot during leaping (Walker 1974; Gebo 1988). However, it is also well known that relative calcaneal length in primates has a strong allometric component, which complicates the simple functional relationship (Walker 1974; Moyà-Solà et al. 2012). Boyer et al. (2013) approached the problem in a systematic way, first by examining the correlation between body size and relative distal calcaneal length to demonstrate that there is a significant negative allometric effect of body size (i.e., larger primates have proportionately smaller calcanei). They then tested for associations between leaping frequency and calcaneal elongation using residuals (the difference between observed and expected values for an individual of a particular body size). Their results indicated that leaping prosimians do tend to have relatively longer calcanei, but the same is not true for leaping anthropoids.

Boyer et al. (2013) also addressed another confounding factor in functional analysis, that of phylogeny. Most traditional statistical methods assume that the data are drawn from a population at random and are independent of one another. However, this is virtually never the case with biological comparative data. Species are linked by phylogenetic relationships that can be represented as a nested (hierarchical) tree. Consequently, individual species that are more closely related to one another tend to resemble each other more than they resemble more distantly-related species, thus resulting in a non-independent sample (Felsenstein 1985; Nunn 2011). Phylogenetic methods take this statistical non-independence into account and ensure that the assumption of independence is not violated (Felsenstein 1985). Whether or not phylogenetic methods are always necessary is yet another subject of debate (Losos 1999; Nunn 2011). In many instances, they may not be. However, methods that test data for phylogenetic signal are now available and easily accessible, making it possible to quantitatively assess whether or not phylogenetic corrective methods should be employed (Nunn 2011). Boyer et al. (2013) demonstrated that primate calcanei preserve significant phylogenetic signal, making the use of phylogenetic methods appropriate in that particular case. The results of phylogenetically corrected analyses will differ depending on the phylogenetic tree being used, so careful thought should be given to which tree or trees will be employed. More information on the use of phylogenetic methods can be found in Arnold et al. (2010), Nunn (2011), and Garamszegi (2014).

Whereas body size and phylogeny are issues that can be quantified and accounted for statistically, functional morphological analyses have other inherent limitations. A fundamental limitation of studying function in fossils lies in the nature of the fossil record. With few exceptions, only hard tissues such as teeth and bones are preserved, leaving paleontologists with limited material for analysis. Further, most fossil material is fragmentary; relatively few complete skeletons are available for analysis. Available remains can be crushed, distorted, or otherwise have the details of morphology obscured. This can place significant limitations on the types of analyses that can be undertaken for most species. However, despite the incomplete nature of the fossil record, a great deal of inference can still be made because bone tissue preserves indications of function, phylogeny, and other processes such as ontogenetic growth and metabolic processes.

Another limitation is the accuracy and precision of functional interpretations. Models and data used in functional analyses are necessarily simplified models of complex biological systems. In studies of living organisms, it is possible to take many factors into account (soft tissues, kinematics, etc.) that must be estimated or omitted when dealing with fossils. As such, results should be considered rough proxies for behavior (Hildebrand 1985). Additionally, functional inferences from simplified models are only as precise as the modern comparative data, which is far from perfect. Some groups are relatively well-studied (e.g., primates) whereas the behavior of others is relatively poorly known (e.g., elephant shrews). Some animals have been extensively studied in a laboratory setting (e.g., rodents and primates), which has the advantage of providing data on the limits of locomotor and behavioral capabilities: what an animal *can do*, but not necessarily what it *does* in its natural environment (Bock and Wahlert 1965). Further, little is known about intraspecific variation both in morphology and behavior, which is crucial for understanding how interspecific differences evolve. For example, Sockol et al. (2007) examined locomotor energetics in chimpanzees and humans and found not only that human locomotion was more efficient compared to that of chimpanzees, but also that chimpanzees with more human-like limb posture were more efficient when walking bipedally than other chimpanzees, thus providing a mechanistic link to the evolution of more human-like posture. Finally, behaviors are points along a continuum rather than discrete categories, so creating such categories for functional analysis introduces further ambiguity and error. Fortunately, lack of detailed information about function in living animals is a limitation that can be overcome with further research.

Future Prospects

For most of its history, functional morphological analysis has been limited to linear measurements, which are adequate for describing lever arms and calculating mechanical advantage but are of limited utility for describing more complex aspects of shape such as curvature and surface area. Other measurements, such as angles, diameters, and circumferences, have also been difficult to measure, usually necessitating photographs of precisely oriented specimens. Many foreseeable improvements in postcranial functional morphology are linked to advances in technologies such as 3D surface modeling and other multidimensional techniques (such as geometric morphometrics and FEA) that enable more accurate quantification of shapes. Several recent studies provide examples of such methods and are discussed below.

Despite the advantages of 3D methods, challenges persist; the relative sizes of joint surfaces are rather easy to calculate using 3D polygonal models, but measuring angles and curvatures is less straightforward. Tocheri and colleagues (e.g., Tocheri et al. 2005, 2011; Marzke et al. 2010; Dunn et al. 2014) used best-fit planes of joint surfaces to calculate the angles between each articular facet in wrist and ankle bones of apes. Boyer and Seiffert (2013) used a slightly different method to approximate the angle between the talar trochlea and the fibular facet on the astragalus of primates, segmenting out portions of each surface according to a precise positioning protocol and fitting planes to those point clouds. Swartz and Ward (2014) assessed the angles of retroversion or anteversion of the femoral head based on functional axes of the femoral shaft and neck and compared these results to angles calculated according to traditional clinical methods. In a recent study, Hammond et al. (2016) compared abduction angles measured from 3D models of primate hips to those measured from anesthetized living primates; they found that 3D models generate conservative estimates of hip mobility and can reliably predict behavior near the extremes of the locomotor spectrum but are less reliable at differentiating more subtle variation.

3D models also allow a more quantitative treatment of curvature than was previously possible. Boyer et al. (2015b) attempted to quantify the shape of the medial tibial facet of the astragalus of primates by using the ratio of perimeter length (i.e., the "outline") to the surface area of the segmented facet. Rounder facets that more approximated a circle had smaller values for this ratio, and more oval facets had larger ones, allowing an approximate quantification of a trait that was traditionally assessed qualitatively. Another method used to measure joint curvature, outlined in Marzke et al. (2010), fits modeled quadric surfaces to a segmented joint surface. Yapuncich and Boyer (2014) used best-fit spheres to approximate facet curvature, with spheres of larger diameter indicating a flatter surface and those of a smaller diameter one that is more curved.

Although 3D methods facilitate quantitative description of surface shape, they can be expensive and time intensive compared to more traditional methods using linear measurements; whether the increased statistical power they offer makes their use worthwhile is debated (Tocheri 2009; Boyer et al. 2011; Gould 2014). New techniques are becoming available that use automated methods to quantify and analyze shapes, thereby significantly reducing measurement time and permitting larger sample sizes, and many of these perform as well as trained observers (e.g., Boyer et al. 2011, 2015a; Koehl and Hass 2015). These studies offer a glimpse of what is to come in the field of postcranial functional morphology as 3D technologies become less expensive and 3D models become more widely available (e.g., via web sites such as www.morphosource.org). Increased access to these resources will facilitate new and creative methods of quantifying shape and lead to a better understanding of the relationship between form and function in living and fossil organisms alike.

References

Abramoff, M. D., Magalhaes, P. J., & Ram, S. J. (2004). Image Processing with ImageJ. *Biophotonics International, 11,* 36–42.

Alexander, R. M. (2003). *Principles of animal locomotion*. New Jersey: Princeton University Press.

Arnold, C., Matthews, L. J., & Nunn, C. L. (2010). The 10kTrees website: a new online resource for primate phylogeny. *Evolutionary Anthropology, 19,* 114–118.

Ashley-Ross, M. A., & Gillis, G. B. (2002). A brief history of vertebrate functional morphology. *Integrative and Comparative Biology, 42,* 183–189.

Atchley, W. R., Gaskins, C. T., & Anderson, D. (1976). Statistical properties of ratios. I. Empirical results. *Systematic Biology, 25,* 137–148.

Barak, M. M., Lieberman, D. E., Raichlen, D. A., Pontzer, H., Warrener, A. G., & Hublin, J.-J. (2013). Trabecular evidence for a human-like gait in *Australopithecus africanus*. *PLoS ONE, 8,* e77687.

Barr, W. A. (2014). Functional morphology of the bovid astragalus in relation to habitat: controlling phylogenetic signal in ecomorphology. *Journal of Morphology, 275,* 1201–1216.

Barr, W. A. (2018). Ecomorphology. In D. A. Croft, D. F. Su & S. W. Simpson (Eds.), *Methods in paleoecology: Reconstructing Cenozoic terrestrial environments and ecological communities* (pp. 337–347). Cham: Springer.

Barr, W. A., & Dunn, R. H. (2015). A method of analyzing complex joint surfaces in ecomorphology using slope rasters derived from digital elevation models. *American Journal of Physical Anthropology, 156,* 78.

Biewener, A. A. (1990). Biomechanics of mammalian terrestrial locomotion. *Science, 250,* 1097–1103.

Bock, W. J. (1990). From biologische anatomie to ecomorphology. *Netherlands Journal of Zoology, 40,* 254–277.

Bock, W. J., & von Wahlert, G. (1965). Adaptation and the form-function complex. *Evolution, 19,* 269–299.

Boyer, D. M., & Seiffert, E. R. (2013). Patterns of astragalar fibular facet orientation in extant and fossil primates and their evolutionary implications. *American Journal of Physical Anthropology, 151,* 420–447.

Boyer, D. M., Patel, B. A., Larson, S. G., & Stern, J. T., Jr. (2007). Telemetered electromyography of peroneus longus in *Varecia variegata* and *Eulemur rubriventer*: implications for the functional significance of a large peroneal process. *Journal of Human Evolution, 53,* 119–134.

Boyer, D. M., Lipman, Y., St. Clair, E. M., Puente, J., Patel, B. A., Funkhouser, T., et al. (2011). Algorithms to automatically quantify the geometric similarity of anatomical surfaces. *Proceedings of the National Academy of Sciences, USA, 108,* 18221–18226.

Boyer, D. M., Seiffert, E. R., Gladman, J. T., & Bloch, J. I. (2013). Evolution and allometry of calcaneal elongation in living and extinct primates. *PLoS ONE, 8,* e67792.

Boyer, D. M., Puente, J., Gladman, J. T., Glynn, C., Mukherjee, S., Yapuncich, G., et al. (2015a). A new fully automated approach for aligning and comparing shapes. *Anatomical Record, 298,* 249–276.

Boyer, D. M., Yapuncich, G. S., Butler, J. E., Dunn, R. H., & Seiffert, E. R. (2015b). Evolution of postural diversity in primates as reflected by the size and shape of the medial tibial facet of the talus. *American Journal of Physical Anthropology, 157,* 134–177.

Carrano, M. T. (1999). What, if anything, is a cursor? Categories versus continua for determining locomotor habit in mammals and dinosaurs. *Journal of Zoology, London, 247,* 29–42.

Connour, J. R., Glander, K., & Vincent, F. (2000). Postcranial adaptations for leaping in primates. *Journal of Zoology, London, 251,* 79–103.

Croft, D. A., & Anderson, L. C. (2008). Locomotion in the extinct notoungulate *Protypotherium*. *Palaeontologia Electronica, 11.1.1A,* 1–20.

Curran, S. C. (2018). Three-dimensional geometric morphometrics in paleoecology. In D. A. Croft, D. F. Su & S. W. Simpson (Eds.), *Methods in paleoecology: Reconstructing Cenozoic terrestrial environments and ecological communities* (pp. 317–335). Cham: Springer.

Dagosto, M., & Terranova, C. J. (1992). Estimating the body size of Eocene primates: a comparison of results from dental and postcranial variables. *International Journal of Primatology, 13,* 307–344.

Demes, B., Jungers, W. L., & Selpien, K. (1991). Body size, locomotion, and long bone cross-sectional geometry in indriid primates. *American Journal of Physical Anthropology, 86,* 537–547.

Dunn, R. H., & Rasmussen, D. T. (2007). Skeletal morphology and locomotor behavior of *Pseudotomus eugenei* (Rodentia, Paramyinae) from the Uinta Formation, Utah. *Journal of Vertebrate Paleontology, 27,* 987–1006.

Dunn, R. H., Tocheri, M. W., Orr, C. M., & Jungers, W. L. (2014). Ecological divergence and talar morphology in gorillas. *American Journal of Physical Anthropology, 153,* 526–541.

Elissamburu, A., & Vizcaíno, S. F. (2004). Limb proportions and adaptations in caviomorph rodents (Rodentia: Caviomorpha). *Journal of Zoology, London, 262,* 145–159.

Felsenstein, J. (1985). Phylogenies and the comparative method. *American Naturalist, 125,* 1–15.

Fisher, R. A. (1936). The use of multiple measurements in taxonomic problems. *Annals of Eugenics, 7,* 188–198.

Fleagle, J. G. (1985). Size and adaptation in primates. In W. L. Jungers (Ed.), *Size and scaling in primate biology* (pp. 1–19). Boston: Springer.

Garamszegi, L. Z. (Ed.). (2014). *Modern phylogenetic comparative methods and their application to evolutionary biology*. Berlin: Springer.

Gebo, D. L. (1988). Foot morphology and locomotor adaptation in Eocene primates. *Folia Primatologica, 50,* 3–41.

Gebo, D. L., & Rose, K. D. (1993). Skeletal morphology and locomotor adaptation in *Prolimnocyon atavus*, an early Eocene hyaenodontid creodont. *Journal of Vertebrate Paleontology, 13,* 125–144.

Gould, F. D. H. (2014). To 3D or not to 3D, that is the question: do 3D surface analyses improve the ecomorphological power of the distal femur in placental mammals? *PLoS ONE, 9,* e91719.

Gould, S. J. (1966). Allometry and size in ontogeny and phylogeny. *Biological Reviews, 41,* 587–638.

Gregory, W. K. (1912). Notes on the principles of quadrupedal locomotion and on the mechanism of the limbs in hoofed animals. *Annals of the New York Academy of Sciences, 22,* 267–294.

Gunz, P., Mitteroecker, P., & Bookstein, F. L. (2005). Semilandmarks in three dimensions. In D. E. Slice (Ed.), *Modern morphometrics in physical anthropology* (pp. 73–98). New York: Kluwer Academic/ Plenum Publishers.

Hammer, Ø., & Harper, D. (2006). *Paleontological data analysis*. Malden: Blackwell Publishing.

Hammer, Ø., Harper, D., & Ryan, P. D. (2001). PAST: paleontological statistics software package for education and data analysis. *Palaeontologia Electronica, 4.1.4A,* 1–9.

Hammond, A. S., Plavcan, J. M., & Ward, C. V. (2016). A validated method for modeling anthropoid hip abduction in silico. *American Journal of Physical Anthropology, 160,* 529–548.

Hamrick, M. W. (1996). Articular size and curvature as determinants of carpal joint mobility and stability in strepsirrhine primates. *Journal of Morphology, 230,* 113–127.

Heinrich, R. E., & Rose, K. D. (1997). Postcranial morphology and locomotor behaviour of two early Eocene miacoid carnivorans, *Vulpavus* and *Didymictis*. *Palaeontology, 40,* 279–305.

Hildebrand, M. (1985). Walking and running. In M. Hildebrand, D. M. Bramble, K. F. Liem & D. B. Wake (Eds.), *Functional vertebrate morphology* (pp. 38–57). Cambridge: Belknap Press.

Hildebrand, M. (1988). Form and function in vertebrate feeding and locomotion. *American Zoologist, 28,* 727–738.

Hildebrand, M., & Goslow, G. (1998). *Analysis of vertebrate structure* (5th ed.). New York: Wiley.

Hildebrand, M., Bramble, D. M., Liem, K. F., & Wake, D. B. (Eds.). (1985). *Functional vertebrate morphology*. Cambridge: Belknap Press.

Hills, M. (1978). On ratios—a response to Atchley, Gaskins, and Anderson. *Systematic Zoology, 27,* 61–62.

Hogg, R. (2018). Permanent record: the use of dental and bone microstructure to assess life history evolution and ecology. In D. A. Croft, D. F. Su & S. W. Simpson (Eds.), *Methods in paleoecology: Reconstructing Cenozoic terrestrial environments and ecological communities* (pp. 75–98). Cham: Springer.

Hopkins, S. S. B. (2018). Estimation of body size in fossil mammals. In D. A. Croft, D. F. Su & S. W. Simpson (Eds.), *Methods in paleoecology: Reconstructing Cenozoic terrestrial environments and ecological communities* (pp. 7–22). Cham: Springer.

Huynh Nguyen, N., Pahr, D. H., Gross, T., Skinner, M. T., & Kivell, T. L. (2014). Micro-finite element (μFE) modeling of the siamang (*Symphalangus syndactylus*) third proximal phalanx: the functional role of curvature and the flexor sheath ridge. *Journal of Human Evolution, 67,* 60–75.

Jacobs, R. L., Boyer, D. M., & Patel, B. A. (2009). Comparative functional morphology of the primate peroneal process. *Journal of Human Evolution, 57,* 721–731.

Jungers, W. L., Falsetti, A. B., & Wall, C. E. (1995). Shape, relative size, and size-adjustments in morphometrics. *American Journal of Physical Anthropology, 38,* 137–161.

Kappelman, J. (1988). Morphology and locomotor adaptations of the bovid femur in relation to habitat. *Journal of Morphology, 198,* 119–130.

Kappelman, J., Plummer, T. W., Bishop, L. C., Duncan, A., & Appleton, S. (1997). Bovids as indicators of Plio-Pleistocene paleoenvironments in east Africa. *Journal of Human Evolution, 32,* 229–256.

Kivell, T. L. (2016). A review of trabecular bone functional adaptation: what have we learned from trabecular analyses in extant hominoids and what can we apply to fossils? *Journal of Anatomy, 228,* 569–594.

Klingenberg, C. P. (2011). MorphoJ: an integrated software package for geometric morphometrics. *Molecular Ecology Resources, 11,* 353–357.

Koehl, M. A. R. (1996). When does morphology matter? *Annual Review of Ecology and Systematics, 27,* 501–542.

Koehl, P., & Hass, J. (2015). Landmark-free geometric methods in biological shape analysis. *Journal of the Royal Society Interface, 12,* 20150795.

Larson, S. G. (1993). Functional morphology of the shoulder in primates. In D. L. Gebo (Ed.), *Postcranial adaptation in nonhuman primates* (pp. 45–69). De Kalb: Northern Illinois University Press.

Lieberman, D. E. (1997). Making behavioral and phylogenetic inferences from hominid fossils: considering the developmental influence of mechanical forces. *Annual Review of Anthropology, 26,* 185–210.

Lieberman, D. E., Devlin, M. J., & Pearson, O. M. (2001). Articular area responses to mechanical loading: effects of exercise, age, and skeletal location. *American Journal of Physical Anthropology, 116,* 266–277.

Lieberman, D. E., Polk, J. D., & Demes, B. (2004). Predicting long bone loading from cross-sectional geometry. *American Journal of Physical Anthropology, 123,* 156–171.

Losos, J. B. (1999). Uncertainty in the reconstruction of ancestral character states and limitations on the use of phylogenetic comparative methods. *Animal Behavior, 58,* 1319–1324.

MacLatchy, L., & Müller, R. (2002). A comparison of the femoral head and neck trabecular architecture of *Galago* and *Perodicticus* using micro-computed tomography (μCT). *Journal of Human Evolution, 43,* 89–105.

MacLeod, N., & Rose, K. D. (1993). Inferring locomotor behavior in Paleogene mammals via eigenshape analysis. *American Journal of Science, 293,* 300–355.

Marelli, C. A., & Simons, E. L. R. (2014). Microstructure and cross-sectional shape of limb bones in great horned owls and red-tailed hawks: how do these features relate to differences in flight and hunting behavior? *PLoS ONE, 9,* e106094.

Marzke, M. W., Tocheri, M. W., Steinberg, B., Femiani, J. D., Reece, S. P., Lincheid, R. L., et al. (2010). Comparative 3D quantitative analyses of trapeziometacarpal joint surface curvatures among living catarrhines and fossil hominins. *American Journal of Physical Anthropology, 141,* 38–51.

Moyà-Solà, S., Köhler, M., Alba, D. M., & Roig, I. (2012). Calcaneal proportions in primates and locomotor inferences in *Anchomomys* and other Palaeogene euprimates. *Swiss Journal of Palaeontology, 131,* 147–159.

Nunn, C. L. (2011). *The comparative approach in evolutionary anthropology and biology*. Chicago: The University of Chicago Press.

Pohlert, T. (2014). The pairwise multiple comparison of mean ranks package (PMCMR). *R package*. http://CRAN.R-project.org/package=PMCMR.

R Core Team. (2015). *R: A language and environment for statistical computing*. GNU General Public License.

Rose, K. D. (1990). Postcranial skeletal remains and adaptations in early Eocene mammals from the Willwood Formation, Bighorn Basin, Wyoming. *Geological Society of America Special Paper, 243,* 107–133.

Ruff, C. B. (1989). New approaches to structural evolution of limb bones in primates. *Folia Primatologica, 53,* 142–159.

Ruff, C. B. (2003a). Long bone articular and diaphyseal structure in Old World monkeys and apes. II: estimation of body mass. *American Journal of Physical Anthropology, 120,* 16–37.

Ruff, C. B. (2003b). Ontogenetic adaptation to bipedalism: age changes in femoral to humeral length and strength proportions in humans, with a comparison to baboons. *Journal of Human Evolution, 45,* 317–349.

Ruff, C. B. (2009). Relative limb strength and locomotion in *Homo habilis*. *American Journal of Physical Anthropology, 138,* 90–100.

Ruff, C. B., & Runestad, J. A. (1992). Primate limb bone structural adaptations. *Annual Review of Anthropology, 21,* 407–433.

Ruff, C. B., Walker, A., & Trinkaus, E. (1994). Postcranial robusticity in *Homo*. III: ontogeny. *American Journal of Physical Anthropology, 93,* 35–54.

Ruff, C. B., Holt, B., & Trinkaus, E. (2006). Who's afraid of the big bad Wolff?: 'Wolff's law' and bone functional adaptation. *American Journal of Physical Anthropology, 129,* 484–498.

Runestad, J. A. (1997). Postcranial adaptations for climbing in Loridae (Primates). *Journal of Zoology, London, 242,* 261–290.

Ryan, T. M., & Ketcham, R. A. (2002a). The three-dimensional structure of trabecular bone in the femoral head of strepsirrhine primates. *Journal of Human Evolution, 43,* 1–26.

Ryan, T. M., & Ketcham, R. A. (2002b). Femoral head trabecular bone structure in two omomyid primates. *Journal of Human Evolution, 43,* 241–263.

Ryan, T. M., & Walker, A. (2010). Trabecular bone structure in the humeral and femoral heads of anthropoid primates. *Anatomical Record, 293,* 719–729.

Schaeffer, B. (1947). Notes on the origin and function of the artiodactyl tarsus. *American Museum Novitates, 1356,* 1–24.

Scherf, H. (2008). Locomotion-related femoral trabecular architectures in primates—high resolution computed tomographies and their implications for estimations of locomotor preferences of fossil primates. In H. Endo & R. Frey (Eds.), *Anatomical imaging* (pp. 39–59). Tokyo: Springer Japan.

Schmitt, D. (1996). Humeral head shape as an indicator of locomotor behavior in extant strepsirrhines and Eocene adapids. *Folia Primatologica, 67,* 137–151.

Shaw, C. N., & Ryan, T. M. (2011). Does skeletal anatomy reflect adaptation to locomotor patterns? Cortical and trabecular architecture in human and nonhuman anthropoids. *American Journal of Physical Anthropology, 147,* 187–200.

Shockey, B. J., Croft, D. A., & Anaya, F. (2007). Analysis of function in the absence of extant functional homologues: a case study using mesotheriid notoungulates (Mammalia). *Paleobiology, 33,* 227–247.

Smith, J. M., & Savage, R. J. G. (1956). Some locomotory adaptations in mammals. *Journal of Zoology, London, 42,* 603–622.

Smith, R. J. (1980). Rethinking allometry. *Journal of Theoretical Biology, 87,* 97–111.

Smith, R. J. (2005). Relative size versus controlling for size. *Current Anthropology, 46,* 249–273.

Sockol, M. D., Raichlen, D. A., & Pontzer, H. (2007). Chimpanzee locomotor energetics and the origin of human bipedalism. *Proceedings of the National Academy of Sciences, USA, 104,* 12265–12269.

Sokal, R. R., & Rohlf, F. J. (1995). *Biometry* (3rd ed.). New York: W. H. Freeman & Co.

Swartz, S., & Ward, C. V. (2014). Variation of proximal femoral angular orientation in human and non-human primates. *American Journal of Physical Anthropology, 153,* 250.

Swartz, S. M. (1993). Biomechanics of primate limbs. In D. L. Gebo (Ed.), *Postcranial adaptation in nonhuman primates* (pp. 5–42). De Kalb: Northern Illinois University Press.

Tocheri, M. W. (2009). Laser scanning: 3D analysis of biological surfaces. In C. W. Sensen & B. Hallgrímson (Eds.), *Advanced imaging in biology and medicine* (pp. 85–101). Berlin: Springer-Verlag.

Tocheri, M. W., Razdan, A., Williams, R. C., & Marzke, M. W. (2005). A 3D quantitative comparison of trapezium and trapezoid relative articular and nonarticular surface areas in modern humans and great apes. *Journal of Human Evolution, 49,* 570–586.

Tocheri, M. W., Solhan, C. R., Orr, C. M., Femiani, J., Frohlich, B., Groves, C. P., et al. (2011). Ecological divergence and medial cuneiform morphology in gorillas. *Journal of Human Evolution, 60,* 171–184.

Tsubamoto, T. (2014). Estimating body mass from the astragalus in mammals. *Acta Palaeontologica Polonica, 59,* 259–265.

Van Valkenburgh, B. (1994). Ecomorphological analysis of fossil vertebrates and their paleocommunities. In P. C. Wainwright & S. M. Reilly (Eds.), *Ecological morphology* (pp. 140–166). Chicago: The University of Chicago Press.

Walker, A. (1974). Locomotor adaptations in past and present prosimian primates. In F. A. Jenkins Jr. (Ed.), *Primate locomotion* (pp. 349–381). New York: Academic Press.

Warrener, A. G., Lewton, K. L., Pontzer, H., & Lieberman, D. E. (2015). A wider pelvis does not increase locomotor cost in humans, with implications for the evolution of childbirth. *PLoS ONE, 10,* e0118903.

Yalden, D. W. (1970). The functional morphology of the carpal bones in carnivores. *Acta Anatomica, 77,* 481–500.

Yalden, D. W. (1971). The functional morphology of the carpus in ungulate mammals. *Acta Anatomica, 78,* 461–487.

Yalden, D. W. (1972). The form and function of the carpal bones in some arboreally adapted mammals. *Acta Anatomica, 82,* 383–406.

Yapuncich, G. S., & Boyer, D. M. (2014). Talar articular surface curvature decreases allometrically among primates. *American Journal of Physical Anthropology, 153,* 278–279.

Yapuncich, G. S., Gladman, J. T., & Boyer, D. M. (2015). Predicting euarchontan body mass: a comparison of tarsal and dental variables. *American Journal of Physical Anthropology, 157,* 472–506.

Zelditch, M. L., Swiderski, D. L., & Sheets, H. D. (2012). *Geometric morphometrics for biologists* (2nd ed.). Amsterdam: Academic Press.

Chapter 4
Inferring Mammal Dietary Ecology from Dental Morphology

Alistair R. Evans and Silvia Pineda-Munoz

Abstract The teeth of mammals are the key interface between food and animal – where the rubber meets the road. Mammals generally use their teeth for mechanical processing, thereby facilitating and increasing rates of ingestion, digestion and fermentation. The various foods eaten by mammals respond to bite forces in different ways: some foods fracture easily, while others resist cracks propagating through them. In addition, some foods must be broken down to small pieces for effective energy and nutrient extraction; others merely need to be small enough to swallow. The most effective tooth morphology therefore varies with the mechanical properties of the food. Tooth shape can help to determine the typical food sources consumed by mammals at a given fossil locality, which in turn informs the broad environmental conditions and community structure once present at the site. In this chapter, we examine the ways in which mammalian tooth morphology can serve as an indicator of diet and thus past environments by examining the materials science of foods and the functional morphology of mammal teeth.

Keywords Biomechanics • Dental topography • Diet • Dietary mechanical properties • Functional morphology • Performance testing • Teeth

A. R. Evans (✉)
School of Biological Sciences, Monash University, Melbourne, Australia
e-mail: alistair.evans@monash.edu

A. R. Evans
Geosciences, Museums Victoria, Melbourne, Australia

S. Pineda-Munoz
Evolution of Terrestrial Ecosystems (ETE), Smithsonian Institution, Washington, DC, USA
e-mail: Pineda-MunozS@si.edu

Introduction

Teeth break food. This fundamental ability provides access to a veritable smorgasbord of foods by aiding in the acquisition and mechanical processing of foods that would otherwise be too difficult to digest. Food processing helps to fuel the high metabolic rates of mammals and, in some instances, the eventual loss of effective teeth may even put a limit on lifespan or reproductive ability (Laws 1968; Logan and Sanson 2002; King et al. 2005). Mammal teeth come in a variety of distinctive shapes, both along the tooth row and among the great diversity of mammal species, past and present. They can provide unique information about animals such as their behavior, environment, feeding preferences and even important life events. Teeth are the hardest material in the body and tend to fossilize well, making them especially valuable to paleontologists attempting to reconstruct the diet of fossil mammals.

Mammals can only consume what is available to them, and many foods are restricted to certain types of environment. Simplistically, then:

$$\text{Diet} \subset \text{Environment}$$

such that diet is always contained within (or is a subset of) an animal's environment. Determining the diet of a fossil mammal can thus help to identify the type of ecosystems it was once part of and aid in reconstructing past environments.

In the vast majority of mammals, teeth are the main mechanical interface between food and the body, with their shape typically reflecting foods and associated structures to be acquired and processed. Therefore, it is critical to know both how the shape of the tooth influences the application of force onto food and how food items respond to these bite forces through deformation and/or fracture.

Mammals have four classes of teeth: incisors, canines, premolars and molars (Fig. 4.1A). Incisors are generally peg-like or spade-shaped teeth but occasionally have more

D. A. Croft et al. (eds.), *Methods in Paleoecology: Reconstructing Cenozoic Terrestrial Environments and Ecological Communities*, Vertebrate Paleobiology and Paleoanthropology, https://doi.org/10.1007/978-3-319-94265-0_4

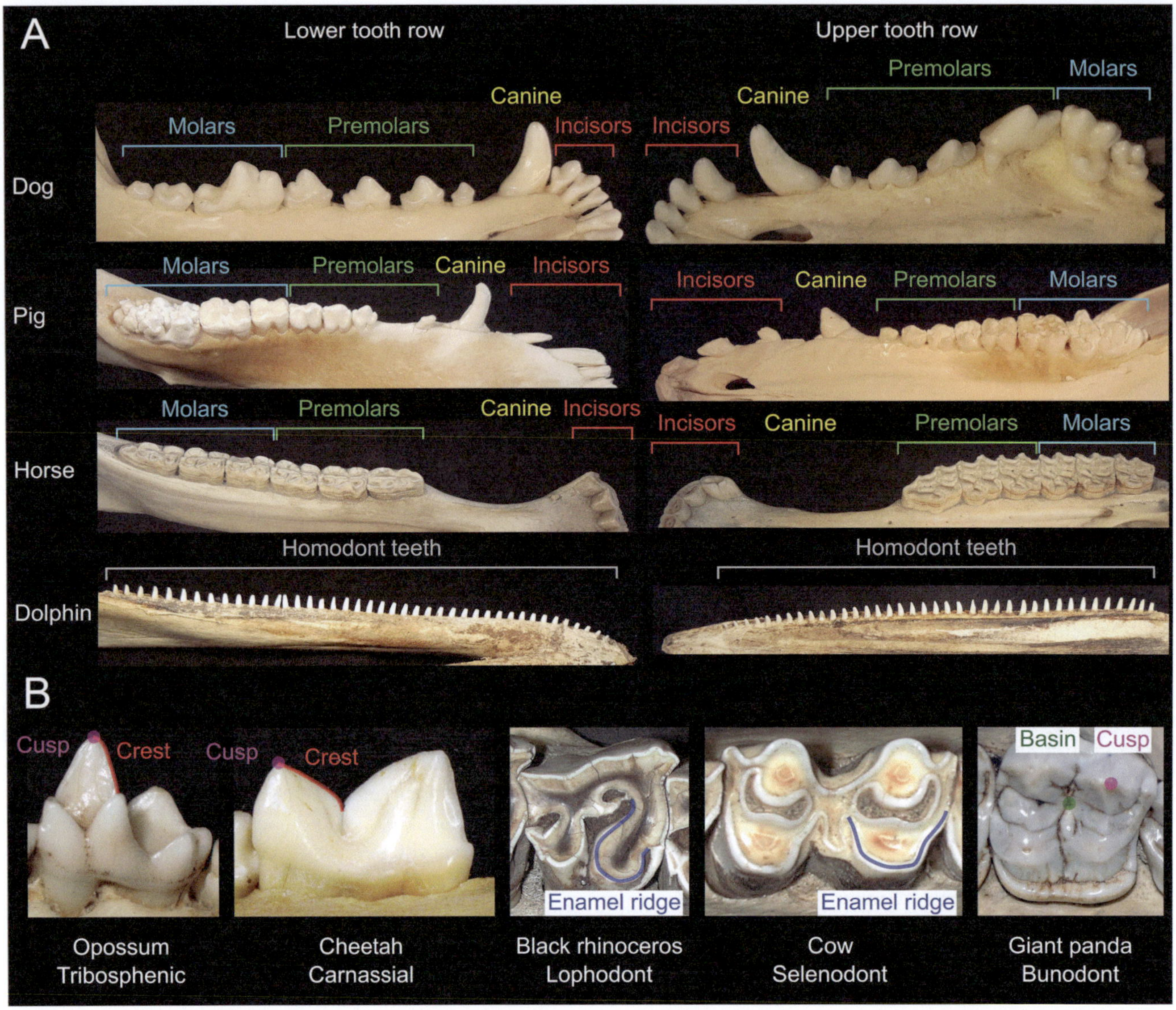

Fig. 4.1 A small sampling of the diversity of tooth morphology in mammals. A, Tooth classes (incisors, canines, premolars and molars, as well as undifferentiated homodont teeth) of lower and upper tooth rows in dog (*Canis lupus familiaris*, MZRC Dog53), wild boar (*Sus scrofa*, MZRC 3304), horse (*Equus ferus*, MZRC 617), and dolphin (*Delphinus* sp., MZRC unregistered). B, Typical features of molars, including crests, enamel ridges, cusps and basins in lower tribosphenic molar of opossum (*Didelphis virginiana*, MZH 1802), lower carnassial molar of cheetah (*Acinonyx jubatus*, MHZ U31), upper lophodont molar of white rhinoceros (*Ceratotherium simum*, ZMB 54402), lower selenodont molar of cow (*Bos taurus*, MZRC 4080), and upper bunodont molar of giant panda (*Ailuropoda melanoleuca*, ZMB 37028). Museum abbreviations: MZH, Helsinki Zoological Museum; MZRC, Monash University Zoology Research Collections; ZMB, Zoological Museum Berlin

complex shapes such as combs. Canines are almost always tall, single-cusped teeth but may have small accessory cusps. These two tooth classes are generally used for capturing, holding, deforming and breaking off fragments small enough to be processed in the mouth. Premolars may also be involved in holding or fragmenting foods. The molars of most mammals are involved in cyclic chewing to fragment foods to small enough particles for fermentation or digestion; when it includes tooth-tooth contact, this is called mastication.

Foods typically undergo both mechanical and chemical breakdown to enable extraction of sufficient nutrients from them. Those foods that are easily broken down and digested in the gut may require very little mechanical processing; mechanical processing merely makes them small enough to swallow. Easily digested foods include many soft animal tissues (muscle and internal organs) and some plant products (pollen, nectar, and sap). Many plant structures represent the greatest processing challenges for mammals feeding on them. The cells of plant structural tissue, including stems, roots, and leaves, contain nutrient-rich cytosol which is encased in relatively indigestible cell walls (Sanson 2006). Herbivores may focus on breaking open the cells to release the contents for digestion or fragmenting the cell walls to increase surface area to ferment the cellulose in them. Mastication is key for both of these steps, with the shape of teeth and precision of occlusion greatly increasing the effectiveness of breaking down plant materials. Teeth may also function in cracking open shells or bones, even when these are not the components to be digested.

Mechanical processing using teeth is either not required or not carried out by some mammals, such as the filter-feeding crab-eater seal *Lobodon carcinophaga* and ant-eating echidna *Tachyglossus aculeatus*. In these animals, the teeth do not reflect the mechanical properties of the diet – making inference of diet from dental morphology unsuccessful – or teeth cannot be studied because they are absent altogether.

Teeth perform many functions in mammals other than capture and processing of food, including grooming, sexual signaling, digging, gouging and gnawing. In general, these additional roles are carried out by the anterior teeth (incisors and canines). Molar teeth are essentially only related to food processing and so generally give stronger dietary signals.

There are many other aspects of teeth that inform about diet and ecology. High tooth height (hypsodonty) can reveal a lot about the diet and environment of a mammal, as discussed by Barr (2018). Processing of food and associated materials usually leaves small physical marks on teeth termed microwear, the patterns of which can vary with diet and environment, as discussed by Green and Croft (2018). Mesowear is another method for interpreting diet from wear-induced shape change and is reviewed in the same chapter.

In this chapter, we review how mammal teeth process particular foods and how this process influences dental morphology. We outline some basic principles of materials science as they relate to food and how it is broken down. Then we discuss how the shape and size of teeth and their component parts are expected to influence their performance and how assessment of performance can be made through computer simulation or physical testing. Through this knowledge, we can make approximations of the diets of fossil mammals based on their gross dental morphology or quantitative assessment of tooth shape. Finally, we explore some recent examples and demonstrate the fundamentals of some of the modern 3D morphology approaches.

Historical Background

General associations among tooth shape, function and diet stretch back to the beginning of the natural sciences, with Aristotle (Physica; Hardie and Gaye 1930) commenting on the differences between cutting incisors and grinding molars. In modern Western science, these themes have been further explored and expanded, such as by Hunter (1771–1803 [1865 post.]), who observed that teeth are used to catch, collect and prepare food for digestion, and that the physical properties of food affect how they are acquired and prepared. Gregory (1922) suggested that tooth shape evolved in order to improve mechanical efficiency for certain foods, and Simpson (1933a) correlated the direction of jaw motion with specific food specializations. Kay (1975) summarized these and other intellectual advances of our knowledge of the functions of teeth through the 18th to 20th centuries.

Consideration of overall tooth shape was a frequently-used approach in early modern paleontology to interpret the likely diet of fossil mammals, and this method was vastly improved by quantification of form. One of the earliest such quantitative studies of dental morphology was that of Crusafont-Pairó and Truyols-Santonja (1956). These authors estimated the degree of carnivory in fossil carnivorans based on angular measurements of the carnassial tooth that quantified the elongation of the carnassial blade and reduction of the talonid basin. The work of Rensberger (1973) deserves highlighting as a landmark in quantitative functional dental morphology, in which he carried out computer simulations of 3D model tooth shapes to examine the effects of enamel band orientation, dentine basin depth and tooth wear on pressure and effectiveness of herbivore dentitions.

From the 1970s onwards, more detailed studies and refined approaches made significant advances in revealing relationships between teeth and diet. First, new methods to quantify tooth shape enabled more precise comparisons among species and correlations with diet (e.g., Kay 1975). Second, several authors undertook more theoretical considerations of how the shape of a tooth affects its ability to fracture food items and what shapes would be expected to most effectively fracture certain foods (e.g., Osborn and Lumsden 1978; Lucas 1982; Lucas and Luke 1984). Third, the crucial role that the mechanical properties of different foodstuffs play in this process began to come to the fore (Kay and Hiiemae 1974; Kay 1975; Rosenberger and Kinzey 1976), along with the necessity to accurately quantify the mechanical properties of the foods (Lucas 1979). These studies began to lay the foundation of a clearer conception of how foods break and the role that tooth shape plays in facilitating breakage.

Approach

We can consider there to be two main approaches for inferring the diet of a mammal from the shape of its teeth: analogy and biomechanics. These approaches have different logical bases and thus vary in the scope and precision of the predictions they are able to make. The choice of methodology often largely depends on the generality of the question and the nature of the available material.

Prediction by Analogy

A straightforward way of inferring what an extinct mammal may have fed on is to compare its dentition to that of extant species with known diets. This is the foundation of much of comparative anatomy and has a long and distinguished history. In its application to the teeth-diet relationship, if the teeth of the extinct and extant species resemble each other, then analogy suggests that both species specialized on similar foodstuffs.

Using analogy for dietary inference from teeth has often been based on tooth type, i.e., the general shape of the teeth (Fig. 4.1A). For instance, tribosphenic tooth shapes are usually interpreted as indicating an insectivorous diet, bunodont a hard food diet, lophodont a plant-based diet, and homodont a piscivorous diet. In a more general sense, a fossil tooth bearing a series of occlusal ridges would resemble the teeth of many extant herbivores and thus may well have belonged to a herbivore. Gregory (1920: 141) used analogy to speculate that the molars of early notharctine primates were adapted to an insectivorous diet, while those of later members of the group were for "vegetable food, especially leaves and fruits".

However, the use of the analogy approach with gross tooth morphology has some major limitations. In particular, there are no clear rules on how similar teeth have to be to justify the assumption of similar diets, which means that assessments of analogy can be highly subjective. Likewise, problems arise when the shape of a fossil tooth does not obviously resemble any living species and therefore cannot be interpreted in a modern context. The phylogenetic history of a species can also substantially influence tooth form. For example, all species of a group may have quite similar dentitions despite varying somewhat in diet (e.g., Chemisquy et al. 2015). This may be because insufficient time has elapsed for sufficient evolutionary adaptation, or the dentition is relatively generalized and can adequately process the range of diets found in the group.

To help address the issue of how to analogize disparate shapes, we can seek ways to quantify shape, allowing more objective assessments of the extent and the manner in which two shapes are similar. Length:width ratios have been the main staple of tooth measurement (e.g., Kieser 1990), but they ignore any detail of tooth morphology beyond the aspect ratio of the whole tooth. Quantification of 2D linear or area measurements of teeth and component parts can give keen insights to some dietary and ecological questions, particularly in taxonomically-limited samples (e.g., Van Valkenburgh 1989).

More sophisticated analyses of 2D shape have sought to hone in on shape characteristics. A pioneer in this field was Schmidt-Kittler (1984), with his work on quantifying 2D parameters of the enamel occlusal surface in herbivores. More recently, Occlusal Enamel Index (OEI) has been developed to quantify the relative length of occlusal enamel bands compared to the tooth area in herbivorous mammals (Famoso et al. 2013; Famoso and Davis 2016). Grazing mammals feeding on more complex and fibrous vegetation typically display higher values of OEI. Fractal Dimensionality, based on fractal theory, computes how the measured length of the enamel ridge changes with measurement resolution (Stone and Telford 2005; Candela et al. 2013). It has been used to study the relationship between morphology and diet and has an advantage over OEI of being more independent of body size (Famoso et al. 2016).

3D linear measurement has been an additional challenge but has been successfully used to differentiate diets based on tooth morphology (e.g., Strait 1993). However, the high degree of variation in mammal tooth shape (Fig. 4.1B) has been a fundamental stumbling block for comparisons across Mammalia.

Detailed shape analysis of the dentition can be captured by geometric morphometric analysis, as detailed by Curran (2018), but the strict requirements of identical number of landmarks means that the use of this technique to compare diverse dental morphologies with differing numbers of cusps or crests is limited.

Prediction Through Biomechanics

Instead of searching for analogues, the biomechanical approach to diet reconstruction borrows concepts from mechanics and materials science to model teeth as tools whose performance under certain conditions can be tested. This functional or performance analysis can consider how teeth fracture food from a bottom-up, theoretical approach by building on materials science concepts, or from a top-down, experimental approach by directly testing real or idealized tooth shapes with food materials. Either way, it is necessary to take into account the properties of both the tool (i.e., the tooth), and the material (i.e., the food) to which it is applied.

Mechanical Properties of Foods

The mechanical properties of a food influence how it can be handled, fractured, and broken into smaller pieces (fragmented) for swallowing and digestion in the gut and are best described using the terms and concepts employed by materials scientists. The concepts are highly detailed and use specific terms, some of which have closely related words in general use with somewhat different meanings, such as 'plastic'. Below, we provide brief descriptions of the most important terms and give relatable examples of everyday foods or materials. A more detailed review is provided by Berthaume (2016). When a force is applied to a material, it is referred to as 'loading' the material, as in a load is being applied.

Materials scientists investigate how a material behaves when it is loaded by mounting it in a force testing machine that pulls (in tension) or pushes (in compression) the material and measures the distance that is applied and the force that results from that movement. When these are plotted against one another, the force-displacement graph (Fig. 4.2 A) can reveal how much force and energy are required to initially deform and then fracture the material. However, it is generally desired that the properties measured be standardized by size. Therefore, we can calculate the relative change in dimensions of the material as the force is applied, called the *strain*, and the standardized force through a cross-section of the material, called the *stress*. The resulting stress-strain graph (Fig. 4.2B) can be used to determine many standard material properties, some of which are described below. Berthaume (2016) explains the details of how to calculate stress and strain.

Strength: The stress that the material can withstand before yielding or fracturing; in practice this means that a stronger material will require a higher force to fracture should all other factors (e.g., shape, size, loading conditions) be held constant. We know that bone is stronger than fingernail or insect exoskeleton, as it takes higher stress to fracture it. For a force-displacement graph, we can measure the breaking strength, or the maximum force that was required to fracture the object (Fig. 4.2A), while the ultimate tensile strength is the maximum engineering stress that a standardized test piece can withstand (Fig. 4.2B).

Elastic modulus (Young's modulus): How much a material deforms when it is loaded while the deformation is still elastic (i.e., before it is permanently deformed). It can be calculated from the slope of the initial, linear part of a stress-strain graph (Fig. 4.2B). Bone has a high elastic modulus, but that of skin is low. In lay terms, this is often considered to be 'stiffness', but materials science often uses 'stiffness' to describe the response of a structure rather than a material therefore, it is scale-dependent.

Toughness: How much energy it takes to propagate a crack through the material (called 'energy release rate' by Berthaume 2016). This can be determined from the area under a stress-strain graph (Fig. 4.2B). Breaking materials such as glass, nut shells, or eggshells can require very little energy because cracks become self-propagating and do not require additional input of energy; more energy must be used to extend a crack in plasticine or muscle tissue.

Brittle: A material that does not deform prior to fracturing (opposite of ductile). The blue line in Fig. 4.2C shows very little strain (deformation) before fracture at high stress. Glass and eggshells are classic examples of brittle materials, as they do not change shape before they fracture – in fact, you can put the pieces back together again, as they have

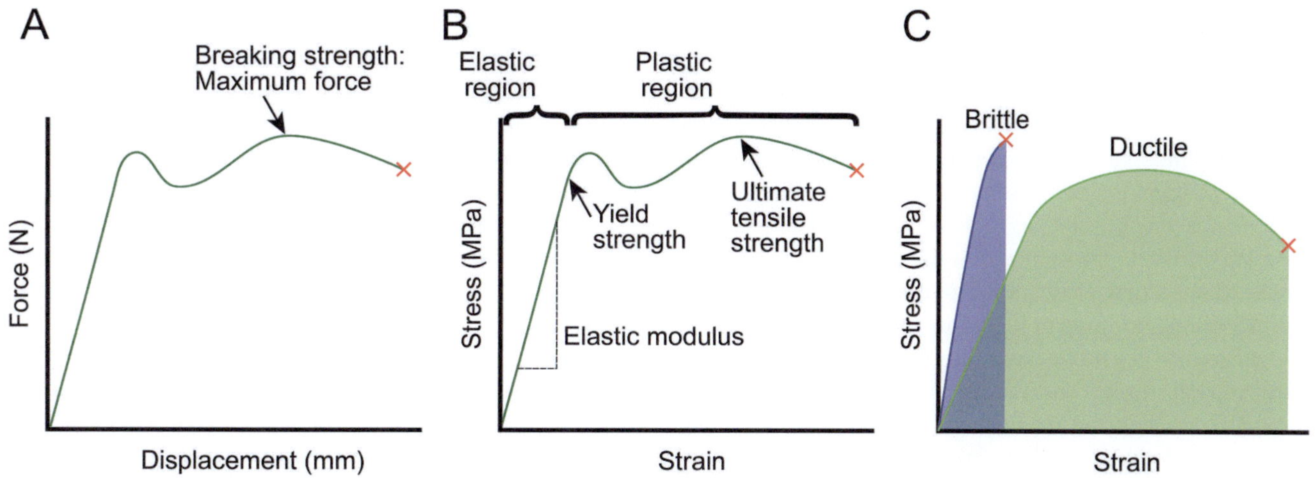

Fig. 4.2 Materials testing graphs illustrating key concepts in materials testing. A, Force (in Newtons, N) exerted by the test piece when the opposing end of the piece is pulled in tension (displacement, in mm). Fracture of the test piece is shown by a red x. B, When force is standardized by cross-sectional area of the test piece, it shows the engineering stress (in megapascals, MPa) in the test piece; engineering strain is the proportional change in the length of the test piece (unitless, or in percentage). The elastic region at the initial part of loading shows a linear change in stress, and the slope of this region is the elastic modulus. The yield stress is the stress at which elastic behaviour ceases and permanent deformation has occurred. The ultimate tensile strength is the highest stress that the material withstands before failure. C, Schematic plots of tensile stress-strain for brittle and ductile metals. Shaded areas under the graphs show that toughness is generally higher for ductile materials than brittle ones. Adapted from Berthaume (2016) and Callister and Rethwisch (2014)

retained their shape; this can also be called elastic fracture.

Ductile (also plastic): A material that deforms permanently before fracturing (opposite of brittle). The green line in Fig. 4.2C shows that substantially more strain (deformation) has occurred before fracture. Plasticine and ripe banana flesh behave in a ductile manner and so are permanently deformed when a force is applied to them.

Elastic: The deformation is not permanent, so the material will spring back to the original shape when the force is removed. Many materials behave in an elastic manner, with a linear stress-strain curve over very small strains (the 'elastic region' in Fig. 4.2B). Truly elastic materials, where the stress-strain curve is linear during moderate strain, are very rare. The clothing material we generally call 'elastic' has the property that, after moderate stretching, allows it to bounce back to its original length when no longer being stretched. However, 'elastic' and rubber behave in a more complex fashion termed hyperelastic. Vertebrate skin and muscle behave in a generally elastic manner.

Viscosity: The behavior of the material varies with time or speed of loading. While this may make you think about liquids with very different viscosities like water and honey, solids can also behave differently when loaded quickly or slowly. Most biological materials combine aspects of elasticity and viscosity and so are *viscoelastic*; the way they respond to loads changes with time (such as speed of loading). 'Silly Putty' is an extreme example of this, as it can bounce or brittle fracture when loaded quickly but flow when the load is applied slowly.

Fracture Mechanics

The surface area of food can be increased by plastic deformation and/or by dividing it into multiple fragments. Very few foods undergo extensive plastic deformation without fracture, perhaps including some soft fruits. To fragment a food, first a crack must be initiated in the material. This crack may then be propagated through the material. Fracture occurs when atomic bonds are broken in the material, usually by pulling atoms apart or sliding them past one another. Real materials generally have tiny cracks throughout, and when loaded, these cracks may elongate or propagate, eventually causing the material to fail. When a tooth first penetrates a food, a crack is initiated. A crack will only propagate when sufficient energy is concentrated at the crack tip, increasing stress and causing more bonds to fail. This is much easier to do when the crack tip is sharp and the material has a higher elastic modulus; both result in a larger percentage of energy being transferred to the crack tip rather than absorbed by the material through plastic deformation.

Cracks are easier to propagate through materials like eggshell that have a low toughness (critical energy release rate) than through materials like plasticine with high toughness. They also propagate much more easily if they are running through a homogeneous material. When a crack hits an interface between different materials, it is usually diverted, which blunts the crack, and energy is wasted in the crack running in alternative directions. Second, the surface energy at the interface between the materials may be higher than the energy needed to continue to propagate the crack through the first material, so the crack cannot initiate and propagate into the second material. Third, large numbers of microcracks in the material cause the energy to be dispersed among them, resulting in a much higher level of energy needed to propagate any single crack. Such crack-stopping mechanisms in structures with plywood-like layering make it less likely that the structure is easily fractured, and increase its toughness.

Tool Shape

Because different foods vary in how they break when force is applied, tooth shape can be optimized to achieve more efficient processing. One way to investigate this optimization process is to test how specific changes in tooth shape alter effectiveness. In such tests, the force and energy required to fracture and fragment a particular type of food are useful indicators of performance (Evans and Sanson 1998). Thus, for a particular tooth shape to be effective, the required force must not exceed the maximum bite force that can be generated by the muscles and lever mechanics of the jaws. Also, the average energy required to process it must be below the amount of energy that can be extracted from the food, such that there is a net gain of energy from all food consumed. A particular food may still be worth consuming even if it requires more energy to process it than is gained if it fulfills other requirements such as providing essential minerals or vitamins.

Teeth come in a great variety of shapes (Fig. 4.1), each of which differs from the next in the number, size, and shape of cusps, crests (including tall blades and low enamel ridges), basins, and valleys that make up the tooth crown. Which of these tools predominate largely depends on the type of food being processed. Lucas and Luke (1984) hypothesized that crack propagation in tough or ductile foods like grass or muscle tends to require an elongate blade, whereas brittle foods like nuts are more efficiently processed by a mortar-and-pestle arrangement. The critical material property requiring blades may actually be fibers that need to be cut rather than high toughness per se (Wright 2005).

Cusps: A cusp is a local topological maximum of the tooth surface, analogous to a hill. Some teeth, such as human canines, have only one cusp, while premolars and molars generally each have two or more cusps, such as the bunodont molars of the giant panda (Fig. 4.1B). The distribution and shape of cusps can vary both within a tooth and across the dentition. Cusps frequently initiate cracks in a food item, sometimes alone and sometimes by occluding into the space between other two cusps or crests. For simple cone- or mound-shaped cusps, there are two main aspects of shape that influence the force and energy required to penetrate and drive through food.

First, the radius of curvature at the tip of the cusp (*tip sharpness*) determines the area of contact between the cusp and food and thus also the pressure (Fig. 4.3A). A smaller radius of curvature (higher tip sharpness) will increase stress in the food for a given bite force (Lucas 1982). Once the tip of the cusp has penetrated the food, the shape of the shaft will determine the force and energy required for the cusp to drive though. This is *cusp sharpness*, which can be quantified as the volume of the cusp at a given distance from the tip (Evans and Sanson 1998). A smaller volume (higher cusp sharpness) will take less force and energy to penetrate most foods.

Most mammal teeth have multiple cusps, and the relative sizes, shapes, and arrangement of all cusps may affect how they interact with foods. Some of these factors have been considered by Berthaume (2014) and are discussed below. When tall cusps have ridges running down their sides, such as many bat canines (Freeman 1988), they may be considered to be crests, and so some of the crest shape characteristics discussed below will also be relevant (Freeman and Lemen 2006).

Crests: A crest is an elongated, blade-like feature with higher curvature in one direction. The newly-erupted tooth crown is referred to as the primary occlusal morphology, and is covered with enamel in most teeth. Prominent enamel-covered crests are found on carnassial and tribosphenic teeth (Fig. 4.1B). Following wear, the underlying dentine becomes exposed and enamel ridges are formed as part of the secondary occlusal morphology (Fortelius 1985). Enamel ridges are found on lophodont and selenodont teeth (Fig. 4.1B).

In a manner similar to cusps, the radius of curvature along the crest edge (*edge sharpness*) will influence how the blade initiates a crack in food (Fig. 4.3B). Occluding blades have one leading or rake surface, and the angle between the rake surface and a line perpendicular to the direction of movement is the *rake angle*. A higher rake angle decreases the contact area between the blade and the food and therefore displaces less material as the blade is driven through the food. Overall, a blade with a high rake angle thus tends to fracture food using less force (Evans and Sanson 2003).

The angle between the long axis of the blade and a line perpendicular to the direction of movement is the *approach angle* (Fig. 4.3C). An approach angle of zero is equivalent to pushing a horizontal knife blade directly down onto food. Increasing the approach angle increases the mechanical advantage of the blade by reducing the amount of material cut per unit distance moved. A higher approach angle therefore tends to reduce force requirements when propagating a crack through food (Evans and Sanson 2003).

Manufactured blades, such as scissors, typically have a small space (called *clearance* or *relief*) behind the edge of the blade (Fig. 4.3B). This reduces friction between blades and also reduces the tendency for material to force occluding blades apart.

Basins and opposing cusps: A basin is a depression in the tooth surface, in some ways the inverse of a cusp (Fig. 4.1B). Some basins, like the talonid basin of a tribosphenic lower molar, accommodate the occlusion of an opposing cusp. Shape characteristics that affect the function of basins are less clear than for the other tool types discussed here. It is possible that simply the area of the basin indicates how much food can be crushed between it and its occluding

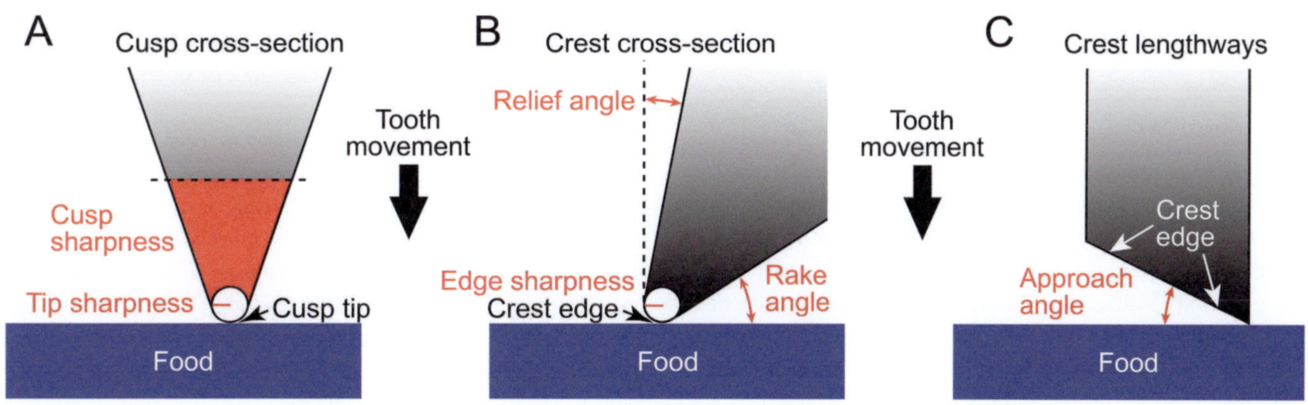

Fig. 4.3 Diagrammatic illustrations of functional parameters for: A, a cusp; B-C, a crest. In all three diagrams, the tooth is moving down from the top of the page to fracture food (blue). The long axis of the crest is perpendicular to the page in B and parallel to it in C. Partly adapted from Evans and Sanson (2003) and Evans (2005)

cusp. Lucas and Luke (1984) predict that some difference in the curvature of the basin and occluding cusp would improve performance. Yamashita (1998) tested this relationship in Malagasy lemurs by hypothesizing that the radius of curvature of the basin should increase relative to the radius of the opposing cusp with increasing food hardness, aiding food fracture by allowing for lateral excursions of the cusp within the basin. However, hard foods were actually correlated with a cusp radius more similar to the corresponding basin radius, creating a tight fit. This may indicate that these hard foods require precise occlusion of the crests surrounding the basin to improve fracture of foods. Basin volume has been measured in chimpanzees through the wear sequence by Klukkert et al. (2012) and in other hominids by Zuccotti et al. (1998) but without specific functional hypotheses.

Quantifying Tooth Shape

One of the significant challenges of investigating the relationships of potential shape parameters with performance and diet is the difficulty of measuring them from real teeth. Parameters such as rake angle are essentially cross-sectional shape measurements. Traditional methods for measuring these characteristics would involve physically sectioning the tooth or a replica, followed by optical projection or imaging for measurement. For example, Freeman (1992) investigated the cross-sectional shape of bat canines using sectioned replicas. Other metrics such as 3D surface area, volume, or surface curvature are even more difficult to assess using physical methods. The challenge is exacerbated by the particularly small teeth of some mammals. Along with the initial application of 3D shape capture methods to teeth (Reed 1997; Zuccotti et al. 1998; Jernvall and Selänne 1999; Evans et al. 2001) came the ability to measure these functional parameters and associate them with performance, diet and tooth wear (Evans 2005; Evans and Sanson 2005; Evans et al. 2006).

Tool Number

Each tooth is comprised of one or more component tools that may vary in type, shape and size. The overall performance of a tooth will depend on the action of all of these tools as they process food. While the precise performance of each component will vary with its shape and mode of occlusion, the number and type of these features will give an indication of how many opportunities there are for food fracture. For instance, when two teeth bearing a single blade occlude, there is one opportunity for the blades to trap a food item between them. However, if each of these teeth had two blades, each of which occluded against the two blades of the opposing tooth, there would be four opportunities for food fracture. Extending this analogy to more complex teeth, one can assume that the total number of food fracture opportunities is likely greater than proportional to the number of component tools.

This approach assumes that all tools act equally in fracturing food, which is unlikely to be the case in most instances. Small features may not interact or have sufficient space to successfully fracture food, while sharper enamel ridges may be more effective than blunter ones. However, for broad comparisons among tooth types and diets, counting the number of available tools may still be informative.

The quantification of tool type and number was pioneered by Jernvall (1995; Jernvall et al. 1996), who classified tooth crowns based on their number of large cusps (buccal and lingual) and crests (longitudinal and transverse). More recent methods have instead used 3D tooth surface data in combination with Geographic Information Systems (GIS) algorithms (Evans et al. 2007). Specifically, the orientation at each point on a digitized 3D surface of a tooth can be expressed in terms of one of eight cardinal compass directions, such as north or south-east (Fig. 4.4D). Adjacent points on the tooth surface with the same direction can be grouped together as a 'patch' or 'clump', with the number of these orientation patches giving a measure of the total number of topographic features and therefore the 'complexity' of the tooth surface. This measure is termed 'orientation patch count' or 'OPC'.

The patches isolated by the OPC method do not uniquely count features such as cusps, crests, or wear facets. Any one feature may vary in how many patches it represents according to its shape and size – a large, conical cusp will have eight patches, while a small one may have only four patches (because of the finite size of the elements representing the surface), and a horizontal blade sometimes only two. Because of the great diversity of component shapes, the OPC method gives an overall estimate of tooth components that have the potential to apply force to food and is expected to be highly correlated with the number of potential breakage sites on the tooth.

The OPC method was first used to compare the cheek tooth row complexity of carnivorans (species in the order Carnivora) and rodents (species in the order Rodentia), two highly divergent groups among mammals that nonetheless have overlapping dietary specializations in their component species (Evans et al. 2007). In that study, tooth row complexity clearly distinguished broad diet types (including hypercarnivory, animal-dominated omnivory, and herbivory), but there was significant overlap between carnivorans and rodents in the same diet class. This indicates some degree of scale- and phylogenetic independence, meaning that dental complexity gives a good indication of general

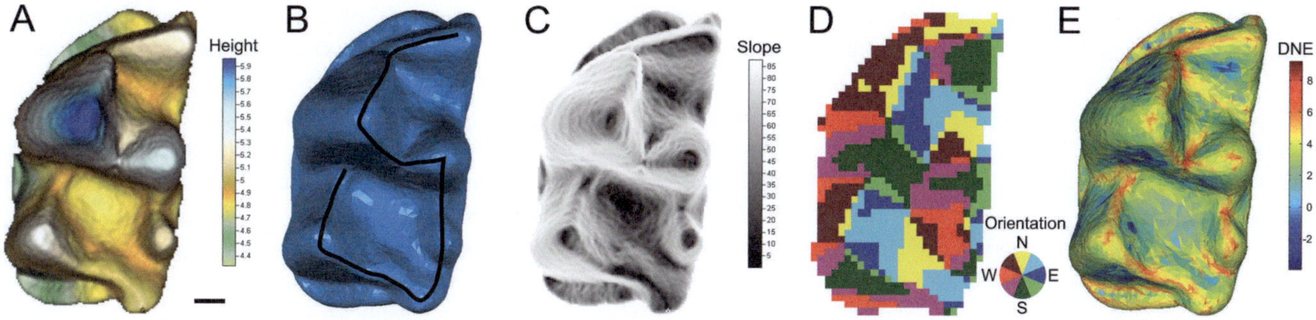

Fig. 4.4 Dental topography and complexity measures in occlusal view, demonstrated using the left lower second molar of the dasyurid marsupial *Antechinus agilis* NMV C12676 (Museums Victoria): A, Height-encoded map (height in mm); B, Shearing edges measured for calculation of shear ratio; C, Slope (in degrees); D, Orientation map for OPC calculation (colour wheel shows orientation and compass directions: north, south, east and west); E. Dirichlet normal energy (DNE). Scale bar = 0.2 mm. Adapted from Evans (2013)

diet regardless of the size or taxonomic affinities of a species. Some difference in the OPC-diet relationship was found between platyrrhines and prosimians by Winchester et al. (2014), but the general patterns of OPC with diet for these groups were similar.

Tooth Topography

Like the number and shape of its component tools, topographic measures of the entire tooth surface can also reveal diet. GIS or topographic algorithms can calculate the slope at each point of the tooth surface (grid point of a GIS map or vertex of a polygonal surface; Fig. 4.4C) with respect to the horizontal occlusal plane (Ungar and Williamson 2000). The mean slope of all points is the *average slope*.

Variation in height over the tooth surface can be quantified using the *relief index*. The relief index is based on the 3D true surface area of the crown (SA) and the 2D projected area of the tooth in occlusal view (PA), either as SA/PA (Ungar and Williamson 2000) or $\ln(\sqrt{SA}/\sqrt{PA})$ (Boyer 2008).

Sharper cusps and crests are likely to improve performance of teeth when fracturing ductile foods. Rather than measuring the sharpness of separate cusps, the average curvature or sharpness of the entire tooth surface can be quantified using Dirichlet normal energy (DNE; Fig. 4.4E; Bunn et al. 2011). DNE measures the degree of bending of the surface (bending energy), in essence giving the average curvature of the tooth surface.

Tool and Tooth Size

The absolute size of a tool or tooth affects its overall performance. Two cusps of the same shape that differ in size will vary in their performance in penetrating food because their interaction with the food occurs at the true size of the tool, not as size-removed shape (Evans and Sanson 1998). The function of a tooth therefore cannot be divorced from its size. In general, a larger tool can process more food with each bite, either through a larger area of contact or because it forms a longer blade. A blade cannot cut food any longer than itself, and a blade of twice the length can, in principle, divide a food item twice as long.

For great apes, Berthaume (2014) showed that cusp radius of curvature is significantly correlated to cusp size, but only weakly. Cusp radius of curvature and the arrangement of multiple cusps interact to influence the optimal tooth shape for a given food size (Berthaume et al. 2014).

Some of the most common dental metrics, such as measures of relative crest length (Kay 1975, 1978), are related to size. In particular, the objective of *shear quotient* and *shear ratio* is to standardize the length of crests relative to the size of the tooth (Fig. 4.4B). Higher shear quotients are associated with foods that require more processing, such as leaves and insects, compared to foods such as fruits (see Bunn et al. 2011 for review). This could be because longer crests are likely to be associated with taller, sharper cusps (Evans and Sanson 1998) or to ensure more food is processed per chew (such as undulating crests in ungulates).

Increasing tool size, and thus the amount of food processed, generally comes at the cost of increased force and energy for a given occlusal stroke, but the overall processing time (such as number of bites or chews) may be reduced. Depending on how much mechanical processing is required, there is likely to be a trade-off between the size and the number of tools present on a given tooth. While "bigger must be better" is often the mantra regarding teeth, as it implies more processing per bite (such as with more potential food fracture events), it also requires a higher bite force. Larger teeth may also increase the chance of missing small foods due to larger gaps between tooth features. Selection may be for minimum number of chews (maximum amount processed per bite), or minimum bite force. If selecting for the former, then current bite force must be adequate or there will be selection pressure for increased bite force.

The total size of a tooth crown has implications for the number of tools (e.g., cusps, crests, basins, etc.) and their relative sizes that can be accommodated. For a fixed tooth size, average tool size must decrease if the number of tools increases, and vice versa. Alternatively, increasing tooth size will increase the number of tools of a given size that can fit on the tooth surface. The overall processing ability (typically of chewing) can be assessed by looking at all occluding teeth, usually the molars and sometimes the premolars. The scaling relationships of tooth to body size in fossil taxa can also be investigated when combined with body size estimation (Hopkins 2018).

Simulation and Performance Testing

Computer simulations can be used to model and test the performance of different tool and tooth shapes under particular circumstances. Finite element analysis (FEA) is based on computer models of the object of interest (e.g., a tooth), with each model comprising many simple triangular or polygonal elements (Rayfield 2007). Simulated loads are applied to the computer model and then used to calculate the distribution of stress and strain. The majority of dental FEA studies so far have focused on the risk of tooth fracture rather than the question of how effectively a given tooth will fracture a given food. For instance, Macho and Spears (1999) used 2D FEA models to relate the differences in occlusal topography in great apes to the stresses in the tooth when exposed to loads required to fracture food items. Recent work by Berthaume et al. (2010) on fossil hominins has greatly extended those early efforts to assess the influence of tooth size and shape along with food size on stress in loaded foods. More precise information on loading, based on simulated tooth contacts using the Occlusal Fingerprint Analyzer, and of the ligamentous and bony support of the tooth leads to greater confidence in interpretation of likely stresses during chewing (Benazzi et al. 2011, 2012, 2014).

Though promising, FEA studies at present are generally limited in that the material properties of foods (e.g., anisotropy and viscoelasticity) are greatly simplified or even ignored (see Skamniotis et al. 2016 for an exception). Likewise, they cannot simulate fracture mechanics, including crack propagation, in a broad array of foods (fibrous foods, leaves, grasses, nuts, seeds, meat, and bone). In addition, crack propagation in enamel and dentine requires further research. As a result, physical testing currently remains the best way to examine such questions further. Abler (1992) was a pioneer in his investigation of the functional consequences of blade shape and serrations in tyrannosaurids and other animals. Anderson and LaBarbera (2008) tested a number of the hypotheses put forward by Abler (1992) and Evans and Sanson (1998), examining the effect of approach angle when fracturing asparagus and shrimp. Berthaume et al. (2010) used steel casts of hominin teeth to load synthetic domes. Other notable examples include Whitenack and Motta (2010) testing performance in shark teeth, and Crofts and Summers (2014) using model morphologies of fish teeth for crushing shells.

Strengths of Approach

Tooth morphology represents the evolved response to many factors, including the mechanical processing of foods and the need to resist fracture and wear. It is obvious that the teeth of any given species of mammal are sufficient for processing relevant foods (where necessary), with the shape of teeth having evolved in response to the mechanical properties of the diet.

Computer simulation and physical testing provide independent sources of information about expected tooth shapes for given foods, as well as the likely performance of both modern and fossil teeth. Control of shape variables of the teeth and mechanical properties of test foods in carefully designed experiments are extremely important in being able to tease apart the influences of these components.

Biases and Shortcomings

The fundamental assumption underlying the use of fossils to reconstruct biology is that there is a tight match between form and function (Bock and Wahlert 1965), although there may be many forms that can produce the same function. Here, we are specifically interested in whether tooth shape truly reflects diet. Given a large enough bite force with a strong, flat tooth surface, probably any food could be consumed, including bone. However, such a system would be over-engineered, more expensive than necessary, and therefore evolutionarily unstable. Nevertheless, current techniques and interpretations of dental morphology cannot reveal all of the foods consumed by an animal.

Limits of Prediction

Foods that require no mechanical processing do not exert a selection pressure for tooth shape, and thus cannot be inferred from tooth morphology. It may be that tooth shape is correlated with the most mechanically challenging foods in a diet. Such foods may be "fallback foods": foods that are only occasionally consumed (e.g., depending on the season or life stage) and that require more or alternative processing effort than the typical diet of the species but that are crucial to the survival of the individual (i.e., under strong natural selection).

While some mammals have evolved an essentially ideal tooth shape for their diet, such as many carnivores and insectivores (Evans and Sanson 2003), dental morphology often represents a trade-off between performance and various constraints, including mechanical (e.g., risk of tooth fracture), developmental, spatial, and phylogenetic. For example, overall tooth shape and number in each class (incisors, canine, premolars and molars) is often conserved across higher taxonomic groups to the point that most postcanine teeth can relatively easily be identified to family level. Similarly, tooth shape may not always entirely reflect the diet of the species in question but rather preserve a 'memory' of the foods consumed by its ancestors. The polar bear *Ursus maritimus* is a dedicated carnivore but retains a relatively complex tooth shape similar to its close relative the brown bear *Ursus arctos*, which is much more omnivorous (Evans et al. 2007). The short divergence time between these two species may have been inadequate to drastically change polar bear molar shape, even though it is visibly less complex. As lineages evolve, there may be a substantial delay between changes in food preferences and subsequent dental morphological adaptations (Davis and Pineda Munoz 2016); however, the rate at which tooth morphology can change has yet to be quantified.

Limitations of Techniques

While the new morphological approaches described above have greatly improved analyses of dental morphology, there are still challenges to establishing consistent, objective and informative techniques. For example, 3D data may be captured via many different scanning systems, such as X-ray computed tomography (CT and microCT), surface scanning (structured light, laser) and photogrammetry. Each imaging method has its limits and biases that must be taken into account. CT enables detailed structural information to be obtained, including from the interior of the object, but requires expensive equipment ($100K–$2M), often based in medical or engineering facilities, and significant data storage (often >2 GB raw data per scan). Surface scanning typically has the advantages of faster data capture and smaller data storage requirements (2–200 MB per scan) but will only scan external, easily visible surfaces. Some laser and structured light scanners can be relatively expensive ($20K–$200K), while size- and resolution-limited machines now exist for less than $5K. Photogrammetry has markedly improved over the last few years, including using mobile phone apps like 123D Catch (Autodesk), but high quality surfaces require sufficient texture on the surface and dozens to hundreds of photographs from many angles followed by hours to days of computer processing.

3D scanning of teeth can be particularly difficult with light-based scanners due to the opalescence and reflectivity of enamel and can require powder-coating to achieve high quality surface reconstructions (e.g., Smith and Strait 2008). Small teeth (less than approximately 4 mm in length) are challenging to scan. Very few surface scanners can successfully scan small, high relief tribosphenic molars, particularly when they are embedded in the jawbone (though Laser Design and Breuckmann can do isolated teeth of this type). MicroCT is the preferred method for these small teeth, as it easily deals with undercuts and closely approximated teeth and bone and can achieve sufficiently high resolution (<10 micron voxel sizes).

Processing of 3D data also includes many complex steps, such as image segmentation, alignment of multiple scan passes, smoothing, downsampling and file format conversion. There remains a need to establish reliable protocols providing consistent results, although a reasonable amount of testing has established boundary conditions for good practice.

Examples of Applications

This chapter reviews a diverse range of approaches and methods to reveal the diets of extinct mammals. Below we provide some examples of how these methods can be applied to specific research questions.

Multituberculates

The typical conception of mammalian evolution has been that mammals did not diversify until after the extinction of the non-avian dinosaurs at the end-Cretaceous mass extinction 66 million years ago. The multituberculates were the most abundant mammals during the Mesozoic and the longest-lived group of mammals, surviving from around 180–35 million years ago. Dental diversity within multituberculates was relatively high, ranging from taxa with a large blade-like lower molars to those with highly complex low-crowned teeth. However, because there are no close extant relatives of multituberculates, their diets have been hotly debated. Wilson et al. (2012) used OPC to study the dental morphology of multituberculate mammals across the Cretaceous-Paleogene boundary. Their findings reveal a major morphological radiation before the mass extinction event, which suggests that mammal diversification had commenced well before the disappearance of non-avian dinosaurs. Many of these new multituberculates appear to have been herbivorous, perhaps feeding on the newly dominant angiosperms. Therefore, mammals may have played a broader ecological role in Mesozoic ecosystems than previously thought.

Marsupials and Placentals

Marsupials and placentals, the two main groups of extant mammals, diverged around the Early Cretaceous and independently radiated on multiple continents throughout the Mesozoic and Cenozoic. Despite multiple reports of morphological convergence in features such as skull shape, body form, and locomotion, it has so far remained unclear whether dental morphology changed with diet in a similar way in both groups. To address this, Pineda-Munoz et al. (2017) developed a new methodology called Multi-Proxy Dental Morphology Analysis (MPDMA), which combines detailed information on diet with multiple quantitative measures of tooth shape described in this review, including OPC, average slope, and relief index. MPDMA was tested for four mammal orders (Carnivora, Diprotodontia, Primates and Rodentia) across eight dietary specializations (herbivory, carnivory, frugivory, granivory, insectivory, fungivory, gumivory, and generalized) and thus offered the possibility to draw more accurate dietary inferences that better represent the multidimensional nature of dental morphology and dietary specializations across all mammals (Davis and Pineda Munoz 2016). The results show that, like carnivorans and rodents (Evans et al. 2007), marsupials and placentals with the same dietary specializations have evolved morphologically similar dentitions, highlighting the tight connection between high-level morphology and diet in tooth evolution.

Large Herbivores and Lophs

Reconstruction of past environments across the globe has been a major challenge for paleoecologists. Important requirements for such a task include taxa that are widespread with fossilized characteristics that reliably indicate ecological variables such as net primary productivity. Liu et al. (2012) correlated dental characteristics of large herbivores (orders Perissodactyla, Artiodactyla, Primates, and Proboscidea) with climate data across the globe to test whether net primary productivity could be estimated from fossil teeth. In large herbivores, the processing capability of a tooth can be estimated by counting the number of longitudinal lophs (LOP) or enamel ridges, with higher LOP values correlating with an increased ability to process plant matter. Likewise, hypsodonty is the relative height of the tooth and is an adaptation to high rates of wear, particularly in arid environments with higher loads of exogenous grit (see Barr 2018). Both correlate with drier environments, where vegetation exercises a higher abrasive effect on tooth surfaces. Liu et al. (2012) observed that the average LOP/hypsodonty ratio of the members of a community is correlated with climate in modern ecosystems and applied this metric to the western Eurasian Sansanian and Pikermian chronofaunas (both Miocene) to interpret their paleoclimates.

Plagiaulacoid Teeth

Bladed teeth with serrations, termed 'plagiaulacoid' by Simpson (1933b), have independently evolved in several groups of mammals: marsupials (several times independently within this group), carpolestid primates, and multituberculates. The importance and likely diet of fossil species with plagiaulacoid teeth has been debated since Simpson's (1933b) first description (Krause 1982; Biknevicius 1986; Dumont et al. 2000). To investigate the functional capabilities of these teeth, Pollock (2016) measured crest shape parameters in marsupials and multituberculates and used the range of values to create 3D printed models of simplified bladed teeth with or without serrations. Each of the models was used in physical force testing with two foods that emulate plant (asparagus) and animal (gelatine) materials. Higher rake angle improved performance for gelatine but not asparagus, while higher approach angle and added serrations did so for asparagus but not gelatine. These observations suggest that a plagiaulacoid tooth shape is likely to be more advantageous for plant feeders and that most extinct plagiaulacoid taxa were probably herbivorous.

Hard Food Feeding in Hominins

During the evolution of the human lineage, several hominin species are thought to have consumed hard foods, including the robust australopith 'Nutcracker Man' *Paranthropus boisei* (Leakey 1959). Molar shape varies among fossil hominins, including the degree of occlusal relief. How could one determine the likely diet of the various hominins based on dental morphology? Berthaume et al. (2010) examined how molar shape affected the fracture of hard foods using both physical testing and finite element analysis. Metal replicas of teeth were used in force tester experiments to fracture acrylic hemispheres (emulating nut shells), where the force and displacement to fracture were measured. Each physical test was directly compared with an equivalent finite element simulation, and these showed reasonable correspondence. While the study did not unambiguously identify dental features that optimize hard food fracture, the approach paved the way for tightly-controlled experiments along with high resolution quantification of dental morphology to resolve these questions.

Methods Demonstration

Below we give a basic outline of the steps required to scan, process, and import a 3D model for calculation of OPC and other dental topographic metrics. Given the variety of available scanning devices and software packages, it is not possible for us to be comprehensive, but this outline will serve as a list and recommendations for requisite operations.

1. Obtain surface or volume scans of tooth surface. Surface scans from laser scanners (NextEngine, Laser Design) or light scanners (Artec, Breuckmann, ATOS); sometimes manual alignment of partial surface models are required (e.g., software Geomagic, SolidWorks). Volume scans from X-ray computed tomography (SkyScan, Xradia) to give image stacks of cross-sections through the specimen; surfaces can be extracted using thresholding or segmentation (software Avizo, Mimics).
2. Process surface model. Remove extraneous surface, undertake smoothing and minor editing.
3. Align surface model to given axis system. For instance, occlusal direction pointing along positive Z axis, anterior/mesial along positive Y axis.
4. Convert to required file format. Software packages for creation and calculations use many different 3D surface file formats including .ply, .obj., .dxf and .stl – some in either binary or ASCII encoding. You will need to ascertain what file formats are imported by your calculation software.
5. Import into calculation software and run analyses. Software packages include Surfer Manipulator (Evans et al. 2007), Teether (Bunn et al. 2011), MorphoTester (Winchester 2016), and molaR in R (Pampush et al. 2016).
6. Interpretations of diet will generally be based on a modern comparative database of species with known diets. Conventional statistical tests can be used to determine whether the fossil population/species differs from extant species (t-test, ANOVA) and discriminant function analysis/canonical variates analysis (DFA/CVA) of extant species can be used to interpret values from focal taxa.

Future Prospects

With the advent of further methods and approaches to quantify tooth shape, we will be entering a new golden age of dental morphology. High-throughput 3D imaging and processing will be key in compiling large datasets of modern and fossil morphologies. In addition, we must continue to strive for a better fundamental understanding of the mechanical properties of foods and their interaction with morphology. We look forward to the application of these advances to better interpret the paleoecology of mammals throughout the world.

Acknowledgements We are very grateful to Darin Croft, Denise Su and Scott Simpson for the invitation to attend the Paleoecology Symposium in Cleveland and to contribute to this volume. We also thank Felix Marx, David Hocking, Gudrun Evans, Darin Croft and three anonymous reviewers for comments and suggestions that greatly improved this manuscript. Partial support for this research was provided by the Australian Research Council Future Fellowship FT130100968 to A.R.E. and NSF-DEB 1257625 and the Evolution of Terrestrial Ecosystems program at NMNH Smithsonian Institution.

References

Abler, W. L. (1992). The serrated teeth of tyrannosaurid dinosaurs, and biting structures in other animals. *Paleobiology, 18,* 161–183.

Anderson, P. S. L., & LaBarbera, M. (2008). Functional consequences of tooth design: effects of blade shape on energetics of cutting. *Journal of Experimental Biology, 211,* 3619–3626.

Barr, W. A. (2018). Ecomorphology. In D. A. Croft, D. F. Su & S. W. Simpson (Eds.), *Methods in paleoecology: Reconstructing Cenozoic terrestrial environments and ecological communities* (pp. 337–347). Cham: Springer.

Benazzi, S., Kullmer, O., Grosse, I. R., & Weber, G. W. (2011). Using occlusal wear information and finite element analysis to investigate stress distributions in human molars. *Journal of Anatomy, 219,* 259–272.

Benazzi, S., Kullmer, O., Grosse, I. R., & Weber, G. W. (2012). Brief communication: comparing loading scenarios in lower first molar supporting bone structure using 3D finite element analysis. *American Journal of Physical Anthropology, 147,* 128–134.

Benazzi, S., Grosse, I. R., Gruppioni, G., Weber, G. W., & Kullmer, O. (2014). Comparison of occlusal loading conditions in a lower second premolar using three-dimensional finite element analysis. *Clinical Oral Investigations, 18,* 369–375.

Berthaume, M. A. (2014). Tooth cusp sharpness as a dietary correlate in great apes. *American Journal of Physical Anthropology, 153,* 226–235.

Berthaume, M. A. (2016). Food mechanical properties and dietary ecology. *American Journal of Physical Anthropology, 159,* 79–104.

Berthaume, M. A., Grosse, I. R., Patel, N. D., Strait, D. S., Wood, S., & Richmond, B. G. (2010). The effect of early hominin occlusal morphology on the fracturing of hard food items. *Anatomical Record, 293,* 594–606.

Berthaume, M. A., Dumont, E. R., Godfrey, L. R., & Grosse, I. R. (2014). The effects of relative food item size on optimal tooth cusp sharpness during brittle food item processing. *Journal of the Royal Society Interface, 11*, 20140965.

Biknevicius, A. R. (1986). Dental function and diet in the Carpolestidae (Primates, Plesiadapiformes). *American Journal of Physical Anthropology, 71,* 157–171.

Bock, W. J., & Wahlert, G. V. (1965). Adaptation and the form-function complex. *Evolution, 19*, 269–299.

Boyer, D. M. (2008). Relief index of second mandibular molars is a correlate of diet among prosimian primates and other euarchontan mammals. *Journal of Human Evolution, 55,* 1118–1137.

Bunn, J. M., Boyer, D. M., Lipman, Y., St Clair, E. M., Jernvall, J., & Daubechies, I. (2011). Comparing Dirichlet normal surface energy of tooth crowns, a new technique of molar shape quantification for dietary inference, with previous methods in isolation and in combination. *American Journal of Physical Anthropology, 145,* 247–261.

Callister, W. D., & Rethwisch, D. G. (2014). *Materials science and engineering: An introduction* (9th ed.). Hoboken, NJ: Wiley.

Candela, A. M., Cassini, G. H., & Nasif, N. L. (2013). Fractal dimension and cheek teeth crown complexity in the giant rodent *Eumegamys paranensis*. *Lethaia, 46,* 369–377.

Chemisquy, M. A., Prevosti, F. J., Martin, G., & Flores, D. A. (2015). Evolution of molar shape in didelphid marsupials (Marsupialia: Didelphidae): analysis of the influence of ecological factors and phylogenetic legacy. *Zoological Journal of the Linnean Society, 173,* 217–235.

Crofts, S. B., & Summers, A. P. (2014). How to best smash a snail: the effect of tooth shape on crushing load. *Journal of the Royal Society Interface, 11*, 20131053.

Crusafont-Pairó, M., & Truyols-Santonja, J. (1956). A biometric study of the evolution of fissiped carnivores. *Evolution, 10,* 314–332.

Curran, S C. (2018). Three-dimensional geometric morphometrics in paleoecology. In D. A. Croft, D. F. Su & S. W. Simpson (Eds.), *Methods in paleoecology: Reconstructing Cenozoic terrestrial environments and ecological communities* (pp. 317–335). Cham: Springer.

Davis, M., & Pineda Munoz, S. (2016). The temporal scale of diet and dietary proxies. *Ecology and Evolution, 6,* 1883–1897.

Dumont, E. R., Strait, S. G., & Friscia, A. R. (2000). Abderitid marsupials from the Miocene of Patagonia: an assessment of form, function, and evolution. *Journal of Paleontology, 74,* 1161–1172.

Evans, A. R. (2005). Connecting morphology, function and tooth wear in microchiropterans. *Biological Journal of the Linnean Society, 85,* 81–96.

Evans, A. R. (2013). Shape descriptors as ecometrics in dental ecology. *Hystrix: The Italian Journal of Mammalogy, 24,* 133–140.

Evans, A. R., & Sanson, G. D. (1998). The effect of tooth shape on the breakdown of insects. *Journal of Zoology, 246,* 391–400.

Evans, A. R., & Sanson, G. D. (2003). The tooth of perfection: functional and spatial constraints on mammalian tooth shape. *Biological Journal of the Linnean Society, 78,* 173–191.

Evans, A. R., & Sanson, G. D. (2005). Correspondence between tooth shape and dietary biomechanical properties in insectivorous microchiropterans. *Evolutionary Ecology Research, 7,* 453–478.

Evans, A. R., Harper, I. S., & Sanson, G. D. (2001). Confocal imaging, visualization and 3-D surface measurement of small mammalian teeth. *Journal of Microscopy, 204,* 108–118.

Evans, A. R., Fortelius, M., Jernvall, J., & Eronen, J. T. (2006). Dental ecomorphology of extant European carnivorans. In E. Żądzińska (Ed.), *Current trends in dental morphology research: Full refereed papers from 13th International Symposium on Dental Morphology* (pp. 223–232, Vol. 3). Łódź: University of Łódź Press.

Evans, A. R., Wilson, G. P., Fortelius, M., & Jernvall, J. (2007). High-level similarity of dentitions in carnivorans and rodents. *Nature, 445,* 78–81.

Famoso, N. A., & Davis, E. B. (2016). On the relationship between enamel band complexity and occlusal surface area in Equids (Mammalia, Perissodactyla). *PeerJ, 4*, e2181.

Famoso, N. A., Feranec, R. S., & Davis, E. B. (2013). Occlusal enamel complexity and its implications for lophodonty, hypsodonty, body mass, and diet in extinct and extant ungulates. *Palaeogeography, Palaeoclimatology, Palaeoecology, 387,* 211–216.

Famoso, N. A., Davis, E. B., Feranec, R. S., Hopkins, S. S. B., & Price, S. A. (2016). Are hypsodonty and occlusal enamel complexity evolutionarily correlated in ungulates? *Journal of Mammalian Evolution, 23,* 43–47.

Fortelius, M. (1985). Ungulate cheek teeth: developmental, functional, and evolutionary interrelations. *Acta Zoologica Fennica, 180,* 1–76.

Freeman, P. W. (1988). Frugivorous and animalivorous bats (Microchiroptera): dental and cranial adaptations. *Biological Journal of the Linnean Society, 33,* 249–272.

Freeman, P. W. (1992). Canine teeth of bats (Microchiroptera): size, shape and role in crack propagation. *Biological Journal of the Linnean Society, 45,* 97–115.

Freeman, P. W., & Lemen, C. (2006). Puncturing ability of idealized canine teeth: edged and non-edged shanks. *Journal of Zoology, 269,* 51–56.

Green, J. L., & Croft, D. A. (2018). Using dental mesowear and microwear for dietary inference: a review of current techniques and applications. In D. A. Croft, D. F. Su & S. W. Simpson (Eds.), *Methods in paleoecology: Reconstructing Cenozoic terrestrial environments and ecological communities* (pp. 53–73). Cham: Springer.

Gregory, W. K. (1920). On the structure and relations of *Notharctus*, an American Eocene primate. *Memoirs of the American Museum of Natural History, New Series, 3,* 49–243.

Gregory, W. K. (1922). *The origin and evolution of the human dentition*. Baltimore: Williams & Wilkins Company.

Hardie, R. P., & Gaye, R. K. (1930). *Physica (Vol. v2, The works of Aristotle)*. Oxford: Clarendon Press.

Hopkins, S. S. B. (2018). Estimation of body size in fossil mammals. In D. A. Croft, D. F. Su & S. W. Simpson (Eds.), *Methods in paleoecology: Reconstructing Cenozoic terrestrial environments and ecological communities* (pp. 7–22). Cham: Springer.

Hunter, J. (1771–1803 [1865 post.]). *The natural history of the human teeth*. London: Robert Hardwicke.

Jernvall, J. (1995). Mammalian molar cusp patterns: developmental mechanisms of diversity. *Acta Zoologica Fennica, 198,* 1–61.

Jernvall, J., Hunter, J. P., & Fortelius, M. (1996). Molar tooth diversity, disparity, and ecology in Cenozoic ungulate radiations. *Science, 274,* 1489–1492.

Jernvall, J., & Selänne, L. (1999). Laser confocal microscopy and geographic information systems in the study of dental morphology. *Palaeontologia Electronica, 2.1.3A*, 12 pp.

Kay, R. F. (1975). The functional adaptations of primate molar teeth. *American Journal of Physical Anthropology, 43,* 195–215.

Kay, R. F. (1978). Molar structure and diet in extant Cercopithecidae. In P. M. Butler & K. A. Joysey (Eds.), *Development, function, and evolution of teeth* (pp. 309–339). London: Academic Press.

Kay, R. F., & Hiiemae, K. M. (1974). Jaw movement and tooth use in recent and fossil primates. *American Journal of Physical Anthropology, 40,* 227–256.

Kieser, J. A. (1990). *Human adult odontometrics*. Cambridge: Cambridge University Press.

King, S. J., Arrigo-Nelson, S. J., Pochron, S. T., Semprebon, G. M., Godfrey, L. R., Wright, P. C., et al. (2005). Dental senescence in a long-lived primate links infant survival to rainfall. *Proceedings of the National Academy of Sciences, USA, 102,* 16579–16583.

Klukkert, Z. S., Teaford, M. F., & Ungar, P. S. (2012). A dental topographic analysis of chimpanzees. *American Journal of Physical Anthropology, 148,* 276–284.

Krause, D. W. (1982). Jaw movement, dental function, and diet in the Paleocene multituberculate *Ptilodus*. *Paleobiology, 8,* 265–281.

Laws, R. M. (1968). Dentition and ageing of the hippopotamus. *East African Wildlife Journal, 6,* 19–52.

Leakey, L. S. B. (1959). A new fossil skull from Olduvai. *Nature, 184,* 491–493.

Liu, L., Puolamaki, K., Eronen, J. T., Ataabadi, M. M., Hernesniemi, E., & Fortelius, M. (2012). Dental functional traits of mammals resolve productivity in terrestrial ecosystems past and present. *Proceedings of the Royal Society B, 279,* 2793–2799.

Logan, M., & Sanson, G. D. (2002). The effects of tooth wear on the activity patterns of free-ranging koalas (*Phascolarctos cinereus* Goldfuss). *Australian Journal of Zoology, 50,* 281–292.

Lucas, P. W. (1979). The dental-dietary adaptations of mammals. *Neues Jahrbuch für Geologie und Paläontologie Monatshefte, 1979,* 486–512.

Lucas, P. W. (1982). Basic principles of tooth design. In B. Kurtén (Ed.), *Teeth: Form, function and evolution* (pp. 154–162). New York: Columbia University Press.

Lucas, P. W., & Luke, D. A. (1984). Chewing it over: basic principles of food breakdown. In D. J. Chivers, B. A. Wood, & A. Bilsborough (Eds.), *Food acquisition and processing in primates* (pp. 283–301). New York: Plenum Press.

Macho, G. A., & Spears, I. R. (1999). Effects of loading on the biomechanical behavior of molars of *Homo*, *Pan*, and *Pongo*. *American Journal of Physical Anthropology, 109,* 211–227.

Osborn, J. W., & Lumsden, A. G. S. (1978). An alternative to "thegosis" and a re-examination of the ways in which mammalian molars work. *Neues Jahrbuch für Geologie und Paläontologie Abhandlungen, 156,* 371–392.

Pampush, J. D., Winchester, J. M., Morse, P. E., Vining, A. Q., Boyer, D. M., & Kay, R. F. (2016). Introducing molaR: a new R package for quantitative topographic analysis of teeth (and other topographic surfaces). *Journal of Mammalian Evolution, 23,* 397–412.

Pineda-Munoz, S., Lazagabaster, I. A., Evans, A. R., & Alroy, J. (2017). Inferring diet from dental morphology in terrestrial mammals. *Methods in Ecology and Evolution, 8,* 481–491.

Pollock, T. I. (2016). *Blade Runner: Revealing the function and dietary implications of blade-like serrated teeth in mammals*. BSc (Hons) Thesis, Monash University.

Rayfield, E. J. (2007). Finite element analysis and understanding the biomechanics and evolution of living and fossil organisms. *Annual Review of Earth and Planetary Sciences, 35,* 541–576.

Reed, D. N. O. (1997). Contour mapping as a new method for interpreting diet from tooth morphology. *American Journal of Physical Anthropology, Suppl, 24,* 194.

Rensberger, J. M. (1973). An occlusal model for mastication and dental wear in herbivorous mammals. *Journal of Paleontology, 47,* 515–528.

Rosenberger, A. L., & Kinzey, W. G. (1976). Functional patterns of molar occlusion in platyrrhine primates. *American Journal of Physical Anthropology, 45,* 281–298.

Sanson, G. (2006). The biomechanics of browsing and grazing. *American Journal of Botany, 93,* 1531–1545.

Schmidt-Kittler, N. (1984). Pattern analysis of occlusal surfaces in hypsodont herbivores and its bearing on morpho-functional studies. *Proceedings of the Koninklijke Nederlandse Akademie Van Wetenschappen series B: palaeontology, geology, physics, chemistry, anthropology, 87,* 453–480.

Simpson, G. G. (1933a). Paleobiology of Jurassic mammals. *Paleobiologica, 5,* 127–158.

Simpson, G. G. (1933b). The "plagiaulacoid" type of mammalian dentition. A study of convergence. *Journal of Mammalogy, 14,* 97–107.

Skamniotis, C. G., Patel, Y., Charalambides, M. N., & Elliott, M. (2016). Fracture investigation in starch-based foods. *Interface Focus, 6,* 20160005.

Smith, N. E., & Strait, S. G. (2008). PaleoView3D: from specimen to online digital model. *Palaeontologia Electronica, 11.2.11A,* 17 pp.

Stone, J., & Telford, M. (2005). Fractal dimensions characterizing mammal teeth: a case study involving Elephantidae. *Mammal Review, 35,* 123–128.

Strait, S. G. (1993). Differences in occlusal morphology and molar size in frugivores and faunivores. *Journal of Human Evolution, 25,* 471–484.

Ungar, P., & Williamson, M. (2000). Exploring the effects of tooth wear on functional morphology: a preliminary study using dental topographic analysis. *Palaeontologia Electronica, 3.1.1A,* 18 pp.

Van Valkenburgh, B. (1989). Carnivore dental adaptations and diet: a study of trophic diversity within guilds. In J. L. Gittleman (Ed.), *Carnivore behavior, ecology, and evolution* (pp. 410–436). Ithaca: Cornell University Press.

Whitenack, L. B., & Motta, P. J. (2010). Performance of shark teeth during puncture and draw: implications for the mechanics of cutting. *Biological Journal of the Linnean Society, 100,* 271–286.

Wilson, G. P., Evans, A. R., Corfe, I. J., Smits, P. D., Fortelius, M., & Jernvall, J. (2012). Adaptive radiation of multituberculates before the extinction of dinosaurs. *Nature, 483,* 457–460.

Winchester, J. M. (2016). MorphoTester: an open source application for morphological topographic analysis. *PLoS ONE, 11,* e0147649.

Winchester, J. M., Boyer, D. M., St Clair, E. M., Gosselin-Ildari, A. D., Cooke, S. B., & Ledogar, J. A. (2014). Dental topography of platyrrhines and prosimians: convergence and contrasts. *American Journal of Physical Anthropology, 153,* 29–44.

Wright, B. W. (2005). Craniodental biomechanics and dietary toughness in the genus *Cebus*. *Journal of Human Evolution, 48,* 473–492.

Yamashita, N. (1998). Functional dental correlates of food properties in five Malagasy lemur species. *American Journal of Physical Anthropology, 106,* 169–188.

Zuccotti, L. F., Williamson, M. D., Limp, W. F., & Ungar, P. S. (1998). Modeling primate occlusal topography using geographic information systems technology. *American Journal of Physical Anthropology, 107,* 137–142.

Chapter 5
Using Dental Mesowear and Microwear for Dietary Inference: A Review of Current Techniques and Applications

Jeremy L. Green and Darin A. Croft

Abstract Mesowear and microwear analyses use data from worn tooth surfaces as proxies for feeding ecology. Mesowear is based on gross dental wear and forms over months to years. The method was originally developed for ungulates but has recently been expanded to other groups, at least preliminarily. Dental microwear has been investigated for well over half a century and continues to be refined. It forms over days to weeks. Wide varieties of techniques are currently used for microwear analysis, all of which require attention to detail. Among these techniques, three-dimensional microwear texture analysis has the greatest potential for accurately reconstructing feeding ecology, yet the "recipe" for analyzing microwear data remains a work in progress. Combining mesowear and microwear with one another and other dietary proxies can permit robust inferences about the feeding ecology of extinct species.

Keywords Dentition • Enamel • Feeding ecology Microscopy • Paleodiet • Tooth wear

Summary

Both dental mesowear and microwear analysis use data from the "damaged" wear surface of a tooth as a proxy for feeding ecology in extant and extinct vertebrates. Mesowear is a relatively new method of dietary interpretation in herbivores that is based on gross dental wear. It measures diet (and, to some extent, habitat) in individuals and populations over the span of months to years. Collecting data is relatively quick and easy. The method was originally developed for ungulates (mostly artiodactyls and perissodactyls) but has recently been expanded to other groups, at least in a preliminary fashion. Dental microwear analysis has been a rich area of investigation for well over half a century and continues to be refined today. It can discriminate not only large-scale dietary differences among taxa but also dietary variation within populations (among individuals). However, all steps of microwear data collection (from cleaning a tooth to analyzing data) require attention to detail and careful planning to ensure the collection of accurate, objective results. Investigators that analyze dental microwear are currently using five distinct techniques: (1) scanning electron microscopy (SEM); (2) stereoscopic low-magnification light microscopy (LM); (3) photomicrographic LM; (4) three-dimensional microwear texture analysis (MTA) with scale-sensitive fractal analysis (SSFA); and (5) MTA with International Organization for Standardization (ISO) 25178-2 parameters. Three-dimensional MTA has the greatest potential for accurately reconstructing feeding ecology, yet the "recipe" for analyzing microwear data remains a work in progress as technology advances and our understanding of the events that create microwear improves. Mesowear and microwear can each be a very powerful tool for paleoecological and environmental reconstruction, but both require strict protocols. Combining these techniques with one another and with other dietary proxies can permit robust inferences about the feeding ecology of extinct species.

Terms

Abrasion = tooth wear caused by interactions between a tooth and exogenous particles (e.g., food, grit).

Attrition = tooth wear that is caused by tooth-on-tooth interactions.

J. L. Green (✉)
Department of Geology, Kent State University at Tuscarawas, 330 University Drive NE, New Philadelphia, OH 44663, USA
e-mail: jgreen72@kent.edu

D. A. Croft
Department of Anatomy, Case Western Reserve University, 10900 Euclid Ave, Cleveland, OH 44106-4930, USA
e-mail: dcroft@case.edu

D. A. Croft et al. (eds.), *Methods in Paleoecology: Reconstructing Cenozoic Terrestrial Environments and Ecological Communities*, Vertebrate Paleobiology and Paleoanthropology, https://doi.org/10.1007/978-3-319-94265-0_5

Browser = an herbivore that feeds primarily (>90%) on leaves, twigs, buds, flowers, and/or fruits.

Cast = an exact replica of a specimen that is produced by filling a mold with epoxy or another material.

Confocal Microscope = a microscope that uses point illumination and a spatial pinhole placed at the confocal plane of the lens to reduce out-of-focus light, creating a three-dimensional image.

Cusp Relief = the relative distance between the cusp tip and base.

Cusp Shape = a measure of the facet development in a cusp.

Erosion = tooth wear that is caused by chemical or acid dissolution.

Facet = a smooth, flat area of enamel on the occlusal surface of a tooth that is formed by wear.

Grazer = an herbivore that feeds primarily (>90%) on grasses.

Mesowear = gross (macroscopic) tooth wear (facet development).

Microwear = microscopic scars that form on a tooth surface from tooth use (use-wear) during life.

Mixed Feeder = an herbivore that feeds on a mixture of leaves, twigs, buds, and grasses.

Mold = a negative copy of a specimen that can be used to produce an exact replica (cast).

Scanning Electron Microscope = a microscope that uses the interaction of a focused beam of electrons with a sample to produce a detailed image; designed for high-magnification use (typically $\geq 100\times$).

Stereomicroscope = a microscope in which light refracting through or reflecting off an object is viewed through two eye pieces, producing a three-dimensional visualization; designed for low-magnification use (typically $\leq 100\times$).

Ungulate = a hoof-bearing mammal; not a natural (monophyletic) group.

Theoretical and Historical Background

Mesowear

Mesowear refers to macroscopic wear on teeth: wear that is visible to the naked eye as a smooth, flat area of enamel on the occlusal surface. Such wear comprises what is known as a facet. Mesowear analysis seeks to describe the degree to which such facets are developed on the teeth of herbivorous mammals and to correlate such patterns with diet. Unlike microwear analysis, mesowear analysis has only been applied to herbivorous mammals. In herbivores that consume relatively soft foods, such as the leaves of dicotyledonous plants, twigs, buds, flowers, and fruits (i.e., browsers), the teeth working against one another is the primary cause of tooth wear; this type of wear is attrition. Attrition creates enamel surfaces that have sharp edges and well-developed facets. In grazing herbivores, which mainly consume grasses and other low-growing vegetation, the food consumed causes more tooth wear than the teeth themselves. This type of wear is termed abrasion, and it tends to obscure facets and round off enamel edges. Mesowear analysis was originally developed for ungulates, particularly artiodactyls and perissodactyls, but it has recently been expanded to other groups including rodents, rabbits, elephants (proboscideans), extinct ungulate-like mammals from South America, and even marsupials (see below).

The term mesowear (meaning "intermediate wear") was coined by Fortelius and Solounias (2000) because it highlighted the intermediate time scale at which mesowear forms: more slowly than microwear, which reflects an animal's diet over days to weeks (see below), but more quickly than changes in the overall structure of a tooth, which only occur over evolutionary time (over many generations). Mesowear develops over an animal's lifetime, or at least at a time scale long enough so that seasonal changes in diet cannot be detected (Rivals et al. 2013). Thus, it reflects "the average diet of a particular species from a particular location in space and time" (Fortelius and Solounias 2000: 2). Different scoring techniques for mesowear exist, but they differ from one another primarily in detail. The greatest advances in mesowear analysis have been in extending the technique to groups of herbivorous mammals beyond artiodactyls and perissodactyls. Collecting mesowear data is generally quick and easy, making the technique highly amenable to large samples (e.g., Mihlbachler et al. 2011).

Microwear

Microwear is the result of microscopic damage to dental tissues (usually enamel) that forms during mastication. This microscopic wear is not a cumulative (long-term) record of chewing (in contrast to mesowear), but rather is continuously recorded and erased with each subsequent feeding event (Teaford and Oyen 1989). As a result, it generally only records an animal's last meal(s), a characteristic dubbed the "Last Supper" phenomenon by Grine (1986). High turnover rates in microwear have been recorded in most vertebrates, ranging from days to weeks (Teaford and Oyen 1989; Baines et al. 2014; Hoffmann et al. 2015). The relative orientations of striations on tooth surfaces is a proxy for direction of jaw movement during chewing (Gordon 1984a), although most microwear studies tend to focus on the relationship between microwear patterns and the texture of particles being consumed (Teaford 1991). Such analyses can target the proportions and densities of discrete features (scratches, pits, gouges) on a wear surface (Teaford and Walker 1984;

Solounias and Semprebon 2002), or, increasingly commonly, the overall texture (roughness) of a chewing surface (Scott et al. 2005). Regardless of the approach, it has long been recognized that microwear patterns in animals with different diets can be qualitatively and quantitatively (even statistically) distinguished (Walker et al. 1978; Teaford and Walker 1984; Solounias and Semprebon 2002; Ungar et al. 2003). This correlation has led to the popularity of microwear as a proxy for reconstructing paleoecology in vertebrates.

Despite the many known correlations between diet and microwear, researchers are still trying to understand the details of these relationships (Ungar 2015). Indeed, a fundamental question that remains unanswered is whether the primary causative agent of tooth wear is endogenous (i.e., a property of the food itself, such as hardness; Erickson 2014; Rabenold and Pearson 2014; Xia et al. 2015) or exogenous (i.e., due to non-biological particles such as dust or sand; Lucas et al. 2013; Hoffmann et al. 2015; Spradley et al. 2016). Data support both interpretations, leading to the inescapable conclusion that a seemingly simple correlation is actually far more complex. The number of microwear studies published each year continues to increase, highlighting its importance in the field of paleoecological reconstruction.

Ungar et al. (2008) provided a thorough review of the deep history of microwear analysis including its origins, philosophical shifts, and methodological innovations. Here, we briefly review its foundations before focusing on the past decade, a time of rapid innovation in the field.

The first studies of microwear were qualitative in nature, using light microscopy to examine the orientations of scratches to determine directionality of chewing motion (Butler 1952; Mills 1955). Subsequent researchers abandoned the light microscope in favor of the high resolution and magnification of scanning electron microscopy (Rensberger 1978; Walker et al. 1978); the quantification of microwear patterns to discern dietary differences soon followed (Gordon 1982; Teaford and Walker 1984). This early research mainly focused on primates; it demonstrated relatively consistent microwear within species (Gordon 1982) and distinct microwear among species of differing diets (Teaford and Walker 1984). Gordon (1984b) recognized the need for standardization of SEM instrumentation parameters, magnification, and tooth sampling loci to control for intra- and inter-individual variation in tooth wear. Quantifying scar features on teeth permitted statistical testing to discriminate between genuine and random microwear patterns, thus leading to a more advanced understanding of the relationship between microwear and feeding ecology (Teaford and Walker 1984; Strait 1993; Ungar and Spencer 1999). In the late 1980s and early 1990s, researchers sought to more accurately quantify microwear using a variety of approaches and variables (e.g., scratch:pit ratios, scratch length/breadth, feature orientation; Grine 1986; Maas 1991; Ungar 1995).

By the early 2000s, it was recognized that observer consistency in microwear feature recognition and measurement on SEM images limited interpretations (Grine et al. 2002) and that SEM analysis could be expensive and time-consuming (Solounias and Semprebon 2002). In this light, Solounias and Semprebon (2002) suggested returning to low-magnification light microscopy (LM) for dental microwear analysis as an effective yet inexpensive (in both time and cost) method. Interest in LM increased rapidly during the subsequent decade, with a concomitant increase in the diversity of non-primate mammals analyzed (e.g., ungulates: Solounias and Semprebon 2002; chalicotheres: Coombs and Semprebon 2005; proboscideans: Green et al. 2005; Rivals et al. 2012; xenarthrans: Green 2009a; rodents: Nelson et al. 2005; Townsend and Croft 2008b; notoungulates: Townsend and Croft 2008a). As with SEM methods, observer consistency was questioned (Ungar et al. 2008). Early studies suggested that intra- and inter-observer error were non-significant in LM (Semprebon et al. 2004b), but later studies have disagreed (Mihlbachler and Beatty 2012; Mihlbachler et al. 2012).

Concerns about observer error and lack of consistency among researchers using SEM and LM fueled the development of an automated three-dimensional technique called "microwear texture analysis" (MTA) (Ungar et al. 2003; Scott et al. 2005). This method differed from previous methods in using computer software to analyze the texture of a wear surface as a whole rather than to quantify discrete features ("scratches" vs. "pits") on a tooth surface. Because this technique analyzes microwear in three dimensions rather than two (as in SEM and some variants of LM; Merceron et al. 2005), it can provide greater power for discriminating subtle differences in feeding ecology, such as seasonal variations in diet (DeSantis et al. 2013).

Microwear researchers presently use all three types of microscopy (listed above) and variations of each are still being proposed. For example, investigators in some recent LM studies have counted features on photomicrographs (under randomized, blind conditions) instead of directly through a microscope in an effort to create a more standardized approach (Fraser et al. 2009; Mihlbachler et al. 2012; Hoffmann et al. 2015); SEM researchers have used randomized, blind conditions for counting features (Green and Resar 2012; Green and Kalthoff 2015); and MTA now uses standardized ISO parameters to characterize textural patterns in addition to scale-sensitive fractal analysis (SSFA) (Purnell et al. 2012; Gill et al. 2014). In all of these cases, it is important to remember that each technique has its limitations and shortcomings. MTA is the most widely used technique today and has greatest promise for the future of microwear research (Calandra and Merceron 2016). Nevertheless, the other two methods can still provide significant insights when properly applied (Green and Resar 2012; Mihlbachler et al. 2012).

Collecting and Analyzing Data

Selecting Specimens

Accurate mesowear and microwear analyses require selecting appropriate comparative specimens. This means using teeth (either fossil or modern) that have well defined, undamaged wear facets. A tooth that has not yet come into wear (i.e., has no obvious wear facet) is unusable, as the processes that result in tooth wear have not yet been initiated. Additionally, overly worn teeth (from older individuals) should be excluded from mesowear analyses (Rivals et al. 2007) because facet development is obscured. Teeth that are *in situ* (in a skull) are ideal since they leave no ambiguity as to: (1) species identification; (2) tooth position and orientation; and (3) ontogenetic age. For extant species, specimens should preferably come from live or recently culled animals whose diet is known and controlled. However, since this is generally not possible, most studies have relied on museum collections, which offer large sample sizes but lack specific (individual) dietary information (Kay and Covert 1983). In such cases, specimens are assigned to the dietary category typical for the species. The lack of precise correspondence between specimens and their presumed diet can be a significant source of error or noise, particularly for species with diets that vary temporally (day to day or seasonally) or spatially (in different parts of their range). Targeting museum specimens associated with detailed collection information (including season of death and longitude-latitude, where available) can help mitigate this error, but parsing specimens into increasingly fine dietary and/or locality bins usually comes at the cost of decreasing sample sizes.

The choice of tooth position and wear facet to sample depends on the question being addressed and the group being studied. It is optimal to have a standardized sampling protocol that targets the same wear facet on the same tooth across all individuals (Gordon 1982). This helps reduce variation in wear that may arise from differences in chewing mechanics across the tooth row (e.g., a posterior or distal tooth will experience less shearing and higher bite forces than a more anterior or mesial one; Gordon 1982). This is usually straightforward when comparing species within particular families or orders. Comparing members of different orders can be problematic if members of the groups under consideration have (or had) different chewing mechanics (Croft and Weinstein 2008; Fraser and Theodor 2010; Saarinen et al. 2015; Ulbricht et al. 2015; Mihlbachler et al. 2016). Studies that have examined intra- and inter-tooth variability have sampled multiple teeth and different surfaces within a single tooth to capture the range of variability that is present (e.g., Kaiser and Fortelius 2003; Krueger et al. 2008; Green 2009b; McAfee and Green 2015).

Extending mesowear and microwear analysis to extinct species poses additional challenges. First, sample sizes are generally smaller due to the rarity of fossil specimens. Second, fossil specimens may not be suitable if they cannot confidently be identified to species and/or if their position in the tooth row cannot be determined. In addition, the process of fossilization itself can significantly alter or erase the original microwear scars on the wear surface (Teaford 1988; King et al. 1999), though this can also be the case for seemingly well-preserved modern specimens (Teaford 1988). A simple comparison of the chewing and non-chewing surfaces of a tooth using a low-magnification microscope can be used to identify fossil teeth that should be discarded. The chewing surface should have visible microwear in regular patterns, whereas the non-occlusal surfaces should be devoid of similar features; if similar scar patterns are observed on chewing and non-chewing surfaces on the same tooth, the specimen has most likely been altered and should be removed from the sample (Teaford 1988). Fortunately, taphonomic processes tend to obliterate microwear patterns rather than mimic genuine scar features (King et al. 1999; Martínez and Pérez-Pérez 2004). Taphonomic breakage is generally obvious for mesowear, as it results in irregular surfaces that lack the smooth polish typical of well-preserved enamel.

Specimen Preparation

Since mesowear is scored on a gross level with the naked eye, no special cleaning of specimens is generally required. It is only necessary that the development of the tooth facets can be observed. This is not true of microwear, as debris (dirt, dust), preservatives, and/or other residues can completely obscure microscopic features (Teaford 1988). Whether to clean the entire tooth or simply a portion of it depends on the size of the tooth and the facet(s) targeted for analysis (described above); the teeth of small mammals (e.g., rodents, primates) can usually be cleaned and molded entirely, whereas only a portion of the tooth of a large mammal (e.g., proboscideans) is usually cleaned. When only part of a tooth is cleaned, the anatomical position and orientation should be recorded so that it is readily available during the analysis phase. Cleaning via physical scrubbing or preparation, especially with metal tools, brushes, etc., should be avoided as this can remove and/or alter scar patterns on the tooth. For fossil specimens, chemical preservatives (e.g., Butvar, Glyptal) are usually dissolved with acetone but may require multiple applications, depending on the stubbornness of the preservative. Researchers should identify islands of pooled preservative on the tooth surface before conducting an analysis. Once glues are removed (or if none were ever applied), applying ethanol with a cotton swab is an excellent technique for physically washing the tooth surface. The tooth should dry completely before proceeding.

Both mesowear and microwear can be scored using replicas (casts) rather than actual specimens. However, since scoring mesowear does not require specialized equipment, making casts is usually unnecessary; original specimens can be scored directly in museum collections. Microwear, by contrast, is frequently scored on casts because microscopes are not easily portable and borrowing large numbers of specimens from a museum for study in a laboratory usually is not possible. It is also much easier to manipulate a tooth cast under a microscope than a mandible or cranium, particularly for medium to large mammals, but original teeth can be analyzed where specimens are available for laboratory study.

The first step in making a cast is creating a mold, which is an impression of the tooth surface. A high quality dental molding compound should be used so that sub-micron level detail is preserved with little (if any) alteration. Polyvinylsiloxane compounds such as President Jet Light Body (low viscosity) and President Jet Regular Body (medium viscosity) consistently produce molds with both high accuracy and precision (Goodall et al. 2015). Other products should be used with caution, and data from studies using different molding compounds should not be compared directly (Goodall et al. 2015). The molding compound should be mixed according to its label, applied to the clean tooth surface in a thin coat, and allowed to dry. A first mold serves as the final cleaning step and should be discarded, as it will be contaminated with any debris remaining on the tooth surface (Solounias and Semprebon 2002). A second mold should be created using a thicker layer that completely covers the cleaned area. The specimen number, species, and any other pertinent information should be kept with this second and final mold. Applying a small tag to the mold as it dries is one way to ensure that the specimen number remains with the mold. Plastic bags or plastic organizer boxes with subdivided compartments are affordable options for transporting and storing molds. Some researchers prefer to analyze microwear directly from molds rather than casts (e.g., Schulz et al. 2010), but this is not feasible for techniques that rely on light refraction or electron beam interaction.

Prior to casting, a dam made from two-part laboratory putty is constructed around each mold to constrain the casting compound. The casts themselves are usually made with a high quality two-part epoxy (resin + hardener) capable of replicating the sub-micron detail captured in the mold. Epo-Tek #301, EpoKwick, and Feropur PR-55 are common choices (Solounias and Semprebon 2002; Galbany et al. 2004). Adjust the ratio of resin to hardener and the mixing time according to manufacturer instructions. Before pouring the mold, excess air bubbles must be removed from the mixture via depressurization in a vacuum chamber and/or by using a centrifuge (poured molds should not be placed under vacuum). Surface bubbles present in the mixture after pouring are removed using a toothpick. Casts should polymerize for at least 48 hours. Up to four casts can be made from a single mold before the replicated microwear pattern is compromised (Galbany et al. 2006).

Mesowear Data Collection

As originally conceived by Fortelius and Solounias (2000), two qualitative variables of a tooth are scored for a mesowear analysis of an ungulate or ungulate-like mammal: occlusal relief (cusp height; scored as high or low), and cusp shape (scored as sharp, rounded, or blunt) (Fig. 5.1A). Occlusal relief can be scored metrically instead of subjectively using digital photographs and pre-determined cutoffs for high versus low (see Croft and Weinstein 2008). In cases where the two buccal cusps differ in sharpness, the sharper of the two is scored. Mesowear is typically scored on an upper cheek tooth in buccal view (usually the second molar), but other upper tooth positions can also be evaluated (Kaiser and Solounias 2003) as can lower teeth, at least in ruminant artiodactyls (Fraser et al. 2014). This is particularly useful for fossil populations, where sample size can be a limiting factor. For both extant and extinct species, at least 10 individuals (preferably 20) should be scored to adequately sample variation within the population. The percentages of specimens (individuals) of each species exhibiting high occlusal relief, blunt cusps, and sharp cusps are used for comparisons among species and for interpreting diets of extinct species. This two-variable method (i.e., using occlusal relief and cusp height) has recently been termed mesowear I (Solounias et al. 2014).

Rivals and Semprebon (2006; see also Rivals et al. 2007) were the first to use a single-variable method for scoring mesowear that combined occlusal relief and cusp height into a single qualitative (categorical) score ranging from zero (high relief and sharp cusps) to three (blunt cusps and essentially no relief). This modified version of mesowear analysis (later dubbed mesowear II; Solounias et al. 2014) was predicated on the observation that occlusal relief and cusp sharpness are correlated with one another and that some morphologies – such as high relief with blunt cusps – are rare or non-existent. Other researchers have used similar approaches, sometimes combined with more traditional scoring (e.g., Croft and Weinstein 2008; Kaiser et al. 2009). Mihlbachler and Solounias (2006) used the percentage of specimens exhibiting blunt or rounded cusps as a single mesowear score, but this approach has not been used in other studies. Mihlbachler et al. (2011) facilitated use of a single, combinatorial variable for mesowear by using a mesowear "ruler" of tooth cusp casts illustrating the two endpoints and

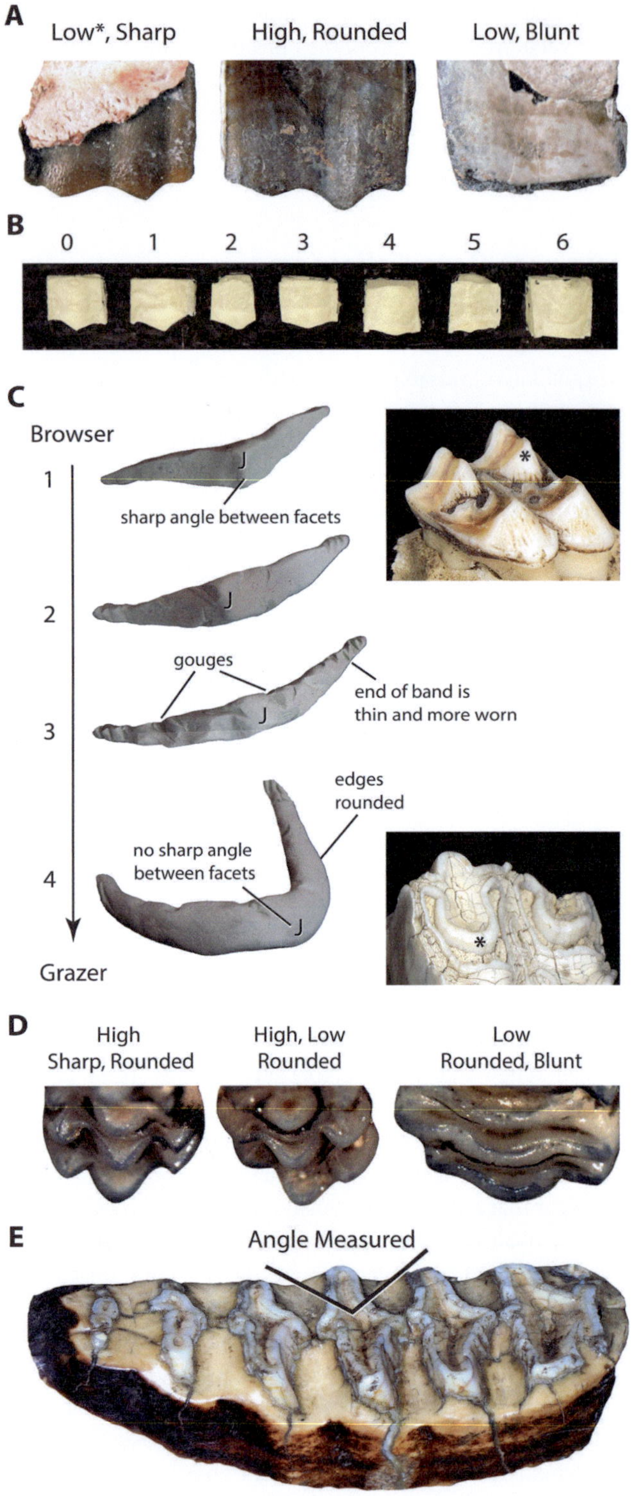

◀ **Fig. 5.1** Examples of mesowear scoring. A, notoungulate right upper second molars in buccal view scored qualitatively for occlusal relief (low or high; asterisk indicates that relief varies depending on metric cutoff used) and cusp shape (blunt, rounded, or sharp) based on Croft and Weinstein (2008); from left to right, *Archaeohyrax suniensis* UF 172969, *A. suniensis* UF 172473, *Trachytherus alloxus* UF 149261; B, mesowear "ruler" of Mihlbachler et al. (2011), digitally reversed so scores read from 0–6; C, inner mesowear (=mesowear III) scoring method of Solounias et al. (2014) and Danowitz et al. (2016); left, clay models (courtesy of N. Solounias) illustrating the four wear stages of the inner enamel band and the junction (J) between its mesial (anterior) and distal (posterior) facets; right, upper molars of a typical browser (above; white-tailed deer, *Odocoileus virginianus*) and a typical grazer (below; domestic cow, *Bos taurus*); the asterisk in each photo indicates the inner enamel band of the paracone; D, left upper molars of the murid *Apodemus flavicollis* in anterior (mesial) view, scored qualitatively for mesowear; relief is scored for both valleys, and shape is scored for all three cusps; modified from Ulbricht et al. (2015: Fig. 3); E, lower molar of an African savanna elephant (*Loxodonta africana* UM R1910) in oblique occlusal view (digitally reversed, anterior to right), illustrating the angle of the worn dentine valley measured by Saarinen et al. (2015) in their study of proboscideans. Abbreviations: UM, University of Michigan Museum of Paleontology; UF, Florida Museum of Natural History, University of Florida

five intermediate stages in this categorical variable (Fig. 5.1 B). Although these authors used a different numerical scale than Rivals and Semprebon (2006) (0–6 versus 0–3, respectively), the number of categories was the same. The major conceptual difference between this strategy and that of Rivals and Semprebon (2006) is that mesowear is scored as a single variable rather than two separate variables that are later combined into a single value. Fraser et al. (2014) also used a single-variable scale for scoring mesowear but included only five stages (two endpoints and three intermediate stages). The "extended mesowear method" of Winkler and Kaiser (2011; see also Taylor et al. 2013) calculates a single mesowear score using cusp shape and a modified scoring scheme for occlusal relief with more than two categories.

Most recently, Solounias et al. (2014) and Danowitz et al. (2016) have developed a new mesowear method for selenodont artiodactyls known as mesowear III or inner mesowear. The latter term highlights the main difference between this mesowear scoring technique and traditional scoring methods (=mesowear I and II), which Danowitz et al. (2016) refer to as outer mesowear: inner mesowear evaluates wear on the inner (more lingual) enamel band of a tooth in occlusal view rather than the outer (external or buccal) enamel band in lateral view. Inner mesowear is evaluated qualitatively using a four-point scale that incorporates facet shape, distinctiveness of facets, and the presence of gouges, among other characteristics (Fig. 5.1C). Three parts of the inner enamel band of an upper tooth cusp (paracone or metacone) are scored separately: (1) the mesial (anterior) facet; (2) the distal (posterior) facet; and (3) the junction between the two facets (J). The resulting inner mesowear data can be analyzed using the same visualization and statistical techniques as outer microwear data (discussed below), and the two can be combined with one another to provide a more nuanced picture of tooth wear (Danowitz et al. 2016).

Mesowear data collection for other groups of mammals varies depending on tooth morphology. Mesowear data for marsupials, murine rodents (mice; Fig. 5.1D), and rabbits (lagomorphs) can be collected using the same variables as for ungulates (Butler et al. 2014; Ulbricht et al. 2015). For proboscideans, which have a very different tooth structure, angular data are obtained from the occlusal surface using a digital angle meter (Saarinen et al. 2015; Fig. 5.1E).

Microwear Data Collection

Light Microscopy (LM) Data Collection: LM analysis uses a stereomicroscope and manipulated light to illuminate microwear features on a clear epoxy cast (Solounias and Semprebon 2002; Semprebon et al. 2004b; Fig. 5.2A). These analyses occur either by counting features directly through the lens (stereoscopic LM; Solounias and Semprebon 2002) or via digital images (photomicrographic LM: Merceron et al. 2005; Fraser et al. 2009; Mihlbachler et al. 2012). Solounias and Semprebon (2002) developed the popular observer-based stereoscopic LM method using a standardized magnification of 35× and a reticle with a 0.4 × 0.4 mm counting square placed in the ocular. Some later studies that focused on smaller mammals have used higher magnifications and/or smaller counting areas (e.g., 70×; Nelson et al. 2005; Townsend and Croft 2008b; Christensen 2014). Directing a light source across the occlusal surface at a shallow angle ("external oblique illumination" sensu Semprebon et al. 2004b) highlights microwear features in the target area.

The original LM variables of Solounias and Semprebon (2002) include: (1) number of scratches; (2) number of pits; (3) presence of >4 large pits; (4) texture of scratches (fine, coarse, or mixed); (5) presence of hypercoarse scratches; (6) presence of >4 cross-scratches; and (7) presence of gouges. Scratches and pits are counted inside the 0.4 × 0.4 mm reticle in two separate (non-overlapping) areas on the target region per individual (specimen), and the values for all individuals of a particular species are averaged to generate a species-specific value. Identifying scratch texture (predominately fine, predominately coarse, or a mixture of fine and coarse), pit size (small vs. large), hypercoarse scratches, and gouges is done qualitatively through the differential refraction of light (Solounias and Semprebon 2002). Deeper features (hypercoarse scratches, large pits) have low refractivity and thus appear darker on a light background, whereas shallower features are highly refractive and shiny on a dark background (Semprebon et al. 2004b; Christensen 2014). The angle of the light source is manipulated in a standardized fashion (e.g., Christensen 2014) to create both bright and dark fields, which helps illuminate the full range of features. Changes in the angle of the light source can directly change the types of features that are visible (Solounias and

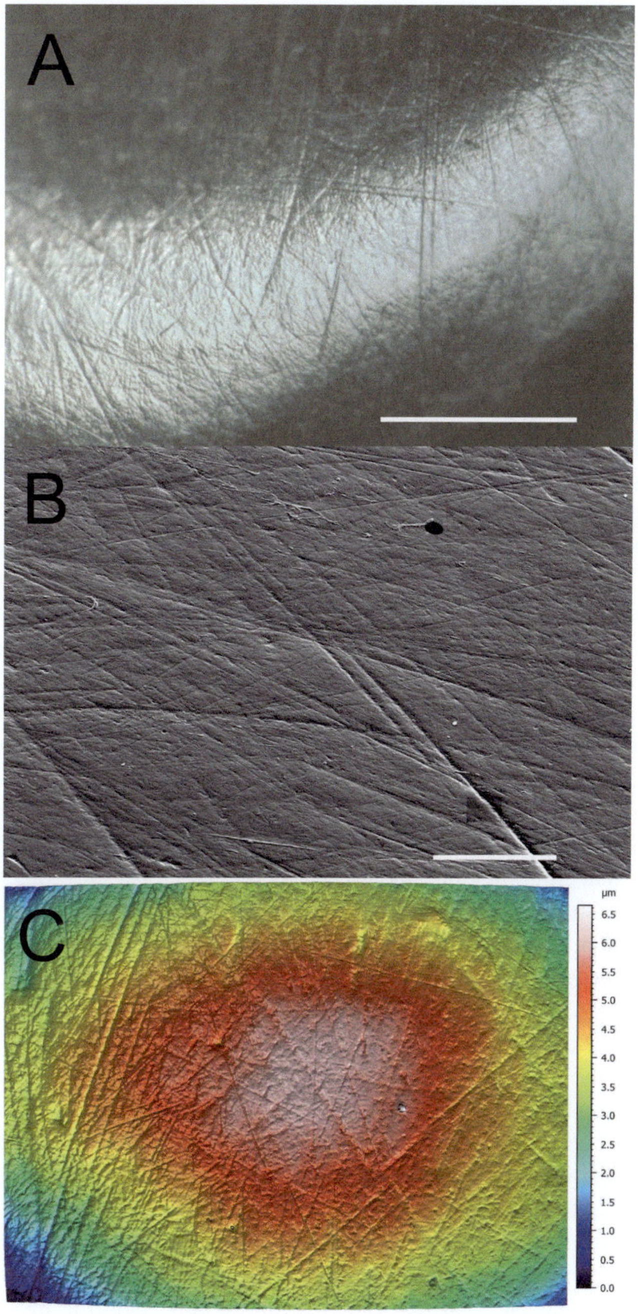

Fig. 5.2 Examples of enamel microwear on the second lower molar of *Mammut americanum* (late Pleistocene, Indiana, USA; INSM 71.3.226.1), imaged via three different microscopes. Images were all captured in the same general region on the central mesial facet of the metalophid. A, light microscope image using external oblique illumination, captured at a resolution of 1199 pixels/mm, scale bar = 400 µm; B, SEM image, captured at 500× magnification, 20 kV, secondary electron image, surface oriented normal to the electron beam, scale bar = 50 µm; C, three-dimensional white-light scanning confocal microscope, image x-y axes equal 276 × 204 µm^2 respectively, colors represent scale of z-axis. INSM = Indiana Museum of Natural History, Indianapolis, IN

Semprebon 2002); thus, it is important that an experienced user teach new individuals interested in stereoscopic LM the technique. For these qualitative variables, the microwear of a particular species is defined as the percentage of individuals showing a particular feature (e.g., hypercoarse scratches). Subsequent studies have introduced additional variables, such as the percent of individuals with a low scratch percentage (Semprebon and Rivals 2007; Rivals et al. 2012). Collectively, these variables provide a quantitative mean microwear signature for each species that can be analyzed statistically.

One limitation of stereoscopic LM is that the full range of features visible through the lens using external oblique lighting cannot be captured as a digital image without loss of resolution (Semprebon et al. 2004b). Alternative photomicrographic LM techniques have been developed to compensate for this problem (Merceron et al. 2005; Fraser et al. 2009; Mihlbachler et al. 2012). These techniques all use a digital camera mounted on a light stereomicroscope but vary in their placement of the light source and other aspects. Merceron et al. (2005) and Mihlbachler et al. (2012) placed the specimen on a glass stage and transmitted light from beneath, whereas Fraser et al. (2009) continued to use external oblique illumination but specified an angle of $\sim 45°$ relative to the surface of the cast. In all photomicrographic LM studies, the surface of the cast is subhorizontal to ensure equal illumination and good focus. Contrast and other variables can be manipulated to produce the best image either before the image is captured (Merceron et al. 2005) or afterwards (using software; Fraser et al. 2009; Mihlbachler et al. 2012). All of these decisions should follow the specifications of the chosen methodology.

In photomicrographic LM, images are analyzed on a computer display, so it is necessary to report digital resolution of both the image (resulting from the combination of pixel density and digital magnification) and the monitor display. It is critical to bear in mind that optical magnification is not the same as digital magnification; since light passing into the camera does not pass through the eyepiece, the magnification of the eyepiece is irrelevant. Confusion with regard to these specifications can lead to a lack of repeatability and inappropriate comparisons among independent microwear analyses (Mihlbachler and Beatty 2012). Digital resolution is reported as pixel density (sensu Mihlbachler and Beatty 2012), and an optimal resolution for photomicrographic LM is chosen on a case-by-case basis, depending on the size of the scars being analyzed (Mihlbachler and Beatty 2012). Highest resolutions can lead to overestimates of feature densities that are difficult to reproduce, as researchers will be tempted to count and analyze the finest scars visible (Mihlbachler and Beatty 2012).

Data collection on digital micrographs is user-dependent yet more controlled than counting directly through the lens (Mihlbachler et al. 2012). Counts should still be restricted to a standardized area, delineated as a digital box or by cropping the image. Users can save images used for analysis and even upload them to open-access online databases (e.g., Dental Microwear Image Library, maintained by NYIT College of Osteopathic Medicine) so that independent researchers can perform counts on the same images to improve long-term repeatability. In addition to counting feature density, photomicrographic analysis allows sizes of scars to be measured either directly (Merceron et al. 2005) or estimated using predefined shapes (Mihlbachler et al. 2012). The lowest levels of observer error and bias in LM (see Microwear Limitations) occur when users have prior experience with microwear analysis and conduct their analyses in a randomized, blind (to taxonomic identity and dietary category) fashion (Mihlbachler et al. 2012). Dimensionless variables (e.g., relative proportions of scratches and pits) may be more robust to repeatability than absolute ones (Mihlbachler and Beatty 2012; Mihlbachler et al. 2012; DeSantis et al. 2013).

Scanning Electron Microscopy (SEM) Data Collection: As mentioned previously, the bulk of dental microwear research prior to the re-introduction of LM or texture analysis was SEM-based, as high resolution micrographs could be captured at high magnification to measure the density and proportions of even the smallest features on a tooth (Gordon 1988; Teaford 1991; Ungar et al. 2008; Fig. 5.2B). This technique requires a conductive specimen to prevent electrostatic charging with the electron beam during imaging (unless an environmental SEM is used; Rose 1983; Galbany et al. 2004). This process is non-reversible, so casts of teeth are usually analyzed rather than actual specimens. The cast is first mounted on a brass disc or aluminum stub (using an adhesive such as term fusible gum, carbon tabs, or PVA) and then a colloidal argent belt (e.g., silver, graphite) is usually applied to the base of the cast and between the cast and stub. The mounted cast is then sputter-coated with a colloidal metal, typically silver or gold. The thickness of the coating can vary (100–400 Å; Gordon 1982; Galbany et al. 2004) but should only be as thick as necessary to produce a quality image with the microscope being used. The specimen is kept in a clean, dust-free container and gently cleaned with compressed air prior to imaging, as dust interferes with electron dispersal during analysis (Galbany et al. 2004).

A mounted specimen is inserted in the SEM and placed under vacuum, following the specified procedure for the instrument being used. A litany of imaging settings can be adjusted, including magnification level, operating voltage, working distance, specimen orientation/tilt, and the type of electrons used. The values used will depend on operator knowledge, the quality of the instrument, and the questions being addressed. However, it cannot be emphasized enough that all of these parameters should be standardized during imaging, as even subtle changes can affect the number, size, and type of scar features that are visible (Gordon 1988; Galbany et al. 2004). Magnification level varies depending on the analysis, with an obvious tradeoff between area sampled and accuracy and precision of measurements (Teaford 1991). Lower magnification levels (100–150×) cover a wider surface area (e.g., 0.4 mm^2 at 150×); as a result, the largest and most representative scars are visible (up to 400 individual features per tooth; Teaford 1991) but finer features are either invisible or too small to permit accurate measurements (Gordon 1988; Fig. 5.3A). Higher magnification levels (300–500×) cover a much smaller area (e.g., 0.03 mm^2 at 500×; Fig. 5.3B), which can be advantageous when working with small teeth and measuring the proportions of very small features. However, larger scars extend beyond the field of view (i.e., their size cannot be accurately measured), and variation in the overall scar pattern is underestimated (Gordon 1988; Fig. 5.3B–C). Thus, data collected from images taken at different magnifications are not directly comparable (e.g., Fig. 5.3), so standardizing this variable in a study is critical (Gordon 1988).

Operating (electron accelerating) voltage and working distance can vary depending on the size of the cast and capability of the instrument. Reported accelerating voltages usually range from 10–25 kV (Galbany et al. 2004; Grine et al. 2006; Green and Resar 2012), while working distances usually range from 10–30 mm (Galbany et al. 2004; Green and Resar 2012). Specimen tilt within the chamber should be consistent, as visual distortion occurs with varying levels of tilt relative to the electron detector, leading to errors in feature measurement (Gordon 1982, 1988). SEM analyses usually employ either back-scattered (e.g., Strait 1993) or secondary electrons (e.g., Ungar and Spencer 1999). The former provides excellent surface relief but may not illuminate scratches aligned with the electron detector (Galbany et al. 2004). Secondary electron imaging is recommended where categories of feature orientation are being considered (Galbany et al. 2004). The final SEM image should be adjusted for brightness and contrast using standardized procedures (e.g., using the "levels" feature in Adobe Photoshop such that the lightest pixel is white and darkest pixel is black; Mihlbachler et al. 2012; McAfee and Green 2015) to maximize feature visibility in a controlled fashion. Digital resolution (in pixel density) of the image and the computer display should be reported. Images are usually either cropped to a standardized size for analysis (with every feature in the cropped image being counted; e.g., Patnaik 2015), or a standardized digital box is constructed around the area to be counted (e.g., Green and Resar 2012).

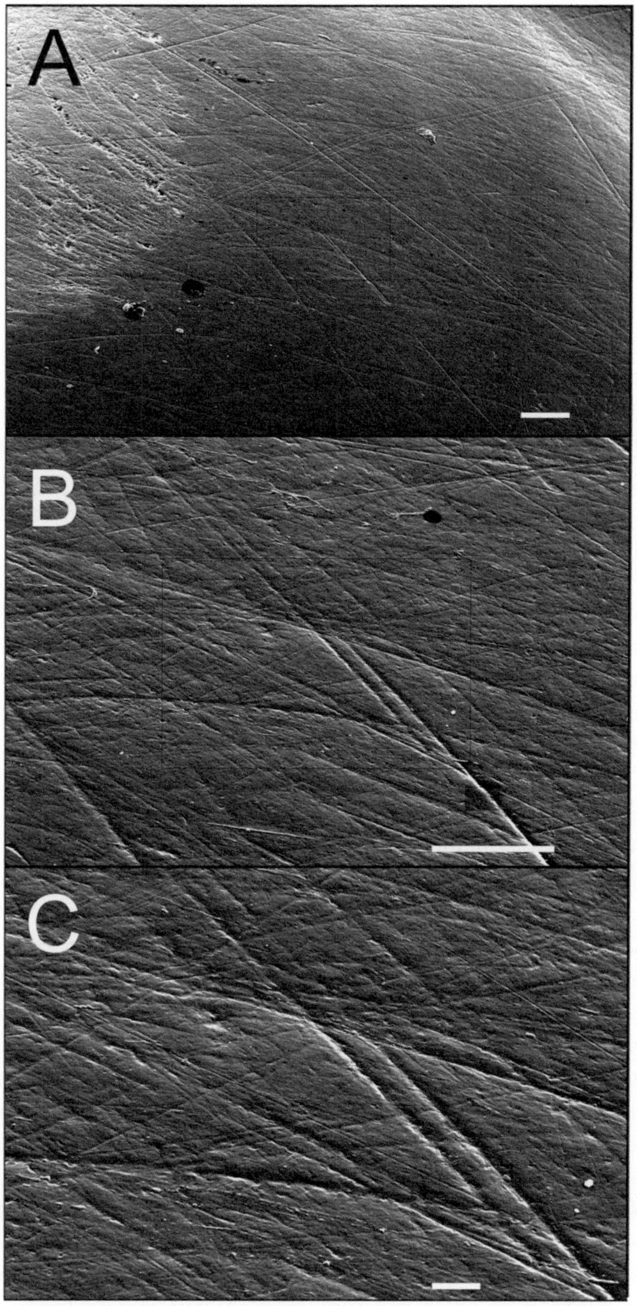

Fig. 5.3 Examples of microwear visible at different SEM magnification scales on the same specimen (*Mammut americanum*, m2, late Pleistocene, Indiana, USA; INSM 71.3.226.1). Images captured on central mesial facet of metalophid, using secondary electrons at 20 kV with surface oriented normal to the electron beam. A, 100× magnification, scale bar = 100 μm, box represents region visible in B; B, 500× magnification, scale bar = 50 μm, box represents region visible in C; C, 1000× magnification, scale bar = 10 μm. INSM = Indiana Museum of Natural History, Indianapolis, IN

Quantification of scar patterns on SEM micrographs has its roots in the early days of microwear research (Gordon 1982), with different techniques varying in magnification, field of view, measurement method, and feature category designation (scratches vs. pits) (Teaford and Walker 1984; Grine 1986; Ungar 1995). Grine et al. (2002) showed that different methods are all prone to inter-observer error (an observer has to identify individual features, an inherently subjective process) and recommended using the semi-automated software package Microware developed by Ungar (1995) to maximize ease of use and consistency. Subsequent SEM studies (Grine et al. 2006; Purnell et al. 2006; Green and Resar 2012; Resar et al. 2013; Green and Kalthoff 2015; Patnaik 2015) have used the latest version of the program (Microware 4.02; Ungar 2002). This software requires that the user identify genuine scars on the image and use mouse-clicks to mark the four endpoints (long axis and short axis) of each feature. Even features with edges that extend past the image are counted and measured. The software automatically generates summary data for each micrograph, including feature density, mean length, mean width, and mean orientation (relative and standard). Scratches are automatically distinguished from pits based on a length:width ratio that can be set by the user (and therefore should be reported); the default setting is a ratio of >4:1 for scratches, which is based on the findings of Grine (1986).

MTA Microscopy Data Collection: The ability to measure microwear in three dimensions (quantifying not only length and width of scars but also depth) adds a level of detail that can discriminate subtle dietary distinctions not obvious using LM or SEM methods (Purnell et al. 2012; DeSantis et al. 2013; Purnell et al. 2013). The advent of MTA (Ungar et al. 2003; Scott et al. 2005, 2006) achieved this result by using a white-light scanning confocal microscope with 100× objective to scan tooth casts. Some authors have used interferometry to acquire three-dimensional surface data (Estebaranz et al. 2007; Merceron et al. 2014), yet confocal microscopy remains the most widely applied platform in MTA (Calandra and Merceron 2016). The user chooses the wear facet to analyze and is responsible for positioning the tooth cast such that the appropriate target surface (i.e., enamel vs. dentine; Haupt et al. 2013) is scanned. Four adjacent scans per tooth (a total field of view of $204 \times 276\ \mu m^2$; Fig. 5.2C) are then digitally leveled, and identifiable surface defects and artifacts (dust particles) are manually removed from the image (Scott et al. 2006). Postmortem alteration, either anthropogenic or sedimentary, is easily recognizable under MTA (see Calandra and Merceron 2016 for examples). A data point cloud is generated and analyzed in Toothfrax (Surfact) and SFrax programs using scale-sensitive fractal analysis (SSFA), which characterizes changes in surface texture (length-scale, area-scale, and volume-filling versus scale analyses) at differing scales (Scott et al. 2006). Since the entire pattern on the surface of the tooth is characterized automatically rather than by an individual user, problems associated with observer error are greatly reduced (Ungar et al. 2003). However, selection of the wear facet to be scanned still requires user competency, which can be a source of error for untrained individuals (Calandra and Merceron 2016). Likewise, differences in settings on the confocal imaging system (e.g., resolution, lateral point spacing, source light type) can lead to incomparable data if such factors are not taken into account by researchers using different instruments (Arman et al. 2016). Manually deleting dust particles and casting artifacts from confocal scans (prior to analysis) is another potential source of error, as both excessive and conservative editing can reduce data quality. Some authors choose to discard scans that include dust particles and casting artifacts, rather than digitally edit each scan (Calandra et al. 2012).

Ecologically significant SSFA variables include: (1) complexity; (2) anisotropy; (3) textural fill volume; (4) scale of maximum complexity; and (5) heterogeneity of complexity. Each of these is briefly summarized below (see Scott et al. 2006 for additional details). Complexity measures change of roughness across different scales, with high complexity usually correlated with a surface that has scratches and pits of different sizes that overlie one other. Anisotropy measures the orientation of roughness on a surface, with high anisotropy corresponding to a surface with long, consistently oriented scratches. Textural fill volume describes how much material has been removed from a surface, with high values correlating with the presence of deep features. Scale of maximum complexity records fine-scale limits of complexity, with higher values corresponding to a greater proportion of small features on a surface. Finally, variation in complexity across a surface (heterogeneity) is measured using subdivisions of the analyzed surface (either a 3×3 or 9×9 grid), with low heterogeneity representing surfaces of more uniform texture.

Three-dimensional MTA using SSFA is applied today as it was originally developed by Ungar et al. (2003; see Withnell and Ungar 2014; Hua et al. 2015; Shearer et al. 2015), although alternative techniques employing standardized areal texture parameter data (e.g., ISO 21578-2) are also in use (see Schulz et al. 2010; Calandra et al. 2012; Purnell et al. 2012, 2013; Schulz et al. 2013; Gill et al. 2014). Extracting ISO parameters requires similar pre-processing as SSFA, but the final scan must also be spatially filtered and adjusted for gross tooth form alterations such as curvature (Calandra et al. 2012; Purnell et al. 2013). Generated parameters include up to 30 variables that define basic geometric properties of surface texture (e.g., height, hybrid, volume, spatial), each quantifying a specific attribute. The ability of particular variables to discriminate diet depends on the taxonomic group; diets of primates and bats are discriminated by a unique combination of height, spatial, and volume variables (Calandra et al. 2012; Purnell et al. 2013), whereas fish require height and hybrid combinations (Purnell et al. 2012). Recently, an analytical technique that quantifies three-dimensional microwear on virtual reality models has been developed and tested (Tausch et al. 2015). Preliminary results suggest that this alternative to MTA may be useful for mapping directionality of chewing in molar facets in extinct taxa and may also have clinical applications in humans (Tausch et al. 2015).

Mesowear Data Analysis

In the case of analyses using mesowear I, the variables analyzed are percentages of individuals in a population or species with a particular trait (e.g., % with sharp cusps). Since percentage data are not normally distributed, they should be transformed using the arcsine transformation if standard (parametric) significance tests follow. Alternatively, non-parametric tests could be employed. Mesowear profiles using these variables can be compared using multivariate techniques such as principal components analysis (PCA), discriminant function analysis (DFA), and hierarchical cluster analysis (HCA) (Fortelius and Solounias 2000; Franz-Odendaal et al. 2003; Semprebon et al. 2004a; Croft and Weinstein 2008; Fraser and Theodor 2011; Louys et al. 2011). Mesowear II analyses use univariate statistical tests to compare means (e.g., ANOVA; Rivals et al. 2013) and can examine correlations using regression (Kaiser et al. 2013). Mesowear distributions can be visualized using histograms and box and whisker diagrams.

Statistical tests should only be used if their *a priori* assumptions are not violated. Many such tests assume that data points are statistically independent, and this is not necessarily the case for biological data; a trait common to two species can be the result of phylogeny (inheritance from a common ancestor) as well as ecology (adaptation to a particular lifestyle). Thus, strong phylogenetic signal in the species and traits being analyzed can confound a paleoecological study and lead to type I errors ("false positives"). The degree to which this impacts mesowear analyses is not entirely clear and likely depends on the particular data set being analyzed and the specific question(s) being asked. Nevertheless, assuming phylogeny is a potentially confounding factor is a conservative approach, and several studies have controlled for non-independence of data using techniques such as phylogenetic generalized least squares (PGLS; Kaiser et al. 2013; Fraser et al. 2014) and phylogenetically independent contrasts (PIC; Mihlbachler and Solounias 2006). Münkemiller et al. (2012) provide a useful review of other statistical techniques that can be used to test for phylogenetic signal in ecological data.

Microwear Data Analysis

Plots of mean scratch vs. mean pit counts among sampled taxa reveal broad dietary differences (grazer vs. leaf browser vs. fruit browser) using both LM and SEM techniques. Grazers tend to have high scratch but low pit counts, leaf browsers have low scratch and low or high pit counts (i.e., "dirty" browsers; Semprebon and Rivals 2007), and fruit browsers have moderate scratch and high pit counts. This relative difference in scratch vs. pit values (i.e., the "trophic triangle") is consistently reproduced in stereoscopic and photomicrographic LM studies (e.g., Semprebon et al. 2004b; Mihlbachler et al. 2012; Christensen 2014). Although absolute values of scratch and pit counts vary among observers (see Microwear Limitations), relative differences will be consistent across observers (Mihlbachler et al. 2012). Thus, it may be more useful to look for statistically significant relative trends across a dataset rather than absolute "cut-offs" between dietary morphospaces (e.g., differentiating browsing from grazing ungulates based on number of scratches; Solounias and Semprebon 2002). Dimensionless variables (e.g., Microwear Index = number of scratches/number of pits; MacFadden et al. 1999) that rely on proportional rather than absolute similarities may also be useful (DeSantis et al. 2013). For SSFA variables in MTA, a significant negative relationship exists between complexity and anisotropy among mammalian herbivores (DeSantis et al. 2013), and these two variables are commonly plotted against one another to delineate dietary ecomorphospaces (Ungar et al. 2007; Donohue et al. 2013; Merceron et al. 2014; Shearer et al. 2015). For a more detailed review of how MTA can be used to study mammalian ecology, see Calandra and Merceron (2016) and DeSantis (2016).

Regardless of how data are collected, statistical tests are necessary to distinguish random noise from significant variation among species (Gordon 1988). Tests should be chosen based on: (1) the nature of the sample population; (2) the assumptions of the test; and (3) the hypotheses being tested. Univariate parametric tests (e.g., student's t-test, ANOVA) can be used to compare species means, but should only be used if the assumptions of normality and equal variance are met. Otherwise, (as is common for microwear data; Scott et al. 2006), data should be transformed beforehand (Conover and Iman 1981), or non-parametric tests (e.g., Mann-Whitney U, Kruskal-Wallis test; Kruskal and Wallis 1952) can be applied (e.g., Calandra et al. 2012; Donohue et al. 2013; Purnell et al. 2013; Withnell and Ungar 2014). Multivariate analyses (PCA, DFA, and HCA) amenable to both continuous and discrete variables can provide a robust view of how LM, SEM, or MTA variables change statistically among taxa (e.g., Solounias and Semprebon 2002; Godfrey et al. 2004; Green et al. 2005; Townsend and Croft 2008b; Purnell et al. 2013; Gill et al. 2014; Green and Kalthoff 2015). As discussed above for mesowear, the presence of a strong phylogenetic signal in microwear data can be tested and controlled for using statistical techniques (e.g., Fraser and Theodor 2011). Independent researchers should be well-versed with statistical techniques or collaborate with statistically-knowledgeable colleagues to avoid potentially catastrophic mistakes (Vaux 2012).

Strengths

General Strengths

A major strength of mesowear and microwear as paleodietary proxies is that only well-preserved teeth are required. The need for relatively complete skeletal remains (e.g., skulls, articulated limbs, etc.) is often a major limiting factor for studies of extinct species using functional morphology (e.g., finite element analysis). Teeth are frequently preserved in the fossil record due to the extreme durability of enamel, and in the case of mammals, even isolated teeth usually preserve sufficient information for a generic or specific identification. This means that studies can be based on relatively large sample sizes (at least from a paleontological perspective), engendering greater confidence in interpretations and providing opportunities to examine patterns within and between populations. Both mesowear and microwear are independent of one another, and statistical power for reconstructing diet is higher when they are used in conjunction (Fraser and Theodor 2011). In contrast to isotopic studies (Higgins 2018), mesowear and microwear are non-destructive techniques.

A second major strength of mesowear and microwear analyses is that they appear to be relatively taxon-independent (see Andrews and Hixson 2014). In other words, since the wear being evaluated is due in large part to the particles consumed, similar ingesta are thought to produce similar features in different groups. Nevertheless, differences in enamel microstructure can influence the size and distribution of microwear features (Maas 1991), and a recent study of artiodactyls and perissodactyls concluded that similar diet does not necessarily result in similar microwear (Mihlbachler et al. 2016). Likewise, since mesowear was developed for ungulates with selenodont or lophodont teeth, extending the method to groups with other types of teeth and/or mastication styles has required developing new scoring methods (see below). Thus, scores are not directly comparable among all groups. The degree to which mesowear and microwear data are taxon-independent (i.e., affected by phylogeny) will undoubtedly be a topic of additional research in coming years. If such data are found to be highly correlated with clade membership, it may limit their utility for inferring diets in families and orders of mammals that have no extant representatives and/or that differ significantly in craniodental morphology from their modern relatives (e.g., Schulz et al. 2007; Croft and Weinstein 2008; Townsend and Croft 2008a; Green 2009a, b; Semprebon et al. 2011; Resar et al. 2013; Green and Kalthoff 2015).

Mesowear Strengths

In addition to the general strengths noted above, mesowear has the advantage of being relatively quick and easy to record on original fossil material. The ease of scoring varies to some degree with the specific method employed, but even more complex schemes only require a pair of calipers or a digital camera and laptop computer. As a result, specimens can be scored directly in museum collections, permitting large samples to be obtained for minimal expense (essentially only the cost of visiting the museum). Mesowear scores tend to be more consistent among different observers than microwear scores, and the technique can be learned relatively quickly (Loffredo and DeSantis 2014).

Microwear Strengths

Dental microwear data provide direct evidence of food and chewing mechanics on tooth surfaces if care is taken with specimen selection and preparation (Gordon 1982; Teaford 1988; Ungar et al. 2007). Microwear thus provides direct evidence of how teeth were actually used by an individual animal, not just what they are evolutionarily capable of doing (Teaford 2007). This proxy can provide detailed information on individual feeding behavior based on physical properties (abrasiveness) of ingesta, which can be used to determine variation in food preference within and among populations of a species (e.g., Ungar and Spencer 1999; Grine et al. 2006; Rivals and Solounias 2007; Merceron et al. 2010; Henton et al. 2014), as well as seasonal and chronologic (evolutionary) changes in feeding ecology at the species, clade, or paleocommunity-level (e.g., Merceron et al. 2010; Rivals et al. 2013; Merceron et al. 2014; Patnaik 2015). Hypotheses of the cause-effect relationship between food texture (size, shape, hardness) and dental biomechanics (tooth shape, tissue hardness, bite force, chewing mechanics) can be experimentally tested under controlled laboratory conditions (e.g., Gordon 1984; Teaford and Oyen 1989; Maas 1991; Lucas et al. 2013; Borrero-Lopex et al. 2015; Constantino et al. 2015; Hua et al. 2015; Xia et al. 2015), and ideally, *in vivo* in wild or captive animals (e.g., Teaford and Oyen 1989; Organ et al. 2006; Schulz et al. 2013; Baines et al. 2014; Hoffmann et al. 2015).

Traditionally, microwear analyses were limited to enamel surfaces, as it was assumed that orthodentine was too complex and heterogeneous to preserve regular, measureable microwear (Lucas 2005). However, multiple recent studies into orthodentine microwear (focused on xenarthrans, which lack enamel on adult tooth surfaces) reveal that scar patterns on this tissue can be statistically quantified (Oliveira 2001; Green 2009a, b; Green and Resar 2012; Haupt et al. 2013; Resar et al. 2013; Green and Kalthoff 2015; McAfee and Green 2015). Although orthodentine has less power to discriminate diet than enamel (Haupt et al. 2013), microwear still varies predictably among xenarthrans with distinct diets regardless of the technique applied (Green and Kalthoff 2015). Thus, microwear can now be used as an ecological proxy for species with enameless teeth (i.e., a presumed limitation is now a strength).

Limitations

General Limitations

The temporal scale of any dietary proxy must be taken into consideration when reconstructing feeding ecology. As noted previously, mesowear reflects an animal's diet over months to years (but see Solounias et al. 2014 for a demonstration of more rapid changes in experimental animals). As a consequence, it cannot detect transitions in diet on shorter time scales, such as with the seasons (e.g., Rivals et al. 2013). However, mesowear can detect dietary transitions before such shifts are reflected in skeletal morphology (tooth shape, chewing mechanics, bite force potential), since changes in mesowear take place more rapidly than evolutionary changes in morphology (e.g., Mihlbachler et al. 2011). Microwear reflects diet on an even shorter time scale than mesowear, generally days to weeks. This can be useful for detecting seasonal changes in diet or dietary differences among extinct and extant populations (e.g., Rivals and Solounias 2007; Rivals et al. 2015), but it can also result in misrepresentations of typical diet if the population sampled is not sufficiently representative (Kay and Covert 1983). When comparing mesowear and microwear across major clades, statistical techniques that account for non-independence of data due to phylogeny should be used to limit the potential for type I errors. Carbon ($\delta^{13}C$) and nitrogen ($\delta^{15}N$) isotope ratios preserved in fossilized bones and/or teeth can also provide dietary information (Higgins 2018). This geochemical information is recorded over an interval similar to mesowear (months to years) but reflects ecological conditions when the tissue was mineralizing (see Higgins 2018) rather than the last few months or years of life. Apparently conflicting dietary information from different proxies can often be resolved by considering the specific intervals that each reflects.

Mesowear Limitations

Perhaps the greatest limitation of mesowear analysis is that it only reflects the relative contributions of attrition (tooth-on-tooth wear) and abrasion (food-on-tooth wear) to tooth facet development. In other words, it gauges the toughness or abrasiveness of the food being consumed. The method was originally developed for relatively large-bodied herbivores (primarily folivores or leaf-eaters) with the primary goal of discriminating abrasion-dominated grazers and/or open habitat feeders from attrition-dominated browsers in forested habitats (as well as identifying varying types of mixed feeders; Fortelius and Solounias 2000). This same grazing-browsing/open habitat-closed habitat axis has formed the basis for all subsequent mesowear analyses, even those focusing on non-ungulates (e.g., Butler et al. 2014; Ulbricht et al. 2015). Unlike microwear, it is unclear whether mesowear (relative attrition vs. abrasion) can be used to distinguish among other types of foods. The relative contributions of food (particularly grass) and habitat to mesowear development continue to be debated (e.g., Kaiser and Schulz 2006; Kaiser et al. 2013; Damuth and Janis 2014; Kubo and Yamada 2014).

Another primary limitation of mesowear is that it is scale-dependent; the cusps on the tooth of a large, browsing rhinoceros do not look the same as those of a small, browsing deer, even though each has teeth with high, pointed cusps compared to close relatives that have more abrasive diets. Thus, although the mesowear ruler of Mihlbachler et al. (2011) functions well for members of the horse family (Equidae), the focus of that study, the scale is less useful for species with much smaller or much larger teeth. This limitation can be overcome relatively easily for species that have modern relatives of similar size but is an impediment for extinct herbivores such as South American notoungulates that span four orders of magnitude but have no living representatives (Croft 1999; Croft and Weinstein 2008).

Until recently, it seemed that overall tooth structure and mastication style were a significant limitation for mesowear investigations. Several recent studies have demonstrated that the basic principles of paleodietary reconstruction based on attrition and abrasion can also be applied to groups as diverse as rodents, rabbits, marsupials, and elephants, and that it is possible to develop methods for mesowear analysis in these groups (Fraser and Theodor 2010; Butler et al. 2014; Saarinen et al. 2015; Ulbricht et al. 2015). Thus, tooth structure and mastication style may be a less significant limitation than previously thought.

Microwear Limitations

Teaford (2007) provides a thorough review of the theoretical limits of dental microwear; the highlights of Teaford's discussion are outlined below. As mentioned previously, microwear is only as good as the specimen on which it is preserved. Once scars are altered or destroyed via post-mortem processes, they are lost forever. If all known specimens of a fossil species have been altered, then that animal's diet must be determined using dietary proxies other than microwear. Even in well-preserved teeth, information about an animal's feeding behavior can be invisible to microwear analysis if it does not scar the tooth surface, as can occur with very soft foods. Microwear in fossil populations is also limited by the fact that not all past environments and habitats still exist today, and that such environments may have had unknowable and untestable impacts on microwear. However, these limits will continue to be pushed as technologies used to image and analyze wear surfaces improve along with the ability to experimentally test the direct impacts of particles on tooth surfaces (e.g., Hua et al. 2015).

Dental microwear relies on the calibration of microwear patterns among extant animals (for which diet is known) to test hypotheses of paleodiet in fossil organisms (Teaford 1991). Unless *in vivo* experimental analyses are applied, microwear studies must rely on specimens in museum collections and generalized dietary information for the species (which may itself be scarce, such as for two-toed sloths; Green and Resar 2012) rather than specific information about the foods consumed by a particular individual (Kay and Covert 1983). Lack of associated climate and environmental data can exacerbate this problem (Teaford 2007). Many mammals are opportunistic and change their diet seasonally and/or regionally (e.g., Bowyer et al. 1983; Heiduck 1997; Sidorovich 2000; Stokke and du Toit 2000). As a result, simply pooling individuals of a single species from different regions could produce a skewed microwear signal (Rivals et al. 2010). In both extant and paleontological studies, this can be minimized by only pooling individuals from limited spatial and temporal intervals (e.g., from the same geographic regions or geological formations).

A large sample of individuals is necessary to accurately estimate microwear signatures and identify normal departures from the mean at either the population or specific level (Gordon 1988). Acquiring a sufficient sample depends on not only the researcher's ability to visit museum collections, but also the time required to analyze each tooth. The latter variable is directly correlated with the technique being applied. Stereoscopic LM is the most amenable to analyzing a large number of samples (in some cases, hundreds of teeth; Solounias and Semprebon 2002; Green 2009a) in the shortest amount of time (hours to days). Confocal MTA also permits analysis of large samples over short time frames, although not as quick as stereoscopic LM, as each specimen takes approximately 20–30 minutes to scan. Photomicrographic LM and SEM require the most time per specimen, as the position and instrumentation settings have to be adjusted manually to capture an appropriate image, and then each image has to be analyzed independently by the user. Additionally, SEM and MTA labs may charge for instrumentation use, whereas this is generally not the case for LM setups. The pros and cons of time required to conduct an analysis versus the ability of a technique to address the study's goals should be considered when designing an analysis.

Arguably, the single largest source of error in microwear studies is consistent identification and measurement of discrete features, either through the lens or on images (LM and SEM; Grine et al. 2002; Semprebon et al. 2004b; Galbany et al. 2005; Mihlbachler et al. 2012; DeSantis et al. 2013). Error rates among users examining the same specimens can be high and statistically significant for all user-based techniques (1–21% [9% average] for SEM; 8–45% for light microscopy), with variation depending on user experience, the specific variable being identified, sample size, and number of iterations (Grine et al. 2002; Galbany et al. 2005; Mihlbachler et al. 2012). In general, a significant negative relationship exists between user experience with microwear feature identification and the magnitude of error (Galbany et al. 2005; Mihlbachler et al. 2012). Even though inter-observer error clearly impacts LM and SEM studies (Grine et al. 2002; Galbany et al. 2005; Mihlbachler et al. 2012; DeSantis et al. 2013), a wide variety of user-based microwear analyses have made and continue to make important contributions toward elucidating feeding ecology (e.g., Walker et al. 1978; Gordon 1982; Teaford and Walker 1984; Teaford and Oyen 1989; Strait 1993; Solounias and Semprebon 2002; Purnell et al. 2006; Fraser et al. 2009; Green 2009a; Green and Resar 2012; Resar et al. 2013; Christensen 2014; Strait 2014; Hoffmann et al. 2015). Microwear studies using MTA provide the solution to the problem of significant user error but are limited by the availability of confocal microscopes, which are available to fewer researchers than pre-existing SEMs and stereomicroscopes. Until more researchers have access to such technologies, studies using user-based techniques should follow strict protocols to minimize noise from observer error. For example, researchers should count their own features on each image or cast that is analyzed (Galbany et al. 2005; Mihlbachler et al. 2012) rather than comparing their values to those of others, such as those compiled for comparative microwear databases (e.g., Green et al. 2005; Rivals et al. 2010, 2012). Such practices are becoming more common among user-based studies (e.g., Green 2009a, b; Green and Resar 2012; Christensen 2014; Hoffmann et al. 2015). Although is not possible to retroactively test for the effects of inter-observer error in past LM studies (Mihlbachler et al. 2012), there is no direct evidence that the fundamental conclusions of such studies were invalid. Indeed, most independent studies found similar, predictable microwear trends among species with distinct diets (Green et al. 2005; Townsend and Croft 2008b; Rivals et al. 2010, 2012).

Observers should also always count under blind conditions, with specimen, taxonomic, and dietary information invisible, in one sitting (see Green and Resar 2012 and Mihlbachler et al. 2012 for examples of how to set up such analyses). This ensures that a microwear analysis is not a "self-fulfilling prophecy" (Mihlbachler et al. 2012).

Examples of Applications

Mesowear and microwear have been used in a wide variety of applications, and no single study can stand as an exemplar of their many uses in paleoecology. Therefore, we briefly summarize below recent representative publications that together provide examples of the techniques and approaches discussed in this chapter.

Mesowear

1. **Mesowear and hypsodonty over time**: Mihlbachler, M. C., Rivals, F., Solounias, N., & Semprebon, G. M. (2011). Dietary change and evolution of horses in North America. *Science*, *331*, 1178–1181. Mihlbachler et al. integrate mesowear data (mesowear II) with crown height (hypsodonty) and body size data (as inferred from tooth size) to examine how dietary abrasiveness in horses (family Equidae) changed over a period of some 55 million years. They find a strong correlation between hypsodonty and dietary abrasiveness and were able to correlate both with changes in climate and vegetation. They use the amount of variation in mesowear scores as an indicator of the relative strength of selective pressure, with high variation corresponding to low selective pressure, and note several instances during which particular horse populations were apparently under greatest selective pressure for increased hypsodonty. This study is an example of how mesowear can be used to examine ecological changes in a single group over an extended interval.
2. **Combined mesowear and stereoscopic LM**: Sánchez-Hernández, C., Rivals, F., Blasco, R., & Rosell, J. (2016). Tale of two timescales: combining tooth wear methods with different temporal resolutions to detect seasonality of Paleolithic hominin occupational patterns. *Journal of Archaeological Research: Reports*, https://doi.org/10.1016/j.jasrep.2015.09.0114. Sánchez-Hernández et al. collect mesowear (mesowear II) and microwear data from archaeological cave site samples of red deer (*Cervus elaphus*; a mixed feeder) and horse (*Equus ferus*; a grazer). They find *E. ferus* mesowear and microwear to be consistent with one another and invariant among caves and stratigraphic intervals, as would be expected for a committed grazer. By contrast, they find discrepancies between microwear and mesowear data for *C. elaphus*, as well as stratigraphic variation in microwear (but not mesowear) scores. They interprete these differences as reflecting the differing time scales over which mesowear and microwear form, with mesowear scores reflecting a year-long average diet and microwear scores recording diet during a particular season. This paper is an example of how combining microwear and mesowear data can provide greater insight than analyzing either one separately.
3. **Combined mesowear and MTA**: Stynder, D. D. (2011). Fossil bovid diets indicate a scarcity of grass in the Langebaanweg E Quarry (South Africa) late Miocene/early Pliocene environment. *Paleobiology*, *37*, 126–139. Stynder collects mesowear data (mesowear I) from fossil horses (family Equidae) and compares these data with microwear data from two previous studies (one using traditional microwear, the other using MTA). He interprets most species as browsers based on mesowear scores and infers that previous interpretations of some of these species as grazers based on microwear (using traditional techniques but not MTA) could be due to their feeding on less abrasive C3 (as opposed to C4) grasses during a specific time of year. Based on the diets of these horses, he reconstructs South Africa's southwest coast as being heavily wooded about 5 million years ago. This study provides an example how mesowear and microwear data can be combined to infer vegetational structure.

In Vivo Microwear in Laboratory Animals

1. **SEM**: Baines, D. C., Purnell, M. A., & Hart, P. J. B. (2014). Tooth microwear formation rate in *Gasterosteus aculeatus*. *Journal of Fish Biology*, *84*, 1582–1589. Baines and colleagues undertake a controlled feeding study in a population of tank-reared three-spined stickleback. Prior to the analysis, all fish had an exclusive limnetic feeding regime and were fed bloodworms in the water column in a tank with no abrasive sediment on the bottom. At the start of the study, a group of fish were transferred to a benthic feeding regime, where they fed exclusively on bloodworms on the tank bottom with quartz sediment substrate. On four consecutive days, four fish were sacrificed per day for study. Microwear feature density (scratch and pit counts) increased significantly from day 0 (control) to day 4 within the study group. These results reveal that: (1) microwear turnover rate is

high even in aquatic vertebrates with non-occluding teeth; and (2) an abrasive substrate has a significant influence on microwear formation.

2. **Photomicrographic LM**: Hoffman, J. M., Fraser, D., & Clementz, M. T. (2015). Controlled feeding trials with ungulates: a new application of *in vivo* dental molding to assess the abrasive factors of microwear. *The Journal of Experimental Biology*, *218*, 1538–1547. Hoffmann and colleagues use a novel live-animal sampling technique to test the influence of ingested abrasives on ungulate tooth microwear. Over a one-month period, four pen-raised sheep were fed three dietary regimes: (1) a control diet of clean hay; (2) hay with fine sand; and (3) hay with coarse sand. Sheep were anesthetized and microwear molds taken between feeding treatments. Under blind, randomized conditions, two independent observers consistently found that pit counts were positively correlated with an increase in abrasive size. This study was the first to use ungulates in a repeatable, controlled feeding experiment and to study the effect of different-sized abrasive particles on microwear formation.
3. **MTA**: Schulz, E., Piotrowski, V., Clauss, M., Mau, M., Merceron, G., & Kaiser, T. M. (2013). Dietary abrasiveness is associated with variability of microwear and dental surface texture in rabbits. *PLoS ONE*, *8*, e56167. Schulz and colleagues measure the effect of silica content on microwear formation in four groups of laboratory rabbits. For 25 weeks, each group was fed a different controlled diet, including lucerne (alfalfa), lucerne plus crushed oats, grass meal with crushed oats, and grass meal. The lucerne group had the lowest silica content and was correlated with higher variability in microwear texture parameters, whereas abrasive diets (grass) had more uniform wear patterns. This suggests that variation in a species' microwear texture signature likely reflect changes in abrasive content of consumed food.

Reconstructing Diet Using Microwear

1. **SEM**: Resar, N. A., Green, J. L., & McAfee, R. K. (2013). Reconstructing paleodiet in ground sloths (Mammalia, Xenarthra) using dental microwear analysis. *Kirtlandia*, *58*, 61–72. Resar and colleagues compare orthodentine microwear patterns in three extinct ground sloths with those of extant two-toed and three-toed sloths. Each fossil species was sampled from a single locality, representing one population. Under blind, randomized conditions, two independent observers found that living and extinct forest-dwelling sloths had a low number of predominately coarse scratches, whereas the more open-habitat sloth, *Thinobadistes*, had unusually high numbers of fine scratches. This result supports the efficacy of orthodentine microwear as a proxy for paleoecology and suggests a possible environmental influence in sloth microwear.
2. **Stereoscopic LM**: Christensen, H. B. (2014). Similar associations of tooth microwear and morphology indicate similar diet across marsupial and placental mammals. *PLoS ONE*, *9*, e102789. Using a large sample of 153 extant mammal species, Christensen combines LM with tooth morphology and body mass to produce an accurate method to predict diet in a species with unknown ecology. The author collected all LM data independently rather than relying on comparative data collected by others. The "trophic triangle" (where grazers, browsers, and hard object feeders vary in scratch and pit counts) was recovered, consistent with independent LM studies. This study demonstrates that microwear data used in conjunction with other variables will more accurately predict ecology than microwear data alone.
3. **Photomicrographic LM**: Fraser, D., Mallon, J. C., Furr, R., & Theodor, J. M. (2009). Improving the repeatability of low magnification microwear methods using high dynamic range imaging. *PALAIOS*, *24*, 818–825. Fraser and colleagues introduce a robust method of LM using high dynamic range imaging to produce high-resolution photomicrographs that can be scored manually. All ranges of scratch and pit categories are visible on a single image, which eliminates the need to manually adjust the light angle, as in stereoscopic LM. Using these images, two independent researchers were able to consistently and correctly classify exant species by diet. This method improves the repeatability of LM studies.
4. **MTA**: Donohue, S. L., DeSantis, L. R. G., Schubert, B. W., & Ungar, P. S. (2013). Was the Giant short-faced bear a hyper-scavenger? A new approach to the dietary study of ursids using dental microwear textures. *PLoS ONE*, *8*, e77531. Donohue and colleagues compare the microwear textures of the extinct *Arctodus simus* (whose diet has been strongly debated) with that of five extant bear species. They found that high anisotropy is characteristic of leaf-eating (folivory) in living bears, whereas high complexity is more correlated with feeding on hard objects, such as bone or fruit. *Arctodus simus* had a microwear texture inconsistent with hard-object feeding and therefore did not consume large amounts of bone (i.e., it was not a hyper-scavenger), falsifying at least one paleodietary hypothesis for this extinct species. This study highlights the ability of MTA to reconstruct paleodiet in extinct species via comparisons with living relatives.

Future Prospects

Several recently published studies have highlighted areas in which mesowear research will likely advance over the next decade. These include extending mesowear analysis to non-ungulate groups of mammals (Butler et al. 2014;

Saarinen et al. 2015; Ulbricht et al. 2015) and examining more detailed aspects of tooth morphology to provide additional data on abrasion and attrition (Solounias et al. 2014; Danowitz et al. 2016). Additionally, controlled studies of the relationship between diet and mesowear (Solounias et al. 2014) are vital for understanding the processes underlying observed patterns. Since mesowear is still a relatively new method, it is likely to be used with increasing frequency to study the diets of extinct species.

One area of mesowear analysis that seems ripe for advancement is automated scoring of teeth. Since mesowear I and II ("outer mesowear") analyses focus on the buccal profile of a tooth, it should be possible to create an automated computer program for extracting a simple shape from a photograph and calculating relevant parameters (heights, angles, etc.). This could eliminate all subjectivity in scoring mesowear and permit more refined statistical tests (and perhaps shape-based analyses; see Curran 2018). On the other hand, since ease of data collection is a major strength of mesowear analysis, the benefits of a more sophisticated approach over traditional methods would have to be clearly demonstrated before justifying wider adoption.

In response to the issue of subjectivity in data collection of LM and SEM, MTA was developed as a fully automated system that was free from observer error (Ungar et al. 2003; Scott et al. 2005). However, 10 years later, revisions to the automated procedure (e.g., incorporating ISO data; Purnell et al. 2012) have now been applied, and improvements to instrumentation parameters are still being made (Arman et al. 2016). The more traditional LM and SEM techniques are still used, but their methodological limits are better defined, and this has improved the objectivity of their conclusions. There is every reason to believe that further advances in LM, SEM, and texture analyses will be forthcoming, although more changes are likely for the latter than the two former methods. The greatest promise involves improvements to three-dimensional modeling and automated reconstruction of tooth surfaces, which will directly improve the level of precision and accuracy of microwear and mesowear data.

One major element that has been traditionally lacking is *in vivo* studies of microwear formation. These studies are truly the source of the majority of our knowledge of how microwear forms as a result of food texture. Fortunately, recent attempts to record microwear in laboratory animals (Schulz et al. 2013; Baines et al. 2014; Hoffmann et al. 2015) have significantly improved our knowledge base. As protocols for such experimental research become better defined, it will be increasingly possible to acquire permits and permissions to analyze microwear *in vivo* in both captive and wild animals. This, in turn, will rapidly refine our interpretations of fossil microwear data (Ungar 2015). Combined with technological advances, future investigators will be better equipped to answer important questions that drive current research. How does the abrasiveness of ingested particles directly influence microwear formation (Hoffmann et al. 2015)? Can microwear features of specific sizes and/or shapes be directly correlated with specific particles (Constantino et al. 2015)? How much of an animal's microwear signature is influenced by its ingested diet and how much by its external surroundings (Xia et al. 2015)? Future studies that apply a combined approach of mesowear and microwear analyses to the same samples will provide robust data on dental wear as it relates to ecology (Sánchez-Hernández et al. 2016). True understanding of the complexity of tooth wear as a natural process is still a speck on the scientific horizon, but every study inches us closer to the destination.

References

Andrews, P., & Hixson, S. (2014). Taxon-free methods of palaeoecology. *Annales Zoologici Fennici, 51,* 269–284.

Arman, S. D., Ungar, P. S., Brown, C. A., DeSantis, L. R. G., Schmidt, C., & Prideaux, G. J. (2016). Minimizing inter-microscope variability in dental microwear texture analysis. *Surface Topography: Metrology and Properties, 4,* 024007.

Baines, D. C., Purnell, M. A., & Hart, P. J. B. (2014). Tooth microwear formation rate in *Gaterosteus aculeatus*. *Journal of Fish Biology, 84,* 1582–1589.

Borrero-Lopez, O., Pajares, A., Constantino, P. J., & Lawn, B. R. (2015). Mechanics of microwear traces in tooth enamel. *Acta Biomaterialia, 14,* 146–153.

Bowyer, R. T., McKenna, S. A., & Shea, M. E. (1983). Seasonal changes in coyote food habits as determined by fecal analysis. *The American Midland Naturalist, 109,* 266–273.

Butler, P. M. (1952). The milk molars of Perissodactyla, with remarks on molar occlusion. *Proceedings of the Zoological Society of London, 121,* 777–817.

Butler, K., Louys, J., & Travouillon, K. (2014). Extending dental mesowear analyses to Australian marsupials, with applications to six Plio-Pleistocene kangaroos from southeast Queensland. *Palaeogeography, Palaeoclimatology, Palaeoecology, 408,* 11–25.

Calandra, I., & Merceron, G. (2016). Dental microwear texture analysis in mammalian ecology. *Mammal Review, 46,* 215–228.

Calandra, I., Schulz, E., Pinnow, M., Krohn, S., & Kaiser, T. M. (2012). Teasing apart the contributions of hard dietary items on 3D dental microtextures in primates. *Journal of Human Evolution, 63,* 85–98.

Christensen, H. B. (2014). Similar associations of tooth microwear and morphology indicate similar diet across marsupial and placental mammals. *PLoS ONE, 9,* e102789.

Conover, W. J., & Iman, R. L. (1981). Rank transformations as a bridge between parametric and nonparametric statistics. *The American Statistician, 35,* 124–129.

Constantino, P. J., Borrero-Lopez, O., Pajares, A., & Lawn, B. R. (2015). Simulation of enamel wear for reconstruction of diet and feeding behavior in fossil mammals: a micromechanics approach. *BioEssays, 38,* 89–99.

Coombs, M., & Semprebon, G. (2005). The diet of chalicotheres (Mammalia, Perissodactyla) as indicated by low magnification stereoscopic microwear analysis. *Journal of Vertebrate Paleontology, 25,* 47A.

Croft, D. A. (1999). Placentals: South American ungulates. In R. Singer (Ed.), *Encyclopedia of paleontology* (pp. 890–906). Chicago: Fitzroy-Dearborn Publishers.

Croft, D. A., & Weinstein, D. (2008). The first application of the mesowear method to endemic South American ungulates (Notoungulata). *Palaeogeography, Palaeoclimatology, Palaeoecology, 269,* 103–114.

Curran, S. C. (2018). Three-dimensional geometric morphometrics in paleoecology. In D. A. Croft, D. F. Su & S. W. Simpson (Eds.), *Methods in paleoecology: Reconstructing Cenozoic terrestrial environments and ecological communities* (pp. 317–335). Cham: Springer.

Damuth, J., & Janis, C. M. (2014). A comparison of observed molar wear rates in extant herbivorous mammals. *Annales Zoologici Fennici, 51,* 188–200.

Danowitz, M., Hou, S., Mihlbachler, M., Hastings, V., & Solounias, N. (2016). A combined-mesowear analysis of late Miocene giraffids from North Chinese and Greek localities of the Pikermian Biome. *Palaeogeography, Palaeoclimatology, Palaeoecology, 449,* 194–204.

DeSantis, L. R. G. (2016). Dental microwear textures: reconstructing diets of fossil mammals. *Surface Topography: Metrology and Properties, 4,* 023002.

DeSantis, L. R. G., Scott, J. R., Schubert, B. W., Donohue, S. L., McCray, B. M., Van Stolk, C. A., et al. (2013). Direct comparisons of 2D and 3D dental microwear proxies in extant herbivorous and carnivorous mammals. *PLoS ONE, 8,* e71428.

Donohue, S. L., DeSantis, L. R. G., Schubert, B. W., & Ungar, P. S. (2013). Was the Giant short-faced bear a hyper-scavenger? A new approach to the dietary study of ursids using dental microwear textures. *PLoS ONE, 8,* e77531.

Erickson, K. L. (2014). Prairie grass phytolith hardness and the evolution of ungulate hypsodonty. *Historical Biology: An International Journal of Paleobiology, 26,* 737–744.

Estebaranz, F., Galbany, J., Martínez, L. M., & Pérez-Pérez, A. (2007). 3-D interferometric microscopy applied to the study of buccal enamel microwear. In S. E. Bailey & J.-J. Hublin (Eds.), *Dental perspectives on human evolution* (pp. 391–403). New York: Springer.

Fortelius, M., & Solounias, N. (2000). Functional characterization of ungulate molars using the abrasion-attrition wear gradient: a new method for reconstructing paleodiets. *American Museum Novitates, 3301,* 1–36.

Franz-Odendaal, T. A., Kaiser, T. M., & Bernor, R. L. (2003). Systematics and dietary evaluation of a fossil equid from South Africa. *South African Journal of Science, 99,* 453–459.

Fraser, D., & Theodor, J. M. (2010). The use of gross dental wear in dietary studies of extinct lagomorphs. *Journal of Paleontology, 84,* 720–729.

Fraser, D., & Theodor, J. M. (2011). Comparing ungulate dietary proxies using discriminant function analysis. *Journal of Morphology, 272,* 1513–1526.

Fraser, D., Mallon, J. C., Furr, R., & Theodor, J. M. (2009). Improving the repeatability of low magnification microwear methods using high dynamic range imaging. *PALAIOS, 24,* 818–825.

Fraser, D., Zybutz, T., Lightner, E., & Theodor, J. M. (2014). Ruminant mandibular tooth mesowear: a new scheme for increasing paleoecological sample sizes. *Journal of Zoology, 294,* 41–48.

Galbany, J., Martinez, L. M., & Perez-Perez, A. (2004). Tooth replication techniques, SEM imaging and microwear analysis in primates: methodological obstacles. *Anthropologie, 42,* 5–12.

Galbany, J., Martínez, L. M., López-Amor, H. M., Espurz, V., Hiraldo, O., Romero, A., et al. (2005). Error rates in buccal-dental microwear quantification using scanning electron microscopy. *Scanning, 27,* 23–29.

Galbany, J., Estebaranz, F., Martínez, L. M., Romera, A., De Juan, J., Turbón, D., et al. (2006). Comparative analysis of dental enamel polyvinylsiloxane impression and polyurethane casting methods for SEM research. *Microscopy Research and Technique, 69,* 246–252.

Gill, P. G., Purnell, M. A., Crumpton, N., Brown, K. R., Gostling, N. J., Stampanoni, M., et al. (2014). Dietary specializations and diversity in feeding ecology of the earliest stem mammals. *Nature, 512,* 303–305.

Godfrey, L. R., Semprebon, G. M., Jungers, W. L., Sutherland, M. L., Simons, E. L., & Solounias, N. (2004). Dental use wear in extinct lemurs: evidence of diet and niche differentiation. *Journal of Human Evolution, 47,* 145–169.

Goodall, R. H., Darras, L. P., & Purnell, M. A. (2015). Accuracy and precision of silicon based impression media for quantitative areal texture analysis. *Scientific Reports, 5,* 10800.

Gordon, K. D. (1982). A study of microwear on chimpanzee molars: implications for dental microwear analysis. *American Journal of Physical Anthropology, 59,* 195–215.

Gordon, K. D. (1984a). The assessment of jaw movement direction from dental microwear. *American Journal of Physical Anthropology, 63,* 77–84.

Gordon, K. D. (1984b). Hominoid dental microwear: complications in the use of microwear analysis to detect diet. *Journal of Dental Research, 63,* 1043–1046.

Gordon, K. D. (1988). A review of methodology and quantification in dental microwear analysis. *Scanning Microscopy, 2,* 1139–1147.

Green, J. L. (2009a). Dental microwear in the orthodentine of the Xenarthra (Mammalia) and its use in reconstructing the palaeodiet of extinct taxa: the case study of *Nothrotheriops shastensis* (Xenarthra, Tardigrada, Nothrotheriidae). *Zoological Journal of the Linnean Society, 156,* 201–222.

Green, J. L. (2009b). Intertooth variation of orthodentine microwear in armadillos (Cingulata) and tree sloths (Pilosa). *Journal of Mammalogy, 90,* 768–778.

Green, J. L., & Kalthoff, D. C. (2015). Xenarthran dental microstructure and dental microwear analyses, with new data for *Megatherium americanum* (Megatheriidae). *Journal of Mammalogy, 96,* 645–657.

Green, J. L., & Resar, N. A. (2012). The link between dental microwear and feeding ecology in tree sloths and armadillos (Mammalia: Xenarthra). *Biological Journal of the Linnean Society, 107,* 277–294.

Green, J. L., Semprebon, G. M., & Solounias, N. (2005). Reconstructing the palaeodiet of Florida *Mammut americanum* via low-magnification stereomicroscopy. *Palaeogeography, Palaeoclimatology, Palaeoecology, 223,* 34–48.

Grine, F. E. (1986). Dental evidence for dietary differences in *Australopithecus* and *Paranthropus*: a quantitative analysis of permanent molar microwear. *Journal of Human Evolution, 15,* 783–822.

Grine, F. E., Ungar, P. S., & Teaford, M. F. (2002). Error rates in dental microwear quantification using scanning electron microscopy. *Scanning, 24,* 144–153.

Grine, F. E., Ungar, P. S., Teaford, M. F., & El-Zaatari, S. (2006). Molar microwear in *Praeanthropus afarensis*: evidence for dietary stasis through time and under diverse paleoecological conditions. *Journal of Human Evolution, 51,* 297–319.

Haupt, R. J., DeSantis, L. R. G., Green, J. L., & Ungar, P. S. (2013). Dental microwear texture as a proxy for diet in xenarthrans. *Journal of Mammalogy, 94,* 856–866.

Heiduck, S. (1997). Food choice in Masked titi monkeys (*Callicebus personatus melanochir*): selectivity or opportunism? *International Journal of Primatology, 18,* 487–502.

Henton, E., MCorriston, J., Martin, L., & Oches, E.A. (2014). Seasonal aggregation and ritual slaughter: isotopic and dental microwear for cattle herder mobility in the Arabian Neolithic. *Journal of Anthropological Archaeology, 33*, 119–131.

Higgins, P. (2018). Isotope ecology from biominerals. In D. A. Croft, D. F. Su & S. W. Simpson (Eds.), *Methods in paleoecology: Reconstructing Cenozoic terrestrial environments and ecological communities* (pp. 99–120). Cham: Springer.

Hoffman, J. M., Fraser, D., & Clementz, M. T. (2015). Controlled feeding trials with ungulates: a new application of *in vivo* dental molding to assess the abrasive factors of microwear. *The Journal of Experimental Biology, 218,* 1538–1547.

Hua, L., Brandt, E. T., Meullenet, J., Zhou, Z., & Ungar, P. S. (2015). Technical note: an *in vitro* study of dental microwear formation using the BITE Master II chewing machine. *American Journal of Physical Anthropology, 158,* 769–775.

Kaiser, T. M., & Fortelius, M. (2003). Differential mesowear in occluding upper and lower molars: opening mesowear analysis for lower molars and premolars in hypsodont horses. *Journal of Morphology, 258,* 67–83.

Kaiser, T. M., & Schulz, E. (2006). Tooth wear gradients in zebras as an environmental proxy – a pilot study. *Mitteilungen aus dem Hamburgischen Zoologischen Museum und Institut, 103,* 187–210.

Kaiser, T. M., & Solounias, N. (2003). Extending the tooth mesowear method to extinct and extant equids. *Geodiversitas, 25,* 321–345.

Kaiser, T. M., Brasch, J., Castell, J. C., Schulz, E., & Clauss, M. (2009). Tooth wear in captive wild ruminant species differs from that of free-ranging conspecifics. *Mammalian Biology – Zeitschrift für Säugetierkunde, 74,* 425–437.

Kaiser, T. M., Müller, D. W. H., Fortelius, M., Schulz, E., Codron, D., & Clauss, M. (2013). Hypsodonty and tooth facet development in relation to diet and habitat in herbivorous ungulates: implications for understanding tooth wear. *Mammal Review, 43,* 34–46.

Kay, R. F., & Covert, H. H. (1983). True grit: a microwear experiment. *American Journal of Physical Anthropology, 61,* 33–38.

King, T., Andrews, P., & Boz, B. (1999). Effect of taphonomic processes on dental microwear. *American Journal of Physical Anthropology, 108,* 359–373.

Krueger, K. L., Scott, J. R., Kay, R. F., & Ungar, P. S. (2008). Technical note: dental microwear textures of "Phase I" and "Phase II" facets. *American Journal of Physical Anthropology, 137,* 485–490.

Kruskal, W. H., & Wallis, W. A. (1952). Use of ranks in one-criterion variance analysis. *Journal of the American Statistical Association, 47,* 583–621.

Kubo, M. O., & Yamada, E. (2014). The inter-relationship between dietary and environmental properties and tooth wear: comparisons of mesowear, molar wear rate, and hypsodonty index of extant sika deer populations. *PLoS ONE, 9,* e90745.

Loffredo, L. F., & DeSantis, L. R. G. (2014). Cautionary lessons from assessing dental mesowear observer variability and integrating paleoecological proxies of an extreme generalist *Cormohipparion emsliei*. *Palaeogeography, Palaeoclimatology, Palaeoecology, 395,* 42–52.

Lucas, P. W. (2005). *Dental functional morphology: How teeth work.* New York: Cambridge University Press.

Lucas, P. W., Omar, R., Al-Fadhalah, K., Almusallam, A. S., Henry, A. G., Michael, S., et al. (2013). Mechanisms and causes of wear in tooth enamel: implications for hominin diets. *Journal of the Royal Society Interface, 10,* 20120923.

Maas, M. C. (1991). Enamel structure and microwear: an experimental study of the response of enamel to shearing force. *American Journal of Physical Anthropology, 85,* 31–49.

MacFadden, B. J., Solounias, N., & Cerling, T. E. (1999). Ancient diets, ecology, and extinction in 5-million-year-old horses from Florida. *Science, 283,* 824–827.

Martínez, L. M., & Pérez-Pérez, A. (2004). Post-mortem wear as an indicator of taphonomic processes affecting enamel surfaces on hominin teeth from Laetoli and Olduvai (Tanzania): implications to dietary interpretations. *Anthropologie, 42,* 37–42.

McAfee, R. K., & Green, J. L. (2015). The role of bite force in the formation of orthodentine microwear in tree sloths (Mammalia: Xenarthra: Folivora): implications for feeding ecology. *Archives of Oral Biology, 60,* 181–192.

Merceron, G., Blondel, C., De Bonis, L., Koufos, G. D., & Viriot, L. (2005). A new method of dental microwear analysis: application to extant primates and *Ouranopithecus macedoniensis* (Late Miocene of Greece). *PALAIOS, 20,* 551–561.

Merceron, G., Escarguel, G., Angibault, J., & Verheyden-Tixier, H. (2010). Can dental microwear textures record inter-individual dietary variations? *PLoS ONE, 5,* e9542.

Merceron, G., Hofman-Kamińska, E., & Kowalczyk. (2014). 3D dental microwear texture analysis of feeding habits of sympatric ruminants in the Białowieża Primeval Forest, Poland. *Forest Ecology and Management, 328,* 262–269.

Mihlbachler, M. C., & Beatty, B. L. (2012). Magnification and resolution in dental microwear analysis using light microscopy. *Palaeontologia Electronica, 15,* 25A.

Mihlbachler, M. C., & Solounias, N. (2006). Coevolution of tooth crown height and diet in Oreodonts (Merycoidodontidae, Artiodactyla) examined with phylogenetically independent contrasts. *Journal of Mammalian Evolution, 13,* 11–36.

Mihlbachler, M. C., Rivals, F., Solounias, N., & Semprebon, G. M. (2011). Dietary change and evolution of horses in North America. *Science, 331,* 1178–1181.

Mihlbachler, M. C., Beatty, B. L., Caldera-Siu, A., Chan, D., & Lee, R. (2012). Error rates and observer bias in dental microwear analysis using light microscopy. *Palaeontologia Electronica, 15,* 12A.

Mihlbachler, M. C., Campbell, D., Ayoub, M., Chen, C., & Ghani, I. (2016). Comparative dental microwear of ruminant and perissodactyl molars: implications for paleodietary analysis of rare and extinct ungulate clades. *Paleobiology, 42,* 98–116.

Mills, J. R. E. (1955). Ideal dental occlusion in primates. *Dental Practice, 6,* 47–61.

Münkemüller, T., Lavergne, S., Bzeznik, B., Dray, S., Jombart, T., Schiffers, K., et al. (2012). How to measure and test phylogenetic signal. *Methods in Ecology and Evolution, 3,* 743–756.

Nelson, S., Badgley, C., & Zakem, E. (2005). Microwear in modern squirrels in relation to diet. *Palaeontologia Electronica, 8,* 14A.

Oliveira, E. V. (2001). Micro-desgaste dentario em alguns Dasypodidae (Mammalia, Xenarthra). *Acta Biologica Leopoldensia, 23,* 83–91.

Organ, J. M., Ruff, C. B., Teaford, M. F., & Nisbett, R. A. (2006). Do mandibular cross-sectional properties and dental microwear give similar dietary signals? *American Journal of Physical Anthropology, 130,* 501–507.

Patnaik, R. (2015). Diet and habitat changes among Siwalik herbivorous mammals in response to Neogene and Quaternary climate changes: an appraisal in the light of new data. *Quaternary International, 371,* 232–243.

Purnell, M. A., Crumpton, N., Gill, P. G., Jones, G., & Rayfield, E. J. (2013). Within-guild dietary discrimination from 3-D textural analysis of tooth microwear in insectivorous mammals. *Journal of Zoology, 291,* 249–257.

Purnell, M. A., Hart, P. J. B., Baines, D. C., & Bell, M. A. (2006). Quantitative analysis of dental microwear in threespine stickleback: a new approach to analysis of trophic ecology in aquatic vertebrates. *Journal of Animal Ecology, 75,* 967–977.

Purnell, M., Seehausen, O., & Galis, F. (2012). Quantitative three-dimensional microtextural analyses of tooth wear as a tool for dietary discrimination in fishes. *Journal of the Royal Society Interface, 9,* 2225–2233.

Rabenold, D., & Pearson, O. M. (2014). Scratching the surface: a critique of Lucas et al. (2013)'s conclusion that phytoliths don't abrade enamel. *Journal of Human Evolution, 74,* 130–133.

Rensberger, J. M. (1978). Scanning electron microscopy, of wear and occlusal event in some small herbivores. In P. M. Butler & K. A. Joysey (Eds.), *Development, function, and evolution of teeth* (pp. 415–438). New York: Academic Press.

Resar, N. A., Green, J. L., & McAfee, R. K. (2013). Reconstructing paleodiet in ground sloths (Mammalia, Xenarthra) using dental microwear analysis. *Kirtlandia, 58,* 61–72.

Rivals, F., & Semprebon, G. M. (2006). A comparison of the dietary habits of a large sample of the Pleistocene pronghorn *Stockoceros onusrosagris* from the Papago Springs Cave in Arizona to the modern *Antilocapra americana*. *Journal of Vertebrate Paleontology, 26,* 495–500.

Rivals, F., & Solounias, N. (2007). Differences in tooth microwear in populations of caribou (*Rangifer tarandus*, Ruminatia, Mammalia) and implications to ecology, migration, glaciations and dental evolution. *Journal of Mammalian Evolution, 14,* 182–192.

Rivals, F., Mihlbachler, M. C., & Solounias, N. (2007). Effect of ontogenetic-age distribution in fossil and modern samples on the interpretation of ungulate paleodiets using the mesowear method. *Journal of Vertebrate Paleontology, 27,* 763–767.

Rivals, F., Semprebon, G., & Lister, A. (2012). An examination of dietary diversity patterns in Pleistocene proboscideans (*Mammuthus*, *Palaeoloxodon*, and *Mammut*) from Europe and North America as revealed by dental microwear. *Quaternary International, 255,* 188–195.

Rivals, F., Rindel, D., & Belardi, J. B. (2013). Dietary ecology of extant guanaco (*Lama guanicoe*) from southern Patagonia: seasonal leaf browsing and its archaeological implications. *Journal of Archaeological Science, 40,* 2971–2980.

Rivals, F., Prignano, L., Semprebon, G. M., & Lozano, S. (2015). A tool for determining duration of mortality events in archaeological assemblages using extant ungulate microwear. *Scientific Reports, 5,* 17330.

Rivals, F., Mihlbachler, M. C., Solounias, N., Mol, D., Semprebon, G. M., de Vos, J., et al. (2010). Palaeoecology of the Mammoth Steppe fauna from the late Pleistocene of the North Sea and Alaska: separating species preferences from geographic influence in paleoecological dental wear analysis. *Palaeogeography, Palaeoclimatology, Palaeoecology, 286,* 42–54.

Rose, J. J. (1983). A replication technique for scanning electron microscopy: applications for anthropologists. *American Journal of Physical Anthropology, 62,* 255–261.

Saarinen, J., Karme, A., Cerling, T., Uno, K., Säilä, L., Kasiki, S. et al. (2015). A new tooth wear-based dietary analysis method for Proboscidea (Mammalia). *Journal of Vertebrate Paleontology,* e918546.

Sánchez-Hernández, C., Rivals, F., Blasco, R., & Rosell, J. (2016). Tale of two timescales: combining tooth wear methods with different temporal resolutions to detect seasonality of Paleolithic hominin occupational patterns. *Journal of Archaeological Research: Reports, 6,* 790–797.

Schulz, E., Calandra, I., & Kaiser, T. M. (2010). Applying tribology to teeth of hoofed mammals. *Scanning, 32,* 162–182.

Schulz, E., Fahlke, J. M., Merceron, G., & Kaiser, T. M. (2007). Feeding ecology of the Chalicotheriidae (Mammalia, Perissodactyla, Ancylopoda). Results from dental micro- and mesowear analyses. *Verhandlungen des Naturwissenschaftlichen Vereins in Hamburg, 43,* 5–31.

Schulz, E., Piotrowski, V., Clauss, M., Mau, M., Merceron, G., & Kaiser, T. M. (2013). Dietary abrasiveness is associated with variability of microwear and dental surface texture in rabbits. *PLoS ONE, 8,* e56167.

Scott, R. S., Ungar, P. S., Bergstrom, T. S., Brown, C. A., Grine, F. E., Teaford, M. F., et al. (2005). Dental microwear texture analysis reflects diets of living primates and fossil hominins. *Nature, 436,* 693–695.

Scott, R. S., Ungar, P. S., Bergstrom, T. S., Brown, C. A., Childs, B. E., Teaford, M. F., et al. (2006). Dental microwear texture analysis: technical considerations. *Journal of Human Evolution, 51,* 339–349.

Semprebon, G., & Rivals, F. (2007). Was grass more prevalent in the pronghorn past? An assessment of the dietary adaptations of Miocene and Recent Antilocapridae (Mammalia: Artiodactyla). *Palaeogeography, Palaeoclimatology, Palaeoecology, 253,* 332–347.

Semprebon, G., Janis, C., & Solounias, N. (2004a). The diets of the Dromomerycidae (Mammalia: Artiodactyla) and their response to Miocene vegetational change. *Journal of Vertebrate Paleontology, 24,* 427–444.

Semprebon, G., Godfrey, L. R., Solounias, N., Sutherland, M., & Jungers, W. L. (2004b). Can low-magnification stereomicroscopy reveal diet? *Journal of Human Evolution, 47,* 115–144.

Semprebon, G., Sise, P., & Coombs, M. (2011). Potential bark and fruit browsing as revealed by stereomicrowear analysis of the peculiar clawed herbivores known as chalicotheres (Perissodactyla, Chalicotherioidea). *Journal of Mammalian Evolution, 18,* 33–55.

Shearer, B. M., Ungar, P. S., McNulty, K. P., Harcourt-Smith, W. E. H., Dunsworth, H. M., & Teaford, M. F. (2015). Dental microwear profilometry of African non-cercopithecoid catarrhines of the Early Miocene. *Journal of Human Evolution, 78,* 33–43.

Sidorovich, V. E. (2000). Seasonal variation in the feeding habits of riparian mustelids in river valleys of ME Belarus. *Acta Theriologica, 45,* 233–242.

Solounias, N., & Semprebon, G. (2002). Advances in the reconstruction of ungulate ecomorphology with application to early fossil equids. *American Museum Novitates, 3366,* 1–49.

Solounias, N., Tariq, M., Hou, S., Danowitz, M., & Harrison, M. (2014). A new method of tooth mesowear and a test of it on domestic goats. *Annales Zoologici Fennici, 51,* 111–118.

Spradley, J. P., Glander, K. E., & Kay, R. F. (2016). Dust in the wind: how climate variables and volcanic dust affect rates of tooth wear in Central American howling monkeys. *American Journal of Physical Anthropology, 159,* 210–222.

Stokke, S., & du Toit, J. T. (2000). Sex and size related differences in the dry season feeding patterns of elephants in Chobe National Park, Botswana. *Ecography, 23,* 70–80.

Strait, S. G. (1993). Molar microwear in extant small-bodied faunivorous mammals: an analysis of feature density and pit frequency. *American Journal of Physical Anthropology, 92,* 63–79.

Strait, S. G. (2014). Myrmecophagous microwear: implications for diet in the homin fossil record. *Journal of Human Evolution, 71,* 87–93.

Stynder, D. D. (2011). Fossil bovid diets indicate a scarcity of grass in the Langebaanweg E Quarry (South Africa) late Miocene/early Pliocene environment. *Paleobiology, 37,* 126–139.

Tausch, J., Kullmer, O., & Bromage, T. G. (2015). A new method for determining the 3D spatial orientation of molar microwear. *Scanning, 37,* 446–457.

Taylor, L. A., Kaiser, T. M., Schwitzer, C., Müller, D. W. H., Codron, D., Clauss, M., et al. (2013). Detecting inter-cusp and inter-tooth wear patterns in rhinocerotids. *PLoS ONE, 8,* e80921.

Teaford, M. F. (1988). Scanning electron microscope diagnosis of wear patterns versus artifacts on fossil teeth. *Scanning Microscopy, 2,* 1167–1175.

Teaford, M. F. (1991). Dental microwear: what can it tell us about diet and dental function? In M. A. Kelley & C. S. Larsen (Eds.), *Advances in dental anthropology* (pp. 341–356). New York: Wiley-Liss.

Teaford, M. F. (2007). What do we know and not know about dental microwear and diet? In P. S. Ungar (Ed.), *Evolution of the human diet: The known, the unknown, and the unknowable* (pp. 106–131). Oxford: Oxford University Press.

Teaford, M. F., & Oyen, O. J. (1989). *In vivo* and *in vitro* turnover in dental microwear. *American Journal of Physical Anthropology, 80,* 447–460.

Teaford, M. F., & Walker, A. (1984). Quantitative differences in dental microwear between primate species with different diets and a comment on the presumed diet of *Sivapithecus*. *American Journal of Physical Anthropology, 64,* 191–200.

Townsend, K. E., & Croft, D. A. (2008a). Diets of notoungulates from the Santa Cruz Formation, Argentina: new evidence from enamel microwear. *Journal of Vertebrate Paleontology, 28,* 217–230.

Townsend, K. E., & Croft, D. A. (2008b). Enamel microwear in caviomorph rodents. *Journal of Mammalogy, 89,* 728–742.

Ulbricht, A., Maul, L. C., & Schulz, E. (2015). Can mesowear analysis be applied to small mammals? A pilot-study on leporines and murines. *Mammalian Biology – Zeitschrift für Säugetierkunde, 80,* 14–20.

Ungar, P. S. (1995). A semiautomated image analysis procedure for the quantification of dental microwear II. *Scanning, 17,* 57–59.

Ungar, P. S. (2002). Microware software. Version 4.02. A semiautomated image analysis system for the quantification of dental microwear. Unpublished. Fayetteville.

Ungar, P. S. (2015). Mammalian dental function and wear: a review. *Biosurface and Biotribology, 1,* 25–41.

Ungar, P. S., & Spencer, M. A. (1999). Incisor microwear, diet, and tooth use in three Amerindian populations. *American Journal of Physical Anthropology, 109,* 387–396.

Ungar, P. S., Merceron, G., & Scott, R. S. (2007). Dental microwear texture analysis of Varswater bovids and Early Pliocene palaeoenvironments of Langebaanweg, Western Cape Province, South Africa. *Journal of Mammalian Evolution, 14,* 163–181.

Ungar, P. S., Brown, C. A., Bergstrom, T. S., & Walker, A. (2003). Quantification of dental microwear by tandem scanning confocal microscopy, and scale sensitive fractal analysis. *Scanning, 25,* 185–193.

Ungar, P. S., Scott, R. S., Scott, J. R., & Teaford, M. (2008). Dental microwear analysis: historical perspectives and new approaches. In J. D. Irish & G. C. Nelson (Eds.), *Technique and application in dental anthropology* (pp. 389–425). New York: Cambridge University Press.

Vaux, D. (2012). Research methods: know when your numbers are significant. *Nature, 492,* 180–181.

Walker, A., Hoeck, H. N., & Perez, L. (1978). Microwear of mammalian teeth as an indicator of diet. *Science, 201,* 908–910.

Winkler, D. A., & Kaiser, T. M. (2011). A case study of seasonal, sexual and ontogenetic divergence in the feeding behaviour of the moose (*Alces alces* LINNÉ, 1758). *Verhandlungen des Naturwissenschaftlichen Vereins Hamburg, 46,* 331–348.

Withnell, C. B., & Ungar, P. S. (2014). A preliminary analysis of dental microwear as a proxy for diet and habitat in shrews. *Mammalia, 78,* 409–415.

Xia, J., Zheng, J., Huang, D., Tian, Z. R., Chen, L., Zhou, Z., et al. (2015). New model to explain tooth wear with implications for microwear formation and diet reconstruction. *Proceedings of the National Academy of Sciences, USA, 112,* 10669–10672.

Chapter 6
Permanent Record: The Use of Dental and Bone Microstructure to Assess Life History Evolution and Ecology

Russell Hogg

Abstract In order to study the impact of ecology on life history evolution, paleobiologists have begun to take a cue from dendrochronology, and have increasingly been making use of microstructural growth records available within mineralized tissues: dental enamel, dentin, and bone. These tissues contain regular growth increments whose periodicities range from 24-hour (circadian), to 2-week, to seasonal and annual cycles, and which have allowed us to generate absolute chronologies of growth in teeth and bone. The study of these increments is known as odontochronology with respect to teeth, and skeletochronology with respect to bone. These two related fields allow us to directly quantify how species' growth patterns are correlated with a number of ecological factors over geological time. Moreover, because growth increments record individual stress events that occurred within individual animals' lifetimes, they allow us to observe important ecological pressures that occurred on shorter timescales as well.

Keywords Growth increments • Odontochronology Skeletochronology • Biorhythm • Cross-striation • Stria of Retzius • Von Ebner's line • Andresen line • Lamella LAG

Introduction

The term life history refers to the sum total of an organism's growth, reproduction, and lifespan according to the demands placed by a limited resource base. In other words, with only so many calories available over individual organisms' lives, what is the best way to use those resources to maximize fitness? For example, is it better to devote as many resources to producing as many offspring as one can and die young ('r' strategy), or perhaps live a long time and invest a lot in a few offspring ('K' strategy)? Ecological context is a major force driving the selection of particular life history characteristics, and so in principle we can learn a lot about present and past ecologies by studying the life history profiles of species within them (e.g., Godfrey et al. 2001; Macho and Williamson 2002). For example, increasing the unpredictability of an environment may cause some species to become hypometabolic, delay reproduction, and decrease range size, as is the case for the lemurs of Madagascar (Hogg et al. 2015).

In practice, it has traditionally been hard to apply life history study to paleoecology directly because growth information for extinct animals has not been as easy to obtain as for extant species. This limitation has been increasingly overcome thanks to the study of growth lines that are deposited over specific time intervals within teeth and bone; while the regular periodicity of these growth lines was demonstrated over a series of landmark studies in the first half of the twentieth century (e.g., Asper 1916; Massler and Schour 1941; Okada 1943), only since the 1980's have they been systematically applied to the study of life history. These growth lines can be used to reconstruct an animal's chronology of growth and even to gain metabolic information. Odontochronology, and more recently skeletochronology (the study of growth lines in teeth and bones, respectively), have been increasingly applied to the study of past environments and evolutionary pressures.

Study of growth in teeth and bone is vital to the paleoecologist for a variety of reasons. First, both teeth and bone can reflect environmental stress within their growth patterns. Therefore, studying their growth patterns can provide information about paleoenvironments. On top of this, normal growth patterns in teeth and bones reflect key developmental milestones, for example in primates the timing of M1 eruption is strongly correlated with weaning (Smith 1989; Smith et al. 1994). Likewise, one would expect to see

R. Hogg (✉)
Department of Rehabilitation Sciences, Florida Gulf Coast University, 10501 FGCU Blvd S, Ft Myers, FL 33965, USA
e-mail: rhogg@fgcu.edu

D. A. Croft et al. (eds.), *Methods in Paleoecology: Reconstructing Cenozoic Terrestrial Environments and Ecological Communities*, Vertebrate Paleobiology and Paleoanthropology, https://doi.org/10.1007/978-3-319-94265-0_6

changes in paleoenvironment across large time scales reflected not only in morphology, but also in changes within growth patterns that, in the fossil record, can only be read from bones and teeth.

This chapter is intended as a review of the current state of odonto- and skeletochronology, providing theoretical and historical background of the discipline, discussion of methods of data collection and analysis, their strengths and shortcomings, as well as providing examples of application and future implications for the field of paleoecology.

Theoretical and Historical Background

Theoretical Framework

The fundamental basis for odontochronology and skeletochronology is that mineralized tissues, such as bone, dental enamel, dentin, and cementum, in vertebrates preserve permanent growth records that can be observed using microscopic imaging techniques. These tissues are characterized by the fact that the majority of their matrix consists of crystals of the mineral hydroxyapatite, in combination with water and an organic component that consists of a variety of different proteins such as collagen and amelogenin. They are formed by the secretory action of epithelial (dental enamel) or mesenchymal-derived (all others) cells, which create a protein matrix that is subsequently "filled in" by precipitation of hydroxyapatite crystals out of the interstitial fluid. Mineralized tissues differ from one another in the percentage of mineral and organic components, the types of proteins contained in the matrix, the precise method of secretion, and in both ultrastructure and microstructure, amongst other features; however, they all share the presence of growth lines within their histological structure. These lines record specific, repeated physiological cycles that occur over regular intervals of time (Boyde 1989; Dean 2000). Therefore, these growth lines can be read and interpreted in a manner similar to tree rings.

Enamel and Dentin

In teeth, enamel and dentin are the most frequently studied tissues, since many mammal taxa such as primates have only a small amount of cementum. Teeth grow from a subgingival structure called the dental papilla, in which are two adjacent epithelia: the outer layer contains future enamel-secreting cells, called ameloblasts, and the inner layer is made up of future dentin-secreting cells, called odontoblasts. Odontoblasts march inward to the future pulp cavity and proliferate downward toward the root, whereas ameloblasts march outward to the tooth surface and proliferate downward toward the future cervix of the tooth (Noyes and Thomas 1921; Ferguson et al. 2000). Thus, the enamel formation front, or EFF, continually advances outward and downward. Tooth formation is therefore a product of matrix deposition by individual cells and the initiation of new ameloblasts and odontoblasts. Therefore, increases or decreases in enamel thickness are achieved by increasing the amount of enamel secreted by individual cells (either through rate or duration changes), whereas overall crown size is a function of this as well as of the number of cells eventually recruited. Both enamel and dentin contain 24-hour growth lines, referred to as cross-striations (enamel) and von Ebner's lines (dentin), which run in the tissue parallel to the deposition front formed by the secretory cells (ameloblasts and odontoblasts, respectively; Asper 1916; Gysi 1931; Schour and Poncher 1937; Schour and Massler 1937; Boyde 1979; Shinoda 1984; Boyde 1989; Bromage 1991; Dean 1993, 2000). Figure 6.1 provides a visual depiction of cross-striations and von Ebner's lines within a tooth. For discussions of the physiology underlying the formation of these lines, readers are referred to Okada (1943); Shinoda (1984); and Lacruz and colleagues (2012). Both enamel and dentin also contain longer-period growth lines, known as striae of Retzius (enamel) and Andresen lines (dentin). These growth increments share the same periodicity with one another, as demonstrated empirically by Bromage (1991) and Dean (1995); hereafter, the shared periodicity of these lines will be referred to as the Retzius Periodicity (RP; Hogg et al. 2015) In the first of these two studies, the author demonstrated that counts of cross striations and von Ebner's lines between striae of Retzius and Andresen lines (respectively) in macaque teeth corresponded with each other and also with the number of days between labeling injections of fluorescent dyes. Moreover, Dean (1995) demonstrated in a human sample that the number of striae of Retzius and Andresen lines is equal between corresponding stress lines in enamel and dentin within the same individual. This periodicity varies both among and within species but not within individual organisms (FitzGerald 1998; Shellis 1998; Dean 2000). These lines are visibly distinguishable from the 24-hour (or *circadian*) lines by having a darker, more accentuated appearance in a microscope, with a regular number of lighter circadian lines interspersed between them (Fig. 6.1). The precise physiological causes for the formation of these lines are much less well understood than that of daily increments. Some authors (Boyde 1989; Risnes 1998) have argued that striae of Retzius and (by extension) Andresen lines represent a sort of exaggerated daily increment, wherein all secreting ameloblasts and odontoblasts undergo a greater slowing of matrix secretion. This slowing is then tied to the physico-chemical differences in crystallites and organic molecules between the lines and the surrounding

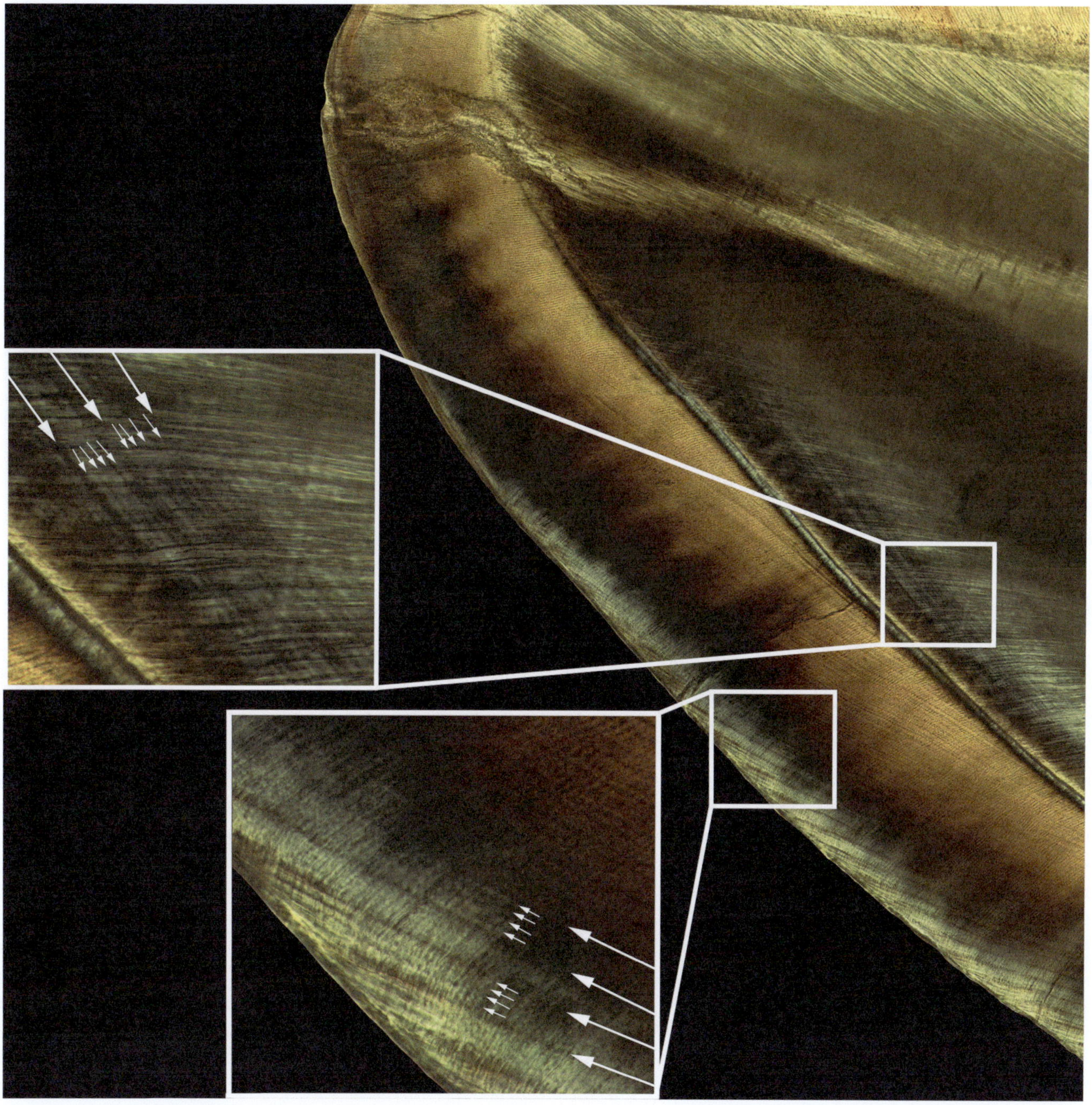

Fig. 6.1 Dental growth increments in a canine tooth from the platyrrhine primate, *Lagothrix lagotricha*. Cusp tip is toward upper left. Upper inset depicts von Ebner lines (24-hour increments, small white arrows) and Andresen lines (large white arrows) in dentin. Lower inset depicts cross-striations (24-hour increments, small white arrows) and striae of Retzius (large white arrows) in enamel. The periodicity of this individual is four, i.e., four 24-hour increments between each set of long-period increments. Field width = 2510 micrometers; top inset = 258 micrometers; bottom inset = 268 micrometers. Image taken in circularly polarized transmitted light microscopy. Specimen courtesy of teaching collection of Dr. Anthony DiFiore

tissue (Risnes 1998). Risnes (1990, 1998) has argued that this slowed secretion is tied to temporary morphological changes of the ameloblasts' Tomes processes, an idea supported by Witzel and colleagues (2006). Others have argued that these lines appear as a result of secretory cells temporarily ceasing their activity altogether over a regular cycle (Newman and Poole 1974), though this is the minority view (e.g., Dean 1987). In addition to these metabolic changes, the long period growth lines may also result from changes in the trajectory of the secretory cell, an option that is not mutually exclusive to the other two (Rose 1979, Ramirez-Rozzi 2002). Irrespective of the details of the local cellular activity, Bromage and colleagues (e.g., 2009, 2012, 2016) have argued that the striae of Retzius and Andresen lines appear in teeth in response to autonomic biorhythms controlled by a biological clock in the hypothalamus, and that the periodicity of the growth lines is a manifestation of the oscillation frequency of this clock, which they refer to as the Havers-Halberg Oscillation (HHO). Striae of Retzius and Andresen lines seem to be formed because the periodic disturbance in the secretory activity of the ameloblasts and odontoblasts affects the orientation and orderliness of enamel crystallites; since all cells are affected at the same time, this creates a visible discontinuity in the enamel or dentin matrix (Boyde 1989; Risnes 1998). There also appear to be chemical differences between the matrix within these growth lines and "ordinary" matrix (as is also the case with circadian increments), as acid-etching tooth sections affects their visibility (Risnes 1998). Wherever striae of Retzius reach the enamel surface, they appear as a series of bands or lines running around the circumference of the tooth, known as perikymata (Bromage and Dean 1985).

Fig. 6.2 Lamellae in a circularly polarized microscopy image of human cortical bone. Lamellae appear as alternating light and dark bands, both as circumferential lamellae in primary bone (indicated by small black arrows), and in secondary bone within osteons (small white arrows). Field width = 430 micrometers. Specimen sourced from teaching collection, Florida Gulf Coast University

Bone

Bone tissue contains lamellae, concentric layers of bone separated by histologically distinguishable growth lines; lamellae have also been shown to have a regular periodicity (Bromage et al. 2009). Lamellae typically appear in primary and secondary cortical bone, rather than in woven or trabecular bone; their presence is a defining feature of what is termed "lamellar" bone (Fig. 6.2). Unlike teeth, bone has no smaller increments between these growth lines. For bone, lamellae (both secondary osteonal and primary circumferential) can be counted in a section of compact bone and their periodicity determined via calibration to circadian and long-period dental growth lines or vital labeling (Bromage et al. 2009; see further discussion in the next section) in order to provide timescales for growth of a particular region of bone. As mentioned above, the daily secretion rate (DSR) can also be quantified by dividing measurements of the breadth between successive lamellae by the lamellar periodicity value (in days) for the individual in question. This allows the researcher to measure the growth rate of a tissue and compare differences in growth rate among differing anatomical regions, individuals, and/or species. This method applies both to growth in primary bone and also to rates of remodeling in secondary bone (Bromage et al. 2009, 2011a, b, 2015).

It is worth mentioning here that the lack of circadian increments in bone begs explanation, since dentin and enamel both contain circadian increments despite having very different embryological origins (i.e., enamel derives from ectoderm-derived epithelium, whereas dentin derives from mesenchymal-derived connective tissue), and since bone lamellae share their periodicity with the long term increments of dentin and enamel (Bromage et al. 2009). The absence of circadian increments in bone is even more puzzling in that solid physiological evidence exists to demonstrate that bone cells undergo circadian oscillation in their activities. For example, serum levels of osteocalcin and bone-specific alkaline phosphatase (bALP) are commonly studied as indicators of metabolism in bone, and have been shown to oscillate over a 24-hour cycle in multiple species (Allen 2008). The same is true for numerous other bone transcription factors (Zvonic et al. 2007). Therefore, it seems that there is something about the deposition process of bone or some particular physico-chemical characteristic of bone matrix that prevents us from noticing daily growth lines in bone. For example, it is possible that the rate of matrix formation is too slow to make daily growth lines visibly stand out in bone within a histological section. DSR for enamel in humans has been identified as ranging between 1–5 micrometers per day (Risnes 1998), with medians ranging between 3.5–6 micrometers per day (Dean 2000); DSR for dentin in humans ranges between 1–5 micrometers per day (Dean 2000). For comparison, mean DSR in platyrrhine primates ranges between 4.3 (*Callithrix jacchus*) and 5.72 (*Cebus capucinus*) micrometers, with a weak correlation to body mass (Hogg 2010). However, for bone, a human male with a known Retzius/lamellar periodicity

(RP) of 8 days demonstrated lamellar growth rates ranging between 4–8 micrometers, translating into a DSR of 0.5–1 micrometer per day (Bromage et al. 2011b). Standard light microscopy resolutions that we target for dental specimens, therefore, may not be useful for observing daily increments in bone. That said, Shinoda and Okada (1988) were able to demonstrate using vital labeling that the DSR for bone in an Asiatic chipmunk is about 3.3 micrometers/day. In this case, circadian lines were not encountered between lamellae, because the authors found that one lamella was deposited every 24-hour period. They also found that for a beagle, one lamella was deposited every four days, demonstrating that lamella formation does vary in a manner similar to that of long period dental growth lines and that there is no true 24-hour growth increment in bone that is analogous to cross-striations. This finding was a foundational element in the formation of the HHO hypothesis of Bromage and colleagues (2009), discussed in the examples section below, which states that RP is correlated with body mass in mammals. However, the DSR for beagle bone is about 3.5 micrometers per day (Shinoda and Okada 1988); this means that low growth rates for bone tissue are not a viable explanation for why bone does not exhibit a circadian growth line between lamellae.

Another possibility for the lack of circadian lines in bone is that any circadian changes that are manifested in bone structure are of a chemical nature that do not affect the crystallite orientation and/or light refraction properties of dentin and enamel, meaning that standard imaging protocols (e.g., light and electron microscopy, microCT) may be improper tools for quantifying daily changes in bone matrix secretion. In this case, chemical or isotopic analyses may be better suited to the task. Also, odontoblasts and ameloblasts are polarized columnar cells that secrete matrix only on one end of the cell (Risnes 1998, Ruch and Lesot 2000), whereas osteoblasts are nonpolarized cells that secrete matrix all around the surface of their plasma membrane (and in so doing encase themselves, transitioning into osteocytes; Hall 2005). Thus enamel and dentin have a more linear fashion of deposition that may lend itself more to the appearance of circadian growth lines, whereas this is not the case in bone. All of these are speculations, as there are few concrete data out there assessing this question to date, and any interpretations are not necessarily mutually exclusive.

Bone does have some other unique characteristics that require special consideration as compared to teeth. First, in teeth the mineralized tissues are not subject to remodeling, so each tooth can in principle contain a complete record of its own growth. However, bone tissue is undergoing constant turnover both during growth and once adult size has been reached. It will usually not be possible to recreate a growth record for an entire bone due to the fact that, as the overall bone expands due to appositional deposits on its outer surface, osteoclast activity is eroding earlier-deposited endosteal bone. Moreover, remodeling of matrix throughout the bone continuously erodes earlier growth and remodeling events throughout individuals' lives. Therefore, the greater the age of the individual in question, the more extensive the bone turnover will have been and the less information will be available from that individual's early development. To date, there are no studies that have been performed to demonstrate over what sort of timescales bones from different taxa and different age individuals can provide a reliable chronology, though Bromage and colleagues (2011a, b) have used lamellar bone to demonstrate variation in skeletal deposition rates in humans that are correlated to known environmental stressors that occurred within a known calendrical timescale over a period of several years. But as a general guideline, if one wishes to access information about the growth of bone during the juvenile period *per se*, juveniles or young adults need to sampled.

Besides bone remodeling, effects of osteocyte density, body size, and variable rates of lamellar formation are factors that likely have an impact on the accuracy of skeletochronological reconstructions, but whose effects on this accuracy remain unknown. For example, a study by Metz and colleagues (2003) provided evidence that osteocytes actively inhibit osteoblast function and therefore inhibit bone refilling in secondary osteon, an idea first advanced by Marotti and colleagues (1995). Metz and colleagues argued that this inhibitory function arises from a need to maintain the morphology of the osteon such that it enables proper delivery of nutrients and oxygen to the osteocyte population (for example, maintaining a large enough diameter for the Haversian canal), and facilitates the role of osteocytes as chemosensors and mechanosensors for the bone. Using tetracycline labels, they demonstrated that within sheep, osteons with higher osteocyte density had a negative correlation with osteon wall width and formation period. One noteworthy detail about this study is that while the statistical significance of the osteocyte density correlations was high ($p < 0.01$ in all cases), the strength of the correlations was low (R^2 values between 0.119 and 0.235). Therefore, this relationship seems to be a highly variable one. Another important consideration is that the study sample consisted of domestic animals from one species, and that it only examined remodeling rather than newly growing bone. Therefore, Metz and colleagues' results suggest that when conducting skeletochronology of secondary osteons within an individual and when comparing individuals within a species, osteocyte density likely needs to be corrected for; if this correction is not made, measurements of growth rate between lamellae may not be able to provide a proper environmental signal. That said, we do not understand how this relationship works across species, so the relationship may or may not hold at higher taxonomic levels.

The model is also not inconsistent with the explanation for high osteocyte densities proposed by the HHO model of Bromage and colleagues (2009, see below), which posits that initial osteocyte densities result from rapid proliferation of osteoblasts that are then incorporated to bone and transition to osteocytes. This is partly due to the fact that we have no evidence yet for a negative relationship between osteocyte density and osteoblast activity as just mentioned, but also because the model of Metz and colleagues only applies once this initial osteocyte state is reached (Metz et al. 2003).

The study of Metz and colleagues (2003) also found a relationship between lamellar formation rate and lamellar diameter within secondary osteons. That is, as successive lamellae are deposited within an osteon over time, this deposition rate slows. This phenomenon was first demonstrated by Manson and Waters (1965), in a vital labeling study of secondary osteons in cats, and likely applies across mammals. This means that to find environmental influence on bone deposition rates within an individual, one most likely needs to account for the decay rate of lamellar breadth within each individual osteon. Similarly, since rates of bone turnover are negatively correlated with body size across species (Bromage et al. 2009, 2012, Christen et al. 2015, Hogg et al. 2017), body mass needs to be accounted for when making comparisons across species.

Correspondences Between Bone and Teeth

In 2009, Bromage and colleagues demonstrated that the periodicity of lamellae in bone corresponds to that of striae of Retzius and Andresen lines in teeth, or RP; in other words, there is a systemic rhythmicity that seems to be expressed across the different mineralized tissues (Bromage et al. 2009). In a tooth one can determine the periodicity (in days) of striae of Retzius or Andresen lines by counting the circadian (24-hour) growth increments between them; however, since bone does not contain circadian increments, determination of the RP in bone lamellae requires that one calibrate the lamellar rhythm either using vital labeling, or using the circadian and longer-period lines of teeth. To repeat, this calibration is possible due to the fact that the lamellar rhythm matches that of striae of Retzius and Andresen lines in teeth, as shown via the vital labeling study of Bromage and colleagues (2009). In this study, the authors examined four species of mammal from different orders, all of which had been administered fluorescent labels during their lifespan on a known calendar. These fluorescent labels are incorporated into growing and remodeling bone shortly after administration, and appear as landmarks when viewed using fluorescence microscopy. When the calendar of administration is known, one can measure the distance between subsequent markers in a bone and divide by the total number of days to reveal a daily secretion rate. Using this calendar, Bromage and colleagues were able to estimate the number of days between secretion of lamellae and demonstrate that not only are lamellae periodic, but that their periodicity has a 1:1 ratio with the periodicity of striae of Retzius and Andresen lines. Therefore, summing the number of circadian growth lines in enamel and/or dentin will provide you the periodicity for striae of Retzius, Andresen lines, and bone lamellae within the same individual. That said, even though the growth increments in all three tissues are a match, the constant turnover of bone tissue throughout individual organisms' lifespans means that one cannot have a complete record of growth for an adult bone; since dental tissues do not undergo turnover, it is possible to reconstruct the entire record of growth for a tooth.

Using Growth Increments to Reconstruct Life History

Knowing the periodicities of growth increments in bones and teeth allows the researcher access to several important pieces of information. The first of these is the daily secretion rate (DSR) of the tissue, in other words the amount of matrix deposited per day by each enamel-, dentin-, or bone-secreting cell. Both bones and teeth are formed by secretory cells that lay down matrix in an appositional manner; since we can determine the number of days it takes for a row of cells to deposit the matrix between two growth lines, using a microscope we can measure the distance between those lines and divide that by the number of days in order to calculate how much material per day the cells were laying down. We can then compare measurements of the DSR across different parts of the same tooth or bone as well as across multiple teeth and bones from the same organism to see if it increases or decreases. We can also compare across taxa and individuals to answer questions about life history variation and evolution.

While enamel thickness is a function of the secretory rate and duration of the individual ameloblasts, increase in crown height (the distance between the apex of the dentinal horn and the cervix) is a function of the total number of ameloblasts located along this axis and the rate at which they are serially recruited to initiate matrix secretion. Growth rate of a tooth crown is not only a function of the secretory rate of ameloblasts and odontoblasts, it is also a function of the number of cells actively secreting matrix at any one time. This latter variable should not be confused with the overall number of cells that will be recruited to manufacture the tooth, as this is more of a function of the final size of the tooth and will be more related to *duration* of tooth formation rather than *rate*. As an analogy, consider a construction crew building a new road. For a given size road (say, 5 km of a 2-lane highway), each worker on my payroll has the potential to work quickly or slowly, building more or fewer feet of road per day, respectively. In turn, faster individual workers translates to a shorter time to completion of my road. This is comparable to variations in the DSR. However, I can also change my time to completion by increasing or decreasing the number of workers operating on any given day. For the same 5 km stretch of road, if I have 20 workers who can each lay 0.5 km per day, I will finish sooner than if I have only 10 who can lay the same amount per worker.

Direct assessment of the number of actively secreting cells requires sacrifice of living juvenile animals for the purpose of histologically examining the cells in their tooth buds. In paleoecology, it is essentially impossible to do this since our target population has long since expired and we are typically dealing with fully complete, fossilized teeth. Luckily, the geometry of striae of Retzius with respect to the enamel-dentin junction (EDJ) and overall enamel thickness gives us indirect insight into this aspect of dental growth rate. This can be demonstrated trigonometrically. Imagine a right triangle superimposed on an enamel section, with corners *x, y,* and *z* (Fig. 6.3). Its hypotenuse (*zx*) is a stria of Retzius, with a second side (*yx*) whose length is equivalent to the enamel thickness at the point at which the Retzius line intersects the enamel surface, and the third side (*zy*) consists of the line of the EDJ from the point at which the Retzius line intersects the EDJ. Length of the last side (*zy*) is proportional to the growth rate contributing to crown height. Based on trigonometry, the more acute the angle *xzy*, then the longer the hypotenuse must be to intersect with the same line *yx*, representing the enamel thickness. So, assuming the same enamel thickness between two specimens, a more acute angle between the striae of Retzius and the EDJ requires that the striae of Retzius will be longer. Recall that the striae of Retzius represent a fossilized remnant of the position of the enamel epithelium at successive instants of time. Therefore, assuming that the ameloblast cells of the enamel epithelium are of similar size at least among closely related taxa, this means that having a longer stria of Retzius requires the presence of more actively secreting cells at the enamel formation front. The longer striae are a direct consequence of having a larger number of enamel secreting cells at one time. Given two individuals with similar-sized teeth, similar enamel thickness, and similar DSR, it follows that the one with more acute angles between the striae of Retzius and EDJ is recruiting more cells to secrete enamel at any one time, and so should finish growing its tooth crown sooner. Hogg and Walker (2011) referred to this angle as the Enamel Formation Front (EFF) angle, and argued that when carefully applied, measurements of EFF angle can provide useful comparative information about the growth rate of teeth (the same metric is referred to as the "D degrees" angle by Beynon and Wood 1986). Figures 6.3 and 6.4 provide a visual demonstration of these concepts; for further discussion see Section "Approaches and Analysis", below.

In teeth, since DSR and EFF angle can be used to provide information about the rate at which dental tissues grow, their study has likely ecological implications. For example, Hogg and Walker (2011) and Hogg (2010) demonstrated that among platyrrhine primates, these two growth rate metrics in enamel may be related to the age of foraging independence within a species. For example, species whose juveniles must process their own food independently at an earlier age, will have relatively faster-growing teeth. Beyond this, knowledge of the periodicity and number of striae of Retzius and/or Andresen lines can be used to estimate with high precision the *duration* of growth for a tooth, both in terms of the tooth crown specifically (crown formation time, or CFT) and the entire tooth including the root (Dean 2000).

Aberrant growth lines are also frequently laid down by tooth- and bone-secreting cells as a result of arrested or abnormal growth, such as due to weaning or birth. These lines represent systemic homeostatic disturbances, and therefore appear simultaneously across forming teeth within a tooth row. Therefore, they can be calibrated across teeth, so that one can potentially calculate the time of formation of an entire tooth row (Boyde 1990; Dean and Beynon 1991; Dirks et al. 2002; Dirks 2003; Schwartz et al. 2006). Such growth records can be compared across individuals and species to hunt for evolutionary patterns of growth (for further information on these methods, see Section "Approaches and Analysis", below).

An aberrant growth line is considered to be a result of a stress event that negatively affects an individual animal's physiology, resulting in a disruption of normal matrix formation that is visible in enamel (accentuated striae of Retzius, a.k.a. Wilson lines), dentin, and bone (Lines of Arrested Growth, or LAGs). Birth, weaning, malnutrition, disease, and social disturbances have all been shown to create such lines in teeth (e.g., Macho et al. 1996; Simpson 1999; Dirks et al. 2002; Schwartz et al. 2006; see Fig. 6.5). These lines also may appear in teeth as a result of seasonal changes in resource availability (Macho et al. 1996). In bone, these aberrant growth lines are hypothesized to result most frequently from annual seasonal cycling. LAGs have mostly been studied in nonmammalian vertebrates, especially archosaurs and therapsids of the Mesozoic (e.g., Chinsamy and Rubidge 1993; Ray et al. 2004; Erickson et al. 2009; Woodward et al. 2011). Teeth and bones can also show signs of stress during an animal's life where aberrant growth lines do not appear, in that stress can affect overall growth rates of the tissues (e.g., Witzel et al. 2006; Bromage et al. 2011a). In principal, we can quantify stress-related decreases in growth rate by measuring the distances between

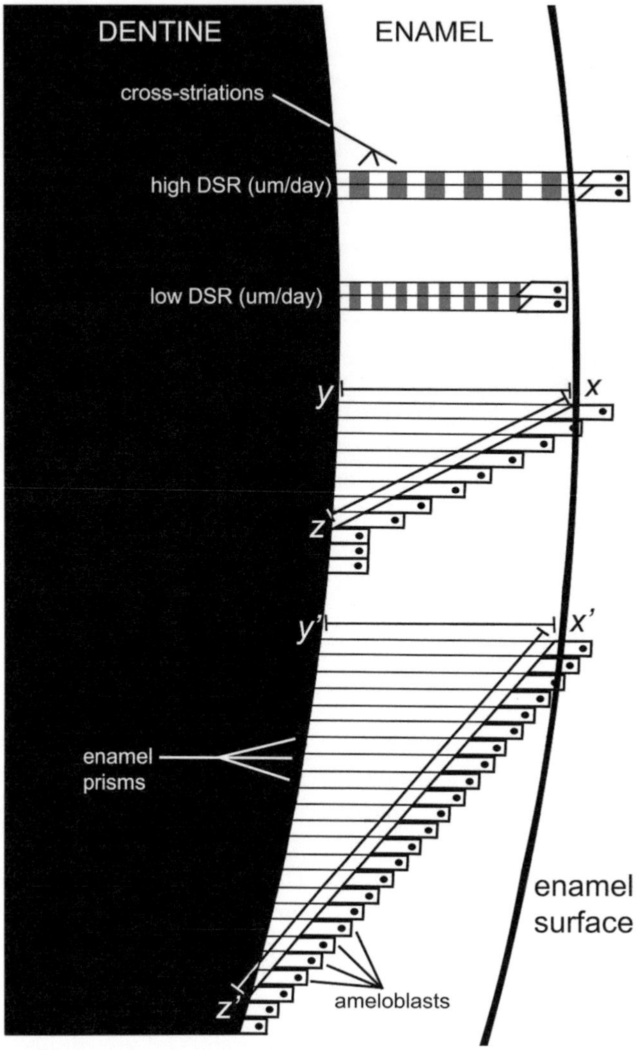

Fig. 6.3 The two components of enamel growth, showing ameloblasts (enamel secreting cells) as they progress from the enamel-dentin junction (EDJ) to the future enamel surface. Top: Growth rate as an effect of the amount of enamel secreted by individual cells each day (DSR). Gray and white bands signify cross-striations; the high DSR cells (=wider bands) are able to secrete more enamel in a lesser amount of time than the low DSR cells, as indicated by the relative number of cross-striations between the two groups. Bottom: Changing the number of actively secreting cells also affects the growth rate. Ameloblasts migrate at an oblique angle as an epithelial sheet from the dentin to the future enamel surface. This sheet of cells is referred to as the enamel formation front (EFF), and it moves outward from the dentin as the cells secrete enamel from their basal surface. Striae of Retzius record the position of the EFF at periodic intervals of time. Based on the geometry of the tooth, a more obtuse angle between the EFF and the EDJ translates to fewer active cells at any given time. This can be demonstrated by constructing a triangle in the enamel section, where a line between corners *z* and *x* is represented by a stria of Retzius, line *yx* represents the thickness of the enamel between the point where the stria of Retzius intersects the enamel surface and the point at which a line perpendicular to the enamel surface intersects the EDJ, and line *zy* represents the length between that point and the point where the stria of Retzius intersects the EDJ. Trigonometrically, given the same length for line *yx*, lengthening line *zx* requires a corresponding decrease in angle *xzy*. Therefore, in the figure, one can see that for the same length *yx*, triangle *x′y′z′* (bottom) has a longer stria of Retzius than triangle xyz (*z′x′* vs. *zx*) and therefore a more acute angle (*x′z′y′* vs. *xzy*). Because the entire stria of Retzius was originally covered with ameloblast cells, if one assumes that ameloblasts are of a similar size, stria of Retzius *z′x′* has a larger number of secretory ameloblasts than *zx*. This can be counted: in the figure, *z′x′* has 22 secretory ameloblasts along its length, whereas *zx* has only eight. Since higher numbers of secretory cells should lead to faster tissue deposition (all else being equal), this relationship enables us to quantify EFF angle to gain information about dental enamel growth rates

successive periodic growth increments. Therefore, growth increments in bones and teeth can also be used to assess the impact of stress on individuals and populations from the fossil record to answer specific ecological questions (e.g., Macho et al. 1996; Macho and Williamson 2002).

However, a study by Kierdorf and colleagues (2013) recently demonstrated experimentally (using vital labeling) that increments in sheep enamel which morphologically resemble cross-striations in primates, are actually lines deposited over a period of about 4.8 hours, whereas the true

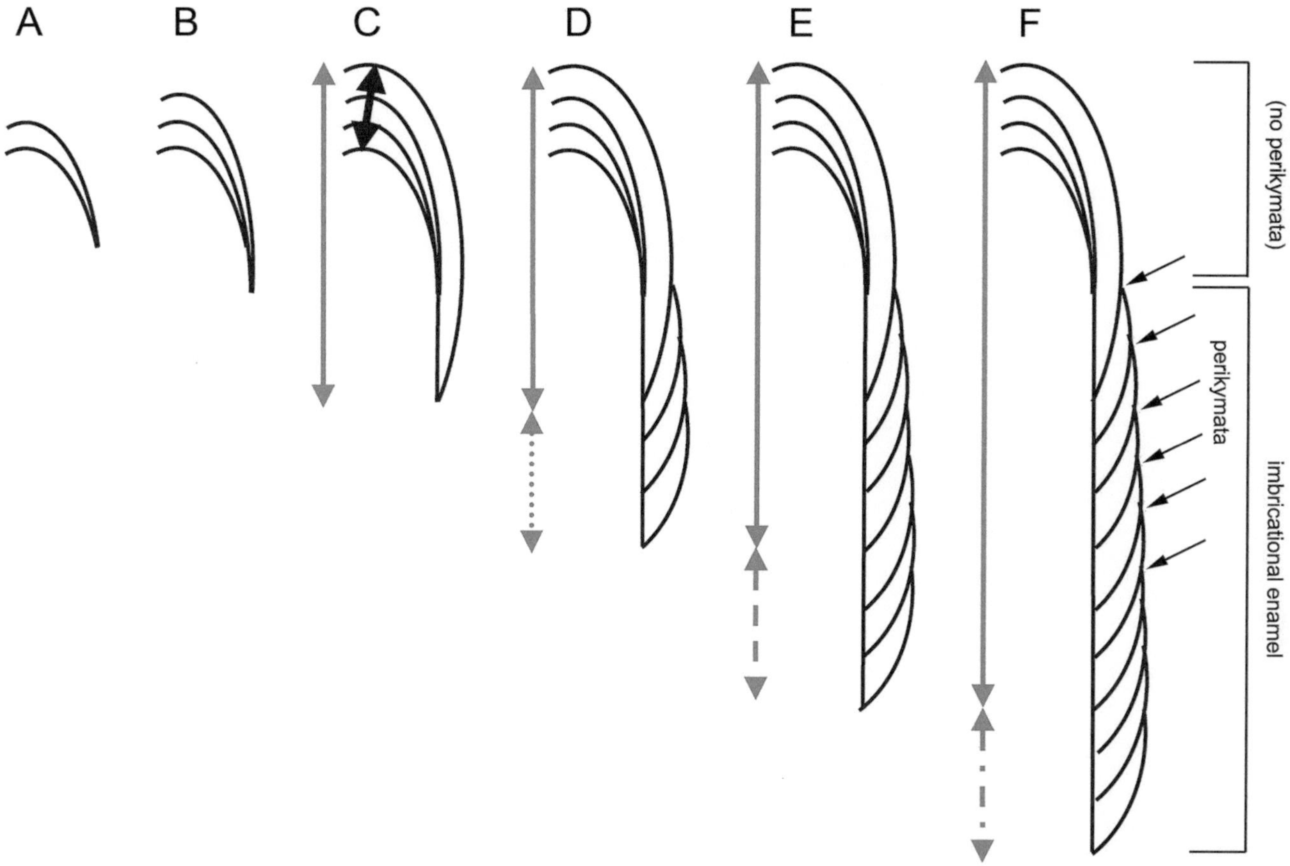

Fig. 6.4 Depiction of tooth crown extension. Letters A–F depict an advancing sequence of tooth growth through time. The enamel formation front (EFF) starts at the cusp of the tooth (A) and advances outward from the dentin (A–C), creating cuspal enamel; black lines represent striae of Retzius, which record the position of the EFF at periodic intervals of time. Once the cusp is complete, proliferation of ameloblast cells drives the EFF downward, creating imbricational enamel while extending the crown in height to the position of the cervix (D–F). Modified from Schwartz and Dean (2001)

daily increments closely resembled striae of Retzius. Moreover, they demonstrated that the DSR for sheep enamel is extremely high compared to primates, between 13.4 micrometers and 17 micrometers. In a later study, Kierdorf and colleagues found similar results for pigs (Kierdorf et al. 2014). If one were using primate morphology as the standard of measurement and analysis, this means that one's odontochronology for sheep will be skewed. Assessments of absolute timescales that incorporate stress lines and periodic growth increments can only be carried out where we have solid experimental evidence demonstrating the timescales that the growth increments represent. As a consequence, we can be most confident about dental growth records reconstructed for primates, which have been very well studied; other mammal taxa need much more sampling and experimentation.

In addition to providing information about schedules and rates of growth, as well as about frequency and impact of stress events, growth increments in bones and teeth have the potential to tell us a great deal about overall physiological processes underlying growth and life history in extinct animals. As noted above, Bromage and colleagues (e.g., 2009, 2012) have recently advanced a hypothesis that in mammals (and possibly other vertebrates), a biological clock within the hypothalamus of the brain acts as a metronome to coordinate various physiological features underlying life history such as the timing and duration of cell proliferation, overall somatic growth and maturation, and metabolic rate. Our hypothesis (Bromage et al. 2009, 2012) postulates that this biological rhythm (the Havers-Halberg oscillation, or HHO) is the physiological mechanism that causes striae of Retzius, Andresen lines, and bone lamellae to be synchronized to one another. This rhythm has a periodicity which occurs as a multiple of whole days. According to the model, changing this periodicity has cascading effects on life history physiology.

For example, a rat has an HHO periodicity of one day (Bromage et al. 2009) and new striae of Retzius and bone lamella are formed each day (Shinoda and Okada 1988). In other words, the RP of rats corresponds with and is a

Fig. 6.5 Accentuated stria of Retzius (stress line) in a canine tooth of the platyrrhine primate *Cebus capucinus*. Cusp tip is toward upper left, enamel is orange-colored, and dentin is a darker brown. The strong dark line running through the enamel (highlighted by white arrows) represents a stress event and is most likely a neonatal line (deposited when enamel growth stops temporarily due to the stress of birth). This line would appear in all teeth being formed at that moment in time and can be used to calibrate growth across teeth. Field width = approximately 750 micrometers. Specimen courtesy of teaching collection of Dr. Alfred Rosenberger

histological manifestation of a physiological oscillation in the hypothalamus, the HHO. Based on certain aspects of sympathetic nervous physiology, we hypothesize that the hypothalamus coordinates the timing of cell division as it appears that most cells depend on the nervous system to tell them how to metabolize calories and when to divide (Brown et al. 2007; Bromage et al. 2012; Hogg et al. 2017). This periodic peak/trough cycling in sympathetic output then triggers a slow-down or "pause" in matrix secretion by osteoblasts, ameloblasts, and odontoblasts that is manifested in the form of a growth line in the tissue; we predict that the growth line in bone appears because osteoblasts are being signaled to proliferate, however there is less clarity into the physiology of the tooth-secreting cells (Hogg et al. 2017). Since the mechanism is proposed to work throughout the organism, and not just in osteoblasts, this means that a rat's cells will receive 14 signals for its cells to proliferate over a two week period. However, elephants have an RP of 14 days (it takes two weeks for a lamella to form, and two weeks for the enamel between striae of Retzius to form). Therefore, in the same time interval, an elephant's cells will receive only one signal to proliferate (Bromage et al. 2009). It follows that the mouse's skeleton would tend to grow at a relatively much greater rate, with regard to the relative amount of matrix secreted. And indeed this is the case: lamellae tend to be of a similar thickness regardless of taxon, however it takes an elephant 2 weeks to form one of its lamellae (based on calibration from striae of Retzius) and only one day for a rat (based both on vital bone labeling and striae of Retzius; Bromage et al. 2009). According to the model, this increased growth rate is tied to differences in osteoblast reproduction rates that are tied to HHO cycling, such that mouse lamellae can appear at the rate they do because mice should have a larger number of secreting osteoblasts and therefore osteocytes per unit bone than larger animals (Bromage et al. 2009). In contrast, for the purposes of the example we can hypothesize from the regression data for mammals uncovered by Bromage and colleagues that the elephant, with its longer growth period, can form a similar amount of bone with relatively fewer cells, by increasing the amount of time available. It follows that smaller-bodied species should have higher osteocyte densities than larger-bodied species, which the data do corroborate (Bromage et al. 2009). It also becomes evident that this greater rate of cell proliferation and activity would be correlated with a more frenetic calorie expenditure per unit time for the mouse, and a greater basal metabolic rate relative to body mass. Thus we can see how mammals of small body mass have rapid cell division, rapid growth, and a relatively fast metabolism, whereas larger animals have the opposite pattern – and all of this can be tied together neatly by a single system in the form of the HHO model.

That said, it is difficult to account for the direct physiological cause underlying stria of Retzius and Andresen line formation. Under the HHO model, lamellae appear because the osteoblasts have received a sympathetic signal to undergo mitosis, and therefore temporarily halt bone secretion until cell division has completed. Since this signal is centrally coordinated, all osteoblasts should in principle be undergoing cell proliferation more or less simultaneously, and this yields a histological aberration in the form of a growth line. In fact, osteoblasts *must* halt matrix secretion when they divide, because mitosis causes the DNA sequences that are necessary for matrix protein formation to be coiled up into chromatid form, where it is inaccessible to RNA transcription (Hogg et al. 2017). Since secretory-stage odontoblasts and ameloblasts are purported not to divide, the current version of the HHO model argues that they are halting secretory activity as a side-effect response to the central nervous signal of the HHO, which is aimed primarily toward other body tissues (Bromage et al. 2009, 2012). However, there is no direct evidence regarding this question currently. In other words, dental increments may be "collateral damage" of a system designed to regulate proliferation and metabolism elsewhere, but it is not possible at this point to say for certain.

With regard to body mass and dental growth, Mahoney and colleagues (2016) recently argued that decreased bone deposition rates in a human sample were correlated to shorter RP, opposite the predictions of the original HHO model, and it is noteworthy that within humans RP may be negatively correlated with body mass, also counter to the original HHO model (Bromage et al. 2011a, b; Mahoney et al. 2016). However, this phenomenon in humans had been earlier observed by Bromage and colleagues (2011a, b), and likely is an intraspecific feature resulting from evolutionary forces unique to humans' own history (Bromage et al. 2011a, b). Regardless, there have been no studies on how best to apply these theoretical findings regarding osteocyte density, variation in lamellar formation, and correlations to body mass directly to the practice of skeletochronology. It should also be noted that Mahoney and colleagues also suggested that RP (and therefore HHO physiology) may differ between deciduous and permanent teeth, which would have implications for changing metabolic rate, etc., over ontogeny (Mahoney et al. 2016). However, while this would be a fascinating component of the puzzle, the data behind this hypothesis relies on identification of RP of 4 and 5 days in deciduous teeth, which is lower than the lowest previously reported value. While this could suggest change, through ontogeny, it could also suggest an abnormal sample (only three individuals in this sample were at 4 or 5 days, the remainder ranged between 6 and 11, which is normal based on prior studies), or more likely it indicates sampling error, as the authors in this study did not follow standard sampling protocols designed to minimize problems in visual identification and measurement of growth increments: it is necessary to use many more than two striations to provide a confident visual assessment of periodicity, and it is necessary to confirm one's visual assessment by dividing Retzius breadth by DSR for that region of the tooth in each case (see Schwartz and Dean 2001; Schwartz et al. 2005, 2006), neither of which safeguard was performed in this study. To properly demonstrate the concept that RP can change, it is necessary to provide solid evidence of changing periodicities within deciduous and permanent teeth of the same individual, following standard sampling procedures.

To date, the evidence for the model is mostly histological and indirect, though Bromage and colleagues (2016) have recently demonstrated that pigs have an oscillation of metabolites in their blood that matches the value of their RP. That being said, the model holds great promise for paleoecology and we have already begun to explore how to use HHO signatures in bones and teeth to inform us about how taxa have evolved in response to shifting ecological demands. For example, we (Hogg et al. 2015) have used striae of Retzius to help understand how the unpredictable environment of Madagascar has impacted life history evolution in lemurs as compared to the anthropoid primates (for further discussion, see Section "Examples of Applications" below).

Foundational Work and Progression of Techniques

The earliest discoveries of growth increments in teeth and bones, and indeed the notion that they encode specific regular intervals in time, go back to the beginning of the field of histology. Fathers of the field Clopton Havers (1691) and Anthony van Leeuwenhoek (1693) gave the earliest descriptions of striae of Retzius as well as lamellae. From Leeuwenhoek:

> I very plainly perceived that … [the] circumference or outside [of the tooth] was composed of a kind of scaly particles laid one on another, and I considered whether each series or layer of these scaly particles might not be the substance or thickness formed in the space of one year (Leeuwenhoek 1693).

Retzius (1837) followed up with a much later description of the striae that (perhaps unfairly, given the above) bear his name. The unfairness is even more striking, considering that Mayer Fraenkel (1835) and Jacob Linderer (1837) had earlier published descriptions of dental growth increments before Retzius. Preiswerk (1895) first recorded the relationship between perikymata and regular striae of Retzius. Asper (1916) first suggested that cross-striations in enamel have a daily time-dependency, an idea corroborated by numerous studies through the 1930's and 1940's for both cross-striations and von Ebner's lines (Gysi 1931; Schour 1936; Schour and Poncher 1937; Hoffman and Schour 1938; Schour and Hoffman 1939; Massler and Schour 1941, 1946; Kawasaki et al. 1979). Gysi (1931) first documented that the periodicity between striae of Retzius in humans was 7 days, setting the stage for future reconstructions of crown growth using both striae of Retzius and perikymata as their external manifestations (e.g., Bromage and Dean 1985). In 1943, it was suggested that the daily rhythm of von Ebner's lines (and by implication also cross-striations) was a result of circadian changes in the pH of blood serum (Okada 1943), providing some physiological basis for understanding growth increments. This idea was only recently placed within the context of a genetic mechanism involving the circadian clock (Lacruz et al. 2012).

Thus, the histological framework for odontochronology was set mostly in the first half of the twentieth century. However, apart from a few early studies by Boyde (e.g., 1963), application of dental increments as a tool to study life history only began to take root in the 1980's with several studies of great apes and fossil hominins (e.g., Boyde and Martin 1982; Martin 1983; Bromage and Dean 1985; Beynon and Dean 1987). Since 1990, we have seen an ever-increasing sophistication in odontochronological studies, as well as an ever-growing sample base and expanding field of researchers who increasingly look toward questions outside of the primate realm (primates being the historical focus of most odontochronological studies). For example, dental growth has now been well sampled across the primates, and dental growth records are being filled in for other mammal orders (e.g., Macho et al. 2003; Hogg et al. 2015; see examples section below for more studies), and we have come to the point where we can begin to calibrate growth records in both teeth and bone to known environmental insults (e.g., Dirks et al. 2002; Dirks 2003; Macho et al. 2003; Bromage et al. 2011a, b; again, see below).

With regard to bone, it has only been in the last few years that we have been able to incorporate lamellae into our studies (Bromage et al. 2009). Therefore, this is still a relatively young area of study and there are only a few papers on lamellar growth rates to date (e.g., Bromage et al. 2011a, b). A more extensive area of research for bone has been the study of LAGs, an area initiated in therapsid and archosaur studies stemming from the 1990's (Chinsamy 1993; Chinsamy and Rubidge 1993; Horner et al. 1999), but only recently applied to mammals (see Section "Examples of Applications" below). The relatively new field of skeletochronology, then, uses growth increments (lamellae and LAGs) to reconstruct growth and remodeling rates over time intervals that are inferred from the periodicity of those growth increments. In this it is like odontochronology. Unlike odontochronology, there are no daily increments, so whereas odontochronology can stand alone, skeletochronology is best supported by calibration of the lamellar rhythm from the daily increments of teeth. Likewise, odontochronology can in principal provide a complete record of the growth of a tooth as an organ, whereas skeletochronology cannot, due to bone remodeling (as discussed above). Therefore, odontochronology is limited in its ability to provide data only during the juvenile period of most mammals (since most mammals do not have indeterminate dental growth), whereas skeletochronology can be used to access physiological information well into adulthood due to the continued presence of bone remodeling. That said, skeletochronology should be distinguished from the process of determining the age and growth patterns of individual mammals based on the appearance and fusion of centers of bony growth at the gross anatomical level.

Since the foundational work, the techniques of odontochronology and skeletochronology have seen much growth. For example, we have better imaging techniques and software to assist the process. We can apply these observations to addressing a variety of evolutionary and ecological problems. We have increased application to the fossil record and are learning about more taxa, such as bovids, cervids, suids, elephantids, canids, and a now-thorough (though not exhaustive) sampling of primates, including many fossil hominins. Moreover, now that we know we can calibrate bone growth schedules from lamellae to absolute timescales using striae of Retzius (Bromage et al. 2009, 2011a, b), we have even more potential questions we can answer. The remaining sections of the chapter will discuss all of these growth areas in more detail.

Approaches and Analysis

Different Approaches

Dental and bone growth increments have a size on the scale of micrometers, so methods of odontochronology and skeletochronology all rely on microscopic analysis of histological specimens. No matter what the imaging method, once photomicrographs or three-dimensional microscopic digital models of the tooth or bone in question are created, the growth increments within them can be measured and counted to provide an accurate assessment of the growth record of that particular tissue.

The microscopy method that is by far the most commonly used is that of transmitted light microscopy, typically using polarized light. Dental and bone increments can be readily visualized at the resolutions provided by typical transmitted light optical systems between 40× and 400×, so the scale is appropriate for optical microscopy. Moreover, the hydroxyapatite crystals in bones and teeth are birefringent in that crystals with different orientations refract polarized light in different ways; this means that when polarized light is passed through them, color differences are created that enhance the contrast of growth increments so that they can be viewed easily without specimen staining (Espirito Santo et al. 2006; Sousa et al. 2006). Additionally, with optical microscopes the imaging process can be readily automated such that large montages of an entire bone or tooth section can be generated, allowing the researcher to study a single image with a fieldwidth in millimeters or centimeters rather than trying to assemble data from many tiny images of their specimen. Figures 6.1 and 6.4 both provide examples of such a montage.

However, the use of polarized light requires that specimens be thin-sectioned and polished, before being mounted on microscope slides. This is an arduous process, and more importantly many collections curators frown on this type of permanent alteration to their specimens. In cases where the preparation time or access to specimens is a limiting factor, confocal light microscopy may be employed. Confocal microscopy can operate at similar resolutions and magnifications as polarized light, but uses a reflected light beam to capture an image from just underneath the enamel surface (Lacruz et al. 2006; Bromage et al. 2007). Thus, an intact specimen with an exposed internal surface (such as a longitudinally fractured tooth or broken bone shaft) can be imaged, without any additional preparation. Therefore, this method is particularly useful in studies of rare fossil specimens (e.g., Lacruz et al. 2006; Bromage et al. 2007). Figure 6.6 provides a comparison of images generated in confocal and polarized light microscopy.

Scanning electron microscopy (SEM, including backscatter SEM) allows much higher-resolution images to be made at much higher magnification, down to the structure of hydroxyapatite crystals themselves. Thin sections do not need to be used – naturally broken surfaces suffice to allow one to visualize growth lines, as is the case with confocal microscopy, however the quality of such images is variable. For example, the frequent absence of a structural signature for striae of Retzius means that they will often not be visible within SEM; daily increments are often difficult to visualize as well (Fig. 6.7). Polishing and etching a cleanly-cut surface provides more consistent viewing of increments (Fig. 6.8), but this process eliminates the advantage of needing less specimen preparation. The advent of newer low-vacuum systems means that specimens can be imaged without being sputter-coated in gold or carbon, so again SEM is a particularly useful method where one is hesitant about creating thin sections or otherwise altering a specimen. However, fields of view are typically much smaller than in optical microscopes, and eroded surfaces still usually require cleaning, polishing, and/or acid etching in order to create a surface that presents clear anatomy. Scanning electron microscopy is also much more expensive than light microscopy. Since SEM inherently captures image of surfaces of objects in three-dimensions, there is also more inherent error in terms of quantification of growth increments, since natural variations in fractured surfaces and any obliquities of specimen surfaces with respect to the electron beam (apart from directly perpendicular) will lead to inaccuracies in measurements from the final two-dimensional measurement on screen. Based on these factors, SEM is not often used for odontochronology and skeletochronology studies, though it is frequently used for studies of enamel structure as it relates to biomechanics (e.g., Martin et al. 2003; Hogg et al. 2011).

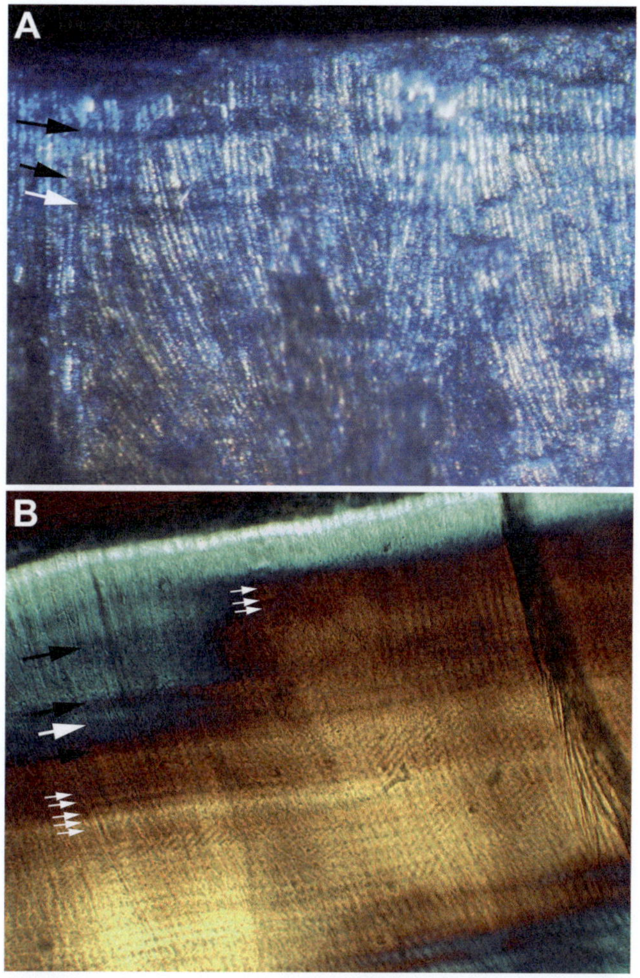

Fig. 6.6 Comparison of white-light confocal microscopy and circularly polarized light microscopy. Both images depict approximately the same region (imbricational enamel on the buccal side of the tooth) of the same canine from an individual of *Canis familiaris* (U. of Missouri mammals collection catalog #3474). In both images, the tooth surface is toward the top of the image and dentin is toward the bottom (not shown). A, Confocal microscopy, image of a cleanly broken surface; B, polarized light microscopy, image of a thin section. Image B presents much better resolution on growth increments. For example, black arrows indicate striae of Retzius; the top two in each image are the same, but in image B, it is possible to see a third stria below the other two that is not visible in A. In B, it is also possible to see numerous cross-striations (small white arrows) that are not clearly visible in A. There are hints of cross-striations in A, but they cannot be confidently measured. However, confocal microscopes can generate more useful images than this with very careful image preparations (see Lacruz et al. 2006). The large white arrow depicts an accentuated stria of Retzius or stress line, visible in both images. Field width = approximately 250 micrometers in both images

Two more recent high-tech approaches to assessing hard tissue growth increments are microCT and synchrotron

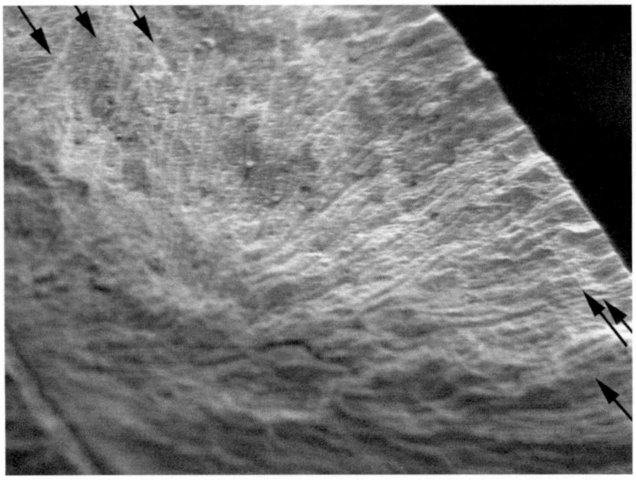

Fig. 6.7 Scanning electron microscope image of enamel in a broken section of a *Cebus apella* premolar (AMNH 133900). Enamel surface toward upper right, dentin in lower left corner (just visible). With no additional etching or other preparation, increments can be faintly seen, as indicated by the black arrows. However, the oblique angles with the camera created by the uneven surface and the low contrast for the growth lines make a reliable quantification of these increments difficult. Field width = 315 micrometers. Institutional abbreviation: AMNH, American Museum of Natural History

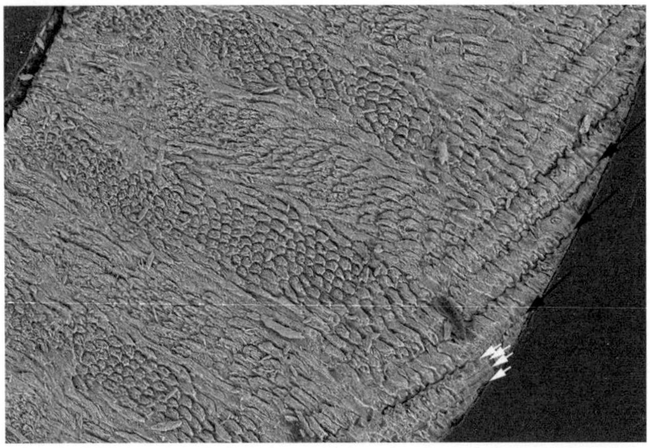

Fig. 6.8 Scanning electron microscope image of a cut and polished longitudinal slice through a *Cebus albifrons* premolar (AMNH 62838). Enamel surface is toward the lower right, dentin is the dark region in the upper left. After being cut, polished, and acid-etched, this specimen provides much better detail of growth increments than the image in Fig. 6.7 but still is not particularly useful for reconstructing a growth chronology for the tooth (for example, striae of Retzius cannot be seen extending all the way to the dentin). Black arrows indicate striae of Retzius, white arrows indicate cross-striations. Field width = 412 micrometers. Institutional abbreviation: AMNH, American Museum of Natural History

"virtual" microscopy (Tafforeau et al. 2007; Smith et al. 2015). In both methods, X-rays are used to create three-dimensional photomicrographs of specimens, at super-high resolution. These images can be manipulated and quantified digitally as virtual histological models (e.g., Tafforeau et al. 2007; Smith et al. 2015). The potential for both of these methods is tremendous, however currently there are multiple limiting factors to their use, including cost and difficulty of access to only a few machines; due to this, only a few research groups currently employ them (see "Future Prospects" for further discussion).

How Does One Conduct an Analysis?

Regardless of the imaging method used, once digital photomicrographs are acquired, the next step is to quantify the growth increments using some form of imaging and measurement software, of which NIH ImageJ is a very serviceable base-line example. The software is free, and is available for download at https://imagej.nih.gov/ij/. Most digital microscopy imaging systems (whether optical or SEM), as well as microCT and synchrotron systems, will also have their own software packages that allow for measurement and manipulation of histological images.

With regard to important measurements, in teeth the breadths between daily lines can be taken to provide DSR (daily secretion rate). As noted above, estimating DSR in bone requires data for both lamellar breadth from the bone and RP from the teeth of the same individual. Enamel formation front (EFF) angle can also be measured across the entire height of the tooth to provide comparative information about the rate of increase in crown height; if multiple measurements are taken, this provides a mean angle for the tooth and also allows one to assess changes in growth rate of the crown through time.

In bone, concentric or circumferential lamellae in a region can be counted and breadth between them measured to provide growth rate information, and their periodicity can be calibrated from tooth sections of the same individual. Likewise, LAGs can be measured to identify seasonal changes in growth (Köhler et al. 2012). Convention calls for sampling from midshaft femur if available (e.g., McFarlin 2006; Bromage et al. 2009, 2011a). However, when dealing with fossils midshaft femur specimens may not always be available, and there is always a possibility that one's particular research questions will make other specimens more appropriate (for example, if one wanted to compare growth rates specifically within the wings of two different bird species). The specimen should have good cortical bone available, since the thickness of cortical bone compared to cancellous bone provides much more anatomy for quantification, and ideally it should not be pathological unless pathology is part of the research question (e.g., it should not be osteoporotic, osteomyelitic, contain neoplasms, etc.). With regard to location within the cortical bone, different studies utilize subperiosteal, midcortical, and perioendosteal bone depending on their specific purpose (e.g., McFarlin 2006; Bromage et al. 2009, 2011a).

Recalling that DSR and EFF (enamel formation front) angle provide estimates of growth *rate* within the tooth, the next task is to determine the overall length of time it took to form the tooth crown (CFT) or the entire tooth. There are two primary methods for doing this. The first was described by Boyde (1963) and Risnes (1986), and is applicable only to the crown; it tracks striae of Retzius from dentin to surface and factors in the DSR to provide a total number of days it took to form the crown (Beynon et al. 1991; Ramirez-Rozzi 1995; Hogg 2010; Jordana and Köhler 2010; Fig. 6.9). In essence, the researcher follows a "zig-zag" path along the tooth crown (Fig. 6.9). Because this method is more labor-intensive, it is now usually reserved for scenarios where a large section of enamel is missing striae of Retzius due to complexity of the enamel structure or degraded specimen quality. The second method entails counting the number of striae of Retzius/Andresen lines and multiplying that by the RP to provide a total number of days for formation (e.g., Schwartz and Dean 2001). This method is more expedient and allows one to

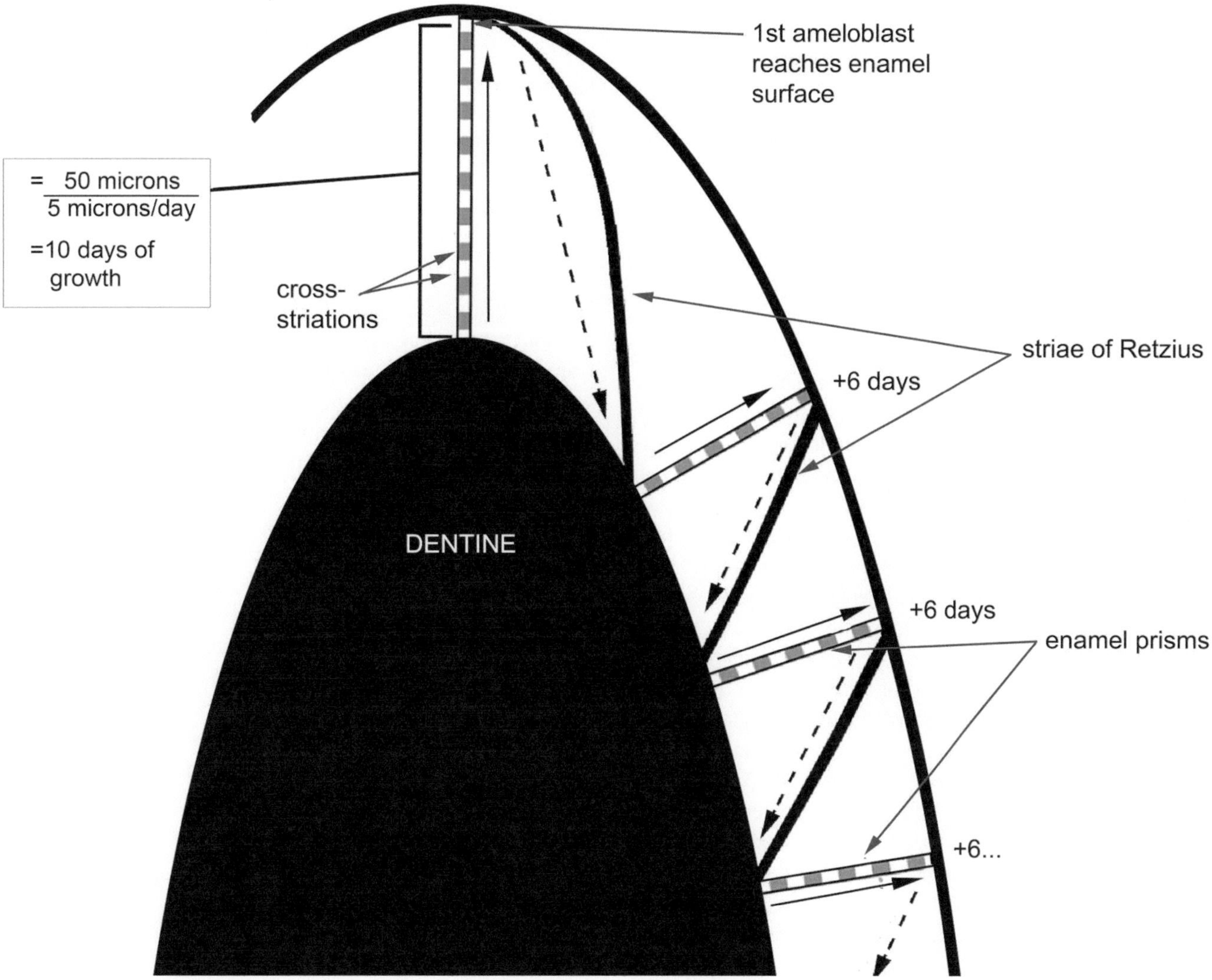

Fig. 6.9 A method used to estimate total crown formation time (CFT). First, an estimate of cuspal enamel formation time is calculated (=10 days in illustration) by measuring the greatest distance between the dentin and enamel surface at the cusp tips parallel to enamel prism paths (50 micrometers in illustration) and dividing this by the known DSR, which is the mean breadth of cross-striations for the tooth (in this case, 5 micrometers/day). When the first cuspal ameloblast reaches the enamel surface, it is finished secreting. The solid arrows proceed out along prisms and represent the number of days it took to form that segment of enamel (i.e., 10 days to form the full thickness at the cusp tip). From this point, the nearest stria of Retzius is traced down to the EDJ (see dashed arrows); at this new location, a new ameloblast has started to secrete at nearly the same time as when the first cuspal ameloblast is finished. The time it takes to complete enamel thickness at this new point can be calculated in the same manner as the first and the process repeated until the cervix of the tooth is reached. Total CFT can be estimated by summing the number of days it took to secrete the enamel thickness at each segment

include dentin (and therefore tooth roots) in the analysis, but it does require a nearly complete record of striae of Retzius or Andresen lines through the entire tooth. Once the growth chronology for a tooth is established, one can use stress lines (which will be deposited in all still-forming teeth simultaneously) as calibration points to reconstruct the growth chronology of potentially an entire tooth row, if all teeth are available for study.

Surface manifestations of striae of Retzius, the perikymata, can also be used to quantify changes in growth and the presence of environmental stress (Guatelli-Steinberg 2001). Given their regular periodicity, perikymata can be summed along a tooth crown to provide estimates of CFT (though such estimates exclude cuspal striae of Retzius, which do not reach the surface as perikymata; Bromage and Dean 1985). Also, disturbances in growth rate can be inferred from differences in breadth between perikymata. When stress events have a major inhibiting factor on crown growth and effect enamel thickness locally, these defects known as enamel hypoplasias appear; these defects typically span multiple perikymata and are a frequently used as stress indicators in anthropology and paleoecology (Hillson 1996; Guatelli-Steinberg 2001; Franz-Odendaal et al. 2004; Kierdorf et al. 2012).

Once all data are in hand, typically they are analyzed using fairly standard statistical approaches: t-tests and ANOVA are commonly used to discern differences among samples, and both bivariate and multivariate regression analyses are frequently used to search for correlations between growth data and predictor variables (e.g., Schwartz and Dean 2001; Bromage et al. 2009). Given large taxonomic sets, phylogenetically controlled versions of such analyses, such as phylogenetic generalized least squares (PGLS), are appropriate (e.g., Hogg and Walker 2011; Hogg et al. 2015). When looking at CFT or tooth formation times, overall ontogenetic trajectories for teeth can also be tracked, such as in studies of sexual dimorphism in extant hominoid primates and the extinct adapiform primate *Cantius* (Schwartz and Dean 2001; Schwartz et al. 2005; Fig. 6.10).

Ontogenetic trajectories can also be generated for bone in cases where substantial remodeling has not occurred, however studies creating ontogenetic trajectories to date have only employed use of LAGs in archosaurs (e.g., Horner and Padian 2004). Lamellae have not been used to generate ontogenetic trajectories, and mammals (as well as other non-archosaur vertebrates) have been little studied in this regard. The reason that archosaurs have received the most attention appears to be that most interest in understanding skeletal tissue growth rates at the histological level started in the dinosaur paleontology community (e.g., Horner et al. 1999), rather than there being any histological obstacle provided by mammals or other non-archosaur vertebrates. However, Bromage and colleagues (2011a, b) have used

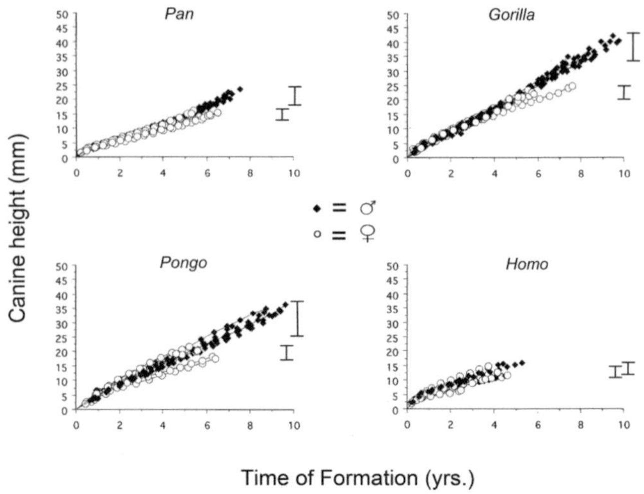

Fig. 6.10 Longitudinal canine growth trajectories for male and female hominoids, taken from Schwartz and Dean (2001). Note that male trajectories are extended further than those for females (most markedly in *Gorilla* and *Pongo*, the most dimorphic hominoids), but that male and female slopes show little difference from one another. This indicates that males achieve larger canine size via longer growth times rather than faster growth rates

lamellae calibrated to the RP to study changes in bone growth rate and remodeling rate in a small sample of adult humans from Malawi. Important for the study of paleoecology, they were able to find a correspondence between bone deposition rates as demonstrated by lamellae, and seasonal effects on rate of bone growth as well as the impact of drought (for more detailed discussion of methods, see examples section below). While the correspondence is currently speculative and needs to be replicated, this type of study provides a tool that is potentially invaluable for paleoecological reconstructions. There are several key differences between this type of study and those of LAGs: (1) when calibrated with growth increments in teeth, lamellae can provide a direct chronology whereas that of LAGs must be inferred; (2) study of lamellae has the potential for elucidating changes in bone deposition as a response to environmental insults that may be too subtle to cause the formation of a LAG; and (3) study of lamellae provides greater resolution with regard to timescales (near-weekly as opposed to seasonally or annually). In principle this type of analysis is transferrable to other mammals, and perhaps other vertebrates as well, however this has not been tested.

Strengths of the Approach

Odontochronology and skeletochronology provide several major advantages to researchers interested in life history study of extinct animals and in paleoecology. First, they can record longitudinal data which are still available for study after the death of the individual animal, and help circumvent difficulties acquiring cross-sectional data that are inherent to paleontology. Longitudinal studies typically track animals across their lifespan, which is obviously not a possibility for paleontology, where the lives of the organisms in question have already ended. Cross-sectional studies use a wide swath of sampling across a population to infer growth and other life history patterns, but this can be difficult in paleontology because (1) sample sizes of available fossils are usually too small for such analyses and (2) cross-sectional studies rely on having a reasonable estimate for the age of the animals in question to make their inferences. Odontochronology and skeletochronology provide an absolute record of longitudinal data for the individual by giving us the ability to study growth across the entire time span of a tooth or dentition (for odontochronology), and/or the spans of time recorded in bone that are contingent to the forces of remodeling (for skeletochronology). These data are useful at the individual level so that species/population-wide estimates of growth are not imposed onto the individual. Moreover, these individual data can contribute to population-wide descriptions that can be applied through deep time in the fossil record. By providing real-time calibration of growth events, tooth and bone increments allow us to track the growth of an individual animal with great precision long after they have died. The toolkit also makes it easier for us to study extant taxa, by reducing the need to perform expensive and time-consuming (potentially over years and decades) studies of live animals. Moreover, the growth chronologies we can create with bones and teeth can be very detailed and precise. Due to the short periodicity of growth lines within teeth and lamellae within bone, we have a resolution down to days or weeks, not just years as is the case with dendrochronology. As noted above, we can also directly track the influence of environment on a past population or even across species by quantifying stress lines and by assessing changes in growth rate. This is a key window to past ecological systems.

Biases and Shortcomings

Of course, odontochronology and skeletochronology are not "magic bullets" that give perfect access to life history information. Several factors affect the quality of data, the first of these being the quality of the specimens and, where applicable, the thin sections themselves. Degraded tooth and bone specimens can preserve little viable microanatomy for study. Likewise, preparing thin sections well or even acid-etching a specimen for SEM are delicate processes that require proper training as well as a cautious hand and eye, supplemented with practice. An unskilled preparator or an error even from an experienced technician can quickly ruin a section, or at the very least make it difficult to extract the appropriate data. Likewise, recognizing growth increments, distinguishing the different types from one another, and distinguishing them from stress lines in a histological section takes a trained, experienced eye. Different increments can look similar, and different species present different challenges for the microscopist. For example, I find that dentin is usually easier to read than enamel in *Canis* due to the microanatomical complexity of enamel in this genus, whereas I find the converse is usually true in primates. Therefore, it is possible for a novice to misread and mismeasure things. This is why it is vitally important to measure specimens multiple times and conduct intra-observer error studies, particularly during the period of training (e.g., Hogg 2010). When imaging with microCT, the size of the specimen also will affect the resolutions that can be achieved, and in synchrotron imaging, the imaging algorithms used can greatly affect image quality (Schwartz, pers. comm.).

Apart from the quality of data itself, additional factors can affect the validity of our conclusions. The most important such factor is the extent of sampling. Collection curators usually are reticent with regard to making tooth and bone specimens available for thin-sectioning, especially in rare or particularly important fossils. On top of this, there are obvious size limitations in the hypodigm of most fossil species. Therefore, odontochronology and skeletochronology typically must make do with small samples that often consist of broken pieces, and this is especially so with regard to fossils. Therefore, we frequently cannot say much about ranges of variation within species, though there are exceptions. MicroCT and synchrotron imaging can assess histology without permanently altering specimens and so can get around this obstacle, but they are available to a limited few researchers. As a consequence, interpretations of growth increments can often be limited in scope.

Another limitation of the approach is that in mammals at least, odontochronology is typically useful to study the life of an organism up to the point when its permanent dentition is complete – that is, it can only be used to study the juvenile period. There are exceptions to this, such as in the ever-growing incisors of rodents (though it can also be difficult to read enamel increments in small-bodied mammals). In animals such as crocodiles that have a continual, indeterminate tooth succession throughout their lives, we also have the potential to study life history processes into adulthood by study of enamel and dentin, however odontochronology has yet to be systematically deployed in these types of animals. Indeed, we cannot say if odontochronological techniques can be systematically employed outside of mammals. Another limitation is that we will have inherently limited interpretations in edentulous taxa such as baleen whales; in these animals we do have bones to study, but we have no teeth with which we can calibrate the periodicity of bone lamellae. With regard to bone tissue, another limitation is that cortical drift and bone remodeling can erase signatures of previous growth periods and life history events, making these periods of life inaccessible for study. Likewise, adjacent osteons may have formed at very different times through ontogeny, meaning that comparison of osteons may be very difficult unless landmarks that span multiple osteons can be found that help create a chronology (as can be performed in dendrochronology to compare multiple trees from the same region). Whether or not this is a feasible task has yet to be explored.

Examples of Applications

After much of the theory and technique of odontochronology was solidified in the 1980's and 1990's and the number of researchers steadily increased, the last twenty years have seen a steady growth in direct ecological applications of the technique. An excellent example of this is a series of studies aimed at understanding the evolution of insular dwarfism (Bromage et al. 2002; Dirks et al. 2011; Jordana and Köhler 2011). Odontochronology lends itself particularly well to this area, since insular dwarfism is as much a life history problem as it is an ecological one. The predominant theory of insular dwarfism is that when large-bodied species migrate to a new island environment, they are forced to adapt to novel changes in competition, the size of their resource base, as well as changes in predation pressure, all of which foster evolution of smaller body sizes (Raia and Meiri 2006). However, evolution of a smaller body size has a host of life history consequences: relative metabolic rate must change, reproduction must likely change, and organisms must "choose" whether they will decrease body mass by either growing more slowly, growing for a shorter period of time, or some combination of the two. In fossil elephantids, odontochronology has demonstrated that the island dwarf species from Cyprus, *Palaeoloxodon cypriotes,* achieved its smaller size via both a slowed growth trajectory and a shorter juvenile period (Bromage et al. 2002; Dirks et al. 2011). Interestingly, *Palaeoloxodon*'s RP is much shorter than other elephant species, at six days as opposed to the more usual two-week rhythm (Bromage et al. 2009). This result seems superficially similar to the findings of Mahoney and colleagues (2016), who recently argued that decreased bone deposition rates in a human sample was correlated to shorter RPs opposite the predictions of the original HHO model. However, as noted above this phenomenon is likely an intraspecific feature resulting from evolutionary forces unique to humans' own history (Bromage et al. 2011a). As such, it is not directly applicable to the case of *Palaeoloxodon*'s evolutionary history as an insular dwarf. It is also worth noting here that there is no experimental evidence of enamel or dentin secretion rates in elephants, and therefore following the implications of Kierdorf and colleauges (2013) the results for elephant RP should be treated with caution.

The overall findings of Bromage, Dirks, and colleagues regarding insular dwarfism contrasts the findings of Jordana and Köhler (2011), who found a much slower growth rate but prolonged growth period in teeth of the Plio-Pleistocene dwarf bovid *Myotragus balearicus* from the Balearic Islands of Spain. It is interesting that these two species should respond so differently to their island conditions yet still be identifiable as "insular dwarfs." Studies of additional species and locations are needed to understand what combination of ecological parameters and evolutionary background might generate these specific types of patterns.

Another series of studies has sought to determine relationships between dental growth, life history, and ecological niche with respect to diet. In a landmark paper, Godfrey and colleagues (2001) first assessed this question in a study of tooth eruption timing in primates, demonstrating that folivorous species complete dental eruption sooner, that correlated with an earlier age at weaning and an earlier age at foraging independence. Dirks (2003) followed this study with an odontochronological assessment of four species of catarrhine primates, using CFT estimates to corroborate the idea that folivores initiate tooth eruption at an earlier age and complete it sooner. Hogg and Walker (2011) examined DSR and EFF angles among platyrrhine primates in a phylogenetically controlled study, and also found evidence that species with an earlier age of foraging independence and weaning exhibit faster dental growth rates.

With regards to studies of environmental stress and its impact on animal species, Macho and colleagues (1996) provided the first targeted study. They examined single teeth from eleven individuals of the extinct gelada relative *Theropithecus oswaldi*, to quantify accentuated striae of Retzius (that is, stress lines). In addition to generating CFT estimates, they demonstrated that the accentuated lines occurred over predictable cycles of approximately 90, 102, and 132 days. The authors argued that this was likely due to cyclical changes in food resource availability caused by seasonality in rainfall. Similarly, anubis and hamadryas baboons have been demonstrated to display accentuated lines in response to their mother's first postpartum estrus cycle, and also develop stress lines in relation to reproductive costs such as menarche and lactation, although these may also have been due to drought (Dirks et al. 2002). Focusing on perikymata, Franz-Odendaal and colleagues also argued that enamel hypoplasias in the Pliocene giraffid *Sivatherium* reflected periodic stress incidents associated with seasonality. In 2002, Macho and Williamson conducted a similar study to that of Dirks and colleagues, using teeth in extant African bovids. They demonstrated that closed-habitat species exhibited relatively fewer disturbances in enamel formation than open-habitat species. This finding has not only ecological implications for the species themselves (e.g., the authors argue that it suggests closed-habitat species suffer less juvenile mortality), it also provides a tool for understanding paleoenvironments in fossil sites containing bovids. A comparison of modern bovids with extinct bovids from the 3.9 Ma Allia Bay site in Kenya demonstrated that stress lines in teeth resulting from seasonal effects were similar in both living and fossil species (Macho et al. 2003). This implies that overall seasonality at 3.9 Ma in Allia Bay is similar to that of modern Kenya, with major implications for evolution of the hominin *Australopithecus anamensis* from this site. It is important to note here, however, that the temporal spacing between growth lines for any mammal taxa outside of primates is really not well-established, given Kierdorf and colleagues' (2012, 2014) demonstration that daily growth increments in ungulates can morphologically resemble striae of Retzius in primates. Therefore, while the methods of odontochronological studies of ungulates are a vital contribution to paleoecology, their results need to be experimentally verified.

Another area where odontochronology has been applied to ecology concerns the way in which environment may drive evolution of RP. As discussed above, the earliest studies of RP with regard to HHO physiology (Bromage et al. 2009; Hogg 2010) showed strong correlations between RP and body mass as well as brain mass in a large sample of anthropoid primates. However, two large subfossil lemur species, *Megaladapis edwardsi* (about the size of a female gorilla) and *Palaeopropithecus ingens* (about the size of a chimpanzee), had a lower RP than predicted by their body mass based on a regression analysis of anthropoid primates. To explore the evolutionary and ecological significance of this, Hogg and colleagues (2015) examined striae of Retzius in a large sample of both extant and extinct lemur species. They showed that lemurs as a group (not just the two species examined before) have very low RPs compared to anthropoid primates. Among anthropoids the range of mean RPs across species ranges from 1–11 days, whereas for lemurs it is limited to 2–4 days – this despite the fact that both taxa share a similar range of body mass variation. Hogg and colleagues interpreted the restricted range of lemur RP variation as being tied to a long period of independent evolution in the environment of Madagascar, which became increasingly unpredictable through geologic time and is highly seasonal today (Hogg et al. 2015). Under this idea, the narrow range of RPs in lemurs is a metabolic adaptation that is part of a suite of distinctive life history adaptations in this group. Importantly, the authors also argued that *Daubentonia madagascariensis* (Aye-aye), the extant lemur species that has arguably the most stable and predictable resource base, has the RP, metabolic rate, and other life history characteristics that most resemble those of anthropoids. These findings demonstrate that ecology could play major roles in determining exactly how HHO physiology coordinates growth and metabolism. Therefore, studies of RP in fossil assemblages also have the potential to inform us about paleoenvironments, though this area needs much more study.

With regard to bone increments, as noted above studies of LAGs have predominantly given attention to Mesozoic taxa, however given the Cenozoic focus of this volume we will discuss the growing literature addressing LAGs in mammalian fauna. While it was at first believed that LAGs characterized bones of ectothermic animals only, recent studies have demonstrated the appearance of LAGs in Euarchontoglires, Laurasiatheria, Afrotheria, and Xenarthra among the placental mammals (Sander and Andrassy 2006; Straehl et al. 2013). A good example is a study of LAGs in Late Pleistocene ungulates, rhinoceratids, and elephantids from Germany, in which the authors speculated that a high occurrence of LAGs in their sample may have been related to the glacial period with which the remains were associated. However, there was no corresponding extant sample with which to test this hypothesis (Sander and Andrassy 2006).

As another case, Köhler and Moya-Sola (2009) examined bones of *Myotragus balearicus* from the Balearic Islands, constructing a growth chronology using LAGs (as well as looking at bone tissue types) to argue that this species evolved a crocodilian-like (i.e., indeterminate) growth pattern, metabolism, and life history as an adaptation to the relatively small resource base of its insular ecosystem. This could potentially explain why this species does not show the reduced RP seen in *Palaeoloxodon*, but more data are needed. More recently, Marin-Moratalla and colleagues (2013) used data from the extant (but endangered) antelope *Addax nasomasculatus* to demonstrate that growth chronologies created using LAGs can provide information about growth rates, time of death (during the annual cycle), and population health. The authors argued that each of these types of data are important for creation of demographic models that could be applied to conservation efforts, but it is just as evident that this type of information could be applied to paleontological specimens and paleoecological problems. One of the most important studies that will allow us to perform such activities with confidence is that of Köhler and colleagues (2012), who directly examined thermometabolic physiology and hormone levels in a sample of ruminants, and compared it against histological specimens generated from the same animals. In so doing, they established a physiological framework within which LAGs can be interpreted with more confidence. For example, they demonstrated that a winter LAG in red deer (*Cervus elaphus*) is associated with a period of hypometabolism and low core body temperature, which seem to have the purpose of lowering thermometabolic costs during the winter season.

With regard to skeletochronology using lamellae, there is only one published study to date since this is a new area. In 2011, Bromage and colleagues (2011a, b) examined lamellae in histological sections of modern humans from Malawi, for each of whom life history information was available (height, weight, age at death, domicile, occupation, etc.). Measuring the lamellae, they found six- to eight-week growth cycles as well as annual growth rhythms in some cases. Importantly, in one individual they discovered a lower growth rate variability (i.e., less difference between low and peak periods of growth). The authors linked this to a period of drought that the individual was known to have experienced. Therefore, this study demonstrates proof-of-concept, in that lamellae can be used to construct useful growth chronologies and assess the impact of stress events on growth.

Future Prospects

The primary improvements in odontochronology and skeletochronology for the near future pertain to technological improvement and sampling. The most important potential improvements come specifically in imaging technology; while confocal and polarized light microscopy are certainly adequate to the tasks of odontochronology and skeletochronology, and have many particular advantages, we have seen that they do have limitations. As noted above, they require either clean broken sections of a specimen or specially prepared histological thin sections, and these conditions do restrict our ability to acquire specimens for study as well as the quality of those specimens. Moreover, these imaging modalities do not allow us to construct three-dimensional representations of an entire specimen. Synchrotron and microCT imaging allow us to circumvent these problems, however these techniques also provide limitations in terms of availability for the former, and resolution for the latter (most microCT scanners have a limit of around five micrometers in resolution, whereas cross-striations in enamel can be as small as one to two micrometers in breadth). That said, microCT machines with appropriate resolution are now becoming available and should become more commonly accessible within the next decade – one exciting prospect is a set of new machines that add microCT capacity to SEM (e.g., Bruker Corporation Micro-CT for SEM, www.bruker.com). Given that synchrotron machines involve large particle accelerators and are prohibitively expensive, a proliferation of microCT in odontochronology and skeletochronology seems much more likely.

Another area of potential technological improvement relates to software. Within both three-dimensional and two-dimensional imaging modalities, most of the work of creating growth chronologies is done by a person quantifying the growth increments visually on a computer. Since skeletal growth increments occur in a given pattern, however, it should be possible to develop software that would automate at least some of the tasks of odontochronology and skeletochronology. Using such a program, a researcher could access a pre-existing image file within the software and, through a few simple commands, get statistics on DSR, EFF angles, CFT, and more.

Sampling is another area that should see great improvement in the near future. As discussed above, odontochronology has been thoroughly applied within the primates, but there are far fewer studies of non-primate mammals. Therefore, understanding how growth of dental tissues is linked to ecology across mammals is limited. Likewise, odontochronology is an almost entirely untapped toolkit with regard to nonmammal vertebrates, and indeed we cannot say for certain if the theory and techniques apply equally well in these taxa. With regard to bone, the converse is true for studies of LAGs, in that mammals are less-well understood. Moreover, given the results of Bromage and colleagues (2011a, b), a whole field of research is available in studying growth using lamellae, and applying these data to paleoecological questions, since only one limited population of one species (modern humans from Malawi) has been examined thus far. That said, bone tissue has received a great deal of attention with regard to other indicators of growth and stress. For example, Harris lines are macroscopic features that purportedly reveal stress episodes and an associated slowing of growth at the epiphyses of mammal long bones, and have received much attention in archaeology and anthropology (e.g., McHenry and Schulz 1976; Mays 1995; Scott and Hoppa 2015). Within these same fields, isotopic analysis has also been used to great advantage to assess environmental stress and pathology (Olsen et al. 2014; Beaumont et al. 2015). By adding in an understanding of crown growth and mineralization to such data, we should be able to use serial sampling within and across dentitions to provide temporal information on cyclical changes in diet or water availability Interestingly, it has recently been demonstrated that within humans, overall dental tissue growth is far less impacted by stress than bone tissue growth is, suggesting that teeth are physiologically insulated from stress more than other body tissues (Elamin and Liversidge 2013). Therefore, there is a storehouse of potential information to be had from bone, integrating histological study of osseous tissues with other methods as well as with studies of teeth.

In sum, odontochronology and skeletochronology are highly valuable tools for the paleoecologist. They provide direct information about how both individual organisms and populations respond to environmental change, while also giving us a window into what large scale environmental changes may have actually occurred in the past. They also allow us to access information about specific ecological stresses that happened on a timescale of days, weeks, and months within the deep geological past. Since these techniques have so much promise with regard to the information they can provide, and so much need in terms of the sampling that needs to be performed, they should prove a fruitful avenue of research for new scholars entering the field for decades to come.

References

Allen, M. J. (2008). Biomechanical markers of bone metabolism in animals: uses and limitations. *Veterinary Clinical Pathology, 32,* 101–113.

Asper, H. (1916). Uber die "Braune Retzius" sche Parallelstreifung im Schmelz der Menschlichen Zahne. *Schweiz. Vierteljahrschrift Zahnheilk, 26,* 277–314.

Beaumont, J., Montgomery, J., Buckberry, J., & Jay, M. (2015). Infant mortality and isotopic complexity: new approaches to stress, maternal health, and weaning. *American Journal of Physical Anthropology, 157,* 441–457.

Beynon, A. D., & Dean, M. C. (1987). Crown-formation time of a fossil hominid premolar tooth. *Archives of Oral Biology, 32,* 773–780.

Beynon, A. D., & Wood, B. A. (1986). Variations in enamel thickness and structure in East African hominids. *American Journal of Physical Anthropology, 70,* 177–195.

Beynon, A. D., Dean, M. C., & Reid, D. J. (1991). Histological study on the chronology of the developing dentition in gorilla and orangutan. *American Journal of Physical Anthropology, 86,* 189–203.

Boyde, A. (1963). Appendix on the teeth of the skeletal remains of a child from Dinas Powys. In D. J. Alcock (Ed.), *Dinas Powys*. Cardiff: University of Wales Press.

Boyde, A. (1979). Carbonate concentration, crystal centers, core dissolution, caries, cross striations, circadian rhythms, and compositional contrasts in the SEM. *Journal of Dental Research, 58 (special issue B),* 981–983.

Boyde, A. (1989). Enamel. In A. Oksche & L. Vollrath (Eds.), *Handbook of microscopic anatomy, Vol. V/6: Teeth* (pp. 309–473). Berlin: Springer.

Boyde, A. (1990). Developmental interpretations of dental microstructure. In C. J. De Rousseau (Ed.), *Primate life history and evolution* (pp. 229–267). New York: Wiley-Liss.

Boyde, A., & Martin, L. B. (1982). Enamel microstructure determination in hominoid and cercopithecoid primates. *Anatomy and Embryology, 165,* 193–212.

Bromage, T. G. (1991). Enamel incremental periodicity in the pig-tailed macaque: a polychrome fluorescent labeling study of dental hard tissues. *American Journal of Physical Anthropology, 86,* 205–214.

Bromage, T. G., & Dean, M. C. (1985). Re-evaluation of the age at death of immature fossil hominids. *Nature, 317,* 525–527.

Bromage, T. G., Dirks, W., Erdjument-Bromage, H., Huck, M., Kulmer, O., Öner, R., et al. (2002). A life history and climate change solution to the evolution and extinction of insular dwarfs: a Cypriot experience. In W. H. Waldren & J. A. Ensenyat (Eds.), *World islands in prehistory: International insular investigations V Deia conference of prehistory. British Archaeological Reports International Series, 105,* 420–427.

Bromage, T. G., Lacruz, R., Perez-Ochoa, A., & Boyde, A. (2007). Portable confocal scanning optical microscopy of *Australopithecus africanus* enamel microstructure. In S. Bailey & J. J. Hublin (Eds.), *Dental perspectives on human evolution: State of the art research in dental paleoanthropology* (pp. 193–209). New York: Springer.

Bromage, T. G., LaCruz, R., Hogg, R., Goldman, H., McFarlin, S. C., Warshaw, J., et al. (2009). Lamellar bone reconciles enamel rhythms, body size, and organismal life history. *Calcified Tissue International, 84,* 388–404.

Bromage, T. G., Juwayeyi, Y. M., Smolyar, I., Hu, B., Gomez, S., & Chisi, J. (2011a). Enamel-calibrated lamellar bone reveals long period growth rate variability in humans. *Cells Tissues Organs, 194,* 124–130.

Bromage, T. G., Juwayeyi, Y. M., Smolyar, I., Hu, B., Gomez, S., Scaring, V. J., et al. (2011b). Signposts ahead: hard tissue signals on rue de Armand de Ricqles. *Comptes Rendus Palevol, 10,* 499–507.

Bromage, T. G., Hogg, R. T., LaCruz, R. S., & Hou, C. (2012). Primate enamel evinces long period biological timing and regulation of life history. *Journal of Theoretical Biology, 305,* 131–144.

Bromage, T. G., Juwayeyi, Y. M., Katris, J. A., Gomez, S., Ovsiy, O., Goldstein, J., et al. (2015). The scaling of human osteocyte lacuna density with body size and metabolism. *Comptes Rendus Palevol, 15,* 32–39.

Bromage, T. G., Idaghdour, Y., Lacruz, R. S., Crenshaw, T. D., Ovsiy, O., Rotter, B., et al. (2016). The swine plasma metabolome chronicles "many days" biological timing and functions linked to growth. *PLoS ONE, 11,* e0145919.

Brown, M. F., Gratton, T. P., & Stuart, J. A. (2007). Metabolic rate does not scale with body mass in cultured mammalian cells. *American Journal of Physiology – Regulatory Integrative, and Comparative Physiology, 292,* R2115–R2121.

Chinsamy, A. (1993). Bone histology and growth trajectory of the prosauropod dinosaur *Massospondylus carinatus* Owen. *Modern Geology, 18,* 319–329.

Chinsamy, A., & Rubidge, B. S. (1993). Dicynodont (Therapsida) bone histology: phylogenetic and physiological implications. *Palaeontologia Africana, 30,* 97–102.

Christen, P., Ito, K., & van Rietbergen, B. (2015). A potential mechanism for allometric trabecular bone scaling in terrestrial mammals. *Journal of Anatomy, 226,* 236–243.

Dean, M. C. (1987). Growth layers and incremental markings in hard tissues: a review of the literature and some preliminary observations about enamel structure in *Paranthropus boisei*. *Journal of Human Evolution, 16,* 157–172.

Dean, M. C. (1993). Daily rates of dentine formation in macaque tooth roots. *International Journal of Osteoarchaeology, 3,* 199–206.

Dean, M. C. (1995). The nature and periodicity of incremental lines in primate dentine and their relationship to periradicular bands in OH 16 (*Homo habilis*). In J. Moggi-Cecchi (Ed.), *Aspects of dental biology: Paleontology, anthropology, and evolution* (pp. 239–265). Florence: Angelo Pontecorboli.

Dean, M. C. (2000). Incremental markings in enamel and dentine: what they can tell us about the way teeth grow. In M. F. Teaford, M. M. Smith & M. W. J. Ferguson (Eds.), *Development, function, and evolution of teeth* (pp. 119–130). Cambridge, UK: Cambridge University Press.

Dean, M. C., & Beynon, A. D. (1991). Histological reconstruction of crown formation times and initial root formation times in a modern human child. *American Journal of Physical Anthropology, 86,* 215–228.

Dirks, W. (2003). Effect of diet on dental development in four species of catarrhine primates. *American Journal of Primatology, 61,* 29–40.

Dirks, W., Reid, D. J., Jolly, C. J., Phillips-Conroy, J. E., & Brett, F. L. (2002). Out of the mouths of baboons: stress, life history, and dental development in the Awash National Park hybrid zone, Ethiopia. *American Journal of Physical Anthropology, 118,* 239–252.

Dirks, W., Bromage, T. G., & Agenbroad, L. D. (2011). The duration and rate of molar plate formation in *Palaeoloxodon cypriotes* and *Mammuthus columbi* from dental histology. *Quaternary International, 255,* 79–85.

Elamin, F., & Liversidge, H. M. (2013). Malnutrition has no effect on the timing of human tooth formation. *PLoS ONE, 8,* e72274.

Erickson, G. M., Rauhut, O. W. M., Zhou, Z., Turner, A. H., Inouye, B. D., Hu, D., et al. (2009). Was dinosaurian physiology inherited by birds? Reconciling slow growth in *Archaeopteryx. PLoS ONE, 4,* e7390.

Ferguson, C. A., Hardcastle, Z., & Sharpe, P. T. (2000). Development and patterning of the dentition. In P. O'Higgins & M. J. Cohn (Eds.), *Development, growth, and evolution: Implications for the study of the hominid skeleton* (pp. 187–205). San Diego: Academic.

FitzGerald, C. M. (1998). Do enamel microstructures have regular time dependency? Conclusions from the literature and a large-scale study. *Journal of Human Evolution, 35,* 371–386.

Fraenkel, L. (1835). *De penitiori dentium humanorum structura observationis.* In Diss Vratislaviae: Pressburg, Hungary.

do Espirito Santo, A. R., Novaes, P. D., Line, S. R. P. (2006). Anisotropic properties of the enamel organic extracellular matrix. *European Journal of Oral Sciences, 114 Supplement*, 333–337.

Franz-Odendaal, T. A., Chinsamy, A., & Lee-Thorp, J. (2004). High prevalence of enamel hypoplasia in an early Pliocene giraffid (*Sivatherium hendeyi*) from South Africa. *Journal of Vertebrate Paleontology, 24,* 235–244.

Godfrey, L. R., Samonds, K. E., Jungers, W. L., & Sutherland, M. R. (2001). Teeth, brains, and primate life histories. *American Journal of Physical Anthropology, 114,* 192–214.

Guatelli-Steinberg, D. (2001). What can developmental defects of enamel reveal about physiological stress in nonhuman primates? *Evolutionary Anthropology, 10,* 138–151.

Gysi, A. (1931). Metabolism in adult enamel. *Dental Digest, 37,* 661–668.

Hall, B. K. (2005). *Bones and cartilage: Developmental and evolutionary skeletal biology.* Amsterdam: Academic Press.

Havers, C. (1691). *Osteologia nova.* London.

Hillson, S. (1996). *Dental anthropology* (2nd ed.). Cambridge: Cambridge University Press.

Hoffman, M. M., & Schour, I. (1938). Rate and gradients of growth in rat molar as demonstrated by injections of alizarin red. *Journal of Dental Research, 17,* 307–308.

Hogg, R. T. (2010). Dental microstructure and growth in the cebid primates. Ph.D. Dissertation, City University of New York.

Hogg, R. T., & Walker, R. (2011). Correlates of ecology in the enamel microstructure of Cebidae (Platyrrhini, Primates). *Anatomical Record, 294,* 2193–2206.

Hogg, R. T., Ravosa, M., Vinyard, C., & Ryan, T. (2011). Functional morphology of the anterior masticatory apparatus of marmosets (*Callithrix*, Cebidae, Primates). *Journal of Morphology, 272,* 833–849.

Hogg, R. T., Godfrey, L. R., Schwartz, G. T., Dirks, W., & Bromage, T. G. (2015). Lemur biorhythms and life history evolution. *PLoS ONE, 10,* e0134210.

Hogg, R. T., Bromage, T. G., Goldman, H. M., Katris, J. A., & Clement, J. G. (2017). The Havers-Halberg oscillation cycle and bone metabolism. In C. J. Percival & J. T. Richtsmeier (Eds.), *Building bone: Early bone development in anthropology* (pp. 254–280). Cambridge: Cambridge University Press.

Horner, J. R., & Padian, K. (2004). Age and growth dynamics of *Tyrannosaurus rex. Proceedings of the Royal Society of London B, 271,* 1875–1880.

Horner, J. R., de Ricqles, A., & Padian, K. (1999). Variation in dinosaur skeletochronology indicators: implications for age assessment and physiology. *Paleobiology, 25,* 295–304.

Jordana, X., & Köhler, M. (2011). Enamel microstructure in the fossil bovid *Myotragus balearicus* (Majorca, Spain): implications for life-history evolution of dwarf mammals in insular ecosystems. *Palaeogeography, Palaeoclimatology, Palaeoecology, 300,* 59–66.

Kawasaki, K., Tanaka, S., & Ishikawa, T. (1979). On the daily incremental lines in human dentine. *Archives of Oral Biology, 24,* 939–943.

Kierdorf, H., Witzel, C., Upex, B., Dobney, K., & Kierdorf, U. (2012). Enamel hypoplasia in molars of sheep and goats, and its relationship to the pattern of tooth crown growth. *Journal of Anatomy, 220,* 484–495.

Kierdorf, H., Kierdorf, U., Frölich, K., & Witzel, C. (2013). Lines of evidence-incremental markings in molar enamel of Soay sheep as revealed by a fluorochrome labeling and backscattered electron imaging study. *PLoS ONE, 8,* e74597.

Kierdorf, H., Breuer, F., Richards, A., & Kierdorf, U. (2014). Characterization of enamel incremental markings and crown growth parameters in minipig molars. *The Anatomical Record, 297,* 1935–1949.

Köhler, M., & Moya-Sola, S. (2009). Physiological and life history strategies of a fossil large mammal in a resource-limited environment. *Proceedings of the National Academy of Sciences, USA, 106,* 20354–20358.

Köhler, M., Marin-Moratalla, N., Jordana, X., & Aanes, R. (2012). Seasonal bone growth and physiology in endotherms shed light on dinosaur physiology. *Nature, 487,* 358–361.

Lacruz, R. S., Ramirez-Rozzi, F., & Bromage, T. G. (2006). Variation in enamel development of South African fossil hominids. *Journal of Human Evolution, 51,* 580–590.

Lacruz, R. S., Hacia, J. G., Bromage, T. G., Boyde, A., Lei, Y., Xu, Y., et al. (2012). The circadian clock modulates enamel development. *Journal of Biological Rhythms, 27,* 237–245.

Leeuwenhoeck, A. (1693). Microscopical observations of the structure of teeth and other bones: made and communicated, in a letter by Mr. Anthony Leeuwenhoeck. *Philosophical Transactions of the Royal Society of London, 12*, 1002–1003.

Linderer, C. J. (1837). *Handbuch der Zahnheilkunde*. Berlin: Schlesinger.

Macho, G. A., & Williamson, D. K. (2002). The effects of ecology on life history strategies and metabolic disturbances during development: an example from the African bovids. *Biological Journal of the Linnean Society, 75,* 271–279.

Macho, G. A., Reid, D. J., Leakey, M. G., Jablonski, N., & Beynon, A. D. (1996). Climatic effects on dental development of *Theropithecus oswaldi* from Koobi Fora and Olorgesailie. *Journal of Human Evolution, 30,* 57–70.

Macho, G. A., Leakey, M. G., Williamson, D. K., & Jiang, Y. (2003). Palaeoenvironmental reconstruction: evidence for seasonality at Allia Bay, Kenya, at 3.9 million years. *Palaeogeography, Palaeoclimatology, Palaeoecology, 199,* 17–30.

Mahoney, P., Miszkiewicz, J. J., Pitfield, R., Schlecht, S. H., Deter, C., & Guatelli-Steinberg, D. (2016). Biorhythms, deciduous enamel thickness, and primary bone growth: a test of the Havers-Halberg Oscillation hypothesis. *Journal of Anatomy, 228,* 919–928.

Manson, J. D., & Waters, N. E. (1965). Observations on the rate of maturation of the cat osteon. *Journal of Anatomy, 99,* 539–549.

Marin-Moratalla, N., Jordana, X., & Köhler, M. (2013). Bone histology as an approach to providing data on certain key life history traits in mammals: implications for conservation biology. *Mammalian Biology – Zeitschrift für Säugetierkunde, 78,* 422–429.

Marrotti, G., Ferretti, M., Remaggi, F., & Palumbo, C. (1995). Quantitative evaluation on osteocyte canalicular density in human secondary osteons. *Bone, 16,* 125–128.

Martin, L. B. (1983). The relationships of the later Miocene Hominoidea. Ph.D. Dissertation, University College London.

Martin, L. B., Olejniczak, A. J., & Maas, M. C. (2003). Enamel thickness and microstructure in pitheciin primates, with comments on dietary adaptations of the middle Miocene hominoid *Kenyapithecus*. *Journal of Human Evolution, 45,* 351–367.

Massler, M., & Schour, I. (1941). Growth potential and appositional life span of the enamel and dentin forming cells in the human deciduous teeth and first permanent molars. *Journal of Dental Research, 20,* 282–283.

Massler, M., & Schour, I. (1946). The appositional lifespan of the enamel and dentine forming cells. *Journal of Dental Research, 25,* 145–50.

Mays, S. (1995). The relationship between Harris lines and other aspects of skeletal development in adults and juveniles. *Journal of Archaeological Science, 22,* 511–520.

McFarlin, S. (2006). Ontogenetic variation in long bone microstructure in catarrhines and its significance for life history. Ph.D. Dissertation, City University of New York.

McHenry, H., & Schulz, P. D. (1976). The association between Harris lines and enamel hypoplasia in prehistoric California Indians. *American Journal of Physical Anthropology, 44,* 507–511.

Metz, L. N., Martin, R. B., & Turner, A. S. (2003). Histomorphometric analysis of the effects of osteocyte density on osteonal morphology and remodeling. *Bone, 33,* 753–759.

Newman, H., & Poole, D. (1974). Observations with scanning and transmission electron microscopy on the structure of human surface enamel. *Archives of Oral Biology, 19,* 1135–1143.

Noyes, F. B., & Thomas, N. G. (1921). *A text-book of dental histology and embryology, Including laboratory directions*. New York: Lea & Febiger.

Okada, M. (1943). Hard tissues of animal body – highly interesting details of Nippon studies in periodic patterns of hard tissue are described. *Shanghai Evening Post, Medical Edition, 43,* 15–31.

Olsen, K. C., White, C. D., Longstaffe, F. J., von Heyking, K., McGlynn, G., Grupe, G., et al. (2014). Intraskeletal isotopic compositions ($\delta^{13}C$, $\delta^{15}N$) of bone collagen: nonpathological and pathological variation. *American Journal of Physical Anthropology, 153,* 598–604.

Preiswerk, G. (1895) *Beitrage zur Kenntniss der Schmelzstructur bei Saugethieren mit Besonderer Berucksichtigung der Ungulaten.* Ph.D. Dissertation, University of Basel.

Raia, P., & Meiri, S. (2006). The island rule in large mammals: paleontology meets ecology. *Evolution, 60,* 1731–1742.

Ramirez-Rozzi, F. V. (1995). Time of crown formation in Plio-Pleistocene Hominid teeth. In J. Moggi-Cecchi (Ed.), *Aspects of dental biology: Palaeontology, anthropology, and evolution* (pp. 217–238). Florence: International Institute for the Study of Man.

Ramirez-Rozzi, F. (2002). Enamel microstructure in hominids: new characteristics for a new paradigm. In N. Minugh-Purvis & K. J. MacNamara (Eds.), *Human evolution through developmental change* (pp. 319–348). Baltimore: Johns Hopkins.

Ray, S., Botha, J., & Chinsamy, A. (2004). Bone histology and growth patterns of some nonmammalian therapsids. *Journal of Vertebrate Paleontology, 24,* 634–648.

Retzius, A. (1837). Bemerkungen uber den innern Bau der Zahne, mit besonderer Rucksicht auf den im Zahnknochen vorkommenden Rohrenbau. In J. Muller (Ed.), *Archiv für Anatomie, Physiologie und Wissenschaftliche Medicin, in Verbindung mit Gelehrten* (pp. 486–566). Berlin: Verlag von W. Thome.

Risnes, S. (1986). Enamel apposition rate and the prism periodicity in human teeth. *European Journal of Oral Sciences, 94,* 294–404.

Risnes, S. (1990). Structural characteristics of staircase-type Retzius lines in human dental enamel analyzed by scanning electron microscopy. *Anatomical Record, 226,* 135–146.

Risnes, S. (1998). Growth tracks in dental enamel. *Journal of Human Evolution, 35,* 331–350.

Rose, J. C. (1979). Morphological variations in growth of enamel prisms within abnormal striae of Retzius. *Human Biology, 51,* 139–151.

Ruch, J. V., & Lesot, H. (2000). Molecules implicated in odontoblast terminal differentiation and dentinogenesis. In M. F. Teaford, M. M. Smith, & M. W. J. Ferguson (Eds.), *Development, function, and evolution of teeth* (pp. 13–21). Cambridge, UK: Cambridge University Press.

Sander, P. M., & Andrassy, P. (2006). Lines of arrested growth and long bone histology in Pleistocene large mammals from Germany: what do they tell us about dinosaur physiology? *Palaeontographica Abteilung A, 277,* 143–159.

Scott, A. B., & Hoppa, R. D. (2015). A re-evaluation of the impact of radiographic orientation on the identification and interpretation of Harris lines. *American Journal of Physical Anthropology, 156,* 141–147.

Schour, I. (1936). The neonatal line in the enamel and dentin of human deciduous teeth and first permanent molar. *Journal of the American Dental Association, 23,* 1946–1955.

Schour, I., & Hoffman, M. M. (1939). Studies in tooth development: I: the 16 microns calcification rhythm in the enamel and dentin from fish to man. *Journal of Dental Research, 18,* 91–102.

Schour, I., & Massler, M. (1937). The rate and gradient of growth in deciduous teeth with special reference to neonatal ring. *Journal of Dental Research, 16,* 349–350.

Schour, I., & Poncher, H. G. (1937). Rate of apposition of enamel and dentin, measured by the effect of acute fluorosis. *American Journal of Diseases of Children, 54,* 757–776.

Schwartz, G. T., & Dean, M. C. (2001). Ontogeny of canine dimorphism in extant hominoids. *American Journal of Physical Anthropology, 115,* 259–288.

Schwartz, G. T., Miller, E. R., & Gunnell, G. F. (2005). Developmental processes and canine dimorphism in primate evolution. *Journal of Human Evolution, 48,* 97–103.

Schwartz, G. T., Reid, D. J., Dean, M. C., & Zihlman, A. L. (2006). A faithful record of stressful life events recorded in the dental development record of a juvenile gorilla. *International Journal of Primatology, 27,* 1201–1219.

Shellis, R. P. (1998). Utilization of periodic markings in enamel to obtain information on tooth growth. *Journal of Human Evolution, 35,* 387–400.

Shinoda, H. (1984). Faithful records of biological rhythms in dental hard tissues. *Chemistry Today, 162,* 40–43.

Shinoda, H., & Okada, M. (1988). Diurnal rhythms in the formation of lamellar bone in young growing animals. *Proceedings of the Japan Academy, 64,* 307–310.

Simpson, S. W. (1999). Reconstructing patterns of growth disruption from enamel microstructure. In R. D. Hoppa & C. M. FitzGerald (Eds.), *Human growth in the past: Studies from bones and teeth* (pp. 241–263). Cambridge, UK: Cambridge University Press.

Smith, B. H. (1989). Dental development as a measure of life history in primates. *Evolution, 43,* 683–688.

Smith, B. H., Crummett, T. L., & Brandt, K. L. (1994). Ages of eruption of primate teeth: a compendium for aging individuals and comparing life histories. *Yearbook of Physical Anthropology, 37,* 177–231.

Smith, T. M., Tafforeau, P., Le Cabec, A., Bonnin, A., Houssaye, A., Pouech, J., et al. (2015). Dental ontogeny in Pliocene and early Pleistocene hominins. *PLoS ONE, 10,* e0118118.

Sousa, F. B., Vianna, S., & Snatos-Magalhaes, N. (2006). A new approach for improving the birefringence analysis of dental enamel mineral content using polarizing microscopy. *Journal of Microscopy, 221,* 79–83.

Straehl, F. R., Scheyer, T. M., Forasiepi, A. M., MacPhee, R. D., & Sanchez-Villagra, M. R. (2013). Evolutionary patterns of bone histology and bone compactness in xenarthran mammal long bones. *PLoS ONE, 8,* e69275.

Tafforeau, P., Bentaleb, I., Jaeger, J. J., & Martin, C. (2007). Nature of laminations and mineralization in rhinoceros enamel using histology and X-ray synchrotron microtomography: potential implications for palaeoenvironmental isotopic studies. *Palaeogeography, Palaeoclimatology, Palaeoecology, 246,* 206–227.

Witzel, C., Kierdorf, U., Dobney, K., Ervynck, A., Vanpoucke, S., & Kierdorf., H. (2006). Reconstructing impairment of secretory ameloblast function in porcine teeth by analysis of morphological alterations in dental enamel. *Journal of Anatomy, 209,* 93–110.

Woodward, H. N., Rich, T. H., Chinsamy, A., & Vickers-Rich, P. (2011). Growth dynamics of Australia's polar dinosaurs. *PLoS ONE, 6,* e23339.

Zvonic, S., Ptitsyn, A. A., Kilroy, G., Wu., Conrad, S. A., Scott, L. K., et al. (2007). Circadian oscillation of gene expression in murine calvarial bone. *Journal of Bone and Mineral Research, 22,* 357–365.

Chapter 7
Isotope Ecology from Biominerals

Pennilyn Higgins

Abstract Stable isotopes of carbon and oxygen can be used to make inferences about ancient climates and habitats. Biological hard tissues, particularly the biominerals of bioapatite, calcite, and aragonite are excellent archives of these isotopes and can be used to explore environments of Earth's deep past.

Keywords Carbon • Oxygen • Bioapatite • Calcite • Aragonite • Mass spectrometry • Stable isotope

Summary

Many tissues in the bodies of organisms record the chemical environment in which the organism lived at the time the tissue was forming. Biominerals, like those that compose bones, teeth, and shells, often retain these chemical signals even after the organism has died and fossilized, whereas other tissues are lost due to decay. Paleobiologists and geochemists can use analytical methods to read the chemical records in fossils and understand more about the lives and habitats of ancient organisms.

Different elements and molecules present in biominerals are affected by different aspects of the organism's life (e.g., metabolism and growth) and environment (e.g., climate, vegetation, and competition), allowing them to be used as a proxy, or a way to estimate, these aspects. There are a multitude of such proxies. This chapter will focus on arguably the most common and easiest to measure of these proxies, isotopes of carbon and of oxygen.

Isotopes are atoms of a single element, such as carbon, that differ only in the number of neutrons in the nucleus resulting in subtle differences in their atomic mass, but not otherwise greatly changing their chemical behavior. Some isotopes of an element are stable, whereas others may be radioactive. For example, in carbon there are three isotopes (^{12}C, ^{13}C, and ^{14}C). Both ^{12}C and ^{13}C are stable and most common, comprising 98.93 and 1.07 percent of the natural abundance of carbon, respectively. Atoms of these two isotopes of carbon will remain unchanged over an infinite period. Unstable or radioactive isotopes, such as ^{14}C (also called radiocarbon), will decay to become other elements while emitting radiation. Such elements do not remain unchanged in molecules over infinite timescales. The rate at which many radioactive isotopes decay is known, which forms the basis for radiometric dating techniques that are used to assign ages to geological specimens. In this chapter, however, we are interested in the stable isotopes only, for these are the isotopes that record and preserve ecological signals.

All isotopes of the same element undergo chemical reactions in the same way; however, the difference in mass between isotopes, though often slight, does have some measurable affect where chemical bonds are made or broken, or when molecules change phase (such as from liquid to gas). The result is that the ratios of different isotopes in the product of the reaction (e.g., biological tissues) will differ from that in the original ingredients in a characteristic fashion. This process is fractionation. Fractionations occur between environmental inputs (typically through diet or water ingestion) and the tissue composition in organisms. In paleontological contexts, the most commonly used hard tissues are teeth (primarily enamel), enameloid in fish scales, and shells of molluscs. When we understand how fractionation works, we can then make interpretations about the habitat and dietary preferences of the organism.

In this chapter, we look at isotopes of carbon (^{12}C and ^{13}C) and of oxygen (^{16}O and ^{18}O) from fossil bones, teeth, and shells. Scientists discuss the ratios of isotopes in materials using the delta notation. For example, $\delta^{13}C$ (said "delta

P. Higgins (✉)
Department of Earth and Environmental Sciences, University of Rochester, 227 Hutchison Hall, Rochester, NY 14627, USA
e-mail: pennilyn.higgins@rochester.edu

D. A. Croft et al. (eds.), *Methods in Paleoecology: Reconstructing Cenozoic Terrestrial Environments and Ecological Communities*, Vertebrate Paleobiology and Paleoanthropology, https://doi.org/10.1007/978-3-319-94265-0_7

thirteen C") is a way to talk about the relative amounts of ^{13}C and ^{12}C in a sample. Likewise, $\delta^{18}O$ is used to discuss the relative amounts of ^{18}O and ^{16}O in a material. Carbon isotopes ($\delta^{13}C$) provide a proxy for metabolic processes and diet as our primary uptake of carbon is from food. The building of sugars in photosynthesis fractionates the carbon isotopes as they are drawn into the plant, and the digestion of plant or animal tissues causes another fractionation. Interpretation of oxygen isotopes is more complex, but $\delta^{18}O$ is used widely as a proxy for temperature, precipitation, and humidity.

The relative amounts of different isotopes of an element are measured using mass spectrometry. Until recently, almost all isotopic analyses from biominerals were performed in essentially the same way, i.e., by converting the powdered mineral to carbon dioxide gas (CO_2) that could be ionized. Molecules of the gas are then separated into different masses (due to differing isotopic composition) and their relative abundances are measured. Newer methods, described in this chapter, utilize lasers either for improved sampling or to improve precision of analyses for better results.

Terms

Aragonite and Calcite

Aragonite and calcite are two minerals with the chemical formula $CaCO_3$ (calcium carbonate). They differ in the way their composing atoms are stacked in the mineral matrix. Aragonite is frequently found in the shells of marine organisms. It is generally unstable on the Earth's surface and will recrystallize to form calcite over time. Calcite also is frequently found in the shells of marine organisms. It is also the dominant mineral in limestone and in soil carbonate nodules.

Bioapatite (Apatite)

Apatite minerals possess the phosphate group (PO_4^{3-}) and are notably resistant to chemical processes that could alter the mineral structure. Bioapatite is any apatite mineral that forms due to biological processes, and is typically a calcium phosphate. Bones and teeth in vertebrates are composed of bioapatite, as are some invertebrate hard parts. In some cases, the phosphate group is replaced with a carbonate group (CO_3^{2-}). Oxygen from the phosphate groups and both carbon and oxygen from carbonate groups can be analyzed for stable isotopic composition.

There are many apatite minerals, defined in part by the content of fluorine, calcium, and hydroxyl, as well as the relative amounts of phosphate and carbonate (CO_3^{2-}) which will substitute for phosphate in the apatite matrix. The specific mineral present depends upon the position of the crystal in the body (bones versus teeth, and even the specific part of the tooth), as well as whether or not the tooth or bone has experienced alteration due to processes following death and fossilization.

Biomineral

Minerals are naturally occurring inorganic solids with crystalline structures and a specified chemical composition. Biominerals are minerals that form due to biological processes. These can include structural elements (bone, teeth, shell, etc.) or could be minerals precipitated due to metabolic processes external to the organism, such as the precipitation of siderite ($FeCO_3$) near iron-reducing bacteria (Mortimer and Coleman 1997; Romanek et al. 2003).

Bulk Analysis

Analysis of a single, representative sample of a specimen. This may be a single tooth, a single shell, or whole ground rock. Typically, this is collected to attempt to get an average value for the sample that includes its entire span of formation.

Calcite

See Aragonite.

Carbonate

A type of mineral that possesses the carbonate functional group CO_3^{2-}. Sometimes refers only to the functional group when substituted into a non-carbonate mineral's crystal matrix.

Delta Value (δ)

Delta (δ) values are the common means of discussing relative abundances of stable isotopes compared to a defined standard, using the following equation:

$$\delta = [(R_{sample} - R_{standard})/(R_{standard})] \times 1000$$

where **R** is the ratio of the less common vs the more common isotopes of an element, $\mathbf{R_{standard}}$ is that ratio for a known material (the standard to which the unknown sample will be compared), and $\mathbf{R}_{\mathbf{sample}}$ is that ratio in an unknown material.

Delta values are not concentrations. Instead, delta values are isotopic abundances relative to international standards with defined delta values. For example, on the V-SMOW scale (see below), SMOW itself has a defined $\delta^{18}O$ value of 0‰. A sample that has a relatively greater amount of ^{18}O ("heavy" oxygen) or relatively less ^{16}O ("light" oxygen) than SMOW would have a positive delta value, whereas a sample that has relatively less ^{18}O would have a negative delta value.

Depletion

Depletion is a relative term describing a trend in the relative abundance of one isotope over another in a sample, most often referring to a loss the heavier isotope (for example, ^{13}C [the 'heavier' isotope] loses its abundance relative to ^{12}C, the 'lighter' isotope). However, it is possible for the 'light' isotope (^{12}C in this example) to become depleted relative to the heavier isotope (^{13}C). Thus, it is important to specify which isotope is depleted when using this term.

The term depleted is often incorrectly applied to delta values. Delta values cannot be depleted or enriched (see below). They can only be more positive or more negative.

Enamel

Enamel is bioapatite that commonly forms the exterior of the teeth of vertebrates, though enamel-like biominerals (enameloid) are also present on many fish scales. Enamel is distinct from other bioapatite skeletal materials such as bone and dentine due to the lack of porosity and the very low organic material content in its matrix. This makes enamel highly resistant to diagenesis, meaning that fossilized enamel is frequently unaltered from its original state.

Enrichment

Enrichment is a relative term describing a trend in the relative abundance of one isotope over another in a sample, most often referring to an increase in the heavier isotope (for example, ^{13}C [the 'heavier' isotope] increases its abundance relative to ^{12}C, the 'lighter' isotope). However, it is possible for the 'light' isotope (^{12}C in this example) to become enriched relative to the heavier isotope (^{13}C). Thus, it is important to specify which isotope is enriched when using this term.

The term enriched is often incorrectly applied to delta values. Delta values cannot be depleted or enriched (see below). They can only be more positive or more negative.

Fractionation

Fractionation is a change in the relative abundances of isotopes of an element from naturally occurring universal ratios due to chemical or physical processes. Examples of such processes are evaporation of water into water vapor, dissolution of carbonate into carbon dioxide gas, the formation of an aragonite shell from carbonate dissolved in water, or the inclusion of dietary carbon into tooth enamel. Many of these fractionations are temperature dependent. This temperature dependence forms the basis of isotopic paleothermometers (McCrea 1950).

Fractionation Factor (α)

An expression of the fractionation of isotopes during reactions or changes of phase. For biominerals, this could be the fractionation of carbon isotopes from food (substance A) to carbon isotopes found in the tooth enamel of large herbivorous mammals (substance B). It is shown as:

$$\alpha_{A-B} = R_A/R_B$$

where $\boldsymbol{\alpha}_{\mathbf{A-B}}$ is the fractionation factor from substance A (food) to substance B (tooth enamel); $\mathbf{R}_{\mathbf{A}}$ is the ratio of the heavier isotope of carbon (^{13}C) to the lighter isotope of carbon (^{12}C) in substance A (food); $\mathbf{R}_{\mathbf{B}}$ is the ratio of the heavier isotope of carbon (^{13}C) to the lighter isotope of carbon (^{12}C) in substance B (enamel).

This may also be presented at an apparent fractionation factor, $\boldsymbol{\alpha}^*$, where:

$$\alpha_{A-B}* = (1000 + \delta_A)/(1000 + \delta_B)$$

where $\boldsymbol{\delta}_{\mathbf{A}}$ and $\boldsymbol{\delta}_{\mathbf{B}}$ are the delta values for the element of interest for substance A and substance B, respectively.

Compare with Enrichment Factor (ε*).

Ion

An ion is an atom of an element that has either gained or lost electrons, thereby imparting a negative or positive charge to the atom.

Isotope

An isotope of an element is an atom that has the same number of electrons and protons of that element (which defines the element), but that differs in the number of neutrons in the nucleus. Changing the number of neutrons alters the mass of the atom, but not its other chemical properties. It is this difference in mass that leads to fractionation.

Isotopic Enrichment (ε*)

An expression of the fractionation of isotopes during reactions or changes of phase. For biominerals, this could be the fractionation of carbon isotopes from food (substance A) to carbon isotopes found in the tooth enamel of large herbivorous mammals (substance B). It is shown as:

$$\varepsilon_{A-B}* = (\alpha_{A-B} * -1)$$

where $\boldsymbol{\alpha_{A-B}}$ is the fractionation factor from substance A (food) to substance B (tooth enamel).

$\varepsilon_{A-B}*$ may also be approximated as:

$$\delta_A - \delta_B$$

where $\boldsymbol{\delta_A}$ and $\boldsymbol{\delta_B}$ are the delta values for the element of interest for substance A and substance B, respectively.

ε* is reported in per mil (‰)

Compare with Fractionation Factor (α).

Limestone

Limestone is a rock composed of calcium carbonate, usually in the form of calcite. Limestones typically form in aqueous environments, sometimes by inorganic precipitation of calcium carbonate directly from the water. Other times, biological processes may be involved.

Limestones can by layered or show little internal structure. They often contain shelly fossils, and may at times be composed almost entirely of fossils.

Mass Spectrometer

Mass spectrometers are devices used to measure the abundance of specific elements, compounds, or isotopes in a sample by using atomic mass. There are many different types of mass spectrometer, depending upon the application and the materials being studied.

Pee-Dee Belemnite (PDB)

A belemnite is the fossilized internal skeleton of a squid-like organism. Belemnites are typically composed of calcium carbonate. The Pee-Dee Belemnite is (or was) a fossil collected from the Pee-Dee Formation of South Carolina. It was originally ground and used as a laboratory standard for isotopes of carbon and oxygen in carbonates, and assigned delta (δ) values of 0‰ (per mil) for both $\delta^{13}C$ and $\delta^{18}O$ (Craig 1957). The original Pee Dee belemnite is gone, replaced by international standards NBS-19 and L-SVEC (Coplen et al. 2006).

Per mil (‰)

The unit of the delta notation (an expression of the relative ratios of isotopes of elements in a material; see the above definition). Delta values are small, often only hundredths or thousandths of a unit. These numbers are multiplied by 1000 to allow discussion to be easier. The per mil (‰) symbol is a reflection of this multiplication.

See Delta Value.

Phosphate

A mineral that possesses the phosphate functional group PO_4^{3-}. Sometimes refers only to the functional group when isolated from the mineral. Apatites are phosphate minerals.

Serial Analysis

A type of analysis in which fossil specimens that show incremental growth, such as teeth or shells, are sampled multiple times along the direction of growth to capture isotopic changes during the growth of the organism. This pattern of change can be used to interpret environmental or metabolic changes during the life of the organism.

Stable Isotope

Isotopes are atoms of elements of different masses, due to variations of the number of neutrons in the nucleus. For some isotopes, the different number of neutrons makes the nucleus unstable, resulting in radioactive decay and the conversion of that atom to a different element, such as $^{14}C \rightarrow ^{14}N$. For others, no decay occurs, and the isotope is called stable (e.g., ^{12}C and ^{13}C). Stable isotopes of elements exist in known and reasonably constant relative abundances (e.g., ^{12}C: ~98.9%; ^{13}C: ~1.1%), but may be fractionated by chemical or physical processes.

Standard

Standards are samples of known isotopic composition. Standards are run alongside unknowns. Because of the nature of isotope ratio mass spectrometry, the measured value (the data directly from the mass spectrometer) of all samples differs from the true value. Standards are used then to correct analyzed values to true values.

Stratigraphy

Stratigraphy is the placement in relative order (oldest to youngest) of geological rock units.

V-PDB

An international scale used for measurement of carbon and oxygen isotopes, typically from carbonate. PDB refers to the Pee Dee Belemnite, which was collected from the Pee Dee Formation in South Carolina in the 1950s. The $\delta^{13}C$ and $\delta^{18}O$ values for the original PDB were assigned values of zero. The supply of the original Pee Dee Belemnite has since been exhausted.

The International Atomic Energy Agency (IAEA) now manages and distributes isotopic standards internationally, including those for carbonates. In 1993, the International Atomic Energy Agency (IAEA) defined values for the V-PDB scale, using the more commonly available standards NBS-19 (a limestone) and L-SVEC (a lithium carbonate). The V-convention is used in reference to the IAEA's headquarters in Vienna, Austria. Today, the international community uses defined values NBS-19 ($\delta^{13}C$ = +1.95‰, $\delta^{18}O$ = −2.20‰; Coplen 1994) and L-SVEC (for carbon only, $\delta^{13}C$ = −46.6‰; Coplen et al. 2006) as the reference points for calibrating laboratory analyses to the V-PDB scale.

V-SMOW

An international scale used for measurement of hydrogen and oxygen isotopes, typically from water. SMOW refers to Standard Mean Ocean Water, a presumed average of the isotopic values of ocean water world-wide.

The International Atomic Energy Agency (IAEA) now manages and distributes isotopic standards internationally, including those for water. In 1993, the International Atomic Energy Agency (IAEA) defined values for the V-SMOW scale, using the commonly available standards SMOW and SLAP (Standard Light Antarctic Precipitation). These standards are stored in large tanks at the IAEA in Vienna, Austria. The V-convention is used in reference to the IAEA's headquarters in Vienna. Today, the international community uses defined values SMOW ($\delta^{2}H$ = 0‰, $\delta^{18}O$ = 0‰) and SLAP ($\delta^{2}H$ = −427.5‰, $\delta^{18}O$ = −55.50‰) as the reference points for calibrating laboratory analyses to the V-SMOW scale.

Principles and History of Stable Isotope Analysis

Theoretical Underpinnings of Stable Isotope Analysis

Isotopes are atoms of the same element (which is defined by a specific number of protons in the atomic nucleus) that differ in the number of neutrons in the atomic nucleus resulting in a different atomic mass. All isotopes of an element possess similar chemical behavior and will substitute for one another in molecules. However, heavier isotopes typically form stronger bonds than lighter isotopes of the same element. Because of this, whenever chemical reactions occur, whether inorganic or biologically mediated, there is a slight change in the ratio of heavy to light isotopes of the same element. This process is called fractionation. The new, altered ratio can be measured from fossils and rocks and can tell us about the conditions and chemical reactions that occurred in the ancient past.

First Isotopic Analyses

Isotopic analysis began with the desire to understand the individual components of matter, what we know now as the elements. Scientists developed techniques to explore what really caused various materials to differ from one another. In the process, isotopes were discovered.

The first mass spectrographs: The original mass spectra were used to learn about the relative sizes of different elements. Much work was done beginning in the late 19th century and early 20th century by researchers like J. J. Thompson (1897) and W. Wein (1898) who developed various methods to determine the masses of individual elements (Budzikiewicz and Grigsby 2006). Much of this was done by ionizing pure gasses into 'cathode rays', which we now know are ion beams with positive charges. Cathode rays were deflected in a vacuum using electric and magnetic fields, where the electric field was held constant and the magnetic field was changed stepwise, resulting in a spectrum. The intensity and position of the resultant rays were recorded as voltages in a Faraday cup or on a screen. Bright spots on photographic film indicated the presence of elements based on their different masses.

Discovery of isotopes: When mass spectrometry first developed, the recorded spots or peaks were broad and poorly focused. With advances in detection and vacuum, peaks became sharper and new details of the overall peak shape were able to be detected. With improvements, Thompson (1911) worked with F. W. Aston to build a mass spectrograph that revealed the mass spectrum for carbon dioxide, with peaks for CO_2, CO, O, and C. Further improvements showed that the cathode ray for neon actually had two peaks, one much smaller than the other (Thompson 1913). We know now that what he observed were two stable isotopes of neon, ^{20}Ne and ^{22}Ne.

Modern mass spectrometry: In 1918, A. J. Dempster developed a new type of mass spectrometer which is very similar to the ones in today's laboratories. Rather than using canal ray guns (which make beams of positive ions) to ionize the gasses, he used thermal desorption or electrical ionization to ionize the study gas. Like modern mass spectrometers, these gas ions were accelerated away from the ion source using controlled electrical potentials. The resultant beams passed through a slit into the analyzer. This new instrument had much higher resolution than the original systems.

F. W. Aston went on to receive the Nobel Prize in Chemistry in 1922 in part for showing that elements can have more than one stable isotope (see Aston 1921), something that was not possible without the higher resolution mass spectrometers like that designed by Dempster (1918). Aston's work in developing the modern mass spectrometer and identifying various isotopes helped lay the ground work for what is now modern mass spectrometry.

Paleothermometry with oxygen isotopes from carbonates: The implications and possible applications of the study of stable isotopes became clear in 1950, when J. M. McCrea published his paper on the relationship between oxygen isotopes and temperature in carbonates (McCrea 1950). He showed that calcium carbonate precipitated from water of a known δ^{18}O and of a particular temperature had a predictable isotopic value. This meant that if we could make assumptions about the δ^{18}O of ancient waters, we could possible determine paleotemperature using isotopic analysis. In this seminal work, McCrea (1950) helped us understand the temperature dependence of fractionation of isotopes, and in the process, he developed the methods that we still use today for isotopic analysis of carbonates.

By 1953, researchers were already using McCrea's (1950) relationship to learn about ancient environments. Among the first papers was one by Emiliani and Edwards (1953) where foraminifera (single-celled aquatic organisms with a calcite shell) were collected from ocean cores and analyzed using McCrea's (1950) methods. This work showed a temperature change over time from which they inferred a change in the mixing of ocean waters during glacial times. This was among the very first times that isotopes were used to examine past climate. Today, isotopes are an essential part of any paleoclimate study.

Daansgard (1964) helped scientists better understand the global distribution of isotopes in water, especially that isotopic composition varied seasonally, and with latitude and altitude. We can apply these principles to isotopic values in carbonate and bioapatite to better interpret temperature and environment. Craig and Gordon (1965) helped us better understand the distribution of hydrogen and oxygen isotopes in the hydrosphere. Rosanski et al. (1993) showed that the amount and timing of precipitation events can also have an effect on isotopic ratios of oxygen in that precipitation, helping scientists further refine their interpretations of oxygen isotopes from carbonate minerals. Zachos and colleagues (2001) showed the overall pattern of marine oxygen and carbon isotopes over the last 65 million years. The variation in oxygen isotope composition helps us understand large-scale global climate patterns, including episodes of extreme warmth and the onset of glaciation. More recently, Bowen and Wilkinson (2002) used new GIS technologies plus thousands of new data points to better map variation in δ^{18}O composition in meteoric water.

Modern study of carbon isotopes: Smith and Epstein (1971) demonstrated that plants can be divided into different isotopic groups. This provides a way to use carbon isotopes to look at the relative composition of different types of vegetation (tropical grasses versus leafy trees and shrubs, etc.) in an ancient environment. O'Leary (1981) took this a step further by showing that the isotopic differences between plant groups (arid region grasses versus most shrubs and trees) are a consequence of their different metabolic pathways. These metabolic pathways can also tell us something about the environment, because particular pathways are adaptively favored in certain climates. Cerling and Harris (1999) showed how carbon isotopes from tooth enamel can reflect dietary preferences in mammals and how they can be used to distinguish open grassland environments from more closed forested environments. Zachos et al. (2001) described the global pattern of carbon isotopic change over the last 65 million years. They concluded that fluctuations in δ^{13}C often correlate with major tectonic, biotic, and climatic events. Because different sources of carbon (decaying plants, methane hydrates, burning of fossil fuels, etc.) have different isotopic values, it is also possible to learn about the causes of climatic or biotic events through the use of carbon isotopes.

Improvements – Reading the Records

Once the modern mass spectrometer was developed (Dempster 1918), improvements have been made to simplify and automate the process. The original methods involved large mass spectrometers with complex vacuum systems that required a great deal of technical skill to operate. Today, with the most advanced mass spectrometers, determining the isotope geochemistry of a sample has become common and relatively easy and inexpensive.

Improvements in sensitivity of mass spectrometers have made it possible to run smaller samples much faster. With this new capability, specimens that were once too small or precious to analyze are now able to be analyzed. Plus, many more samples can be run in a shorter period of time. The result is massive amounts of data (comparatively) and opportunities increase the resolution of our analyses. Increased stability of systems have meant an increase in precision of analyses, making it possible to draw interpretations from very small variations in isotopic ratios that would have in the past been lost in the error of the measurement. Newer automated systems have increased the number of samples that can be analyzed. Originally, sample preparation was a long and tedious process fraught with opportunities for error, including potential loss or contamination of precious samples. Today, the potential for sample loss is greatly reduced using computerized automated systems.

All of this, alongside advances in the environmental and biological sciences, have helped us better understand the how variations in our environment, such as temperature, timing and amount of precipitation, or the abundance of carbon dioxide in the atmosphere, can affect isotopic ratios, and provide us with new methods for exploring ancient ecology from fossils and the rocks that contain them.

Working with Isotopes from Biominerals

Commonly Studied Biominerals

In general, those working with isotopes from biominerals have focused mostly on carbonate (CO_3^{2-}) or phosphate (PO_4^{3-}) groups in minerals. Carbonates are a favorite to work with, as they are relatively easy to analyze and provide records of carbon and oxygen. Calcite and aragonite are most commonly analyzed. Phosphates are more challenging, and can only provide records of oxygen, but are generally less likely to be altered during fossilization, making them appealing.

Not every mineral containing carbonate that preserves an environmental signal is a biomineral. Calcite is a mineral common in soils. Isotopic analysis of this calcite can be used as a proxy for overall environmental conditions (see Berke 2018). Other carbonates like siderite (iron carbonate) can provide information about the presence or absence of abundant free oxygen in an environment as well as the degree of microbial activity (Mortimer and Coleman 1997).

Bioapatite – Teeth and Bones: Bioapatite is the mineral component of teeth and bones. The mineral is a member of the apatite group, one of many minerals that possess the phosphate (PO_4^{3-}) functional group, and has the general formula $Ca_5(PO_4)_3(OH, F, Cl)$. There are many apatite minerals, defined in part by the content of fluorine, calcium, and hydroxyl, as well as the relative amounts of phosphate and carbonate (CO_3^{2-}) which will substitute for phosphate in the apatite matrix during mineral formation. Rather than attempt to correctly identify which species of apatite is present in biological tissues, it is much easier to simply refer to all biogenic apatite as bioapatite. In bioapatites, either the carbonate (which has substituted for phosphate and includes both carbon and oxygen) may be analyzed or the oxygen from the phosphate itself may be isolated and analyzed.

Whether from carbonate or phosphate, the oxygen isotopic value from bioapatite reflects the animal's environment. If the animal is terrestrial, it gets much of its oxygen from water it drinks, though some oxygen also comes from its food and from the air it breathes. Information about temperature, humidity, and rainfall are contained with $\delta^{18}O$, because evaporation, amount of rainfall, and humidity have profound effects on the $\delta^{18}O$ of surface water. For marine or freshwater organisms, oxygen from the water the animals swim in is incorporated into the skeleton. In this case, oxygen isotopes typically reflect temperature and salinity of the water.

Carbon can only be measured from the carbonate component of bioapatite. It can provide information about the diet of the animal, which then can be used to interpret plant availability, climate, and niche partitioning (the sharing and utilization of resources by different species in a given area to reduce direct competition between species).

Bioapatite crystals from different body tissues form at different times. Bone tends to be modified continuously during the life of the animal (which is why broken bones can heal). Because of this, isotopes from a sample of bone give us a time-averaged record of a large part of an animal's life, rather than a snapshot of a narrower period of time. As bone is also porous to allow blood vessels and connective tissues pass through, mineral precipitation or infiltration into these spaces during fossilization may over-print the original isotopic signal.

The enamel and dentine that form teeth differ from bone as after these tissues are formed, they do not remodel throughout their lifespan. One caveat is the outermost layer of enamel can have minor changes in mineral composition due to the nature of dietary and water composition either during life – we add fluorine to our drinking water to modify slightly the composition of our enamel to better resist tooth decay (Petersen and Lennon 2004) – or through diagenesis. Thus, enamel is a better candidate for long-term preservation of original isotopic signals.

Tooth crown formation begins by the formation of a poorly mineralized matrix underlain by the dentin core. This matrix is subsequently mineralized forming the hardened enamel cap (Passey and Cerling 2002). Mineralization occurs from the tip of the tooth toward the root providing a potential time-series of isotopic change (Passey and Cerling 2002). What is most important is that teeth, and tooth enamel in particular, are heavily mineralized with little porosity and are most likely to survive fossilization with minimal alteration of the original isotopic signal. Dentin is porous (though not nearly as much as bone) with a greater organic content and may be altered where enamel on the same tooth may be close to pristine.

Calcite and Aragonite – Shells: Calcite and aragonite are common minerals that form skeletons of organisms. They share the same chemical formula, calcium carbonate ($CaCO_3$), but differ in their crystalline structure. Calcite is a common material making up the shells of organisms like snails and clams. Ostracods, a type of shelled crustacean, also have a calcite shell. Calcite also forms the supporting skeleton of corals and of echinoderms like sea urchins and sea stars. Many single-celled organisms such as foraminifera have tiny shells (called tests) made of calcite.

Some pelecypods (the group that includes clams) and gastropods (snails and their kin) have shells made of aragonite. Aragonite is less stable than calcite on the Earth's surface, so often shells made of aragonite convert to calcite during fossilization. This process can alter the isotopic values preserved in the shell. If you find a fossil shell and can demonstrate that it is still aragonite, you can be more confident that you have an unaltered signal preserved.

The carbonate portion of the mineral is analyzed, providing ratios for oxygen and carbon. Oxygen, as in bioapatite, provides information about the environment. As most organisms that have calcite or aragonite skeletons are aquatic, oxygen typically informs us about the temperature and salinity of the water in which they live (Abell and Hoelzmann 2000). Terrestrial organisms that possess shells, such as land snails, record information about precipitation, evaporation, and humidity, similar to how it is recorded in the teeth of terrestrial mammals (Prendergrast et al. 2015; Yanes et al. 2014).

Carbon is more challenging. It provides information about carbon sources for the animal, which includes dissolved inorganic carbon (DIC – typically carbonic acid) as well as some proportion of the animal's own metabolic carbon – basically what it breathes out. DIC can come from the atmosphere, but also from the decay products of plants and animals that may wash into the water. Whatever the source of dissolved inorganic carbon, the organism further fractionates the signal with its own metabolic processes while mineralizing the shell.

Other Carbonates – Rocks: The metabolic processes of some organisms can result in the additional formation of minerals external to the body of the organism. In most cases, the mineral that forms is a carbonate, frequently calcite (Andrews et al. 1997). For example, stromatolites, which are algal mounds that build up layers of limestone, preserve water and metabolic conditions in the calcite that forms due to their respiration (Schidlowski 2001; Andres et al. 2006). As with skeletal calcite, oxygen provides environmental information. Carbon isotopes provide evidence of metabolic processes and for very old rocks, isotopes may be the only evidence for the existence of life. Carbon can be used to understand what types of organisms were present.

Carbonate from rock even when not precipitated by biological processes can be helpful. Carbonate often forms in soils as nodules and can be preserved in the rock record as fossil soils or paleosols. In paleosol carbonates, carbon provides a record of the types of vegetation present, for example whether C3 or C4 plants grew in that location in the distant past, and can potentially provide insight into water stress on those plants (Cerling 1984; Cerling et al. 1989). Oxygen in paleosol carbonates is a proxy for temperature, humidity, and precipitation (Dworkin et al. 2005). Lake sediments also frequently preserve carbonate minerals from which isotopes can provide clues to the lake's history, such as episodes of high evaporation (Leng and Marshall 2004).

Specimen Collection, Preparation, and Analysis

Biominerals are all solids. Somehow, these solids must be broken down into the pure gasses that can be analyzed in the mass spectrometer. This section will cover the step by step the process for most typical samples.

Collecting Samples: Our first step is to collect a sufficient amount of powdered biomineral for analysis. For pure carbonates like shells or limestones, a milligram of raw powder may be sufficient to provide reliable data. For bioapatites, like enamel or bone, you need substantially more than that, usually at least 10 mg.

There are plenty of ways to get this sample. One is to simply crush up a whole shell or tooth, especially if it is small. However, there are good reasons why you would not want to do this. Different parts of the tooth or shell preserve different portions of the animal's life. There are also significant differences in the sensitivity to diagenesis depending on the part of the tooth or shell. For example, in teeth, the analysis should be limited to enamel if at all possible, because this is least likely to be diagenetically altered.

We always start by cleaning up the specimen. Dirt clinging to surfaces can contain its own carbonate that could contaminate the sample. This can be done with soap and water, a dental pick, or with a dental drill at slow speed with dental burrs. Sometimes it is necessary to wrap the tooth in an inert material like teflon tape to prevent surface dirt from other parts of the specimen to contaminate the desired sample.

Once the fossil specimen is clean, the samples are most often collected using a slow-speed drill with dental burrs, either carbide or diamond. Samples can be collected in one of two ways, either bulk or serial.

Bulk samples are collected to get a general sense of the average isotopic value for the whole specimen. If growth lines are evident, bulk samples are collected by drilling a line perpendicular to the growth lines along the length of the shell or tooth. If the growth rate of the shell or tooth is known, it is possible to limit the bulk sample line so as to collect only a known period's worth of growth. Most of the time this is not known, or the fossil is broken in some way to make this impossible, so we just collect a bulk sample along the maximum length of the specimen.

Serial samples are collected from fossils that are known to grow incrementally. Often such fossils show growth lines. These lines are not necessarily yearly lines, but like tree rings, show the general stages of growth. A series of samples can be taken tracing these lines and then can be used to reconstruct a life history for the organism. Depending on the size and type of fossil, many samples (>50) can be collected from a single specimen.

Each sample is collected onto weighing paper and placed into a carefully labeled microcentrifuge vial. Once all the samples are collected, we can move onto the next step.

Preparing Biominerals for Analysis: One goal in the sampling is to ensure that what we put into the mass spectrometer for analysis is the purest material possible. It is likely, however, that the sample has unwanted extra contaminants. For pure carbonates and bioapatites, these are typically organic contaminants, such as the oils from our hands. Pretreatments of the powders eliminate these contaminants.

Getting rid of organic contaminants is typically done with either hydrogen peroxide or sodium hypochloride, although hydrazine hydrate might also be used (Pellegrini and Snoeck 2016; Snoeck and Pellegrini 2015). Bioapatites frequently undergo a second pretreatment with acetic acid to eliminate any extra carbonate (that which is not in the mineral matrix). Each pretreatment step, and the choice of chemical used, has the potential to alter the isotopic values and thus are used with caution (see Pellegrini and Snoeck 2016; Snoeck and Pellegrini 2015).

Once pretreated and dry, the samples are ready for analysis of isotopes of carbon and oxygen from carbonate. Depending on the particular mass spectrometer and its inlet systems, the samples might be weighed into vials, reaction vessels, or boats for analysis.

If the research goal is to analyze the phosphate oxygen from bioapatite, additional steps must be taken to first break the phosphate-oxygen bond then precipitate silver phosphate crystals that can be analyzed using a different kind of inlet system than used for carbonates. Preparing silver phosphate requires the use of hydrofluoric acid (HF) that requires a lot more time than just running the carbonate component of bioapatite. However, the phosphate group in bioapatite is less likely to have been altered by the process of fossilization, and could be advantageous in certain cases.

Mass Spectrometry: In order for a biomineral to be analyzed for its isotopic composition, it must first be converted to a gas. How this is done is dependent upon the mass spectrometer being used and on what component of the biomineral is being studied. The most routine type of analysis is for carbon and oxygen isotopes from carbonate minerals, or from substituted carbonate in bioapatite. In both cases, the carbonate is formed by reacting the powdered sample with pure phosphoric acid. This results in carbon dioxide gas that can be analyzed for heavy carbon and heavy oxygen.

Phosphate oxygen can also be analyzed, but not through simple reaction with acid. Instead, the silver phosphate that was prepared earlier has to be put through an oven heated to about 1400°C and reacted with glassy carbon. This yields carbon monoxide, which can then be analyzed for its $\delta^{18}O$.

When the gas (either carbon dioxide or carbon monoxide), is initially formed it is generally contaminated with other molecules. Sometimes sulfur or nitrogen appears as the result of impurities in the samples. To ensure that the cleanest sample goes into the mass spectrometer, the gasses are cleaned using various means including fluid-cooled cold finger condenser and gas chromatography columns.

The purified gas is then introduced into the ion source of the mass spectrometer. An electron beam ionizes the gas and it is accelerated and focused out of the source by a series of electrified plates. The resultant ion beam then passes through a magnetic field and is bent. Ions of different mass bend different amounts. For carbon dioxide, the most common form, with light oxygen and light carbon (mass 44), bends the most. Carbon dioxide of mass 45 has one heavy carbon atom and bends slightly less. Carbon dioxide that has a heavy oxygen atom (mass 46) bends the least. Thus, after going past the magnet, what was one beam is now three, representing three different masses of carbon dioxide.

Mass spectrometers have collector cups set up such that the three beams each hit a different cup. Because each beam is composed of ions, striking the collector generates an electrical current. This electrical current is measured from each cup, then ratios are calculated between the different cups. The ratio of mass 44 to mass 45 contains information about the carbon isotopic ratios. The ratio of mass 44 to mass 46 provides oxygen isotopic data.

It is from the ratios of generated voltages or currents that the isotopic ratios can be calculated by the software specific to the mass spectrometer. This generates raw data that can then be normalized to international scales for research.

Delta Values and Standards for Comparison: Isotopic ratios, especially for light isotopes like H, O, C, and N, are reported in delta values. $\delta^{13}C$ is our symbol for the delta value (in this case, for carbon). We also talk about $\delta^{18}O$, $\delta^{15}N$, $\delta^{2}H$ (or sometimes δD [deuterium]). These delta values give us information about the amounts of different isotopes in a substance. $\delta^{13}C$ compares ^{13}C with ^{12}C. $\delta^{18}O$ compares ^{16}O with ^{18}O.

The delta value is a ratio and is always in the units of per mil (‰). It is a comparison between a substance of known isotopic composition (such as NBS-19 for carbonate and SMOW for water) and the substance for which we want to know the composition. The delta value is calculated this way:

$$[(R_{sample} - R_{standard})/(R_{standard})] \times 1000 = \delta_{sample-standard}$$

where, **R** is the ratio between the heavy isotope and the light isotope. This number is usually very, very small, because heavy isotopes are rare. This is a ratio, so there are no units.

$\mathbf{R_{standard}}$ is the ratio measured from a standard – something for which we know the relative amounts of the heavy and light isotopes, or have assigned a value, such as NBS-19 or SMOW.

$\mathbf{R_{sample}}$ is the ratio measured from our 'unknown' – the sample we want to know the delta value for.

The delta (δ) value is therefore a way to express the difference between Rstandard and Rsample.

The ‰ part comes in when we multiply by 1000. We multiply by 1000 because the number is very small and makes it easier for us to talk about the results, as in "1.34 per mil" is easier to say than "0.00134."

What a delta value tells you is how much more or less the unknown sample has of the heavy isotope than the standard. If it has relatively more of the heavy isotope, the delta values are positive, and the more positive (larger) the number, the more of the heavy isotope there is. In this case, the unknown is said to be 'enriched' in the heavy isotope. If the unknown has less of the heavy isotope than the standard, then the delta value is negative, and the unknown is called 'depleted' in the heavy isotope.

Communicating results using international scales: The first results are currents on collector cups. The mass spectrometer software converts these currents to delta values representing the relative amounts of heavy and light atoms of the same element. This raw value is generally relative to a reference gas on the mass spectrometer, one that is injected with every analysis to act as a zero line for comparison between samples.

The raw delta value from the mass spectrometer is not on any absolute scale. It is like measuring temperature with a thermometer that has divisions on it, but none is labeled. We know one sample is two per mil higher than another, but we can't relate this to any other samples run on any other mass spectrometer.

To relate our delta values to international reference scales, we analyze several samples of known delta value alongside our unknowns. These standards are calibrated to the two commonly used scales for carbonates, V-PDB (for carbon and oxygen) and V-SMOW (for oxygen only). You can think of these like Fahrenheit and Celsius, going back to the thermometer analogy.

When all the samples in a particular set are run, a least squares regression is completed on the samples of known isotopic value. This gives us a conversion equation that we can then apply to the samples of unknown isotopic value. This process is referred to as normalization or reduction. Once this is done, the values for the unknowns are now on an international reference scale and can be compared with results from other laboratories all over the world.

Why Isotopes from Biominerals are Informative

Biominerals provide valuable clues to ancient ecology. Through interactions with their environment (e.g., diet, drinking, interactions with the hydrosphere), the isotopic signature of an organism's surroundings is incorporated into the its hard tissues with known or predictable fractionations. Biominerals become incorporated into the rock record through their interactions with the biosphere providing a mineral record of the paleoenvironment as long as the rock exists.

Skeletal minerals can survive the processes that turn sediments into rock, while soft tissues tend to be lost. Phosphate minerals, like those of teeth and bones, are especially easy to preserve, but calcite also often survives the process of fossilization.

Analysis of carbonates (or those minerals that contain the carbonate group) is relatively easy and routine in some laboratories. Preparation of samples can be completed at institutions that lack mass spectrometry capability and then sent off to contract laboratories for analysis. For these reasons, biominerals that contain carbonate are commonly used and powerful tools to explore ancient ecology.

Caveats: Why Isotopic Analysis of Biominerals Can Be Problematic

Fractionation

The metabolic processes in organisms are complex, and are quite different from group to group, especially for distantly related groups. While a mussel and a clam might look similar, the way in which they mineralize their shells is likely to be different. It has been shown in ostracods, which are small shelled crustaceans, that their local ecology and metabolism strongly affects fractionation in their shells (Von Grafenstein et al. 1999). While it is generally agreed that horses and bison incorporate carbon from their diets into tooth enamel similarly ($\varepsilon^* \sim 14.1‰$, Cerling and Harris 1999), one must be cautious when extending that relationship to manatees ($\varepsilon^* \sim 14‰$, MacFadden et al. 2004) or rodents ($\varepsilon^* = 9.5–9.7‰$, DeNiro and Epstein; 1978; $\varepsilon^* = 9.5–10.3‰$, Ambrose and Norr 1993). Current research shows that differences between mammals exist in how they fractionate carbon (Passey et al. 2005).

Though this is improving, it is currently time consuming and costly to conduct the research to determine exactly how isotopes fractionate in the variety of organisms that we might see in the fossil record. For invertebrates, this typically means growing populations of animals in controlled isotopic and environmental conditions, then studying the shells as the animals grow, meaning that the animals must be killed in the process (see Goodwin et al. 2003; Balakrishnan et al. 2005; Geist et al. 2005; Carroll et al. 2006; Gillikin et al. 2007: McConnaughey and Gillikin 2008). Similarly, with vertebrates, the organisms' environmental inputs need to be controlled, and biominerals (e.g., bones and teeth) collected after the death of the animal (see Bryant et al. 1996a; Fricke and O'Neil 1996; Kohn et al. 1998; Cerling and Harris 1999; Sponheimer, and Lee-Thorp 1999; MacFadden et al. 2004).

Mineralization

The process of forming a biomineral in the body of a living organism is complex and not completely understood in all organisms. This results in a certain number of assumptions to be made while interpreting the data. For example, it is often assumed that visible growth lines are daily or yearly, more like tree rings, but that might not be the case (Jones and Quitmyer 1996; Dettman et al. 1999). Also, unlike tree rings, the mineralization of biominerals is not instantaneous. It may take several months for one 'growth band' in a mammal tooth to fully mineralize (Balasse 2002; Passey and Cerling 2002; Zazzo et al. 2005). The rate of growth, or elongation rate, might also vary along the length on the tooth or the shell (Hoppe et al. 2004). To complicate matters, these growth lines are seldom straight, nor are they conveniently perpendicular to the outer surface of the tooth or shell (Passey and Cerling 2002; Ivany et al. 2004). The result of all of this is that any single sample is likely to be an average of some amount of time, which varies from animal to animal. The implications of this are that small or temporary changes in an organism's diet, behavior, or environment may be masked in a time-averaged sample.

Many of our interpretations are also dependent upon the mineral we sample. We understand how fractionation works for calcite, but it might not be the same for aragonite, both of which are common minerals in shells (Ivany et al. 2004). Bioapatite in teeth and bones is variable in composition (Bryant et al. 1996a, b), and to date, the exact fractionations by mineral or by temperature are not fully worked out. This means we need to use caution when collecting samples and when interpreting results.

Diagenesis

The changes that occur during fossilization are collectively known as diagenesis. As scientists use fossils to interpret past environments, it is important to be able to assess diagenesis in the fossils to be confident that the interpretations are accurate. Diagenesis can change the shape or form of the minerals in a process called recrystallization. For example, a common mineral constituent of mollusk shells is aragonite. However, aragonite is unstable under the conditions of the Earth's surface and often converts to calcite during fossilization. During this conversion, oxygen and carbon from the environment can be exchanged, altering the isotopic ratios in the fossil shell. As calcite is such a common mineral in sedimentary environments, calcite or aragonite shells can also undergo dissolution and recrystallization, so that the original shape of the shell is preserved, but the original mineral has been completely replaced.

In bioapatites, the individual crystallites that make up bones and teeth can grow bigger during diagenesis (Keenan and Engel 2017). In fact, the crystal matrices are often further altered by the introduction of new ions that distort the crystal. Fossilized bioapatite often has an increased concentration of fluorine and rare earth elements (Jacques et al. 2008; Tütken et al. 2008). Uranium is also far more abundant in vertebrate fossils than during the life of the animal, so much so that scintillation meters detecting radioactive decay of uranium have been used to detect fossils remotely (Jones and Burge 1998). These chemical processes that alter bioapatite during fossilization can potentially (and often do) alter the preserved isotopic ratios, resulting in the possibility that what is being measured are the conditions of fossilization and not those experienced by the living animal.

With shells, we can use relatively simple spectrographic methods, like Fourier Transform Infrared Spectroscopy (FTIR) to assess whether the mineral is aragonite or calcite. If our fossil shells are aragonite (and we assume that they were originally aragonite) we can be confident that diagenetic alteration was minimal. FTIR can also be used with bioapatites to measure relative concentrations of phosphate and carbonate, as well as roughly determine the size of crystallites in the matrix. Another method, proposed by Iacumin et al. (1996) suggested measuring $\delta^{18}O$ from both the carbonate and the phosphate components in bioapatite. Though each component has a slightly different origin, the isotopic values, if diagenesis is minimal, should be around 8–10‰ (Iacumin et al. 1996). Whether this is a valid method has been debated (Martin et al. 2008) and some new methods to detect diagenesis have been proposed (Labs-Hochstein and MacFadden 1996).

Standards and Normalization

Isotopic data, specifically those for oxygen and carbon, are often reported in the delta (δ) notation, which is a measure of how different the sample is from a defined standard (see the section on delta values). Isotopic values for most biominerals are measured against the V-PDB (Vienna – Pee Dee Belemnite) scale, which was defined originally by taking a fossil belemnite and assigning its carbon and oxygen isotopes to a value of zero. Today, the original Pee Dee Belemnite is gone, and all analyses are calibrated against other international standards that are assigned values on the V-PDB scale. Specifically, NBS-19, a calcium carbonate with defined values for $\delta^{18}O$ and $\delta^{13}C$, is commonly used as a standard for biomineral analyses (Friedman, et al. 1982). NBS-18, another calcium carbonate, is also commonly used. L-SVEC is a lithium carbonate with a defined value for $\delta^{13}C$ but not for $\delta^{18}O$.

Almost all laboratories have some NBS-19 for their calibrations. Most also use NBS-18, but not nearly so many use L-SVEC. Because of the differences in international standards that laboratories might use, measured values for the same sample might also differ from lab to lab, due to mathematical differences in the normalization process.

There is no international standard for $\delta^{13}C$ or $\delta^{18}O$ of apatite, so most laboratories use the same carbonate standards above. However, because we do not fully understand the temperature dependence of fractionation when bioapatites are dissolved in acid to produce the CO_2 to measure, we can't be completely confident that we can make a direct comparison between results from bioapatite that are analyzed using different temperatures. This is why it is important to match the substrate of the standard with that of the sample analyzed.

V-SMOW (Vienna – Standard Mean Ocean Water) is another scale used for reporting $\delta^{18}O$ from samples. Typically, this scale is used only for water samples, but there is a mathematical conversion available to change measurements in V-PDB to V-SMOW. Since there are a couple of equations one could choose, the results may differ in subtle ways between labs or publications.

There is also substantial difference between labs as to how they correct analyses to reflect the international standards. As supplies of the international standards are limited, most laboratories do not have enough to run alongside unknowns on a daily basis. Laboratories instead calibrate internal standards, those run only in their own lab but in quantities large enough to be run every day, to the international standards. What these standards are and how many are run is as variable as the number of international standards available for any substrate. Some laboratories will calibrate a tank of carbon dioxide to use as a standard. This can be especially problematic as the isotopic ratios of gas in a cylinder can change rapidly as the volume in the tank runs low.

Whatever standards are used, whether they are international or internal standards, after the analyses are completed, the results must be normalized. The data are corrected, often using a least squares regression so that the analyzed standards give the 'correct' answer and the rest of the samples are adjusted accordingly. The resultant regression equation will be different if only one standard is used, or if multiple standards are used. It is generally agreed that multiple standards are preferred (Coplen 1994; Coplen et al. 2006), but not all labs use multiple standards, nor are there always multiple international standards available. However, differences in the standards used and the normalization method applied, can result in differences in values for the same sample when measured in different laboratories.

Analytical Techniques

Differences in results can also arise from the handling of samples before and during analysis. As noted earlier, pretreatments of samples can affect the isotopic values of the biomineral. Pellegrini and Snoeck (2015) and Snoeck and Pellegrini (2015) discuss the potential problems for bioapatites. For carbonates, pretreatments may or may not be done, depending on the preference of the laboratory and that of the researcher.

The mass spectrometer itself can also pose challenges. Even two of the same model of mass spectrometer can behave differently which can cause some variation. The systems by which we get our powdered solid samples converted to a gas and then drawn into the mass spectrometer are highly variable, in terms of analysis temperature, presence or absence of vacuum, pressure of the analyzed gas and the pressure of reference gasses. These little differences can add up. The use of international standards as noted in the previous section takes care of some, but not all of this analytically-based variation.

Examples of Applications

Carbon and Interpretation of Ancient Niche Partitioning

Carbon isotopes measured from tooth enamel of mammals are known to reflect the animal's diet. For large terrestrial herbivorous mammals, the fractionation from carbon isotopes in their food (plants) and the carbon isotopes in tooth enamel is approximately 14‰ (Cerling and Harris 1999). Plants fall into only a few of isotopic categories due to metabolic differences. Most important of these are plants that utilize the C3 or the C4 photosynthetic pathway (Teeri and Stowe 1976). C3 plants, generally cool season grasses as well as most trees and shrubs, have a carbon isotopic value around −27‰. C4 plants, typically grasses associated with arid environments, have carbon values around −14‰ (Bender 1971; Smith and Epstein 1971). In mammal tooth enamel, due to fractionation, a C3-feeder has an isotopic ratio of around −13‰, and a C4-feeder has a ratio around 0‰ (Cerling and Harris 1999) (Fig. 7.1).

In environments where both C3 and C4 plants co-exist, an animal that eats primarily C3 vegetation is usually a browser, that is, a mammal that prefers leaves from trees and shrubs in preference over grasses. In that same environment, a C4-specialist is usually a grazer. Some mammals, like horses, can make use of browse or graze depending on what is available as well as potential competition with other mammals (Feranec and MacFadden 2006; Feranec et al. 2009).

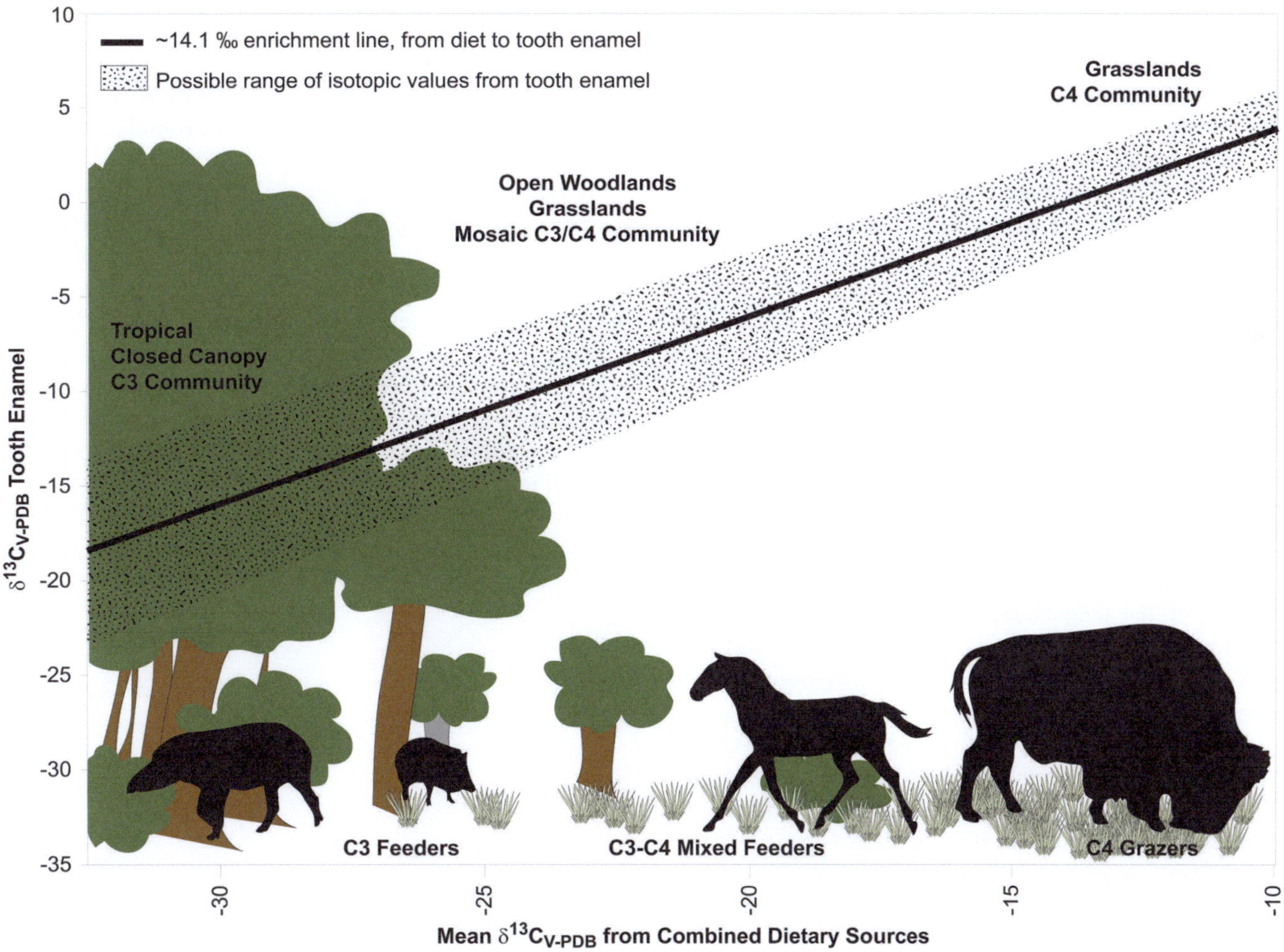

Fig. 7.1 Figure illustrating the relationship between ecological niche and carbon isotopic values in both the plant material (X-axis) and tooth enamel (Y-axis), in a hypothetical environment where both C3 and C4 plants coexist. The slope of this line is approximately 1, based on the simple fractionation from food to enamel of about −14‰ (Cerling and Harris 1999), where the $\delta^{13}C$ values from food are approximately 14‰ more negative than those measured from tooth enamel

This is called niche partitioning, where animals living in the same place make use of different parts of that environment.

We can do better than just distinguishing browsers from grazers. Some regions are heavily forested, and others are more open. The carbon isotopes of plants growing in these environments reflect differences in 'openness' as well as water stress. C3 plants growing in a closed canopy forest may have carbon isotopic values of less than −30‰ (van der Merwe and Medina 1991). This is called the 'canopy effect'. C3 plants living in open and relatively dry areas tend to have more positive values, perhaps as high as −22‰. This high value reflects the metabolic effects of water stress (Farquhar et al. 1989).

Knowing this, we can ask questions about the environments present in the remote past, as well as learn something of mammals with no modern analogue. MacFadden and Higgins (2004) conducted a study of Miocene mammals in Panama. At that time in geological history, there were no C4 grasses available, so MacFadden and Higgins (2004) knew they were dealing with a pure C3 environment. They were particularly interested in learning if a closed-canopy forest (similar to today's) existed in the past. They had specimens from six mammal taxa, including an oreodont, a camel-like protoceratid artiodactyl, two horses, and two rhinos. The $\delta^{13}C$ isotopic ratios measured from bulk samples of tooth enamel for these taxa ranged from −15.9 to −10.1‰. This suggests environments that ranged from forested to open woodland, to an open environment where plants may have experienced some water stress. One thing that is certain is that there was no fully closed canopy. Furthermore, MacFadden and Higgins (2004) showed that the oreodont preferred the forest, whereas the artiodactyl fed in more open environments.

Oxygen and Patterns of Annual Temperature and Precipitation

Because of the temperature dependence of isotopic fractionation, stable isotopes can be used as a paleothermometer (McCrea 1950; Emiliani and Edwards 1953; Goodwin et al. 2003; Dworkin et al. 2005). Oxygen isotopes especially lend themselves to use as a paleothermometer, as oxygen is a major component of water and water is abundant. Fractionation effects due to evaporation and precipitation are large, easy to measure, and heavily related to temperature. The oxygen isotopes in the fractionated water then become incorporated into biominerals of organisms. For mammals, the best proxy material is tooth enamel.

A single mammal tooth preserves an incremental record of some part of the mammal's early life. For example, horse and bison third molars mineralize over about 18 months starting just after the animal is weaned (Hoppe 2004). While the tooth is mineralizing, its carbon and oxygen are in isotopic equilibrium with the animal's environment, thus preserving a chemical record of what the environment was like for that animal at that time. Collecting and analyzing sequential or serial samples along the growth axis of the tooth can document a sequential record of that animal's experience. We can use these records from single or multiple individuals to learn about the overall climate in a region in the remote past.

The oxygen isotope paleothermometer is not without challenges, most importantly because the initial oxygen isotopic value of the original water must be known in order to know how large the temperature-related isotopic shift is (e.g., Eberle et al. 2010). There are many ways to estimate the original isotopic value of ancient water, including studies of modern water systems in regions of different temperatures, or through the use of proxy samples that are other materials from which oxygen isotopes can be measured for which the fractionation of oxygen with water and the relationship between the sample and the unaltered water is well understood (Craig and Gordon 1965; Zhou et al. 2008). Often, the fractionations are so complex that it is impossible to measure absolute temperatures in degrees (but see the section on clumped isotopes below), but overall patterns can be determined.

Temperature varies with season (summer is warmer than winter), latitude (the poles are colder than the equator), and altitude (higher altitudes are colder than lower altitudes). Depending upon the biomineral studied, more positive oxygen values can represent either warmer or colder temperatures. For tooth enamel of large terrestrial mammals, animals that must drink water from ponds or streams, more positive values generally indicate warmer temperatures due to evaporation (Higgins and MacFadden 2004; Kohn et al. 1998) (Fig. 7.2).

There are additional complications for interpretation of oxygen data in mammal teeth. In particular, the Amount Effect (Higgins and MacFadden 2004), wherein episodes of

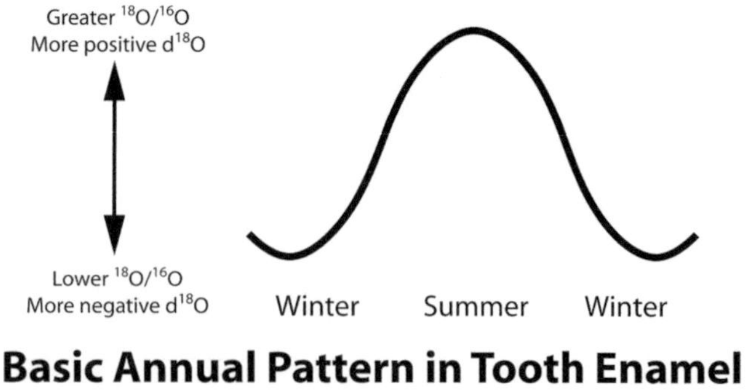

Fig. 7.2 The basic pattern of oxygen isotopic variation due to seasonal temperature change as seen in tooth enamel. Cool conditions (winter) are represented by lower $\delta^{18}O$ values than warm conditions (summer)

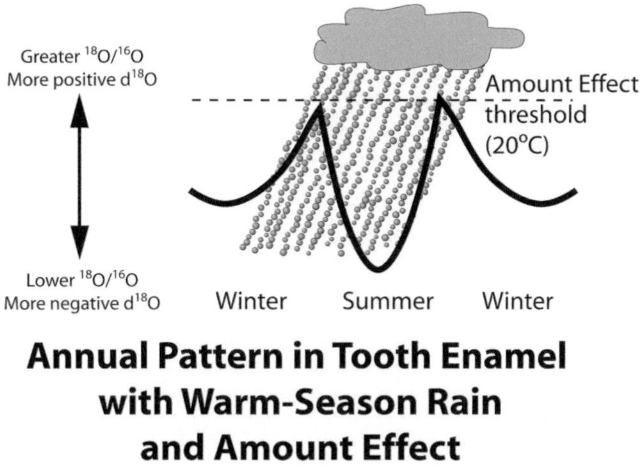

Fig. 7.3 Predicted pattern of oxygen isotopic values in regions where temperatures rise above 20°C (the Amount Effect threshold) and there is significant precipitation. Precipitation during times where temperatures are above the Amount Effect threshold causes a decrease in the isotopic values of precipitation, mimicking cooler temperatures

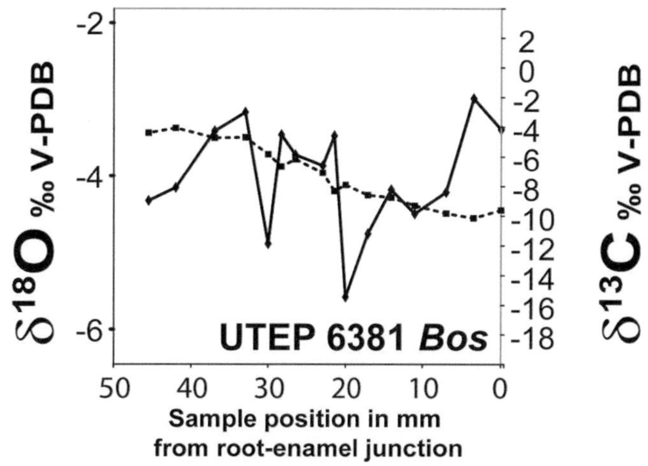

Fig. 7.5 Serial isotopic data from a modern cow (*Bos*) mandibular third molar (m3) collected near El Paso, Texas, USA. Solid line shows oxygen values; dashed line shows carbon values. Compare the oxygen isotopic pattern with that for precipitation in Fig. 7.4

heavy rain occurring at temperatures above 20°C can cause a depression in the oxygen isotopic curves that mimic winter-time cool temperatures (Fig. 7.3).

The amount effect is evident in data from the Global Network of Isotopes in Precipitation (GNIP). Figure 7.4 shows the relationship between temperature, precipitation amount, and the isotopic values of precipitation as seen at a GNIP station near Waco, Texas, USA.

Higgins and MacFadden (2004) used serial isotopic data from modern, feral mammals that lived near GNIP stations to show that the signal from the amount effect was indeed preserved in enamel. Figure 7.5 shows data from a molar of a cow (*Bos*) collected near El Paso, Texas, USA. It was expected that the isotopic signal from this cow would be similar to that from Waco, with two downward turns in the oxygen curve during the summer due to greater amounts of precipitation at the beginning and end of the summer.

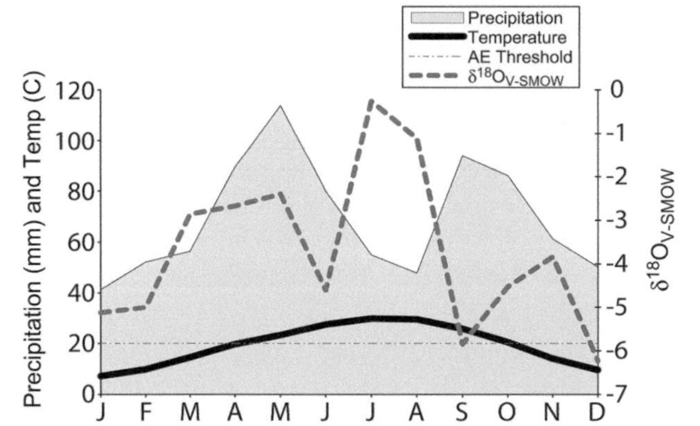

Fig. 7.4 Average monthly values of temperature, amount of precipitation, and δ^{18}OV-SMOW for Waco, Texas, USA on the Global Network of Isotopes in Precipitation (http://isohis.iaea.org; IAEA/WMO 2001). The Amount Effect threshold (20°C) is shown as a dashed line

Higgins and MacFadden (2009) compared serial isotopic data of carbon and oxygen with predicted temperature and precipitation patterns from several Pleistocene sites throughout North America. Patterns of annual temperature and precipitation were modeled from modern data using a computer program (Bryson and DeWall 2007). Using these modeled patterns, patterns of variation in oxygen isotopes were predicted and compared with isotopic values measured from fossil teeth. They showed that predicted patterns were preserved in the fossilized teeth.

Oxygen and Paleoaltitude

Using oxygen isotopes from biominerals then can allow us to not only understand weather patterns (as noted in the previous section), but can also give us information about the ancient elevation of sites. This can be important if one wishes to know, for example, the rate and timing of uplift of mountain ranges. Because temperature varies with altitude, one can use oxygen isotopes from teeth to estimate temperature, and from this elevation.

There is particular interest in the uplift of the Earth's high altitude plateaux, the Tibetan Plateau and the Altiplano. As elevation increases, one would expect temperatures to decrease, and therefore δ^{18}O values of rain and surface water to likewise decrease. Bershaw et al. (2010)

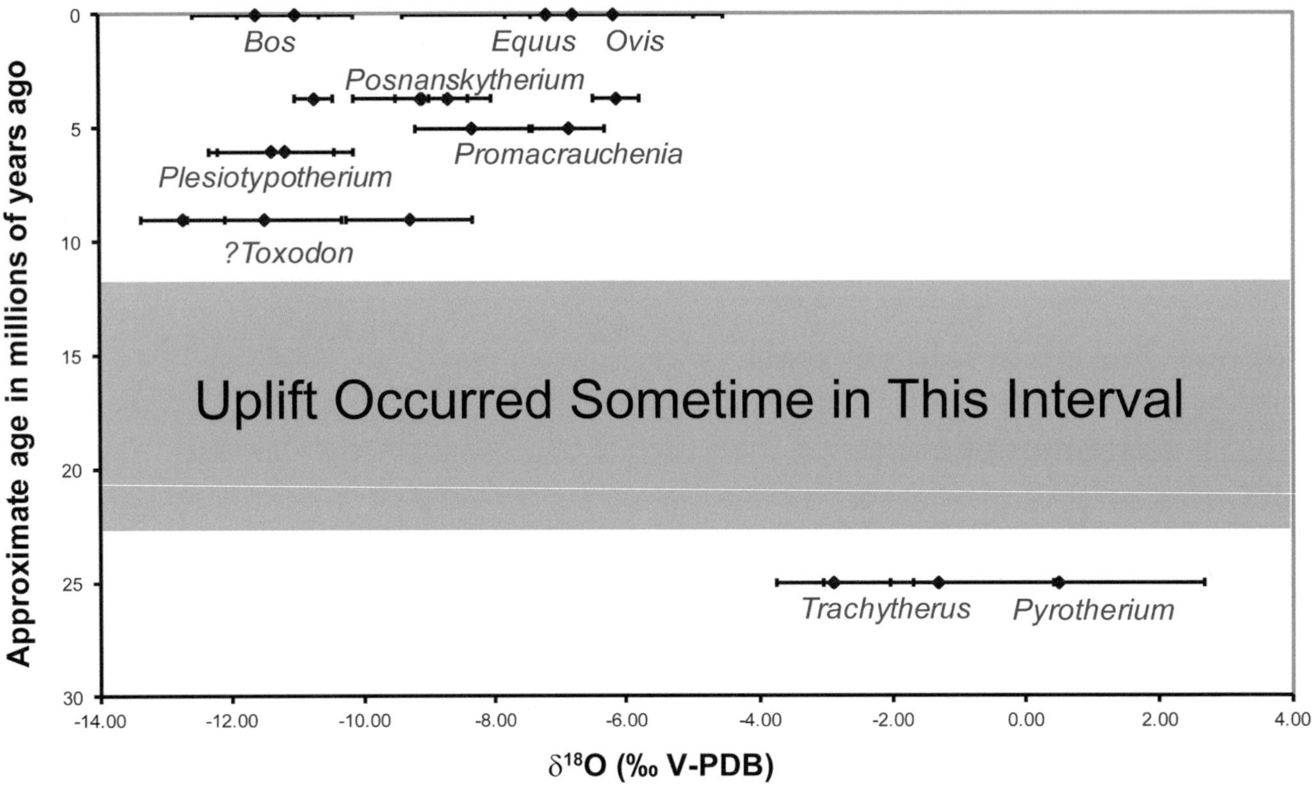

Fig. 7.6 Delta values for oxygen from multiple species of South American ungulate mammals plotted against time. Symbols represent mean value of all serial samples from a single tooth. Bars represent one standard deviation

used oxygen isotopic data from a series of hoofed mammals, including modern horses and cattle, and members of the extinct South American group of large hoofed mammals called the notoungulates, to constrain the timing of uplift of the Bolivian Altiplano, that has otherwise been investigated using isotopes of soils (see Garzione et al. 2006). The $\delta^{18}O$ of tooth enamel from large mammals reflects the $\delta^{18}O$ of the surface water that the mammal drank. Potential temperature-related fractionations due to body temperature in the mammals can be ignored because most large mammals have a constant body temperature. Thus, any trend in $\delta^{18}O$ values over time is likely due to temperature. Bershaw et al. (2010) observed a negative shift in $\delta^{18}O$ values over time that correlates with the uplift of the Altiplano (Fig. 7.6).

While it is not necessarily possible to calculate the exact elevation at any point, it is possible to infer that elevation increased dramatically between 25 and 8 million years ago based on these data alone.

Aridity Index

Aridity is a general term dryness (or lack of precipitation) in an area. This can be expressed at the difference (in mm/y) between water lost to the environment (through evaporation or transpiration) and the water gained through precipitation in a specific area. Where there is relatively more water lost than gained (commonly due to high rates of evaporation), $\delta^{18}O$ values of the remaining water tend to decrease relative to the original source water (typically rain). Because of this, it is possible to estimate how arid a region is by comparing the $\delta^{18}O$ value of the source water (the original precipitation) with that of waters that have experienced evaporation (Blumenthal et al. 2017).

Direct measurement of water isotopes is impossible in fossils. However, $\delta^{18}O$ can be measured from biominerals. Some organisms are not sensitive to evaporation effects in their $\delta^{18}O$. These are called 'evaporation insensitive' and labeled EI (Levin et al. 2006). The $\delta^{18}O$ measured from EI organisms reflects the $\delta^{18}O$ of the original source water. Other organisms are affected by evaporation, and are called 'evaporation sensitive' (ES). The $\delta^{18}O$ of ES organisms reflects the $\delta^{18}O$ of evaporated waters. The comparison between $\delta^{18}O$ values of EI organisms versus those of ES organisms is termed the aridity index (Levin et al. 2006; Blumenthal et al. 2017).

The use of aridity index is limited to species that are known to be either ES or EI. This is particularly useful for fossil mammals for which there are modern analogues. For example, Blumenthal et al. (2017) showed that members of the Hippopotamidae, the Elephantidae, and the Rhinocerotidae are EI, whereas members of the Giraffidae, the Tragelaphini (spiral horned antelopes that include kudus), and the Hippotragini ("horse" antelopes such as oryx) are ES. For each of these taxa, the aridity index differs (Levin et al. 2006; Blumenthal et al. 2017), so it is essential to have well characterized EI and ES organisms, and to have a full understanding of the index as it applies to each of those organisms.

Recent Advances

Working with Smaller Samples

Advances in instrumentation, including increased long-term stability, greater sensitivity, and methods for 'focusing' the sample gasses make it possible to run very small samples. For example, early analyses of tooth enamel required large (10+ mg) samples to be collected from a tooth. Today, samples as small as 1 mg may be used (and even smaller with laser ablation, see below). This makes it possible to collect multiple samples from a single tooth, making serial analysis possible.

Most new instruments also have automated systems for getting the gas generated by the samples into the mass spectrometer. In most cases, carbon dioxide gas destined for the mass spectrometer is generated by reacting powdered sample with phosphoric acid in a vacuum. Early in the history of these sorts of analyses, this gas had to be trapped in glass breakseal tubes using a series of vacuum lines, cold baths to isolate the desired gas, and a torch to seal the gas into the tube for analysis, each sample taking several hours to a day to react then 10 to 15 minutes to extract. Then, about 10 or 20 tubes per day could be loaded into a peripheral device on the mass spectrometer and were then sequentially broken open, releasing the gas to be analyzed.

Today, we have automated systems, where one only needs to weigh the correct amount of sample into a vial, boat, or reaction vessel, and the reaction and extraction are automated and take 20 minutes or less. With automated systems, what used to take a year or more to analyze, can be done in a month or less, and can be accomplished with less training and greater safety than required to use vacuum lines and prepare breakseal tubes.

Laser Ablation

The ability to analyze samples of 1 mg or smaller of tooth enamel, or 50 micrograms if dealing with pure carbonate, has improved the resolution and precision of analysis, especially where sequential analyses of single teeth, shells, or thin rock layers is needed. However, now we can go even smaller with greater precision using laser methods.

Laser ablation (LA) is a method by which a laser is aimed at a specific spot on a specimen, producing a fine aerosol of dust for analysis. This method is currently most often used for heavier elements, like lead, iron, and uranium, and is not without its problems for lighter isotopes like oxygen and carbon. However, several authors (starting with Cerling and Sharp 1996 and Sharp and Cerling 1996) have worked to develop methods of LA mass spectrometry that would work. Gas Chromatography (GC) (Passey and Cerling 2006) and direct laser fluorination (Grimes et al. 2008) have emerged as the most common methods.

While both methods are promising there are yet problems. In tooth enamel, there are two sites for oxygen, in the phosphate or the substituted carbonate. Each of these has a different history and a different fractionation from body water, and these signals may be mixed when ablation occurs. Thus, oxygen results from tooth enamel may be difficult to interpret (Passey and Cerling 2006). Newer methods, such as laser fluorination, may help alleviate this problem, but are still in testing (Linders et al. 2001; Grimes et al 2008). This is not so much a problem in shells, where the oxygen resides only in the carbonate phase.

Carbon in both teeth and shells resides only in the carbonate phase, meaning that laser methods can be effective and yield easily interpretable results. With a laser, it is possible to get a very detailed time series even from the smallest of teeth such as rodents (see Hynek et al. 2012; Kimura et al. 2013; Blumenthal et al. 2014).

Laser Absorption

Lasers can do more than just collect small samples. In the last few years, systems have been developed that use the absorption of laser light to determine the isotopic ratios in gasses. Gas vapors are drawn into a chamber into which a laser of a specific range of wavelengths is fired (Wahl et al. 2006; Crosson 2008). This laser bounces back and forth in the chamber between mirrors on each end, slowly fading away. Not every wavelength fades at the same rate, and these relative differences can be used to determine the isotopic values of the gas. Depending upon the interest of the user, these analyses can be done every few minutes and can be completely automated.

What is particularly important about these systems is their size and ruggedness. Systems can be run out of a car or on a boat with no problem. They do not need special electrical needs and can be plugged into a regular wall socket. And they can be attached to intake systems that sample rivers, oceans, or air at intervals that are essentially continuous. Current systems are designed to measure isotopes of hydrogen and oxygen from water and also carbon dioxide gas. Such systems are now deployed globally to look at environmental changes (Chen et al. 2010) but can be modified by adding specialized inlet systems to do the same sorts of analyses we do now with standard mass spectrometry.

There are two points of interest here. One is that the system can analyze samples more rapidly that a classic mass spectrometer. The other is that these tend to cost significantly less. Laser systems typically cost half or less than a traditional mass spectrometer and have much lower demands for consumables and electricity. Traditional mass spectrometers are also large and incredibly sensitive to environmental conditions. However, laser systems still require similar inlet systems like the larger mass spectrometers to take solid samples and convert them to gasses for analysis. Systems are being designed now to support these needs.

Clumped Isotope Geochemistry

The main thrust of this chapter is the discussion of how carbon and oxygen from the carbonate phase in minerals can provide information about ancient environments. One of our challenges is that these are only proxy measures and are not able to quantify environmental conditions. For example, from oxygen isotopes we can infer times of relative warmth or aridity, or the seasonality of rain, but we can't define a temperature without making many, many assumptions. Clumped isotope geochemistry provides a way to quantify temperature using isotopic methods.

Carbonate (CO_3^{2-}) contains both carbon and oxygen. We assume in most of our analyses that only one or the other of these is a rare isotope. That is, if carbonate has a heavy carbon, the oxygen atoms are all the lighter isotope ($^{13}C^{16}O_3^{2-}$). Or if there is one heavy oxygen, the remaining atoms are all the common, lighter isotope ($^{12}C^{16}O_2^{18}O^{2-}$). However, there are instances where carbonate molecules exist with both a heavy carbon and one heavy oxygen ($^{13}C^{16}O_2^{18}O^{2-}$). These are called clumped isotopes (Eiler 2007). We also know that clumping in carbonates decreases with increasing temperature at which the carbonate formed (Ghosh et al. 2006 and 2007). If we can measure the amount of clumping, we can calculate the temperature at which a carbonate formed. Depending upon the material analyzed, this could be the temperature at which a soil carbonate mineralized (using inorganically precipitated calcite), the average temperature of a lake (using precipitated limestone or shells of organisms living in the lake), or body temperature in extinct organisms (using bioapatite from teeth). The use of clumped isotopes makes it possible to ask questions about a wide range of significant issues such as body temperature in dinosaurs (Eagle et al. 2010, 2011) and surface temperatures in a rising mountain range (Ghosh et al. 2006).

In standard mass spectrometry, when carbon dioxide is produced from carbonate, we get carbon dioxide of three masses: 44 (all light atoms); 45 (one heavy carbon); 46 (one heavy oxygen). However, other masses of carbon dioxide do exist that possess the clumped isotopes. Of particular interest is mass 47, with one heavy oxygen and a heavy carbon. It is because of this interest in mass 47, you will often hear this method colloquially referred to as Delta-47 ($\Delta 47$). Carbonate with an atomic mass of 47 is exceedingly rare, with only about 0.11% of all carbonates having this mass ($^{13}C^{16}O_2^{18}O^{2-}$; Ghosh et al. 2006). To measure it, a mass spectrometer with very high voltage sources, extra collectors, and a long time is required.

This method is still in its infancy. There are now several labs running such analyses, but there are challenges, including developing the necessary calibration curves by growing carbonates at specific temperatures then analyzing them (see Ghosh et al. 2007). It is also a challenge as the sample size needed is extremely large, especially for bioapatites. Whereas a typical bioapatite analysis for isotopes of carbon and oxygen requires about 2 mg of powdered mineral, a clumped isotope analysis requires about a gram. With continued work and better instrumentation, the time and amount of sample required will be reduced.

Triple Oxygen Analysis

In almost all cases, stable isotope analysis of oxygen focuses on the most common isotopes, ^{16}O and ^{18}O. Another stable isotope of oxygen is ^{17}O. It is much rarer than ^{16}O and ^{18}O, making it particularly challenging to analyze. Improved methods using either standard gas-source mass spectroscopy or laser absorption allow for measurement of this rare element. Analyses that measure $\delta^{18}O$ and $\delta^{17}O$ are called "triple oxygen analyses" because three stable isotopes of oxygen are being measured.

Potential applications of $\delta^{17}O$ analysis are described in detail by Passey et al. (2014). In natural waters, like rain, rivers, oceans, and clouds, there is a set relationship between $\delta^{18}O$ and $\delta^{17}O$. However, as this water is fractionated through evaporation, intake into leaves, being eaten (as leaf water) or drunk (like surface water), the relationship between $\delta^{18}O$ and $\delta^{17}O$ changes, such that there is a different amount of ^{17}O than would otherwise be expected (Passey et al. 2014). This difference between average, expected values, and the actual measured value is called ^{17}O-excess, and is denoted as $\delta^{17}O$. The relative abundance of ^{17}O is also affected by the ratio of atmospheric O_2 to CO_2, meaning that triple oxygen isotope analyses may make it possible to measure past concentrations of atmospheric CO_2 (Passey et al. 2014).

Among the great challenges for the triple oxygen analysis is to be able to distinguish between how much of the measured $\delta^{17}O$ is due to evaporation or to oxygen concentrations in the atmosphere. Confounding this further is the additional fractionations caused by the biologically mediated mineralization of skeletal tissues (bones, teeth, shells) that might be used for triple oxygen analysis (Passey et al. 2014). Active research is underway to solve these problems.

Chapter Summary

Analyses of stable isotopes from skeletal biominerals like calcite and apatite (important constituents of shells, bones, and teeth) changed the way paleobiologists study ancient environment and ecosystems. In this chapter, the focus was primarily on isotopes of carbon and oxygen, which provide insights into ancient vegetation, precipitation, and temperature.

Carbon is particularly useful in understanding food sources, as ultimately carbon incorporated in biominerals comes from the diet of the organism. In particular, for land-dwelling herbivores, carbon isotopes can be used to distinguish not only between two major types of carbon fixation in plants (C3 and C4 carbon fixation pathways), but also the relative amounts of plant using the different pathways, which can provide insight into the overall environment.

Oxygen is more complex, but isotopic ratios of oxygen can ultimately be tied back to precipitation, humidity, and temperature. Teasing these things apart is a challenge, but newer methods, such as clumped isotope paleothermometry makes it possible to know the temperature to better understand the relationship between temperature and humidity.

Most biominerals preserve both carbon and oxygen isotopes, making it possible to use both isotope systems in concert for a fuller understanding of overall environment and of the life history of a specific individual organism.

References

Abell, P. I., & Hoelzmann, P. (2000). Holocene palaeoclimates in northwestern Sudan: stable isotope studies on molluscs. *Global and Planetary Change, 26,* 1–12.

Ambrose, S. H., & Norr, L. (1993). Experimental evidence for the relationship of the carbon isotope ratios of whole diet and dietary protein to those of bone collagen and carbonate. In J. Lambert & G. Grupe (Eds.), *Prehistoric human bone, archaeology at the molecular level* (pp. 1–31). Berlin: Springer.

Andres, M. S., Sumner, D. Y., Reid, R. P., & Swart, P. K. (2006). Isotopic fingerprints of microbial respiration in aragonite from Bahamian stromatolites. *Geology, 34,* 973–976.

Andrews, J. E., Riding, R., & Dennis, P. F. (1997). The stable isotope record of environmental and climatic signals in modern terrestrial microbial carbonates in Europe. *Palaeogeography, Palaeoclimatology, Palaeoecology, 128,* 171–189.

Aston, F. W. (1921). Isotopes and atomic weights. *Nature, 107,* 334–338.

Balakrishnan, M., Yapp, C. J., Theler, J. L., Carter, B. J., & Wyckoff, D. G. (2005). Environmental significance of $^{13}C/^{12}C$ and $^{18}O/^{16}O$ ratios of modern land-snail shells from the southern great plains of North America. *Quaternary Research, 63,* 15–30.

Balasse, M. (2002). Reconstructing dietary and environmental history form enamel isotopic analysis: time resolution of intra-tooth sequential sampling. *International Journal of Osteoarcheology, 12,* 155–165.

Bender, M. M. (1971). Variations in the $^{13}C/^{12}C$ ratios of plants in relation to the pathway of photosynthetic carbon dioxide fixation. *Phytochemistry, 10,* 1239–1244.

Berke, M. A. (2018). Reconstructing terrestrial paleoenvironments using sedimentary organic biomarkers. In D. A. Croft, D. F. Su & S. W. Simpson (Eds.), *Methods in paleoecology: Reconstructing Cenozoic terrestrial environments and ecological communities* (pp. 121–149). Cham: Springer.

Bershaw, J., Garzione, C. N., Higgins, P., MacFadden, B. J., Anaya, F., & Alvarenga, H. (2010). Spatial-temporal changes in Andean plateau climate and elevation from stable isotopes of mammal teeth. *Earth and Planetary Science Letters, 289,* 530–538.

Blumenthal, S. A., Cerling, T. E., Chritz, K. L., Bromage, T. G., Kozdon, R., & Valley, J. W. (2014). Stable isotope time-series in mammalian teeth: *in situ* $\delta^{18}O$ from the innermost enamel layer. *Geochimica et Cosmochimica Acta, 124,* 223–236.

Blumenthal, S. A., Levin, N. E., Brown, F. H., Brugal, J.-P., Chritz, K. L., Harris, J. M., et al. (2017). Aridity and hominin environments. *Proceedings of the National Academy of Sciences, USA, 114,* 7331–7336.

Bowen, G. J., & Wilkinson, B. (2002). Spatial distribution of $\delta^{18}O$ in meteoric precipitation. *Geology, 30,* 315–318.

Bryant, J. D., Koch, P. L., Froelich, P. N., Showers, W. J., & Genna, B. J. (1996a). Oxygen isotope partitioning between phosphate and carbonate in mammalian apatites. *Geochimica et Cosmochimica Acta, 60,* 5145–5148.

Bryant, J. D., Froelich, P. N., Showers, W. J., & Genna, B. J. (1996b). Biological and climatic signals in the oxygen isotopic composition of Eocene-Oligocene equid enamel phosphate. *Palaeogeography, Palaeoclimatology, Palaeoecology, 126,* 75–89.

Bryson, R. A., & DeWall, K. M. (2007). An introduction to the archaeoclimatology macrophysical climate model. In R. A. Bryson & K. M. DeWall (Eds.), *A Paleoclimatology workbook: High resolution, site-specific macrophysical climate modeling* (pp. 3–10). The Mammoth Site of Hot Springs, SD, Inc.

Budzikiewicz, H., & Grigsby, R. D. (2006). Mass spectrometry and isotopes: a century of research and discussion. *Mass Spectrometry Reviews, 25,* 146–157.

Carroll, M., Romanek, C., & Paddock, L. (2006). The relationship between the hydrogen and oxygen isotopes of freshwater bivalve shells and their home streams. *Chemical Geology, 234,* 211–222.

Cerling, T. E. (1984). The stable isotopic composition of modern soil carbonate and its relationship to climate. *Earth and Planetary Science Letters, 71,* 229–240.

Cerling, T. E., Quade, J., Wang, Y., & Bowman, J. R. (1989). Carbon isotopes in soils and palaeosols as ecology and palaeoecology indicators. *Nature, 341,* 138–139.

Cerling, T. E., & Harris, J. M. (1999). Carbon isotope fractionation between diet and bioapatite in ungulate mammals and implications for ecological and paleoecological studies. *Oecologia, 120,* 347–363.

Cerling, T. E., & Sharp, Z. D. (1996). Stable carbon and oxygen isotope analysis of fossil tooth enamel using laser ablation. *Palaeogeography, Palaeoclimatology, Palaeoecology, 126,* 173–186.

Chen, H., Winderlich, J., Gerbig, C., Rella, C. W., Crosson, E. R., Van Pelt, A. D., et al. (2010). High-accuracy continuous airborne measurements of greenhouse gasses (CO_2 and CH_4) using the cavity ring-down spectroscopy (CRDS) technique. *Atmospheric Measurement Techniques, 3,* 375–386.

Coplen, T. B. (1994). Reporting of stable hydrogen, carbon, and oxygen isotopic abundances. *Pure and Applied Chemistry, 66,* 273–276.

Coplen, T. B., Brand, W. A., Gehre, M., Gröning, M., Meijer, H. A. J., Toman, B., et al. (2006). New guidelines for $\delta^{13}C$ measurements. *Analytical Chemistry, 78,* 2439–2441.

Craig, H. (1957). Isotopic standards for carbon and oxygen and correction factors for mass-spectrometric analysis of carbon dioxide. *Geochemica et Cosmochimica Acta, 12,* 133–149.

Craig, H., & Gordon, L. I. (1965). Deuterium and oxygen 18 variations in the ocean and the marine atmosphere. In E. Tongiogi (Ed.), *Stable isotopes in oceanographic studies and paleotemperatures* (pp. 9–130). Pisa: CNR-Laboratorio di Geologia Nucleare.

Crosson, E. R. (2008). A cavity ring-down analyzer for measuring atmospheric levels of methane, carbon dioxide, and water vapor. *Applied Physics B, 92,* 403–408.

Dansgaard, W. (1964). Stable isotopes in precipitation. *Tellus, 16,* 436–468.

Dempster, A. J. (1918). A new method of positive ray analysis. *Physics Reviews (Series 6), 11,* 316–325.

DeNiro, M. J., & Epstein, S. (1978). Influence of diet on the distribution of carbon isotopes in animals. *Geochimica et Cosmochimica Acta, 45,* 341–351.

Dettman, D. L., Reische, A. K., & Lohmann, K. C. (1999). Controls on the stable isotope composition of seasonal growth bands in aragonitic fresh-water bivalves (Unionidae). *Geochimica et Cosmochimica Acta, 63,* 1049–1057.

Dworkin, S. I., Nordt, L., & Atchley, S. (2005). Determining terrestrial paleotemperatures using the oxygen isotopic composition of pedogenic carbonate. *Earth and Planetary Science Letters, 237,* 56–68.

Eagle, R. A., Schauble, E. A., Tripati, A. K., Tutken, T., Hulbert, R. C., & Eiler, J. M. (2010). Body temperatures of modern and extinct vertebrates from ^{13}C-^{18}O bond abundances in bioapatite. *Proceedings of the National Academy of Sciences, USA, 107,* 10377–10382.

Eagle, R. A., Tutken, T., Marton, T. S., Tripati, A. K., Fricke, H. C., Connely, M., et al. (2011). Dinosaur body temperatures determined from isotopic (^{13}C-^{18}O) ordering in fossil biominerals. *Science, 333,* 443–445.

Eberle, J. J., Fricke, H. C., Humphrey, J. D., Hackett, L., Newbrey, M. G., & Hutchison, J. H. (2010). Seasonal variability in Arctic temperatures during early Eocene time. *Earth and Planetary Science Letters, 296,* 481–486.

Eiler, J. M. (2007). "Clumped-isotope" geochemistry – the study of naturally-occurring, multiply-substituted isotopologues. *Earth and Planetary Science Letters, 262,* 309–327.

Emiliani, C., & Edwards, G. (1953). Tertiary ocean bottom temperatures. *Nature, 4359,* 887–888.

Farquhar, G. D., Ehleringer, J. R., & Hubick, K. T. (1989). Carbon isotope discrimination and photosynthesis. *Annual Reviews of Plant Physiology and Plant Molecular Biology, 40,* 503–537.

Feranec, R. S., & MacFadden, B. J. (2006). Isotopic discrimination of resource partitioning among ungulates in C_3-dominated communities from the Miocene of Florida and California. *Paleobiology, 32,* 191–205.

Feranec, R. S., Hadly, E. A., & Paytan, A. (2009). Stable isotopes reveal seasonal competition for resources between late Pleistocene bison (*Bison*) and horse (*Equus*) from Rancho La Brea, southern California. *Palaeogeography, Palaeoclimatology, Palaeoecology, 271,* 153–160.

Fricke, H. C., & O'Neil, J. R. (1996). Inter- and intra-tooth variation in the oxygen isotope composition of mammalian tooth enamel phosphate: implications for palaeoclimatological and palaeobiological research. *Palaeogeography, Palaeoclimatology, Palaeoecology, 126,* 91–99.

Garzione, C. N., Molnar, P., Libarkin, J. C., & MacFadden, B. J. (2006). Rapid late Miocene rise of the Bolivian Altiplano: evidence for removal of mantle lithosphere. *Earth and Planetary Science Letters, 241,* 543–556.

Geist, J., Auerswald, K., & Boom, A. (2005). Stable carbon isotopes in freshwater mussel shells: environmental record or marker for metabolic activity? *Geochimica et Cosmochimica Acta, 69,* 3545–3554.

Ghosh, P., Eiler, J., Campana, S. E., & Feeney, R. F. (2007). Calibration of the carbonate 'clumped isotope' paleothermometer for otoliths. *Geochimica et Cosmochimica Acta, 71,* 2736–2744.

Ghosh, P., Garzione, C. N., & Eiler, J. M. (2006). Rapid uplift of the Altiplano revealed through ^{13}C-^{18}O bonds in paleosol carbonates. *Science, 311,* 511–515.

Gillikin, D. P., Lorrain, A., Meng, L., & Dehairs, F. (2007). A large metabolic carbon contribution to the $\delta^{13}C$ record in marine aragonitic bivalve shells. *Geochimica et Cosmochimica Acta, 71,* 2936–2946.

Goodwin, D. H., Schone, B. R., & Dettman, D. L. (2003). Resolution and fidelity of oxygen isotopes as paleotemperature proxies in bivalve mollusk shells: models and observations. *PALAIOS, 18,* 110–125.

Grimes, S. T., Collinson, M. E., Hooker, J. J., & Mattey, D. P. (2008). Is small beautiful? A review of the advantages and limitations of using small mammal teeth and the direct laser fluorination analysis technique in the isotope reconstruction of past continental climate change. *Palaeogeography, Palaeoclimatology, Palaeoecology, 266,* 39–50.

Higgins, P., & MacFadden, B. J. (2004). "Amount effect" recorded in oxygen isotopes of Late Glacial horse (*Equus*) and bison (*Bison*) teeth from the Sonoran and Chihuahuan deserts, southwestern United States. *Palaeogeography, Palaeoclimatology, Palaeoecology, 206,* 337–353.

Higgins, P., & MacFadden, B. J. (2009). Seasonal and geographic climate variabilities during the Last Glacial Maximum in North America: applying isotopic analysis and macrophysical climate models. *Palaeogeography, Palaeoclimatology, Palaeoecology, 283,* 15–27.

Hoppe, K. A., Stover, S. M., Pascoe, J. R., & Amundson, R. (2004). Tooth enamel biomineralization in extant horses: implications for isotopic microsampling. *Palaeogeography, Palaeoclimatology, Palaeoecology, 206,* 355–365.

Hynek, S. A., Passey, B. H., Prado, J. L., Brown, F. H., Cerling, T. E., & Quade, J. (2012). Small mammal carbon isotope ecology across the Miocene-Pliocene boundary, northwestern Argentina. *Earth and Planetary Science Letters, 321–322,* 177–188.

Iacumin, P., Bocherens, H., Mariotti, A., & Longinelli, A. (1996). Oxygen isotope analyses of co-existing carbonate and phosphate in biogeneic apatite: a way to monitor diagenetic alteration of bone phosphate? *Earth and Planetary Science Letters, 142,* 1–6.

IAEA/WMO, (2001). Global Network of Isotopes in Precipitation. The GNIP Database.

Ivany, L. C., Wilkinson, B. H., Lohmann, K. C., Johnson, E. R., McElroy, B. J., & Cohen, G. J. (2004). Intra-annual isotopic variation in Venericardia bivalves: implications for the early Eocene termperatue, seasonality, and salinity on the U.S. gulf coast. *Journal of Sedimentary Research, 74,* 7–19.

Jacques, L., Ogle, N., Moussa, I., Kalin, R., Vignaud, P., Brunet, M., et al. (2008). Implications of diagenesis for the isotopic analysis of Upper Miocene large mammalian herbivore tooth enamel from Chad. *Palaeogeography, Palaeoclimatology, Palaeoecology, 266,* 200–210.

Jones, D. S., & Quitmyer, I. R. (1996). Marking time with bivalve shells: oxygen isotopes and season of annual increment formation. *PALAIOS, 11,* 340–346.

Jones, R., & Burge, J. (1998). The successful application of radiological survey instruments to map bone locations and aid excavation at a Lower Cretaceous, Cedar Mountain Formation, dinosaur quarry. In S. G. Lucas, J. I. Kirkland & J. W. Estep (Eds.) *Lower and middle cretaceous terrestrial ecosystems* (pp. 310–329). New Mexico Museum of Natural History and Science Buletin No. 14.

Keenan, S. W., & Engel, A. S. (2017). Early diagenesis and recrystallization of bone. *Geochimica et Cosmochimica Acta, 96,* 209–223.

Kimura, Y., Jacobs, L. L., Cerling, T. E., Uno, K. T., Ferguson, K. M., Flynn, L. J., et al. (2013). Fossil mice and rats show isotopic evidence of niche partitioning and change in dental ecomorphology related to dietary shift in late miocene of Pakistan. *PLoS ONE, 8*(8), e69308.

Kohn, M. J., Schoeninger, M. J., & Valley, J. W. (1998). Variability in oxygen isotope compositions of herbivore teeth: reflections of seasonality or developmental physiology? *Chemical Geology, 152,* 97–112.

Labs-Hochstein, J., & MacFadden, B. J. (2006). Quantification of diagenesis in Cenozoic sharks: elemental and mineralogical changes. *Geochimica et Cosmochimica Acta, 70,* 4921–4932.

Leng, M. J., & Marshall, J. D. (2004). Palaeoclimate interpretation of stable isotope data from lake sediment archives. *Quaternary Science Reviews, 23,* 811–831.

Levin, N. E., Cerling, T. E., Passey, B. H., Harris, J. M., & Ehleringer, J. R. (2006). A stable isotope aridity index for terrestrial environments. *Proceedings of the National Academy of Sciences, USA, 103,* 11201–11205.

Lindars, E. S., Grimes, S. T., Mattey, D. P., Collinson, M. E., Hooker, J. J., & Jones, T. P. (2001). Phosphate $\delta^{18}O$ determination of modern rodent teeth by direct laser fluorination: an appraisal of methodology and potential application to palaeoclimate reconstruction. *Geochimica et Cosmochimica Acta, 65,* 2535–2548.

MacFadden, B. J., & Higgins, P. (2004). Ancient ecology of 15-million-year-old browsing mammals within C3 plant communities from Panama. *Oecologia, 140,* 169–182.

MacFadden, B. J., Higgins, P., Clementz, M. T., & Jones, D. S. (2004). Diets, habitat preferences, and niche differentiation of Cenozoic sirenians from Florida: evidence from stable isotopes. *Paleobiology, 20,* 297–324.

Martin, C., Bentaleb, I., Kaandorp, R., Iacumin, P., & Chatri, K. (2008). Intra-tooth study of modern rhinoceros enamel $\delta^{18}O$: is the difference between phosphate and carbonate $\delta^{18}O$ a sound diagenetic test? *Palaeogeography, Palaeoclimatology, Palaeoecology, 266,* 183–189.

McConnaughey, T. A., & Gillikin, D. P. (2008). Carbon isotopes in mollusk shell carbonates. *Geo-Marine Letters, 28,* 287–299.

McCrea, J. M. (1950). On the isotopic chemistry of carbonates and a paleotemperature scale. *The Journal of Chemical Physics, 18,* 849–857.

Mortimer, R. J. G., & Coleman, M. L. (1997). Microbial influence on the ocysgne isotopic composition of diagenetic siderite. *Geochimica et Cosmochimica Acta, 61,* 1705–1711.

O'Leary, M. H. (1981). Carbon isotope fractionation in plants. *Phytochemistry, 20,* 553–567.

Passey, B. H., & Cerling, T. E. (2002). Tooth enamel mineralization in ungulates: implications for recovering a primary isotopic time-series. *Geochimica et Cosmochimica Acta, 66,* 3225–3234.

Passey, B. H., Robinson, T. F., Ayliffe, L. K., Cerling, T. E., Sponheimer, M., Dearing, M. D., et al. (2005). Carbon isotope fractionation between diet, breath CO_2, and bioapatite in different mammals. *Journal of Archaeological Science, 32,* 1459–1470.

Passey, B. H., & Cerling, T. E. (2006). *In situ* stable isotope analysis ($\delta^{13}C$, $\delta^{18}O$) of very small teeth using laser ablation GC/IRMS. *Chemical Geology, 235,* 238–249.

Passey, B. H., Hu, H., Ji, H., Montanari, S., Li, S., Henkes, G. A., et al. (2014). Triple oxygen isotopes in biogenic and sedimentary carbonates. *Geochimica et Cosmochimica Acta, 141,* 1–25.

Pellegrini, M., & Snoeck, C. (2015). Comparing bioapatite carbonate pre-treatments for isotopic measurements: Part 2 – Impact on carbon and oxygen isotope compositions. *Chemical Geology, 420,* 88–96.

Petersen, P. E., & Lennon, M. A. (2004). Effective use of fluorides for the prevention of dental caries in the 21st century: the WHO approach. *Community Dentistry and Oral Epidemiology, 32,* 319–321.

Prendergast, A. L., Stevens, R. E., Barker, G., & O'Connell, T. C. (2015). Oxygen isotope signatures from land snail (Helix melanostoma) shells and body fluid: proxies for reconstructing Mediterranean and North African rainfall. *Chemical Geology, 409,* 87–98.

Romanek, C. S., Zhang, C. L., Li, Y., Horita, J., Vali, H., Cole, D. R., et al. (2003). Carbon and hydrogen isotope fractionations associated with dissimilatory iron-reducing bacteria. *Chemical Geology, 195,* 5–16.

Rozanski, K., Araguas-Araguas, L., & Gonfiantini, R. (1993). Isotopic Patterns in Modern Global Precipitation. In P. K. Swart, K. C. Lohmann, J. McKenzie & S. Savin (Eds.) *Climate change in continental isotopic records* (pp. 1–36). AGU Geophysical Monograph 78.

Schidlowski, M. (2001). Carbon isotopes as biogeochemical recorders of life over 3.8 Ga of Earth history: evolution of a concept. *Precambrian Research, 106,* 117–134.

Sharp, Z. D., & Cerling, T. E. (1996). A laser GC-IRMS technique for *in situ* stable isotope analyses of carbonates and phosphates. *Geochimica et Cosmochimica Acta, 60,* 2909–2916.

Smith, B. N., & Epstein, S. (1971). Two Categories of $^{13}C/^{12}C$ ratios for higher plants. *Plant Physiology, 47,* 380–384.

Snoeck, C., & Pellegrini, M. (2015). Comparing bioapatite carbonate pre-treatments for isotopic measurements: Part 1 – Impact on structure and chemical composition. *Chemical Geology, 417,* 394–403.

Sponheimer, M., & Lee-Thorp, J. A. (1999). Oxygen isotopes in enamel carbonate and their ecological significance. *Journal of Archaeological Science, 26,* 723–728.

Teeri, J. A., & Stowe, L. G. (1976). Climatic patterns and the distribution of C_4 grasses in North America. *Oecologia, 23,* 1–12.

Thomson, J. J. (1897). Cathode rays. *Philosophical Magazine (Series 5), 44,* 293–316.

Thomson, J. J. (1911). Rays of positive electricity. *Philosophical Magazine (Series 2), 21,* 225–249.

Thomson, J. J. (1913). Rays of positive electricity. *Proceedings of the Royal Society of London (Series A), 89,* 1–20.

Tütken, T., Vennemann, T. W., & Pfretzschner, H.-U. (2008). Early diagenesis of bone and tooth apatite in fluvial and marine settings: constraints from combined oxygen isotope, nitrogen and REE analysis. *Palaeogeography, Palaeoclimatology, Palaeoecology, 266,* 254–268.

van der Merwe, N. J., & Medina, E. (1991). The canopy effect, carbon isotope ratios and foodwebs in Amazonia. *Journal of Archaeological Science, 18,* 249–259.

Von Grafenstein, U., Erlernkeuser, H., & Trimborn, P. (1999). Oxygen and carbon isotopes in modern fresh-water ostracod valves: assessing vital offsets and autecological effects of interest for palaeoclimate studies. *Palaeogeography, Palaeoclimatology, Palaeoecology, 148,* 133–152.

Wahl, E. H., Fidric, B., Rella, C. W., Koulikov, S., Kharlomov, B., Tan, S., et al. (2006). Applications of cavity ring-down spectroscopy to high precision isotope ratio measurement of $^{13}C/^{12}C$ in carbon dioxide. *Isotopes in Environmental and Health Studies, 42,* 21–35.

Wien, W. (1898). Untersuchungen über die electrische Entladung in verdünnten Gasen. *Annals of Physical Chemistry (neue Folge), 65,* 440–452.

Yanes, Y., Izeta, A. D., Cattaneo, R., Costa, T., & Gordillo, S. (2014). Holocene ($\sim$4.5–1.7 cal. kyr BP) paleoenvironmental conditions in central Argentina inferred from entire-shell and intra-shell stable isotope composition of terrestrial gastropods. *The Holocene, 24,* 1193–1205.

Zachos, J., Pagani, M., Sloan, L., Thomas, E., & Billups, K. (2001). Trends, rhythms, and aberrations in global climate 65 Ma to present. *Science, 292,* 686–693.

Zazzo, A., Balasse, M., & Patterson, W. P. (2005). High-resolution $\delta^{13}C$ intratooth profiles in bovine enamel: implications for mineralization pattern and isotopic attenuation. *Geochimica et Cosmochimica Acta, 69,* 3632–3642.

Zhou, J., Poulsen, C. J., Pollard, D., & White, T. S. (2008). Simulation of modern and middle Cretaceous marine $\delta^{18}O$ with an ocean-atmosphere general circulation model. *Paleoceanography, 23,* PA3223.

Chapter 8
Reconstructing Terrestrial Paleoenvironments Using Sedimentary Organic Biomarkers

Melissa A. Berke

Abstract Organic geochemical biomarkers are an increasingly utilized set of tools for reconstructing past terrestrial conditions. These biomarkers can be preserved from all life, representing Archaea, Eukarya, and Bacteria, and are deposited in sediments and soils, and yet they can be separated in the laboratory for analysis of individual compounds from the same sample. Often well preserved in the geologic record, new proxies based on these "molecular fossils" extracted from sedimentary archives have allowed for the reconstruction of a wide array of paleoenvironmental and paleoclimatic conditions, including but not limited to past temperature, vegetation, and hydroclimate. This review provides a brief overview of those molecular proxies that might be most applicable for terrestrial environmental conditions.

Keywords Molecular proxies • Compound specific stable isotopes • Paleoclimate • Biomarkers • Organic geochemistry

Introduction

Organic molecules are used as recorders of past climate and environmental conditions in terrestrial and aquatic settings. These "molecular fossils," or biomarkers as they are commonly called, can survive for millions of years and can be linked to specific source organisms or groups of organisms, or particular biogeochemical processes, such as methane production, anoxia, or photosynthesis (Brocks and Summons 2003; Pancost and Boot 2004; Peters et al. 2005). The biomarker concept is based on the separation of various aquatic, microbial, and terrestrial biomarkers in a heterogeneous mixture of organic compounds within the sediments. These compounds are insoluble in water and often are chemically inert with low volatility, making many of them long-lived in the environment and well-suited for studies of the past where low thermal maturation of the host lithology has occurred (Eglinton and Eglinton 2008). Sedimentary archives often contain a diverse mixture of these compounds, including compounds from organisms with poor preservation potential, which can be used to answer questions about the climate and environment over a wide range of spatial and temporal scales (Peters et al. 2005). Sediments contain a complex aggregate of molecular biomarkers that can be chemically separated and identified in the laboratory to help determine the origin and abundance of the organic matter. Therefore, these compounds can be more useful for the examination of specific environmental conditions, such as using compound specific stable isotopes extracted from sediments rather than bulk sediment organic matter isotope analyses that may include a large mixture of sources (Castañeda and Schouten 2011). Accumulation and preservation of biomarkers most commonly occurs in the soils and sediments of lakes, rivers, and oceans, but they have also been found preserved in speleothems (Blyth et al. 2008), rodent middens (Carr et al. 2010; Chase et al. 2012), ooids (Summons et al. 2013), and rock concretions (Kiriakoulakis et al. 2000). Biomarker distributions, abundances, and stable isotopic values are all used to determine past environmental conditions. Specifically, advances in stable isotope analysis at the compound level (focusing here on $\delta^{13}C$ and δD) are now used to address new and unique paleoenvironmental questions. This chapter is aimed at students and non-specialists with an interest in applying molecular proxy techniques for terrestrial reconstruction. Biomarkers and methods most useful for reconstructing Cenozoic terrestrial environments are given preference here, but excellent reviews of both shorter-lived biomarkers and those specific for unique aquatic conditions are available elsewhere (e.g., Brocks and Summons 2003;

M. A. Berke (✉)
Department of Civil and Environmental Engineering and Earth Sciences, University of Notre Dame, Notre Dame, IN 46556, USA
e-mail: mberke@nd.edu

D. A. Croft et al. (eds.), *Methods in Paleoecology: Reconstructing Cenozoic Terrestrial Environments and Ecological Communities*, Vertebrate Paleobiology and Paleoanthropology, https://doi.org/10.1007/978-3-319-94265-0_8

Castañeda and Schouten 2011; Grice and Eiserbeck 2014; Meyers 1997a; Pancost and Boot 2004; Schouten et al. 2013; Sinninghe Damsté and Schouten 2006).This review focuses on the reconstruction of terrestrial vegetation, hydroclimate (broadly defined as water cycle variability linked to climate shifts), and temperatures preserved in soil, lake, and marine settings.

The connection between biomarkers and their biological precursor has been greatly improved upon in recent times. Through modern sampling, laboratory growth experiments, and culturing, we have established the relationship (in many cases in the form of a mathematical calibration) between many carbon-hydrogen structures preserved in geologic materials and their modern counterpart and identified their original function and source. However, the concept of an organic source for preserved hydrocarbons was not always supported. In the late nineteenth century, multmacromolecular carboniple theories for the abiotic origin of oil were presented. Alfred Treibs (1899–1983), a noted organic geochemist of the time, was the first to support the organic link between molecules from rocks and biota, based on a comparison of rock-extracted hydrocarbons to the structure of plant and algal pigments. In 1934, Treibs observed the striking similarity between a vanadyl-porphyrin separated from black shale and modern eukaryote chlorophyll *a*. He later outlined the breakdown of chlorophyll *a* into the observed sedimentary products (Treibs 1936), and as such became the "Father of Organic Geochemistry" (Kvenvolden 2006). Another early pioneer and contemporary of Treibs, Parker Trask (1899–1961), was a geologist who also made significant early contributions to the field. In an effort to better understand oil reservoirs, Trask made comparisons of organic matter in sedimentary units to source beds of petroleum. The field of organic geochemistry was thus born, yet it wasn't until the 1970s that the term biomarker was coined. An amalgamation of 'biological marker,' it is thought that the term was first used by Wolfgang Seifert, one of Treibs' students employed by Chevron (Gaines et al. 2008). Seifert's early work was on steroids in oils and analytical fingerprinting methods utilizing hopanes as an index of thermal maturity (an effect of high heat and pressure on sedimentary hydrocarbons). The late 1970s and early 1980s became a time of moving away from solely using biomarkers to uncover the biological origin of petroleum products and towards a more complete understanding of the formation and preservation of hydrocarbons. Ultimately, the window these biomarkers provided into Earth's history become more apparent. Significant instrumentation advances (see further discussion below) have rapidly moved the field of organic geochemistry forward, with the ability to measure smaller samples and more complex molecules with greater precision and accuracy.

Organic Matter Preservation

Accumulation and preservation of organic matter occurs overwhelmingly in aquatic environments, both terrestrial (often lacustrine) and marine settings. In fact, terrestrial biomarkers are thought to be preserved preferentially in marine sediments compared to marine lipids (Prahl et al. 2003; Sinninghe Damsté et al. 2002b). However, the vast majority of organic matter does not get preserved at all and is ultimately remineralized and respired as CO_2 (Canfield 1994; Hedges and Keil 1995). The rates and location of organic matter production and its composition can help predict the likelihood of preservation. The oceans cover >70% of the Earth's surface area, but only account for a fraction of the long-term burial of organic matter compared to terrestrial aquatic settings (Dean and Gorham 1998). Furthermore, continental margins, including deltas and continental shelves, which make up $\sim 10\%$ of the ocean's surface area, account for >90% of the organic matter buried in marine settings (Berner 1989; Killops and Killops 2005). Regions with low burial efficiency, such as the pelagic or abyssal zones of the ocean, experience almost full remineralization of organic carbon and thus have sediments with low total organic carbon and low reactivity (Burdige 2007).

Although much emphasis has been placed on carbon cycling in marine settings, the sink of carbon in lakes and wetlands significantly outweighs that of the oceans (Dean and Gorham 1998). Lacustrine sediments commonly contain 10% or greater total organic carbon in sediments, even though they cover <4% of earth's surface area (Carroll and Bohacs 2001; Dean and Gorham 1998; Peters et al. 2005; Tranvik et al. 2009). These terrestrial sites collect and sequester large amounts of carbon from the surrounding watersheds, with the majority accumulating in these basin and not continuing out to sea (Cole et al. 2007), making them an important long-term repository for carbon.

The presence of oxygen is one of the important determinants in preservation of organic matter in aquatic environments. The reduction or complete absence of O_2 can occur in the water column during increased phytoplankton activity when algae consume O_2 during respiration and zooplankton consume O_2 while breaking down organic matter. These sediments favor the preservation of organic matter over oxic environments because low dissolved oxygen (<0.1% dissolved O_2) inhibits the bioturbation, resuspension, and exposure of freshly buried organic matter to aerobic degradation by microbes. Although conditions are less favorable for the decomposition of organic matter, microorganisms in anoxic sediments are able to utilize alternative terminal electron acceptors (e.g., NO^{3-}, Fe^{3+}, SO_4^{2-}), and emphasize the importance of anaerobic oxidation in the subsurface (Calvert and Pedersen 1993). In fact, new research suggests that deep subsurface microbes may have a larger role in the presence of organic matter, particularly at extreme depths and high water temperatures than previously thought. Anaerobic microbes have been found living at deeper depths and in higher temperatures than previously thought possible (Hallmann et al. 2008; Inagaki et al. 2015; Pikuta et al. 2007), which has implications for long-term organic matter burial and preservation. For example, Inagaki et al. (2015) found microbial communities at ~1.5 to 2.5 km below the seafloor living at temperatures between ~40° to 60°C in the Pacific Ocean off the coast of Japan. These microbial communities are thought to have survived living on coal and hydrogen from the coal-bed they were found associated with at depth (Inagaki et al. 2015).

For the most part, degradation of the organic biomarkers deposited in sediments occurs quickly following burial, within a few hundred years as observed in lacustrine deposits (Cranwell 1981; Meyers et al. 1980). However, the importance of anaerobic oxidation in the subsurface has become evident in light of new analytical methods in recent years. In fact, new research suggests that deep subsurface microbes may have a larger role in the presence of deep ocean organic matter than previously thought. Anaerobic microbes have been found living at deeper depths and in higher temperatures than previously thought possible (Hallmann et al. 2008; Inagaki et al. 2015; Pikuta et al. 2007), which has implications for long-term organic matter burial and preservation. For example, Inagaki et al. (2015) found microbial communities at ~1.5 to 2.5 km below the seafloor living at temperatures between ~40° to 60°C in the Pacific Ocean off the coast of Japan. The microbial communities are thought to have survived living on coal and hydrogen from the coal-bed they were found associated with at depth (Inagaki et al. 2015).

Organic matter that survives scavenging and microbial degradation in the water column and sediments is separated into two classes: bitumen and kerogen, both characterized by their relative resistance to significant further microbial alteration (Wakeham et al. 1997). Biomarkers can be found in kerogen, highly chemically varied organic matter that is insoluble with organic solvents and is often polymerized into large macromolecules (De Leeuw and Largeau 1993; Tegelaar et al. 1989b; Vandenbroucke and Largeau 2007). Bitumen is organic matter defined by its ability to be extracted from geologic materials using organic solvents (Killops and Killops 2005), and contains the majority of molecules used for reconstructing terrestrial paleoenvironments. Excessive heating and pressure during burial thermally alters the organic compounds in both the bitumen and kerogen components. When kerogen breaks down, some of its degradation byproducts form bitumen. This thermal alteration can continue to occur within the deposit, but biomarkers are no longer preserved at temperatures greater than ~150–250°C (Killops and Killops 2005; Peters et al. 2005).

Ultimately, some biomarkers can be preserved for billions of years (Peters et al. 2005). Although lipids often lose functional groups during diagenesis, the remaining hydrocarbon skeleton is then often stable and long-lasting. Some biomarkers, such as deoxyribonucleic acid (DNA) (Coolen and Overmann 1988; D'Andrea et al. 2006) and certain sterols (D'Anjou et al. 2012) are relatively short-lived, surviving only up to a few thousand years and often in unique chemical or temperature settings allowing enhanced preservation potential. Molecules that are much less labile, including lipids and plant structural components such as lignin may still experience some degree of chemical diagenesis (Killops and Killops 2005). The diagenetic fate of some of these lipids will be discussed further later in this chapter.

Organic matter preservation can also occur with inorganic association within the water column, during transportation, or in the upper sediment layers following burial. In these instances, organic matter is afforded a level of protection that allows it to persist longer than exposed organic matter would (Burdige 2007). This may involve actual physical protection mechanisms, such as sheltering organic matter and preventing remineralization or chemical attack (Burdige 2007). For instance, organic matter preservation was higher when molecules were associated with fine grained, rather than coarser grained sediments (Keil et al. 1994; Mayer 1994), providing a link between mineral surface area and preservation (Hedges and Keil 1995). Mayer (1994) noted a proportional relationship between surface area of sediment grains and organic matter abundance, and proposed an adsorption mechanism, termed "mesopore protection," whereby, bacterial enzymes are excluded due to size restriction of pores. Similarly, the monolayer hypothesis postulates that mineral surfaces are coated with the equivalent of a monolayer of carbon which creates a size restriction and enables protection (Hedges and Keil 1995; Mayer 1994, 1999). Although research has uncovered a connection between mineral surface area and organic matter preservation, we still do not have a complete understanding of the mechanisms involved in organic matter protection and burial.

Sampling Locations

The use of organic biomarkers for Cenozoic environmental reconstruction is limited by the availability of sedimentary units for study. Escaping thermal overprinting is a part of this preservation, but aspects of post-burial exposure and outcrop location are important to preservation as well. Outcrop samples are often available and the easiest to obtain, however, these samples pose recovery challenges for biomarkers because of their exposure. Alteration of organic matter and more specifically biomarkers can occur in outcrop due to oxidation, microbial degradation, and percolation of meteoric water exposing organic matter to further decay (Marynowski et al. 2011; Petsch et al. 2000, 2001a, b; Sageman et al. 2003). Biomarker studies from outcrop require extra care in sampling and interpretation. In a study of a shale weathering profile, Petsch et al. (2001a) used phospholipid fatty acid biomarkers and found that modern microbial communities were subsisting on exposed and ancient ($\sim$365 Ma) macromolecular carbon in shale. This implies that exposed organic matter thought to be refractory is actively being assimilated, and thus changed, in outcrops. Active degradation of biomarkers in exposed outcrops is a possibility and that care should be taken at these sampling locations. Indicators of poor preservation or contamination of biomarkers are further discussed below.

Marine or lacustrine sediment core samples are commonly used for terrestrial environment reconstruction. Recovery from deep drill holes alleviates the post-burial weathering and oxidation that outcrop samples experience and thus they often contain higher overall abundances of biomarkers (Peters et al. 2005). However, there are other considerations that must be made with regards to biomarker sampling from cores. Information regarding coring operations (e.g., recovery and disturbances) are critical when sampling for paleoecological reconstructions. Storage of cores for many years could result in oxidation or microbial degradation of biomarkers (Brittingham et al. 2017). These issues can be prevented with advanced storage techniques such as sealing sediment cores in plastic and archiving in cold, dark storage (e.g., Berke et al. 2012b, 2014).

Leaf Waxes of Past Vegetation

Plant leaf waxes lipids are highly refractory in the sediments and preferentially accumulated in aquatic settings prior to deposition compared to other lipids (Cranwell 1981; Schimmelmann et al. 1999; Sinninghe Damsté et al. 2002b; Wakeham et al. 1997), making them a robust and valuable tool for environmental paleoreconstructions (Eglinton and Eglinton 2008). Waxes are a part of the leaf cuticle, acting as a barrier between soft plant tissues and environmental stressors (Shepherd and Griffiths 2006). This epicuticular wax serves to protect the plant from water loss, disease, and ultraviolet radiation (Shepherd and Griffiths 2006).

Vascular plant biomarkers can be used for a wide range of applications in past environment and climate reconstruction. *n*-Alkanes, *n*-alkanols, *n*-alkanoic acids, and wax esters, collectively termed *n*-alkyl lipids (Fig. 8.1), make up leaf waxes and can be produced by both terrestrial and aquatic organisms. However, the number of carbon atoms in these lipids can be used as a general fingerprint for the source organism (Cranwell 1973; Eglinton and Hamilton 1963, 1967). Long-chain *n*-alkyl lipids make up a significant component of the epicuticular coating on vascular plant leaves, and exhibit a strong predominance of odd instead of even number of carbon atoms in a chain (*n*-alkanes) or even over odd carbon number chain length predominance (*n*-alkanoic acids) (Eglinton and Hamilton 1967). While long carbon chain lengths are commonly more dominant in terrestrial higher plants (C_{29}-C_{35} for *n*-alkanes and C_{28}-C_{34} in *n*-alcohols and *n*-alkanoic acids), aquatic algae and microbes are often predominantly composed of shorter chain lengths (C_{17}-C_{21} for *n*-alkanes and C_{16}-C_{22} in *n*-alkanoic acids) (Blumer et al. 1971; Cranwell et al. 1987). The mid-chain length homologues C_{23}-C_{27} (*n*-alkanes) and C_{24}-C_{26} (*n*-alkanoic acids), often produced in smaller quantities by terrestrial higher plants, make up the majority of leaf waxes of aquatic macrophytes (Cranwell 1984; Ficken et al. 2000).

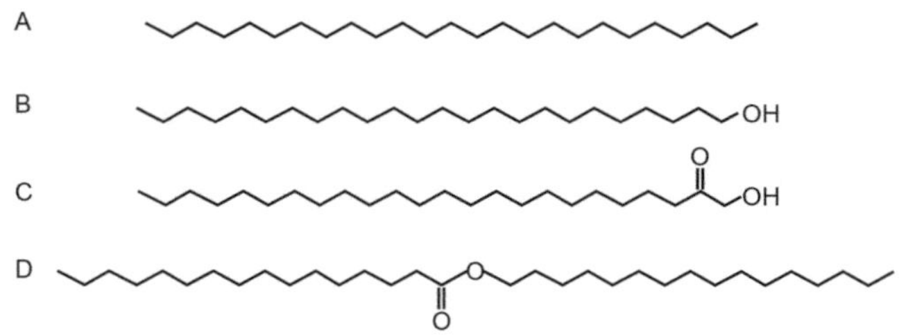

Fig. 8.1 *n*-Alkyl lipid structures found in plants including: A, *n*-alkanes (C_{25} shown); B, *n*-alkanols (C_{24} shown); C, *n*-alkanoic acids (C_{24} shown); and D, wax esters

Though these general guidelines often hold true, there are exceptions that complicate environmental interpretations based upon carbon chain length. For example, Aichner et al. (2010) observed that C_{29} *n*-alkanes, linked most commonly to terrestrial plants, are the dominant *n*-alkane chain length measured from submerged macrophytes in a Tibetan lake while terrestrial plants, such as *Betula* sp., *Sphagnum* moss, and *Salix glauca*, produce an abundance of C_{23} and C_{25} alkanes (Berke et al. 2015b; Bush and McInerney 2013; Sachse et al. 2006). Long-chain alkanes have been linked to aquatic algal sources (Lichtfouse et al. 1994; Metzger et al. 1985; Volkman et al. 1980). Exceptions to the commonly held ideas of plant production suggest that interpretation of *n*-alkyl carbon chain lengths for source identification must proceed with caution.

Angiosperms contribute more *n*-alkanes to the sedimentary record than do gymnosperms (Bush and McInerney 2013; Diefendorf et al. 2011). Diefendorf et al. (2011) found in a study quantifying modern plant lipids that *n*-alkanes are often too low in quantity to be measured in gymnosperms, while angiosperms contained abundant *n*-alkanes, a finding that has been supported with studies from soils in angiosperm forests (Otto and Simpson 2005; Sachse et al. 2006). In landscapes dominated by gymnosperm forests, other compounds may serve as better biomarkers, though fewer paleoreconstruction studies have been done using these biomarkers (Bastow et al. 2001; Hautevelle et al. 2006; Schouten et al. 2007c). For example, terpenoids, defense compounds produced by gymnosperms and angiosperms to ward off pests, pathogens, and herbivory (Coley 1983; Langenheim 1994; Otto and Wilde 2001), may serve as better gymnosperm biomarkers (Diefendorf et al. 2011; Otto and Simoneit 2001, 2002; Schouten et al. 2007c). These aromatic hydrocarbon compounds were first utilized for terrestrial environmental reconstruction from Jurassic marine sediments off the coast of Australia (van Aarssen et al. 2000) and more studies are exploring the use of aromatic biomarkers and their diagenetic precursors (Otto and Wilde 2001). Recent modern exploration of these biomarkers (Diefendorf et al. 2012) also aids in our understanding of terpenoid formation for paleoreconstruction.

While leaf wax lipids are known to be fairly robust in the sedimentary record, it is clear that degradation via microbial oxidation (e.g., Cranwell et al. 1987) and heating associated with burial (e.g., Chikaraishi and Naraoka 2007) can occur. Observations suggest that *n*-alkyl lipids are not affected equally by degradation. Specifically, *n*-alkanes withstand sedimentary degradation better than *n*-alkanoic acids or *n*-alkanols (Cranwell 1981). Additionally, sedimentary *n*-alkanoic acids and *n*-alkanols can convert to *n*-alkanes as deposits approach and move into the oil generation window (increasing thermal maturity via heating and time since burial), ultimately changing composition and abundances of the leaf wax lipids found in the sedimentary record (Collister et al. 1994a; Gupta et al. 2006, 2007). In addition to changing the lipid found in these sediments, thermal maturity also alters carbon chain length distributions, providing an alteration signature of carbon chain lengths and not original distribution (Bray and Evans 1961). These alterations make paleoenvironmental reconstructions based on *n*-alkyl lipid abundances and carbon chain length distributions in ancient deposits that have undergone some thermal maturity difficult. Though leaf wax lipid distribution and carbon chain lengths are altered with thermal maturity, recent research suggests that stable carbon isotopic values likely remain relatively unaltered. Using hydrous pyrolysis experiments of various modern plants, Diefendorf et al. (2015) found little alteration of $\delta^{13}C$ values (ca. 1‰) prior to or within the early temperatures of the oil generation window (temperatures <250°C).

n-*Alkane Distribution*

Plants produce different *n*-alkyl lipid types and while the chemical composition is similar, abundances and distribution may vary with respect to environmental conditions, abundances, thicknesses, chain length distribution (Bush and McInerney 2013). Ratios of *n*-alkane carbon chain lengths can be used to provide general information about changing environmental conditions and the relative contributions of plant communities to the lipid pool (Meyers 1997a). One such ratio, P_{aq}, is used to determine non-emergent aquatic macrophyte input to a lake system, differentiated from emergent macrophytes and terrestrial plants. P_{aq} is calculated using the abundance of mid-chain *n*-alkanes (C_{23} and C_{25}) relative to the sum of these mid-chain lengths and longer C_{29} and C_{31} homologues (Ficken et al. 2000). Similarly, the ratio of long, terrestrially-derived *n*-alkanes to short chain, aquatically-derived *n*-alkanes, referred to as the terrestrial to aquatic ratio (TAR) (Meyers 1997b, 2003), measures the abundance of C_{27}, C_{29}, and C_{31} compared to C_{15}, C_{17}, and C_{19}.

Another ratio commonly applied to *n*-alkane homologues with the purpose of gaining plant and environmental information across sampled sites or through time is known as the average chain length (ACL) (Poynter and Eglinton 1990). The ACL is the weighted average of the carbon chain lengths defined as:

$$ACL_{m-n} = \sum_{i=m}^{n} \frac{i[C_i]}{C_i}$$

The m and n are the shortest and longest chain lengths, respectively, and i is the number of carbon atoms in each homologue. For *n*-alkanes, the odd-numbered homologues are used and for *n*-alkanoic acids, the even-numbered homologues are used in this equation. It is used as an approximation of integrated local vegetation and changing vegetation types through time (e.g., Eglinton and Hamilton 1967). However, beyond general tracking of changes in carbon homologue distribution in sedimentary archives, it is still unclear what plant ACL variability reflects. ACL values do not remain constant as leaves develop (Sachse et al. 2009; Tipple et al. 2013) and also are not consistent across all plant groups or functional types, particularly within woody vegetation (Diefendorf et al. 2011). Studies have found that grasses and deciduous trees exhibit homologue preferences (e.g., Cranwell 1973; Rommerskirchen et al. 2006; Vogts et al. 2009). C_4 grasses (described below) often produce longer homologue lengths than do C_3 grasses, with studies suggesting an abundance of longer chain lengths (higher ACL) being a characteristic of C_4 grasses (Vogts et al. 2012). Furthermore, warmer climates are often associated with longer ACLs (Sachse et al. 2006; Simoneit 1977), which may correspond to a prevalence of C_4 grasses and aridity in some regions (Vogts et al. 2012). Though it is likely that environmental conditions help shape ACL values (e.g., Bush and McInerney 2015; Rommerskirchen et al. 2003), caution must be taken with interpretations that rely on these ratios because there is an incomplete understanding of what drives homologue abundances.

Another commonly used lipid ratio is the carbon preference index (CPI), which looks at the extent of odd to even dominance in *n*-alkanes (Bray and Evans 1961; Marzi et al. 1993). The updated calculation of the original CPI definition (Bray and Evans 1961) by Marzi et al. (1993) is:

$$CPI = \frac{\left(\sum_{odd} C_{17-33}\right) + \left(\sum_{odd} C_{19-35}\right)}{2 \cdot \left(\sum_{even} C_{18-34}\right)}$$

CPI values are often used to distinguish the biological origin and the possible influence of thermal maturity on *n*-alkanes (Bray and Evans 1961; Peters et al. 2005; Tegelaar et al. 1989a). Deposits with high thermal maturity, an abundance of algae or bacteria, degradation of higher carbon chain lengths, or input of petroleum contamination all tend to yield low CPI values, ≤ 1 (Peters et al. 2005). High CPI values have also been found in modern plants (Bush and McInerney 2013; Collister et al. 1994b) and recent sediments (Cranwell 1981).

Leaf Wax $\delta^{13}C$ for Vegetation Reconstruction

All higher plants fall into one of three photosynthetic classifications: C_3 (those that use the Calvin-Benson pathway, $\sim$90% of all plants, including common trees and bushes), C_4 (those that use the Hatch-Slack pathway, many tropical grasses and sedges), or Crassulacean acid metabolism (CAM, those that use a mixture of pathways, most commonly succulents) plants. These pathways describe uptake of atmospheric CO_2 by plants and the efficiency and isotopic fractionation associated with the diffusion and incorporation of their carbon source (O'Leary 1981).

The evolution of C_4 photosynthesis is thought to have developed to help plants cope with the environmental pressure of high aridity and excessively warm conditions that can lead to desiccation in C_3 plants (Ehleringer and Monson 1993; Monson 1989). Interestingly, this metabolic pathway appears to have arisen multiple times through geologic history in over 50 separate evolutionary lineages (Sage 2004; Sage et al. 2011). While the timing of emergence and expansion are debated, multiple lines of evidence indicate that the initial rise of C_4 plants occurred during the Cenozoic, with grasses emerging prior to sedges (Besnard et al. 2009; Sage et al. 2011). Evidence of C_4 photosynthesis in the Carboniferous has not been definitively identified (Cerling 1999), even though low CO_2 and high O_2 environments might have been ideally suited for the initial development of this metabolic pathway (Wright and Vanstone 1991). An expansion of C_4 photosynthesis during the Miocene does not appear to be globally synchronous, however, with evidence suggesting C_4 dominated landscapes occurred earlier in Africa then elsewhere, attributed to some combination of low pCO_2, arid conditions, or shifts in fire regimes (Tipple and Pagani 2007).

The isotopic fractionation between atmospheric carbon from CO_2 and bulk leaf tissue stable carbon isotope ($\delta^{13}C$) values (Farquhar et al. 1989) and compound specific $\delta^{13}C$ values of leaf waxes (Diefendorf and Freimuth 2017; Diefendorf et al. 2010) has been well documented. Carbon isotope use in long-chain leaf wax compounds relates to plant physiology, regional environmental parameters, and large-scale climate and atmospheric controls (Ehleringer et al. 1997; Farquhar et al. 1989). C_3 and C_4 plants both fix carbon using the Calvin-Benson pathway (biochemical reactions within the plants), where CO_2 and a 5-carbon sugar (ribulose bisphosphate, known as RuBP) are converted into 3-phosphoglycerate. This reaction converting CO_2 into sugars is catalyzed by the enzyme ribulose bisphosphate carboxylase oxygenase (RuBisCO), which creates a large fractionation against ^{13}C, 29‰ for higher plants (Farquhar et al. 1989), accounting for lower $\delta^{13}C$ values for plants than $\delta^{13}C$ of CO_2 in the atmosphere.

In C_4 plants, prior to the Calvin-Benson fixation, CO_2 is fixed via carboxylation of phosphoenolpyruvate (PEP) by PEP-carboxylase (PEP-C). C_4 efficiency in CO_2 uptake begins with the enzyme PEP-C, which has a strong preference for CO_2 and not O_2. The 4-carbon acid (oxaloacetate) that is produced is then reduced and moved from the outer cells (mesophyll cells) to the inner cells (bundle sheath cells). CO_2 is then released by decarboxylation, and then Calvin-Benson fixation begins. When temperatures are higher or atmospheric CO_2 concentrations are lower, RuBisCO can react with O_2, the process known as photorespiration, wasting some of the energy generated from photosynthesis. Today, this reduction by photorespiration is 30–40% of the photosynthetic rate (Ehleringer and Monson 1993; Pearcy and Ehleringer 1984). C_4 plants have a physical separation of cells (bundle and mesophyll cells) that keeps CO_2 stomatal uptake and CO_2 fixation separate. Thus, there is a separation of cells where CO_2 is concentrated, keeping nearby cells saturated in CO_2 and ensuring

RuBisCO uses CO_2 instead of O_2, eliminating the energy loss through photorespiration.

In low pCO_2 environments, C_4 plants have an advantage over C_3 plants due to the PEP-C enzyme and CO_2 concentrating mechanism (Ehleringer et al. 1997; Sage et al. 1999; Tipple and Pagani 2007), which leads to ^{13}C enrichment of their tissues compared to C_3 plants (tissues that have less negative $\delta^{13}C$ values because they have more ^{13}C and less ^{12}C). The isotopic difference is related to fractionation associated with RuBisCO discrimination and slow leak of enriched CO_2 from the leaf vascular tissue. C_4 plants decrease their stomatal opening, reducing transpiration while maintaining an equivalent amount of carbon fixation than C_3 plants could in similar environments. Ultimately, this results in higher water use efficiency in C_4 plants than C_3 plants, making C_4 plants better adapted to warmer, more arid conditions (Cerling et al. 1997; Ehleringer et al. 1997; Rommerskirchen et al. 2006). Therefore, C_4 vegetation dominates locations where these adaptations allow survival over C_3 vegetation, such as tropical and subtropical ecosystems (Sage 2004; Sage et al. 2011). However, in higher pCO_2 environments, C_4 vegetation is not favored due to the higher energy requirement necessary for this photosynthetic pathway (Ehleringer et al. 1997). Physiological advantages of C_3 and C_4 plants in different environments and variability of pCO_2 likely played a significant role in vegetation distribution over geologic intervals. For example, it is thought that C_4 vegetation (such as many tropical grasses) dominated the tropics during glacial intervals (Ehleringer et al. 1997). Additionally, within a given environment, C_3 plants seem to respond to the availability of water by producing lower $\delta^{13}C$ values in wetter conditions (Marshall et al. 2007). This may have an impact on vegetation $\delta^{13}C$ reconstructions during wetter and more arid intervals, even if vegetation type does not vary significantly.

The environmental and climate preferences of C_3 and C_4 vegetation allow for reconstruction of environments using stable isotopes. In sediments, this requires compound specific isotopes to distinguish terrestrial vegetation isotope variability from aquatic or bacterial inputs. Variations in $\delta^{13}C$ values have been used to reconstruct past landscape vegetation variability, by means of photosynthetic pathway variability of C_3 plants and C_4 plants. On average, the isotopic signature of plants utilizing these photosynthetic pathways is distinct (Cerling et al. 1997; Dawson et al. 2002; Farquhar et al. 1989). Modern bulk plant isotope values for C_3 are more depleted in ^{13}C (ranging from −19 to −35‰) relative to C_4 plants (ranging from −10 to −16‰) (Cerling and Harris 1999). These isotope values are also reflected in the isotopic values of individual compounds from plants. For example, plants using the C_4 pathway are ^{13}C enriched (more positive $\delta^{13}C$ values) relative to C_3 plants, with an idealized endmember value of ~−21.4‰ (for the C_{29} *n*-alkane) (Castañeda et al. 2009a). Plants utilizing the C_3 pathway are more ^{13}C depleted (more negative $\delta^{13}C$ values) on average than C_4 plants, with a cited endmember of −34.7‰ (C_{29} *n*-alkane) (Castañeda et al. 2009a).

CAM plants, the third photosynthetic pathway, keep stomata shut during the day but open at night, and isotopically resemble an intermediate of C_3 and C_4 $\delta^{13}C$ values (Deines 1980; Farquhar et al. 1989; O'Leary 1981; Osmond et al. 1973). This method of carbon-fixation is found primarily in arid desert plants, especially succulents. A large input of CAM plant waxes to the sediments would complicate the interpretation of C_3/C_4 vegetation reconstruction, as their intermediate isotope values are difficult to distinguish from these other plants (Osmond et al. 1973). However, overall *n*-alkyl production of CAM plants is low relative to C_3 and C_4 vegetation (Black Jr 1973), and thus reconstructions are expected to be unbiased from the insignificant contributions of CAM *n*-alkyl lipids with intermediate $\delta^{13}C$ values.

Simple two-member C_3/C_4 mixing models have been used to assess the relative contributions of vegetation types to sedimentary sequences (e.g., Berke et al. 2014; Schefuß et al. 2003; Sinninghe Damsté et al. 2011b). These estimates were more recently updated to include a ±20% uncertainty estimate based on a small dataset that showed the variability in endmember values (Castañeda et al. 2009a). A recent paper by Diefendorf et al. (2010) found a large range of $\delta^{13}C$ values for globally-distributed C_3 vegetation. Using >3,000 leaf measurements from trees and shrubs at 105 locations on 5 continents, the authors observed values that range from −34.9‰ to −21‰ for C_3 plants (Diefendorf et al. 2010). This finding suggests that a simple binary model cannot be used to assign a relative percentage of C_3/C_4 vegetation from $\delta^{13}C$ values unless there is other information to constrain vegetation type in a specific environment (Dawson et al. 2002; Diefendorf et al. 2010). While the range of variability of C_3 vegetation precludes the use of a single isotopic endmember, it does not prevent the use of $\delta^{13}C$ values to indicate relative trends in C_3/C_4 vegetation change on the landscape. For example, Cerling et al. (2011), showed that there was a nonlinear relationship between woody cover and $\delta^{13}C$ values of soil organic matter, due to C_3 plants being found in open or wooded environments. A large compilation of $\delta^{13}C$ values indicated a continuum between woody to herbaceous plants and C_3 to C_4 vegetation (Cerling et al. 2011). Following this, Magill et al. (2013a, b) compiled $\delta^{13}C$ values for soil organic matter, bulk leaves and leaf *n*-alkanes for 64 plant species and 288 soils from 289 subtropical and tropical geographic sites. The authors measured lipid-leaf fractionation factors and compared them to soil organic matter-leaf fractionation factors to establish relative C_3 and C_4 contribution shifts through time for hominin sites in northern East Africa (Fig. 8.2). For these sites, the study found the same fractionation between soil organic matter $\delta^{13}C$ (with respect to bulk leaf $\delta^{13}C$) and $\delta^{13}C$ of C_{31}

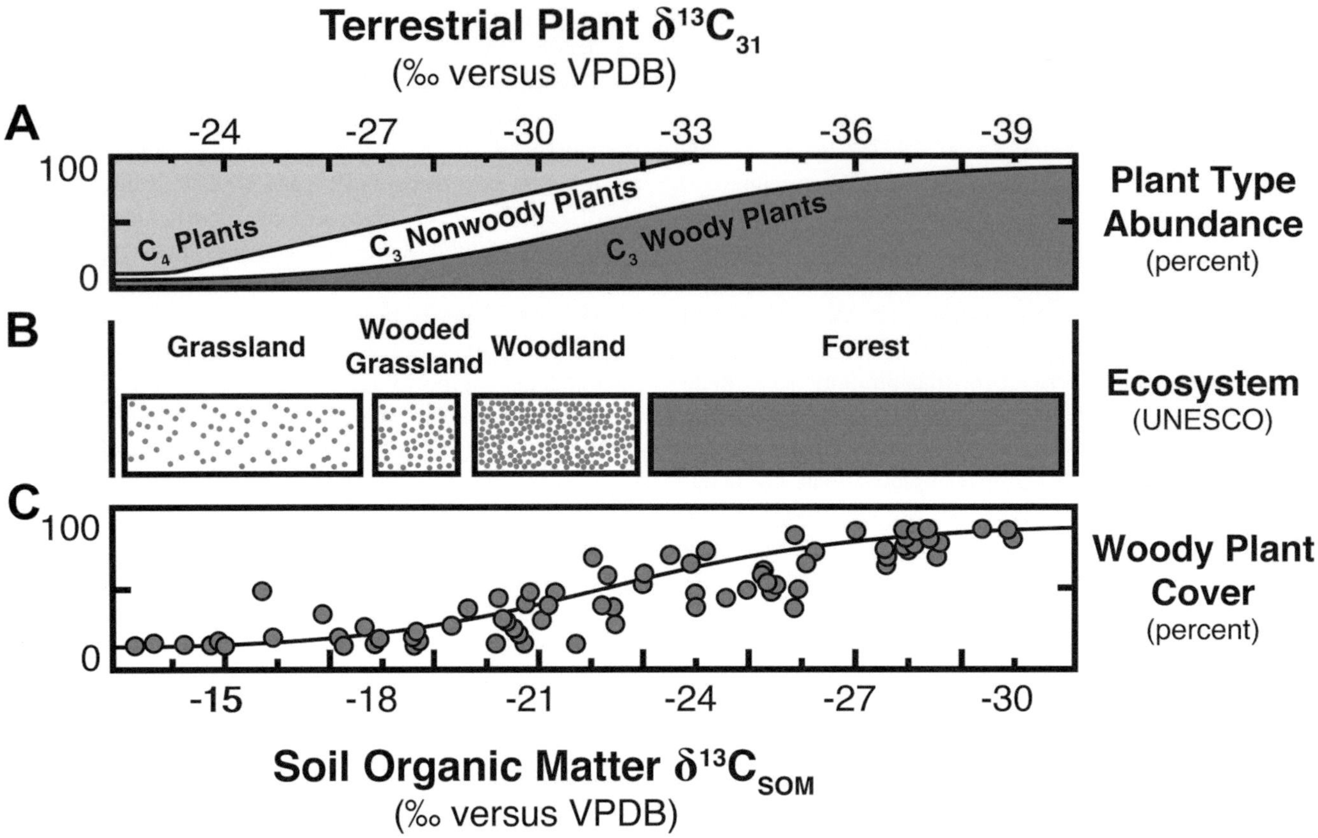

Fig. 8.2 A relationship for the Olduvai region of eastern Africa between $\delta^{13}C$ of soil organic matter (lower axis) and $\delta^{13}C$ of C_{31} plant wax alkane (upper axis) compared to: A, plant type (C_4 plants, C_3 nonwoody plants, and C_3 woody plants); B, UNESCO ecosystem designation; and C, percentage plant woody cover. The isotopic offset (fractionation) for C_3 and C_4 plants relative to soil organic matter were both 9‰, which allowed for this comparison. Reproduced from Magill et al. (2013a)

n-alkane (with respect to bulk leaf $\delta^{13}C$) for both C_3 plants and soils and C_4 plants and soils (9‰). This allows for the estimation of $\delta^{13}C$ of soil organic matter from leaf wax $\delta^{13}C$ and the prediction of woody cover and C_3/C_4 vegetation change from this region using the scheme shown in Fig. 8.2.

Another approach for C_3/C_4 vegetation estimation involved a mixing model but with a constrained third end-member, specifically considering pine (or general evergreen) waxes, for reconstructing vegetation shifts for ~62,000 years from a sedimentary record from Lake Tulane, Florida (Huang et al. 2006). In this case, pollen was used to assist in the reconstruction of vegetation shifts between C_3 and C_4. Huang et al. (2006) showed a 40% increase in C_4 vegetation during the Last Glacial Maximum (LGM, ~21,000 cal years B.P.) compared to the Holocene and attributed the increase to increased regional aridity. Although there is low production of *n*-alkanes from evergreens (Diefendorf et al. 2011), Huang et al. (2006) also showed an isotopic excursion of 4–5‰ interpreted as an increase in pine contribution to the system from the Holocene to present.

In another example, $\delta^{13}C$ was used to reconstruct vegetation variations across North America during the Miocene. In sediments from the Gulf of Mexico (DSDP Site 94), $\delta^{13}C$ of *n*-alkanes illustrated the possible evidence of C_4 vegetation prior to the Middle Miocene, with a significant rise in C_4 vegetation concurrent with the Middle Miocene Climatic Optimum (~14 Ma). Tipple et al. (2010) corrected $\delta^{13}C$ values of sedimentary wax compounds for shifting value of atmospheric CO_2 during the Miocene using reconstructed $\delta^{13}C_{CO2}$ from a benthic foraminifera reconstruction (Tipple et al. 2010; Tipple and Pagani 2010). C_{29} ranged between −26.2‰ and −30.9‰ and C_{31} ranged between −24.8 and −30.2‰. These authors posited that CO_2 wasn't necessarily the driver of C_4 vegetation expansion and that hydroclimate helped grassland expansion in North America (Tipple and Pagani 2010).

Recent research has shown that integration of multiple proxies, comparing biomarker stable isotopes with stable isotopes from other archives, can provide a powerful complimentary dataset of terrestrial changes from both marine and continental archives. Uno et al. (2016) observed $\delta^{13}C$ values from terrestrial leaf wax *n*-alkanes (C_{31} and C_{35}) extracted from a marine sediment core off the east coast of Somalia (in the Somali Basin) with a ^{13}C enrichment beginning at 10 Ma (Fig. 8.3). This was used as evidence for

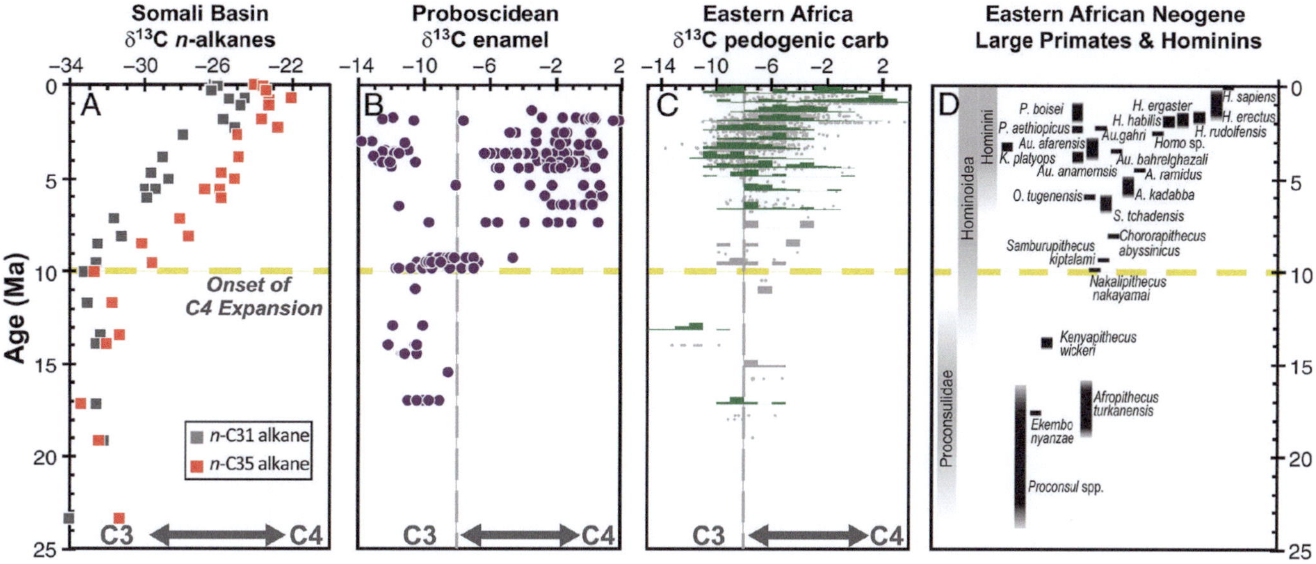

Fig. 8.3 Multiproxy data from eastern Africa showing comparison between leaf wax compound and: A, specific stable isotopes ($\delta^{13}C$ of C_{35} and C_{31} alkanes); B, tooth enamel stable isotopes ($\delta^{13}C$ eastern and central African proboscideans); C, soil carbonate stable isotopes ($\delta^{13}C$ from pedogenic carbonate in eastern Africa); and D, age ranges of Neogene primate taxa from eastern Africa. Reproduced from Uno et al. (2016). Data show good integration of terrestrial lipid isotopes, sourced from leaf wax lipids from marine sediment cores recovered from the Somali Basin, and terrestrial soil and fossil isotopes, sourced from around eastern and central Africa. Uno et al. (2016) interpreted these data as indicating the onset of C_4 vegetation expansion in the region, at 10 Ma, observed in ^{13}C enrichment (3‰) in leaf wax lipids and diet shift to inclusion of C_4 vegetation by proboscideans at this time

the expansion of C_4 vegetation in eastern Africa, alongside data from proboscidean fossil tooth $\delta^{13}C$ values (indicative of a diet shift from C_3 to some C_4 plants), and soil carbonate $\delta^{13}C$ values (a proxy of landscape vegetation). The agreement between these records strengthens the interpretations made and provides evidence of the way that biomarker proxies can complement other geochemical analyses for paleoenvironmental reconstruction.

Reconstructing Patterns of Aridity Using $\delta^{13}C$

In many carbon isotope studies from the tropics and subtropics, $\delta^{13}C$ values of leaf wax compounds have been used as a general indicator of aridity, due the adaptations of C_4 plants to hot, arid conditions (Sage et al. 1999). In these cases, leaf wax $\delta^{13}C$ is used as a general proxy for changes in landscape aridity based on the physiological advantage of the C_4 photosynthesis. C_3 plants dominate the global natural landscape productivity, but C_4 plants play an important role in specific ecosystems (Ehleringer et al. 1997). In the tropics and subtropics, two-thirds of all grasses are C_4 vegetation (Sage et al. 1999). Tropical savannas are also overwhelmingly dominated by C_4 plants, in this case composing more than 90% of the vegetation (Sage et al. 1999). These environmental advantages and skewed landscape distribution allow stable carbon isotopes, specifically proportions of C_4 vegetation, to be used as general proxies of aridity.

Leaf wax compound specific $\delta^{13}C$ is commonly used to reconstruct patterns of aridity across East Africa. For example, Castañeda et al. (2007) and Johnson et al. (2016) used changes in $\delta^{13}C$ of *n*-alkanes from Lake Malawi sediments to reveal aridity patterns in southeastern Africa. A record of $\delta^{13}C$ of the last 23 ka indicated that Lake Malawi was arid during the Last Glacial Maximum (~23–21 ka), Younger Dryas (~13–11.5 ka), early Holocene (11–7.7 ka) and Little Ice Age (~1300–1850 AD) (Castañeda et al. 2007). Extending the Lake Malawi record back to 1.3 Ma, Johnson et al. (2016) observed a progressive ^{13}C depletion that matched the calcium XRF (X-ray Fluorescence) data, with concentrations of calcium a proxy for open/closed status of the basin (Fig 8.4). Together, these data suggest an increase in wetter conditions in southeastern Africa for the last 1.3 Ma, punctuated by abrupt "mega-droughts" seen throughout the record in both proxies of aridity (Johnson et al. 2016). These records would suggest that during drier intervals, the proportion of C_4 vegetation was greater, using the C_3/C_4 relationship for East Africa proposed by Magill et al. (2013a) shown in Fig. 8.2.

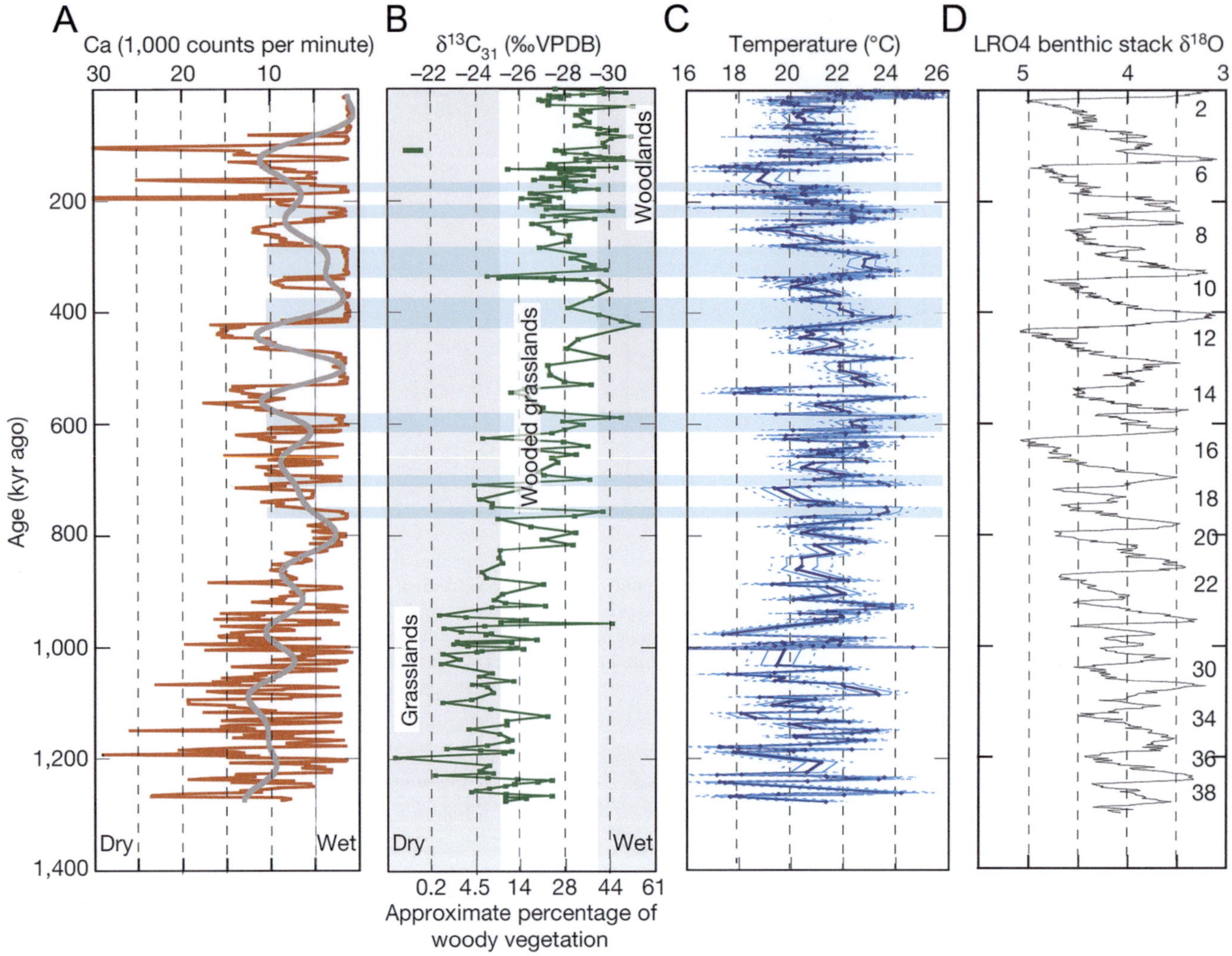

Fig. 8.4 Multiproxy records from Lake Malawi, East Africa, were interpreted to show a change from more arid conditions, to overall wetter climate with long periods of generally stable conditions in hydroclimate over the last 1.3 Ma (reproduced from Johnson et al. 2016). A, calcium concentrations by XRF analysis, with greater variability and higher amplitude shifts between low and high calcium abundances prior to 900,000 years ago indicating wetter and more arid conditions, respectively; B, $\delta^{13}C$ of C_{31} alkane with a trend towards wetter conditions and increased woodlands; C, application of TEX_{86} temperature reconstruction (with 1σ and 2σ uncertainties from analytical and lapse rate corrections in solid and dashed lines, respectively) from a lacustrine setting, with shaded horizontal bars across data panes indicating alignment of warmer, wetter conditions during some of the interglacial periods; D, LR04 stack of Lisiecki and Raymo (2005) showing benthic oxygen isotope stages (glacial proxy indicator) for comparison

δD of Leaf Wax Compounds for Hydroclimate

Evidence of dynamic water processes are preserved in the hydrogen isotopic composition of organic compounds from sediments. Water used in terrestrial plant photosynthesis can be traced back to the composition of meteoric water (Gat 1996), and thus can serve as a paleohydroclimate proxy. The δD values of long-chain, terrestrially-derived sedimentary leaf wax compounds show a strong linear relationship to precipitation δD values across a wide range of environments (Hou et al. 2008; Huang et al. 2004; Sachse et al. 2004). Meteoric water is the ultimate source of the hydrogen values of this organic matter, the isotopic composition of water is significantly modified by biological and physical processes (Sachse et al. 2012). Understanding these processes is critical in the interpretation of δD leaf wax values (Fig. 8.5).

The use of hydrogen isotopes of long-chain leaf compounds avoids complications associated with bulk organic matter hydrogen isotopes analyses. Specifically, establishing the relationship between measured δD of leaf wax compounds and meteoric water may be easier in regions with more well constrained vegetation types or similar fractionations. Long-chain leaf wax hydrogen isotope values (δD) reflect the plant source water with a certain offset that is

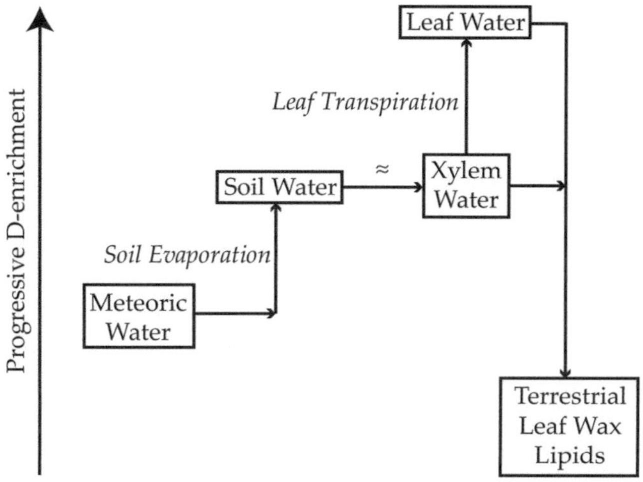

Fig. 8.5 Schematic figure illustrating potential fractionations between plant source water (precipitation) and leaf wax lipids from terrestrial plants used to reconstruct past vegetation and hydroclimate (not drawn to scale). Processes such as soil evaporation and leaf transpiration can act to alter hydrogen isotope values in leaf water, the ultimate source of water used in leaf wax lipid synthesis. Xylem water is often used as a proxy of soil water utilized by the plant due to a lack of fractionation in this uptake. Diagram modified after Sachse et al. (2006) and Smith and Freeman (2006)

dependent on physiological and environmental variables (Chikaraishi and Naraoka 2003; Sachse et al. 2004; Sauer et al. 2001; Sessions et al. 1999). However, even while constraining biosynthetic fractionation by focusing on specific lipids and homologues, there is additional complexity in reconstructing past precipitation based on lipid isotopes (Fig. 8.5). While applications for reconstructing past hydroclimate stress the importance of δD of long-chain leaf waxes as a powerful tool (e.g., Hou et al. 2008; Huang et al. 2004; Sauer et al. 2001), new examinations of modern plants have helped refine the relationship between source water and lipid δD in order to create more quantitative reconstructions of past hydroclimate variability (Sachse et al. 2012).

Variability in environmental water δD values reflects one of the largest determinates of δD values of leaf waxes. The geographic variability in source water can be summarized by the Rayleigh distillation processes (Craig 1961; Dansgaard 1964). Rayleigh distillation fractionates isotopes, with values more negative farther from the source of precipitation. The effects of atmospheric temperature, precipitation amount, latitude, and continentality on meteoric water isotope values (Dansgaard 1964) are reflected in δD values of leaf wax lipids. The temperature effect, predominantly seen outside of tropical regions, results in a strong correlation of isotopic variability and the associated temperature of rainout, such that there is a larger fractionation associated with colder temperatures. The amount effect is most evident in tropical regions where temperature associated fractionation is minor, shows strong correlation between D-depletion and higher rainout rates. Lastly, continentality, or distance the air mass travels from its source, results in a progressive shift towards lower δD values as movement of air masses travel inland. Across a wide range of geographic scales, there is a D-depletion of precipitation as latitude increases (Bowen and Revenaugh 2003; Craig and Gordon 1965).

Hydrogen isotope values of soil water used by plants can also be D-enriched relative to soil water through evaporative effects (Allison et al. 1983; Barnes and Allison 1983), and this process can be expressed in leaf wax lipids (Sachse et al. 2004; Sauer et al. 2001). Effects of soil evaporation are most evident in the uppermost sections of the soil, where the influence of evaporation is the major control on isotopic value beyond source water value. Initial studies of plant tissues found evidence of the influence of evaporation on plant water δD, particularly in desert plants with shallow rooting depths (Ehleringer and Dawson 1992; Ehleringer et al. 1991; Williams and Ehleringer 2000). Plants with deeper rooting depths, such as those in arid environments that are trying to tap into groundwater may not be affected by surface evaporation (Dawson 1993) or may switch between sources depending on the regularity of precipitation events (White et al. 1985). However, it is also important to note that deeply stored groundwater is often an integrated signal of precipitation (Berke et al. 2015a; Ehleringer et al. 1991).

Transpiration can also produce higher δD values in leaf waters relative to xylem and source waters (Feakins and Sessions 2010; Sachse et al. 2006; Smith and Freeman 2006). Fractionation of water isotopes has not been observed during uptake of soil water by roots in most plant types, confirming that xylem water can be used as a proxy for source water (Ehleringer and Dawson 1992; Ehleringer et al. 1991, 1998). However, after initial uptake, xylem water may be modified by water transpiration in the leaves from leaf stomata. The degree of isotope enrichment in plant leaves is controlled by physiological and environmental factors including vapor isotopic composition, relative humidity, and temperature. The relative importance of these processes is still debated, with studies interpreting the majority of meteoric water D-enrichment is from soil evaporation (Hou et al. 2008) or transpiration (Feakins and Sessions 2010; Yang et al. 2009).

In order to quantify the influence of environmental or physiological factors on leaf wax δD, fractionations between source water and lipids are often discussed. Net fractionation (also called apparent fractionation) between source water and plant wax lipid (Fig. 8.5) accounts for evapotranspiration fractionations and any biosynthetic fractionations (Sachse et al. 2006; Smith and Freeman 2006). Sedimentary basins are likely to integrate plant lipids of multiple plant life forms, with a mixture of different biotic fractionations. The differences in isotope fractionation among these plants in some cases may be exacerbated by climate variability across the integrating region. Significant variability in δD fractionation have been described within a single species and single chain length, from −204‰ to −34‰) (Sachse et al. 2012). These differences can be attributed to some combination of environmental and climate factors, including, but not limited to, microclimate, seasonality of wax synthesis, and canopy position of leaves (Sachse et al. 2012). It is important to compare δD values within a compound class and chain length because significant variability within a pathway has been observed due to large enzymatic effects (Chikaraishi et al. 2004, 2009).

Applications of δD for Hydroclimate Reconstruction

Fractionations in modern environments are understudied, leading to an incomplete understanding of the likely variations between source water and wax δD. In order to improve quantitative reconstruction of past hydroclimate, studies have utilized other means of constraining environmental and physiological variables, with independent vegetation reconstruction using pollen or carbonate analyses (Feakins 2013; Magill et al. 2013b).

While there are still many outstanding questions regarding the environmental and physiological factors influencing δD values of leaf wax compounds, the use of this proxy for qualitative hydroclimate variability continues to grow. For example, applications of δD for hydroclimate reconstruction in the tropics have been used to describe changes in the amount of precipitation through time. One of the earliest records to provide insights into hydroclimate of tropical Africa using δD values of leaf wax compounds was a marine sediment core off the coast of the Congolese River catchment (Schefuß et al. 2005a). This core spanned the last 20,000 years and contained waxes integrated from a large catchment covering much of central Africa. Schefuß et al. (2005b) interpreted the lower δD values of leaf wax compounds as indicating wetter conditions and higher δD values representing drier conditions in the Congolese basin. The authors concluded that the largest influence on the δD values was from the precipitation amount affect. The predominant control on the hydroclimate of central Africa for the last 20,000 years was shifting trade winds that would bring monsoonal moisture to the region and lower δD values (Schefuß et al. 2005a).

In another location from Africa, the variability in a sedimentary δD record from Lake Malawi was the result of shifting sources of region precipitation (Konecky et al. 2011). The 140,000 year record showed changes in atmospheric circulation over southeastern Africa, with lower δD values associated with Lake Malawi moisture derived from the descending limb of the proximal Hadley circulation (Konecky et al. 2011). Lake level proxies and δD values prior to 56,000 years ago show extreme variations in hydroclimate, and then show a shift to more stable, wetter conditions with progressive D-enriched leaf wax *n*-alkanoic acid values.

In high latitude settings, δD values have the potential to inform about both the isotopic composition of rainfall and temperature, because temperature and rain-out history are highly correlated (Dansgaard 1964; Jouzel et al. 1997). Modern measurements constrain the relationship between temperature and δD of precipitation and allowed Feakins et al. (2012) to estimate the isotopic value of paleoprecipitation (~−50‰) in the middle Miocene from 20 to 15.5 Ma on the margins of Antarctica using long-chain (C_{28}) *n*-alkanoic acids from sediment cores. Their reconstructions suggested that vegetation coverage on the landscape peaked during the warmest intervals during the Middle Miocene, from 16.4 to 15.7 Ma.

Plant Lignin

Lignin is an abundant component of wood and terrestrial plant tissues (Hedges and Mann 1979) and is well preserved in older sediments. Environmental reconstruction studies in need of additional vegetation differentiation and estimates of continental material transportation can be supplemented with lignin data. Of particular importance for paleoreconstruction is the relative abundances of three phenolic polymers and their ratios to one another: syringyl, vanillyl, and cinnamyl (Goni and Thomas 2000) (Fig. 8.6). Cinnamyl phenols are only found in non-woody tissues, in both gymnosperms and angiosperms (Hedges and Mann 1979). Syringyl phenols are overwhelmingly found in nonwoody and woody angiosperms tissues (Hedges and Mann 1979). Vanillyl phenols are the most general, found in all land plants. However, vanillyl phenols are the only lignin phenol found within woody gymnosperm plant tissues. As such, the ratio of cinnamyl/vanillyl phenols can be used for reconstructions of non-woody to woody vascular plant materials. The ratio of syringyl/vanillyl phenols can be used to look at the relative contributions of angiosperm and gymnosperm plants. These ratios have been used to distinguish between woody/non-woody, angiosperm/gymnosperm, and are particularly powerful when used in combination with $\delta^{13}C$ to distinguish between relative shifts in C_3 and C_4 vegetation. However, studies have suggested that while generally resistant to degradation, syringyl and cinnamyl are less resistant to oxic degradation than vanillyl (Goni et al. 1993). Any differential degradation would skew paleoreconstructions, and thus caution should be exercised where oxic degradation may be possible. Additionally, the degree to which this degradation has affected samples can be assessed by studying the ratio of acid phenols to aldehyde phenols (Hedges and Van Green 1982).

A

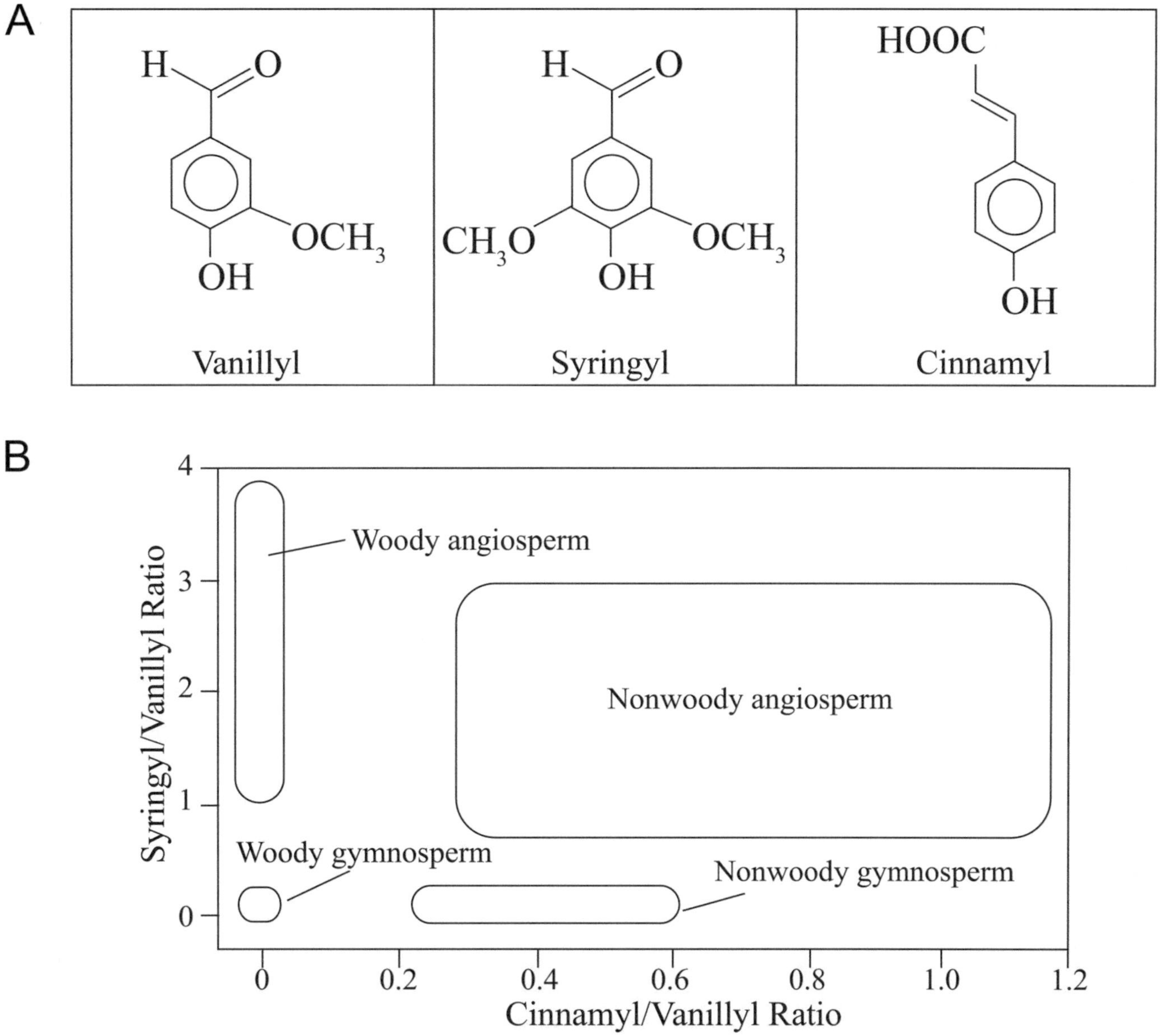

Fig. 8.6 A. Phenolic monomer structures from vascular plant tissues: cinnamyl, syringyl, and vanillyl. B. Lignin phenol crossplot of syringyl/vanillyl ratio to cinnamyl/vanillyl ratio. The crossplot of lignin phenol ratios can be useful in interpreting angiosperm or gymnosperm terrestrial vegetation as well as woody or nonwoody source vegetation. Modified from Hedges and Mann (1979)

Lignin Applications

Early studies of lignin often used phenol abundances to look at the transport of organic matter across the landscape. In the Columbia River drainage basin, lignin phenols were used to examine the accumulation of terrestrial plant matter gathering in the river sediments and study how closely the sedimentary tissues matched with vegetation for the local landscape (Hedges et al. 1984). The authors found large quantities of nonwoody angiosperm tissues in the Columbia River sediments, with vascular land plants composing ~90% of the organic matter found in coastal sediments. For Hu et al. (1999), interpretations of fossil pollen and lignin provided additional information for reconstruction. Their research of Alaskan lake sediments showed a strong similarity between both vegetation proxies; however, as specific pollen types are rich in only certain phenols they suggested that there may be a bias in the lignin record due to preservation. Castañeda et al. (2009b) used $\delta^{13}C$ and distributions of phenols to reconstruct paleovegetation of Lake Malawi sediments since the Late Pleistocene. These authors interpreted the Lake Malawi sedimentary record of cinnamyl/vanillyl phenols as primarily due to inputs of woody C_3 trees to C_4 non-woody grasses. Lignin results supported the $\delta^{13}C$ values from *n*-alkanes, which showed an increase of C_4 vegetation during dry intervals of the Last Glacial Maximum and Younger Dryas during the Holocene.

Proxies For Terrestrial Paleotemperature

Long-Chain Alkenones

Long-chain alkenones are one of the longest standing organic geochemical proxies for past temperature, but they are not found in all lacustrine settings and so are primarily applied for marine surface temperature reconstructions. Therefore, this review will cover fewer details of this proxy. The reader is encouraged to seek out specific reviews of alkenone temperature calibrations and reconstructions for further information (e.g., Herbert 2001).

The $U_{37}^{K'}$ Index is based on long-chain alkenones found in haptophyte algae that are comprised of C_{37}–C_{39} di- and tri-unsaturated ketones, where the degree of unsaturation has a strong positive relationship with growth temperature (Prahl and Wakeham 1987). The proxy is based on the relative proportions of these unsaturations, which are produced by haptophyte algae, primarily *Emiliania huxleyi*, but also by *Geophyrocapsa oceanica*, *Isochryss galbana*, and *Chrysotila lamellosa*. The $U_{37}^{K'}$ Index is defined as the ratio of abundances of C37:2/(C37:2 + C37:3), and is used to estimate surface water temperatures (Prahl et al. 1988) using a core top calibration that is based on modern sediment $U_{37}^{K'}$ Index values and modern water temperatures. It has been observed that $C_{37:4}$ and C_{38} homologues are the most abundant for lacustrine systems (Pearson et al. 2008; Zink et al. 2001), but more research is necessary to determine if this is a characteristic feature of all lacustrine basins. Marine temperature calibrations cannot be applied to lacustrine alkenone records because of the abundance of $C_{37:4}$ to $C_{37:2}$ provides temperature estimates that are too cold (Zink et al. 2001).

The limited presence of alkenones in lakes regionally as well as through time (Zink et al. 2001) and poor correlation with temperature in many terrestrial locations because of the lack of a broadly applicable calibration (Toney et al. 2010) have made this proxy of somewhat limited use in terrestrial (lacustrine) settings. Additionally, in both marine and lacustrine settings there are regions where it is difficult to apply the $U_{37}^{K'}$ Index because global temperatures are at the upper limit of the temperature range that can be measured using this proxy (~27–28°C) (Herbert 2001). However, terrestrial settings where alkenones have been found and used to reconstruct past temperature continue to increase (e.g., Innes et al. 1998; Jaraula et al. 2010; Toney et al. 2010). Where there is good correlation between lake surface temperatures and mean annual or some seasonal air temperature, alkenones can be used to reconstruct air temperature. This is likely to occur in systems where the haptophyte algae live in the uppermost water column or in systems where the basin is fairly homogeneous in temperature (even if only seasonally). Successful application of lacustrine alkenones is often done after determining the source organism and careful measurement of environmental parameters to establish a unique calibration for the site (D'Andrea et al. 2006, 2011).

Glycerol Dialkyl Glycerol Tetraethers (GDGTs)

Glycerol dialkyl glycerol tetraethers (GDGTs) are a major part of membrane lipids from specific marine and lacustrine Archaea and bacteria. The degree of cyclization (0–3 cyclopentane rings, GDGT I – III) increases with increasing growth temperature of the organisms, which can then be used to reconstruct temperatures for sites where they are found (Schouten et al. 2013 and references therein) (Fig. 8.7). Studies suggest that in hyperthermal environments many different GDGTs are formed from diverse Archaeal populations that do not have the same relationship between cyclization and temperature (Pearson et al. 2004; Schouten et al. 2007b). The phylum Thaumarchaeota and Euryarchaeota (Karner et al. 2001; Lipp et al. 2008) dominate in non-hyperthermal settings. Distinguishing between hyperthermophiles or methanogenic/methanotrophic Euryarchaeota (which do not show this temperature relationship) (Schouten et al. 2007b) and Thaumarchaeota is thought to be possible by looking at GDGT IV (often referred to as crenarchaeol), a GDGT suggested to be specific for Thaumarchaeota (Pitcher et al. 2011). The first application of GDGTs for paleoreconstruction using the TEX_{86} proxy described below was done in marine settings (Schouten et al. 2002), but subsequent confirmation of the presence of GDGTs with a correlation to air temperature in lakes was completed soon after (Blaga et al. 2009; Powers et al. 2004, 2005).

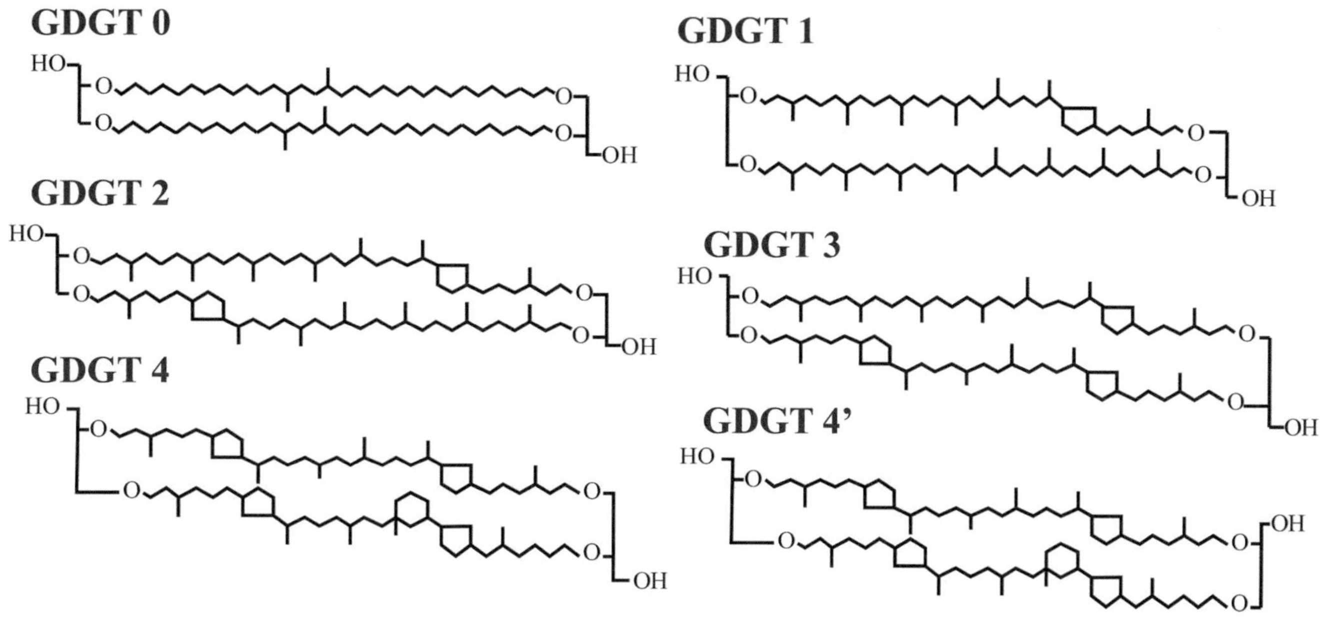

Fig. 8.7 Isoprenoid glycerol dialkyl glycerol tetraether (GDGTs) structures. GDGT 4 is crenarchaeol and GDGT 4' is crenarchaeol region-isomer

GDGTs, like leaf wax compounds, are relatively resistant to degradation, though there is evidence that they can undergo diagenetic alteration with time. Sedimentary evidence suggests that while GDGTs can degrade (Huguet et al. 2008), the TEX_{86} value and therefore the reconstructed temperatures, remain relatively unchanged within proxy error in sediments that have not undergone thermal maturity (Huguet et al. 2009; Schouten et al. 2004; Sinninghe Damsté et al. 2002a). Experimental research shows that more thermally mature sediments (>240°C) would have degraded GDGTs, with inhomogeneous alteration of GDGT structures and an overall decrease in TEX_{86} values (Schouten et al. 2004). Nevertheless, use of TEX_{86} as a temperature proxy has been widely successful since its discovery in 2002, supporting other temperature proxies where there is overlap in analyses, and with wider geographic and time application possibilities than many other paleoclimate proxies (Schouten et al. 2013; Spang et al. 2010).

TEX_{86} Paleotemperature Proxy

The proxy that utilizes the relative abundance of non-thermophilic, aquatically produced GDGTs by Thaumarchaeota is called the TetraEther Index of TetraEthers with 86 carbon atoms (TEX_{86}) (Schouten et al. 2002) (Fig. 8.7). TEX_{86} is an index that was mathematically defined by Schouten et al. (2002) as:

$$TEX_{86} = \frac{[\text{GDGT } 2 + \text{GDGT } 3 + \text{GDGT } 4']}{[\text{GDGT } 1 + \text{GDGT } 2 + \text{GDGT } 3 + \text{GDGT } 4']}$$

The strong linear relationship between measured water and air surface temperatures suggest they are related and that the former can serve as a proxy for the latter because similar aspects of heating and cooling affect both (Powers et al. 2005). There are a series of different empirical calibrations available, all of which use measured sediment core-tops to determine a regional TEX_{86} value compared to measured water temperatures. One of the more recent global marine core-top calibrations (Kim et al. 2008) utilized 223 core-top sediment samples to create the formula:

$$T = -10.78 + 56.2 \times \text{TEX}_{86} \left(r^2 = 0.94\right)$$

Recent marine calibrations have used specific calibrations for low and high temperature waters, which have been difficult to capture in previous calibrations (Kim et al. 2010). The newest calibration efforts utilize a Bayesian statistical approach (BAYSPAR) to integrate global core-top data (Tierney and Tingley 2014; Tierney and Tingley 2015).

While there is significantly less data in lacustrine TEX_{86} calibrations, the TEX_{86} values from certain large lakes also show good correlation with surface water temperatures. The linear calibration of core-top lake data from Powers et al. (2010a) was based on 12 globally distributed large lakes (r^2 = 0.86) while Tierney et al. (2010a) was based on 13 globally distributed lakes, removing cold water lakes and focusing only on lakes with temperatures today >10°C (r^2 = 0.92). However, when these lacustrine calibrations are combined and cold sites are included (Castañeda and Schouten 2011), a total of 19 lacustrine core-top calibrations yields a good correlation as well in the calibration:

$$T = 54.889 \times TEX_{86} - 13.363 (r^2 = 0.89)$$

In lakes with small watersheds relative to surface area, TEX_{86} appears to reconstruct temperatures well. In many smaller lakes, high amounts of soil or *in situ* bacterial production of GDGTs can result in a poor correlation to water temperature, as seen in a European survey of lakes (Blaga et al. 2009).

There is still uncertainty surrounding the source organisms generating GDGTs within the phylum Thaumarchaeota. Specifically, a lack of cultures of the source organisms has prevented a rigorous test of the correlation versus causation of temperature to increases in cyclopentane moieties. For example, new pure cultures of marine ammonia-oxidizing Archaea (AOA), members of Archaea that show GDGT relationship with temperature is not only strain-specific, but is influenced by concentrations of O_2 (Qin et al. 2015). This ecophysiological response means that for certain marine AOA, O_2 limitation can drive TEX_{86} inferred temperatures warmer than measured temperatures. Additionally, where these organisms live may be an added complication. Studies have suggested there may be small quantities of soil-produced GDGTs that have a different relationship to temperature that can get transported to marine or lacustrine sediments (Weijers et al. 2006b). However, there are additional proxies (i.e., the BIT Index) that may help constrain the possibility of soil isoprenoid influence.

Perhaps a larger concern is aquatic Thaumarchaeota occurrence or movement within the water column. Thought to be primarily nitrifiers, Thaumarchaeota have been noted to occupy a niche in timing during the year that may be related to the presence of ammonium in the water column (Konneke et al. 2005; Wuchter et al. 2006). This can then manifest itself as a seasonal cycle of Thaumarchaeota GDGTs, which may only capture a specific seasonal range of temperature and not an integrated annual temperature (Blaga et al. 2010; Sinninghe Damsté et al. 2009; Woltering 2011). Recent work has suggested that in addition to a seasonal bias, Thaumarchaeota may not restrict their movements to the upper photic zone, and thus represent temperatures not associated with surface waters. In fact, a wide range of studies have observed Thaumarchaeota spread throughout the water column (e.g., Castañeda et al. 2009a; Huguet et al. 2007; Woltering et al. 2012; Woltering 2011). However, a study by Wuchter et al. (2005) using sediment trap data found that GDGTs in sediments are sourced primarily from the upper water column, even though Thaumarchaeota live in much deeper waters. This may help explain how TEX_{86} temperatures are so well-correlated to surface water temperatures. Additional studies are necessary to provide confidence in the correspondence between surface temperatures and GDGT location of lipid generation.

TEX_{86} Applications

Even though there are a significant number of complications surrounding the use of TEX_{86} as a paleothermometer, there have been many successful climate reconstructions across the Cenozoic using this proxy based on comparisons with independent proxies of temperature where available. Many TEX_{86} records have been generated from the East African Rift Lakes, such as from Lakes Malawi (Fig. 8.4), Turkana, Victoria, Albert, Challa, and Tanganyika (e.g., Berke et al. 2012a, b, 2014; Johnson et al. 2016; Powers et al. 2005; Sinninghe Damsté et al. 2012; Tierney et al. 2008; Woltering et al. 2011). These records show some similarities to one another, such as the timing of warming following the Last Glacial Maximum (Sinninghe Damsté et al. 2012), but show variability among lakes including amplitude of temperature shifts, which may be related to regional climate complexities or calibration complexities. The oldest reconstructions using TEX_{86} are from Late Jurassic marine sedimentary deposits (Carrillo-Hernandez et al. 2003; Jenkyns et al. 2012). Further extension into the past is limited by the preservation of the GDGT lipids and likely not due to the presence of Archaea generating these lipids. One interval that has undergone the highest amount of scrutiny is the extreme, abrupt warmth at the Paleocene–Eocene Thermal Maximum (PETM) ($\sim$55 Ma). A significant warming of $\sim$5°C was described using the TEX_{86} paleothermometer at the PETM (Zachos et al. 2006), supported by more recent modeling and TEX_{86} calibrations (Dunkley Jones et al. 2013), which also coincides with a similar warming revealed using $\delta^{18}O$ of planktonic foraminifera. Contemporaneous Arctic Ocean warming of <5°C (Sluijs et al. 2006) and Southern Ocean warming (Sluijs et al. 2011) of $\sim$7°C has also been described using the TEX_{86} paleothermometer. In both high latitude cases, absolute temperature values have been difficult to capture with climate models (Sluijs et al. 2006).

Where available in marine settings, comparisons with other available temperature reconstructions such as $\delta^{18}O$ and Mg/Ca from foraminifera and $U^{K'}_{37}$ match well. Examples of excellently preserved GDGTs have allowed multiproxy temperature comparisons including TEX_{86} and $U^{K'}_{37}$ to be generated from another critical climate interval, the Eocene-Oligocene boundary. A significant 3–5°C cooling was described from globally-distributed marine sediments, with similar trends and magnitudes between TEX_{86} and $U^{K'}_{37}$ (Liu et al. 2009). Further, an Eocene-spanning TEX_{86} record from the Southern Ocean (Bijl et al. 2009) showed a similar record to the global ocean benthic $\delta^{18}O$ stack of Zachos et al. (2001).

Branched GDGTs

BIT Index: The branched to isoprenoid ratio of tetraethers, called BIT Index (Hopmans et al. 2004) relies on the relative abundances of branched GDGTs (brGDGTs) and crenarchaeol to estimate the ratio of soil to aquatic derived material entering a system (Hopmans et al. 2004) (Fig. 8.8). As such, the BIT Index has been used to estimate water runoff and landscape variability associated with water runoff through time. As previously described, crenarchaeol was thought to be produced solely by aquatic Thaumarchaeota, while brGDGTs are thought to be a marker for soil bacteria (Weijers et al. 2006b). However, our understanding of the physiological relationship of brGDGT proxies to the producer is incomplete as the majority of producing organisms have not been identified or cultured in a laboratory setting. The BIT Index (Hopmans et al. 2004) utilizes the equation:

$$BIT = \frac{[\text{GDGT I} + \text{GDGT II} + \text{GDGT III}]}{[\text{crenarchaeol} + \text{GDGT I} + \text{GDGT II} + \text{GDGT III}]}$$

BIT Index numbers therefore range from 0 (solely aquatically produced) to 1 (solely terrestrially/soil produced). In marine settings, BIT is often very low, unless samples are taken close to terrestrial sources (a river outflow, for example). This has allowed the BIT Index to be used for reconstruction of terrestrial runoff into sedimentary systems. In lacustrine systems, BIT values can often be high. In fact, in surveys of lake systems, >75% were found to have high BIT values, in this case defined as >0.5 (Blaga et al. 2010; Powers et al. 2010). Most of the lakes with lower BIT values are large lakes with smaller watersheds compared to their surface area (Blaga et al. 2010; Powers et al. 2010).

In addition to soil runoff into lakes leading to high BIT values, recent studies have found that brGDGTs are likely also produced within the water column. Suggestions of *in situ* production have come from varied global lakes (e.g., Blaga et al. 2009; Sinninghe Damsté et al. 2009; Tierney and Russell 2009). Observations of brGDGT abundances in deep waters that surpass those found in surface waters (Sinninghe Damsté et al. 2009) may support the idea that brGDGTs are being generated *in situ* by a yet-to-be-determined organism. In this case, care must be taken when using the BIT Index either as a proxy for sediment transport or as an indication of the validity of TEX_{86} temperatures.

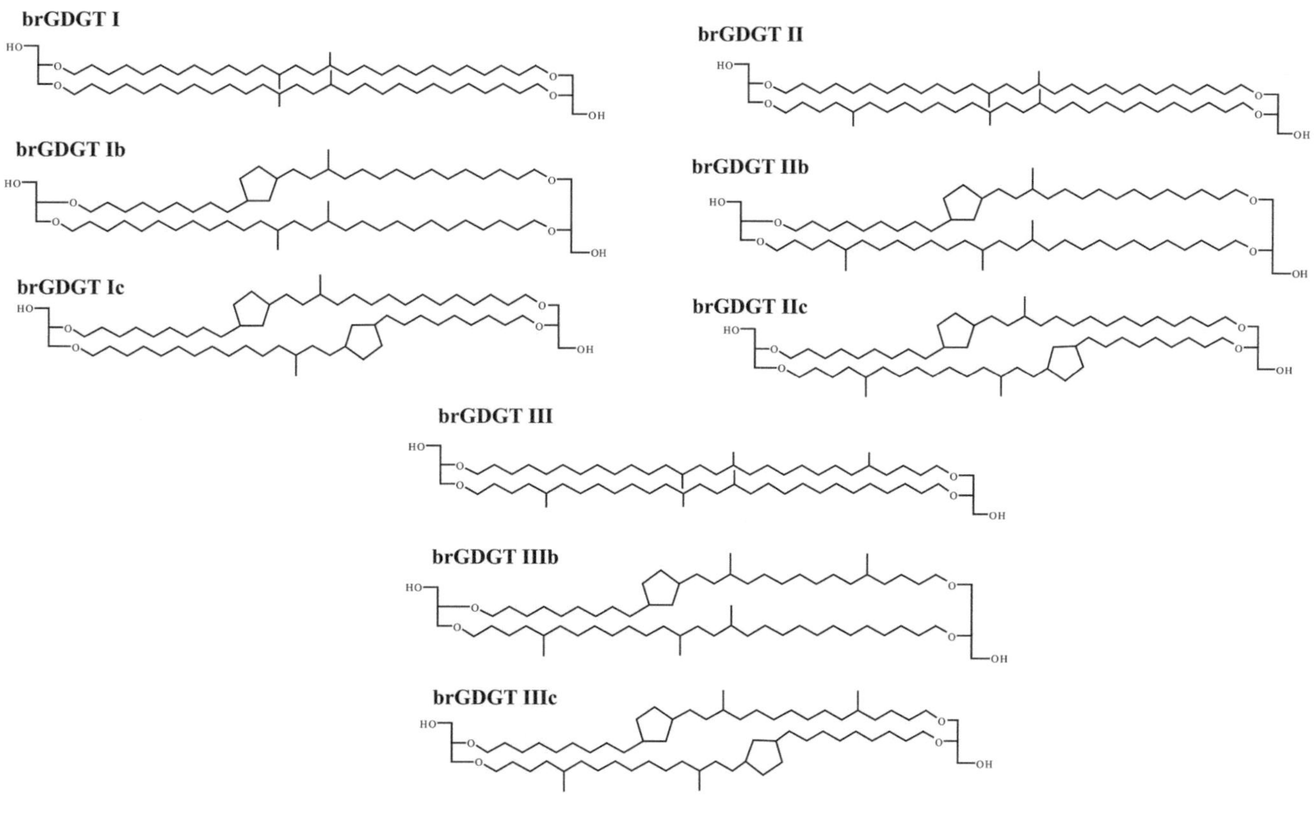

Fig. 8.8 Branched glycerol dialkyl glycerol tetraethers (brGDGT) structures

MBT/CBT: There are two additional proxies that utilize brGDGTs: the methylation of branched tetraethers (MBT) and cyclization of branched tetraethers (CBT) indices (Weijers et al. 2007). The brGDGTs have differing numbers of methyl branches and as many as two cyclopentane structures (Sinninghe Damsté et al. 2000; Weijers et al. 2006a). Initial findings showed that brGDGT distributions had a mathematical relationship to environmental parameters where the soils were collected: MBT providing a proxy of mean annual soil temperature and CBT providing a proxy of pH. Early study of these distributions found that MBT is correlated to soil temperature and also somewhat to soil pH (Weijers et al. 2007). Therefore, CBT, which can be used to reconstruct soil pH, is calculated first, based on the finding that there is a negative, exponential correlation of soil pH with the relative number of cyclopentane ring structures of brGDGTs (Weijers et al. 2007). This can then be used to correct for the effect of soil pH on soil temperature. Soil temperatures often reflect air temperatures, and thus MBT is often used as a proxy for mean annual air temperature (MAAT). The MBT and CBT indices are determined from the following equations, with roman numerals referring to structures found in Fig. 8.8:

$$MBT = \frac{[\text{GDGT I} + \text{GDGT Ib} + \text{GDGT Ic}]}{[\text{GDGT I} + \text{GDGT Ib} + \text{GDGT Ic} + \text{GDGT II} + \text{GDGT IIb} + \text{IIc} + \text{GDGT III} + \text{GDGT IIIb} + \text{GDGT IIIc}]}$$

$$CBT = -\log\left(\frac{(\text{GDGT Ib} + \text{GDGT IIb})}{(\text{GDGT IIb} + \text{GDGT IIc})}\right)$$

Based on a global soil dataset of 134 soils, the uncertainty for the MAAT calculation is ~ 5°C. The CBT and MBT Indices were defined by Weijers et al. (2007) as:

$$CBT = (3.33 - 0.38 \times \text{pH})$$

$$MBT = (0.122 + 0.187 \times \text{CBT} + 0.20 \times \text{MAAT}).$$

While brGDGTs are known to be bacterial in origin (Weijers et al. 2006a), it is unclear what bacteria are making these structures (Sinninghe Damsté et al. 2011a), which can lead to production uncertainties and difficulties in interpretation of results. Specifically, MBT/CBT looked to be a promising method to reconstruct air temperatures in small lake systems where TEX_{86} did not work because soil input was high. Observations from individual lakes support *in situ* production as well as evidence of transported soils. Blaga et al. (2009) found indication of *in situ* brGDGT production within lakes of varying sizes in Europe. Measuring lacustrine sediments and surrounding catchment soils, Tierney and Russell (2009) found that the pH reconstructed using the CBT index from the sediments of an Indonesian lake reflected soil pH values of the watershed. This site showed abundances of brGDGTs that were higher in lake sediments than in the surrounding catchment soils (Tierney and Russell 2009). This likely resulted in colder reconstructed MBT temperatures than MAAT would predict. Nevertheless, brGDGTs from soil and lake production may still produce reasonable temperature reconstructions. In lakes with water temperatures similar to air temperatures and fairly homogeneous water column temperatures, a mixture of aquatically and soil produced brGDGTs may complicate reconstruction but may still provide a good estimate of temperatures of the past. Studies have found that this mixed signal tends to result in an underestimation of MAAT (Blaga et al. 2010; Sun et al. 2011; Tierney et al. 2010b).

An outstanding question in the use of MBT/CBT is whether global or more regional calibrations are the key to achieve the most accurate paleotemperature reconstruction (Peterse et al. 2009; Sinninghe Damsté et al. 2008). For example, a calibration of East African lakes for MBT/CBT used 46 different sites to establish a regional calibration of brGDGTs to MAAT with an r^2 of 0.94, showing a significant improvement over the global calibration for this region (Tierney et al. 2010b). These authors observe temperature and pH control of brGDGTs in the sediments of East African lakes. However, Peterse et al. (2012), in redefining the global calibration for MBT/CBT, tested whether a local calibration is necessary to improve the MBT/CBT global calibration. Using residual temperatures (measured – reconstructed MAAT) for soils in the global dataset, these authors found a lack of systematic variation associated with specific regions, suggesting that a local calibration would not benefit MBT/CBT reconstructions (Peterse et al. 2012). The updated calibration of Peterse et al. (2012) utilizes the initial dataset with a total of 278 global soils and resulted in new equations with a lower correlation for MBT to MAAT ($r^2 = 0.59$ instead of $r^2 = 0.77$ described in Weijers et al. (2007). The CBT was then defined as: pH = 7.90–1.97 × CBT and MBT (now called MBT' and based on only 7 most abundant brGDGTs): MAT = 0.81–5.67 × CBT + 31.0 × MBT' (Peterse et al. 2012).

Recently, new analytical advances have improved chromatography and have helped to further refine the calibration of brGDGTs and provided much needed information about the brGDGT structures (De Jonge et al. 2013; Weber et al. 2015; Yang et al. 2015). Importantly, De Jonge et al. (2013) was able to quantify the 5-methyl and 6-methyl isomers of the penta-methylated and hexa-methylated brGDGTs and found that while the 6-methyl isomers were highly correlated with pH, the 5-methyl isomers were not. Thus, the MBT temperature calibration has been freed from the dependency on pH while improving the calibration (De Jonge et al. 2013). Additionally, abundances of the different 6-methyl brGDGTs are correlated positively to one another, suggesting a common biological source in terrestrial deposits (De Jonge et al. 2014), an important discovery in identifying biological controls and origins in addition to appropriate calibrations. Similarly, the presence of 6-methyl brGDGTs lowered the correlation with MAAT in the original brGDGT MBT/CBT calibration, and this improved understanding of structural information and calibrations has reduced error associated with these reconstructions of temperature, with a reduction of the root mean square error (RMSE) from 6.2 to 4.8°C (De Jonge et al. 2014).

MBT/CBT Applications: Using brGDGTs for independent temperature reconstructions and comparing these results to microfossil or pollen generated temperatures resulted in strong correlation between proxies. Bijl et al. (2013) used both MBT and pollen to reconstruct temperatures from the Early Eocene Climatic Optimum (52–50 Ma) to the continental-scale glaciation of Antarctica beginning around 34 Ma. Using marine sediment cores that span the early to middle Eocene from the Wilkes Land Margin, East Antarctica, reconstructed sea-surface temperatures, determined by TEX_{86}, cooled 2−4°C and air temperatures, determined by MBT and corroborated by pollen assemblages, decreased by 4°C (Bijl et al. 2013) during the opening of the Tasmanian Gateway. The cooling suggested by the MBT reconstruction coincides with the termination of the Early Eocene Climatic Optimum (EECO) at ~49−50 Ma and through-flow of the westbound Antarctic Counter Current, supporting the interpretation that the opening of this seaway was influential in the climate events of this transition.

PETM deposits were an initial target of the MBT paleotemperature proxy, but early reconstructions appeared too warm (Weijers et al. 2007). A newer calibration by Peterse et al. (2012) decreased the absolute temperature of these reconstructed temperatures by ~2°C, while the trend remained the same. The PETM MBT temperatures of Peterse et al. (2012) agree with previous estimates of MAAT based on fossil leaf margins (Wing et al. 2005), pollen assemblages, and $\delta^{18}O$ of fossil bone material from the Arctic (Eberle et al. 2010).

Lastly, an example from Pleistocene lacustrine sediments from the American southwest (Valles Caldera, New Mexico) showed significant glacial/interglacial temperature variability utilizing MBT reconstructions of MAAT (Fawcett et al. 2011). MAAT reconstructions from MIS 10−14 (~368−552 ka) showed swings between warmer interstadials and cooler stadials, which corresponded to pollen assemblage types (such as an increase in *Quercus* and *Juniperus* pollen in warm intervals) and closed/open basin conditions, as illustrated by scanning XRF calcium records (with closed conditions indicated by high calcium) (Fawcett et al. 2011). The alignment of these records and agreed upon interpretation between proxies lends additional credence to the use of MBT for MAAT.

Analytical Approach

Preparation and Extraction

The analytical sections of this chapter are far from a complete guide to all aspects of preparation, analysis, and interpretation of lipid biomarkers. Instead, this section introduces some of the most important analytical concepts with further details available in Killops and Killops (2005) and Peters et al. (2005), essential guides to the field of organic geochemistry.

All analyses described here require that laboratory materials in contact with biomarker archives (e.g., sediments, soils, rocks) must either be purchased at the highest purity grade possible (such as HPLC-grade solvents), or be pre-extracted or distilled (lower solvent grade with initial possible contaminants). Following rinsing (acid wash is not necessary), all glass materials must be placed in a muffle furnace at 400−500°C for at least 4 hours, until possible organic contaminants are removed. Glass and metal materials in contact with sediments can be solvent rinsed with decreasing polarity solvents (for example, methanol followed by hexane). Aluminum foil can either be cleaned of its organic coating by placing in the muffle furnace as above or by solvent rinsing. Chromatographic column gels that can withstand high temperatures (such as alumina and silica) can either be placed in the muffle furnace or solvent rinsed prior to use. No plastics can be utilized with procedures described, though polytetrafluorethylene (PTFE/Teflon) tubing or squirt bottles are often used with satisfactory results. Cutting rock samples or cores must be done with no petroleum based lubricants, relying solely on clean water for blade lubrication. Depending on the preference of the laboratory, nitrile gloves are worn to protect samples from contamination.

However, nitrile gloves are also a source of contamination, especially if wetted with organic solvents, and so care must be taken whether gloves are worn or not to minimize contamination. Sampling from cores should be done near the middle of the core, to avoid possible contamination. This contamination could be from a core liner, especially if plastic, or potentially from drilling fluids (Brocks et al. 2008). This includes drilling fluids recirculating around the cores. However, the middle of the core may be uncontaminated depending on lithology permeability.

Extraction of molecular biomarkers from geologic matrices can be completed using a few different methods. Here the focus will be on methods most often used for general extraction of sediments and soils, with reviews for more specialized extractions, such as intact polar membrane lipids (e.g., Lengger et al. 2012), found elsewhere. Sediments are dried using a freeze dryer to ease grinding and solvent penetration into the material and then homogenized using a solvent-cleaned mortar and pestle or rock pulverizing equipment. A total lipid extract can be isolated from sediment using soxhlet extraction, ultrasonication, accelerated solvent extraction (ASE) or microwave extraction devices. ASE and microwave extraction techniques are both rapid, highly automated, and use much smaller quantities of solvent than more traditional extraction methods such as soxhlet (Kornilova and Rosell-Melé 2003). Often these methods are noted to have higher recovery of analytes (Kornilova and Rosell-Melé 2003; Pastor et al. 1997). However, these methods of lipid extraction involve an initial large cost purchasing these devices, and time required for post-extraction disassembly and cleaning of parts. Additionally some more labile compounds cannot be efficiently extracted using these more "harsh" extraction techniques (e.g., bacteriohopanepolyols (Talbot and Farrimond 2007) and membrane lipids of GDGTs with intact polar head groups (Huguet et al. 2010; Lengger et al. 2012; Pitcher et al. 2009). Soxhlet extraction is slow (inhibited by the number of soxhlet refluxing stations, and the 24–72 hour refluxing time) and require more solvents than other methods described, but is in general a more gentle extraction method. Ultrasonication requires multiple intervals in an ultrasonic bath followed by centrifuging and decanting. The ultimate extraction efficiency of each method is best tested by adding internal standard to samples prior to extraction. In all extraction methods, a mixture of polar and less polar solvents should be used. An extraction blank should be run and checked for contamination for each set of extracted samples.

Lipid Extract Separations

Following organic solvent extraction, a series of chromatographic columns can be run to separate the lipid extract into compound classes based on polarity. Chromatographic columns are often run utilizing gravity and a stationary phase, such as silica or alumina, and organic solvents of increasing polarity, following the principle that only compound classes that can be dissolved in that polarity solvent will be moved from the column bed. Compound classes or groups of similar polarities are then eluted into separate vials (Wakeham and Pease 1992). Analysis of total lipid extracts from geologic materials without column chromatography runs the risk of compound co-elution or a lack of baseline separation during gas chromatography, and thus, poor data quality.

Gas Chromatography

Analysis via a gas chromatograph (GC) works by separating complex mixtures of organic compounds through the interaction with the stationary phase of the fused silica GC column. This necessitates that compounds be GC-amenable, namely, that they are volatile and thermally stable for analysis. Compounds that are not volatile (too polar or containing functional groups) must be chemically altered to be analyzed, usually by derivatization (Meier-Augenstein 2004).

Following column chromatography in the laboratory, there is still a complicated mixture of compounds that can be separated based on different boiling points and polarity. Samples are dissolved in a solvent, commonly hexane or ethyl acetate, and are introduced to the system via syringe. This solvent is then vaporized at the injector. Compounds carried by this vapor elute following their travel and interaction through the column. Retention time depends on the flow rate of the carrier gas (usually H_2 or He), the temperature program, and the physical properties of the GC column

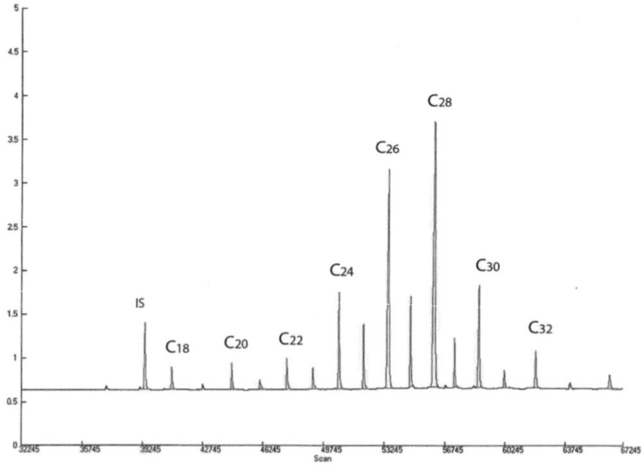

Fig. 8.9 Partial gas chromatogram example showing leaf wax *n*-alkanoic acid, showing even/odd chain length dominance). A known quantity and concentration of internal standard (IS) is added to help quantify compounds by comparison of curve areas

(coating thickness and chemistry, column length, and inner diameter). Compound elution is signified by peaks emerging from the detector, as a plot of time versus intensity, the total series of which is called a chromatogram (Fig. 8.9). Peak height or elution time has no direct relationship to amount of compound or identity of compound without standard or additional analyses. Ideally, increasing column length, slowing flow rate, or otherwise altering the program method so there is more time for compounds to interact with the column may help baseline separation. However, there is a limit to instrumental changes that can be made and still preserve the chromatography.

The GC can be coupled to a series of different detectors. The most common of these is the Flame Ionization Detector (FID). The GC-FID can be used to determine the amount of hydrocarbons exiting the column through pyrolysis by a flame. The resulting area under the peak can be integrated and compared to a known amount of internal standard or standard calibration curves for quantification. GC-FID analysis is highly sensitive with low detection limits. Unlike mass spectrometry (described below), there is little variation of the detector due to compound class, though linearity cannot be assumed.

A GC coupled to a mass spectrometer (GC-MS) allows for the identification of individual compounds of interest. After traveling through the GC column, vaporized compounds are ionized by electron bombardment. This ionization fragments the molecules of the compounds into smaller charged pieces which are then sorted according to their mass to charge ratio using a magnetic field. Often an electron multiplier is necessary to detect these fragments, but once detected, they can be identified based on the way in which they fragmented (charged ions), which serves as a characteristic fingerprint of the molecule in question. A total ion current (TIC) is generated from a full scan spectra of the relative abundance of detected ions as a function of the mass-to-charge ratio. This is similar to an FID trace where all ions are plotted versus time.

One of the most commonly utilized methods for determining the compound specific stable isotopes described in this chapter (δD or $\delta^{13}C$) requires a GC coupled to an Isotope Ratio Mass Spectrometer (GC-IRMS) (Brand 2004; Matthews and Hayes 1978; Sessions 2006). This continuous flow method of compound measurement for light stable isotopes has allowed larger numbers of samples to be run with high precision and accuracy and allows different measurements to be made on the same compound (currently requiring multiple injections and switching between cups and analysis modes as described below). The IRMS contains an electron ionization source, magnetic analyzer, Faraday collector cup array, and acquisition system (Sessions 2006). After passing through the capillary column of the GC, organic compounds enter either enter the combustion or pyrolysis reactor. The combustion reactor is an alumina tube containing Cu, Ni, and Pt wires and is often operated ~900°C. Under oxidizing conditions in the combustion reactor, organic compounds are converted to CO_2 gas. The pyrolysis reactor (also known as thermal conversion) is an open alumina tube and is operated at higher temperatures, often between 1400−1450°C. Organic compounds pass through this reducing reactor and are converted to H_2 gas. After the CO_2 or H_2 passes through the ion source, the resulting gas is then separated by the magnetic analyzer and detected by the Faraday cups. Care must be taken not to fractionate during analysis, which can occur at the inlet if heavy and light isotopes are not moved through the inlet uniformly, or with poor chromatography or peak integration. Typical precision for $^{2}H/^{1}H$ analysis is ~2−5‰ and for $^{13}C/^{12}C$ analysis is 0.1−0.3‰ (Sessions 2006).

In order to allow for comparison between systems and across laboratories, measured values must be compared to international, universal standards, Vienna Pee Dee Belemnite (VPDB) for carbon and Vienna Standard Mean Ocean Water (VSMOW) for hydrogen (Coplen 1988; Coplen 2011).

High Performance Liquid Chromatography-Mass Spectrometry

As mentioned, compounds must be amenable or made amenable through derivatization or methylation procedures to be analyzed on a GC. For those compounds that cannot be vaporized at GC inlet temperatures below ~350°C, usually complex and very polar molecules such as phospholipids and GDGTs, high performance liquid chromatography coupled to an MS (HPLC-MS) can be used (Sturt et al. 2004). These less volatile compounds can be carried by a mobile phase that is injected onto the HPLC stainless steel column. This column is filled with fine silica particles (the stationary phase) that help to separate compounds through their interaction (similar to GC columns described previously). Eluting components from the column are introduced to the MS via an interface, most commonly either via electrospray ionization or atmospheric pressure chemical ionization. During electrospray ionization, the analyte solution has a charge applied to it, forcing it from the electrospray needle. Smaller droplets are formed and continue to travel away from the needle until being detected by the MS, which is aided by the charge the analytes carry. Atmospheric pressure chemical ionization, commonly used for GDGT analysis (Schouten et al. 2007a), is not appropriate for thermally labile compounds, instead being used for low mass compounds. At atmospheric pressure, ionization occurs differently from the electrospray technique described above by the introduction of a pneumatic nebulizer that helps to create ions from the analytes introduced. Solvent molecules in vapor form and initial ions undergo repeated collision, eventually forming analyte ions with little to no fragmentation.

Conclusions

The abundance, diversity, and refractory nature of biomarkers and archives in which they are found make it clear why these chemical fossils have become an important fixture in paleoenvironmental reconstructions. With continual new proxy developments, refined understanding and calibration of existing proxies, and recent analytical advances, biomarker reconstructions of Cenozoic terrestrial environments are likely to continue to serve a critical role in our understanding of past climates and environments. Exciting new developments in biomarker research today are built on decades of foundational exploration, starting with exploration of an organic origin for sedimentary compounds, continuing to the pioneering early work on terrestrial plant wax proxies in the mid-twentieth century, and continuing with ongoing instrumentation and calibration advances, such as those in the realm of the terrestrial temperature proxies of TEX_{86} and MBT. In many cases, there is still much research to be done to uncover linkages between biological sources and environmental response, and what information is necessarily recorded in the lipids that are used for reconstruction. Nevertheless, preserved sedimentary biomarkers provide robust tools to understand Earth's past changes and can be used individually or in multiproxy investigations to provide essential records of the past.

In this chapter, we examined the origins of the term biomarker and the discipline looking to link sedimentary molecules to specific organisms or processes. We explored the wider topic of organic matter preservation, where sediments are likely to contain a record of past environmental variability and the conditions and locations that allow for preferential organic matter preservation and minimal molecular degradation. The chapter discussed some of the biomarker proxies used to reconstruct terrestrial paleoenvironments, and focused on their robustness in the geologic record, first use of the proxy, original and current calibrations, and potential pitfalls in usage. A few applications of each proxy were described to provide examples of how these biomarker proxies have been applied. Reconstructions using leaf wax lipid carbon chain lengths and carbon and hydrogen stable isotopes were described as proxies of landscape change and preservation, aridity and vegetation changes, and hydroclimate variability, respectively. Plant lignin was introduced as a proxy for woody/non-woody and angiosperm/gymnosperm variations. Temperature reconstructions using alkenones, GDGTs, and brGDGTs from haptophyte algae, archaea, and bacteria, respectively, were discussed. Lastly, analytical considerations and instrumentation were discussed as they related to the proxies covered in the chapter.

References

Aichner, B., Herzschuh, U., & Wilkes, H. (2010). Influence of aquatic macrophytes on the stable carbon isotopic signatures of sedimentary organic matter in lakes on the Tibetan Plateau. *Organic Geochemistry, 41,* 706–718.

Allison, G. B., Barnes, C. J., & Hughes, M. W. (1983). The distribution of deuterium and ^{18}O in dry soils 2. Experimental. *Journal of Hydrology, 64,* 377–397.

Barnes, C. J., & Allison, G. B. (1983). The distribution of deuterium and ^{18}O in dry soils: 1. Theory. *Journal of Hydrology, 60,* 141–156.

Bastow, T. P., Singh, R. K., van Aarssen, B. G. K., Alexander, R., & Kagi, R. I. (2001). 2-Methylretene in sedimentary material: a new higher plant biomarker. *Organic Geochemistry, 32,* 1211–1217.

Berke, M., Tipple, B., Hambach, B., & Ehleringer, J. (2015a). Life form-specific gradients in compound-specific hydrogen isotope ratios of modern leaf waxes along a North American monsoonal transect. *Oecologia, 179,* 1–17.

Berke, M. A., Bush, R. T., Cartagena-Sierra, A., Cheah, D., Costello, C., Muldoon, T., & Tillema, M. (2015b). Environmental and physiological controls on plant leaf wax δD from western greenland. *Proceedings American Geophysical Union,* Abstract PP33A-2282.

Berke, M. A., Johnson, T. C., Werne, J. P., Grice, K., Schouten, S., & Sinninghe Damsté, J. S. (2012a). Molecular records of climate variability and vegetation response since the Late Pleistocene in the Lake Victoria basin, East Africa. *Quaternary Science Reviews, 55,* 59–74.

Berke, M. A., Johnson, T. C., Werne, J. P., Livingstone, D. A., Grice, K., Schouten, S., et al. (2014). Characterization of the last deglacial transition in tropical East Africa: insights from Lake Albert. *Palaeogeography, Palaeoclimatology, Palaeoecology, 409,* 1–8.

Berke, M. A., Johnson, T. C., Werne, J. P., Schouten, S., & Sinninghe Damsté, J. S. (2012b). A mid-Holocene thermal maximum at the end of the African Humid Period. *Earth and Planetary Science Letters, 351–352,* 95–104.

Berner, R. A. (1989). Biogeochemical cycles of carbon and sulfur and their effect on atmospheric oxygen over phanerozoic time. *Global and Planetary Change, 1,* 97–122.

Besnard, G., Muasya, A. M., Russier, F., Roalson, E. H., Salamin, N., & Christin, P.-A. (2009). Phylogenomics of C4 photosynthesis in sedges (Cyperaceae): multiple appearances and genetic convergence. *Molecular Biology and Evolution, 26,* 1909–1919.

Bijl, P. K., Bendle, J. A. P., Bohaty, S. M., Pross, J., Schouten, S., Taux, L., Stickley, C. E., McKay, R. M., Röhl, U., Olney, M., Sluijs, A., Escutial, C., Brinkhuis, H., & Expedition 318 Scientists, (2013). Eocene cooling linked to early flow across the Tasmanian Gateway. *Proceedings of the National Academy of Sciences, USA, 110,* 9645–9650.

Bijl, P. K., Schouten, S., Brinkhuis, H., Sluijs, A., Reichart, G. J., & Zachos, J. C. (2009). Early Palaeogene temperature evolution of the Southwest Pacific Ocean. *Nature, 461,* 776–779.

Black Jr, C. C. (1973). Photosynthetic carbon fixation in relation to net CO_2 uptake. *Annual Review of Plant Physiology, 24,* 253–286.

Blaga, C., Reichart, G.-J., Heiri, O., & Sinninghe Damsté, J. (2009). Tetraether membrane lipid distributions in water-column particulate matter and sediments: a study of 47 European lakes along a north–south transect. *Journal of Paleolimnology, 41,* 523–540.

Blaga, C. I., Reichart, G.-J., Schouten, S., Lotter, A. F., Werne, J. P., Kosten, S., et al. (2010). Branched glycerol dialkyl glycerol tetraethers in lake sediments: can they be used as temperature and pH proxies? *Organic Geochemistry, 41,* 1225–1234.

Blumer, M., Guillard, R. R. L., & Chase, T. (1971). Hydrocarbons of marine plankton. *Marine Biology, 8,* 183–189.

Blyth, A. J., Baker, A., Collins, M. J., Penkman, K. E. H., Gilmour, M. A., Moss, J. S., et al. (2008). Molecular organic matter in speleothems and its potential as an environmental proxy. *Quaternary Science Reviews, 27,* 905–921.

Bowen, G. J., & Revenaugh, J. (2003). Interpolating the isotopic composition of modern meteoric precipitation. *Water Resources Research, 39,* 1299.

Brand, W. A. (2004). In P. A. de Groot (Ed.), *Handbook of stable isotope analytical techniques* (pp. 835–856). Amsterdam: Elsevier.

Bray, E. E., & Evans, E. D. (1961). Distribution of n-paraffins as a clue to recognition of source beds. *Geochimica et Cosmochimica Acta, 22,* 2–15.

Brittingham, A., Hren, M., & Hartman, G. (2017). Microbial alteration of the hydrogen and carbon isotopic composition of n-alkanes in sediments. *Organic Geochemistry, 107,* 1–8.

Brocks, J. J., Grosjean, E., & Logan, G. A. (2008). Assessing biomarker syngeneity using branched alkanes with quaternary carbon (BAQCs) and other plastic contaminants. *Geochimica et Cosmochimica Acta, 72,* 871–888.

Brocks, J. J., & Summons, R. E. (2003). Sedimentary hydrocarbons, biomarkers for early life. In W. Schlesinger (Ed.), *Treatise on geochemistry 8,* (pp. 63–115). Amsterdam: Elsevier.

Burdige, D. J. (2007). Preservation of organic matter in marine sediments: controls, mechanisms, and an imbalance in sediment organic carbon budgets? *Chemical Reviews, 107,* 467–485.

Bush, R. T., & McInerney, F. A. (2013). Leaf wax n-alkane distributions in and across modern plants: implications for paleoecology and chemotaxonomy. *Geochimica et Cosmochimica Acta, 117,* 161–179.

Bush, R. T., & McInerney, F. A. (2015). Influence of temperature and C4 abundance on n-alkane chain length distributions across the central USA. *Organic Geochemistry, 79,* 65–73.

Calvert, S. E., & Pedersen, T. F. (1993). Geochemistry of recent oxic and anoxic marine sediments: implications for the geological record. *Marine Geology, 113,* 67–88.

Canfield, D. E. (1994). Factors influencing organic carbon preservation in marine sediments. *Chemical Geology, 114,* 315–329.

Carr, A. S., Boom, A., & Chase, B. M. (2010). The potential of plant biomarker evidence derived from rock hyrax middens as an indicator of palaeoenvironmental change. *Palaeogeography, Palaeoclimatology, Palaeoecology, 285,* 321–330.

Carrillo-Hernandez, T., Schaeffer, P., Adam, P., Albrecht, P., Derenne, S., & Largeau, C. (2003). Remarkably well-preserved archaeal and bacterial membrane lipids in 140 million year old sediments from the Russian platform (Kashpir Oil Shales, Upper Jurassic). In *Proceedings, 21st International Meeting on Organic Geochemistry* (pp. 77–78). Krakow.

Carroll, A. R., & Bohacs, K. M. (2001). Lake-type controls on petroleum source rock potential in nonmarine basins. *AAPG Bulletin, 85,* 1033–1053.

Castañeda, I. S., Mulitza, S., Schefuß, E., Lopes dos Santos, R. A., Sinninghe Damsté, J. S., & Schouten, S. (2009a). Wet phases in the Sahara/Sahel region and human migration patterns in North Africa. *Proceedings of the National Academy of Sciences, USA, 106,* 20159–20163.

Castañeda, I. S., & Schouten, S. (2011). A review of molecular organic proxies for examining modern and ancient lacustrine environments. *Quaternary Science Reviews, 30,* 2851–2891.

Castañeda, I. S., Werne, J., & Johnson, T. C. (2007). Wet and arid phases in the southeast African tropics since the Last Glacial Maximum. *Geology, 35,* 823–826.

Castañeda, I. S., Werne, J. P., Johnson, T. C., & Filley, T. R. (2009b). Late Quaternary vegetation history of southeast Africa: the molecular isotopic record from Lake Malawi. *Palaeogeography, Palaeoclimatology, Palaeoecology, 275,* 100–112.

Cerling, T. E. (1999). Paleorecords of C_4 plants and ecosystems. In R. F. Sage & R. K. Monson (Eds.), *C_4 plant biology*. San Diego, CA, USA: Academic Press.

Cerling, T. E., & Harris, J. M. (1999). Carbon isotope fractionation between diet and bioapatite in ungulate mammals and implications for ecological and paleoecological studies. *Oecologia, 120,* 347–363.

Cerling, T. E., Harris, J. M., MacFadden, B. J., Leakey, M. G., Quade, J., Eisenmann, V., et al. (1997). Global vegetation change through the Miocene/Pliocene boundary. *Nature, 389,* 153–158.

Cerling, T. E., Wynn, J. G., Andanje, S. A., Bird, M. I., Korir, D. K., Levin, N. E., et al. (2011). Woody cover and hominin environments in the past 6 million years. *Nature, 476,* 51–56.

Chase, B. M., Scott, L., Meadows, M. E., Gil-Romera, G., Boom, A., Carr, A. S., et al. (2012). Rock hyrax middens: a palaeoenvironmental archive for southern African drylands. *Quaternary Science Reviews, 56,* 107–125.

Chikaraishi, Y., & Naraoka, H. (2003). Compound-specific δD-$\delta^{13}C$ analyses of *n*-alkanes extracted from terrestrial and aquatic plants. *Phytochemistry, 63,* 361–371.

Chikaraishi, Y., & Naraoka, H. (2007). $\delta^{13}C$ and δD relationships among three *n*-alkyl compound classes (*n*-alkanoic acid, *n*-alkane and *n*-alkanol) of terrestrial higher plants. *Organic Geochemistry, 38,* 198–215.

Chikaraishi, Y., Naraoka, H., & Poulson, S. R. (2004). Hydrogen and carbon isotopic fractionations of lipid biosynthesis among terrestrial (C3, C4 and CAM) and aquatic plants. *Phytochemistry, 65,* 1369–1381.

Chikaraishi, Y., Tanaka, R., Tanaka, A., & Ohkouchi, N. (2009). Fractionation of hydrogen isotopes during phytol biosynthesis. *Organic Geochemistry, 40,* 569–573.

Cole, J. J., Prairie, Y. T., Caraco, N. F., McDowell, W. H., Tranvik, L. J., Striegl, R. G., et al. (2007). Plumbing the global carbon cycle: integrating inland waters into the terrestrial carbon budget. *Ecosystems, 10,* 172–185.

Coley, P. (1983). Herbivory and defensive characteristics of tree species in a lowland tropical forest. *Ecological Monographs, 53,* 209–233.

Collister, J. W., Lichtfouse, E., Hieshima, G., & Hayes, J. M. (1994a). Partial resolution of sources of n-alkanes in the saline portion of the Parachute Creek Member, Green River Formation (Piceance Creek Basin, Colorado). *Organic Geochemistry, 21,* 645–659.

Collister, J. W., Rieley, G., Stern, B., Eglinton, G., & Fry, B. (1994b). Compound-specific $\delta^{13}C$ analyses of leaf lipids from plants with differing carbon dioxide metabolisms. *Organic Geochemistry, 21,* 619–627.

Coolen, M. J. L., & Overmann, J. (1988). Analysis of subfossil molecular remains of purple sulfur bacteria in a lake sediment. *Applied and Environmental Microbiology, 64,* 4513–4521.

Coplen, T. B. (1988). Normalization of oxygen and hydrogen isotope data. *Chemical Geology: Isotope Geoscience Section, 72,* 293–297.

Coplen, T. B. (2011). Guidelines and recommended terms for expression of stable-isotope-ratio and gas-ratio measurement results. *Rapid Communications in Mass Spectrometry, 25,* 2538–2560.

Craig, H. (1961). Isotopic variations in meteoric waters. *Science, 133,* 1702–1703.

Craig, H., & Gordon, L. I. (1965). Deuterium and oxygen 18 variations in the ocean and marine atmosphere. In E. Tongiogi (Ed.), *Proceedings of a conference on stable isotopes in oceanographic studies and paleotemperatures* (pp. 9–130) Spoleto, Italy: V. Lishi e F.

Cranwell, P. A. (1973). Chain-length distribution of n-alkanes from lake sediments in relation to post-glacial environmental change. *Freshwater Biology, 3,* 259–265.

Cranwell, P. A. (1981). Diagenesis of free and bound lipids in terrestrial detritus deposited in a lacustrine sediment. *Organic Geochemistry, 3,* 79–89.

Cranwell, P. A. (1984). Lipid geochemistry of sediments from Upton Broad, a small productive lake. *Organic Geochemistry, 7,* 25–37.

Cranwell, P. A., Eglinton, G., & Robinson, N. (1987). Lipids of aquatic organisms as potential contributors to lacustrine sediments. *Organic Geochemistry, 11,* 513–527.

D'Andrea, W. J., Huang, Y., Fritz, S. C., & Anderson, N. J. (2011). Abrupt Holocene climate change as an important factor for human migration in West Greenland. *Proceedings of the National Academy of Sciences, USA, 108,* 9765–9769.

D'Andrea, W. J., Lage, M., Martiny, J. B. H., Laatsch, A. D., Amaral-Zettler, L. A., Sogin, M. L., et al. (2006). Alkenone producers inferred from well-preserved 18S rDNA in Greenland lake sediments. *Journal of Geophysical Research-Biogeosciences, 111,* G03013.

D'Anjou, R. M., Bradley, R. S., Balascio, N. L., & Finkelstein, D. B. (2012). Climate impacts on human settlement and agricultural activities in northern Norway revealed through sediment biogeochemistry. *Proceedings of the National Academy of Sciences, USA, 109,* 20332–20337.

Dansgaard, W. (1964). Stable isotopes in precipitation. *Tellus, 16,* 436–468.

Dawson, T. E. (1993). Water sources of plants as determined from xylem-water isotopic composition: perspectives on plant competition, distribution, and water relations. In J. R. Ehleringer, A. E. Hall & G. D. Farquhar (Eds.), *Stable isotopes and plant carbon-water relations* (pp. 465–496) San Diego: Academic Press.

Dawson, T. E., Mambelli, S., Plamboeck, A. H., Templer, P. H., & Tu, K. P. (2002). Stable isotopes in plant ecology. *Annual Review of Ecology and Systematics, 33,* 507–559.

De Jonge, C., Hopmans, E. C., Stadnitskaia, A., Rijpstra, W. I. C., Hofland, R., Tegelaar, E., et al. (2013). Identification of novel penta- and hexamethylated branched glycerol dialkyl glycerol tetraethers in peat using HPLC–MS, GC–MS and GC–SMB-MS. *Organic Geochemistry, 54,* 78–82.

De Jonge, C., Hopmans, E. C., Zell, C. I., Kim, J.-H., Schouten, S., & Sinninghe Damsté, J. S. (2014). Occurrence and abundance of 6-methyl branched glycerol dialkyl glycerol tetraethers in soils: implications for palaeoclimate reconstruction. *Geochimica et Cosmochimica Acta, 141,* 97–112.

De Leeuw, J. W., & Largeau, C. (1993). A review of macromolecular organic compounds that comprise living organisms and their role in kerogen, coal, & petroleum formation. In M. Engel & S. Macko (Eds.), *Organic geochemistry* (Vol. 11, pp. 23–72). US: Springer.

Dean, W. E., & Gorham, E. (1998). Magnitude and significance of carbon burial in lakes, reservoirs, and peatlands. *Geology, 26,* 535–538.

Deines, P. (1980). The isotopic composition of reduced organic carbon. In P. Fritz & J. C. Fontes (Eds.), *Handbook of environmental geochemistry: The terrestrial environment* (Vol. 1, pp. 329–406). Amsterdam: Elsevier.

Diefendorf, A. F., Freeman, K. H., & Wing, S. L. (2012). Distribution and carbon isotope patterns of diterpenoids and triterpenoids in modern temperate C_3 trees and their geochemical significance. *Geochimica et Cosmochimica Acta, 85,* 342–356.

Diefendorf, A. F., Freeman, K. H., Wing, S. L., & Graham, H. V. (2011). Production of *n*-alkyl lipids in living plants and implications for the geologic past. *Geochimica et Cosmochimica Acta, 75,* 7472–7485.

Diefendorf, A. F., & Freimuth, E. J. (2017). Extracting the most from terrestrial plant-derived n-alkyl lipids and their carbon isotopes from the sedimentary record: a review. *Organic Geochemistry, 103,* 1–21.

Diefendorf, A. F., Mueller, K. E., Wing, S. L., Koch, P. L., & Freeman, K. H. (2010). Global patterns in leaf 13C discrimination and implications for studies of past and future climate. *Proceedings of the National Academy of Sciences, USA, 107,* 5738–5743.

Diefendorf, A. F., Sberna, D. T., & Taylor, D. W. (2015). Effect of thermal maturation on plant-derived terpenoids and leaf wax n-alkyl components. *Organic Geochemistry, 89–90,* 61–70.

Dunkley Jones, T., Lunt, D. J., Schmidt, D. N., Ridgwell, A., Sluijs, A., Valdes, P. J., et al. (2013). Climate model and proxy data constraints on ocean warming across the Paleocene-Eocene Thermal Maximum. *Earth-Science Reviews, 125,* 123–145.

Eberle, J. J., Fricke, H. C., Humphrey, J. D., Hackett, L., Newbrey, M. G., & Hutchison, J. H. (2010). Seasonal variability in Arctic temperatures during early Eocene time. *Earth and Planetary Science Letters, 296,* 481–486.

Eglinton, G., & Hamilton, R. J. (1963). The distribution of alkanes. In T. Swaine (Ed.), *Chemical plant taxonomy* (pp. 87–217). US: Academic.

Eglinton, G., & Hamilton, R. J. (1967). Leaf epicuticular waxes. *Science, 156,* 1322–1335.

Eglinton, T. I., & Eglinton, G. (2008). Molecular proxies for paleoclimatology. *Earth and Planetary Science Letters, 275,* 1–16.

Ehleringer, J. R., Cerling, T. E., & Helliker, B. R. (1997). C_4 photosynthesis, atmospheric CO_2, and climate. *Oecologia, 112,* 285–299.

Ehleringer, J. R., & Dawson, T. E. (1992). Water uptake by plants: perspectives from stable isotopes. *Plant, Cell and Environment, 15,* 1073–1082.

Ehleringer, J. R., Evans, R. D., & Williams, D. (1998). Assessing sensitivity to change in desert ecosystems—A stable isotope approach. In H. Griffiths (Ed.), *Stable isotopes: Integration of biological, ecological, and geochemical processes* (pp. 223–237). Oxford: BIOS Scientific Publishing.

Ehleringer, J. R., & Monson, R. K. (1993). Evolutionary and ecological aspects of photosynthetic pathway variation. *Annual Reviews of Ecological Systematics, 24,* 411–439.

Ehleringer, J. R., Phillips, S. L., Schuster, W. S. F., & Sandquist, D. R. (1991). Differential utilization of summer rains by desert plants. *Oecologia, 88,* 430–434.

Farquhar, G. D., Ehleringer, J. R., & Hubick, K. T. (1989). Carbon isotope discrimination and photosynthesis. *Annual Review of Plant Physiology and Plant Molecular Biology, 40,* 503–537.

Fawcett, P. J., Werne, J. P., Anderson, R. S., Heikoop, J. M., Brown, E. T., Berke, M. A., et al. (2011). Extended megadroughts in the southwestern United States during Pleistocene interglacials. *Nature, 470,* 518–521.

Feakins, S. J. (2013). Pollen-corrected leaf wax D/H reconstructions of northeast African hydrological changes during the late Miocene. *Palaeogeography, Palaeoclimatology, Palaeoecology, 374,* 62–71.

Feakins, S. J., & Sessions, A. L. (2010). Controls on the D/H ratios of plant leaf waxes in an arid ecosystem. *Geochimica et Cosmochimica Acta, 74,* 2128–2141.

Feakins, S. J., Warny, S., & Lee, J.-E. (2012). Hydrologic cycling over Antarctica during the middle Miocene warming. *Nature Geoscience, 5,* 557–560.

Ficken, K. J., Li, B., Swain, D. L., & Eglinton, G. (2000). An n-alkane proxy for the sedimentary input of submerged/floating freshwater aquatic macrophytes. *Organic Geochemistry, 31,* 745–749.

Gaines, S. M., Eglinton, G. & Rullkotter, J. (2008). *Echoes of life: What fossil molecules reveal about Earth history.* Oxford University Press.

Gat, J. R. (1996). Oxygen and hydrogen isotopes in the hydrologic cycle. *Annual Review of Earth and Planetary Sciences, 24,* 225–262.

Goni, M. A., Nelson, B., & Blanchette, R. A. (1993). Fungal degradation of wood lignins: geochemical perspectives from CuO-derived phenolic dimers and monomers. *Geochimica et Cosmochimica Acta, 57,* 3985–4002.

Goni, M. A., & Thomas, K. A. (2000). Sources and transformations of organic matter in surface soils and sediments from a tidal estuary (North Inlet, South Carolina, USA). *Estuaries, 23,* 548–564.

Grice, K., & Eiserbeck, C. (2014). The analysis and application of biomarkers. In H. Holland & K. Turekian (Eds.), *Treatise on geochemistry* (pp. 47–77). Oxford: Elsevier.

Gupta, N. S., Collinson, M. E., Briggs, D. E. G., Evershed, R. P., & Pancost, R. D. (2006). Reinvestigation of the occurrence of cutan in plants: implications for the leaf fossil record. *Paleobiology, 32,* 432–449.

Gupta, N. S., Michels, R., Briggs, D. E. G., Collinson, M. E., Evershed, R. P., & Pancost, R. D. (2007). Experimental evidence for the formation of geomacromolecules from plant leaf lipids. *Organic Geochemistry, 38,* 28–36.

Hallmann, C., Schwark, L., & Grice, K. (2008). Community dynamics of anaerobic bacteria in deep petroleum reservoirs. *Nature Geoscience, 1,* 588–591.

Hautevelle, Y., Michels, R., Malartre, F., & Trouiller, A. (2006). Vascular plant biomarkers as proxies for palaeoflora and palaeoclimatic changes at the Dogger/Malm transition of the Paris Basin (France). *Organic Geochemistry, 37,* 610–625.

Hedges, J. I., & Keil, R. G. (1995). Sedimentary organic matter preservation: an assessment and speculative synthesis. *Marine Chemistry, 49,* 81.

Hedges, J. I., & Mann, D. C. (1979). The characterization of plant tissues by their lignin oxidation products. *Geochimica et Cosmochimica Acta, 43,* 1803–1807.

Hedges, J. I., Turin, H. J., & Ertel, J. R. (1984). Sources and distributions of sedimentary organic matter in the Columbia River drainage basin, Washington and Oregon. *Limnology and Oceanography, 29,* 35–46.

Hedges, J. I., & Van Green, A. (1982). A comparison of lignin and stable carbon isotope compositions in Quaternary marine sediments. *Geochimica et Cosmochimica Acta, 11,* 43–54.

Herbert, T. D. (2001). Review of alkenone calibrations (culture, water column, and sediments). *Geochemistry, Geophysics, and Geosystems, 2,* 1055.

Hopmans, E. C., Weijers, J. W. H., Schefuß, E., Herfort, L., & Sinninghe Damsté, J. S. (2004). A novel proxy for terrestrial organic matter in sediments based on branched and isoprenoid tetraether lipids. *Earth and Planetary Science Letters, 224,* 107–116.

Hou, J., D'Andrea, W. J., & Huang, Y. (2008). Can sedimentary leaf waxes record D/H ratios of continental precipitation? Field, model, & experimental assessments. *Geochimica et Cosmochimica Acta, 72,* 3503–3517.

Hu, F. S., Hedges, J. I., Gordon, E. S., & Brubaker, L. B. (1999). Lignin biomarkers and pollen in postglacial sediments of an Alaskan lake. *Geochimica et Cosmochimica Acta, 63,* 1421–1430.

Huang, Y., Shuman, B., Wang, Y., & Webb III, T. (2004). Hydrogen isotope ratios of individual lipids in lake sediments as novel tracers of climatic and environmental change: a surface sediment test. *Journal of Paleolimnology, 31,* 363–375.

Huang, Y., Shuman, B., Wang, Y., Webb III, T., Grimm, E. C., & Jacobson, J. (2006). Climatic and environmental controls on the variations of C_3 and C_4 plant abundances in central Florida for the past 62,000 years. *Palaeogeography, Palaeoclimatology, Palaeoecology, 237,* 428–435.

Huguet, C., de Lange, G. J., Gustafsson, Ö., Middelburg, J. J., Sinninghe Damsté, J. S., & Schouten, S. (2008). Selective preservation of soil organic matter in oxidized marine sediments (Madeira Abyssal Plain). *Geochimica et Cosmochimica Acta, 72,* 6061.

Huguet, C., Kim, J.-H., de Lange, G. J., Sinninghe Damsté, J. S., & Schouten, S. (2009). Effects of long term oxic degradation on the UK' 491 37, TEX86 and BIT organic proxies. *Organic Geochemistry, 40,* 1188–1194.

Huguet, C., Martens-Habbena, W., Urakawa, H., Stahl, D. A., & Ingalls, A. E. (2010). Comparison of extraction methods for quantitative analysis of core and intact polar glycerol tetraethers (GDGTs) in environmental samples. *Limnology and Oceanography: Methods, 8,* 127–145.

Huguet, C., Schimmelmann, A., Thunell, R., Lourens, L. J., Sinninghe Damsté, J. S., & Schouten, S. (2007). A study of the TEX86 paleothermometer in the water column and sediments of the Santa Barbara Basin, California. *Paleoceanography*, *22*, PA3203.

Inagaki, F., Hinrichs, K.-U., Kubo, Y., Bowles, M. W., Heuer, V. B., Hong, W.-L., et al. (2015). Exploring deep microbial life in coal-bearing sediment down to ~2.5 km below the ocean floor. *Science, 349,* 420–424.

Innes, H. E., Bishop, A. N., Fox, P. A., Head, I. M., & Farrimond, P. (1998). Early diagenesis of bacteriohopanoids in recent sediments of Lake Pollen, Norway. *Organic Geochemistry, 29,* 1285–1295.

Jaraula, C. M. B., Brassell, S. C., Morgan-Kiss, R. M., Doran, P. T., & Kenig, F. (2010). Origin and tentative identification of tri to pentaunsaturated ketones in sediments from Lake Fryxell, East Antarctica. *Organic Geochemistry, 41,* 386–397.

Jenkyns, H. C., Huybers-Schouten, L., Schouten, S., & Sinninghe Damsté, J. S. (2012). Middle Jurassic-Early Cretaceous high-latitude sea-surface temperatures from the Southern Ocean. *Climate of the Past, 8,* 215–226.

Johnson, T. C., Werne, J. P., Brown, E. T., Abbott, A., Berke, M., Steinman, B. A., et al. (2016). A progressively wetter climate in southern East Africa over the past 1.3 million years. *Nature, 537,* 220–224.

Jouzel, J., Alley, R. B., Cuffey, K. M., Dansgaard, W., Grootes, P., Hoffmann, G., et al. (1997). Validity of the temperature reconstruction from water isotopes in ice cores. *Journal of Geophysical Research, 102,* 26471–26487.

Karner, M., DeLong, E. F., & Karl, D. M. (2001). Archaeal dominance in the mesopelagic zone of the Pacific Ocean. *Nature, 409,* 507–510.

Keil, R. G., Tsamakis, E., Fuh, C. B., Giddings, C., & Hedges, J. I. (1994). Mineralogical and textural controls on the organic composition of coastal marine sediments: hydrodynamic separation using SPLITT-fractionation. *Geochimica et Cosmochimica Acta, 58,* 879–893.

Killops, S. & Killops, V. (2005). *Introduction to organic geochemistry.* Blackwell Publishing.

Kim, J.-H., Schouten, S., Hopmans, E. C., Donner, B., & Sinninghe Damsté, J. S. (2008). Global sediment core-top calibration of the TEX_{86} paleothermometer in the ocean. *Geochimica et Cosmochimica Acta, 72,* 1154.

Kim, J.-H., van der Meer, J., Schouten, S., Helmke, P., Willmott, V., Sangiorgi, F., et al. (2010). New indices and calibrations derived from the distribution of crenarchaeal isoprenoid tetraether lipids:

implications for past sea surface temperature reconstructions. *Geochimica et Cosmochimica Acta, 74,* 4639–4654.

Kiriakoulakis, K., Marshall, J. D., & Wolff, G. A. (2000). Biomarkers in a Lower Jurassic concretion from Dorset (UK). *Journal of the Geological Society, 157,* 207–220.

Konecky, B. L., Russell, J. M., Johnson, T. C., Brown, E. T., Berke, M. A., Werne, J. P., et al. (2011). Atmospheric circulation patterns during late Pleistocene climate changes at Lake Malawi, Africa. *Earth and Planetary Science Letters, 312,* 318–326.

Konneke, M., Bernhard, A. E., de la Torre, J. R., Walker, C. B., Waterbury, J. B., & Stahl, D. A. (2005). Isolation of an autotrophic ammonia-oxidizing marine archaeon. *Nature, 437,* 543–546.

Kornilova, O., & Rosell-Melé, A. (2003). Application of microwave-assisted extraction to the analysis of biomarker climate proxies in marine sediments. *Organic Geochemistry, 34,* 1517–1523.

Kvenvolden, K. A. (2006). Organic geochemistry – A retrospective of its first 70 years. *Organic Geochemistry, 37,* 1–11.

Langenheim, J. H. (1994). Higher-plant terpenoids – a phytocentric overview of their ecological roles. *Journal of Chemical Ecology, 20,* 1223–1280.

Lengger, S. K., Hopmans, E. C., Sinninghe Damsté, J. S., & Schouten, S. (2012). Comparison of extraction and work up techniques for analysis of core and intact polar tetraether lipids from sedimentary environments. *Organic Geochemistry, 47,* 34–40.

Lichtfouse, É., Derenne, S., Mariotti, A., & Largeau, C. (1994). Possible algal origin of long chain odd n-alkanes in immature sediments as revealed by distributions and carbon isotope ratios. *Organic Geochemistry, 22,* 1023–1027.

Lipp, J. S., Morono, Y., Inagaki, F., & Hinrichs, K.-U. (2008). Significant contribution of Archaea to extant biomass in marine subsurface sediments. *Nature, 454,* 991–994.

Lisiecki, L. E., & Raymo, M. E. (2005). A Pliocene-Pleistocene stack of 57 globally distributed benthic $\delta^{18}O$ records. *Paleooceanography, 20,* PA1003.

Liu, Z., Pagani, M., Zinniker, D., DeConto, R., Huber, M., Brinkhuis, H., et al. (2009). Global cooling during the Eocene-Oligocene climate transition. *Science, 323,* 1187–1190.

Magill, C. R., Ashley, G. M., & Freeman, K. H. (2013a). Ecosystem variability and early human habitats in eastern Africa. *Proceedings of the National Academy of Sciences, USA, 110,* 1167–1174.

Magill, C. R., Ashley, G. M., & Freeman, K. H. (2013b). Water, plants, and early human habitats in eastern Africa. *Proceedings of the National Academy of Sciences, USA, 110,* 1175–1180.

Marshall, J. D., Brooks, J. R., & Lajtha, K. (2007). Sources of variation in the stable isotopic composition of plants. *Stable Isotopes in Ecology and Environmental Science, 2,* 22–60.

Marynowski, L., Kurkiewicz, S., Rakociski, M., & Simoneit, B. R. T. (2011). Effects of weathering on organic matter: I. Changes in molecular composition of extractable organic compounds caused by paleoweathering of a Lower Carboniferous (Tournaisian) marine black shale. *Chemical Geology, 285,* 144–156.

Marzi, R., Torkelson, B. E., & Olson, R. K. (1993). A revised carbon preference index. *Organic Geochemistry, 20,* 1303–1306.

Matthews, D. E., & Hayes, J. M. (1978). Isotope-ratio-monitoring gas chromatography-mass spectrometry. *Analytical Chemistry, 50,* 1465–1473.

Mayer, L. M. (1994). Surface area control of organic carbon accumulation in continental shelf sediments. *Geochimica et Cosmochimica Acta, 58,* 1271–1284.

Mayer, L. M. (1999). Extent of coverage of mineral surfaces by organic matter in marine sediments. *Geochimica et Cosmochimica Acta, 63,* 207–215.

Meier-Augenstein, W. (2004). In P. A. de Groot (Ed.), *Handbook of stable isotope analytical techniques* (pp. 154–176). Amsterdam: Elsevier.

Metzger, P., Casadevall, E., Pouet, M. J., & Pouet, Y. (1985). Structures of some botryococcenes: branched hydrocarbons from the b-race of green alga *Botryococcus braunii. Phytochemistry, 24,* 2995–3002.

Meyers, P. (1997a). Organic geochemical proxies for paleoceanographic, paleolimnologic and paleoclimatic processes. *Organic Geochemistry, 27,* 213–250.

Meyers, P. A. (1997b). Organic geochemical proxies of paleoceanographic, paleolimnologic, and paleoclimatic processes. *Organic Geochemistry, 27,* 213–250.

Meyers, P. A. (2003). Applications of organic geochemistry to paleolimnological reconstructions: a summary of examples from the Laurentian Great Lakes. *Organic Geochemistry, 34,* 261–289.

Meyers, P. A., Bourbonniere, R. A., & Takeuchi, N. (1980). Hydrocarbons and fatty acids in two cores of Lake Huron sediments. *Geochimica et Cosmochimica Acta, 44,* 1215–1221.

Monson, R. K. (1989). On the evolutionary pathways resulting in C_4 photosynthesis and Crassulacean acid metabolism, *19,* 57–110.

O'Leary, M. H. (1981). Carbon isotope fractionation in plants. *Phytochemistry, 20,* 553–567.

Osmond, C. B., Allaway, W. G., Sutton, B. G., Troughton, J. H., Queiroz, O., Luttge, U., et al. (1973). Carbon isotope discrimination in photosynthesis of CAM plants. *Nature, 246,* 41–42.

Otto, A., & Simoneit, B. R. T. (2001). Chemosystematics and diagenesis of terpenoids in fossil conifer species and sediment from the Eocene Zeitz formation, Saxony Germany. *Geochimica et Cosmochimica Acta, 65,* 3505–3527.

Otto, A., & Simoneit, B. R. T. (2002). Biomarkers of Holocene buried conifer logs from Bella Coola and north Vancouver, British Columbia, Canada. *Organic Geochemistry, 33,* 1241–1251.

Otto, A., & Simpson, M. J. (2005). Degradation and preservation of vascular plant-derived biomarkers in grassland and forest soils from Western Canada. *Biogeochemistry, 74,* 377–409.

Otto, A., & Wilde, (2001). Sesqui-, di-, and triterpenoids as chemosystematic markers in extant conifers - A review. *The Botanical Review, 67,* 141–238.

Pancost, R. D., & Boot, C. S. (2004). The palaeoclimatic utility of terrestrial biomarkers in marine sediments. *Marine Chemistry, 92,* 239–261.

Pastor, A., Vazquez, E., Ciscar, R., & de la Guardia, M. (1997). Efficiency of microwave-assisted extraction of hysrocarbons and pesticides from sediments. *Analytica Chimica Acta, 344,* 149–241.

Pearcy, R. W., & Ehleringer, J. (1984). Comparative ecophysiology of C3 and C4 plants. *Plant, Cell & Environment, 7,* 1–13.

Pearson, A., Huang, Z., Ingalls, A. E., Romanek, C. S., Wiegel, J., Freeman, K. H., et al. (2004). Nonmarine crenarchaeol in Nevada Hot Springs. *Applied and Environmental Microbiology, 70,* 5229–5237.

Pearson, E. J., Juggins, S., & Farrimond, P. (2008). Distribution and significance of long chain alkenones as salinity and temperature indicators in Spanish saline lake sediments. *Geochimica et Cosmochimica Acta, 72,* 4035–4046.

Peters, K. E., Walters, C. C., & Moldowan, J. M. (2005). *The biomarker guide: Biomarkers and isotopes in petroleum exploration and Earth history*. New York: Cambridge University Press.

Peterse, F., Kim, J.-H., Schouten, S., Kristensen, D. K., Koç, N., & Sinninghe Damsté, J. S. (2009). Constraints on the application of the MBT/CBT palaeothermometer at high latitude environments (Svalbard, Norway). *Organic Geochemistry, 40,* 692–699.

Peterse, F., van der Meer, M., Schouten, S., Weijers, J., Fierer, N., Jackson, R. B., et al. (2012). Revised calibration of the MBT-CBT paleotemperature proxy based on branched tetraether membrane lipids in surface soils. *Geochimica et Cosmochimica Acta, 96,* 215–229.

Petsch, S. T., Berner, R. A., & Eglinton, T. I. (2000). A field study of the chemical weathering of ancient sedimentary organic matter. *Organic Geochemistry, 31,* 475–487.

Petsch, S. T., Eglinton, T. I., & Edwards, K. J. (2001a). ^{14}C-dead living biomass: evidence for microbial assimilation of ancient organic carbon during shale weathering. *Science, 292,* 1127–1131.

Petsch, S. T., Smernik, R. J., Eglinton, T. I., & Oades, J. M. (2001b). A solid state ^{13}C-NMR study of kerogen degradation during black shale weathering. *Geochimica et Cosmochimica Acta, 65,* 1867–1882.

Pikuta, E. V., Hoover, R. B., & Tang, J. (2007). Microbial extremophiles at the limits of life. *Critical Reviews in Microbiology, 33,* 183–209.

Pitcher, A., Hopmans, E. C., Mosier, A. C., Park, S.-J., Rhee, S.-K., Francis, C. A., et al. (2011). Core and intact polar glycerol dibiphytanyl glycerol tetraether lipids of ammonia-oxidizing Archaea enriched from marine and estuarine sediments. *Applied and Environmental Microbiology, 77,* 3468–3477.

Pitcher, A., Hopmans, E. C., Schouten, S., & Sinninghe Damsté, J. S. (2009). Separation of core and intact polar archaeal tetraether lipids using silica columns: insights into living and fossil biomass contributions. *Organic Geochemistry, 40,* 12–19.

Powers, L. A., Johnson, T. C., Werne, J. P., Castañeda, I., Hopmans, E. C., Sinninghe Damsté, J. S., & Schouten, S. (2005). Large temperature variability in the southern African tropics since the last glacial maximum. *Geophysical Research Letters, 32,* L08706.

Powers, L. A., Werne, J. P., Johnson, T. C., Hopmans, E. C., Sinninghe Damsté, J. S., & Schouten, S. (2004). Crenarchaeotal membrane lipids in lake sediments: a new paleotemperature proxy for continental paleoclimate reconstruction? *Geology, 32,* 613–616.

Powers, L. A., Werne, J. P., Van der Woude, A. J., Sinninghe Damsté, J. S., Hopmans, E. C., & Schouten, S. (2010). Applicability and calibration of the TEX_{86} paleothermometer in lakes. *Organic Geochemistry, 41,* 404–413.

Poynter, J. G., & Eglinton, G. (1990). Molecular composition of three sediments from hole 717C: the Bengal Fan. In J. R. Cochran, D. A. V. Stow et al. (Eds.), *Proceedings of the Ocean Drilling Program, Scientific Results, 116,* 155–161.

Prahl, F. G., Muehlhausen, L. A., & Zahnle, D. L. (1988). Further evaluation of long-chain alkenones as indicators of paleoceanographic conditions. *Geochimica et Cosmochimica Acta, 52*(9), 2303–2310.

Prahl, F. G., Cowie, G. L., De Lange, G. J., & Sparrow, M. A. (2003). Selective organic matter preservation in "burn-down" turbidites on the Madeira Abyssal Plain. *Paleoceanography, 18,* 1052.

Prahl, F. G., & Wakeham, S. G. (1987). Calibration of unsaturation patterns in long-chain ketone compositions for palaeotemperature assessment. *Nature, 330,* 367–369.

Qin, W., Carlson, L. T., Armbrust, E. V., Devol, A. H., Moffett, J. W., Stahl, D. A., et al. (2015). Confounding effects of oxygen and temperature on the TEX_{86} signature of marine Thaumarchaeota. *Proceedings of the National Academy of Sciences, USA, 112,* 10979–10984.

Rommerskirchen, F., Eglinton, G., Dupont, L., Güntner, U., Wenzel, C., & Rullkötter, J. (2003). A north to south transect of Holocene southeast Atlantic continental margin sediments: relationship between aerosol transport and compound-specific $\delta^{13}C$ land plant biomarker and pollen records. *Geochemistry Geophysics Geosystems, 4,* 1101.

Rommerskirchen, F., Plader, A., Eglinton, G., Chikaraishi, Y., & Rullkötter, J. (2006). Chemotaxonomic significance of distribution and stable carbon isotopic composition of long-chain alkanes and alkan-1-ols in C4 grass waxes. *Organic Geochemistry, 37,* 1303–1332.

Sachse, D., Billault, I., Bowen, G. J., Chikaraishi, Y., Dawson, T. E., Feakins, S. J., et al. (2012). Molecular paleohydrology: interpreting the hydrogen-isotopic composition of lipid biomarkers from photosynthesizing organisms. *Annual Review of Earth and Planetary Sciences, 40,* 221–249.

Sachse, D., Kahmen, A., & Gleixner, G. (2009). Significant seasonal variation in the hydrogen isotopic composition of leaf-wax lipids for two deciduous tree ecosystems (*Fagus sylvativa* and *Acerpseudoplatanus*). *Organic Geochemistry, 40,* 732–742.

Sachse, D., Radke, J., & Gleixner, G. (2004). Hydrogen isotope ratios of recent lacustrine sedimentary *n*-alkanes record modern climate variability. *Geochimica et Cosmochimica Acta, 68,* 4877–4889.

Sachse, D., Radke, J., & Gleixner, G. (2006). δD values of individual *n*-alkanes from terrestrial plants along a climatic gradient – Implications for the sedimentary biomarker record. *Organic Geochemistry, 37,* 469–483.

Sage, R. F. (2004). The evolution of C_4 photosynthesis. *New Phytologist, 161,* 341–370.

Sage, R. F., Christin, P.-A., & Edwards, E. J. (2011). The C4 plant lineages of planet Earth. *Journal of Experimental Botany, 62,* 3155–3169.

Sage, R. F., Wedin, D. A., & Li, M. (1999). The biogeography of C_4 photosynthesis: patterns and controlling factors. In R. F. Sage & R. K. Monson (Eds.), *C_4 plant biology* (pp. 313–373). San Diego: Academic Press.

Sageman, B. B., Murphy, A. E., Werne, J. P., Ver Straeten, C. A., Hollander, D. J., & Lyons, T. W. (2003). A tale of shales: the relative roles of production, decomposition, and dilution in the accumulation of organic-rich strata, Middle Upper Devonian, Appalachian Basin. *Chemical Geology, 195,* 229–273.

Sauer, P. E., Eglinton, T. I., Hayes, J. M., Schimmelmann, A., & Sessions, A. L. (2001). Compound-specific D/H ratios of lipid biomarkers from sediments as a proxy for environmental and climatic conditions. *Geochimica et Cosmochimica Acta, 65,* 213–222.

Schefuß, E., Jansen, J. H. F., & Sinninghe Damsté, J. S. (2005a). Tropical environmental changes at the mid-Pleistocene transition: insights from lipid biomarkers. In M. J. Head & P. L. Gibbard (Eds.), *Early–middle Pleistocene transitions: The land–ocean evidence* (pp. 35–63). Geological Society of London Special Publications.

Schefuß, E., Schouten, S., & Schneider, R. R. (2005b). Central African hydrologic changes during the past 20,000 years. *Nature, 437,* 1003–1006.

Schefuß, E., Schouten, S., Jansen, J. H. F., & Sinninghe Damsté, J. S. (2003). African vegetation controlled by tropical sea surface temperatures in the mid-Pleistocene period. *Nature, 422,* 418–421.

Schimmelmann, A., Lewan, M. D., & Wintsch, R. P. (1999). D/H isotope ratios of kerogen, bitumen, oil, and water in hydrous pyrolysis of source rocks containing kerogen types I, II, IIS, and III. *Geochimica et Cosmochimica Acta, 63,* 3751–3766.

Schouten, S., Hopmans, E. C., Schefuß, E., & Sinninghe Damsté, J. S. (2002). Distributional variations in marine crenarchaeotal membrane lipids: a new tool for reconstructing ancient sea water temperatures? *Earth and Planetary Science Letters, 204,* 265–274.

Schouten, S., Hopmans, E. C., & Sinninghe Damsté, J. S. (2004). The effect of maturity and depositional redox conditions on archaeal tetraether lipid palaeothermometry. *Organic Geochemistry, 35,* 567–571.

Schouten, S., Hopmans, E. C., & Sinninghe Damsté, J. S. (2013). The organic geochemistry of glycerol dialkyl glycerol tetraether lipids: a review. *Organic Geochemistry, 54,* 19–61.

Schouten, S., Huguet, C., Hopmans, E. C., Kienhuis, M. V. M., & Sinninghe Damsté, J. S. (2007a). Analytical methodology for TEX_{86} paleothermometry by high-performance liquid chromatography/

atmospheric pressure chemical ionization-mass spectrometry. *Analytical Chemistry, 79,* 2940–2944.
Schouten, S., van der Meer, M. T. J., Hopmans, E. C., Rijpstra, W. I. C., Reysenbach, A.-L., Ward, D. M., et al. (2007b). Archaeal and bacterial glycerol dialkyl glycerol tetraether lipids in hot springs of Yellowstone National Park. *Applied and Environmental Microbiology, 73,* 6181–6191.
Schouten, S., Woltering, M., Rijpstra, W. I. C., Sluijs, A., Brinkhuis, H., & Sinninghe Damsté, J. S. (2007c). The Paleocene–Eocene carbon isotope excursion in higher plant organic matter: differential fractionation of angiosperms and conifers in the Arctic. *Earth and Planetary Science Letters, 258,* 581–592.
Sessions, A. L. (2006). Isotope-ratio detection for gas chromatography. *Journal of Separation Science, 29,* 1946–1961.
Sessions, A. L., Burgoyne, T. W., Schimmelmann, A., & Hayes, J. M. (1999). Fractionation of hydrogen isotopes in lipid biosynthesis. *Organic Geochemistry, 30,* 1193–1200.
Shepherd, T., & Griffiths, D. W. (2006). The effects of stress on plant cuticular waxes. *New Phytologist, 171,* 469–499.
Simoneit, B. R. T. (1977). Biogenic lipids in particulates from the lower atmosphere over the eastern Atlantic. *Nature, 267,* 682–685.
Sinninghe Damsté, J. S., Hopmans, E. C., Pancost, R. D., Schouten, S., & Geenevasen, J. A. J. (2000). Newly discovered non-isoprenoid glycerol dialkyl glycerol tetraether lipids in sediments. *Chemical Communications, 2000,* 1683–1684.
Sinninghe Damsté, J. S., Hopmans, E. C., Schouten, S., van Duin, A. C. T., & Geenevasen, J. A. J. (2002a). Crenarchaeol: the characteristic core glycerol dibiphytanyl glycerol tetraether membrane lipid of cosmopolitan pelagic Crenarchaeota. *Journal of Lipid Research, 43,* 1641–1651.
Sinninghe Damsté, J. S., Ossebaar, J., Abbas, B., Schouten, S., & Verschuren, D. (2009). Fluxes and distribution of tetraether lipids in an equatorial African lake: constraints on the application of the TEX_{86} palaeothermometer and branched tetraether lipids in lacustrine settings. *Geochimica et Cosmochimica Acta, 73,* 4232–4249.
Sinninghe Damsté, J. S., Ossebaar, J., Schouten, S., & Verschuren, D. (2008). Altitudinal shifts in the branched tetraether lipid distribution in soil from Mt. Kilimanjaro (Tanzania): implications for the MBT/CGT continental palaeothermometer. *Organic Geochemistry, 39,* 1072–1076.
Sinninghe Damsté, J. S., Ossebaar, J., Schouten, S., & Verschuren, D. (2012). Distribution of tetraether lipids in the 25-ka sedimentary record of Lake Challa: extracting reliable TEX_{86} and MBT/CBT palaeotemperatures from an equatorial African lake. *Quaternary Science Reviews, 50,* 43–54.
Sinninghe Damsté, J. S., Rijpstra, W. I. C., Hopmans, E. C., Weijers, J. W. H., Foesel, B. U., Overmann, J., et al. (2011a). 13,16-Dimethyl octacosanedioic acid (iso-diabolic acid): a common membrane-spanning lipid of Acidobacteria subdivisions 1 and 3. *Applied and Environmental Microbiology, 77,* 4147–4154.
Sinninghe Damsté, J. S., Rijpstra, W. I. C., & Reichart, G.-J. (2002b). The influence of oxic degradation on the sedimentary biomarker record II. Evidence from Arabian Sea sediments. *Geochimica et Cosmochimica Acta, 66,* 2737–2754.
Sinninghe Damsté, J. S., & Schouten, S. (2006). Biological markers for anoxia in the photic zone of the water column. In J. Volkman (Ed.), *Marine organic matter: Biomarkers, isotopes and DNA* (Vol. 2N, pp. 127–163). Berlin Heidelberg: Springer.
Sinninghe Damsté, J. S., Verschuren, D., Ossebaar, J., Blokker, J., van Houten, R., van der Meer, M. T. J., et al. (2011b). A 25,000-year record of climate-induced changes in lowland vegetation of eastern equatorial Africa revealed by the stable carbon-isotopic composition of fossil plant leaf waxes. *Earth and Planetary Science Letters, 302,* 236–246.
Sluijs, A., Bijl, P. K., Schouten, S., Röhl, U., Reichart, G. J., & Brinkhuis, H. (2011). Southern Ocean warming, sea level and hydrological change during the Paleocene-Eocene thermal maximum. *Climate of the Past, 7,* 47–61.
Sluijs, A., Schouten, S., Pagani, M., Woltering, M., Brinkhuis, H., Sinninghe Damsté, J. S., et al. (2006). Subtropical Arctic Ocean temperatures during the Palaeocene-Eocene thermal maximum. *Nature, 441,* 610–613.
Smith, F. A., & Freeman, K. H. (2006). Influence of physiology and climate on δD of leaf wax *n*-alkanes from C_3 and C_4 grasses. *Geochimica et Cosmochimica Acta, 70,* 1172–1187.
Spang, A., Hatzenpichler, R., Brochier-Armanet, C., Rattei, T., Tischler, P., Spieck, E., et al. (2010). Distinct gene set in two different lineages of ammonia-oxidizing Archaea supports the phylum Thaumarchaeota. *Trends in Microbiology, 18,* 331–340.
Sturt, H. F., Summons, R. E., Smith, K., Elvert, M., & Hinrichs, K.-U. (2004). Intact polar membrane lipids in prokaryotes and sediments decipered by high performance liquid chromatography/electrospray ionization multistage mass spectrometry—New biomarkers for biogeochemistry and microbial ecology. *Rapid Communications in Mass Spectrometry, 18,* 617–628.
Summons, R. E., Bird, L. R., Gillespie, A. L., Pruss, S. B., Roberts, M., & Sessions, A. L. (2013). Lipid biomarkers in ooids from different locations and ages: evidence for a common bacterial flora. *Geobiology, 11,* 420–436.
Sun, Q., Chu, G., Liu, M., Xie, M., Li, S., Ling, Y., et al. (2011). Distributions and temperature dependence of branched glycerol dialkyl glycerol tetraethers in recent lacustrine sediments from China and Nepal. *Journal of Geophysical Research: Biogeosciences, 116,* G01008.
Talbot, H. M., & Farrimond, P. (2007). Bacterial populations recorded in diverse sedimentary biohopanoid distributions. *Organic Geochemistry, 38,* 1212–1225.
Tegelaar, E., Matthezing, R., Jansen, J., Horsfield, B., & DeLeeuw, J. (1989a). Possible origin of normal-alkanes in high-wax crude oils. *Nature, 342,* 529–531.
Tegelaar, E. W., de Leeuw, J. W., Derenne, S., & Largeau, C. (1989b). A reappraisal of kerogen formation. *Geochimica et Cosmochimica Acta, 53,* 3103–3106.
Tierney, J. E., & Russell, J. M. (2009). Distributions of branched GDGTs in a tropical lake system: implications for lacustrine application of the MBT/CBT paleoproxy. *Organic Geochemistry, 40,* 1032–1036.
Tierney, J. E., Mayes, M. T., Meyer, N., Johnson, C., Swarzenski, P. W., Cohen, A. S., et al. (2010a). Late twentieth-century warming in Lake Tanganiyika unprecedented since AD 500. *Nature Geoscience, 3,* 422–425.
Tierney, J. E., Russell, J. M., Eggermont, H., Hopmans, E. C., Verschuren, D., & Sinninghe Damsté, J. S. (2010b). Environmental controls on branched tetraether lipid distributions in tropical East African lake sediments. *Geochimica et Cosmochimica Acta, 74,* 4902–4918.
Tierney, J. E., Russell, J. M., Huang, Y., Sinninghe Damsté, J. S., Hopmans, E. C., & Cohen, A. S. (2008). Northern Hemisphere controls on tropical southeast African climate during the past 60,000 years. *Science, 322,* 252–255.
Tierney, J. E., & Tingley, M. P. (2014). A Bayesian, spatially-varying calibration model for the TEX86 proxy. *Geochimica et Cosmochimica Acta, 127,* 83–106.
Tierney, J. E., & Tingley, M. P. (2015). A TEX_{86} surface sediment database and extended Bayesian calibration. *Scientific Data, 2,* 150029.
Tipple, B. J., Berke, M. A., Doman, C. E., Khachaturyan, S., & Ehleringer, J. R. (2013). Leaf-wax *n*-alkanes record the plant–water

environment at leaf flush. *Proceedings of the National Academy of Sciences, USA, 110,* 2659–2664.

Tipple, B. J., Meyers, S. R., & Pagani, M. (2010). Carbon isotope ratio of Cenozoic CO_2: a comparative evaluation of available geochemical proxies. *Paleoceanography, 25,* PA3202.

Tipple, B. J., & Pagani, M. (2007). The early origins of terrestrial C_4 photosynthesis. *Annual Review of Earth and Planetary Sciences, 35,* 435–461.

Tipple, B. J., & Pagani, M. (2010). A 35 Myr North American leaf-wax compound-specific carbon and hydrogen isotope record: implications for C_4 grasslands and hydrologic cycle dynamics. *Earth and Planetary Science Letters, 299,* 250–262.

Toney, J. L., Huang, Y. S., Fritz, S. C., Baker, P. A., Grimm, E., & Nyren, P. (2010). Climatic and environmental controls on the occurrence and distributions of long chain alkenones in lakes of the interior United States. *Geochimica et Cosmochimica Acta, 74,* 1563–1578.

Tranvik, L. J., Downing, J. A., Cotner, J. B., Loiselle, S. A., Striegl, R. G., Ballatore, T. J., et al. (2009). Lakes and reservoirs as regulators of carbon cycling and climate. *Limnology and Oceanography, 54,* 2298–2314.

Treibs, A. (1936). Chlorophyll- und Häminderivate in organischen Mineralstoffen. *Angewandte Chemie, 49,* 682–686.

Uno, K. T., Polissar, P. J., Jackson, K. E., & deMenocal, P. B. (2016). Neogene biomarker record of vegetation change in eastern Africa. *Proceedings of the National Academy of Sciences, USA, 113,* 6355–6363.

van Aarssen, B. G. K., Alexander, R., & Kagi, R. I. (2000). Higher plant biomarkers reflect palaeovegetation changes during Jurassic times. *Geochimica et Cosmochimica Acta, 64,* 1417–1424.

Vandenbroucke, M., & Largeau, C. (2007). Kerogen origin, evolution and structure. *Organic Geochemistry, 38,* 719–833.

Vogts, A., Moossen, H., Rommerskirchen, F., & Rullkötter, J. (2009). Distribution patterns and stable carbon isotopic composition of alkanes and alkan-1-ols from plant waxes of African rain forest and savanna C3 species. *Organic Geochemistry, 40,* 1037–1054.

Vogts, A., Schefuß, E., Badewien, T., & Rullkötter, J. (2012). n-Alkane parameters from a deep sea sediment transect off southwest Africa reflect continental vegetation and climate conditions. *Organic Geochemistry, 47,* 109–119.

Volkman, J. K., Eglinton, G., Corner, E. D. S., & Forsberg, T. E. (1980). Long-chain alkenes and alkenones in the marine coccolithophorid *Emiliania huxleyi*. *Phytochemistry, 19,* 619–2622.

Wakeham, S. G., Lee, C., Hedges, J. I., Hernes, P. J., & Peterson, M. J. (1997). Molecular indicators of diagenetic status in marine organic matter. *Geochimica et Cosmochimica Acta, 61,* 5363–5369.

Wakeham, S. G., & Pease, T. K. (1992). *Lipid analysis in marine particles and sediment samples*. Savannah: Skidaway Institute of Oceanography.

Weber, Y., De Jonge, C., Rijpstra, W. I. C., Hopmans, E. C., Stadnitskaia, A., Schubert, C. J., et al. (2015). Identification and carbon isotope composition of a novel branched GDGT isomer in lake sediments: evidence for lacustrine branched GDGT production. *Geochimica et Cosmochimica Acta, 154,* 118–129.

Weijers, J. W. H., Schouten, S., Donker, v. d., Jurgen C., Hopmans, E. C., & Sinninghe Damsté, J. S. (2007). Environmental controls on bacterial tetraether membrane lipid distribution in soils. *Geochimica et Cosmochimica Acta, 71,* 703–713.

Weijers, J. W. H., Schouten, S., Hopmans, E. C., Geenevasen, J. A. J., David, O. R. P., Coleman, J. M., et al. (2006a). Membrane lipids of mesophilica anaerobic bacteria thriving in peats have typical archaeal traits. *Environmental Microbiology, 8,* 648–657.

Weijers, J. W. H., Schouten, S., Spaargaren, O. C., & Damste, J. S. S. (2006b). Occurrence and distribution of tetraether membrane lipids in soils: implications for the use of the TEX_{86} proxy and the BIT index. *Organic Geochemistry, 37,* 1680–1693.

White, J. W. C., Cook, E. R., Lawrence, J. R., & Wallace, S. B. (1985). The D/H ratios of sap in trees: implications for water sources and tree ring D/H ratios. *Geochimica et Cosmochimica Acta, 49,* 237–246.

Williams, D. G., & Ehleringer, J. R. (2000). Intra- and interspecific variation for summer precipitation use in pinyon-juniper woodlands. *Ecological Monographs, 70,* 517–537.

Wing, S. L., Harrington, G. J., Smith, F. A., Bloch, J. I., Boyer, D. M., & Freeman, K. H. (2005). Transient floral change and rapid global warming at the Paleocene-Eocene boundary. *Science, 310,* 993–996.

Woltering, M., Johnson, T. C., Werne, J. P., Schouten, S., & Sinninghe Damsté, J. S. (2011). Late Pleistocene temperature history of Southeast Africa: a TEX_{86} temperature record from Lake Malawi. *Palaeogeography, Palaeoclimatology, Palaeoecology, 303,* 93–102.

Woltering, M., Werne, J. P., Kish, J. L., Hicks, R., Sinninghe Damsté, J. S., & Schouten, S. (2012). Vertical and temporal variability in concentration and distribution of thaumarchaeotal tetraether lipids in Lake Superior and the implications for the application of the TEX_{86} temperature proxy. *Geochimica et Cosmochimica Acta, 87,* 136–153.

Woltering, M. L. (2011). Thaumarchaeota distribution in the water columns of Lakes Superior and Malawi: Implications for the TEX_{86} lacustrine temperature proxy. Ph.D. Dissertation, University of Minnesota.

Wright, V. P., & Vanstone, S. D. (1991). Assessing the carbon dioxide content of ancient atmospheres using paleocretes: theoretical and empirical constraints. *Journal of the Geological Society, 148,* 945–947.

Wuchter, C., Abbas, B., Coolen, M. J. L., Herfort, L., van Bleijswijk, J., Timmers, P., et al. (2006). Archaeal nitrification in the ocean. *Proceedings of the National Academy of Sciences, USA, 103,* 12317–12322.

Wuchter, C., Schouten, S., Wakeham, S. G., & Sinninghe Damsté, J. S. (2005). Temporal and spatial variation in tetraether membrane lipids of marine Crenarchaeota in particulate organic matter: implications for TEX_{86} paleothermometry. *Paleoceanography, 20,* PA3013.

Yang, H., Lü, X., Ding, W., Lei, Y., Dang, X., & Xie, S. (2015). The 6-methyl branched tetraethers significantly affect the performance of the methylation index (MBT′) in soils from an altitudinal transect at Mount Shennongjia. *Organic Geochemistry, 82,* 42–53.

Yang, H., Pagani, M., Briggs, D. E. G., Equiza, M. A., Jagels, R., et al. (2009). Carbon and hydrogen isotope fractionation under continuous light: implications for paleoenvironmental interpretations of the High Arctic during Paleogene warming. *Oecologia, 160,* 461–470.

Zachos, J., Pagani, M., Sloan, L., Thomas, E., & Billups, K. (2001). Trends, rhythms, and aberrations in global climate 65 Ma to present. *Science, 292,* 686–693.

Zachos, J. C., Schouten, S., Bohaty, S., Quattlebaum, T., Sluijs, A., Brinkhuis, H., et al. (2006). Extreme warming of mid-latitude coastal ocean during the Paleocene-Eocene Thermal Maximum: inferences from TEX_{86} and isotope data. *Geology, 34,* 737–740.

Zink, K.-G., Leythaeuser, D., Melkonian, M., & Schwark, L. (2001). Temperature dependency of long-chain alkenone distributions in recent to fossil limnic sediments and in lake waters. *Geochimica et Cosmochimica Acta, 65,* 253–265.

Chapter 9
Paleopedology as a Tool for Reconstructing Paleoenvironments and Paleoecology

Emily J. Beverly, William E. Lukens, and Gary E. Stinchcomb

Abstract Soils form as a product of physical, chemical, and biological activity at the outermost veneer of Earth's surface. Once buried and incorporated into the sedimentary record, these soils, now paleosols, preserve archives of ancient climates, ecosystems, and sedimentary systems. Paleopedology, the study of paleosols, includes qualitative interpretation of physical characteristics and quantitative analysis of geochemical and mineralogical assays. In this chapter, the paleosol macroscopic, micromorphological, mineralogical, and geochemical indicators of paleoecology are discussed with emphasis on basic analytical and interpretative techniques. These data can reveal a breadth of site-specific interpretations of vegetation, sedimentary processes, climatic variables, and durations of landscape stability. The well-known soil-forming factors are presented as a theoretical framework for understanding landscape-scale soil evolution through time. Vertical and lateral patterns of stacked paleosols that appear in the rock record are discussed in order to address practical approaches to identifying and describing paleosols in the field. This chapter emphasizes a robust multi-proxy approach to paleopedology that combines soil stratigraphy, morphology, mineralogy, biology, and chemistry to provide an in-depth understanding of paleoecology.

Keywords Paleo-Critical Zone • Soil • Paleosol • Paleocatena • Paleoenvironments • Human evolution

Theoretical Background

Introduction

Soils are the product of organic matter accumulation and mineral weathering on the terrestrial surface, integrating the dynamic physical, biological, and chemical interactions that take place over time. Soils should not be confused with sediment, a material product of weathering and erosion occurring elsewhere and subject to transport by wind, water, or gravity. In contrast, soils form *in situ* and are the alteration products of weathered sediment, rock, and/or organic components. Dramatic changes in mineralogy and porosity accompany pedogenesis, which, along with bioturbation, act to disrupt primary depositional bedding. These processes impart new structural and chemical characteristics that are recognizable using field and lab techniques. Such patterns are preserved as soil profiles and can be buried and incorporated into the stratigraphic record (Fig. 9.1A). The resulting paleosols appear in many terrestrial depositional successions as intervals with variegated color, surface texture, and recessed erosional relief in outcrop.

State-Factor Approach

Paleopedology uses concepts and tools of Earth systems science to understand paleosols in the Quaternary and pre-Quaternary sedimentary record (Yaalon 1971; Retallack 2001; Sheldon and Tabor 2009; Tabor and Myers 2015).

E. J. Beverly
Department of Geosciences, Georgia State University, Atlanta, GA, USA

E. J. Beverly · W. E. Lukens
Terrestrial Paleoclimatology Group, Department of Geology, Baylor University, Waco, 76798 TX, USA
e-mail: Bill_Lukens@baylor.edu

E. J. Beverly (✉)
Earth and Environmental Sciences, University of Michigan, Ann Arbor, 48109 MI, USA
e-mail: ebeverly@umich.edu

G. E. Stinchcomb
Department of Geosciences and Watershed Studies Institute, Murray State University, Murray, 42071 KY, USA
e-mail: gstinchcomb@murraystate.edu

D. A. Croft et al. (eds.), *Methods in Paleoecology: Reconstructing Cenozoic Terrestrial Environments and Ecological Communities*, Vertebrate Paleobiology and Paleoanthropology, https://doi.org/10.1007/978-3-319-94265-0_9

The conceptual framework of most paleosol-based paleoenvironmental studies relies upon the fundamental state-factor theory for modern soils (Jenny 1941):

$$S\, or\, s = f(cl, o, r, p, t, \ldots), \quad (1)$$

where S is the soil and s is any soil property, and cl is climate, o is organisms, r is relief, p is parent material, and t is time. The series of dots, or ellipsis, represent other possible unaccounted for factors influencing S. For modern soils, Eq. 1 shows that the five primary factors (cl, o, r, p, t) influence the direction and magnitude of pedogenesis, S, or in some cases seem to predict S (Jenny 1941). For example, modern climosequence relationships that relate S to cl are applied to paleosols using an environmental-factor approach (*sensu* Retallack 1994), where s is measured in the paleosol using a variety of bulk and isotope geochemical, magnetic mineral, and physical proxies (Maynard 1992; Maher and Thompson 1995; Maher 1998; Ludvigson et al. 1998; Birkeland 1999; Retallack 2001; Sheldon et al. 2002; Holliday 2004; Driese et al. 2005; Dworkin et al. 2005; Retallack 2005; Nordt et al. 2006; Sheldon 2006; Sheldon and Tabor 2009; Nordt and Driese 2010a, b; Passey et al. 2010; Retallack and Huang 2010; Gulbranson et al. 2011; Gallagher and Sheldon 2013; Ludvigson et al. 2013; Hyland et al. 2015). The results of such soil-sequence studies allow for the development of empirical, quantitative relationships between soil properties and the state-factor studied (e.g., climate). The reader is directed to Rasmussen et al. (2005) and Rasmussen and Tabor (2007) and references therein for an example of alternative, quantitative frameworks for pedogenesis.

A Brief History of Paleosols in Paleoenvironmental Studies

The application of paleopedology as a tool for paleoecological reconstruction originated in the mid- to late 19th century, when geologists used buried forest beds to correlate glacial stratigraphy in the Midwestern USA, and the early 20th century, when many studies in the USA were focused on identifying the environments of "early man", (i.e., Paleoindian culture) (Leighton 1937; Holliday 2004, 2006). The accumulated knowledge of the early to middle 20th century likely led to the formal designation of the field as paleopedology at International Union for Quaternary Science (INQUA) meetings, led by pioneering paleopedologists Ruhe and Yaalon, which eventually led to two highly influential volumes (Ruhe 1965; Yaalon 1971).

The study of paleosols in the past 50 years has seen advancements through both pre-Quaternary and Quaternary disciplines. Some of the first pre-Quaternary investigations into paleoecology using paleosols can be traced to Retallack's (1983) systematic study of Eocene-Oligocene paleosols in the Badlands of South Dakota, USA. Retallack's (1983) work and successive studies ushered in an era of paleopedological research in a wide range of geologic settings (e.g., Bown and Kraus 1987; Bestland et al. 1997; Terry 2001). The science of paleopedology gradually evolved to become more quantitative with an increasing number of numerical proxies for climate, vegetation, and edaphic soil properties (Retallack 1994; An and Porter 1997; Maher 1998; Sheldon et al. 2002; Driese and Ober 2005; Dworkin et al. 2005; Retallack 2005; Prochnow et al. 2006; Kraus and Hasiotis 2006; Nordt and Driese 2010a, b; Retallack and Huang 2010; Gulbranson et al. 2011; Gallagher and Sheldon 2013; Trendell et al. 2013a, b; Hyland et al. 2015).

Paleosols are important paleoecological archives because they are records of Critical Zones of the Earth's past, preserving the *in situ* substrate of ancient landscapes (Amundson et al. 2007; Leopold et al. 2011; Nordt et al. 2012, 2013; Marin-Spiotta et al. 2014). This approach to studying paleo-Critical Zones (paleo-CZs) continues to gain traction as a multi-disciplinary method for understanding the Earth's environments and processes. Recent work has focused on unraveling and quantifying the biogeochemical processes operating within a paleosol system (Nordt and Driese 2010b; Nordt et al. 2012, 2013), comparable with modern Critical Zone studies (Fig. 9.1A; Amundson et al. 2007; Brantley et al. 2007). In light of these advances, this chapter establishes a theoretical framework for the field of paleopedology through the lens of a paleo-CZ and is applicable throughout geologic time.

This chapter does not review isotope geochemical methods in paleopedology and paleoenvironmental reconstruction, which can be found elsewhere (Nordt 2001; Sheldon and Tabor 2009; Cerling 2014; Levin 2015; Tabor and Myers 2015; Zamanian et al. 2016; Berke 2017). Much of the information compiled in this chapter was adapted from journal articles and books published by pioneers in the field of Quaternary and pre-Quaternary paleopedology (e.g., Yaalon 1971; Brimhall and Dietrich 1987; Birkeland 1999; Retallack 2001; Anderson et al. 2002; Holliday 2004; Sheldon and Tabor 2009; Nordt et al. 2013; Schaetzl and Thompson 2015; Tabor and Myers 2015). The reader is encouraged to consult these publications, which contain conceptual and methodological frameworks for analyzing soils and paleosols.

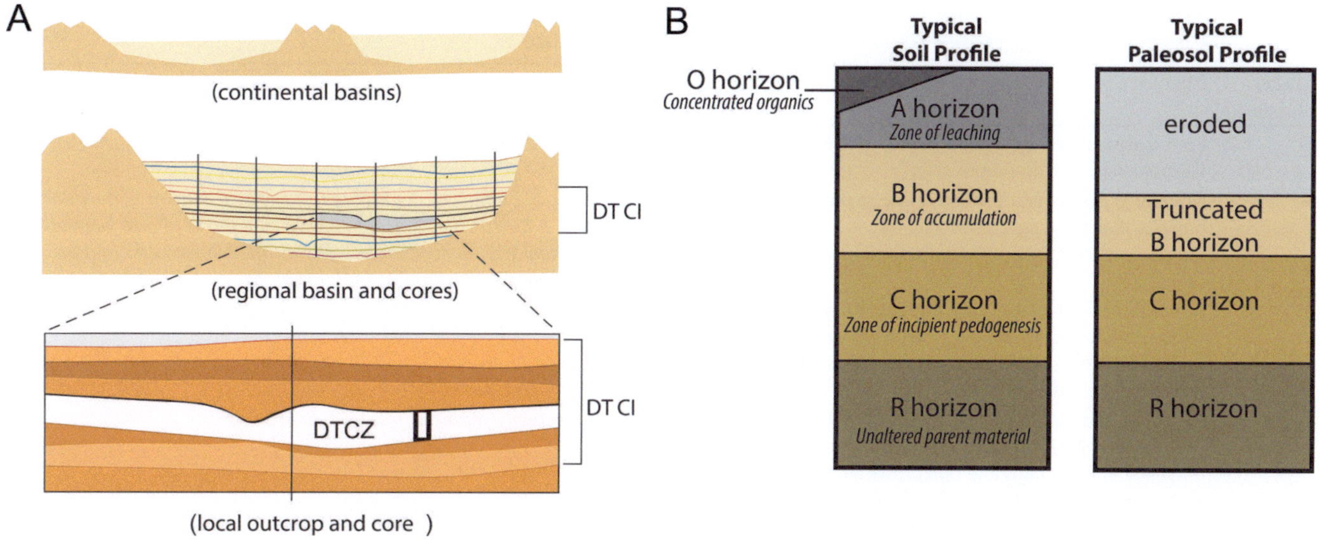

Fig. 9.1 A, Schematic cross-section of a paleo-Critical Zone reproduced from Nordt and Driese (2013). Top Panel: Alluvial deposits in a continental-scale basin. Middle Panel: Regional basin with paleo-Critical Zones illustrated by colored lines that could be studied in outcrop or using sediment cores indicated by vertical black lines. Lower Panel: Local outcrop of an alluvial succession with a paleo-Critical Zones and paleosol. DTCI = Deep time critical interval. The black box indicates the location of Fig. 9.1B. B, A typical soil profile and typical paleosol profile preserved after burial and diagenesis. Modified from Tabor and Myers (2015)

Approaches

Macroscale Physical Characterization

Field observation is the first step for identifying and describing paleosols. This step relies upon observing paleosol morphological features with unassisted vision or a hand lens with ~10× magnification. Much like modern soils, paleosols are identified in the field using color, texture, and weathering features within peds or along former voids (i.e., ped/void features). A number of features may suggest the presence of a paleosol: authigenic mineralization, rhizoliths, coloration, bioturbation, presence of soil aggregates (peds), redistribution of primary/authigenic minerals, shrink-swell features, disruption of primary sedimentary bedding, and deflation surfaces or biomantles as indicated by stone lines. After the paleosol(s) has been identified, a trench should be excavated to a depth that exposes sediment or rock unaffected by modern processes.

Field description and horizon delineation: The description of some paleosols differs from soils due to the inhibiting effects of compaction and lithification on the measurement of many properties. We advocate for describing paleosols in as much detail as the geologic setting allows, as these data will aid in more informed interpretations of paleolandscapes, paleoclimate, and paleovegetation. The reader is directed to the Illustrated Guide to Soil Taxonomy (Soil Survey Staff 2014a) and Keys to Soil Taxonomy (Soil Survey Staff 2006) for standard protocols for field description and horizon designation.

Paleosol horizons should be described using a standardized, published descriptive nomenclature, such as U.S. Department of Agriculture (USDA) soil nomenclature (Tables 9.1 and 9.2; Schoeneberger et al. 2012). A typical soil profile includes the O, A, B, C and R master horizons, where O is the zone of concentrated organics, A is the zone of leaching, B is the zone of accumulation, C is unaltered or unconsolidated parent material, and R is consolidated parent material. The A and O horizons are typically not preserved or recognized in paleosols due to erosion or oxidation of organic matter during the burial process (Fig. 9.1B). Less common master horizons are described in Table 9.1. Stratigraphic successions preserved in the rock record consist of aggrading sedimentary deposits, and therefore many paleosols do not have R horizons.

Subordinate indicators are added as lowercase suffixes to master horizons to identify the presence of pedogenic minerals, organic complexes, or other distinct features (Table 9.2). For example, common types of B horizons include Bk (accumulation of carbonate) and Bt (accumulation of translocated clay; Fig. 9.2B, H). The naming of soil horizons begins with field description but is frequently modified after laboratory and micromorphologic analysis (Birkeland 1999).

Detailed field descriptions of paleosol profiles include texture, peds, mineralizations, color, the presence of fossil roots, trace fossils and any other notable features (e.g., Schoeneberger et al. 2012). Paleosol texture is equivalent to grain size in sedimentology and is described in the field or measured in the laboratory. Peds should be identified and described with respect to size, shape, and grade (Fig. 9.3;

Table 9.1 Master soil horizons that may be observed in paleosols and their paleoecological implications. Modified from Guthrie and Witty (1982), Soil Classification Working Group (1998), and Schaetzl and Thompson (2015)

Master horizon	Characteristics	Paleoecological implications if present in paleosols
O	Layers dominated by organic material (litter and humus) in various stages of decomposition	*Forested environment with high biomass input*
A	Mineral horizons that formed at the surface or below an O horizon that are characterized by the accumulation of humified organic matter mixed with the mineral fraction	*Zone of bioturbation and highest organic content in the mineral soil. A horizons may best preserve signatures of past vegetation, particularly as root traces, phytoliths, biomarkers, and/or isotopic composition of bulk organics*
V	Mineral horizons that formed at the soil surface, or below a layer of rock fragments, e.g., desert pavement, or a physical or biological crust, and which are characterized by the predominance of vesicular pores	*Desert ecosystem, scrub vegetation likely*
E	Light-colored mineral horizons in which the main feature is loss of weatherable minerals. The result is often a concentration of uncoated quartz grains and other resistant minerals and poor soil structure	*Commonly found in forested environments*
B	Subsurface mineral horizons dominated by: (1) illuvial accumulation of clay, iron, aluminum, humus, etc.; (2) removal, addition, or transformation of primary carbonates or gypsum; (3) residual concentrations of sesquioxides; (4) loss of all or most sedimentary structure; (5) brittleness, and/or (6) gleying	*Paleoecological implications vary depending on the horizon suffix* (Table 9.2)
C	Mineral horizons that retain parent material properties and have thus been less affected by pedogenesis	*If C horizon is composed of allogenic parent material (e.g., alluvium, dust), then the organic matter may also be allogenic (detrital) and hold a basin-wide signature*
R	Consolidated hard, continuous bedrock	*N/A*

Table 9.3; Schoeneberger et al. 2012). Ped types are related to parent materials, intra-profile water cycling, biotic activity, and degree of pedogenic alteration, all of which have net paleoecological implications (Table 9.3). For example, columnar peds, which are vertically oriented with round tops, form as a result of clay flocculation in saline soils of arid and semi-arid environments (Fig. 9.3). Such soils tend to exclude most plant types due to the combined effect of low porosity, which inhibits root growth, and nutrient limitation (Qadir and Shubert 2002).

Recognition of the pedogenic redistribution of primary minerals and precipitation of secondary minerals can offer support for interpretations of climate, redox conditions, or soil maturity. Clays are translocated downward with soil water from eluvial (leaching) to illuvial (accumulating) horizons during wetting events. Subsequent drying causes the suspended clays to adhere to void features, which include the surfaces of peds, slickensides, root traces, and tubules. Calcium carbonate precipitates along root traces as rhizoliths or as discrete nodules in the matrix of Bk horizons. Similarly, Fe–Mn masses precipitate along voids or as hard matrix nodules as a result of alternating redox conditions. Salts, such as barite or gypsum, typically accumulate as discrete masses in the matrix of soils with extreme water stress. Geochemical and mineralogical approaches to the characterization of pedogenic minerals are addressed below.

The color of the paleosol matrix, mineralizations, and other notable features are characterized using Munsell® soil color charts. Matrix color can be used to identify soil organic matter (SOM) input and iron speciation associated with redox states but should only be used as an initial tool for description before further laboratory analysis (Holliday 2004). Darker colors, such as the dark browns or dark grayish browns found in buried A horizons, may suggest higher SOM production. However, dark colors can also reflect parent materials, especially for Mg-smectite clays (Fig. 9.2D; Beverly et al. 2014, 2015) or mafic igneous rocks.

The oxidation state and mineralogy of Fe imparts distinct color characteristics on soils and paleosols (Schwertmann 1993; Birkeland 1999) and may directly reflect local environmental conditions such as drainage, fire, temperature, and duration of soil development (Table 9.4; Holliday 2004). Green, blue, or gray colors often suggest poorly-drained, anaerobic conditions (Vepraskas 1992, 2001). Prolonged saturation reduces the amount of available oxygen for soil respiration, which leads to microbially-mediated reduction of Fe and Mn species and an overall gray appearance (Vepraskas and Faulkner 2001). Re-precipitation of Fe and Mn in localized oxygen-rich zones creates a mottled appearance (Fig. 9.2C). These processes frequently occur in wetland or lake-margin paleosols proximal to paleo-lakes or on paleo-floodplains (Retallack 1994; Ufnar et al. 2004;

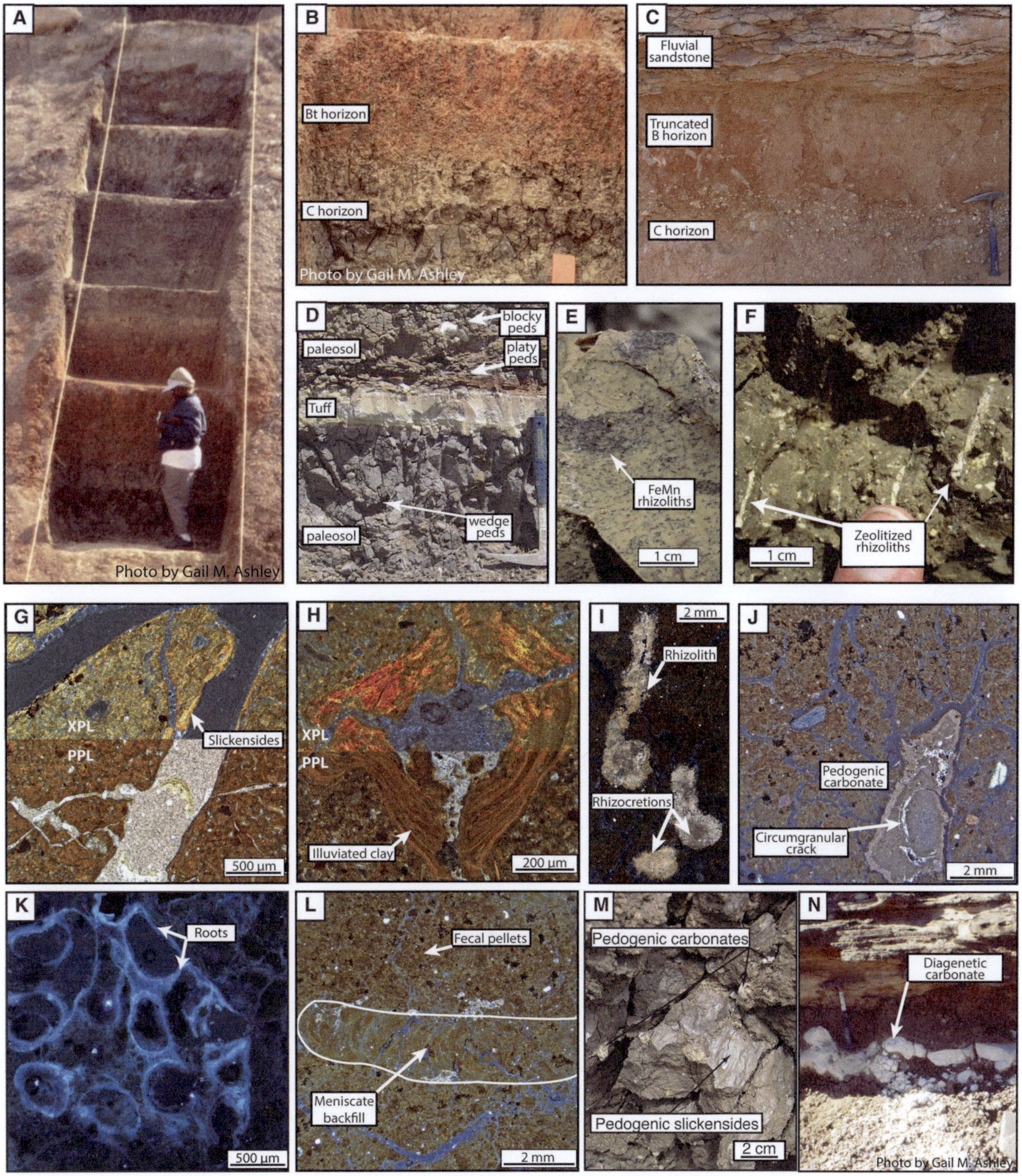

Fig. 9.2 Photographs of pedogenic features in the field and in micromorphology. A, Oxidized and volcaniclastic-rich paleo-Aridisol from Olduvai Gorge, Tanzania. B, Close-up of lowermost step in trench in Fig. 9.2A, which has excellent horizonation, clay illuviation, and ped structure (angular blocky and columnar). C, Fining upward, coarse-grained alluvial paleosol from the Chalbi Desert, Kenya. D, Stacked paleo-Vertisols from Olduvai Gorge separated by a tuff with examples of blocky, platy, and wedge peds. E, FeMn filled rhizoliths from Olduvai Gorge (Reproduced from Beverly et al. 2014). F, Zeolitized rhizoliths from Olduvai Gorge (Reproduced from Beverly et al. 2014). G, Pedogenic B-fabric (Reproduced from Beverly et al. 2015a). H, Illuviated clay in a pore (Reproduced from Beverly et al. 2015a). I, Carbonate rhizolith and rhizocretions in thin section from Karungu, Kenya. J, Pedogenic carbonate in thin section with circumgranular cracking (Reproduced from Beverly et al. 2015a). K, Preserved root structures identified in a paleosol using UVf (Reproduce from Beverly et al. 2015b). L, Fecal pellets and meniscate backfill (Reproduced from Beverly et al. 2015a). M, Pedogenic slickensides in a paleo-Vertisol from Karungu, Kenya, and pedogenic carbonate nodule. N, Non-pedogenic carbonate from Olduvai Gorge. Groundwater carbonate precipitates after primary deposition. Note rock hammer for scale

Table 9.2 Suffix symbols used with master soil horizons and their possible paleoecological implications when observed in paleosols. Modified from Guthrie and Witty (1982), Soil Classification Working Group (1998), Soil Survey Staff (2010), and Schaetzl and Thompson (2015)

Suffix	Used with master horizons	Characteristics/comments	Paleoecological implications
p	O, A	Plowed, tilled, or otherwise disturbed surface layer	*Humans likely played significant role in shaping local ecosystem*
c	B, C	Presence of concretions or nodules, usually of Fe, Al, Mn, or Ti	*Water table fluctuations*
h	B	Accumulation of amorphous and dispersable organic materials, low Munsell value and chroma	*Acidic soil environment; frequently observed in coniferous or boreal forests; Observed in cool-damp shrub environments; acid parent materials (e.g., quartz sand) may also influence development*
j	B, C	Accumulation of jarosite (iron hydroxyl sulfate), as either ped coatings or nodules	*Poorly-drained environment, frequently observed in tidal, swamp, or marsh settings*
k	B, C	Accumulation of visible pedogenic calcium carbonate (<50% by volume) in filaments, soft masses, nodules, pendants, and finely disseminated in matrix	*May occur in arid or semi-arid regions with seasonal wetness; Also known to occur in poorly-drained alkaline environments*
kk	B, C	Engulfment of horizon by pedogenic calcium carbonate ($\geq 50\%$ by volume); coating particles and filling pores	*May occur in arid or semi-arid regions; Indicate protracted surface stability; Often confused with calcium carbonate layers that precipitated in or near the phreatic zone*
n	B, C	Accumulation of exchangeable sodium	*Frequently observed in arid or semi-arid regions*
q	B, C	Accumulation of secondary silica as concretions, opal, etc.	*May occur in arid or semi-arid regions with seasonal wetness; commonly associated with volcanic materials*
s	B	Illuvial accumulation of amorphous, dispersible sesquioxides (Fe- and Al-oxides) and organic matter	*Associated with humid forests*
t	B, C	Illuvial accumulation of clay with clay coatings, bridges, or lamellae	*Well-drained soil where water moves through profile to transport clay with periodic drying of soil when clay adheres to pores; Indicate protracted surface stability*
y	B	Accumulation of pedogenic gypsum (<50% by volume)	*Frequently observed in arid or playa-like environments; Shallow pedogenic gypsum more associated with arid environments; may also results from precipitation near/in the phreatic zone*
yy	B	Horizon dominated by pedogenic gypum where crystals disrupt pedogenic or sedimentary features ($\geq 50\%$ by volume)	*Same as y suffix*
z	B	Accumulation of salts more soluble than gypsum such as NaCl	*May occur in arid or semi-arid regions*
w	B	Weakly expressed color or structural development; used when no other suffix is applicable	*Weak pedogenesis*
x	E, B	Horizon that is firm, brittle, and physically root restrictive	*Restrictive of water movement through the soil*
o	B, C	Residual accumulation of sesquioxides (Fe- and Al-oxides)	*May occur in humid tropical or subtropical environments; Typified by intense weathering and often associated with 1:1 clays (e.g., kaolinite)*
ss	A, B, C	Presence of slickensides	*May occur in seasonally humid environments. Seasonal water stress occurring in smectite-dominated soils; Strong contrast in water input through drought or flooding is frequently observed*
g	A, E, B, C	Strong gleying and Munsell Chroma of ≤ 2	*Fe reduction and loss due to saturation and anaerobic conditions*
r	C	Weathered or soft bedrock	N/A
m	B	Root-restrictive, pedogenically cemented horizon with >90% carbonate, Fe, silica, gypsum, or other salts	*May occur in arid or semi-arid regions*
u	Any	Presences of artifacts	*Evidence of hominins*
b	A, E, B	Buried horizon that exhibits past pedogenesis	*If the succession of measured paleosols includes the surface soil, then denoting buried soil with "b" is useful. However, many paleosol successions consist entirely of buried, fossilized, soils; thus, the "b" is not useful in this scenario and typically excluded*
^	Any	Horizons that are composed of human-transported material	*Evidence of hominins*

Table 9.3 Classification of soil peds or aggregates. Modified from Retallack (2001) and Brady and Weil (2008). Examples are shown in Fig. 9.3

Ped type	Description	Usual Horizon	Paleoecological Implications
Platy	Tabular and horizontal to land surface	Common in C, but may occur in any horizon	*Weathering of relict bedding, soil compaction, or impeded water flow*
Prismatic	Vertically oriented peds with flat, angular top	Usually found in B horizons	*Commonly occur in soils with good drainage and can be armored with translocated clays in humid environments. Can occur near the surface in arid and semi-arid environments*
Columnar	Vertically oriented peds with rounded top	Usually found in B horizons	*Commonly occur in soils affected by shrinking and swelling clays and high in sodium in arid and semi-arid environments*
Angular Blocky	Irregular and roughly cube-shaped with sharp edges	Usually found in B horizons, but may occur in A horizons	*Form as a result of rooting, burrowing, and wetting and drying of the soil*
Subangular Blocky	Irregular and roughly cube-shaped with rounded edges	Usually found in B horizons, but may occur in A horizons	*Form as a result of rooting, burrowing, and wetting and drying of the soil*
Granular	Spheroidal in shape with slightly interlocking edges, porous	Common in A horizons	*Common in grassland soils due to earthworms, high in organic matter, poorly preserved in the rock record due to erosion*
Crumb	Spheroidal in shape, very porous	Common in A horizons	*Common in grassland soils due to earthworms, high in organic matter, poorly preserved in the rock record due to erosion and oxidation of organic matter during lithification of paleosol*
Wedge	Triangular in shape	Common in Bss horizons	*Commonly occur in Vertisols affected by shrinking and swelling clays*

Fig. 9.3 Photographs of common peds (soil aggregates) identified in paleosols. See Table 9.3 for explanation of these ped types

Table 9.4 Paleoecological implications of soil color. Modified from Schwertmann (1993) and Holliday (2004)

Iron oxide	Munsell color	Soils	Paleoecological Implications
Goethite (FeOOH)	7.5YR-2.5Y	All soils with Fe released	*Wherever weathering takes place*
Hematite (Fe_2O_3)	7.5R-5YR	Aerobic soils of the tropics and subtropics with dry seasons	*High soil temperature, low water activity, rapid biomass turnover, high Fe release rate from rocks*
Lepidocrocite (FeOOH)	5YR-7.5YR (Value ≥ 6)	Aquic subgroups in temperate regions (Pseudogleys)	*Anaerobic/aerobic systems, noncalcareous*
Ferrihydrite ($Fe_5HO_8 \cdot 4H_2O$)	5YR-7.5YR (Value ≤ 6)	Gleys, Podzolic B horizons	*Rapid oxidation in humic environments*
Maghemite (Fe_2O_3)	2.5YR-5YR	Mainly tropical and subtropical soils	*Usually a product of fires*

Tabor et al. 2007; Rosenau et al. 2013; Ashley et al. 2013, 2014; Beverly et al. 2014; Driese and Ashley 2016).

Red or reddish-brown horizons usually indicate well-drained conditions or higher temperature soil-forming environments (Fig. 9.2A–C; Ashley and Driese 2000; Driese et al. 2016). Well-drained conditions promote Fe oxidation in O_2-rich soils, which leads to the development of goethite and hematite that impart a reddish-brown to red color in the soil (Fig. 9.2A–C; Schwertmann 1993; Birkeland 1999; Holliday 2004). Increasing temperatures can facilitate chemical reactions, and this is thought to influence the reddening of the soil. Fire-altered soil often shows reddening (Schwertmann 1993; Mentzer 2014) like those affected by heating from lavas. However, careful observation of pedogenic features and geochemical trends can disentangle the competing roles of pedogenesis versus thermal alteration in causing soil redness. For example, Sheldon (2003) showed evidence that rubified paleosols between Columbia River basalt flows appear red primarily due to oxidation and clay production during pedogenesis rather than thermal alteration. Paleosols can also inherit their color from parent materials, so care must be taken when interpreting the paleoclimatic or paleoenvironmental significance of red paleosols (Gile 1979; Sheldon 2005). Finally, Maxbauer et al. (2016b) and Clyde et al. (2013) recently showed that oxidation has altered paleosol color to depths of 20 to 30 m below the modern outcrop surface in core from the Bighorn Basin Coring Project. In the unweathered portion of the core, the colors are much more muted in comparison to the weathered paleosols in outcrop and in the upper 25 m of the core.

Sampling of paleosol horizons: After paleosol horizons are identified and their thicknesses and properties have been described, bulk soil, sediment, or paleosol samples should be collected by horizon. It is important to collect "unweathered" rock sample and avoid modern fractures and weathering (see Maxbauer et al. 2016b). Soil micromorphology (see below) can be used to identify pedogenic processes, evidence of overprinting, and mineralogical composition. Undisturbed, oriented thin-section samples should be collected from all representative horizons, air-dried, epoxy impregnated, and then prepared for in-house or commercial fabrication. In unlithified paleosols, epoxy impregnation may be necessary in the field for removal of oriented samples because of their unconsolidated state.

Bulk samples can also be collected using an arbitrary sampling interval (e.g., every 10 cm), depending on the research question(s) that are being addressed. The bulk samples can be analyzed for their particle size (depending on degree of lithification), bulk and isotopic geochemistry, and bulk and clay mineralogy. Samples of unique features should also be collected if time and space permit, e.g., root traces, mineralizations, diagnostic ped/void features (see below), organic mats, and trace fossils. Soil trace fossils can be used as a complementary data set to paleosol analysis. For further discussion of ichnology see Hasiotis (2007), Hasiotis et al. (2007a, b), and Hembree et al. (2014).

Bulk samples for geochemical analysis should be collected from uppermost B horizons (ideally 25–100 cm depth), as required for some paleoclimate reconstructions (Sheldon et al. 2002; Nordt and Driese 2010a, b; Stinchcomb et al. 2016). Bulk density of paleosols is required for constitutive characterization (e.g., mass-balance geochemistry; see section on Constitutive Mass-Balance). The bulk density of soil and sediment samples can be calculated using the clod method on peds or core method if unconsolidated (Blake and Hartge 1986). If this is not possible, bulk density can be estimated using pedotransfer functions based on the organic content and texture (Rawls 1983) or SiO_2 concentration in Vertisols (Nordt and Driese 2010b).

Micromorphological Characterization

Soil micromorphological features can be studied using a polarized light microscope and described using a variety of terms developed for micromorphological observation of soils. Brewer (1976) is the seminal work regarding the theoretical framework and application of soil micromorphology for interpreting soil development. Later, simpler terminology was developed by Bullock et al. (1985), Fitzpatrick (1993), and

Stoops (2003). Micromorphological terminology can be extensively subdivided and modified depending on the data needed to answer specific research questions. An analysis of the microstructure often begins with a description of the mineral soil matrix by describing the coarse to fine ratio (c/f) and oriented birefringent fabric (b-fabric). A description of weathering of primary minerals can be useful for determining the degree of pedogenic development (Bullock et al. 1985). B-fabric within the soil matrix is often described and can be indicative of the paleo-hydrology of the system. For further discussion of b-fabrics see Stoops (2003) and Stoops et al. (2010). Pedogenic slickensides can be identified from illuviated clay by a lack of laminations and an identical grain size between the matrix and oriented clay (Fig. 9.2G). Oriented b-fabric may indicate that repetitive wetting and drying caused clays to shrink and swell, leading to inclined orientation due to the confining pressure of overlying soil (Blokhuis et al. 1990). It could also indicate bioturbation and reorganization of the soil matrix by other means creating grano-, channel, or poro-striated b-fabric (Stoops et al. 2010).

Pedofeatures, which are >20 μm and are distinguishable from the soil matrix, should also be described in detail (Stoops 2003). In thin section, burrows and root voids can be distinguished by analyzing the textures around the void (Stoops 2003; Stoops et al. 2010). Coatings (lining voids), hypocoatings (impregnated into matrix around pore), quasicoatings (impregnated into matrix surrounding pore but at some distance from feature), and infillings are also important to describe. A variety of materials such as Fe–Mn (Fig. 9.2E), illuviated clay or silt (Fig. 9.2H), or carbonate can fill these voids. Both silt and clay can be translocated down profile via the vertical drainage of water (Stoops et al. 2010) and indicate that the water table was low enough for drainage, but that there was enough water to translocate clay or silt. The illuviation of silt in thin section is sometimes used to infer the physical translocation of freshly deposited silt after a flood or dust accumulation (Kubiena 1970; Fedoroff and Goldberg 1982; Kemp 1999).

The accumulation of clay coatings observed in thin section has been related to the residence time (duration of weathering in the system) of soils using a chronosequence of deeply weathered, sub-tropical soils from Mississippi (Ufnar 2007). It is possible that the Ufnar (2007) chronofunction may overestimate the length of pedogenesis in certain climates (i.e., monsoonal), but this technique can be a useful approximation (Beverly et al. 2015a). Although clay illuviation has been inferred to represent warm-wet climate conditions in some studies (e.g., Bronger and Heinkele 1989), later work showed that Bt horizons can form in semi-arid grassland landscapes when enough time has elapsed (Han et al. 1998). A corollary of the Han et al. (1998) findings is that clay coatings could result from windblown dust deposits weathered and translocated down from silty A horizons.

Fe–Mn concentrations and depletions can be used to infer drainage (Fitzpatrick 1984; Bullock et al. 1985) and more specifically to infer the relative influence of epi- versus endo-saturation in the paleosol. Fe oxides coating peds with reduced interiors are indicative of endosaturation, where water is wicked upward through micropores within aggregate interiors. Oxidizing conditions remain in macropores, where the spacing between grains is too great for water to adhere. Reduction present along the exterior of peds that contain oxidized interiors suggests periodic episaturation (Fitzpatrick 1984; Bullock et al. 1985). Water saturation from surficial ponding will fill macropores and inhibit the diffusion of atmospheric oxygen into pore waters. Consumption of oxygen by microbes will result in the reduction of Fe on aggregate exteriors, which will penetrate into aggregate interiors with time. If the soil becomes drained of ponded water before the anoxic pore waters completely infiltrate aggregate interiors, Fe within aggregates will remain oxidized (Fitzpatrick 1984; Bullock et al. 1985).

Nodules are three-dimensional features not associated with a void or pore feature and are important as they can reveal information about calcification or redox conditions (Stoops 2003). The most commonly identified nodules are pedogenic carbonate, but Fe–Mn nodules are also important (Stiles et al. 2001). The accumulation of $CaCO_3$ in soils frequently indicates an arid or semi-arid soil-forming environment (Gile et al. 1966; Machete 1985). Cathodoluminescence (CL) is also a useful tool for determining the redoximorphic environment of carbonate identified in a paleosol, assessing diagenetic alteration (Barnaby and Rimstidt 1989; Machel et al. 1991), or identifying parent material contamination by marine carbonates (Michel et al. 2013). Carbonate forming in a shallow reducing environment with higher manganese to iron ratios will luminesce; whereas, carbonate forming in an oxidizing environment will have no luminescence (Barnaby and Rimstidt 1989; Machel et al. 1991). The importance of pedogenic carbonate is further discussed below.

An ultraviolet fluorescence (UVf) microscope attachment causes organic matter to fluoresce, facilitating the identification of organics that would be otherwise unobservable. UVf can reveal features such as root cell structure not visible in cross-polarized or plane light (Fig. 9.2K; Beverly et al. 2014, 2015b). Thin sections are often point counted to quantify mineralogy, organic content, or other morphological features that elucidate pedogenic processes (e.g., Stinchcomb et al. 2014).

Identifying features created by soil fauna often yields critical paleoecological information. Faunal features include: fecal pellets, channels or burrows, chambers, coatings, and infillings (Stoops 2003). Faunal features are not always readily identifiable but can yield significant information about paleohydrology. Termites, for example, do not live below the water table, and thus the depth of their burrowing can indicate the height of the water table associated with their colonization surface. Soil-dwelling crayfish can be of similar use in identifying the water table because they require standing water at the base of their burrows (Hasiotis and Honey 2000). Earthworm fecal pellets and burrows are often identified in paleosols and provide macropores favorable to root growth and microbial activity and stabilize organic carbon in the soil (Fig. 9.2L; Stoops et al. 2010). For further discussion of faunal features in micromorphology see Ch. 18 of Stoops et al. (2010).

Characterizing Pedogenic Minerals

Bulk mineralogy of soils and paleosols can be characterized using X-ray diffraction (XRD) (Poppe et al. 2001). Common pedogenic minerals and their paleoecological implications are shown in Table 9.5 (Bullock et al. 1985). Harris and White (2008) provide a summary of how to use XRD to identify common minerals in soils.

Clay Mineralogical Characterization: Clay mineralogical analysis is a powerful tool for recognizing weathering products versus inherited materials, which ultimately aids in understanding the weathering pathway(s) that occurred in paleosols (e.g., Tabor and Montañez 2002; Rosenau et al. 2013). The composition of the clay-sized fraction (<2 μm) is analyzed using XRD and should be prepared following standard methods (Moore and Reynolds 1997). When preparing samples for clay mineralogy, multiple aliquots may be needed to help diagnose 2:1 clay minerals using the following treatments: (1) air-dry, (2) K saturation, (3) Mg saturation by ethylene glycol solvation, and (4) subsequent 550°C heat treatment of the K-saturated sample. Tabor and Myers (2015) address the characterization and interpretation of clay minerals for paleoclimatic purposes.

Pedogenic Carbonate: Pedogenic calcium carbonates ($CaCO_3$ – limestone; $(Ca,Mg)(CO_3)_2$) – dolomite) are an important paleoecological indicator as they typically form in dry environments where precipitation is exceeded by evapotranspiration (Fig. 9.2K), though carbonate precipitation can occur in some poorly-drained alluvial back-swamp soils (Aslan and Autin 1998; Mintz et al. 2011). Carbonates initially precipitate in the soil matrix as soft masses with diffuse boundaries and micritic crystalline texture (Nordt et al. 2004). Zones of carbonate can develop septarian shrinkage cracks (Fig. 9.2J), multi-generational carbonate overgrowths, and coatings of other minerals, such as Fe and Mn oxides (Bullock et al. 1985; Stoops 2003). Such alteration converts the soft masses to hard nodules, which has been interpreted to occur outside the zone of active carbonate precipitation (Nordt et al. 2004). Carbonates also precipitate locally around roots, where CO_2 is high due to respiration, forming filaments and nodules that converge to form rhizoliths. If vertic soil processes caused by shrinking and swelling of clays during the wet and dry seasons are present, these rhizoliths can be broken up to form rhizocretions (Fig. 9.2I).

The degree to which carbonates precipitate in the soil matrix is primarily a function of time (provided that the other soil forming factors are held constant) and can be described

Table 9.5 Common pedogenic minerals and their paleocological implications. Modified from Bullock et al. (1985)

Mineral	Composition	Potential paleoecological implications
Pyrite	FeS_2	*Potentially acid sulphate soils (i.e., waterlogged conditions)*
Hematite	FeS_2O_3	*Potentially tropical or subtropical soils*
Lepidocrocite	$FeO(OH)$	*Gleyed soils (i.e., waterlogged conditions)*
Goethite	$FeO(OH)$	*Temperate and tropical soils*
Gibbsite	$Al(OH)_3$	*Tropical or subtropical soils, Andisols*
Halite	$NaCl$	*Saline soils*
Calcite	$CaCO_3$	*Arid to semi-arid soils or from calcareous parent material*
Siderite	$FeCO_3$	*Marshy soils*
Natron	$Na_2O_3*10H_2O$	*Saline soils*
Trona	$Na_3H(CO_3)_2*2H_2O$	*Saline soils*
Thenardite	Na_2SO_4	*Saline soils*
Barite	$BaSO_4$	*Soils with saline groundwater*
Celestite	$SrSO_4$	*Gypsiferous soils*
Anhydrite	$CaSO_4$	*Arid soils*
Gypsum	$CaSO_4*2H_2O$	*Semi-arid soils; acid sulphate soils*
Jarosite	$KFe_3(OH)_6(SO_4)_2$	*Acid sulphate soils (i.e., waterlogged conditions)*
Mirabilite	$Na_2SO_4*10H_2O$	*Saline soils*
Vivianite	$Fe_3(PO_4)_2*8H_2O$	*Marshy soils*
Bassanite	$CaSO_4*1/2H_2O$	*Gypsiferous soils*

Table 9.6 Stages of calcium formation in calcic soils and pedogenic calcretes. Modified from Gile et al. (1966) and Machete (1985)

Type	Stage	Gravel content	Diagnostic morphology	$CaCO_3$ distribution	Max. $CaCO_3$ content
Calcic soil	I	High	Thin, discontinuous coatings on pebbles, usually on undersides	Coatings sparse to common	tr – 2
		Low	A few filaments in soil or faint coatings on ped faces	Filaments sparse to common	tr – 4
	II	High	Continuous, thin to thick coatings on tops and undersides of pebbles	Coatings common, some carbonate in matrix, but matrix still loose	2–10
		Low	Nodules, soft, 0.5 to 4 cm in diameter	Nodules common, matrix generally noncalcareous to slightly calcareous	4–20
	III	High	Massive accumulations between clasts, becomes cemented in advanced form	Essentially continuous dispersion in matrix	10–25
		Low	Many coalesced nodules, matrix is firmly to moderately cemented		20–60
Pedogenic calcrete (indurated calcic soil)	IV	Any	Thin (<0.2 cm) to moderately thick (1 cm) laminae in upper part of K horizon. Thin laminae may drape over fractured surfaces	Cemented platy to weak tabular structure and indurated laminae. K horizon is 0.5 – 1 m thick	>25 in high gravel content; >60 in low gravel content
	V	Any	Thick laminae (>1 cm) and thin to thick pisolites. Vertical faces and fractures are coated with laminated carbonate (case-hardened surfaces)	Indurated dense, strong platy to tabular structure. K horizon is 1–2 m thick	>50 in high gravel content; >75 in low gravel content
	VI	Any	Multiple generations of laminae, breccia, and pisolites; cemented. Many case-hardened surfaces	Indurated and dense, thick strong tabular structure. K horizon is commonly >2 m thick	>75 in all gravel contents

High: >50% gravel content; Low: <20% gravel content; Tr – trace

in stages of accumulation (Table 9.6; Gile et al. 1966; Machete 1985). Soil carbonates can also be inherited from the parent material, particularly when materials are sourced from marine limestone (Michel et al. 2013). When describing carbonates in the field, an effervescence test is useful for describing the degree of effervescence in the paleosol matrix and/or in secondary carbonate features, which may indicate the relative concentration of carbonate in the paleosol horizon (Schoeneberger et al. 2012).

Stable isotopes of carbon and oxygen from pedogenic carbonates have been used extensively to interpret paleoecology (Cerling and Hay 1986; Cerling et al. 1988), where long sedimentary records have been used to understand the changing climate and environment of the past several million years. Carbon isotopes in pedogenic carbonates are of particular importance because they can be used to interpret open and closed environments by using the enrichment or depletion of ^{13}C relative to ^{12}C, which can result from changes in biomass utilizing disparate photosynthetic pathways (C_3 vs C_4) or from water stress in C_3 plants (Cerling and Quade 1993). Pedogenic carbonates should be collected from at least 50 cm depth to ensure that the soil CO_2 reflects the ^{13}C composition of respired organic C in soil CO_2 rather than infiltrated atmospheric CO_2 (Cerling and Quade 1993). A discussion of biases associated with pedogenic carbonate can be found in **Biases and Shortcomings**.

Pedogenic Siderite: Siderite (Fe(II)CO_3) crystallizes in the reducing phreatic waters of poorly drained soils, commonly around the roots of hydrophilic vegetation (Ludvigson et al. 1998, 2013). Spherulitic masses of siderite are detectable in thin-section as typically mm-scale crystals with radial extinction. In addition to its use as a drainage and redox indicator, siderite has been used to reconstruct the $\delta^{18}O$ composition of ancient meteoric water (Ludvigson et al. 2013, and references therein).

Geochemical Characterization

In contrast to mineralogical characterization, which reveals the organization of elements in soils, geochemical analysis measures the concentrations of elements in a sample. Paleosols should be geochemically analyzed to determine parent material uniformity, the presence and magnitude of pedogenic processes (Birkeland 1999), and to estimate paleoenvironments or paleoclimate variables (Sheldon and Tabor

2009; Tabor and Myers 2015). Bulk geochemical samples should be prepared in a similar fashion to modern soils. Samples are air-dried, broken up with a mortar and pestle, and macroscopic minerals, such as mineral nodules or salts, are discarded so that only the soil matrix is analyzed (Soil Survey Staff 2014b). Treatment with HCl is not necessary and will bias quantitative geochemical analysis if Ca^{2+} in the matrix is removed with carbonate (Myers et al. 2014; Tabor and Myers 2015). Samples are then powdered (typically with >85% of sample finer than 100 μm) and analyzed using inductively-coupled plasma (ICP) – mass spectrometry (MS), – optical emission spectrometry (OES), – atomic emission spectrometry (AES), or X-ray fluorescence (XRF). The results are usually reported in elemental or oxide weight percent. Several applications of geochemical characterization are described here, but for a more thorough review, see Sheldon and Tabor (2009) and Tabor and Myers (2015).

Molecular Weathering Ratios: Molecular weathering ratios can be used as a qualitative approximation to understand weathering in the paleosol and to evaluate relative changes down profile (Retallack 2001), but more quantitative methods such as constitutive mass-balance are now preferred by paleopedologists (see below). Oxides are by convention converted from weight percent to molar percent (dividing by molecular weight of each oxide) before calculating molecular weathering ratios. A discussion of the inferences that can be drawn from molecular weathering ratios can be found in Retallack (2001). The chemical index of alteration (CIA) and chemical index of alteration minus potassium (CIA-K) can be used as a proxy for weathering intensity in a paleosol (Table 9.7; Nesbitt and Young 1982; Maynard 1992; Sheldon et al. 2002). These weathering indices measure clay formation and base loss associated with feldspar weathering and base retention and carbonate precipitation in arid environments (Maynard 1992; Sheldon et al. 2002; Lukens et al. 2018). CIA-K is the preferred weathering ratio for pre-Quaternary paleosols as it reduces digenetic effects of potassium metasomatism (Maynard 1992; Sheldon et al. 2002).

Constitutive Mass-Balance: Mass-balance is a more powerful tool than molecular weathering ratios because it quantitatively compares geochemical and volumetric changes in soil horizons relative to their parent material (Brimhall and Dietrich 1987; Brimhall et al. 1991a, b). Due to variability in parent material compositions, the first step in the mass-balance approach is to evaluate parent material

Table 9.7 Common MAP proxies

Proxy	Soil Orders	Equation for MAP (mm yr^{-1})	Error (mm yr^{-1})	Range (mm yr^{-1})	Reference
Depth to Bk	Any soil with a Bk horizon	$MAP = 137.24 + 6.45D - 0.013D^2$ D = depth to Bk (cm)	±147	≤ 1000	Retallack (2005)
Depth to Bk	Only Vertisol microlows	$MAP = 2.81 \times 10^{-4}D^2 - 0.4025D + 217.35$ (1%) $MAP = 1.41 \times 10^{-4}D^2 - 0.1255D + 109.58$ (2%) $MAP = 6.00 \times 10^{-5}D^2 - 0.0328D + 40.16$ (5%) D = depth to Bk (cm)	±420 ±230 ±230	700–1400	Nordt et al. (2006)
Depth to Bk	Only Vertisol microhighs	$MAP = 4.42 \times 10^{-4}D^2 - 0.9437D + 51252$ (1%) $MAP = 9.48 \times 10^{-4}D^2 - 2.0687D + 1141.12$ (2%) $MAP = 1.52 \times 10^{-3}D^2 - 3.35D + 1853.2$ (5%) D = depth to Bk (cm)	±480 ±490 ±370	700–1400	Nordt et al. (2006)
Depth to By	Any soil with gypsic horizon	$MAP = 87.593e^{0.0209D}$ D = depth to By (cm)	±129	≤ 1000	Retallack and Huang (2010)
CIA-K	Any soil without gleying or with ≤ 5 wt% carbonate as well as Oxisols, Entisols, Andisols, and Vertisols	$MAP = 14.3x - 37.6$ $x = Al_2O_3/(Al_2O_3 + CaO + Na_2O) \times 100$	±172	200–1600	Maynard (1992), Sheldon et al. (2002a), Tabor and Myers (2015)
Bases/ Alumina	Same as CIA-K	$MAP = -259.3 \ln(x) + 759$ $x = (CaO + MgO + Na_2O + K_2O)/Al_2O_3$		200–1600	Sheldon et al. (2002)
CALMAG	Only Vertisols	$MAP = 22.69x - 435.8$ $x = Al_2O_3/(Al_2O_3 + CaO + MgO)$	±108	200–1400	Nordt and Driese (2010a, b)
Goethite/ Hematite	All except Vertisols, Andisols, Entisols, and Gelisols	$MAP = 1003.5[G/H] + 49.3$ G/H = goethite-hematite ratio	±157	100–3300	Hyland et al. (2015)
Fe_{TOT}	Only Vertisols	$MAP = 31.52Fe_{TOT} + 654.38$ Fe_{TOT} = total Fe content of FeMn nodules (wt%)		850–1300	Stiles et al. (2001)

composition using immobile and chemically inert major, trace, and rare earth elements (e.g., Ti, Zr, Al, Nb, La, Ce). The most commonly used immobile constituents are Ti, Zr, and Al, as the deviation from the parent material tends to be relatively small. The concentration of other elements (e.g., Na, Mg, K, Ca, Si, Fe(II), Mn(II), C, P) is weighed against the immobile element to elucidate additions and/or losses relative to the parent material. For these methods, selection of an appropriate, homogeneous, and unweathered parent material is best, although some studies have found that using the lowermost, least-weathered horizon for each paleosol provides acceptable results (e.g., Driese et al. 2016). If the parent material is approximated, it is best to discuss the mass-balance results in a conservative manner (e.g., minimum apparent loss or gain of an element).

Vertical changes in grain size in soil profiles are common and induce heterogeneity in starting mineralogy, which can influence the results of mass balance calculations. Likewise, additions of sedimentary material during pedogenesis can produce variability in weathering trends toward the top of profiles. To account for this issue, parent material uniformity can be examined using geochemical reconstructions applied to buried soils and paleosols following Maynard (1992). The ratio of TiO_2/Zr for each candidate parent material and the ratio TiO_2/Zr of the average composition of the paleosol can be used to determine percent deviation from the parent material (Maynard 1992):

$$\text{Percent Deviation} = \left(\frac{\text{Ti}_{\text{soil}}}{\text{Zr}_{\text{soil}}} - \frac{\text{Ti}_{\text{parent}}}{\text{Zr}_{\text{parent}}}\right) \Big/ \left(\frac{\text{Ti}_{\text{parent}}}{\text{Zr}_{\text{parent}}}\right) \times 100 \tag{2}$$

Cross plots of immobile elements are also useful for determining parent material such as $Zr/(Zr/TiO_2)$ vs. $TiO_2/(TiO_2/Zr)$ (e.g. Ashley and Driese 2000; Driese et al. 2000; Beverly et al. 2014). The selection of Ti and Zr as immobile index elements has also been discussed by Stiles et al. (2003a, b), who found that Zr can be introduced as windblown silt in semi-arid environments.

The following calculations for constitutive mass-balance analysis are derived from Brimhall and Dietrich (1987), Brimhall et al. (1991a, 1991b), Chadwick et al. (1991) and Anderson et al. (2002). Closed system weathering conserves mass and volume, such that:

$$V_p \rho_p C_{i,p} = V_w \rho_w C_{i,w} \tag{3}$$

where the volume (V), bulk density (ρ) and concentration (C) of an immobile element (i) in the parent material (p) are equal to those of a weathered horizon (w).

In most soils, weathering is an open system process, wherein constituents are added or removed from the system. In such cases, volumetric change is tracked as strain (ε):

$$\varepsilon_{i,w} = \frac{\rho_p C_{i,p}}{\rho_w C_{i,w}} - 1 \tag{4}$$

Note that if losses of an element in a weathered horizon are offset by changes in volume, strain is zero. Elemental additions and losses are calculated using the open-system mass-transport function (τ), which incorporates volumetric changes:

$$\tau_{j,w} = \frac{\rho_w C_{j,w}}{\rho_p C_{j,p}} \left(\varepsilon_{i,w} + 1\right) - 1 \tag{5}$$

where $C_{j,w}$ is the concentration of an element of interest (j) in weathered material and $C_{j,p}$ is the concentration of an element of interest (j) in the parent material. When $\tau_{j,w} = -1$, 100% of the element j has been removed in a weathered horizon compared the parent material. Likewise, when $\tau_{j,w} = 1$, 100% of element j has been added to a weathered horizon compared the parent material starting composition. Definitions of all variables in the above equations are summarized in Table 9.8.

Interpretation of mass-balance calculations should be formed in terms of soil-forming process, which may vary depending on the environmental setting or parent material type. Translocation of carbonate is identified by simultaneous changes in Ca, Mg, and Sr. Clay or zeolite accumulation is indicated by increases in Na, Al, Si, and Mg (Ashley and Driese 2000; Beverly et al. 2014). In contrast, Mg fluxes have also been found to closely track clay-rich, Mg-smectite parent material rather than carbonate (Beverly et al. 2014). Losses of the redox-sensitive elements (e.g., V, Fe, Cu, and Mn) may indicate prolonged waterlogging followed by drainage. Cu is also important in the formation organic ligands and can be indicative of changes in organic matter content (Brantley et al. 2007). Likewise, phosphorus

Table 9.8 Glossary of variables modified from Brimhall and Dietrich (1987)

$C_{i,w}$	= concentration of immobile element (i) in weathered material
$C_{i,p}$	= concentration of immobile element (i) in parent material
ρ_p	= bulk density of parent material
ρ_w	= bulk density of weathered material
$\varepsilon_{i,w}$	= strain or volumetric change of immobile element (i) in weathered material
$\tau_{j,w}$	= translocation of element of interest (j) in weathered material
$C_{j,w}$	= concentration of element of interest (j) in weathered material
$C_{j,p}$	= concentration of element of interest (j) in parent material

enrichment at the soil surface can indicate biocycling (Brantley et al. 2007). Human remains and debris can also be enriched in P relative to surrounding material, but additional evidence would be needed to support this interpretation (Eidt 1985; Holliday 2004; Holliday and Gartner 2007).

Pedotransfer Functions: Pedotransfer functions relate bulk geochemistry of soils to properties that are not directly measureable in paleosols. For example, modern soils are studied using a suite of standard characterization properties, which are measured on unconsolidated material to assess physical and chemical properties operating at the colloidal scale (Soil Survey Staff 2014b). Lithification and compaction of paleosols precludes such measurements, but the relationship between characterization properties and geochemical indices allows for detailed pedologic study of paleosols (Nordt and Driese 2010b). Soil characterization properties can yield quantitative paleoecological information on paleosol fertility and hydrology unavailable by any other means (Table 9.9; Nordt et al. 2011). A detailed explanation of modern soil properties can be found in Soil Survey Staff (2014b). Currently, most pedotransfer functions are restricted to paleo-Vertisols (Nordt and Driese 2010b), which are commonly found in the geologic record (Driese and Forman 1992; Driese et al. 1993; Prochnow et al. 2006; Cleveland et al. 2007; Rosenau et al. 2013; Torres and Gaines 2013; Driese et al. 2016). Efforts to develop more broadly applicable pedotransfer functions for soil characterization variables are ongoing and currently limited to soil pH (Lukens et al. 2018). If a paleosol is unconsolidated, soil properties could theoretically be directly measured using established methods (Nordt et al. 2013; Soil Survey Staff 2014b). However, it is not clear to what extent characterization properties are affected in buried paleosols that have not experienced macroscopic evidence of diagenesis.

Table 9.9 Pedotransfer functions for paleo-Vertisols modified from Nordt and Driese (2010b)

Soil Property	Soil Horizon	Equation	Error	Paleoecological importance
Total clay (%)	Noncalcareous	$y = 3.965x + 3.681$ $x = Al_2O_3$ (%)	±4	*Porosity and water holding capacity*
	All ≤5% CaO	$y = 3.118x + 15.22$ $x = Al_2O_3$ (%)	±5	
Fine clay (%)	All ≤10% CaO	$y = 9.974 \ln(x) + 23.68$ $x = Al_2O_3/(CaO + MgO)$*	±7	*Translocation of pedogenic clay*
COLE Coefficient of linear extensibility (cm cm^{-1})	Noncalcareous	$y = 1.345x + 0.077$ $x = CaO + MgO$*	±0.019	*Shrink-swell behavior*
	All ≤10% CaO	$y = 0.00238x - 0.02965$ $x = SiO_2$ (%)	±0.025	
Bulk density (g cm^{-3})	Noncalcareous	$y = 0.011x + 0.457$ $x = SiO_2$ (%)	±0.10	*Porosity and aeration of soil*
CEC_7 ($cmol_c$ kg^{-1}) Cation exchange capacity	Noncalcareous	$y = 585.8x + 14.66$ $x = CaO + MgO$*	±4	*Soil fertility (plant available nutrients)*
	All ≤10% CaO	$y = -8.203x + 44$ $x = (CaO + MgO)/Al_2O_3$*	±8	
pH (H_2O)	All	$y = 0.87 \ln(x) + 9.648$ $x = CaO + MgO$*	±0.6	*Soil fertility (acidity)*
BS Base saturation (%)	All	$y = 14.69 \ln(x) + 134.0$ $x = CaO + MgO$*	±8	*Soil fertility (alkalinity)*
$CaCO_3$ (%)	All	$y = 1.436x + 0.025$ $x = CaO$ (%)	±4	*Buffers soil pH*
Fe_d (%) Crystalline Iron Oxide	All	$y = 0.085x^2 - 0.352x + 0.658$ $x = Fe_2O_3$ (%)	±0.4	*Oxidizing conditions*
Fe_o (%) Amorphous Iron Oxide	All	$y = 0.018e^{570.3x}$ $x = Ni$ (%)	± 0.05	*Poorly drained conditions*
ESP (%) Exchangeable Sodium Percentage	All ≤10% CaO	$y = 110.4x - 2.935$ $x = Na_2O/Al_2O_3$*	±4	*Soil fertility (sodicity)*
EC (dS m^{-1}) Electrical conductivity	All ≤10% CaO	$y = 290.7x^2 - 24.33x + 1.145$ $x = Na_2O/Al_2O_3$*	±3	*Soil fertility (salinity)*

% calculated using weight percent; *converted to molar before calculation

Mean Annual Precipitation and Temperature Proxies

Mean annual precipitation (MAP) can be estimated from paleosols by either physical or geochemical techniques (Table 9.7). The depth to calcic horizon (Bk) and depth to gypsic horizon (By) in soil profiles increases as a function of increasing MAP (Retallack 2005; Nordt et al. 2006; Retallack and Huang 2010). Nordt et al. (2006) used a "family of curves" approach to derive significant relationships between depth to nodular Bk horizons based on the concentration of nodules (1, 2, and 5%) for each of the microhigh and microlow portion of Vertisol profiles. The limitations for applying these depth-to-mineral functions are: (1) the tops of paleosols are commonly eroded (Myers et al. 2014), which results in MAP underestimates, (2) carbonate and gypsum must be nodular in morphology, (3) carbonate and gypsum must be analyzed in thin section to demonstrate pedogenic origin, and (4) burial compaction must be accounted for (Sheldon and Retallack 2001).

The bulk geochemistry of paleosol B horizons is a more reliable method of reconstructing MAP because it does not require the original thickness of the paleosol to be preserved. Typically, the uppermost B horizon is used as the "control interval" (25–100 cm depth in profile; Nordt and Driese 2010a). Bulk geochemistry is related to MAP through two processes: (1) hydrolysis of silicate minerals and progressive leaching of base cations at high MAP, and (2) precipitation of carbonates and retention of base cations at low MAP. Sheldon et al. (2002) found the weathering ratio CIA-K (Sect. 4.5.1) to be the most useful predictor of MAP across a range of soil types, which increases as CaO and Na_2O are preferentially leached relative to Al_2O_3. The acid neutralization capacity of secondary carbonates precipitated at low MAP (*sensu* Chadwick and Chorover 2001) results in retention of structural CaO and Na_2O and correspondingly low CIA-K values. K_2O is excluded from CIA-K because of potential diagenetic addition of K^+ during illitization (Fedo et al. 1995; Sheldon et al. 2002; Lukens et al. 2018).

More recently, attempts have been made to decrease standard error in MAP functions by developing proxies specific to more narrow ranges of soil environments (i.e., *cl, o, r, p, t;* Sect. 3.1). CALMAG (Nordt and Driese 2010a) was developed specifically for clay-rich Vertisols, which are soils dominated by vertic features such as gilgai topography, pedogenic slickensides, and wedge peds that develop in soils that experience seasonal water deficit. Vertisols tend to be developed on pre-weathered parent material where little hydrolysis occurs, and therefore, tracking the flux of CaO and MgO (relative to Al_2O_3) as carbonates or exchangeable cations within the system results in improved MAP prediction (Nordt and Driese 2010a).

It should be noted that climofunctions using weathering indices that relate base cations to Al_2O_3 are only useful for estimating MAP up to a maximum of ~1600 mm (Sheldon et al. 2002; Nordt and Driese 2010a). Above 1600 mm, kaolinite is the dominant clay mineral and any weatherable bases are either completely depleted – yielding an effective ratio of alumina:alumina that does not change with increasing MAP – or bases have yet to be leached and the soil is not in equilibrium with climatic conditions. A few novel methods to estimate MAP utilize Fe in Fe–Mn nodules (Stiles et al. 2001) or the relative abundance of goethite to hematite (Hyland et al. 2015). The weight percent of Fe in Fe–Mn nodules found in Vertisols increases directly with MAP (Stiles et al. 2001) and can be used in conjunction with other methods. Most recently the goethite to hematite ratio (G:H) in the B horizon of a global dataset of soils is shown to correlate with MAP (Hyland et al. 2015). The ratio of G:H can be measured using either X-ray diffraction (XRD) or isothermal remnant magnetization (IRM) acquisition curves, but IRM is preferred due to the higher sensitivity in low concentrations of magnetic minerals in soils and paleosols (Hyland et al. 2015). G:H is the first MAP proxy to extend beyond the barrier of 1600 mm, with a predictive range of 100–3300 mm (Hyland et al. 2015). However, Maxbauer et al. (2016a) note that goethite is difficult to measure using IRM methods and easily recrystallizes during diagenesis, which would preclude measurements of original G:H ratios for most paleosols.

Mean annual temperature (MAT) proxies for paleosols have also improved over the last decade. Early attempts relied on bulk geochemical proxies such as the salinization index (Sheldon et al. 2002). An alternative approach to predicting paleo-MAT uses the oxygen stable isotopic composition of carbonates ($\delta^{18}O$), which is dependent on both water ($\delta^{18}O$) composition and the temperature at which carbonates precipitated (i.e., soil temperature) (Cerling 1984). Dworkin et al. (2005) found a possible way to side-step original water composition by combining two equations that relate (1) the fractionation of oxygen from meteoric water into carbonate, and (2) the spatial relationship between MAT and the $\delta^{18}O$ of meteoric water. However, it is possible that past time intervals had systematic differences in the spatial distribution of meteoric water $\delta^{18}O$ and variation in mean annual temperature, which would preclude the application of this proxy (e.g., Lukens et al. 2017a).

Recent advances in the analytical resolution of isotopes of carbon and oxygen have allowed for the measurement of clumped isotopes (Δ_{47}). In this technique, the abundance of a doubly substituted CO_2 molecule of mass 47 ($^{13}C^{18}O^{16}O$) produced when carbonate is digested during measurement is proportional to soil temperature, without assumptions based on original $\delta^{18}O$ composition of soil waters (Ghosh et al. 2006; Passey et al. 2010; Quade et al. 2013). These carbonates are probably recording the warm, dry season temperature when carbonate is more likely to form and therefore do not represent a mean annual temperature (Passey et al. 2010; Gallagher and Sheldon 2016). The Δ_{47} temperature estimates provide a unique view into critical paleoclimatic variables recovered from paleosols (e.g., Snell et al. 2013) when burial temperatures do not exceed 100°C (Henkes et al. 2014).

Classifying Paleosols

Several perspectives exist on the taxonomic classification of paleosols (Mack et al. 1993; Retallack et al. 1993; Nettleton et al. 2000), which stem from the fact that some distinctions in modern soil taxonomy rely on soil and climate properties (e.g., moisture and temperature regimes) that are immeasurable in paleosols. Each system has strengths and weaknesses (i.e., Retallack et al. 1993, comment and reply), and to date no taxonomy is uniformly accepted among paleopedologists. The two most commonly applied paleosol classification systems are Mack et al. (1993) and Retallack et al. (1993). The Mack et al. (1993) system classifies paleosols without the use of modern soil properties such as bulk density or base saturation, which are not measurable in most paleosols, and creates new names for some paleosol orders. The Retallack et al. (1993) classification system simplifies USDA Soil Taxonomy to allow for paleosol classification by omitting properties that cannot be directly measured in paleosols but retains the naming conventions of Soil Taxonomy. The authors prefer to adapt Soil Taxonomy where possible for comparisons with modern soils and to encourage collaboration with modern soil scientists. The Illustrated Guide to Soil Taxonomy (Soil Survey Staff 2014a, b) is a useful resource for those unfamiliar with the complexities of Soil Taxonomy and is practical for most paleopedological classification purposes.

Rates of Pedogenesis: Paleosols and paleosol successions can also be categorized based on the varying degrees of pedogenesis and sedimentation. Paleosols in alluvial environments are broadly split into three different categories: compound, composite, and cumulative (Fig. 9.4). Compound soils are weakly developed with vertically stacked profiles that occur when sedimentation rate is not steady and there is little erosion. In composite or welded paleosols the rate of pedogenesis exceeds that of sedimentation, and with little erosion creates vertically stacked profiles that partially overlap. Cumulative paleosols have slow but overall constant sedimentation rates where pedogenesis can modify the sediment as it is deposited (Kraus 1999). Cumulative soils tend to preserve remains, such as artifacts or fossils, in discrete depth ranges (Holliday 2004, and references therein). With erosion, these three categories can be further split into compound truncated set, composite truncated set, and cumulative truncated set (Kraus 1999). A careful study of root traces can also help classify rates of pedogenesis. Root mats can help identify surfaces within the paleosol but can also form at a lithologic boundary or fragipan (x horizon suffix; Table 9.3) preventing growth. Clay coatings, redoximorphic features, carbonate, Fe–Mn, and zeolites can all preserve roots features (Fig. 9.2E–F; Ashley and Driese 2000; Stoops et al. 2010; Beverly et al. 2014, 2015a). For further discussion of rhizoliths and their paleoecological implications see Kraus and Hasiotis (2006).

Pedofacies and Pedotypes: Paleosols in stratigraphic successions often contain similar attributes, which relate to their depositional environment. Such clustering can reveal vertical and lateral trends that are the result of geologic and climatic phenomena. Changes in paleosol maturity and drainage are typically assessed using qualitative indices (Table 9.10). Pedofacies use a combination of sedimentological and pedological features to categorize paleosols (Bown and Kraus 1987; Kraus 1987). The pedofacies approach was developed in alluvial settings and relies on the inverse relationship between deposition and soil development – grain size and flooding frequency decrease with distance from sediment source, which is accompanied by increased soil maturity and thicker individual soil profiles.

Retallack (1994) developed a more universal model of paleosol categorization, in which paleosols with similar features are grouped as pedotypes. A pedotype is a representative paleosol and reference to the type profile for each pedotype allows for a non-genetic classification of each paleosol and is indicative of one set of soil forming (*cl, o, r, p, t*) conditions (Retallack 1994). Analysis of pedofacies and pedotypes in vertical stratigraphic successions can reveal long-term trends in climate or basin subsidence (Atchley et al. 2004; Cleveland et al. 2007; Aziz et al. 2008; Atchley et al. 2013). By modifying concepts originally developed in marine sequences, Atchley et al. (2004, 2013) used combinations of grain size and paleosol thickness, maturity, and drainage to delineate fluvial aggradational cycles (FACs) where fining upward successions are overlain by paleosols (e.g., Figure 9.2C). See Atchley et al. (2013) for a review of paleosol stacking pattern analysis.

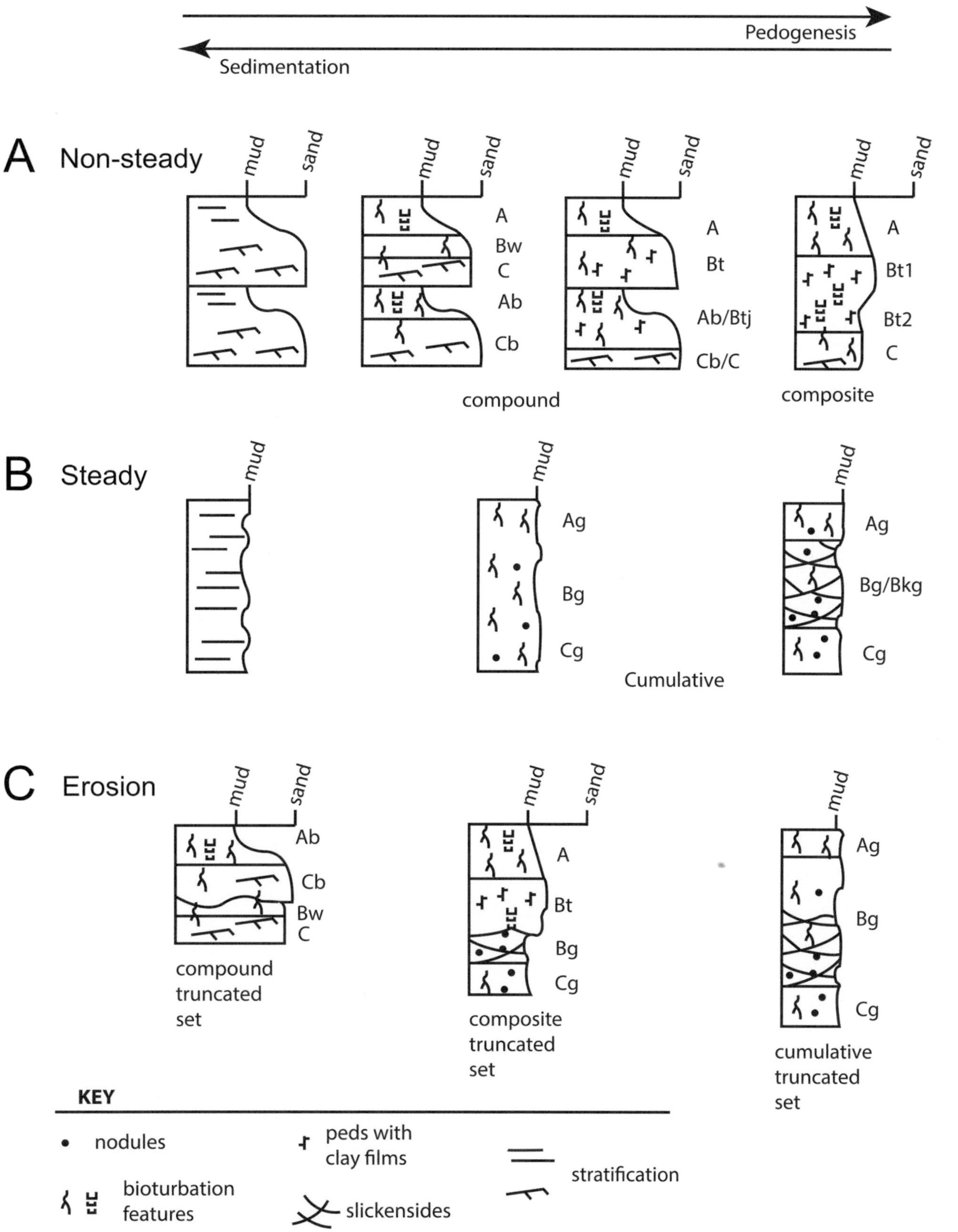

Fig. 9.4 Reproduction of Fig. 9.2 from Kraus (1999) illustrating the pedofacies method of categorizing paleosols. Abbreviations: Ag – gleyed A horizon; Bg – gleyed B horizon; Bw – B horizon showing color or structure development but little if any illuvial accumulation; Bt – B horizon showing accumulation of clays; Btj – incipient development of a Bt horizon; Cg – gleyed C horizon. Reprinted with permission from Elsevier

Table 9.10 Paleosol maturity and drainage indices. Modified from Atchley et al. (2013)

	Rank	Features
Maturity		
1	Very weakly developed	Weak mineral surface horizon (A); no subsurface horizon; massive to faint bedding; root traces
2	Weakly developed	Mineral surface horizon (A); incipient subsurface Bk horizons with less than 5% carbonate nodules, not qualifying as a Calcic diagnostic horizon
3	Moderately developed	Mineral surface horizon (A); subsurface Bk horizon with more than 5% carbonate nodules qualifying as a Calcic diagnostic horizon
4	Strongly developed	Mineral surface horizon (A) and E horizon; red or purple subsurface Bt horizon with sufficient clay enrichment to qualify as an Argillic diagnostic horizon
Drainage		
1	Poorly drained	Gray of blue green surface and subsurface horizons
2	Moderately drained	Yellow or brown subsurface horizons
3	Well drained	Red or purple subsurface horizons

Strengths of Approaches

Because paleosols are by their nature an unconformable surface, they have been ignored in the past in favor of sediment cores from lakes or oceans to understand the effect of long-term climate changes on paleoecology. However, paleosols provide regional, site-based, environmental reconstructions and are thus very useful for understanding paleoecological variability in the immediate site area. Collecting paleosol samples for analysis using the variety of methods previously mentioned is fairly simple. When used properly to reconstruct climate or vegetative cover, paleosols act as *in situ* archives that record environmental conditions on moderate time scales.

With an increase in the detail to which paleoecological data can be extracted from individual specimens and bonebeds as a whole, it is essential to understand the context of fossil accumulations. Where paleosols contain fossil remains, they can be used to infer the method in which fossils were incorporated into the stratigraphic record. The relative degree of paleosol development at or near an archaeological or paleontological site can also provide information on the state of artifact and bone preservation (Holliday 2004). Understanding the depositional and post-depositional context of fossil and archeological sites is a key component to unraveling the taphonomy of the remains to be studied, the evolution of landscapes between depositional events, and can inform researchers on potential biases of preservation for a given locality.

Measuring and analyzing multiple paleosol profiles along a particular buried or exhumed paleo-CZ or paleocatena is a popular approach for reconstructing the paleoecosystem associated with an archaeological site. A catena study assesses the variability of soils across a landscape caused by changes in topography due to changes in hydrology (Morrison 1967; Birkeland 1999; Holliday 2004). Paleolandscapes can be studied at different scales depending on the research question being addressed. At the local scale, the relationship between depositional processes, weathering styles, drainage, and artifact accumulation can be examined. On the basin-scale, amalgamated strata representing a multitude of paleoenvironments can be used to understand long-term changes in climate and subsidence (Kraus 1999; Atchley et al. 2004; Cleveland et al. 2007, 2008; Aziz et al. 2008; Nordt et al. 2012; Atchley et al. 2013; Nordt and Driese 2013). An example of this method at Olduvai Gorge in Tanzania, where correlative surfaces buried by volcanic ashes have been used to reconstruct aspects of paleo-CZs from paleocatena successions (see description in **Applications**; Ashley and Driese 2000; Sikes and Ashley 2007; Ashley et al. 2014; Beverly et al. 2014, 2015a; Driese and Ashley 2016).

Biases and Shortcomings

In the spirit of Vance Holliday's (2004) insightful review of paleosols as paleoenvironmental indicators, we also quote Valentine and Dalrymple (1976, p. 218), "Although soil science is under great pressure to furnish environmental evidence, it is debatable whether we understand the interaction of the soil-forming processes with the site and environmental factors well enough yet to make confident extrapolations." This point alludes to the fact that many researchers make interpretive leaps to paleoenvironmental reconstructions without detailed understanding of the soil-forming and post-burial (diagenetic) processes that may have altered the paleosol system. This point should not deter future paleosol work. Rather, with a detailed analysis and an understanding of the biases discussed here, paleosol research using multiple proxies can yield a more nuanced interpretation of the paleoenvironment.

Disequilibrium with Climate or Environment

Soils are the product of local conditions, which may be considered an advantage in some paleoecological studies using paleosols; however, soils may not reflect the regional conditions due to the local hydrology, which may mimic changes in climate (e.g., Driese and Ashley 2016). It has been shown that multiple pedogenic pathways can yield the same soil property (Holliday 2004). For example, lateral changes in the development of a subsoil argillic (Bt) horizon from clay illuviation may be due to drainage or the duration of soil formation. This phenomenon, known as equifinality, suggests that in some cases a single paleoecological interpretation may not be possible when relying solely on the paleosol record. Thus, it may be difficult to extrapolate environmental and climate reconstructions beyond the site or pedon (Holliday 2004), but if correlation using precise dating methods or correlative tephras is possible, this issue can be avoided by studying the site as a paleocatena or paleo-CZ. By looking at the variations in a paleocatena across the whole landscape rather than focusing solely on the archaeological or fossil site, diachronous soil development can be identified and help researchers better interpret the paleoecology from paleosols. With this detailed method, the changes in local hydrology will be understood within the regional context, and paleoenvironmental and paleoclimatic interpretations can be made.

Pedogenesis on any substrate is a time-transgressive process and may not reach steady state with respect to climate and other environmental factors for upwards of 10^3–10^4 years (Yaalon 1971; Huggett 1998, and references therein). Therefore, less than centennial-scale changes may not be detectable in the soil or paleosol record and may reflect a sum of the environmental changes over the course of formation (Holliday 2004; Richter and Yaalon 2012). The relatively long durations of time required for soils to equilibrate with environmental conditions stands in stark contrast to the time required for the accumulation of artifacts or fossils, which can be instantaneous. Thus, a disparity of time resolution exists between the presence or burial of fossil remains and our ability to reconstruct the environment from which they are derived. This may result in disagreement between results of pedogenic and paleoecological or archeological proxies for environments. Young soils can have morphological and geochemical properties present in other (often drier) climatic settings. For example, excess base cations in immature soils will yield artificially low MAP and MAT estimates using climate proxies, resulting in "mock-aridity" or "mock-frigidity." This should be kept in mind when interpreting climate change in stratigraphic sequences using paleosols of varying maturity (Stinchcomb et al. 2016).

Misuse of Proxies

Many paleosol proxies for climate have inherent limitations and are therefore subject to misapplication. For example, the depth to Bk and By proxies (Retallack 2005; Retallack and Huang 2010) predicts MAP based on the depth from the surface of a soil to the zone of nodular carbonate precipitation. As mentioned previously, many paleosols in the rock record are truncated by overlying fluvial deposits such that A horizons are a rarity in alluvial paleosols (Fig. 9.2C; Myers et al. 2014). This limits the depth to carbonate proxy to relatively few paleosols and must be applied with justification that A horizons were present or that the MAP value is an under estimation (Nordt et al. 2006; Prochnow et al. 2006). Early applications of the depth to carbonate proxy were performed without assessment of the origin of the carbonate, presence of surface horizons, or taxonomic interpretation of paleosols (e.g., Retallack 1997; Sheldon and Retallack 2004). It is now recognized that depth to carbonate in Vertisols follows a unique function (Nordt et al. 2006; Prochnow et al. 2006) that deviates from the more universal equation of Retallack (2005), which should caution researchers in application of such proxies in the absence of detailed paleosol description. Bulk geochemical proxies for MAP and MAT can be easily misused as well. Most paleosol proxies for climate are limited to specific soil orders or depositional conditions (Table 9.9). Misclassification of paleosols by horizons or soil order can therefore result in misapplication of climofunctions.

Perhaps a more widespread abuse of paleosols in climate studies is that very few researchers verify that the discrete geochemical values of paleosol samples fall within the range of values used to generate climate proxies. Because the relationship between a geochemical index and climate (e.g., CIA-K and MAP) is derived through a training data set, any applications of the regression equation must be confined to data that are similar to the training data. For example, unusually high concentrations of MgO (up to 17%) are present in paleosols at Olduvai Gorge because the parent material is an authigenic lacustrine Mg-rich smectite (Beverly et al. 2014). This is significantly more MgO than the dataset on which the CALMAG precipitation proxy was developed, which is only applicable in paleosols with <3% MgO (Nordt and Driese 2010a, b). This does not mean that these proxies are ineffective at predicting precipitation in diverse settings, but that care must be taken to correctly apply the proxies.

Stable isotopes of pedogenic carbonates are also commonly used as paleoecological indicators, but an understanding of soils and soil formation is necessary to apply these proxies. Researchers should interpret the presence of carbonate nodules or pedogenic carbonates in general with caution as several studies have shown that these features can be influenced by soil hydraulic properties (porosity, permeability), texture, clay content, and slope position (Holliday 2004). Also, the identification of carbonate in a paleosol does not mean that it is pedogenic in origin. For example, carbonate concretions are commonly found at Olduvai Gorge up to 20 or 30 cm in diameter and are associated with post-depositional groundwater movement through some paleosols (Fig. 9.2N), as indicated by their large size, irregular morphology, and sparry texture.

Reworking and Inheritance

Paleosols are components of sedimentary systems and are therefore subject to processes of erosion and deposition. The parent material of a soil reflects regional geology, which can include soils actively forming under variegated environmental (e.g., *cl, o, r, p, t)* conditions, bedrock of various compositions and windblown sediment. Any material removed from these reservoirs can be deposited on or incorporated into a soil. Geochemical proxies for climate (e.g., CIA-K) and vegetation (e.g., phytoliths, $\delta^{13}C$ of organic matter) can be altered or biased if any constituent is included from reworked sediment.

Pedotransfer functions for climate that rely on paleosol bulk geochemistry will provide underestimates if sediment is added through any means of deposition. Geochemical climofunctions should not be applied to cumulative paleosols (Sheldon et al. 2002; Sheldon and Tabor 2009), which can form in eolian, fluvial, or mixed-source environments. Nordt and Driese (2010a) note that the smectite on which most Vertisols form is inherited from the surrounding catchment as alluvial clays rather than being authigenically manufactured in the soil profile. Other reworked pedogenic minerals, such as kaolinite or hematite, can make relatively immature paleosols appear to be mature, or can influence the color of a paleosol. Cross-referencing mineralogy with field and micromorphologic observations can mitigate misapplication or misinterpretation of paleoclimate in such settings.

Reworked organic matter and pedogenic carbonates are also a potential problem in alluvial deposits. Because paleosols typically have very low organic C concentrations (Wynn 2007), the relative abundance of *in situ* versus inherited organic matter can be difficult to discern. Rounded pedogenic carbonates found with coarser grains at the base of alluvial deposits are clear evidence for reworking (Lukens et al. 2017b). Comparison between SOM $\delta^{13}C$, pedogenic carbonate $\delta^{13}C$, and phytoliths has been a useful tool to reveal biases in the pedogenic carbonate vegetation signal (Cotton et al. 2012; Chen et al. 2015; Garrett et al. 2015). The difference between the $\delta^{13}C$ of co-occurring organic matter and pedogenic carbonate is 14–17‰ in modern systems (Cerling and Quade 1993) but can potentially shift beyond that range given changes in atmospheric pCO_2 in the past (e.g., Mora et al. 1996) or isotopic fractionation of SOM during decomposition (Wynn 2007).

Diagenesis

Any alteration of paleosols after burial is referred to as diagenesis, and since the study of paleosols inherently involves analyzing soils that have left the *cl, o, r, p, t* conditions in which they were formed, one must carefully differentiate physical and geochemical features formed during pedogenesis versus those formed through diagenesis. Lithification is perhaps the most readily recognizable product of diagenesis and involves compaction and cementation. Compaction reduces pore volume, which may sufficiently inhibit fluid flow in clay-rich paleosols, but is unlikely to prevent groundwater transmission through paleosols with sandier textures. Compaction can also lead to deformation of mineral grains and reduction on primary porosity (Pettijohn et al. 1987).

Post burial cementation – not to be confused with pedogenic calcretes or fragipans – occurs in groundwater and plugs void space with a range of cementing agents. Calcite, silica, iron oxides, and clays are the most common cements in sedimentary rocks and are recognizable due to their occurrence within primary voids (e.g., root traces), along features of textural contrast (e.g., burrows infilled with sediment of an overlying deposit), or as overgrowths on mineral grains and pedofeatures (Stoops 2003). Calcareous paleosols typically show a mix of carbonate morphologies due to both pedogenic and diagenetic remobilization of primary pedogenic carbonates. The current rule of thumb in sampling pedogenic carbonate is to target micrite and avoid any material with sparry or megacrystalline texture (Sheldon and Tabor 2009; Tabor and Meyers 2015), though soils formed on coarse-grained sediments can contain pedogenic carbonates with sparry or microspar texture (Weider and Yaalon 1982). Any researcher attempting to study pedogenic carbonate for paleovegetation or paleoclimate reconstruction should first analyze carbonates in thin section and micro-drill targeted zones to avoid diagenetic overgrowths. Cathodoluminescence is a useful technique to identify histories of carbonate cementation (Barnaby and Rimstidt 1989; Machel et al. 1991; Michel et al. 2013) and should be considered before isotopic sampling is performed.

Diagenetic mineralization can alter the color of a paleosol to reflect the color of cementing agents or as a result of the dissolution of unstable minerals. Burial reddening occurs due to hematite cementation in oxidizing groundwater (Retallack 2001), whereas calcite cements can impart a white to pale color. Decomposition of organic matter in paleo-surface horizons will remove the dark colors evident in modern O and A horizons, leaving more uniform coloration to preserved paleosol profiles. If organic matter is decomposed by anaerobes after the paleosol is buried below the water table, blue-green drab halos will form around individual root traces or laterally along paleo-root mats (Retallack 2001). These colors can be differentiated from gleying due to saturation by identification of pedogenic features indicative of well-drained conditions, such as ped structure, translocation of clay or Fe, and precipitation of pedogenic calcite at depth.

Paleosols flooded by lacustrine or marine waters can exhibit a drab, reduced zone near their upper boundary (e.g., Driese and Ober 2005). This process is called pseudogley because it imparts gleyed colors typically observed in saturated soils. Pedogenic features indicative of episaturation should be present if gleying were due to periodic standing water on the soil surface, such as lateral-branching root traces, Fe-redox halos around roots and voids, pedogenic siderite, pyrite or jarosite, or accumulations of organic matter. Pseudogley has also been attributed to flushing of alkaline groundwater through buried paleosols (Pimentel et al. 1996).

Potassium is added during diagenetic conversion of smectite to illite (Fedo et al. 1995), which is the primary reason for excluding K_2O from pedotransfer functions and climate proxies (Harnois 1988; Maynard 1992). However, the physical pedogenic features indicative of smectite-rich Vertisols – including slickensides, sepic-plasmic clay fabrics and wedge peds – will remain after illitization has taken place (e.g., Driese and Foreman 1992). Current practice in the paleopedology community is to avoid using bulk geochemical data for climofunctions or pedotransfer functions if paleosols have experienced illitization or significant diagenetic mineralization (Sheldon and Tabor 2009; Driese et al. 2007; Medaris et al. 2017).

Pedogenic Mixing of Artifacts or Fossils

Pedogenic mixing processes can obscure discrete artifact layers and widen the depth range (Ferring 1992; Waters 1992; Cremeens et al. 1998). This can occur with aggressive bioturbation by soil fauna, through pedogenic processes such as cryoturbation from freeze-thaw cycles or due to vertic processes causing movement of artifacts or bones. Vertisol movement can also break and polish artifacts (Quade et al. 2004). However, even in Vertisols, which have considerable internal mobility of mass due to shrink-swell processes, artifacts can be placed in relative stratigraphic order with careful study (Waters et al. 2011).

Applications

The paleopedological techniques discussed in this chapter have been applied throughout geologic time, including diverse settings such as Neoarchean paleoweathering surfaces (Driese et al. 2011), late Ordovician and Silurian Appalachian Basin paleosols (Driese and Foreman 1992; Driese et al. 1992), Triassic paleosols at Petrified Forest, Arizona (Cleveland et al. 2007, 2008; Trendell et al. 2013a, b), paleosols spanning the Paleocene-Eocene Thermal Maximum in the Bighorn Basin, Wyoming (Kraus 1997; Kraus and Brown 1988; Kraus and Hasiotis 2006; Aziz et al. 2008; Snell et al. 2013) and Pleistocene-Holocene paleosols in eastern North America (Driese et al. 2005). Here we present examples from East Africa where paleosols have been used to understand the paleoecology at paleontological and archaeological sites throughout the Miocene and Quaternary.

There is a long history of paleosol research in East Africa that began with stable isotopes of C and O from pedogenic carbonates in the Turkana Basin (Cerling et al. 1988) and Olduvai Gorge (Cerling and Hay 1986). These extensive, well-dated paleosol records showed that climate became warmer and drier over the past 2 Ma (Cerling and Hay 1986; Cerling et al. 1988; Levin et al. 2011). This research ushered in a wave of studies focusing on stable isotopes of pedogenic carbonates to understand the paleo-vegetation and paleo-rainfall history at archaeological sites (Cerling et al. 1988; Wynn and Feibel 1995; Sikes et al. 1999; Wynn 2000; Levin et al. 2004; Quade et al. 2004; Wynn 2004; Lepre et al. 2007; Quinn et al. 2007; Campisano and Feibel 2008; White et al. 2009; WoldeGabriel et al. 2009; Cerling et al. 2010; Passey et al. 2010; Cerling et al. 2011; Levin et al. 2011; Quinn et al. 2013; Garrett et al. 2015; Lüdecke et al. 2016). For summaries of the use of stable isotopes in paleoenvironmental reconstructions at hominin sites see reviews by Cerling (2014) and Levin (2015).

This focus on stable isotopes of pedogenic carbonate has yielded long-term records during the Pleistocene for comparison with marine records but has also ignored the wealth of paleoclimatic and paleoenvironmental information that can be gained from the breadth of methods discussed in this chapter. Much remains to be discovered regarding the historical and process-based details of how hominins interacted with their surrounding ecosystem (National Research Council 2010). In this section, we focus paleosol applications at Olduvai Gorge to demonstrate the process and utility of paleosol research (Fig. 9.5). Other similar studies have recently been completed at significant fossil sites in western Kenya at Rusinga Island (Michel et al. 2014) and Karungu (Beverly et al. 2015a; Driese et al. 2016; Lukens et al. 2017b).

Olduvai Gorge is a well-dated archeological site that has been studied for over 50 years, but only recently have the paleosols been recognized for their paleoenvironmental potential (Ashley and Driese 2000; Hover and Ashley 2003; Sikes and Ashley 2007; Magill et al. 2013a, b; Ashley et al. 2014; Beverly et al. 2014; Driese and Ashley 2016). Using well-dated and laterally correlative tephras, Olduvai Gorge can be reconstructed as a series of paleocatenas or paleo-CZs at multiple scales of detail (Fig. 9.6; Ashley et al. 2014; Beverly et al. 2014).

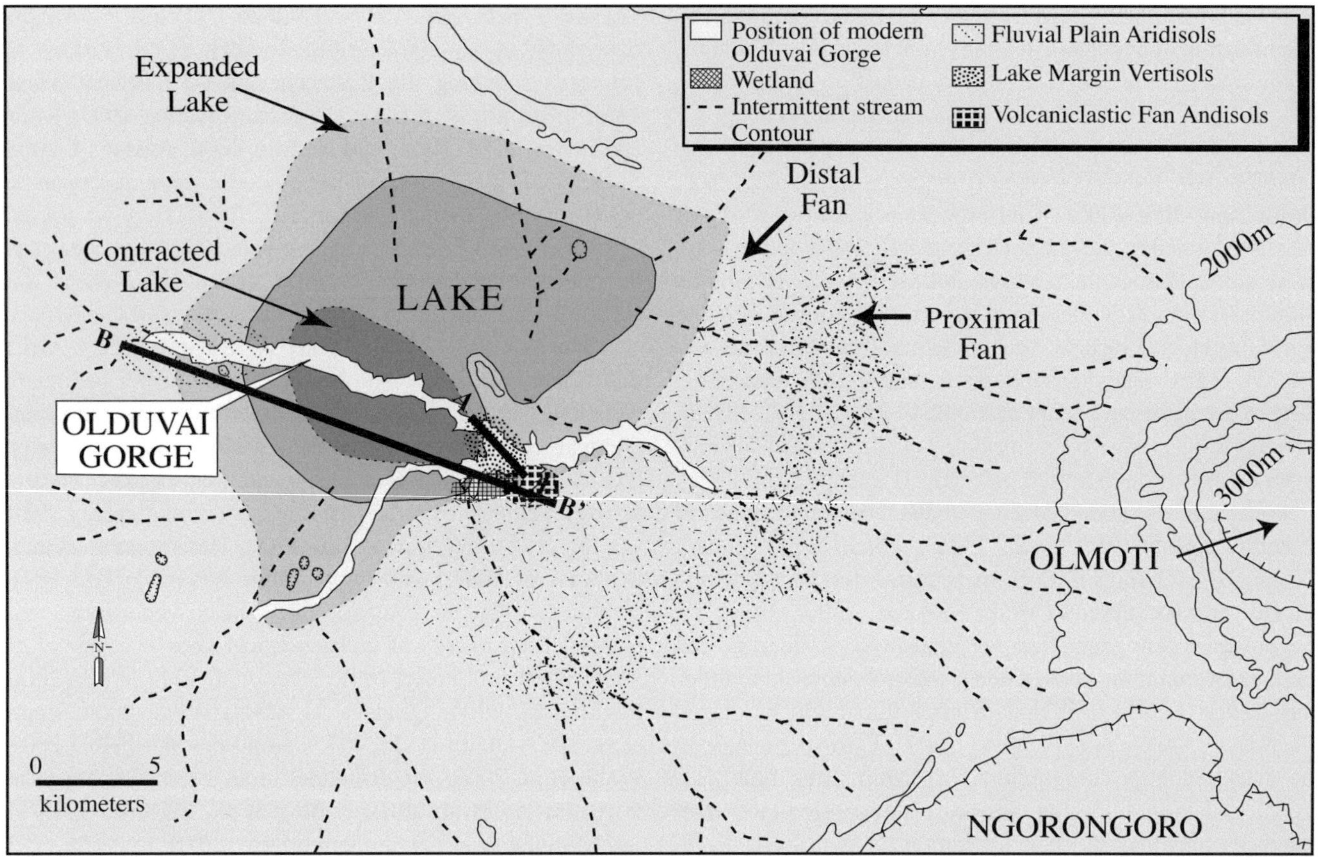

Fig. 9.5 Paleogeographic map of Olduvai Gorge modified from Ashley et al. (2010) and Beverly et al. (2014) showing paleo-Lake Olduvai and the locations of soil-forming environments. Cross-sections A and B are shown in Fig. 9.6

Near the margin of paleo-Lake Olduvai a series of time-equivalent, stacked paleosols were identified and correlated using field relationships and tephrostratigraphy. Four trenches were described by horizon and classified as paleo-Vertisols and then sampled every 10 cm for bulk geochemistry. Oriented samples for micromorphology were also collected. The combination of micromorphology and bulk geochemistry proved to be a powerful tool in describing a paleocatena where minor differences in paleotopography due to small faults and distance from a fault-controlled spring affected the amount and type of pedogenesis. Close to the Zinj fault and a related spring on the downthrown fault block, stacked paleosols developed (Figs. 9.3D and 9.6A), but further away cumulative soils developed (Fig. 9.6A). Using micromorphology, the degree of pedogenesis could be assessed, and with UVf, the amount of organic matter could be visually estimated. Closer to the spring, weakly developed b-fabrics (monostriated), Fe–Mn filled rhizoliths, and high abundances of organic matter were observed. Pedogenesis was weaker due to a higher water table that did not allow for movement of water through the soil (Fig. 9.6A). Further from the spring, organic matter abundance decreased, but b-fabrics were better developed (parallel striated, cross-striated, and granostriated), volcaniclastic grains were weathered and pitted, and Fe–Mn and zeolites formed hypocoatings along ped and pore boundaries. When combined with the phytolith and pollen record from these paleosols (Barboni et al. 2010), a reconstruction of the paleocatena is complete with a groundwater forest closest to the spring and a more open environment with palms and grasses further from the spring (Fig. 9.6A).

Because the carbonates in these paleo-Vertisols were interpreted to be diagenetic precipitates based on their large, irregular size and sparry crystalline texture, bulk geochemical methods were utilized to reconstruct paleoclimatic conditions. An example of mass-balance from the bulk geochemistry is presented here using Na, Al, Si, and K, which are typically associated with clay translocation, but because this paleo-Vertisol formed on Mg-rich smectites, Mg is also included (Fig. 9.7A–B). Results of mass-balance calculations indicate that greater translocation occurs up-section, which is best illustrated with a box and whisker plot using Na as an example (Fig. 9.7C). The increased translocation overlaps with a precession cycle that has been previously identified in uppermost Bed I using stable isotopes, fauna, flora and lithostratigraphy (Cerling and Hay

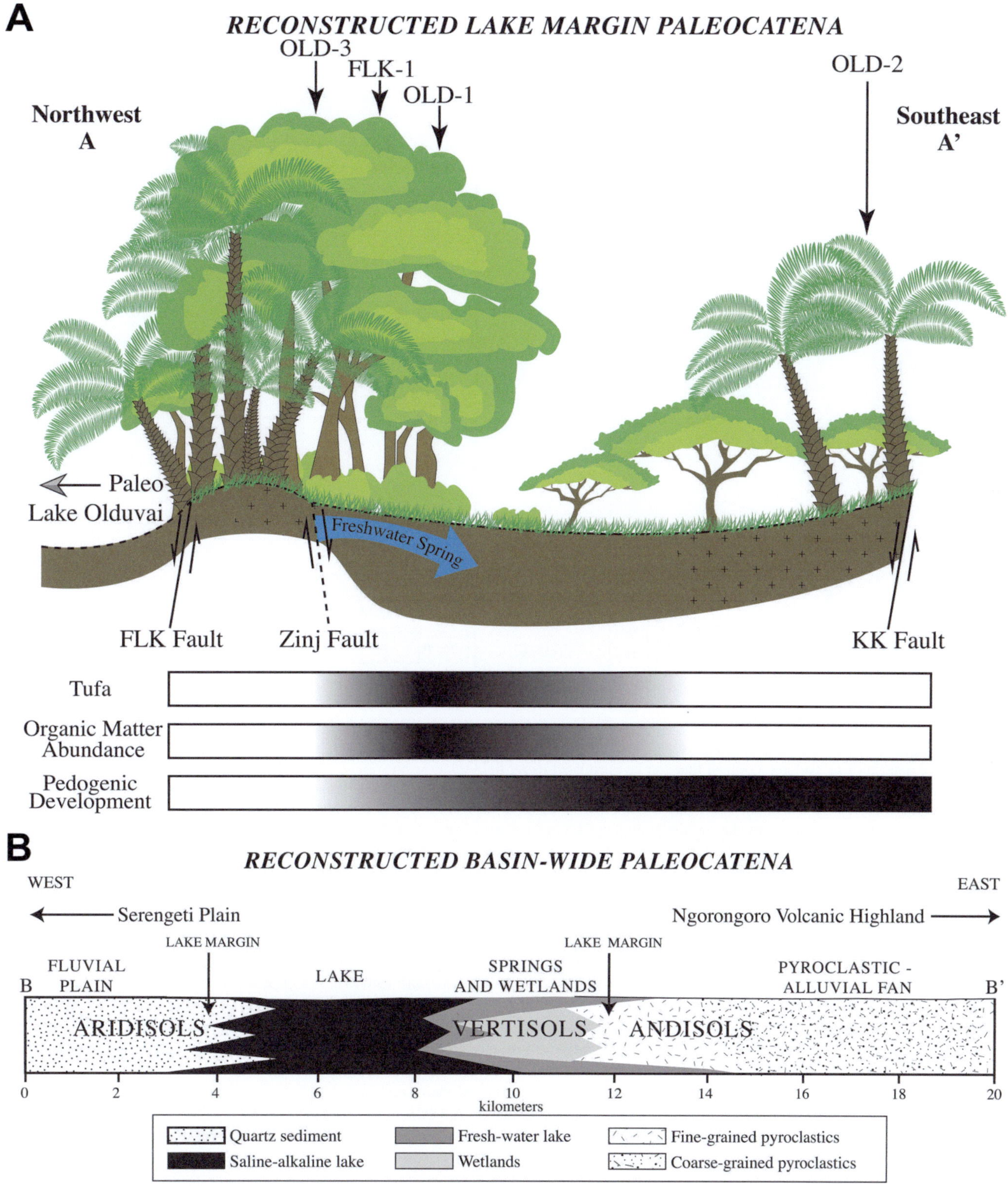

Fig. 9.6 Reconstruction of a paleocatena or paleo-Critical Zone from Olduvai Gorge at two different scales reproduced from Beverly et al. (2014). The locations of cross-sections A and B are shown in Fig. 9.4. A. The local scale reconstructs a paleocatena along the margin of paleo-Lake Olduvai using field description of paleosols, micromorphology, and mass-balance of bulk geochemistry where very minor differences in topography and distance from the spring affected pedogenesis. B. The basin scale reconstruction is a summary of a decade of paleosol research and combines the use of bulk geochemistry, stable isotopes, mineralogy, and detailed field descriptions

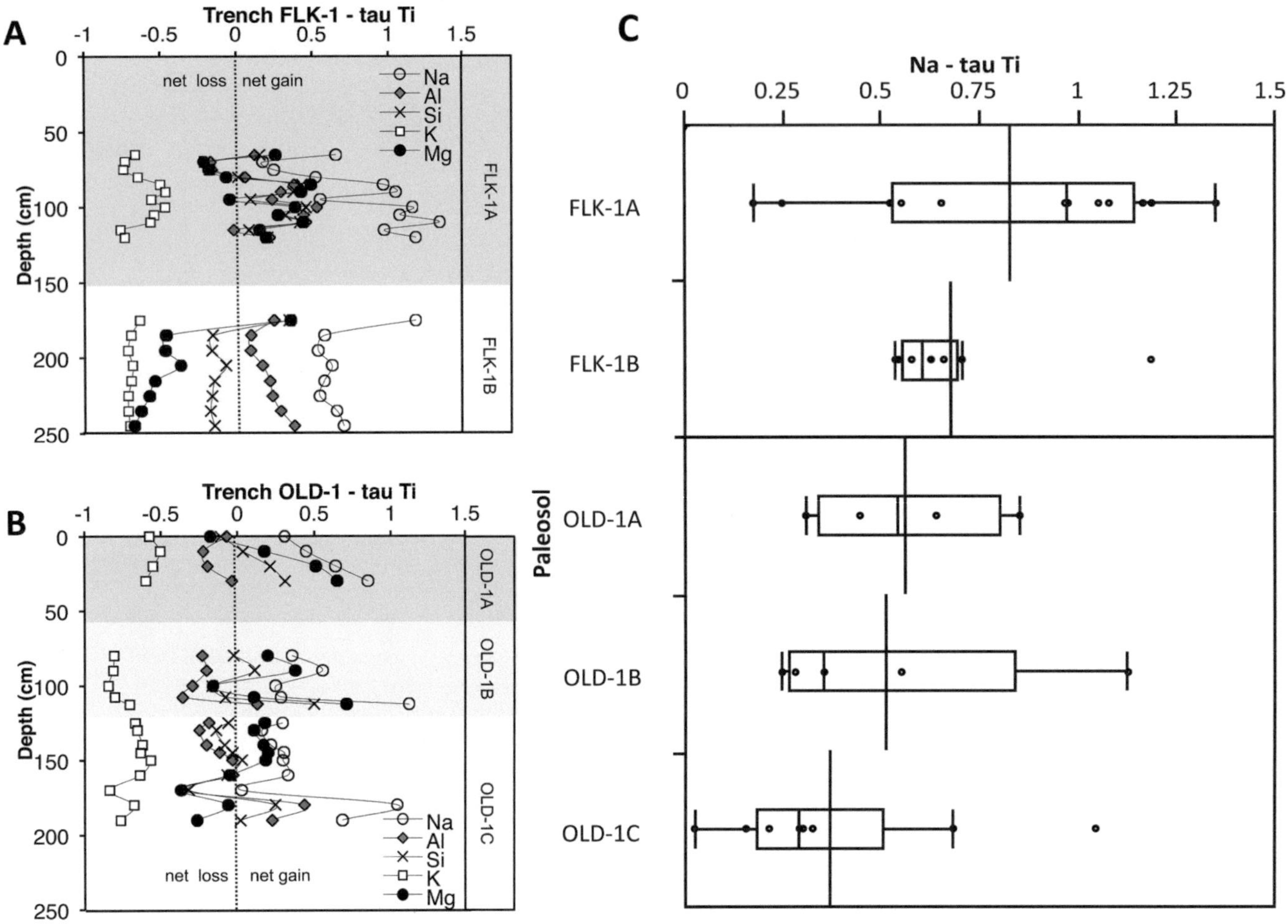

Fig. 9.7 Application of mass-balance to Olduvai paleosols. Reproduced from Beverly et al. (2014). A and B, Time-equivalent stacked paleosols near the lake margin of paleo-Lake Olduvai that can be correlated using field relationships and tephrostratigraphy. This example shows elements grouped by association with clay or zeolite accumulation. Ti is held constant as the immobile element (Tau Ti) and positive values indicate a net gain and negative values indicate a net loss relative to the parent material. Multiply by 100 to convert to percent gains or losses. C, The box and whisker plot of Tau – Na using Ti as the immobile element for each paleosol indicates that the amount of translocation increases up section

Table 9.11 Summary of Olduvai paleosol characteristics modified from Ashley et al. (2014)

Paleosol Characteristics	Fluvial Plain Paleo-Aridisol	Lake Clay Paleo-Vertisol	Pyroclastic Fan Paleo-Andisol
Parent material	Quartzo-feldspathic sediments	Dominantly Mg-smectite lake clay with some tephra influence	Volcaniclastic sediments and tephra
Texture	Sand and silt	Clay dominant	Very coarse sand to silt
Drainage	Well-drained	Dominantly well-drained to poorly drained during lake highstands	Dominantly well-drained but poorly drained proximal to springs
Immobile elements	Zr > Ti: dominated by Serengeti Plain	Ti ≈> Zr: mixed Serengeti Plain and basalts	Ti > Zr: dominated by basalts
Calcium Carbonate	High, includes calcrete	Low	Low, except for detrital grains
Zeolites	Generally low	Abundant, filling root pores and channels, replacing tephra grains	Very abundant, replacing tephra grains and filling root pores
Shrink-swell (vertic) features	Absent	Abundant, slickensides and wedge peds	Absent
Redoximorphic features	Absent	Rare	Rare on fan, abundant near springs

1986; Ashley 2007; Barboni et al. 2010; Beverly et al. 2014, and references therein).

These methods can also be applied to a basin-wide reconstruction of the paleocatena or paleo-CZ (Figs. 9.5 and 9.6B). By analyzing these paleosols in a more holistic method that includes field relationships and careful description of paleosol features such as ped structure or redoximorphic features, stable isotopes, micromorphology, bulk geochemistry, and grain size (Table 9.11), a detailed reconstruction of the paleoenvironment in which hominins were living can be created (Fig. 9.6B; Ashley et al. 2014). Differences in parent material and topography have a great effect on paleosol development at Olduvai Gorge irrespective of climate. Paleo-Aridisols formed in the quartzo-feldspathic sediment on the fluvial plain to the west of paleo-Lake Olduvai (Fig. 9.2A–B) in contrast to the paleo-Vertisols formed in the Mg-rich smectitic clay along the lake margin (Fig. 9.2D–F). The topography of the basin has a great effect on the water table, and further from the lake, soil drainage increased and soils were generally redder in color due to oxidation of Fe (Ashley and Driese 2000; Sikes and Ashley 2007; Ashley et al. 2014). This example illustrates profoundly variegated soils that result from variations in climate, organisms, relief, parent material, and time (*cl, o, r, p, t*). Without careful study and correlation using tephrostratigraphy, the changes in paleosol types could easily be attributed to changes in one specific soil-forming factor such as climate – the paleo-Vertisol high in OM would have formed a wetter climate and carbonate-rich paleo-Aridisols would have formed in a drier climate. In actuality, these paleosols formed coevally on the landscape and the changes in relief and parent material were predominant controls on soil formation. With careful study, climate can be teased out of this complex signal, but climate cannot be the assumed cause of paleosol heterogeneity on the landscape (Figs. 9.5, 9.6A–B; Ashley and Driese 2000; Sikes and Ashley 2007; Ashley et al. 2014; Beverly et al. 2014).

Future Prospects

The addition and refinement of paleosol proxies for climate and environment will advance paleopedology-based paleoecological reconstructions in the near future. The recent development of instrumentation capable measuring of clumped isotopes (Δ_{47}) in paleosol carbonates has allowed for great advances in reconstructing paleotemperature using paleosols. Ongoing research suggests that clumped isotopes may be seasonally biased and sensitive to soil hydroclimate (Passey et al. 2010; Gallagher and Sheldon 2016), and presents opportunities for future research. In addition, the development of the capability to measure ^{17}O in addition to ^{16}O and ^{18}O presents new and interesting prospects in understanding the relationship between evaporation and precipitation using triple oxygen isotopes (Passey et al. 2014). New microanalytical techniques on pedogenic carbonate using U-series dating and laser ablation ICP-MS and ion microprobe to measure variations in $\delta^{18}O$ and $\delta^{13}C$ are also pushing forward the field of paleopedology (Oerter et al. 2016).

Larger modern soil data sets with more quantitative soil attributes will yield new models for predicting both environmental factors and soil properties. This may improve both climate prediction and the ability to more rigorously classify paleosols using Soil Taxonomy (Nordt and Driese 2013). Larger datasets may also provide a means for developing multiple models designed for specific soil characteristics (e.g., soil order, drainage, parent material). Using a data-driven, multivariate statistical approach applied to a large soil dataset, Stinchcomb et al. (2016) showed that both MAP and MAT can be predicted simultaneously for a diverse range of soils forming across a large range of climatic conditions. This is especially useful when the paleopedologist has little environmental data at the outset of a project. Greater effort in estimating fundamental soil properties also has a bright future in the field of paleopedology. Soil pH and Eh proxies were used to construct a "paleo-Pourbaix diagram" used in the assessment of soil nutrient cycling and ecosystem response from Late Cretaceous and early Paleocene paleosols in the Western Interior of North America (Nordt et al. 2011). Similar efforts from East African and other locations could yield new insight into soil-plant-water interactions accompanying evolution.

Key Terms

Catena – A series of closely-related soils that manifest unique properties due to lateral changes in soil-forming conditions that vary as a function of slope. The catena concept has been applied to paleosols as their position relates to valley bottom features such as springs and channels (see pedofacies).

Critical Zone (CZ) – Near-surface environment in which complex interactions involving rock, soil, water, air, and living organisms regulate the natural habitat and determine the availability of life-sustaining resources (National Research Council 2001).

Edaphic – Influenced, produced, or related to soil, typically in reference to mineralogical or chemical characteristics that affect biota.

Eluviation – The transport and removal of soil solutes or material (mineral or organic) from an upper soil horizon. The material is commonly clay to silt size or dissolved in pore water.

Illuviation – The introduction of soil solutes or material (mineral or organic) into a soil horizon from an overlying horizon. The material is commonly clay to silt size.
Lithologic discontinuity – The contact between two different sediment types or genetically unrelated sediments that is usually marked by a distinct surface. When present, time may be unrepresented in the stratigraphic record.
Master horizon – A dominant soil horizon within the soil profile. Master horizons include: O, A, E, B, C, R, and K. The B and C master horizons are the most common in the sedimentary record. Master horizons also may have subordinate horizon modifiers describing secondary characteristics within that soil horizon.
Paleo-Critical Zone (paleo-CZ) – The deep-time equivalent of a modern CZ that is primarily studied using proxies for reconstructing past biogeochemical cycles related to environmental factors. The paleo-CZ is also known as a Deep Time Critical Zone (DTCZ). The DTCZ concept was originally confined to pre-Quaternary rocks and sediments (Nordt and Driese 2013). The concept is extended to the Quaternary (<2.6 Ma) in this chapter.
Paleoecology – The study of animal and/or plant interactions within ancient environments. Studies may incorporate direct fossil evidence or rely on proxy methods.
Paleopedology – The study of paleosols.
Paleosol – A soil that formed on a landscape of the past or under soil-forming conditions no longer operating at a given locality. Paleosols can occur as buried, exhumed, lithified, and/or relict soils. Paleosols are also known as fossil soils.
Parent material – The material (sediment or rock) from which a soil forms.
Pedogenic – Adjective used for soil, soil-forming processes, or products of soil formation (e.g., pedogenic minerals)
Pedology – A subdiscipline of soil science that emphasizes the study of soil formation, especially as it relates to environmental factors (e.g., climate, organisms, relief, parent material, and time).
Soil – a depth profile consisting of distinct layers that result from the physical, chemical, and biological alteration (i.e., weathering) of pre-existing parent materials into a more stable form.
Soil facies or pedofacies – lateral variation in soil development caused by variation in environmental factors of soil formation (i.e., topography).
Soil horizon – A layer within the soil profile that is different from the overlying and underlying layers, usually based on physical features such as color and texture.
Surface horizon – The zone within a soil profile dominated by organic matter inputs from the vegetative cover (e.g., A horizon).
Subordinate horizon indicators – Lower-case letters that are appended to master horizon designations that add distinction to the master horizon (see Table 9.2 for a list of subordinate horizon indicators).
Subsurface horizon – Also called the mineral soil, it is a zone where chemical reactions and physical actions cause the mobilization of ions and fine-grained particles, including clays, metals, and organo-metallic complexes (e.g., B horizon).
Translocation – The movement (usually downward) of chemical and physical constituents through the percolation of water in pores or via biotic agents.

Key References

1. *Soils in Archaeological Research* (Holliday 2004) is devoted to soil and paleosol investigations as they relate to the archaeological record. This textbook is geared towards Quaternary paleopedologists and geoarchaeologists and has a chapter devoted to soils in paleoenvironmental reconstructions.
2. *The Nature and Properties of Soils* (Brady and Weil 2008) is a comprehensive introduction to modern soil science with an emphasis on chemical, physical, and mineralogical characterization of weathering as it pertains to the modern U.S. Soil Survey program.
3. *Soils and Geomorphology* (Birkeland 1999) is a well-known textbook on the dynamics of soil formation related to the five environmental factors: climate, organisms, relief, parent material, and time. This book is frequently used by modern and paleopedologists.
4. *Guidelines for analysis and description of soil and regolith thin sections* (Stoops 2003) is widely available text that serves as a step-by-step guide for analyzing and describing soil thin-sections and includes examples and a CD with images. This book is partially based on Bullock et al. (1985), which is more difficult to find.
5. *Interpretation of Micromorphological Features of Soils and Regoliths* (Stoops et al. 2010) is a textbook with excellent color images illustrating features that can be identified in soils and paleosols.
6. *Stable isotope evidence for hominin environments in Africa* (Cerling 2014) is a chapter summarizing how stable isotopes can be used to reconstruct hominin environments.
7. *Environment and Climate of Early Human Evolution* (Levin 2015) is another summary of the use of stable isotopes to reconstruct hominin paleoenvironments.
8. *Quantitative paleoenvironmental and paleoclimatic reconstruction using paleosols* (Sheldon and Tabor 2009) is an extensive summary of the state of paleosol research and is good place to start for those interested in utilizing paleosols for paleoenvironmental and paleoclimatic research.

9. *Paleosols as Indicators of Paleoenvironment and Paleoclimate* (Tabor and Myers 2015) is an update to the Sheldon and Tabor (2009) summary of paleosol research.
10. *Revisitation of Concepts in Paleopedology: Transactions of the Second International Symposium on Paleopedology* (Follmer 1998) is a collection of abstracts, essays and reports by various authors on numerous facets of paleosols, following the first paleopedology symposium (Yaalon 1971).
11. *Weathering, Soils & Paleosols* (Martini and Chesworth 1992) is an edited volume that is a blend of pedology, classic geology (especially mineralogy and geochemistry) and geomorphology that discusses present and past weathering processes. Although this book offers an intriguing and unique view of paleopedology, it contains little content on paleosols as paleoecological indicators.
12. *A Colour Guide to Paleosols* (Retallack 1997) is a hard-to-find introductory book on paleopedology that contains remarkable paleosol images and accessible text on the basics of the field, including paleosol identification, description and interpretation.
13. *Soils of the Past: An introduction to paleopedology* (Retallack 2001) is a paleopedology textbook that introduces concepts and techniques in the field and provides a chapter-by-chapter overview of paleosols recorded through Earth history and their paleoenvironmental importance.

References

Amundson, R., Richter, D. D., Humphreys, G. S., Jobbagy, E. G., & Gaillardet, J. (2007). Coupling between biota and Earth materials in the critical zone. *Elements, 3,* 327–332.

An, Z. S., & Porter, S. C. (1997). Millennial-scale climatic oscillations during the last interglaciation in central China. *Geology, 25,* 603–606.

Anderson, S. P., Dietrich, W. E., & Brimhall, G. H. (2002). Weathering profiles, mass-balance analysis, and rates of solute loss: linkages between weathering and erosion in a small, steep catchment. *Geological Society of America Bulletin, 114,* 1143–1158.

Ashley, G. M. (2007). Orbital rhythms, monsoons, and playa lake response, Olduvai Basin, equatorial East Africa (ca. 1.85–1.74 Ma). *Geology, 35,* 1091–1094.

Ashley, G. M., & Driese, S. G. (2000). Paleopedology and paleohydrology of a volcaniclastic paleosol interval: implications for Early Pleistocene stratigraphy and paleoclimate record: Olduvai Gorge, Tanzania. *Journal of Sedimentary Research, 70,* 1065–1080.

Ashley, G. M., Barboni, D., Domínguez-Rodrigo, M., Bunn, H. T., Mabulla, A. Z. P., Diez-Martin, F., et al. (2010). Paleoenvironmental and paleoecological reconstruction of a freshwater oasis in savannah grassland at FLK North, Olduvai Gorge, Tanzania. *Quaternary Research, 74,* 333–343.

Ashley, G. M., Deocampo, D. M., Kahmann-Robinson, J. A., & Driese, S. G. (2013). Groundwater-fed wetland sediments and paleosols: it's all about water table. In S. G. Driese & L. C. Nordt (Eds.), *New frontiers in paleopedology and terrestrial paleoclimatology* (SEPM Special Publication No. 104) (pp. 47–62). Tulsa: SEPM.

Ashley, G. M., Beverly, E. J., Sikes, N. E., & Driese, S. G. (2014). Paleosol diversity in the Olduvai Basin, Tanzania: effects of geomorphology, parent material, depositional environment, and groundwater on soil development. *Quaternary International, 322,* 66–77.

Aslan, A., & Autin, W. J. (1998). Holocene flood-plain soil formation in the southern lower Mississippi Valley: implications for interpreting alluvial paleosols. *Geological Society of America Bulletin, 110,* 433–449.

Atchley, S. C., Nordt, L. C., Dworkin, S. I., Ramezani, J., Parker, W. G., Ash, S. R., et al. (2013). A linkage among Pangean tectonism, cyclic alluviation, climate change, and biologic turnover in the Late Triassic: the record from the Chinle Formation, Southwestern United States. *Journal of Sedimentary Research, 83,* 1147–1161.

Atchley, S., Nordt, L., & Dworkin, S. I. (2004). Eustatic controls on alluvial sequence stratigraphy: a possible example from the Cretaceous-Tertiary transition of the Tornillo Basin, Big Bend National Park, West Texas, U.S.A. *Journal of Sedimentary Research, 74,* 391–404.

Aziz, H. A., Hilgen, F. J., van Luijk, G. M., Sluijs, A., Kraus, M. J., Pares, J. M., et al. (2008). Astronomical climate control on paleosol stacking patterns in the upper Paleocene-lower Eocene Willwood Formation, Bighorn Basin, Wyoming. *Geology, 36,* 531–534.

Bae, C. J. (2013). Archaic Homo sapiens. *Nature Education Knowledge, 4,* 4.

Barboni, D., Ashley, G. M., Domínguez-Rodrigo, M., Bunn, H. T., Mabulla, A., & Baquedano, E. (2010). Phytoliths infer dense and heterogeneous paleovegetation at FLK North and surrounding localities during upper Bed I time, Olduvai Gorge, Tanzania. *Quaternary Research, 74,* 344–354.

Barnaby, R. J., & Rimstidt, J. D. (1989). Redox conditions of calcite cementation interpreted from Mn and Fe contents of authigenic calcites. *Geological Society of America Bulletin, 101,* 795–804.

Berke, M. A. (2017). Reconstructing Terrestrial Paleoenvironments Using Sedimentary Organic Biomarkers. In D. A. Croft, S. W. Simpson, & D. F. Su (Eds.), *Methods in paleoecology: Reconstructing Cenozoic terrestrial environments and ecological communities* (pp. 121–149). Cham: Springer.

Berke, M. A. (2018). Reconstructing terrestrial paleoenvironments using sedimentary organic biomarkers. In D. A. Croft, D. F. Su & S. W. Simpson (Eds.), *Methods in paleoecology: Reconstructing Cenozoic terrestrial environments and ecological communities* (pp. 121–149). Cham: Springer.

Beverly, E. J., Ashley, G. M., & Driese, S. G. (2014). Reconstruction of a Pleistocene paleocatena using micromorphology and geochemistry of lake margin paleo-Vertisols, Olduvai Gorge, Tanzania. *Quaternary International, 322–323,* 78–94.

Beverly, E. J., Driese, S. G., Peppe, D. J., Arellano, L. N., Blegen, N., Faith, J. T., et al. (2015a). Reconstruction of a semi-arid Late Pleistocene paleocatena from the Lake Victoria region, Kenya. *Quaternary Research, 84,* 368–381.

Beverly, E. J., Driese, S. G., Peppe, D. J., Johnson, C. R., Michel, L. A., Faith, J. T., et al. (2015b). Recurrent spring-fed rivers in a Middle to Late Pleistocene semi-arid grassland: implications for environments of early humans in the Lake Victoria Basin, Kenya. *Sedimentology, 62,* 1611–1635.

Bestland, E. A., & Retallack, G. J. (1993). Volcanically influenced calcareous paleosols from the Kiahera Formation, Rusinga Island, Kenya. *Journal of the Geological Society of London, 148,* 1067–1078.

Bestland, E. A., Retallack, G. J., & Swisher III, C. C. (1997). Stepwise climate change recorded in Eocene-Oligocene paleosol sequences from central Oregon, *The Journal of Geology, 105,* 153–172.

Birkeland, P. W. (1999). *Soils and geomorphology*. Oxford: Oxford University Press.

Blake, G. R., & Hartge, K. H. (1986). Bulk Density. In A. Klute (Ed.), *Methods of soil analysis: Part I. physical and mineralogical methods* (pp. 363–375). Madison: Soil Science Society of America Inc.

Blokhuis, W. A., Kooistra, M. J., & Wilding, L. P. (1990). Micromorphology of cracking clayey soils (Vertisols). In L. A. Douglas (Ed.), Soil micromorphology: a basic and applied science, *Developments in Soil Science, 19*, 123–148.

Bown, T. M., & Kraus, M. J. (1987). Integration of channel and floodplain suites in aggrading fluvial systems, I. Developmental sequence and lateral relations of lower Eocene alluvial paleosols, Willwood Formation, Bighorn Basin, Wyoming. *Journal of Sedimentary Petrology, 57*, 587–601.

Brady, N. C., & Weil, R. R. (2008). *The nature and properties of soil* (14th ed).

Brantley, S. L., Goldhaber, M. B., & Ragnarsdottir, K. V. (2007). Crossing disciplines and scales to understand the Critical Zone. *Elements, 3*, 307–314.

Brimhall, G. H., & Dietrich, W. E. (1987). Constitutive mass balance relations between chemical composition, volume, density, porosity, and strain in metasomatic hydrochemical systems: results on weathering and pedogenesis. *Geochimica et Cosmochimica Acta, 51*, 567–587.

Brimhall, G. H., Chadwick, O. A., Lewis, C. J., Compston, W., Williams, I. S., Danti, K. J., et al. (1991a). Deformational mass in invasive processes transport and soil evolution. *Science, 255*, 695–702.

Brimhall, G. H., Lewis, C. J., Ford, C., Bratt, J., Taylor, G., & Warin, O. (1991b). Quantitative geochemical approach to pedogenesis: importance of parent material reduction, volumetric expansion, and eolian influx in laterization. *Geoderma, 51*, 51–91.

Bradley, R. S. (1999). *Paleoclimatology: Reconstructing climates of the quaternary*. San Diego: Academic Press.

Brewer, R. (1976). *Fabric and mineral analysis of soils*. New York: Robert E. Krieger Publishing Company.

Bronger, A., & Heinkele, T. (1989). Micromorphology and genesis of paleosols in the Luochuan loess section, China: pedostratigraphic and environmental implications. *Geoderma, 45*, 123–143.

Bullock, P., FeDoroff, N., Jungerius, A., Stoops, G., Tursina, T., & Babel, U. (1985). *Handbook for soil thin section description*. Wolverhampton: Waine Research Publications.

Campisano, C. J., & Feibel, C. S. (2008). Depositional environments and stratigraphic summary of the Pliocene Hadar Formation at Hadar, Afar Depression, Ethiopia. In J. Quade & J. G. Wynn (Eds.), The geology of early humans in the Horn of Africa. *Geological Society of America Special Paper, 446*, 179–201.

Cerling, T. E. (1984). The stable isotopic composition of modern soil carbonate and its relationship to climate. *Earth and Planetary Science Letters, 71*, 229–240.

Cerling, T. E. (2014). Stable isotope evidence for hominin environments in Africa. In T. E. Cerling (Ed.), *Treatise on geochemistry, Vol. 14: Archaeology and anthropology* (pp. 157–67). Oxford: Pergamon.

Cerling, T., & Quade, J. (1993). Stable carbon and oxygen isotopes in soil carbonates. *Geophysical Monograph Series 78.*

Cerling, T. E., & Hay, R. L. (1986). An isotopic study of paleosol carbonates from Olduvai Gorge. *Quaternary Research, 25*, 63–78.

Cerling, T. E., Bowman, J. R., & O'Neil, J. R. (1988). An isotopic study of a fluvial-lacustrine sequence: the Plio-Pleistocene Koobi fora sequence, East Africa. *Palaeogeography, Palaeoclimatology, Palaeoecology, 63*, 335–356.

Cerling, T. E., Harris, J. M., & Passey, B. H. (2003). Diets of East African bovidae based on stable isotope analysis. *Journal of Mammalogy, 84*, 456–470.

Cerling, T. E., Levin, N. E., Quade, J., Wynn, J. G., Fox, D. L., Kingston, J. D., et al. (2010). Comment on the paleoenvironment of *Ardipithecus ramidus*. *Science, 328*, 1105.

Cerling, T. E., Wynn, J. G., Andanje, Sa, Bird, M. I., Korir, D. K., Levin, N. E., et al. (2011). Woody cover and hominin environments in the past 6 million years. *Nature, 476*, 51–56.

Chadwick, O. A., & Chorover, J. (2001). The chemistry of pedogenic thresholds. *Geoderma, 100*, 231–353.

Chadwick, O. A., Brimhall, G. H., & Hendricks, D. M. (1991). From a black to a gray box—a mass balance interpretation of pedogenesis. *Geomorphology, 3*, 369–390.

Chen, S. T., Smith, S. Y., Sheldon, N. D., & Strömberg, C. (2015). Regional-scale variability in the spread of grasslands in the late Miocene. *Palaeogeography, Palaeoclimatology, Palaeoecology, 437*, 42–52.

Cleveland, D. M., Atchley, S. C., & Nordt, L. C. (2007). Continental sequence stratigraphy of the Upper Triassic (Norian Rhaetian) Chinle strata, northern New Mexico, U.S.A.: allocyclic and autocyclic origins of paleosol-bearing alluvial successions. *Journal of Sedimentary Research, 77*, 909–924.

Cleveland, D. M., Nordt, L. C., Dworkin, S. I., & Atchley, S. C. (2008). Pedogenic carbonate isotopes as evidence for extreme climatic events preceding the Triassic-Jurassic boundary: implications for the biotic crisis? *Geological Society of America Bulletin, 120*, 1408–1415.

Clyde, W. C., Gingerich, P. D., Wing, S. L., Rohl, U., Westerhold, T., Bowen, G., et al. (2013). Bighorn Basin Coring Project (BBCP): a continental perspective on early Paleogene hyperthermals. *Scientific Drilling*, 21–31.

Cotton, J. M., & Sheldon, N. D. (2012). High-resolution isotopic record of C_4 photosynthesis in a Miocene grassland. Palaeogeography, Palaeoclimatology, Palaeoecology, 337–338, 88–98.

Cremeens, D. L., Hart, J. P., & Darmody, R. G. (1998). Complex pedostratigraphy of a terrace fragipan at the Memorial Park site, central Pennsylvania. *Geoarchaeology, 13*, 339–359.

Driese, S., & Foreman, J. L. (1992). Paleopedology and paleoclimatic implications of Late Ordovician vertic paleosols, Juniata Formation, southern Appalachians. *Journal of Sedimentology Research, 62*, 71–83.

Driese, S. G., & Mora, C. I. (1993). Physio-chemical environment of pedogenic carbonate formation in Devonian vertic paleosols, central Appalachians, USA. *Sedimentology, 40*, 199–216.

Driese, S. G., Mora, C. I., Cotter, E., & Foreman, J. L. (1992). Paleopedology and stable isotope chemistry of Late Silurian vertic Paleosols, Bloomsburg formation, central Pennsylvania. *Journal of Sedimentary Research, 62*(5), 825–841.

Driese, S. G., Mora, C. I., Stiles, C. A., Joeckel, R. M., & Nordt, L. C. (2000). Mass-balance reconstruction of a modern Vertisol: implications for interpreting the geochemistry and burial alteration of paleo-Vertisols. *Geoderma, 95*, 179–204.

Driese, S. G., Li, Z.-H., & Horn, S. P. (2005). Late Pleistocene and Holocene climate and geomorphic histories as interpreted from a 23,000 ^{14}C yr B.P. paleosol and floodplain soils, southeastern West Virginia, USA. *Quaternary Research, 63*, 136–149.

Driese, S. G., Peppe, D. J., Beverly, E. J., DiPietro, L. M., Arellano, L. N., & Lehmann, T. (2016). Paleosols and paleoenvironments of the early Miocene deposits near Karungu, Lake Victoria, Kenya. *Palaeogeography, Palaeoclimatology, Palaeoecology, 443,* 167–182.

Driese, S. G., & Ober, E. G. (2005). Paleopedologic and paleohydrologic records of precipitation seasonality from early Pennsylvanian "underclay" Paleosols, USA. *Journal of Sedimentary Research, 75,* 997–1010.

Driese, S. G., Medaris Jr, L. G., Ren, M., Runkel, A. C., & Langford, R. P. (2007). Differentiating pedogenesis from diagenesis in early terrestrial paleoweathering surfaces formed on granitic composition parent materials. *The Journal of Geology, 115*(4), 387–406.

Driese, S. G., Jirsa, M. A., Ren, M., Brantley, S. L., Sheldon, N. D., Parker, D., et al. (2011). Neoarchean paleoweathering of tonalite and metabasalt: implications for reconstructions of 2.69 Ga early terrestrial ecosystems and paleoatmospheric chemistry. *Precambrian Research, 189,* 1–17.

Driese, S. G., & Ashley, G. M. (2016). Paleoenvironmental reconstruction of a paleosol catena, the Zinj archeological level, Olduvai Gorge, Tanzania. *Quaternary Research, 85,* 133–146.

Dworkin, S. I., Nordt, L., & Atchley, S. (2005). Determining terrestrial paleotemperatures using the oxygen isotopic composition of pedogenic carbonate. *Earth and Planetary Science Letters, 237,* 56–68.

Eidt, R. C. (1985). Theoretical and practical considerations in the analysis of anthrosols. In G. Rapp & J. A. Gifford (Eds.), *Archaeological geology* (pp. 155–190). New Haven: Yale University Press.

Fedo, C. M., Nesbitt, H. W., & Young, G. M. (1995). Unraveling the effects of potassium metasomatism in sedimentary rocks and paleosols, with implications for paleoweathering conditions and provenance. *Geology, 23,* 921–924.

Fedoroff, N., & Goldberg, P. (1982). Comparative micromorphology of two late Pleistocene paleosols (in the Paris Basin). *Catena, 9,* 227–251.

Fitzpatrick, E. A. (1984). *Micromorphology of soils*. London and New York: Chapman and Hall.

Fitzpatrick, E. A. (1993). *Soil microscopy and micromorphology*. New York: Wiley.

Ferring, C. R. (1992). Alluvial pedology and geoarchaeological research. In V. T. Holliday (Ed.), *Soils in archaeology* (pp. 1–40). Washington D. C., Smithsonian Institution Press.

Follmer, L. R. (1998). Preface. In L. R. Follmer, D. L. Johnson, & J. A. Catt (Eds), *Revisitation of concepts in paleopedology: Transactions of the Second International Symposium on Paleopedology. Quaternary International* (vol. 51/52, pp. 1–3).

Gallagher, T. M., & Sheldon, N. D. (2013). A new paleothermometer for forest paleosols and its implications for Cenozoic climate. *Geology, 41,* 647–650.

Gallagher, T. M., & Sheldon, N. D. (2016). Combining soil water balance and clumped isotopes to understand the nature and timing of pedogenic carbonate formation. *Chemical Geology 435,* 79–91.

Garrett, N. D., Fox, D. L., McNulty, K. P., Tryon, C. A., Faith, J. T., Peppe, D. J., et al. (2015). Stable isotope paleoecology of Late Pleistocene middle stone age humans from the equatorial East Africa, Lake Victoria basin, Kenya. *Journal of Human Evolution, 82,* 1–14.

Ghosh, P., Adkins, J., Affek, H., Balta, B., Guo, W., Schauble, E., et al. (2006). 13C–18O bonds in carbonate minerals: a new kind of paleothermometer. *Geochimica et Cosmochimica Acta, 70,* 1439–1456.

Gile, L. H. (1979). Holocene soils in eolian sediments of Baily County, Texas. *Soil Science Society of America Journal, 43,* 994–1003.

Gile, L. H., Peterson, F. F., & Grossman, R. B. (1966). Morphological and genetic sequences of carbonate accumulation in desert soils. *Soil Science, 101.*

Gulbranson, E. L., Montanez, I. P., & Tabor, N. J. (2011). A proxy for humidity and floral province from paleosols. *Journal of Geology, 119,* 559–573.

Guthrie, R. L., & Witty, J. E. (1982). New designations for soil horizons and layers and the new Soil Survey Manual. *Soil Science Society of America Journal, 46*(2), 443–444.

Han, J. T., Fyfe, W. S., & Longstaffe, F. J. (1998). Climatic implications of the S5 paleosol complex on the southernmost Chinese Loess Plateau. *Quaternary Research, 50,* 21–33.

Harnois, L. (1988). The CIW index: a new chemical index of weathering. *Sedimentary Geology, 55,* 319–322.

Harris, W., & White, N. (2008). X-ray diffraction techniques for soil mineral identification. *Methods of soil analysis part 5—Mineralogical methods*. Madison: Soil Science Society of America.

Hasiotis, S. T. (2007). Continental ichnology: fundamental processes and controls on trace-fossil distribution. In W. Miller III (Ed.), *Trace fossils—Concepts, problems, prospects* (pp. 268–284). Elsevier Press.

Hasiotis, S. T., Platt, B. F., Hembree, D.I., & Everhart, M. (2007a). The trace-fossil record of vertebrates. In W. Miller III (Ed.), *Trace fossils—Concepts, problems, prospects* (pp. 196–218). Elsevier Press.

Hasiotis, S. T., Kraus, M. J., & Demko, T. M. (2007b). Climate controls on continental trace fossils. In W. Miller III (Ed.), *Trace fossils—Concepts, problems, prospects* (pp. 172–195). Elsevier Press.

Hasiotis, S. T., & Honey, J. G. (2000). Paleohydrologic and stratigraphic significant of crayfish burrows in continental deposits: examples from several Paleocene Laramide basins in the Rocky Mountains. *Journal of Sedimentary Research, 70,* 127–139.

Hembree, D. I., Platt, B. F. & Smith, J. J. (Eds.) (2014). *Experimental approaches to understanding fossil organisms: Lessons from the Living*. Netherlands: Springer.

Henkes, G. A., Passey, B. H., Grossman, E. L., Shenton, B. J., Perez-Huerta, A., & Yancey, T. E. (2014). Temperature limits for preservation of primary calcite clumped isotope paleotemperatures. *Geochimica et Cosmochimica Acta, 139,* 362–382.

Holliday, V. T. (2004). *Soils in archaeological research*. Oxford: Oxford University Press.

Holliday, V. T. (2006). A history of soil geomorphology in the United States. In B. P. Warkentin (Ed.), *Footprints in the soil: People and ideas in soil history* (pp. 187–254). Amsterdam: Elsevier.

Holliday, V. T., & Gartner, W. G. (2007). Soil phosphorus and archaeology: a review and comparison of methods. *Journal of Archaeological Science, 34,* 301–333.

Holloway, R. L., Broadfield, D. C., & Yuan, M. S. (2004). *The human fossil record Vol. 3*. Wiley.

Hover, V. C., & Ashley, G. M. (2003). Geochemical signatures of paleodepositional and diagenetic environments: a STEM/AEM study of authigenic clay minerals from an arid rift basin, Olduvai Gorge, Tanzania. *Clays and Clay Minerals, 51,* 231–251.

Huggett, R. J. (1998). Soil chronosequences, soil development, and soil evolution: a critical review. *Catena, 32,* 155–172.

Hyland, E. G., Sheldon, N. D., Van der Voo, R., Badgley, C., & Abrajevitch, A. (2015). A new paleoprecipitation proxy based on soil magnetic properties: implications for expanding paleoclimate reconstructions. *Geological Society of America Bulletin.*

Jenny, H. (1941). *Factors of soil formation: a system of quantitative pedology*. McGraw-Hill book company, Inc.

Jenny, H. (1980). *The soil resource. Origin and behavior.* Berlin: Springer.

Kemp, R. A. (1999). Micromorphology of loess-paleosol sequences: a record of paleoenvironmental change. *Catena, 35,* 179–196.

Kraus, M. J. (1987). Integration of channel and floodplain suites in aggrading fluvial systems, II. Vertical relations of alluvial paleosols. *Journal of Sedimentary Petrology, 57*(4), 602–612.

Kraus, M. J. (1997). Lower Eocene alluvial paleosols: pedogenic development, stratigraphic relationships, and paleosol/landscape associations. *Palaeogeography, Palaeoclimatology, Palaeoecology, 129,* 387–406.

Kraus, M. J. (1999). Paleosols in clastic sedimentary rocks: their geologic applications. *Earth-Science Review, 47,* 41–70.

Kraus, M. J., & Brown, T. M. (1988). Pedofacies analysis; a new approach to reconstructing ancient fluvial sequences. In J. Reinhard & W. R. Sigleo (Eds.), *Paleosols and weathering through geologic time: Principles and applications* (Geological Society of America Special Paper 216) (pp. 143–152). Denver: Geological Society of America.

Kraus, M. J., & Hasiotis, S. T. (2006). Significance of different modes of rhizolith preservation to interpreting paleoenvironmental and paleohydrologic settings: examples from Paleogene paleosols, Bighorn Basin, Wyoming, U.S.A. *Journal of Sedimentary Research, 76,* 633–646.

Kubiena, W. (1970). Micromorphology of polygenetic soils and paleosoils in polar regions. *Annales de Edafologia y Abrobiologia,* 845–856.

Leighton, M. M. (1937). The significance of profiles of weathering in stratigraphic archaeology. In G. G. MacCurdy (Ed.), *Early Man* (pp. 163–172). New York: Lippincott.

Leopold, M., Völkel, J., Dethier, D., Huber, J., & Steffens, M. (2011). Characteristics of a paleosol and its implication for the Critical Zone development, Rocky Mountain Front Range of Colorado, USA. *Applied Geochemistry, 26,* S72–S75.

Lepre, C. J., Quinn, R. L., Joordens, J. C. a, Swisher, C. C., & Feibel, C. S. (2007). Plio-Pleistocene facies environments from the KBS Member, Koobi Fora Formation: implications for climate controls on the development of lake-margin hominin habitats in the northeast Turkana Basin (northwest Kenya). *Journal of Human Evolution, 53,* 504–514.

Levin, N. E. (2015). Environment and climate of early human evolution. *Annual Review of Earth and Planetary Sciences, 43,* 405–429.

Levin, N. E., Cerling, T. E., Passey, B. H., Harris, J. M., & Ehleringer, J. R. (2006). A stable isotope aridity index for terrestrial environments. *Proceedings of the National Academy of Sciences, USA, 103,* 11201–11205.

Levin, N. E., Quade, J., Simpson, S. W., Semaw, S., & Rogers, M. J. (2004). Isotopic evidence for Plio-Pleistocene environmental change at Gona, Ethiopia. *Earth and Planetary Science Letters, 219,* 93–110.

Levin, N., Brown, F. H., Behrensmeyer, A. K., Bobe, R., & Cerling, T. E. (2011). Paleosol carbonates from the Omo Group: isotopic records of local and regional environmental change in East Africa. *Palaeogeography, Palaeoclimatology, Palaeoecology, 307,* 75–89.

Lüdecke, T., Schrenk, F., Thiemeyer, H., Kullmer, O., Bromage, T. G., Sandrock, O., et al. (2016). Persistent C3 vegetation accompanied Plio-Pleistocene hominin evolution in the Malawi Rift (Chiwondo Beds, Malawi). *Journal of Human Evolution, 90,* 163–175.

Ludvigson, G. A., Gonzalez, L. A., Fowle, D. A., Roberts, J. A., Driese, S. G., Villarreal, M. A., et al. (2013). Paleoclimatic applications and modern process studies of pedogenic siderite. In S. G. Driese & L. C. Nordt (Eds.), *New frontiers in paleopedology and terrestrial paleoclimatology* (SEPM Special Publication No. 104) (pp. 47–62). Tulsa: SEPM.

Ludvigson, G. A., Gonzalez, L. A., Metzger, R. A., Witzke, B. J., Brenner, R. L., Murillo, A. P., et al. (1998). Meteoric sphaerosiderite lines and their use for paleohydrology and paleoclimatology. *Geology, 26,* 1039–1042.

Lüdecke, T., Schrenk, F., Thiemeyer, H., Kullmer, O., Bromage, T. G., Sandrock, O., et al. (2016b). Persistent C3 vegetation accompanied Plio-Pleistocene hominin evolution in the Malawi Rift (Chiwondo Beds, Malawi). *Journal of Human Evolution, 90,* 163–175.

Lukens, W. E., Driese, S. G., Peppe, D. J., & Loudermilk, M. (2017a). Sedimentology, stratigraphy, and paleoclimate at the late Miocene Coffee Ranch fossil site in the Texas Panhandle. *Palaeogeography, Palaeoclimatology, Palaeoecology, 485,* 361–376.

Lukens, W. E., Lehmann, T., Peppe, D. J., Fox, D. L., Driese, S. G., & McNulty, K. P. (2017b). The early Miocene Critical Zone at Karungu, Western Kenya: an equatorial, open habitat with few primate remains. *Frontiers in Earth Science, 5,* 87.

Lukens, W. E., Nordt, L. C., Stinchcomb, G. E., Driese, S. G., & Tubbs, J. D. (2018). Reconstructing pH of Paleosols Using Geochemical Proxies. *The Journal of Geology, 126*(4), 427–449.

Mack, G. H., James, W. C., & Monger, H. C. (1993). Classification of paleosols. *Geological Society of America Bulletin, 105,* 129–136.

Machette, M. N. (1985). Calcic soils of the southwestern United States. In *Geological Society of America Special Paper 203* (pp. 1–21).

Machel, H. G., Mason, R. A., Mariano, A. N., & Mucci, A. (1991). Causes and emission of luminescence in calcite and dolomite. In C. E. Barker and O. C. Kopp (Eds.), *Luminescence microscopy and spectroscopy: Quantitative and qualitative applications (SC25)* (pp. 9–25). Tulsa: SEPM.

Magill, C. R., Ashley, G. M., & Freeman, K. H. (2013a). Ecosystem variability and early human habitats in eastern Africa. *Proceedings of the National Academy of Sciences, USA, 110,* 1167–1174.

Magill, C. R., Ashley, G. M., & Freeman, K. H. (2013b). Water, plants, and early human habitats in eastern Africa. *Proceedings of the National Academy of Sciences, USA, 110,* 1175–1180.

Maher, B. A. (1998). Magnetic properties of modern soils and Quaternary loessic paleosols: paleoclimatic implications. *Palaeogeography, Palaeoclimatology, Palaeoecology, 137,* 25–54.

Maher, B. A., & Thompson, R. (1995). Paleorainfall reconstructions from pedogenic magnetic susceptibility variations in the Chinese loess and paleosols. *Quaternary Research, 44,* 383–391.

Marin-Spiotta, E., Chaopricha, N. T., Plante, A. F., Diefendorf, A. F., Mueller, C. W., Grandy, A. S., et al. (2014). Long-term stabilization of deep soil carbon by fire and burial during early Holocene climate change. *Nature Geosci, 7,* 428–432.

Maxbauer, D. P., Feinberg, J. M., & Fox, D. L. (2016a). Magnetic mineral assemblages in soils and paleosols as the basis for paleoprecipitation proxies: a review of magnetic methods and challenges. *Earth-Science Review, 155,* 28–48.

Maxbauer, D. P., Feinberg, J. M., Fox, D. L., & Clyde, W. C. (2016b). Magnetic minerals as recorders of weathering, diagenesis, and paleoclimate: a core-outcrop comparison of Paleocene-Eocene paleosols in the Bighorn Basin, WY, USA. *Earth and Planetary Science Letters, 452,* 15–26.

Maynard, J. B. (1992). Chemistry of modern soils as a guide to interpreting Precambrian paleosols. *Journal of Geology, 100,* 279–289.

Medaris Jr, L. G., Driese, S. G., & Stinchcomb, G. E. (2017). The Paleoproterozoic Baraboo paleosol revisited: quantifying mass fluxes of weathering and metasomatism, chemical climofunctions, and atmospheric pCO_2 in a chemically heterogeneous protolith. *Precambrian Research, 301,* 179–194.

Mentzer, S. M. (2014). Microarchaeological approaches to the identification and interpretation of combustion features in prehistoric archaeological sites. *Journal of Archaeoogical Method and Theory, 21,* 616–668.

Michel, L. A., Driese, S. G., Nordt, L. C., Breecker, D. O., Labotka, D. M., & Dworkin, S. I. (2013). Stable-Isotope geochemistry of

Vertisols formed on marine limestone and implications for deep-time paleoenvironmental reconstructions. *Journal of Sedimentary Research, 83,* 300–308.

Michel, L. A., Peppe, D. J., Lutz, Ja, Driese, S. G., Dunsworth, H. M., Harcourt-Smith, W. E. H., et al. (2014). Remnants of an ancient forest provide ecological context for Early Miocene fossil apes. *Nature Communications, 5,* 1–9.

Mintz, J. S., Driese, S. G., Breecker, D. O., & Ludvigson, G. A. (2011). Influence of changing hydrology on pedogenic calcite precipitation in Vertisols, Dance Bayou, Brazoria County, Texas, U.S.A.: implications for estimating paleoatmospheric pCO_2. *Journal of Sedimentary Research, 81,* 394–400.

Moore, D. M., & Reynolds, R. C. (1997). *X-Ray diffraction and the identification and analysis of clay minerals*. New York: Oxford University Press.

Mora, C. I., Driese, S. G., & Colarusso, L. A. (1996). Middle to Late Paleozoic atmospheric CO_2 levels from soil carbonate and organic matter. *Science, 271,* 1105–1107.

Morrison, R. B. (1967). Principles of Quaternary soil stratigraphy. In R. B. Morrison & H. E. Wright (Eds.), *Quaternary soils* (pp. 1–69). Reno: University of Nevada Desert Research Institute, Center for Water Resources Research.

Myers, T. S., Tabor, N. J., & Rosenau, N. A. (2014). Multiproxy approach reveals evidence of highly variable paleoprecipitation in the Upper Jurassic Morrison Formation (western United States). *Geological Society of American Bulletin, 126,* 1105–1116.

National Research Council. (2001). *Basic research opportunities in Earth science*. Washington, D. C.: National Academies Press.

National Research Council. (2010). *Understanding climate's influence on human evolution*. Washington, D. C.: National Academies Press.

Nettleton, W. D., Olson, C. G., & Wysocki, D. A. (2000). Paleosol classification: problems and solutions. *Catena, 41,* 61–92.

Nesbitt, H. W., & Young, G. M. (1982). Earth Proterozoic climates and plate motions inferred from major element chemistry of lutites. *Nature, 299,* 715–717.

Nordt, L. C. (2001). Stable C and O isotopes in soils: applications for archaeological research. In P. Goldberg, V. Holliday & R. Ferring (Eds.), *Earth-sciences and archaeology* (pp. 419–445).

Nordt, L. C., & Driese, S. D. (2010a). New weathering index improves paleorainfall estimates from Vertisols. *Geology, 38,* 407–410.

Nordt, L. C., & Driese, S. G. (2010b). A modern soil characterization approach to reconstructing physical and chemical properties of paleo-Vertisols. *American Journal of Science, 310,* 37–64.

Nordt, L. C., & Driese, S. G. (2013). Application of the Critical Zone concept to the deep-time sedimentary record. *The Sedimentary Record, 11,* 4–9.

Nordt, L. C., Dworkin, S. I., & Atchley, S. C. (2011). Ecosystem response to soil biogeochemical behavior during the Late Cretaceous and early Paleocene within the western interior of North America. *Geological Society of America Bulletin, 123,* 1745–1762.

Nordt, L. C., Hallmark, C. T., Driese, S. G., Dworkin, S. I., & Atchley, S. C. (2012). Biogeochemical characterization of a lithified paleosol: implications for the interpretation of ancient Critical Zones. *Geochimica et Cosmochimica Acta, 87,* 267–282.

Nordt, L. C., Hallmark, C. T., Driese, S. G., & Dworkin, S. I. (2013). Multianalytical pedosystem approach to characterizing and interpreting the fossil record of soils. In S. G. Driese & L. C. Nordt (Eds.), *New frontiers in paleopedology and terrestrial paleoclimatology* (SEPM Special Publication No. 104) (pp. 47–62, 89–108). Tulsa: SEPM.

Nordt, L., Orosz, M., Driese, S., & Tubbs, J. (2006). Vertisol carbonate properties in relation to mean annual precipitation: implications for paleoprecipitation estimates. *Journal of Geology, 114,* 501–510.

Nordt, L. C., Wilding, L. P., Lynn, W. C., & Crawford, C. C. (2004). Vertisol genesis in a humid climate of the coastal plain of Texas, U.S.A. *Geoderma, 122,* 83–102.

Oerter, E. J., Sharp, W. D., Oster, J. L., Ebeling, A., Valley, J. W., Kozdon, R., et al. (2016). Pedothem carbonates reveal anomalous North American atmospheric circulation 70,000–55,000 years ago. *Proceedings of the National Academy of Sciences, USA, 113,* 919–924.

Passey, B. H., Levin, N. E., Cerling, T. E., Brown, F. H., & Eiler, J. M. (2010). High-temperature environments of human evolution in East Africa based on bond ordering in paleosol carbonates. *Proceedings of the National Academy of Sciences, USA, 107,* 11245–11249.

Passey, B. H., Hu, H., Ji, H., Montanari, S., Li, S., Henkes, G. A., et al. (2014). Triple oxygen isotopes in biogenic and sedimentary carbonates. *Geochimica et Cosmochimica Acta, 141,* 1–25.

Pettijohn, F. J., Potter, P. E., & Siever, R. (1987). *Sand and sandstone* (2nd ed.). Springer-Verlag.

Poppe, L. J., Paskevich, V. F., Hathaway, J. C., & Blackwood, D. S. (2001). A laboratory manual for X-ray powder diffraction. *US Geological Survey Open-File Report, 1*(041), 1–88.

Prochnow, S. J., Nordt, L. C., Atchley, S. C., & Hudec, M. R. (2006). Multi-proxy paleosol evidence for Middle and Late Triassic climate trends in eastern Utah. *Palaeogeography, Palaeoclimatology, Palaeoecology, 232,* 53–72.

Pimentel, N. L., Wright, V. P., & Azevedo, T. M. (1996). Distinguishing early groundwater alteration effects from pedogensis in ancient alluvial basins: examples form the Palaeogene of southern Portugal. *Sedimentary Geology, 105,* 1–10.

Qadir, M., & Schubert, S. (2002). Degradation processes and nutrient constraints in sodic soils. *Land Degradation & Development, 13*(4), 275–294.

Quade, J., Levin, N., Semaw, S., Stout, D., Renne, P., Rogers, M., et al. (2004). Paleoenvironments of the earliest stone toolmakers, Gona, Ethiopia. *Geological Society of America Bulletin, 116,* 1529.

Quade, J., Eiler, J., Daëron, M., Achyuthan, H. (2013). The clumped isotope geothermometer in soil and paleosol carbonate. *Geochimica Cosmochimica Acta, 105,* 92–107.

Quinn, R. L., Lepre, C. J., Wright, J. D., & Feibel, C. S. (2007). Paleogeographic variations of pedogenic carbonate d13C values from Koobi Fora, Kenya: implications for floral compositions of Plio-Pleistocene hominin environments. *Journal of Human Evolution, 53,* 560–573.

Quinn, R. L., Lepre, C. J., Feibel, C. S., Wright, J. D., Mortlock, R. A., Harmand, S., et al. (2013). Pedogenic carbonate stable isotopic evidence for wooded habitat preference of early Pleistocene tool makers in the Turkana Basin. *Journal of Human Evolution, 65,* 65–78.

Rasmussen, C., & Tabor, N. J. (2007). Applying a quantitative pedogenic energy model across a range of environmental gradients. *Soil Science Society of America Journal, 71*(6), 1719–1729.

Rasmussen, C., Southard, R. J., & Horwath, W. R. (2005). Modeling energy inputs to predict pedogenic environments using regional environmental databases. *Soil Science Society of America Journal, 69*(4), 1266–1274.

Rawls, W. J. (1983). Estimating soil bulk density from particle size analyses and organic matter content. *Soil Science, 135,* 123–125.

Retallack, G. J. (1983). Late Eocene and Oligocene paleosols from Badlands National Park, South Dakota. *Geological Society of America. Special Papers, 193,* 82.

Retallack, G. J. (1997). Early forest soils and their role in Devonian global change. *Science, 276,* 583–585.

Retallack, G. J., Wynn, J. G., Benefit, B. R., & Mccrossin, M. L. (2002). Paleosols and paleoenvironments of the middle Miocene,

Maboko Formation, Kenya. *Journal of Human Evolution, 42*(6), 659–703.
Retallack, G. J. (1994). The environmental-factor approach to the interpretation of paleosols. In R. J. Luxmoore & J. M. Bartels (Eds.), *Factors of soil formation: A fiftieth anniversary retrospective* (pp. 31–64). Madison: Soil Science Society of America.
Retallack, G. J. (2001). *Soils of the past: An introduction to paleopedology* (2nd ed.). Oxford: Blackwell Science Ltd.
Retallack, G. J. (2005). Pedogenic carbonate proxies for amount and seasonality of precipitation in paleosols. *Geology, 33,* 333–336.
Retallack, G. J., James, W. C., Mack, G. H., & Monger, H. C. (1993). Classification of paleosols: discussion and reply. *Geological Society of America Bulletin, 105,* 1635–1637.
Retallack, G. J., Orr, W. N., Prothero, D. R., Duncan, R. A., Kester, P. R., & Ambers, C. P. (2004). Eocene-Oligocene extinction and paleoclimatic change near Eugene. *Oregon. Geological Society of America Bulletin, 116,* 817.
Retallack, G. J., & Huang, C. (2010). Depth to gypsic horizon as a proxy for paleoprecipitation in paleosols of sedimentary environments. *Geology, 38,* 403–406.
Richter, D. deB, & Yaalon, D. H. (2012). "The changing model of soil" revisited. *Soil Science Society of America Journal, 76,* 766–778.
Rosenau, N. A., Tabor, N. J., Elrick, S. D., & Nelson, W. J. (2013). Polygenetic history of paleosols in Middle-Upper Pennsylvanian cyclothems of the Illinois Basin, U.S.A.: Part I. Characterization of paleosol types and interpretations of pedogenic processes. *Journal of Sedimentary Research, 83,* 606–636.
Ruhe, R. V. (1965). Quaternary paleopedology. In H. E. Wright, & D. G. Frey (Eds.), The Quaternary of the United States (pp. 755–764). Princeton, NJ: Princeton University Press.
Saxton, K. E., & Rawls, W. J. (2006). Soil water characteristic estimates by texture and organic matter for hydrologic solutions. *Soil Science Society of America Journal, 70,* 1569–1578.
Schaetzl, R. J., & Thompson, M. L. (2015). *Soils*: Cambridge University Press.
Schoeneberger, P. J., Wysocki, D. A., Benham, E. C., & Broderson, W. D. (2012). *Field book for describing and sampling soils, version 3.0.* Lincoln: Natural Resources Conservation Service, National Soil Survey Center.
Schwertmann, U. (1993). Relations between iron oxides, soil color, and soil formation. In J. M. Bigham & E. J. Ciolkosz (Eds.), *Soil color* (Special Publication No. 31) (pp. 51–69). Madison: Soil Science Society of America.
Sheldon, N. (2003). Pedogenesis and geochemical alteration of the Picture Group subgroup, Columbia River basalt, Oregon. *Geological Society of America Bulletin, 115,* 1377–1387.
Sheldon, N. D. (2005). Do red beds indicate paleoclimatic conditions?: a Permian case study. *Palaeogeography, Palaeoclimatology, Palaeoecology, 228,* 305–319.
Sheldon, N. D. (2006). Precambrian paleosols and atmospheric CO_2 levels. *Precambrian Research, 147,* 148–155.
Sheldon, N. D., & Retallack, G. J. (2001). Equation for compaction of paleosols due to burial. *Geology, 29,* 247–250.
Sheldon, N. D., & Retallack, G. J. (2004). Regional paleoprecipitation records from the late eocene and oligocene of North America. *The Journal of Geology, 112,* 487–494.
Sheldon, N. D., & Tabor, N. J. (2009). Quantitative paleoenvironmental and paleoclimatic reconstruction using paleosols. *Earth-Science Review, 95,* 1–52.
Sheldon, N. D., Retallack, G. J., & Tanaka, S. (2002). Geochemical climofunctions from North American soils and application to paleosols across the Eocene-Oligocene boundary in Oregon. *Journal of Geology, 110,* 687–696.
Sikes, N. E., Potts, R., & Behrensmeyer, A. K. (1999). Early Pleistocene habitat in member 1 Olorgesailie based on paleosol stable isotopes. *Journal of Human Evolution, 37,* 721–746.
Sikes, N. E., & Ashley, G. M. (2007). Stable isotopes of pedogenic carbonates as indicators of paleoecology in the Plio-Pleistocene (upper Bed I), western margin of the Olduvai Basin, Tanzania. *Journal of Human Evolution, 53,* 574–594.
Snell, K. E., Thrasher, B. L., Eiler, J. M., Koch, P. L., Sloan, L. C., & Tabor, N. J. (2013). Hot summers in the Bighorn Basin during the early Paleogene. *Geology, 41,* 55–58.
Soil Classification Working Group. (1998). *The Canadian system of soil classification* (3rd ed., Agriculture and Agri-Food Canada Publication 1646). Ottawa: NRC Research Press.
Soil Survey Staff. (2006). *Keys to soil taxonomy* (10th ed.). Washington, D. C.: United States Department of Agriculture Natural Resources Conservation Service.
Soil Survey Staff. (2014a). *Illustrated guide to soil taxonomy* (1.0 ed.). Lincoln: United States Department of Agriculture Natural Resources Conservation Service.
Soil Survey Staff. (2014b). *Kellogg soil survey laboratory methods manual, Version 5.0.* (Soil Survey Investigations Report No. 42). Lincoln: United States Department of Agriculture Natural Resources Conservation Service.
Stiles, C. A., Mora, C. I., & Driese, S. G. (2001). Pedogenic iron-manganese nodules in Vertisols: a new proxy for paleoprecipitation? *Geology, 29,* 943–946.
Stiles, C. A., Mora, C. I., & Driese, S. G. (2003a). Pedogenic processes and domain boundaries in a Vertisol climosequence: evidence from titanium and zirconium distribution and morphology. *Geoderma, 116,* 279–299.
Stiles, C. A., Mora, C. I., Driese, S. G., & Robinson, A. C. (2003b). Distinguishing climate and time in the soil record: mass-balance trends in Vertisols from the Texas coastal prairie. *Geology, 31,* 331–334.
Stinchcomb, G. E., Driese, S. G., Nordt, L. C., DiPietro, L., & Messner, T. C. (2014). Early Holocene soil cryoturbation in northeastern USA: implications for archaeological site formation. *Quaternary International, 342,* 186–198.
Stinchcomb, G. E., Nordt, L. C., Driese, S. G., Lukens, W. E., Williamson, F. C., & Tubbs, J. D. (2016). A data-driven spline model designed to predict paleoclimate using paleosol geochemistry. *American Journal of Science, 316,* 746–777.
Stoops, G. (2003). Guidelines for analysis and description of soil and regolith thin sections. Madison: Soil Science Society of America, Inc.
Stoops, G., Marcelino, V., & Mees, F. (Eds.) (2010). *Interpretation of micromorphological features of soils and regoliths, 1st ed.* Netherlands: Elsevier.
Tabor, N. J., & Montañez, I. P. (2002). Shifts in late Paleozoic atmospheric circulation over western equatorial Pangea: insights from pedogenic mineral $\delta^{18}O$ compositions. *Geology, 30,* 1127–1130.
Tabor, N. J., & Myers, T. S. (2015). Paleosols as indicators of paleoenvironment and paleoclimate. *Annual Review of Earth and Planetary Science, 43,* 11.1–11.29.
Tabor, N. J., Montañez, I. P., Steiner, M. B., & Schwindt, D. (2007). $\delta^{13}C$ values of carbonate nodules across the Permian-Triassic boundary in the Karoo Supergroup (South Africa) reflect a stinking sulfurous swamp, not atmospheric CO_2. *Palaeogeography, Palaeoclimatology, Palaeoecology, 252,* 370–381.
Terry, D. O. (2001). Paleopedology of the Chadron Formation of northwestern Nebraska: implications for paleoclimatic change in the North America midcontinent across the Eocene-Oligocene boundary. *Palaeogeography, Palaeoclimatology, Palaeoecology, 168,* 1–38.
Torres, M. A., & Gaines, R. R. (2013). Paleoenvironmental and paleoclimatic interpretations of the late Paleocene Golder

Formations, southern California, U.S.A., based on paleosol geochemistry. *Journal of Sedimentary Research, 83,* 591–605.

Trendell, A. M., Nordt, L. C., Atchley, S. C., LeBlanc, S. L., & Dworkin, S. I. (2013a). Determining floodplain plant distributions and populations using paleopedology and fossil root traces: upper Triassic Sonsela Member of the Chinle Formation at Petrified Forest National Park, Arizona. *PALAIOS, 28,* 471–490.

Trendell, A. M., Atchley, S. C., & Nordt, L. C. (2013b). Facies analysis of a probable large-fluvial-fan depositional system: the Upper Triassic Chinle Formation at Petrified Forest National Park, Arizona, U.S.A. *Journal of Sedimentary Research, 83,* 873–895.

Ufnar, D. F., Ludvigson, G. A., González, L. A., Brenner, R. L., & Witzke, B. J. (2004). High latitude meteoric $\delta^{18}O$ compositions: Paleosol siderite in the Middle Cretaceous Nanushuk Formation, North Slope. *Alaska. Geological Society of America Bulletin, 116* (3/4), 463–473.

Ufnar, D. (2007). Clay coatings from a modern soil chronosequence: a tool for estimating the relative age of well-drained paleosols. *Geoderma, 141,* 181–200.

Valentine, K. W. G., & Dalrymple, J. B. (1976). Quaternary buried paleosols: a critical review. *Quaternary Research, 6,* 209–222.

Vepraskas, M. J. (1992). Redoximorphic features for identifying aquic conditions. *North Carolina State University Technical Bulletin 301.* Raleigh: North Carolina Agricultural Research Service.

Vepraskas, M. J. (2001). Morphological features of seasonally reduced soils. *Wetland soils: Genesis, hydrology, landscapes, and classification* (pp. 163–182). New York: Lewis Publishers.

Vepraskas, M. J., & Faulkner, S. P. (2001). Redox chemistry of hydric soils. In J. L. Richardson & M. J. Vepraskas (Eds.), *Wetland soils: Genesis, hydrology, landscapes, and classification* (pp. 85–105). New York: Lewis Publishers.

Waters, M. R. (1992). *Principles of geoarchaeology: A North American perspective.* Tucson: University of Arizona Press.

Waters, M. R., Forman, S. L., Jennings, T. A., Nordt, L. C., Driese, S. G., Feinberg, J. M., et al. (2011). The Buttermilk Creek Complex and the origins of Clovis at the Debra L. Friedkin Site. *Texas. Science, 331,* 1599–1603.

Weider, M., & Yaalon, D. H. (1982). Micromorphological fabrics and developmental stages of carbonate nodular forms related to soil characteristics. *Geoderma, 28,* 203–220.

White, T. D., Asfaw, B., Beyene, Y., Haile-Selassie, Y., Lovejoy, C. O., Suwa, G., et al. (2009). *Ardipithecus ramidus* and the paleobiology of early hominids. *Science, 326*(64), 75–86.

WoldeGabriel, G., Ambrose, S. H., Barboni, D., Bonnefille, R., Bremond, L., Currie, B., et al. (2009). The geological, isotopic, botanical, invertebrate, and lower vertebrate surroundings of *Ardipithecus ramidus. Science, 326,* 65, 65e1–65e5.

Wynn, J. G. (2000). Paleosols, stable carbon isotopes, and paleoenvironmental interpretation of Kanapoi, Northern Kenya. *Journal of Human Evolution, 39,* 411–432.

Wynn, J. G. (2004). Influence of Plio-Pleistocene aridification on human evolution: evidence from paleosols of the Turkana Basin, Kenya. *American Journal of Physical Anthropoloy, 123,* 106–118.

Wynn, J. G. (2007). Carbon isotope fractionation during decomposition of organic matter in soils and paleosols: implications for paleoecological interpretations of paleosols. *Palaeogeography, Palaeoclimatology, Palaeoecology, 251,* 437–448.

Wynn, J. G., & Feibel, C. S. (1995). Paleoclimatic implications of Vertisols within the Koobi Fora Formation, Turkana Basin, Northern Kenya. *Journal of Undergraduate Research, 6,* 34–42.

Yaalon, D.H., International Society of Soil Science, International Union for Quaternary Research. (1971). *Paleopedology: origin, nature, and dating of paleosols.* Jerusalem: International Society of Soil Science.

Zamanian, K., Pustovoytov, K., & Kuzyakov, Y. (2016). Pedogenic carbonates: forms and formation processes. *Earth-Science Reviews, 157,* 1–17.

Chapter 10
The Role of Continental Trace Fossils in Cenozoic Paleoenvironmental and Paleoecological Reconstructions

Daniel Hembree

Abstract Trace fossils are the preserved remains of burrows, tracks, trails, nests, borings, or any other record of the behavioral interaction between an organism and a substrate. This chapter reviews the significance of trace fossils to paleoenvironmental and paleoecological studies with an emphasis on their application to the interpretation of Cenozoic continental deposits. Topics include the behavioral and environmental significance of trace fossils, methods utilized in studying trace fossils in outcrop and core, as well as analogous traces of extant organisms, and the strengths, biases, and limitations of trace-fossil data and analyses. The utility of continental trace fossils to paleoecological, paleoenvironmental, and paleoclimatic interpretations are illustrated by two case studies of continental depositional environments from the Eocene-Oligocene and Miocene of North America. Future areas of research with trace fossils then are discussed including experimental and field research with modern organisms and digitization of both modern and ancient traces.

Keywords Ichnofossils • Paleoecology • Ichnology

Introduction

Ichnology is an interdisciplinary science that incorporates paleontology, sedimentology, stratigraphy, and geochemistry to interpret ancient ecosystems and depositional environments through the study of biologically produced sedimentary structures, or trace fossils (Seilacher 2007). Trace fossils are the preserved remains of burrows, tracks, trails, nests, borings, or any other record of the interaction between an organism and a substrate or medium (Frey 1973, 1975; Bromley 1996). The substrate most often consists of loose to firm sediment or soil, but it can also include rock, wood, leaves, bones, shells, or even individual mineral grains depending on the size of the tracemaker (Bromley 1996; MacEachern et al. 2007a). Trace fossils provide a unique view of ancient ecosystems. They often represent groups of organisms not typically preserved as body fossils due to a lack of mineralized hard parts or their presence in environments not conducive to preservation (Bromley 1996; Gingras et al. 2007a). In many settings, such as ancient soils, trace fossils may provide the only evidence of life (Bromley 1996; Hasiotis 2007).

Trace fossils are described and named based on their morphology, typically a set of features, termed ichnotaxobases, that include overall shape, architecture, orientation, internal structure, lining, and surface features (Bromley 1996; Bertling et al. 2006). Trace fossils are given ichnogenus and ichnospecies names following guidelines of the International Code of Zoological Nomenclature. Most studies require diagnosis of trace fossils to the ichnogenus level since the behavioral significance is similar for most ichnospecies of the same ichnogenus. Importantly, however, trace fossils are named based on the morphology of the trace alone and not to the interpreted behavior, tracemaker, or environmental significance (Bertling et al. 2006; Rindsberg 2012).

Trace fossils are especially important to paleontologists because they provide a direct record of behavior, an interpretation generally inferred only indirectly from body morphology (Seilacher 1953; Frey 1978; Bromley 1996). Behavior, however, is also sensitive to environmental fluctuations. Organisms behave in response to biotic and environmental factors, producing distinct traces diagnostic of behavior and various environmental limiting factors (Frey 1975; Seilacher 1978; MacEachern et al. 2007b). The traces produced can provide information about the environment not recorded by other physical evidence, such as salinity, temperature, or nutrient content (MacEachern et al. 2007b). As a

D. Hembree (✉)
Department of Geological Sciences, 316 Clippinger Laboratories, Ohio University, Athens, OH 45701, USA
e-mail: hembree@ohio.edu

D. A. Croft et al. (eds.), *Methods in Paleoecology: Reconstructing Cenozoic Terrestrial Environments and Ecological Communities*, Vertebrate Paleobiology and Paleoanthropology, https://doi.org/10.1007/978-3-319-94265-0_10

result, trace fossils are critical for interpreting sedimentary environments and facies analysis, as well as complete paleoecological reconstructions.

Trace fossil research has been conducted on rocks indicative of almost every environment on Earth and even those considered analogous to environments on other planets. Whereas most of the early research on trace fossils was conducted in ancient and modern marine environments (e.g., Frey 1973; Seilacher 1978, 1986; Bromley 1996), recent work has revealed an equivalently diverse and complex trace-fossil record preserved in continental deposits (e.g., Hasiotis 2002; Buatois and Mángano 2004; Genise et al. 2004; Hasiotis 2007). Detailed trace fossil studies include the identification and characterization of specific ichnotaxa, behavioral and trophic types, population strategies, ichnodiversity, degree of bioturbation, and tiering structure (Frey 1975; Bromley 1996). Many methods have been utilized to better reconstruct and interpret trace fossils including investigation at the scale of outcrops, cores, and thin sections, casting of modern biogenic structures, and the use of various visualization techniques including X-radiography, X-ray computed tomography (CT) scanning, magnetic resonance image (MRI) scanning, and three-dimensional digitization (Gingras et al. 2007a; Knaust 2012a). The application of new technology including visualization techniques has greatly improved our ability to see trace fossils that were once hidden from view.

The purpose of this chapter is to present the paleontological, sedimentological, and stratigraphic significance of trace fossils and their application to the interpretation of Cenozoic continental deposits. This chapter begins with an introduction to the behavioral and environmental significance of trace fossils with an emphasis on continental trace fossils. This is followed by a description of the various methods and techniques for studying trace fossils in outcrop and core, as well as traces produced by extant organisms in modern environments. The strengths, biases, and limitations of trace-fossil data and analyses are then presented to place interpretations based on trace fossils into context with other approaches. In order to illustrate the utility of continental trace fossils to paleoecological, paleoenvironmental, and paleoclimatic interpretations, case studies of two continental depositional environments from the Eocene-Oligocene and Miocene of North America are presented. Finally, future prospects of the use of trace fossils in the reconstruction of Cenozoic paleoecosystems are discussed.

The Utility of Trace Fossils

Trace fossils are the product of a combination of the effects of tracemaker morphology, behavior, and environmental conditions (e.g., Frey 1978; Seilacher 1978, 2007; Bromley 1996; MacEachern et al. 2007b). The morphology of a trace fossil may be very distinctive and characteristic of a specific type of organism, behavior, or environmental setting (Bromley 1996). The overlap of these three influences, however, can result in a wide array of different morphologies. In some instances, the morphology of trace fossils produced by a single tracemaker may be very different from one another due to variations in behavior or environmental conditions (Bromley 1996; Pemberton et al. 2001). In others, diverse organisms may produce almost identical trace fossils due to similarities in the behaviors being expressed (Bromley 1996; Pemberton et al. 2001; Gingras et al. 2007a). Despite these potential issues, ichnologists have developed an understanding of the major controls on trace-fossil morphology which allow their use in reconstructing biodiversity, behaviors, and environments.

Trace Fossils as Records of Behavior

Generally, the most easily obtained, and often most useful, information that can be interpreted from trace fossil morphology is the behavior that was engaged in to produce the structure (Frey 1978; Seilacher 1986, 2007; Bromley 1996; Pemberton et al. 2001; Hasiotis 2007). Whereas body fossils allow paleontologists to determine the physical form of ancient organisms, trace fossils provide evidence for their life activities. In addition, understanding the biological, ecological, and environmental significance of trace fossils requires understanding the behavior they represent (Frey 1975).

Trace fossils are often classified based on their interpreted behavioral significance (Fig. 10.1A). Seilacher (1953) proposed a classification system based on a series of five broad behavioral categories including resting, locomotion, grazing, feeding, and dwelling. These major types of behaviors are expressed throughout the Phanerozoic. Seilacher's system has been refined and expanded since its establishment with the addition of more specialized, temporally to environmentally restricted behaviors such as equilibrium, escape, farming, predation, nesting, and pupation (Frey 1973; Ekdale et al. 1984; Ekdale 1985; Genise and Bown 1994; Bromley 1996; Genise et al. 2007). It is understood, however, that most trace fossils likely represent more than one behavior (Seilacher 1953). A burrow constructed to serve as a dwelling, for example, may also be involved in feeding (e.g., stationary deposit feeders or ambush predators) or a feeding trace may also result from locomotion (e.g., mobile deposit feeders or grazers). Trace fossils, therefore, tend to be classified according to the dominate behavior being expressed.

Different types of behavior influence and modify substrates in a variety of ways to produce discrete structures (Bromley 1996). The simplest may involve an appendage

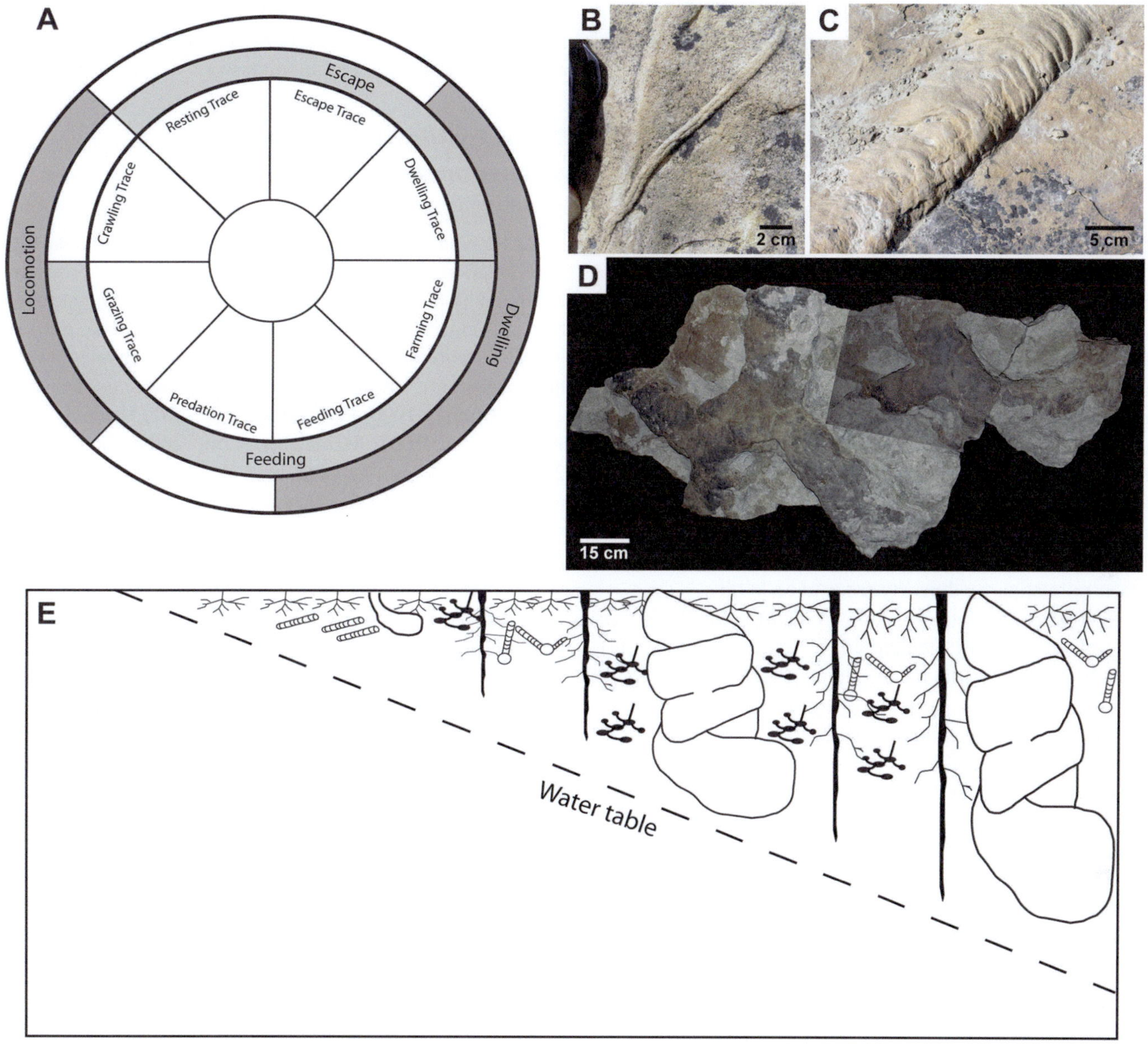

Fig. 10.1 Trace fossils are useful for interpreting organism behavior and environment. A, Typical behaviors and ethological categories represented by trace fossils. The outer two rings represent the basic types of overlapping behaviors and the inner circle represents the standard ethological categories (modified from Frey and Pemberton 1985). B, Natural cast of a bilobed trail representing locomotion. C, Backfilled burrow representing both feeding and locomotion. D, Composite photograph of a large, complex dwelling burrow with multiple, interconnected and branching tunnels. E, Relationship between trace fossil depth and environmental conditions, such as depth to the water table

impacting a soft substrate surface to produce a track representing locomotion. Animals may also move by walking or dragging their bodies along a soft sediment surface to produce shallow to deep trails (Fig. 10.1B). More complex behaviors such as deposit feeding may involve sorting through sediment particles within a substrate with mouthparts or appendages, removing organic material, and redepositing the remaining material; these backfilled burrows are easily recognizable by an interconnected series of meniscate packets of sediment (Fig. 10.1C). Dwelling behaviors typically result in open, maintained burrows or borings excavated into soft to hard substrates. These may consist of a simple tunnel or a complex array of interconnected tunnels and chambers (Fig. 10.1D).

Trace Fossils as Records of Biodiversity

Many of the organisms that produce trace fossils in terrestrial deposits are not normally preserved as body fossils

(Frey 1975; Bromley 1996). This is especially true of soil-dwelling organisms, which are generally soft-bodied and inhabit environments that are not conducive for preservation (Hasiotis 2007). The bias against soft-bodied organisms is problematic as they typically represent a significant part of the biomass in any given community. The ability to interpret the identity of tracemakers to specific trace fossils, therefore, has the potential to greatly improve paleoecological reconstructions.

Ichnodiversity differs from biodiversity, however, in that it represents the range of trace fossil morphologies (typically ichnogenera) which are expressions of behavioral interactions with the substrate rather than different kinds of organisms. A single species, for example, may produce several different ichnogenera if it engages in many behaviors (Bromley 1996). In this case ichnodiversity will likely be higher than biodiversity. Conversely, many different organisms may produce the same ichnogenus while engaged in a similar behavior (Bromley 1996); here ichnodiversity will be lower than biodiversity. This difference must be kept in mind when estimating biodiversity from trace fossils.

Interpretation of tracemakers from trace fossils is typically not done below the family level unless a body fossil is directly associated with the trace (i.e., preserved at the end of a trail or inside a burrow) (Frey 1975; Osgood 1975; Bromley 1996). Definitive connections with tracemakers often rely on impressions of the underside of the body made upon a substrate surface such as resting traces of sea stars and sea anemones (Frey 1975; Osgood 1975; Bromley 1996; Seilacher 2007). Bioglyphs, or surface ornaments, produced by animals during trace production or occupation may also provide insight into the tracemaker (Osgood 1975; Seilacher 2007). These bioglyphs can include scratches made by appendages, digits, and teeth or impressions made by various body parts including limbs, tails, and heads. Tracemaker identity may also be inferred from the behavior(s) being expressed and the environment in which it was produced (Frey 1975; Osgood 1975; Bromley 1996; Seilacher 2007). These may help to eliminate possible tracemakers, leaving a subset of potential organisms.

Integration of trace fossil evidence of life with body fossil evidence is critical to making complete paleoecological reconstructions. In terrestrial paleoenvironments, the soil biota represent the foundation of the surface ecosystem and include primary producers, decomposers, shredders, grazers, and predators (Aber and Melillo 2001; Lavelle and Spain 2001; Hasiotis 2002). These include bacteria, protists, fungi, plants, and animals. The majority of these organisms are not preserved as body fossils; however, the trace fossils produced by their life activities provide a record of their presence (Hasiotis 2002, 2007). As a result, trace fossils allow improved reconstructions of ancient food webs, community structure, and biodiversity.

Trace Fossils as Indicators of Environment

Trace fossil analysis is an important part of facies analysis and interpretation of sedimentary environments. The behaviors exhibited by trace-making organisms are generally controlled by environmental conditions, including environmental stability, turbidity, salinity, oxygenation, moisture content, water table depth, nutrient levels, and temperature (e.g., Seilacher 1978; Bromley 1996; Pemberton et al. 2001; Hasiotis 2007; Hasiotis et al. 2007; MacEachern et al. 2007a, b; Seilacher 2007). Burrowing, in particular, is often undertaken in response to external environmental factors. Importantly, many of the environmental factors that affect organism behavior and control trace production are not recorded by other physical sedimentary evidence (MacEachern et al. 2007b). Examples from terrestrial environments include temperature, moisture content, and nutrient levels (Hasiotis 2007; Hasiotis et al. 2007). This makes trace fossil analysis vital to fully understanding paleoenvironments. It is important to understand, however, that organisms typically respond to fluctuations in multiple environmental variables at any given time, for example, to interrelated changes in both temperature and moisture during dry summers. This can complicate attempts to link a behavioral response to a single environmental cause.

Morphological properties of specific types of trace fossils, particularly burrows, which can be altered by environmental conditions generally include depth and diameter (Bromley 1996; Hasiotis 2007; MacEachern et al. 2007b). Trace-making organisms may alter their depth of penetration of the substrate in response to variations in available oxygen, water salinity, nutrient content, surface temperature, or depth to the water table (Bromley 1996; Hasiotis 2007; Hasiotis et al. 2007; MacEachern et al. 2007b). These various limiting factors may result in burrow assemblages that are shallower or deeper than average (Fig. 10.1E). In stressed conditions, burrowing animals also tend to be smaller than in ideal settings, resulting in burrows with correspondingly smaller sizes (MacEachern et al. 2007b). In addition to morphological variation, specific behaviors may be excluded by environmental conditions. In turbid environments, suspension feeding may be disadvantageous, whereas in a high-energy setting with a constantly shifting substrate, near-surface deposit feeding would be impossible. The total diversity of trace fossils may also be impacted by stressful environmental conditions, resulting in monospecific suites. The occurrence of large numbers of a single ichnogenus often indicates conditions such as high energy, high salinity, low oxygen, or low nutrient availability (MacEachern et al. 2007b).

Methods and Approaches

The analysis of trace fossils can be approached in a number of ways, but three are most commonly employed: analysis of assemblages of ichnotaxa, analysis of ichnofabric, and analysis of modern burrowing organisms or neoichnology. The approach that is used is generally related to the type of material being studied and the questions being asked. Each method has its own strengths and, more likely than not, an ichnologist employs two if not all three methods in a study.

Trace Fossil Assemblages, Ichnocoenoses, and Ichnofacies

Trace fossils are rarely useful individually. A single trace fossil may provide evidence of a certain behavior or organism, but in order to be paleoenvironmentally or paleoecologically significant, trace fossils should be considered in terms of groups of multiple specimens within a single unit (Frey 1975; Seilacher 1978; Bromley 1996). The nature of these groupings can be quite variable in terms of occurrence, distribution, and significance; these include assemblages, ichnocoenoses, and ichnofacies. The type of grouping that is used in a trace fossil analysis depends on the nature of the exposure, the intensity of bioturbation, and the questions being asked about the trace-bearing units.

Trace fossil assemblage: An assemblage is a basic grouping that includes all of the trace fossils within a lithologic unit (Fig. 10.2A) (Bromley 1996). There are no interpretations regarding the origin of the assemblage or the age relationships of the trace fossils; it is simply a collection of all of the identifiable ichnogenera or ichnospecies present. The trace fossils may have been produced simultaneously, in a few discrete intervals (i.e., storm beds), or over a prolonged period of tens to hundreds of thousands of years of continuous sedimentation, resulting in intense overprinting (Howard 1978). Assemblages are useful when all that is needed is a basic characterization of the types of behaviors being expressed in a location over an unspecified period of time.

Ichnocoenosis: An ichnocoenosis is a group of trace fossils hypothesized to have been simultaneously produced by a single community of organisms (Fig. 10.2B) (Bromley 1996). The trace fossils are, therefore, temporally and ecologically related and the trace-making organisms should have been interacting at some level. A trace fossil assemblage may be divided into several different ichnocoenoses each representing a distinct community active for a short period of time. This older community may then be replaced by a new one, the resulting traces crosscutting and overprinting those of the previous ichnocoenosis. Recognition of ichnocoenoses in modern substrates is relatively simple; doing so in the fossil record is more complex.

The succession of communities within the substrate tends to be quite rapid, especially in environments with low sedimentation rates, resulting in pronounced overprinting of individual, ecologically related groups of traces (Bromley 1996). In these situations, cross-cutting relationships may be used to show which trace fossils are linked based on relative ages. Ichnocoenoses are best recognized, however, in environments that are prone to frequent disturbance by rapid depositional events (Bromley 1996). Regular, deep burial of the substrate surface fills and preserves the underlying community and ideally prevents its overprinting by the community that colonizes the new surface. If properly identified, an ichnocoenosis allows the interpretation of individual communities of burrowing organisms and provides the best view of the functioning paleoecosystem.

Ichnofacies: Ichnofacies are groups of trace fossils that reflect one or more community's response to environmental conditions (Fig. 10.2C) (Seilacher 1978; Bromley 1996; MacEachern et al. 2007a). An ichnofacies consists of a defined assemblage of ichnogenera representing a group of related behaviors. The grouping of behaviors is the result of different limiting factors in the environment which makes those behaviors advantageous and others disadvantageous. Ichnofacies are temporally and geographically recurrent, meaning they should not be restricted in geologic time and should occur globally (Seilacher 1978; MacEachern et al. 2007a). Instead, ichnofacies are facies controlled; they occur within a small range of lithologies indicative of a specific environment of deposition. In this way, ichnofacies can be compared across time and space to evaluate differences in similar environments whether they are in the Cambrian of North America or in the Neogene of Australia. Ichnofacies are not described from a single location; rather ichnofacies are laterally continuous just as lithofacies (Seilacher 1978; MacEachern et al. 2007a). As a result, establishment of ichnofacies should lead to mappable constructs, showing the lateral variations in an environment on local to regional scales. Like lithofacies, ichnofacies should also have predictable vertical successions, showing changes in environment through time as a result of variations in autogenic to allogenic processes (Seilacher 1978; MacEachern et al. 2007a).

Three major ichnofacies have been established that correspond to major types of environments. Five softground ichnofacies (*Psilonichnus, Skolithos, Cruziana, Zoophycos, Nereites*) occur in loose sediment generally, but not always, in marine environments ranging from shorelines to the deep sea

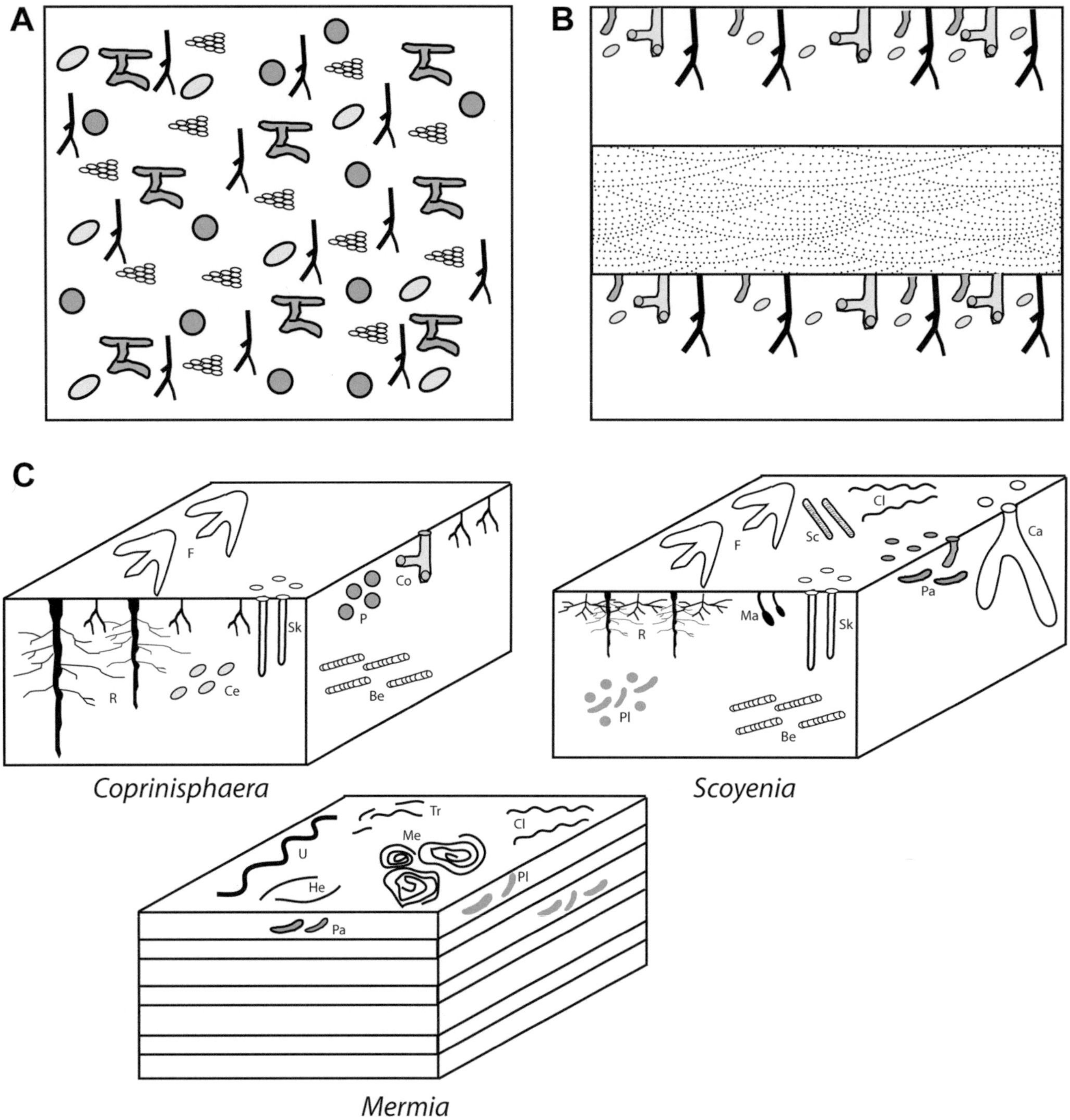

Fig. 10.2 A, Diagram of a trace-fossil assemblage. B, Diagram of two ichnocoenoses separated by undisturbed sediment. C, Diagram illustrating the three continental ichnofacies controlled primarily by substrate moisture (modified from MacEachern et al. 2007a). *Coprinisphaera* ichnofacies (low moisture): Tetrapod foootprints (F), rhizoliths (R), *Beaconites* (B), *Celliforma* (Ce), *Coprinisphaera* (Co), *Pallichnus* (P), *Skolithos* (Sk). *Scoyenia* ichnofacies (moderate moisture): Tetrapod foootprints (F), rhizoliths (R), *Beaconites* (B), *Camborygma* (Ca), *Cochlichnus* (Cl), *Macanopsis* (Ma), *Palaeophycus* (Pa), *Planolites* (Pl), *Scoyenia* (Sc), *Skolithos* (Sk). *Mermia* ichnofacies (high moisture): *Cochlichnus* (Cl), *Helminthoidichnites* (He), *Mermia* (Me), *Palaeophycus* (Pa), *Planolites* (Pl), *Treptichnus* (Tr), *Undichna* (U)

(Seilacher 1964, 1967; Frey and Pemberton 1987; MacEachern et al. 2007a). A wide range of environmental variables control trace-making organisms and behaviors, including energy level, oxygen content, food resource types, depositional rates, substrate consistency, temperature, and salinity (MacEachern et al. 2007a). Three substrate-controlled ichnofacies (*Glossifungites, Trypanites, Teredolites*) occur in specialized substrates such as firm, consolidated mud, rock, or wood (MacEachern et al. 2007a). In these ichnofacies, the nature of the substrate has the greatest control on the types of burrowing organisms present and the behaviors expressed. Three continental ichnofacies (*Mermia, Scoyenia, Coprinisphaera*) (Fig. 10.2C) are confined to freshwater and subaerial environments (MacEachern et al. 2007a). In these settings, the amount of moisture present in the substrate is the greatest limiting factor ranging from saturated (*Mermia*) to dry (*Coprinisphaera*).

Ichnofacies are most useful for paleoenvironmental interpretations. The standard, or archetypal, ichnofacies provide a basis for comparison; a standard representation of behaviors exhibited under similar environmental conditions. Similar suites of trace fossils may then be compared to these established ichnofacies to ascertain the dominant environmental conditions. These also allow the recognition of departures from normal conditions, or environmental stresses such as low oxygen or nutrient content, which may result in deviations from established ichnofacies (MacEachern et al. 2007a, b).

Ichnofabric

The ichnofabric approach is a comprehensive means of evaluating bioturbated deposits (Taylor et al. 2003). Ichnofabric includes all aspects of the texture of the internal structure of a substrate that result from bioturbation or bioerosion at any scale (Bromley and Ekdale 1986; Bromley 1996). In analyzing ichnofabric, both discrete, identifiable trace fossils, as well as diffuse, mottled textures are considered. As a result, the ichnofabric approach is largely utilized when observing trace-fossil assemblages in cross-section, particularly in core. Cross-sections provide the most complete view of bioturbation, including sediment homogenization, disruption features, and bedding modification.

Ichnofabrics are divided into those that are simple and composite (Fig. 10.3A) (Bromley and Ekdale 1986). Simple ichnofabrics result from the activity of a single community whose traces are produced at the same time (analogous to ichnocoenoses) (Bromley and Ekdale 1986). These are often the result of colonization events by opportunistic organisms such as after storm or flood deposition. Composite ichnofabrics are more common and are produced by either the upward replacement of successive communities or the upward movement of a single tiered community (Bromley and Ekdale 1986). Community replacement occurs as a result of environmental change and colonization by a different trace-making community. Upward migration of the same tiered community, however, results from gradual sediment accretion in the same environment causing the overprinting of newer, deeper traces over older, buried shallow traces.

Ichnofabric is most commonly used to assess the total amount of bioturbation in a deposit and the vertical partitioning of the substrate by a community, or tiering (Bromley 1996). The quantification of bioturbation largely involves estimating the percentage of the substrate affected by burrowing organisms. The ichnofabric index (ii) of Droser and Bottjer (1986) and bioturbation index (BI) of Taylor and Goldring (1993) use different grades of bioturbation ranging from 1–5 and 0–6, respectively (Fig. 10.3B). In these models, total bioturbation (ii = 5 or BI = 6) indicates that the rate of biogenic reworking exceeded sedimentation. This generally suggests that there was sufficient time to completely mix the substrate and the environment was favorable for colonization (Bromley 1996). The actual environmental conditions necessary to achieve total bioturbation are quite variable and depend on the type of organisms and the behaviors being expressed. Incomplete bioturbation (ii = 2–4 or BI = 1–5), however, indicates that some stress factor prevented total reworking of the substrate (Bromley 1996). This could be the result of insufficient time of exposure, low nutrient content, low oxygen content, or many others. The type and quality of the trace fossil assemblage must be assessed to determine the causal limiting factor(s). Complete absence of bioturbation (ii = 1 or BI = 0) is more problematic. It can be used to indicate lack of burrowing activity; however, it is often a result of taphonomy (Bromley 1996). In environments with periodic high-energy conditions, for example, reworking of the sediment can destroy biogenic structures. This can occur along stormy coasts or along rivers prone to frequent, high-energy floods.

The substrate can be divided into different vertical zones or tiers based on environmental gradients (Fig. 10.3C) (Bromley 1996). Tiering is controlled by physical, chemical, and biological parameters. In general, the shear strength of the substrate increases downward due to compaction, whereas nutrient and oxygen content decreases (Bromley 1996). As a result, most animals are restricted to the upper 2–4 cm of the substrate. This shallow tier has a low energy cost to move through and a greater concentration of resources. Due to this abundance of organisms, the shallow tiers tend to be represented by homogenized sediment with few discrete traces. Deeper tiers tend to be occupied by fewer organisms due to lower availability of resources and difficulty of movement, but they also provide safety from physical stress, predation, and disturbance by episodic events (Bromley 1996). Deeper tiers are, therefore, characterized by more distinct traces with fewer instances of overlap with increasing depth. Recognizing the tier in which different trace fossils were produced provides critical information about tracemakers and their behavior. Over time, however, environmental change or simple accretion could cause trace fossils in different tiers to overlap. The power of the ichnofabric approach is that it permits the integration of

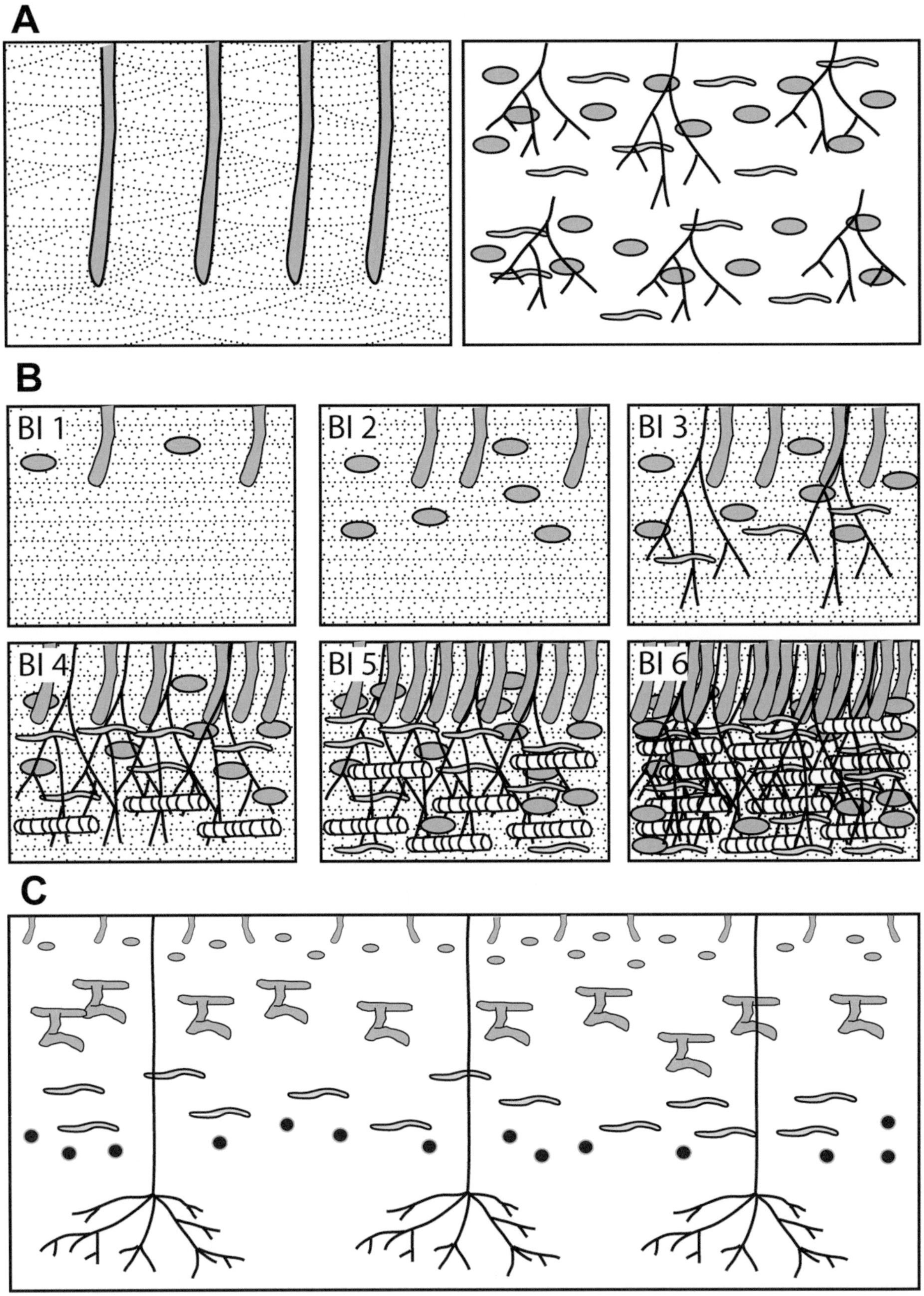

Fig. 10.3 Ichnofabrics. A, Diagrams of simple (left) vs. composite (right) ichnofabrics (modified from Bromley and Ekdale 1986). B, Diagrams of the ichnofabric indices (BI 1–6) of Taylor and Goldring (1993). BI 0 is characterized by no bioturbation. BI 1 (1–4%) has a few discrete traces and sparse bioturbation. BI 2 (5–30%) has low bioturbation with preserved primary sedimentary structures. BI 3 (31–60%) has discrete trace fossils, moderate bioturbation, and some distinguishable bedding planes. BI 4 (61–90%) has intense bioturbation with high density of overlapping trace fossils. BI 5 (91–99%) has completely bioturbated sediment. BI 6 (100%) has completely bioturbated sediment and repeated overprinting of trace fossils (modified from Taylor and Goldring 1993). C, Model of tiering in a single community (modified from Bromley 1996)

all aspects of bioturbation to interpret the sequence of events that produced a final preserved trace fossil assemblage.

Neoichnology

Neoichnology is the study of the trace-making behaviors and resulting traces of modern organisms (Frey 1975). The interpretation of trace fossils is difficult without first understanding the biogenic structures produced by modern organisms. Neoichnological studies are invaluable to provide the data used to interpret the behaviors, burrowing methods, and tracemakers represented by trace fossils (Fig. 10.4). It was only through neoichnologic experiments performed with marine organisms, including annelids, arthropods, echinoderms, and bivalves, that the environmental factors that affect trace-fossil morphology and distribution were accurately determined (Frey 1968, 1970; Schäfer 1972; Bromley and Asgaard 1975; Ekdale and Berger 1978; Frey et al. 1984; Pemberton and Frey 1985; Atkinson 1986; Kanazawa 1992; Seilacher and Seilacher 1994; Gingras et al. 2002a, 2004; Needham et al. 2005; Dashtgard and Gingras 2005; Uchman and Pervesler 2006; Gingras et al. 2007b). These studies were designed to determine the morphological variations in biogenic structures created by specific organisms engaged in known

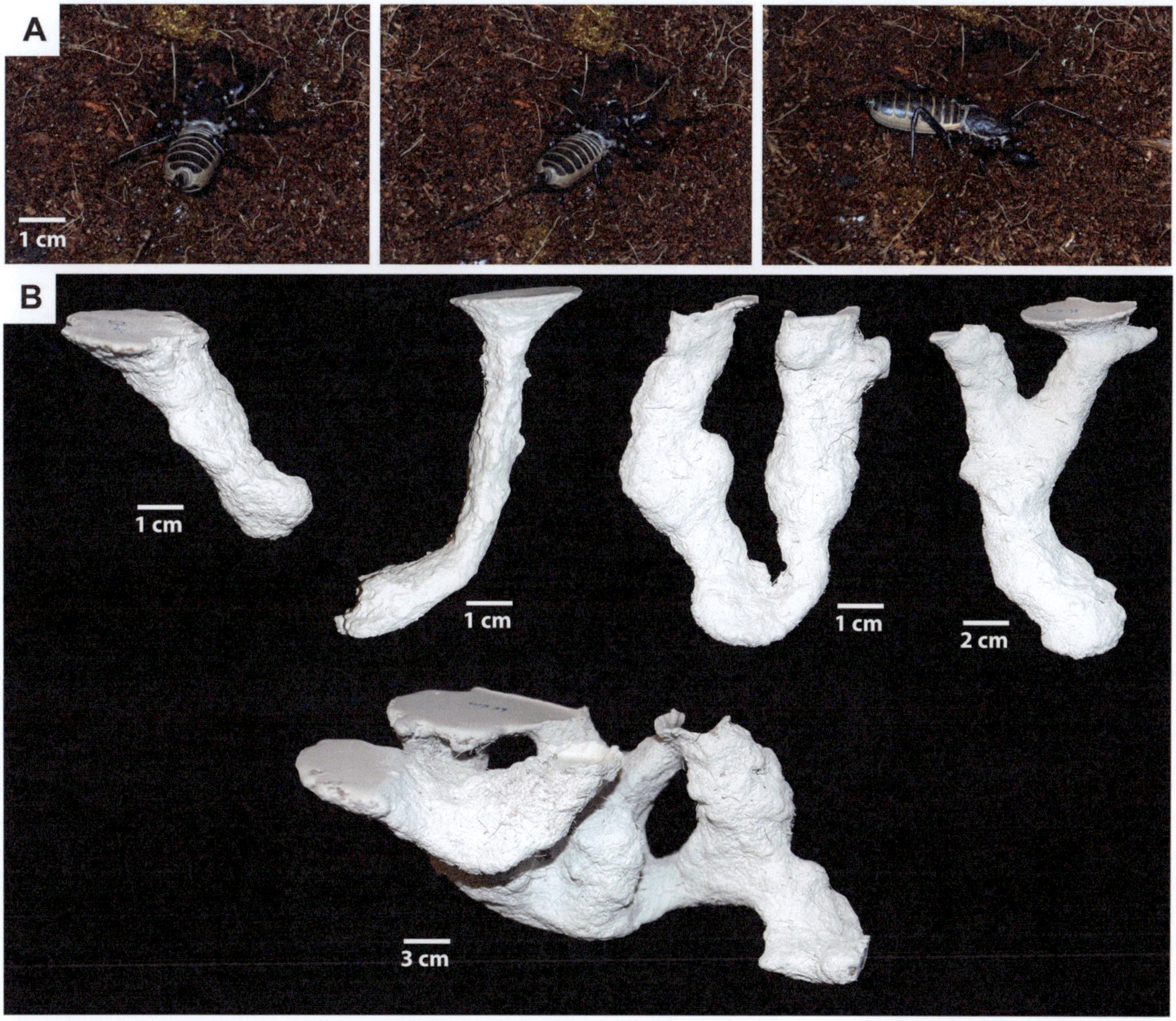

Fig. 10.4 Neoichnological experiments are conducted with modern animals to determine the behaviors associated with the production of biogenic structures and the morphology of the structures (modified from Hembree 2013). A, *Mastigoproctus giganteus* (giant vinegaroon) excavating a burrow in a sediment-filled terrarium. B, Assemblage of burrow morphologies produced by specimens of *M. giganteus* including (from left to right) a subvertical shaft, J-shaped burrow, U-shaped burrow, Y-shaped burrow, and mazework

behaviors under controlled environmental conditions, which could then be used to interpret the paleoecological and paleoenvironmental significance of many marine trace fossils. For example, studies of modern crustacean burrows revealed that the ichnospecies *Ophiomorpha nodosa* often loses its wall lining and becomes *Thalassinoides suevicus* due to changes in substrate consistency (Bromley and Frey 1974; Frey et al. 1978). Similar studies of modern heart urchins (echinoids) have indicated that these organisms produce different trace morphologies depending on their behavior: the ichnospecies *Scolicia prisca* while grazing and *Cardioichnus planus* when resting (Smith and Crimes 1983).

Neoichnologic studies may be conducted in the laboratory or the field. Typically, specific organisms or all of the tracemakers in a specific environment are selected for study. The advantage of laboratory studies is the ability to control the environment so that different parameters may be modified to determine the tracemakers' behavioral response and resulting trace morphologies. Field studies, on the other hand, are largely uncontrolled but do allow the observation of tracemakers in natural settings where responses to other organisms affect expressed behaviors. In addition, the size of some animals and the scale of their traces preclude their study in a laboratory setting and require investigation in the field. Both types of studies, however, are important for understanding the controls on trace fossil morphology and the connection of trace fossils to potential tracemakers.

Strengths, Biases, and Limitations of Trace Fossil Analyses

Strengths

Trace fossil analyses have several strengths that make them invaluable to paleoenvironmental and paleoecological reconstructions. Most importantly, trace fossils record the behavior of organisms rather than simply their morphology, providing a view of the life activities of organisms (Frey 1975, 1978; Bromley 1996). When used in combination with body fossils of known trace makers, a better understanding of organisms as functioning entities can be obtained. In addition, trace fossils are essentially biologically-induced sedimentary structures and are, therefore, easily preserved in conditions that normally lead to the destruction of body fossils (Bromley 1996; Pemberton et al. 2001). Trace fossils may even be enhanced by diagenetic processes that prevent the preservation of body fossils (Gingras et al. 2007a). Under such conditions, trace fossils may provide the only evidence of life. Furthermore, the majority of organisms are soft-bodied or have poorly mineralized hard parts and generally not preserved as fossils except under exceptional circumstances. These soft-bodied components of most ecosystems often represent a greater biomass than those with mineralized hard parts and their absence in the fossil record severely biases paleoecological reconstructions (Bromley 1996). The presence of these organisms, however, may be recorded in the physical structures produced by their interactions with the sediment and other types of media. Recognizing the biological significance of trace fossils helps to fill these gaps in ancient ecosystems (Osgood 1975). Trace fossils are also preserved *in situ*; they are rarely transported from where they were produced and indicate where organisms actually lived (Frey 1975; Bromley 1996). As such they record environmental conditions at the site of preservation and are, therefore, ideal for paleoenvironmental reconstruction. Finally, most ichnogenera have long temporal ranges allowing the comparison of similar ecosystems and sedimentary environments across long spans of geologic time (Bromley 1996).

Biases and Limitations

Despite these strengths, several issues bias or limit the use of trace fossils in paleoenvironmental and paleoecological studies. First, the behaviors and kinds of traces produced by many modern organisms are poorly understood, making the interpretation of trace fossils difficult (Bromley 1996). A major problem faced by neoichnologists is that the study of three-dimensional structures within the sediment under natural conditions is inherently challenging. In addition, cataloguing the biogenic structures produced by all possible organisms engaged in all types of behaviors is a significant undertaking and it is unlikely that a complete understanding of burrowing organisms through geologic time will ever be obtained.

Through the study of modern trace-making organisms, several important patterns have been discovered that can complicate the interpretation of trace fossils. First, individual organisms can produce several different types of traces when engaged in different behaviors or when in different environmental conditions (Bromley 1996; Pemberton et al. 2001). This can easily lead to the misinterpretation of greater biodiversity based on the presence of multiple ichnogenera single locality, when in reality, a single species may have produced all of them. Second, different organisms may produce similar traces when engaged in similar types of behaviors (Bromley 1996; Pemberton et al. 2001). This may result in an underestimation of true biodiversity where several different types of organisms may produce identical trace morphologies (for example, a vertical tube). Third, a single organism can produce a structure having multiple parts with different morphologies (i.e., a compound structure) or multiple organisms can produce a single structure composed of multiple distinct elements (i.e., a composite structure) (Bromley 1996). These are easy to recognize in modern substrates, but require careful interpretation in the fossil record.

Finally, the long-term preservation of trace fossils may alter their original appearance, making direct comparison to modern analogs difficult (Bromley 1996). When studying modern traces, neoichnologists only see organism activity over short time scales (usually less than one year). But biogenic structures such as burrows may be occupied for long periods by multiple generations of organisms that continuously modify the structure while shifting its position. This process can result in trace fossils that are aggregates of multiple copies of the same structure (i.e., cumulative structures) (Bromley 1996). As such, there may be no modern equivalent to the complete trace fossil and the recognition of a single, core component is required. During preservation, processes of filling, cementation, and diagenesis may make the trace fossil larger, more amorphous, and mask important surficial features (Gingras et al. 2007a). Many trace fossils, especially those produced deep in the substrate, can only be viewed as two-dimensional exposures (in outcrop or core) of three-dimensional structures. This limits proper visualization of the complete trace, sometimes revealing only cross-sectional shapes and diameters (Bromley 1996). This loss or lack of access to critical information again makes it challenging to compare trace fossils to modern traces and make proper interpretations.

Cenozoic Case Studies

Trace fossils allow improved interpretations of continental settings by filling in data gaps related to environments and ecosystems. Trace fossils can be used to recognize unconformities and other surfaces of prolonged subaerial exposure, details related to the complex lateral heterogeneity of terrestrial landscapes, conditions of the soil environment, and short-term temporal variations in environmental conditions. Trace fossils also provide evidence of the presence of organisms not preserved as body fossils, improving the assessment of biodiversity, and they lead to a better understanding of the complex interactions of soils, flora, and fauna. In the last two decades, trace fossils have become important tools in the reconstruction of Cenozoic terrestrial environments of Argentina, including the early Eocene Gran Salitral Formation (Melchor et al. 2002), the middle Eocene to early Miocene Sarmiento Formation (Bellosi and González 2010; Bellosi et al. 2010; Sánchez et al. 2010), the early Miocene Pinturas Formation (Bown and Laza 1990; Genise and Bown 1994), the late early Miocene Santa Cruz Formation (Krapovikas 2012), and the Miocene Toro Negro Formation (Krapovickas et al. 2009), Spain, including the Oligocene Noguera Lacustrine System (de Gibert and Sáez 2009), the Miocene Jaraba Formation (Mueller and McCann 2014), and the upper Miocene Torre Formation (Neto de Carvalho et al. 2015), and the United States, including the Paleocene-Eocene Willwood Formation of Wyoming (Kraus and Hasiotis 2006; Smith et al. 2008a, b), the late Eocene-early Oligocene White River Formation (Hembree and Hasiotis 2007a, b), the late Oligocene to early Miocene John Day Formation of Oregon (Retallack 2004), and the middle Miocene Pawnee Creek Formation (Hembree and Hasiotis 2008).

To illustrate the utility of trace fossils to paleoecological and paleoenvironmental reconstructions, two case studies from the Paleogene and Neogene of the North American midcontinent are presented in detail. The case studies involve sites in northeastern Colorado, U.S.A., within the late Eocene-early Oligocene White River Formation (Hembree and Hasiotis 2007a, b) and the middle Miocene Pawnee Creek Formation (Hembree and Hasiotis 2008). Both sites consist of deposits of alluvial systems dominated by localized channel conglomerate and sandstone and abundant, widespread paleosols. Despite their general similarities, the two units were deposited under very different climatic conditions with distinct landscapes and ecosystems. Much of this information is recorded in the trace fossils preserved within the paleosols.

Eocene-Oligocene White River Formation

Background: The late Eocene to early Oligocene was the beginning of modern climatic conditions and involved major changes in climates and biota (Prothero and Berggren 1992; Prothero et al. 2003). The Eocene was characterized by tropical climates and minimal latitudinal temperature gradients, while the Oligocene had generally cooler, more seasonal climates with steep temperature gradients across latitudes (Fluegeman 2003; Hickman 2003; Myers 2003; Oleinik and Marincovich 2003; Yancy et al. 2003). Whereas the Eocene-Oligocene transition is well documented in the marine record (Miller et al. 1987; Miller 1992; Ivany et al. 2003), details of the transition in continental environments are limited by the scattered distribution of sedimentary basins and the difficulty of correlating continental and marine strata from this interval of time. Terrestrial plant- and vertebrate-fossil assemblages of North America suggest the development of lower mean annual temperature and precipitation in the late Eocene and early Oligocene (Hutchinson 1992; Leopold et al. 1992; Retallack 1992; Stucky 1992; Wolfe 1992; Sullivan and Holman 1996; Retallack et al. 2004). The paleosols in and on which these organisms lived, however, are well preserved and abundant. These paleosols contain the trace fossils of the above- and below-ground soil biota, including plants, invertebrates, and vertebrates The incorporation of this trace fossil evidence with the rest of the continental record allows for a more complete interpretation of the continental landscapes and climate during the late Paleogene climate transition.

The White River Formation is an Eocene-Oligocene unit of the central Rocky Mountains and Great Plains of North America composed primarily of 5–460 m of fine-grained, tuffaceous mudstone (Seeland 1985; Swinehart et al. 1985). The lower members of the White River Formation are estimated to be 36–32 Ma based on mammal assemblages, whereas upper members are 32–28 Ma (Galbreath 1953; Prothero and Whittlesey 1998). In this study, a 25 km long and 75–175 m thick exposure of the White River Formation in northeastern Colorado, U.S.A. was investigated (Fig. 10.5A). Exposed strata consisted of fine-grained, alluvial overbank deposits, as well as sandstone and conglomerate (Galbreath 1953; Seeland 1985). The units are broken into three members, the Horsetail Creek, Cedar Creek, and Vista.

Methods Utilized: Analysis of the White River Formation included a detailed study of the paleosols and associated trace fossils in each member. The paleosols were recognized by such diagnostic pedogenic features as disrupted primary sedimentary structures, blocky textures, root traces, gradational contacts, burrows, tracks, and desiccation cracks. The trace fossil assemblages from each paleosol were described and bioturbation indices (BI) were determined to assess the total diversity and abundance of trace fossils. Neoichnological studies from the literature were used to interpret the different kinds of trace fossils, including the behaviors represented, potential tracemakers, and their environmental significance. The assemblages were then subdivided into ichnocoenoses where possible, based on the repeated grouping of certain kinds of trace fossils and their occurrence in paleosols diagnostic of limited times of formation (tens to hundreds of years) and rapid burial. The ichnocoenoses were used as the basic unit for environmental interpretation. Ichnocoenoses were compared to modern analogs from the literature to evaluate the ecological significance of these assemblages. Trace fossil assemblages were not assigned to ichnofacies due to the low amount of variability between individual assemblages.

Results: The paleosols of the White River Formation formed within a dynamic alluvial system. The paleosols fall into four major types, Types I–IV, based on differences in lithology, degree of pedogenic development, and trace fossil assemblages (Fig. 10.5B–E). Type I and II paleosols occur at the base of the studied section in the Horsetail Creek and Cedar Creek members, whereas Type III and IV paleosols occur in the upper Cedar Creek and Vista members. The paleosols and trace fossils record a change in the environment and associated soil ecosystem from the late Eocene to early Oligocene in the basin.

Type I paleosols: Type I paleosols occur as stacked series of 10–15 cm thick, poorly developed paleosols (blocky mudstone) with clearly defined upper surfaces and separated by unmodified, scour-based crossbedded sandstone (Fig. 10.5B). Type I paleosols formed as a result of periodic, rapid sedimentation events that buried the previous landscape surface, resulting in the formation of a new soil (e.g., Kraus 1999). The poorly developed nature of Type I paleosols indicates that these events occurred frequently (e.g., Schaetzl and Anderson 2009). Ichnofossils associated with Type I paleosols include small diameter, calcareous root traces (rhizoliths), brood burrows of ground nesting bees (*Celliforma* isp.), burrows of adult and larval beetles (Type A and B *Planolites* isp.), simple brood burrows of dung beetles consisting of a single tunnel and dung ball (*Macanopsis* isp.–*Pallichnus* isp.), and vertebrate carnivore coprolites (Fig. 10.6A–F). These trace fossils occur as recurring assemblages beneath the buried surfaces of poorly developed paleosols and are interpreted as two different ichnocoenoses. Bioturbation is moderate, and Type I paleosols have a BI of 3.

The rhizoliths are 0.5–1.0 mm diameter, filamentous structures up to 30 cm long with small lateral branches that extend throughout the upper 10 cm of the paleosol profiles to form dense networks (Fig. 10.6A). The small size, shallow depth, and high density of the rhizoliths are similar to modern shallow-rooted grasses present in poorly drained soils with a high water table (e.g., Lavelle and Spain 2001). In alluvial settings, these conditions tend to occur near, but not adjacent to active channels (Birkeland 1999; Kraus 1999).

Burrows in Type I paleosols are interpreted as products of two categories of soil invertebrates, those living in poorly drained, moist soils and those living in well-drained soils. This difference in inferred moisture levels supports the interpretation that Type I paleosols formed near active channels. In modern floodplains, such alternating moisture conditions occur on proximal crevasse-splay and overbank deposits. Soils developing in these environments tend to be water saturated during and immediately after flooding and dry out over time (Birkeland 1999). The resulting ichnologic pattern would be the overprinting of traces of organisms living in wet or poorly drained soil by traces of organisms living in dry or well-drained soil; each would constitute a single ichnocoenosis (Fig. 10.7A). In Type I paleosols, *Planolites* occur as two forms, each within a different ichnocoenosis. Type A *Planolites* are near-surface, horizontal burrows that cut primary sedimentary structures, are associated with soft-sediment deformation features, and are overprinted by pedogenic features (Fig. 10.6B). Type B *Planolites* are 20–30 cm deep, vertical to horizontal, sharply

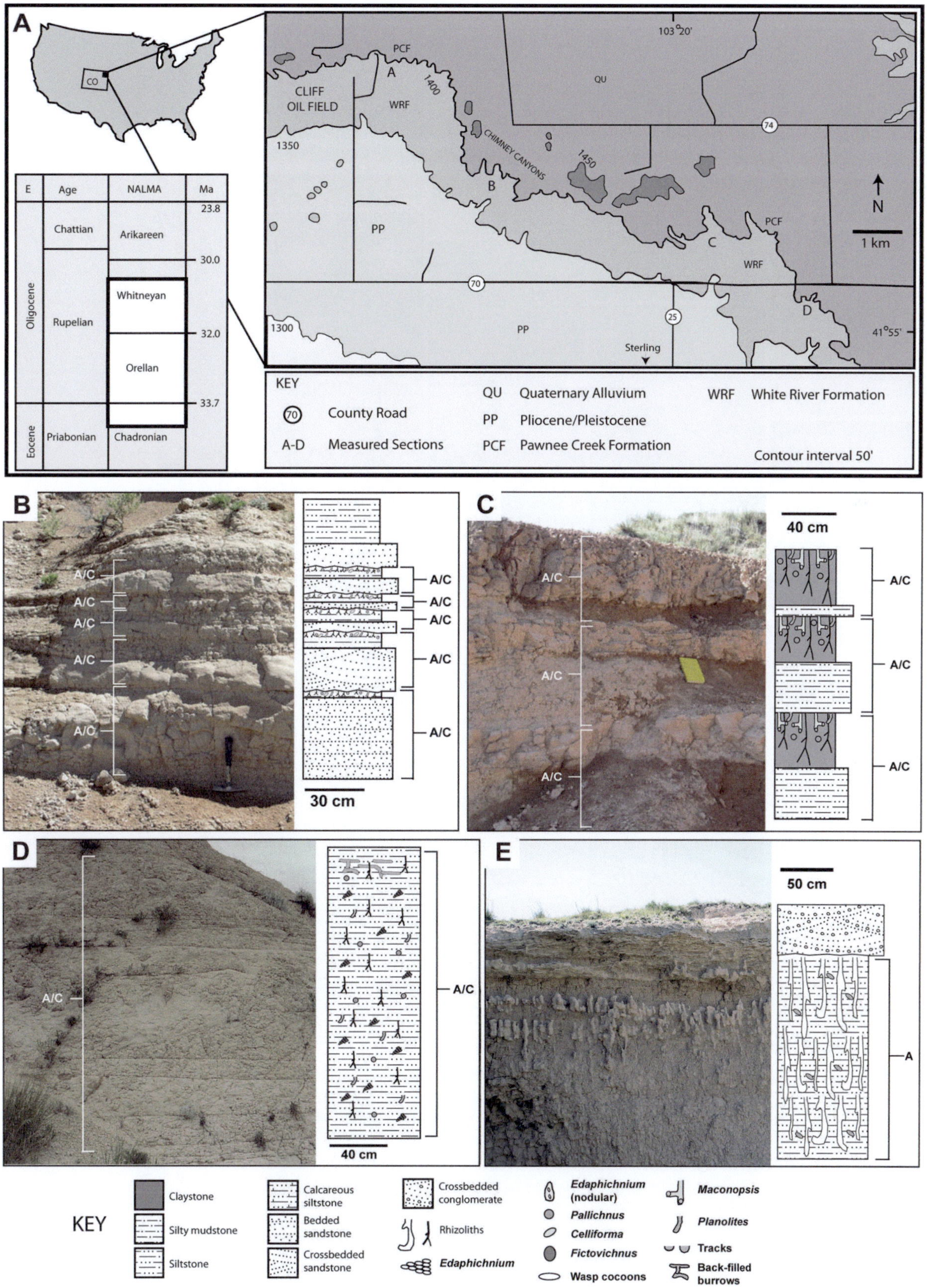

Fig. 10.5 Late Eocene-early Oligocene White River Formation, northeastern Colorado (modified from Hembree and Hasiotis 2007a, b). A, Locality map of the study area. B, Type I paleosol. C, Type II paleosol. D, Type III paleosol. E, Type IV paleosol

Fig. 10.6 Trace fossil assemblages of the White River Formation paleosols. A–F, Type I paleosol trace fossils: A, small diameter, calcareous rhizoliths (at arrows and inset); B, Type A *Planolites* isp. (outlined); C, Type B *Planolites* (at arrow); D, *Celliforma* isp.; E, Simple *Macanopsis* isp.–*Pallichnus* isp.; F, Vertebrate carnivore coprolites. G–K, Type II paleosol trace fossils: G, Root cast (at arrow); H, Type A *Planolites* (outlined); I, Simple *Macanopsis–Pallichnus*; J, Complex *Macanopsis–Pallichnus* (preserved dung balls at arrows); K, Trampled ground. L–R). Type III paleosol trace fossils: L, Root casts; M, Type A *Planolites;* N, Type B *Planolites;* O, *Parowanichnus* isp; P, *Edaphichnium* isp. (with close-up inset); Q, Simple *Macanopsis–Pallichnus*; R, *Beaconites* isp. (with close-up inset); S–X, Type IV paleosol trace fossils: S–T, Large diameter root casts; U, Calcareous rhizoliths; V, *Edaphichium*; W, *Fictovichnus parvus*; X, Wasp cocoons

defined burrows that cut through the massive, blocky pedogenic texture and some Type A *Planolites* (Fig. 10.6C) Based on these properties, the Type A *Planolites* were produced when the sediment was very moist, most likely due to recent flooding events – an interpretation supported by the presence of Type A *Planolites* on surfaces of rippled, fine-grained sandstone and sandy siltstone beds. Type B *Planolites* were produced in the developing soil when it was more cohesive, better drained, and with low soil moisture near the surface and higher moisture at depth.

Insect brood burrows such as *Celliforma ficoides*, *Macanopsis*, and *Pallichnus* are also indicative of a well-drained, dry, upper paleosol profile (Fig. 10.6D–E) (e.g., Sakagami and Michener 1962; Halffter and Matthews 1966; Hasiotis 2002) and occur in an ichnocoenosis with Type B *Planolites*. Brood burrows require that Type I paleosols remained unsaturated for at least one to five months of the year to allow for the successful incubation of the insect eggs (e.g., Retallack 1984; Cambefort and Hanski 1991; Hasiotis 2002). Based on comparison with modern dung beetle burrows, deep *Macanopsis* tunnels associated with *Pallichnus* suggest that they were constructed directly beneath pats of dung rather than being transported from elsewhere (Halffter and Matthews 1966). This implies that herbivores and plant material were abundant in the environment during paleosol development. Preservation of carnivore coprolites and not herbivore coprolites further supports well-drained, oxidizing conditions since organic-rich herbivore dung is preserved in wet, reducing environments, but rapidly decomposes in dry settings (Fig. 10.6F) (Hunt et al. 1994). The absence of herbivore coprolites may also be a result of dung consumption by invertebrates.

Type II Paleosols: Type II paleosols occur as stacked series of 40–60 cm thick, moderately well-developed paleosols (blocky mudstone) with clearly defined upper surfaces and separated by unmodified, massive siltstone (Fig. 10.5C). Type II paleosols formed as a result of periodic, rapid sedimentation events that buried the previous landscape surface resulting in the formation of a new soil (e.g., Kraus 1999). The better development of Type II paleosols suggests that these events were less frequent than in Type I paleosols, allowing more time for pedogenesis (e.g., Schaetzl and Anderson 2009). Ichnofossils associated with Type II paleosols include medium-diameter, vertically elongate root casts, Type A *Planolites,* large numbers of complex, branching burrow networks of *Macanopsis–Pallichnus,* and trampled ground consisting of large numbers of overlapping, indistinct tracks (Fig. 10.6G–K). The trampled ground is seldom associated with other trace fossils, likely due to the disruption of the underlying sediment. Bioturbation is intense with some overlapping trace fossils, and Type II paleosols have a BI of 4.

The 5–20 cm long root casts are interpreted as taproots based on their morphology and indicate that Type II paleosols were well-drained with a deeper water table than where Type I paleosols formed (Fig. 10.6G). In general, taproots extend into the upper vadose zone where soil moisture is high, but not to the point of saturation (Aber and Melillo 2001). In addition, soils containing plants with taproots tend to be better developed due to the resulting increase of downward transportation and translocation of organic material, detrital sediment, and minerals by water (Buol et al. 2003).

The burrows are suggestive of two separate ichnocoenoses produced immediately following deposition while soil moisture was high (Type A *Planolites*) and again after prolonged subaerial exposure when soil moisture was low (*Macanopsis–Pallichnus*) (Fig. 10.7B). The Type A *Planolites* are sinuous and disrupt relict lamination; laminae are either broken into segments or ductilely deformed (Fig. 10.6H). This morphology requires relatively high soil moisture at the time of construction. The presence of trampled ground further suggests that there were periods when the paleosol was saturated, either immediately after flooding or as a result of high rainfall events (Fig. 10.6K). The concentration and complexity of *Macanopsis–Pallichnus* suggest a large population of dung beetles and successive generations of beetles reusing the same soil as a breeding site (e.g., Cambefort and Hanski 1991). Complex and abundant *Macanopsis–Pallichnus* networks, some containing 20–30 dung balls (Fig. 10.6J), also imply the presence of a large population of herbivores to provide the necessary dung, despite the lack of complete skeletal material of such animals in the paleosols (e.g., Hasiotis 2004).

Type III Paleosols: Type III paleosols occur as generally uniform, 30-m thick sequences of moderately well-developed paleosols (blocky silty mudstone) that lack clearly defined upper surfaces (Fig. 10.5D). Type III paleosols formed as a result of slow but continuous aggradation of the landscape surface resulting in a uniform, over-thickened soil (Kraus 1999). Type III paleosols are characterized by a vertically continuous distribution of diverse ichnofossils including millimetric, calcareous rhizoliths, centimetric, elongate root casts, Type A and B *Planolites,* ant nests (*Parowanichnus* isp.), earthworm burrows and fecal pellets (*Edaphichnium* isp.), simple *Macanopsis–Pallichnus*, and backfilled burrows (*Beaconites* isp.) produced by the movement and feeding of larval soil bugs and beetles (Fig. 10.6L–R). Bioturbation is moderate to intense, and Type III paleosols have a BI of 4.

The ichnofossil assemblage of Type III paleosols is suggestive of an organic-rich, well-drained, moist soil allowing permanent occupation of the subsurface by air-breathing animals including ants, soil bugs, beetles, and earthworms. All but Type A *Planolites* occur in the same ichnocoenosis (Fig. 10.7C).

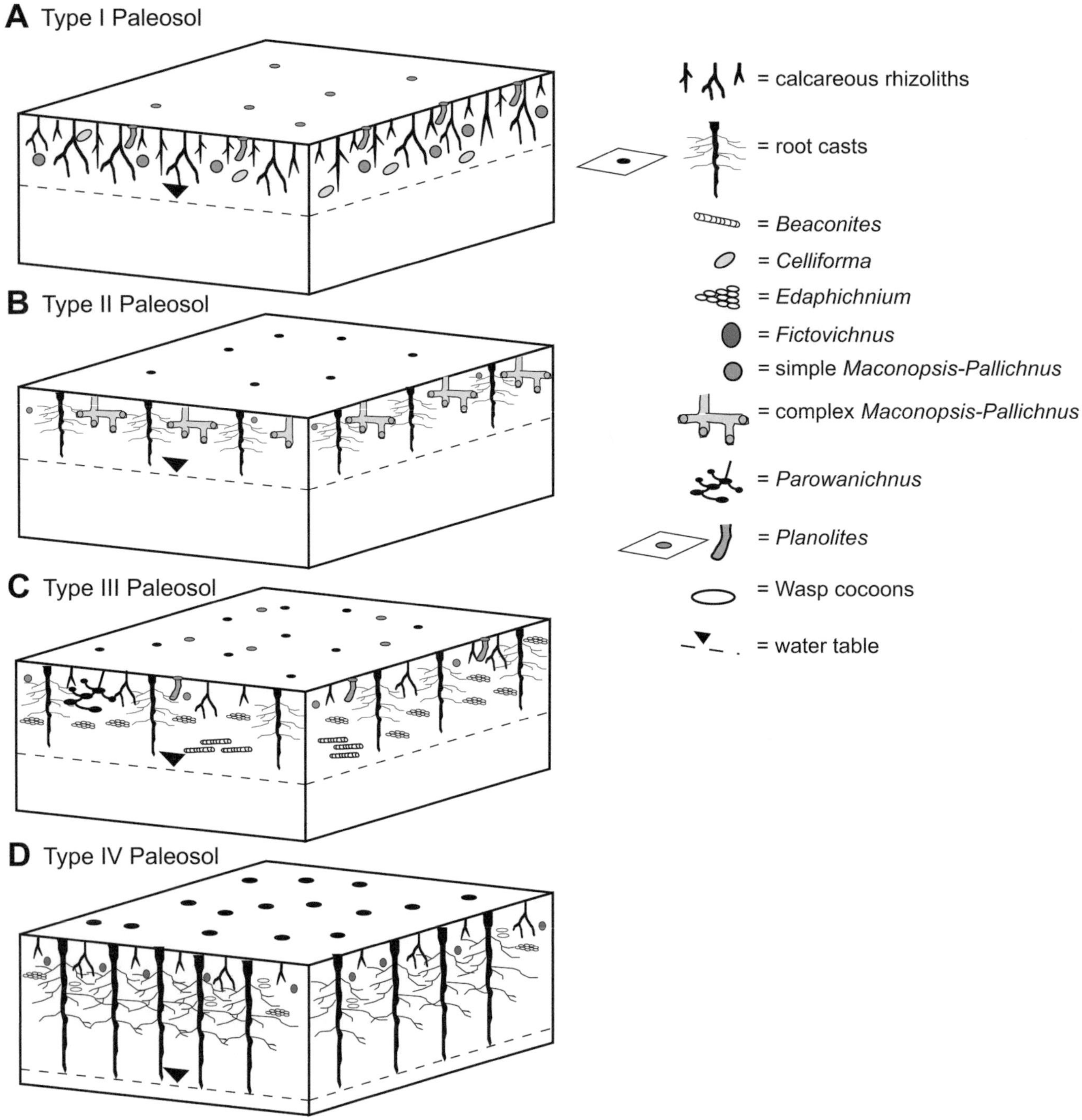

Fig. 10.7 Ichnocoenoses of the White River Formation paleosols: A, Type I paleosol; B, Type II paleosol; C, Type III paleosol; D, Type IV paleosol

The occurrence both of millimetric rhizoliths concentrated near soil surfaces and centimetric root casts penetrating deep into the soil (Fig. 10.6L) suggests variable surface vegetation including grasses and shrubs (e.g., Lavelle and Spain 2001). Roots responded to differences in the depth of available soil water, likely as a result of seasonal changes in precipitation. Type B *Planolites*, simple *Macanopsis–Pallichnus,* and *Parowanichnus* indicate that the paleosol was relatively dry and consolidated for long periods of its development (Fig. 10.6N, P–Q). The complexity of the preserved ant nests in particular suggests that the paleosols were stable with a low sedimentation rate (Fig. 10.6O) (e.g., Sudd 1967; Elmes 1991). Moderate soil-moisture levels are also indicated by abundant *Edaphichnium* (Fig. 10.6P); the presence of earthworms requires that the soil was moist enough to prevent dehydration of these soft-bodied animals but possessed open pore spaces that permitted respiration (e.g., Hasiotis 2002). *Edaphichnium* and *Beaconites* also indicate that Type III paleosols were organic-rich despite the absence of preserved organic matter (Fig. 10.6P, R). The tracemakers of both burrows consumed plant debris and other organic matter in the soil; their abundance indicates a large supply of food resources. Additional organic matter was likely concentrated in the upper horizons by the incorporation of dung and plant material in the soil by burrowing beetles and ants (e.g., Hole 1981; Elmes 1991).

The presence of Type A *Planolites* indicates that changes in the soil consistency and moisture did occur during pedogenesis, from consolidated and dry conditions to unconsolidated and moist conditions. These changes likely occurred in relation to flooding events in which layers of new sediment were deposited on the soil surface.

Type IV Paleosols: Type IV paleosols occur as stacked series of 40–60 cm thick, overlapping, moderately well-developed paleosols (blocky siltstone) with poorly defined upper surfaces (Fig. 10.5E). The upper surface of each underlying paleosol in the sequence is partially overlapped and overprinted by the overlying paleosol. Evidence for this overlap largely derives from the extension of large-scale root casts that extend from the top of one paleosol down into an underlying paleosol with its own discrete assemblage of root traces (Fig. 10.6S–T). Type IV paleosols formed as a result of periodic sedimentation events that buried the previous landscape surface, but where the layer of sediment was not thick enough to prevent the modification of the buried paleosol by the newly developing soil at the surface (e.g., Kraus 1999). Trace fossils associated with Type IV paleosols include 10–15 cm diameter, $\sim$1 m long root casts, centimetric, elongate root casts, millimetric calcareous rhizoliths, *Edaphichium,* beetle cocoons (*Fictovichnus parvus*), and wasp cocoons (Fig. 10.6S–X). Bioturbation is intense, and Type IV paleosols have a BI of 5.

The abundance of laterally and vertically extensive, large diameter and elongate root casts indicate that Type IV paleosols were uniformly well-drained across large distances with a water table at least 2 m below the surface (Fig. 10.7D) (e.g., Lavelle and Spain 2001). The absence of plant fossils in Type IV paleosols makes the identification of the plants represented by the rhizoliths difficult; however, the rhizoliths do provide evidence of considerable vegetative cover on the landscape surface despite the lack of body fossils. In addition, the density and lateral extent of the large rhizoliths suggests an increase in the scale of vegetative cover in the area by the early Oligocene (Fig. 10.6S–T). The large number of beetle and wasp cocoons represent multiple seasons of nest-site reoccupation (Fig. 10.6W–X). Based on the ecology of extant beetles and wasps, these ichnofossils also indicate that these paleosols were long-lived, well drained, and represent thick, preserved A horizons (e.g., Lavelle and Spain 2001). Whereas *Edaphichnium* indicates high soil organic content, the lower quantity and density of these trace fossils relative to those of Type III paleosols suggests a decrease in the earthworm population (Fig. 10.6V). This decrease may have been due to a lowering in either soil moisture or organic matter availability (e.g., Lee 1985). The abundance of large plants could lower total organic matter content, with a greater percentage being recycled by plants rather than accumulating in the soil (e.g., Lavelle and Spain 2001). The smaller average size of the individual *Edaphichnium* pellets is likely also the result of changes in vegetation; large earthworm pellets are typical of modern grasslands and savannas, whereas those of woodlands and forests are 25–50% smaller (Lee 1985).

Conclusions: Trace fossils are essential tools for understanding the complex local to regional processes affecting terrestrial environments and ecosystems during the Eocene-Oligocene transition. The transition from sparsely vegetated, poorly developed paleosols in the late Eocene Horsetail Creek Member to densely vegetated, well-developed paleosols in the early Oligocene Vista Member suggests a change in ecosystems from open grassland to wooded grassland. This ecosystem shift is different from that interpreted from paleosols of South Dakota and central Oregon where the Oligocene is characterized by reduced vegetation and poorly developed paleosols in comparison to the late Eocene. As a result, soils and soil ecosystems appear to have been largely controlled by autogenic processes rather than driven uniformly by broad allogenic processes. In particular, the distance from fluvial channels was the main control on the types of paleosols and the associated trace-fossil assemblages that developed; the ichnocoenoses were controlled by the amount of available soil moisture, surface stability, and organic content. Total ichnodiversity remained nearly stable up section and the few changes in trace-fossil assemblages through time appear to relate to changes in the soil biota as regulated by local hydrology and landscape stability.

Miocene Pawnee Creek Formation

Background: The Miocene was another interval of transition in the Earth's climate and ecosystems. The Miocene was a time of warmer global climates than the Oligocene and the Pliocene (Zachos et al. 2001). Oceanic paleotemperature curves indicate that the early Miocene was the warmest time during the Cenozoic since the Eocene (Zachos et al. 2001). In North America, uplift of the Rocky Mountains through to the middle Miocene resulted in a seasonally dry midcontinental climate characterized by low average precipitation and frequent droughts (Leopold and Denton 1987). By the late Miocene, uplift of the Sierra Nevada and Cascade mountain ranges also resulted in aridity in regions west of the Rocky Mountains (Leopold and Denton 1987). In the continental interior, grassland ecosystems expanded whereas forests contracted in response to the drying (Retallack 2001a). The changing environments resulted in changes in populations of herbivorous mammals, with forest browsers giving way to prairie browsers (Behrensmeyer et al. 1992). The changes in the climate and vegetation also had significant effects on soil properties and soil ecosystems. Insects, especially ants and termites, constituted a large part of the animal biomass in the new open vegetation habitats that developed in the late Cenozoic (Behrensmeyer et al. 1992).

The Pawnee Creek Formation (PCF) of northeastern Colorado is middle Miocene in age (Ar/Ar date of 14.5 ± 0.09 Ma and 14.3 ± 0.02 Ma) (Fig. 10.8A) (Dold and Prothero 2003). The PCF consists of a series of valley fills composed of siltstone, sandstone, and conglomerate that cover most of the Oligocene paleosurface (Galbreath 1953). Paleosols occur within the valley-fill successions and record periods of depositional hiatus and landscape stability. This study described two long-lived, stacked Miocene paleosols that are recognized primarily by their possession of distinct assemblages of small (2-mm-diameter) to large (800-mm-diameter) burrows and root traces. Burrow morphology was used to determine potential tracemakers and behaviors, as well as to reconstruct details of the environment and ecosystem represented by the paleosols of the PCF.

Methods Utilized: Two paleosols of the Pawnee Creek Formation were described along with their associated trace fossil assemblages. Recognizable trace fossils were described to assess the total diversity and abundance of trace fossils. Less distinct trace fossils were also included to estimate bioturbation indices (BI). Previous neoichnological studies were used to interpret the distinct trace fossils including the behaviors represented, potential tracemakers, and their environmental significance. Trace-fossil assemblages of each paleosol were then divided into ichnocoenoses rather than ichnofacies based on the repeated grouping of certain kinds of trace fossils and similarity in environmental signatures. The ichnocoenoses were used to evaluate the ecological significance of these assemblages based on modern analogs.

Results: The high degree of bioturbation by terrestrial organisms in PCF paleosols provides primary evidence that two surfaces of prolonged subaerial exposure (paleosols) are preserved, representing intervals of low sedimentation separated by a period of rapid deposition. The trace fossils of the PCF are interpreted as temporary to permanent dwellings of terrestrial invertebrates and vertebrates that served as refuges from warm and dry climatic conditions. Many of the trace fossils were produced by plants and animals that are not preserved as body fossils in the PCF; therefore, the trace fossils provide a better estimate of the true biodiversity than the body fossils.

Paleosols: Paleosol 1 is 2–3 m thick and consists of an upper silicified, calcareous mudstone and a lower calcareous silty mudstone (Fig. 10.8B). Pedogenic features include angular blocky peds, absence of primary sedimentary structures, a well-developed pedogenic carbonate horizon near the paleosurface, extensive vertical and horizontal branching root traces, and numerous simple burrows and complex burrow systems (Fig. 10.8D). Bioturbation of Paleosol 1 is complete, with a BI of 5.

Paleosol 2 unconformably overlies Paleosol 1 and forms the upper surface of the outcrop in the field area. Paleosol 2 is 3–4 m thick and consists of an upper silicified, calcareous siltstone and a lower sandy siltstone (Fig. 10.8C). Pedogenic features of Paleosol 2 include an upper calcic horizon, blocky structure, absence of primary sedimentary structures, pedogenic carbonate nodules, dense concentrations of vertical and horizontal branching calcareous rhizoliths, and small- to large-diameter burrows and burrow systems (Fig. 10.8D). Bioturbation of Paleosol 2 is complete with repeated overprinting of trace fossils and assigned a BI of 6.

Trace fossils: Trace fossils described from the two paleosols include a variety of architectures. Bioturbation patterns consist mainly of dense concentrations of root traces and small diameter shafts, tunnels, and chambers. Where bioturbation density is highest, architectural diversity is low. Larger burrows are less common but have a greater architectural diversity. The density of the large burrows relative to their average burrow diameter is high.

Root traces are vertical to horizontal and fall into three main size categories, 1 mm, 5 mm, and 10+ mm diameter, that reach depths of 750 mm, 1750 mm, and 3000 mm, respectively (Fig. 10.9A–D). The root traces represent the roots of a mix of grasses, shrubs, and trees based on their size and architecture.

Trace fossils associated with invertebrates include: (A) simple backfilled burrows covered with short, linear scratches (*Beaconites knestosichnus*) produced by adult beetles and soil bugs engaged in locomotion and feeding behaviors (Fig. 10.9E–F); (B) backfilled burrows with chambers and cocoons (*Beaconites kytosichnus*) produced by larval stages of beetles, soil bugs, and other insects engaged in locomotion, feeding, and pupation (Fig. 10.9G–H); (C) complex networks of interconnected, small diameter tunnels, shafts, and chambers (*Parowanichnus* isp.) occurring around large root traces and interpreted as ant nests that represent complex social dwelling behaviors (e.g., Hasiotis 2003) (Fig. 10.9I–J).

Trace fossils associated with vertebrates include: (A) horizontal networks of cylindrical, sinuous tunnels with Y-shaped branches connected to the surface by short,

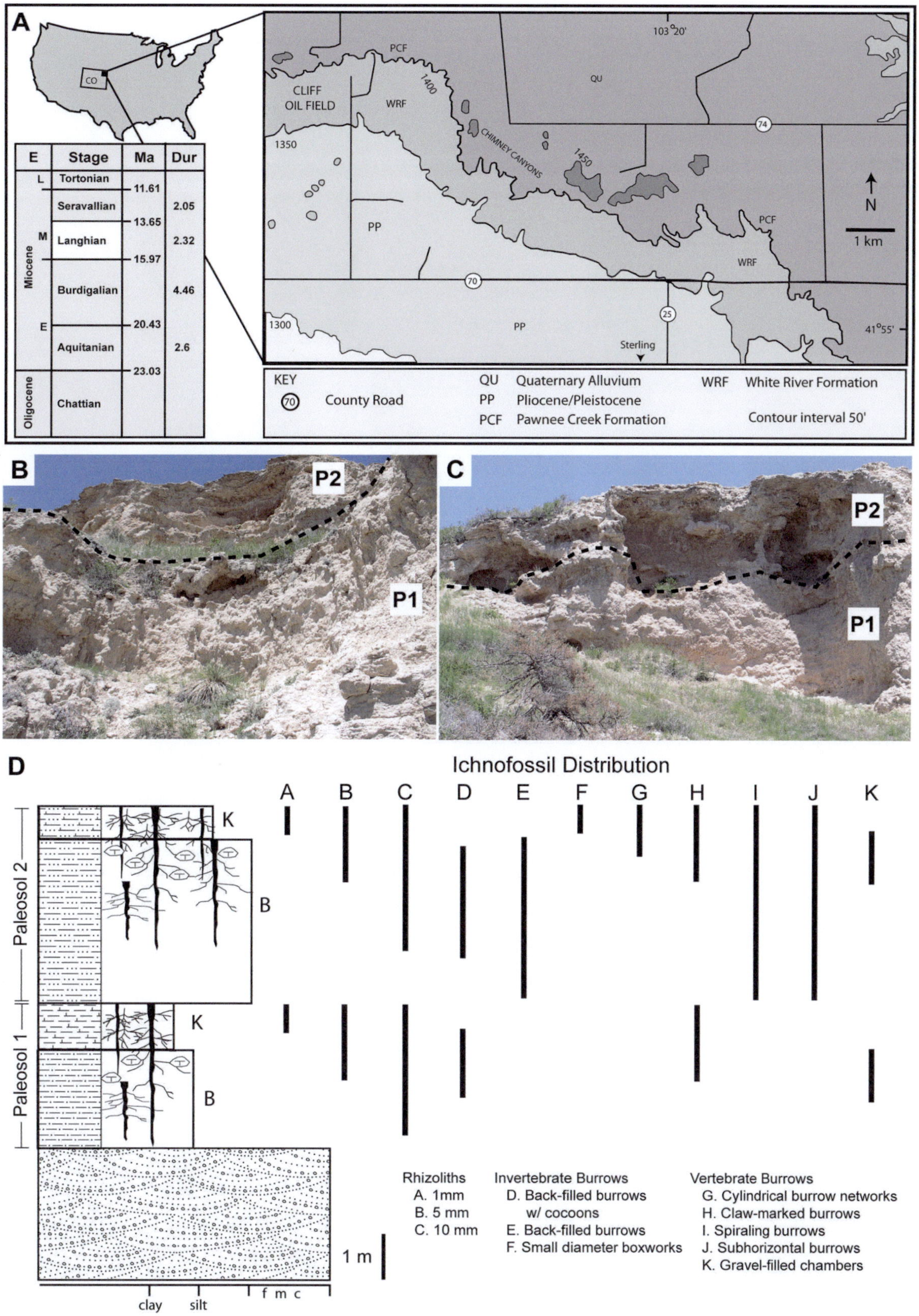

Fig. 10.8 Middle Miocene Pawnee Creek Formation, northeastern Colorado (modified from Hembree and Hasiotis 2008). A, Locality map of the study area. B, Paleosol 1 (P1) is 2–3 m thick with an upper silicified, calcareous mudstone and a lower calcareous silty mudstone. The upper surface is recognized by the presence of a dense mat of rhizoliths and large, scratch-covered burrows. C, Paleosol 2 (P2) is 3–4 m thick with an upper silicified, calcareous siltstone and a lower sandy siltstone. D, Composite stratigraphic section of the Pawnee Creek Formation paleosols. Ichnofossils are tiered from shallow, intermediate, and deep levels within the paleosols

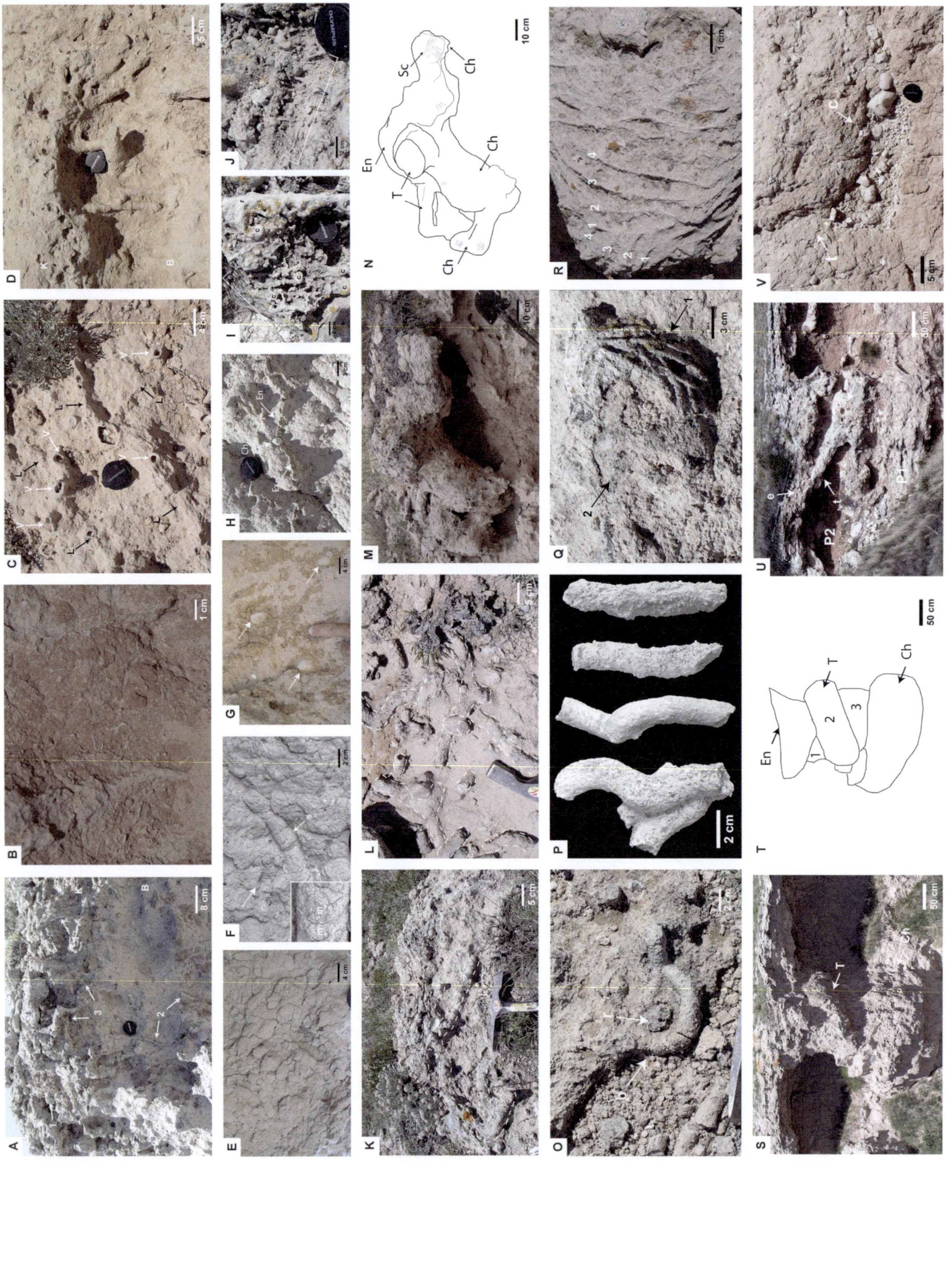

A
B
C
D
E
F
G
H
I
J
K
L
M
N
O
P
Q
R
S
T
U
V
En
T
Ch
Sc
10 cm
1
2
3
50 cm
P1
P2

◀ **Fig. 10.9** Trace fossils of the Pawnee Creek Formation paleosols: A, The upper meter of Paleosol 2 containing 1 mm (1), 5 mm (2), and 10 mm (3) diameter rhizoliths; B, Fine, 1 mm diameter rhizoliths branch laterally and form a dense mat in the upper meter of the paleosols; C, Plan view of vertical (V) and lateral (L) branches of 5–10 mm diameter rhizoliths in the upper meter of Paleosol 2; D, Cross-sectional view of vertical tap root and lateral branches of the large, 5–10 mm diameter rhizoliths in Paleosol 1; E, Dense clusters of backfilled burrows (*Beaconites knestosichnus*); F, *Beaconites knestosichnus* showing backfill menisci (m) (see inset photo) and scratch marks (s) on the burrow wall; G, *Beaconites kytosichnus* with associated closely spaced, filled chambers containing ovoid casts; H, A backfilled tunnel (*B. kytosichnus*) leads to (En) and exits from (Ex) an ovoid cast-bearing chamber (Ch); I, *Parowanichnus* isp. with multiple chambers (c) connected by horizontally and vertically branching tunnels associated with a 10 mm rhizolith (r); J, *Parowanichnus* surrounding a lateral branch of a large diameter rhizolith (r); K, Oblique view of *Kladosystemites* isp. network (outlined) in the upper calcareous horizon of Paleosol 2; L, Interconnected tunnels of *Kladosystemites*; M, Large *Polychoredrites* isp. complex in the upper profile of Paleosol 1; N, Interpretive drawing of *Polychoredrites* from M showing the position of the burrow entrance (En), tunnels (T), chambers (Ch), and scratch marks (Sc); O, Plan view of a U-shaped branch of *Kladosystemites*; P, Sinuous and straight tunnel segments of *Kladosystemites*; Q, Terminal chamber of *Polychoredrites* showing the cemented outer layer (1) and inner friable siltstone fill (2); R, Scratch marks on the walls of *Polychoredrites* consist of multiple sets of three to four 50–120 mm parallel striations; S, Vertically oriented *Daimonelix* isp. with spiraling tunnel (T) and lower chamber (Ch); T, Interpretive drawing of *Daimonelix* in S showing the entrance (En), helical tunnel (T) and terminal chamber (Ch); numbers 1–3 represent successive coils of the tunnel; U, Large subhorizontal *Katarrhedrites* isp. with an upper entrance (e), inclined tunnel (t), and a terminal chamber (c); V, Small, gravel-filled *Katarrhedrites* with tunnel (t) and chamber (c)

subvertical shafts (*Kladosystemites* isp.) produced by small, possibly limbless vertebrates (reptiles or mammals) as dwelling burrows which were likely associated with prey capture (Fig. 10.9K–L, O–P); (B) interconnected complexes of subhorizontal to subvertical tunnels and chambers covered by sets of four parallel, straight scratches (*Polychoredrites* isp.) produced by small- to medium-sized mammals, likely rodents or mustelids, as permanent dwellings possibly for multiple individuals (Fig. 10.9M–N, Q–R); (C) large, vertical, helical burrows with a terminal chamber (*Daimonelix* isp.) produced by large mammals as dwellings, perhaps for hibernation or aestivation as the coiled architecture is typically related to microclimate control (Fig. 10.9S–T); (D) small to large, subhorizontal tunnel with a terminal chamber (*Katarrhedrites* isp.) most likely produced by small mammals to reptiles or large mammals, likely carnivores, as a temporary to permanent dwelling (Fig. 10.9U–V).

Ichnocoenoses: PCF trace fossil assemblages represent at least three distinct ichnocoenoses (Fig. 10.10). The *Kladosystemites* ichnocoenosis is dominated by *Kladosystemites* and also contains *Polychoredrites*, *Parowanichnus*, shallowly penetrating 1-mm-diameter rhizoliths, deeply penetrating 5-mm-diameter rhizoliths, and deeply penetrating 10-mm-diameter rhizoliths (Fig. 10.1A). Trace fossils of the *Kladosystemites* ichnocoenosis were concentrated in the A and upper B horizons of well-drained, consolidated, and nutrient-rich soils with dense concentrations of small to large plants at the surface. The *Beaconites* ichnocoenosis is dominated by *Beaconites* isp. and also contains *Katarrhedrites*, deeply penetrating 5-mm-diameter rhizoliths, and deeply penetrating 10-mm-diameter rhizoliths (Fig. 10.1B). Trace fossils of the *Beaconites* ichnocoenosis are concentrated in the lower B horizons of soils characterized by seasonally variable, saturated and unsaturated conditions, as well as a relatively high organic content. The *Daimonelix* ichnocoenosis is dominated by *Daimonelix* and also contains dense concentrations of *Beaconites*, small to large diameter *Katarrhedrites*, deeply penetrating 5-mm-diameter rhizoliths, and deeply penetrating 10-mm-diameter rhizoliths (Fig. 10.1C). Trace fossils of the *Daimonelix* ichnocoenosis are also concentrated in the lower B horizons of soils characterized by seasonally variable, saturated and unsaturated conditions, but with a higher organic content.

Paleoecology: Trace fossils in the PCF paleosols indicate a diverse biota despite the low abundance of body fossils. The fauna included a large biomass of soft-bodied and poorly mineralized taxa including various beetles and soil bugs that are only known from the dense concentrations of backfilled burrows and cocoon chambers. The larger burrows indicate a vertebrate fauna that likely included burrowing reptiles (tortoises, lizards, and amphisbaenians), mammalian carnivores and omnivores (canids, ursids, mustelids, and possibly felids), and mammalian herbivores (lagomorphs and rodents). Plant fossils are also poorly preserved in the PCF, but the aboveground ecosystem can be interpreted from the paleosols and the root trace assemblages. The paleosol properties and the size and distribution of the rhizoliths indicate that a shrubland and grassland ecosystem for Paleosols 1 and 2, respectively (e.g., Lavelle and Spain 2001; Retallack 2001b; Buol et al. 2003). Shrubland vegetation includes patchy, low-growing bushes and grasses with areas of bare ground. The roots of this vegetation tend to be sparse and densely packed beneath vegetated areas. The upper soil profile may have been encrusted with salt or cemented with calcareous nodules (e.g., Birkeland 1999). Grasslands have soils characterized by well-structured, organic-rich surface horizons with dense networks of fine (<2 mm) roots and an abundant infauna. In regions with semiarid climates, such as the short grass prairie

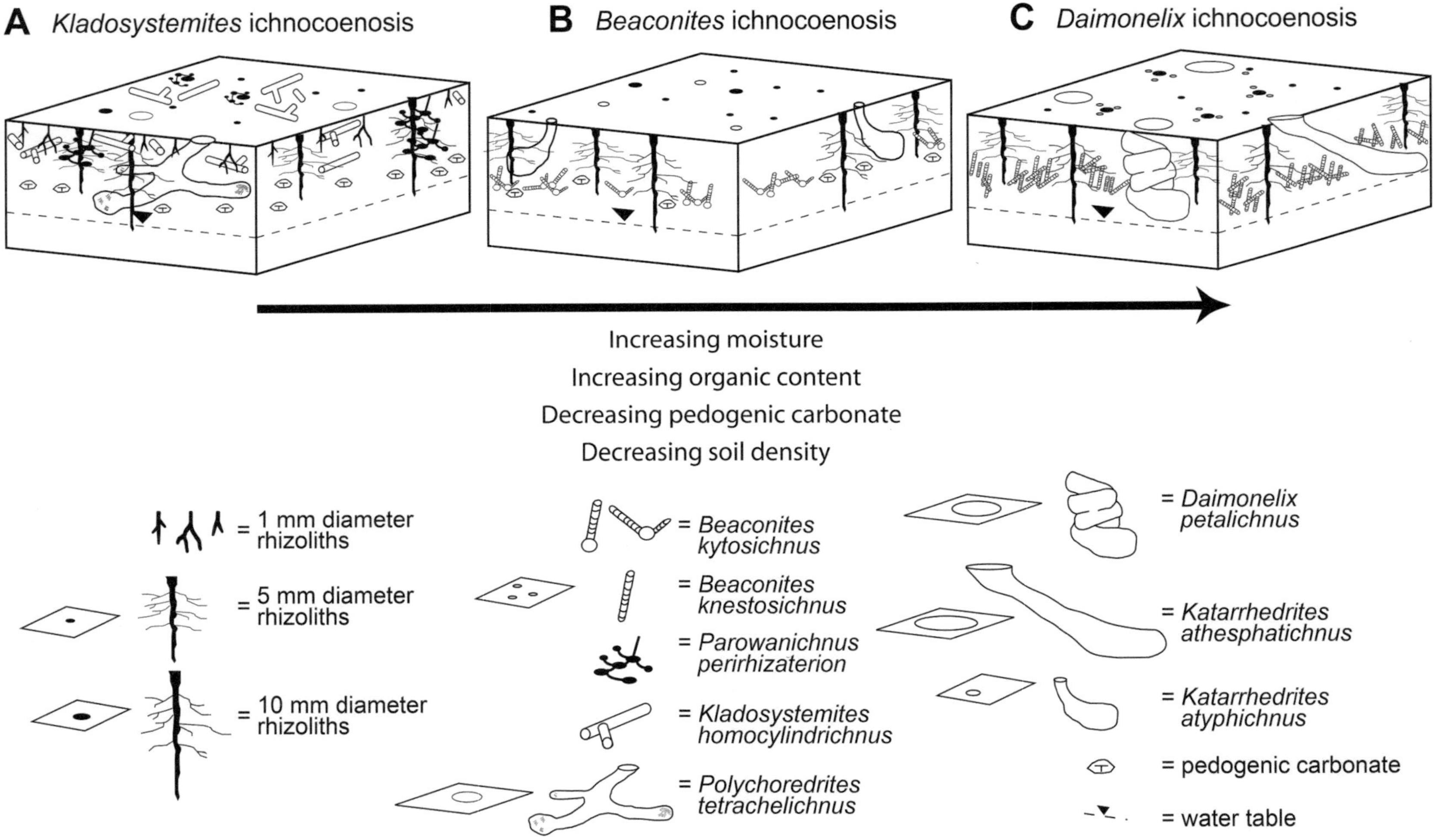

Fig. 10.10 Ichnocoenoses of the Pawnee Creek Formation paleosols (modified from Hembree and Hasiotis 2008): A, *Kladosystemites* ichnocoenosis; B, *Beaconites* ichnocoenosis; C, *Daimonelix* ichnocoenosis

of the western Great Plains, U.S.A., this horizon tends to be shallow and may include an underlying zone of calcareous nodules (Birkeland 1999).

Paleoenvironment: The trace fossils present in the PCF paleosols record the details of the surface and soil environment, as well as the local paleohydrologic setting. The size and depth of rhizoliths are indicative of local precipitation, soil-water content, soil-nutrient content, and soil maturity (Weaver 1919; Hasiotis 2002). Roots of most modern plants, for example, occur only in the upper to lower portion of the vadose zone (e.g., Weaver 1919; van Donselaar-ten Bokkel Huinink 1966). The depth distribution of specific plants does depend on the rooting strategy of the individual species, but typically shallow roots are present in poorly drained soils with a high water table or in arid to semiarid regions with little precipitation. Deep roots, on the other hand, occur in regions with deeply penetrating precipitation in well-drained soils or areas with a low water table (Weaver 1919; van Donselaar-ten Bokkel Huinink 1966; Glinski and Lipiec 1990; Aber and Melillo 2001). When both types of rhizoliths occur in the same paleosol, as in the paleosols of the PCF, the shallower rhizoliths record the depth of moisture similar to the upper vadose zone, whereas the deeper rhizoliths indicate the depth of the intermediate vadose zone (Hasiotis 2002; Kraus and Hasiotis 2006).

The position of burrows of larval beetles and other arthropods also indicates the portion of the soil profile that was moist but above the water table (e.g., Willis and Roth 1962; Hasiotis and Bown 1992; Villani et al. 1999; Hasiotis 2007). Modern larval beetles and soil bugs are typically present in the A and upper B horizons within the upper vadose zone with a moisture of $\sim$7–37% and are most common in grasslands with dry to seasonally wet climates (e.g., Willis and Roth 1962; Villani et al. 1999; O'Geen and Busacca 2001; Hasiotis 2002; O'Geen et al. 2002; Hasiotis 2007). In addition, preserved pupal chambers suggest a lack of extreme fluctuations in water content and a relatively stable soil environment. The presence of linings on the pupal chambers, however, suggests dry, relatively loose soil (e.g., Villani et al. 1999). Dense concentrations of backfilled burrows indicate areas where the soil was rich in roots, nutrients, and other organics even at depth and likely correspond to areas that were more heavily vegetated at the surface. The activity of beetles and soil bugs also altered the soils, increasing their drainage by excavating more permeable burrows and by cycling nutrients and organic material within the profile through deposit-feeding and excretion behaviors (e.g., Lavelle and Spain 2001).

The occurrence of complex burrow systems such as ant nests suggests a firm and cohesive soil required for the long-term maintenance of an open network of tunnels and chambers. The nests are also indicative of a stable landscape with low sedimentation and a soil with good drainage, moderate moisture content, and deep water table. Similarly, the simple to complex, large-diameter burrows interpreted as dwellings of mammals and reptiles require a firm and cohesive soil to support multiple, large, unlined void spaces, as well as a relatively deep water table with minimal fluctuations in its depth to prevent the accumulation of ground water in the lower chambers (e.g., Butler 1995). Therefore, paleosols containing these burrows can be interpreted as having a water table deeper than the deepest burrow. In the PCF, paleosols contains burrows that extend 1–3 m down from the paleosurface. The maintenance required of permanent burrow networks open to the surface also suggests a setting with low sedimentation and erosion rates throughout the time of occupation.

Paleoclimate: Local to regional environmental and climatic conditions played an important role in the type and distribution of soil biota of the PCF. Terrestrial burrows offer a refuge from harsh or severely fluctuating surface conditions and are common in modern semiarid and arid regions, as well as temperate regions with seasonal extremes in temperature and precipitation. The middle to late Miocene was a time of alternating cool and warm climate (Frakes et al. 1992). During the formation of PCF paleosols, ice sheets built up on East Antarctica, resulting in global cooling and an increase in latitudinal temperature gradients (Frakes et al. 1992). The high abundance and diversity of burrowing animals and burrow architectures in the PCF may be best explained by this climatic event. The soil environment provides an important refuge for many terrestrial animals, particularly those inhabiting arid to semiarid regions where they are exposed to environmental extremes that can impose severe physiological stress (Kinlaw 1999; Jackson 2000). To counter this, many species have developed specialized activity patterns and refuge strategies, including the excavation of underground burrows with their own more moderate microclimate. The major advantage of this is protection from extremes in seasonal temperatures and humidity. Surface air temperature may fluctuate by 30°C over a 24-hour period. Subsurface temperatures, however, remain relatively stable over the same period, changing only 10°C at 10 cm depth or 1–2°C at 50 cm depth (Lavelle and Spain 2001). The number of burrows attributed to vertebrates within a single PCF paleosol is a dramatic increase over paleosols from earlier Cenozoic deposits in North America (Retallack 1983; Kraus 1997; Retallack et al. 2000; Retallack 2004; Hembree and Hasiotis 2007a, b). Fossorial vertebrate burrows may then provide excellent first-order indicators of local to regional climate change.

Conclusions: The trace fossils of the Pawnee Creek Formation are invaluable for understanding the physical and biological processes of soil ecosystems in the midcontinent of North America in the middle Miocene. Trace fossils allow the recognition of continental unconformities and surfaces of prolonged subaerial exposure that would otherwise be overlooked. Well-developed paleosols with high densities of terrestrial trace fossils, therefore, have sequence-stratigraphic significance defining surfaces that indicate low sedimentation rate, base level, and accommodation. The architectural and surficial morphology of PCF trace fossils are also informative with regard to the soil environment, including the relative amount and fluctuation of sediment consistency, soil moisture, water-table depth, and precipitation. When combined with paleosol data, the distribution, abundance, and diversity of trace fossils produced by arthropods, reptiles, and mammals indicate that PCF paleosols formed under a seasonal climate subject to wide variation in temperature and precipitation. The unique and diverse assemblages of burrow morphologies demonstrate the resulting adaptation of the local fauna to these environmental extremes. The response of soil ecosystems to short- and long-term changes in climate is an important consideration in any study of the effects of climate change in the geologic record.

Future Prospects

The field of ichnology has advanced considerably in recent years, becoming increasingly interdisciplinary and relevant to a variety of other geological fields including sedimentology, sequence stratigraphy, geomorphology, and geochemistry in addition to other fields such as archaeology, botany, ecology, and zoology. Recent work on ichnotaxonomy has sought to better characterize and define the classification of trace fossils with the goal of making it simpler for nonspecialists to penetrate the complex and often chaotic series of ichnogenera and ichnospecies (Knaust 2012b). In the last decade, field and experimental research on modern tracemaking organisms, particularly those in continental environments, has expanded, increasing our knowledge of how and why many different types of organisms interact with different substrates. Importantly, much research has been published recently in both the geological and biological literature making the need for communication and collaboration between ichnologists and ecologists vital. New technology has also been increasingly applied to trace-fossil analysis, including digitization techniques which allow three-dimensional reconstruction of trace fossils and more direct comparison between modern and ancient traces.

Experimental and Field Research

The modern science of invertebrate ichnology began almost 200 years ago with experiments conducted on terrestrial and marine invertebrates in the field and laboratory (Cadée and Goldring 2007; Baucon et al. 2012). Early studies were conducted along the coast of the Wadden Sea in the Netherlands, Germany, and Denmark. It was through these early investigations of modern burrowing animals that trace fossils were recognized as record of behavior rather than as body fossils of algae, plants, or animals (Cadée and Goldring 2007). The continued expansion of neoichnological studies has improved the interpretation of known trace fossils, especially in continental settings where relatively limited work has been conducted. Laboratory experiments with a variety of terrestrial animals, including annelids, diplopods, arachnids, insects, amphibians, reptiles, and mammals, have provided a suite of trace morphologies that can be connected to specific types of tracemakers, behaviors, and environments (e.g., Tschinkel 2003; Gobetz 2005; Hembree and Hasiotis 2006, 2007a, b; Rodríguez-Tovar 2007; Smith and Hasiotis 2008; Counts and Hasiotis 2009; Hembree 2009, 2013, 2014; Halfen and Hasiotis 2010; Hembree et al. 2012; Platt and Hasiotis 2012; Bowen and Hembree 2014; Catena and Hembree 2014; Dzenowski and Hembree 2014; Hils and Hembree 2015; Johnson and Hembree 2015). An incredible diversity of organisms interacts with substrates in a manner that may produce structures with a high preservation potential and a tremendous amount of study remains to be done. In addition, field research in modern environments reveals many new types of biogenic structures that can be sought out in the fossil record. Field studies of modern continental tracemakers are especially vital since they allow the investigation of organism response to specific environmental conditions, both physical and biological (e.g., Williams 1966; Ahlbrandt et al. 1978; Ratcliffe and Fagerstrom 1980; Shorthouse and Marples 1980; Reynolds and Wakkinen 1987; Doonan and Stout 1994; Tschinkel 2003; Gobetz 2005; Mora et al. 2005; Lawfield and Pickerill 2006; M'rabet et al. 2007; Melchor et al. 2012; Genise et al. 2013; Mikus and Uchman 2013; Doody et al. 2014, 2015; Sarzetti et al. 2014; Cantil et al. 2015). The improved understanding of the biogenic structures produced by modern soil animals has been particularly critical to the interpretation of paleosols and changing soil ecosystems through geologic time.

Digitization of Modern and Ancient Biogenic Structures

New technology related to the two- and three-dimensional imaging of rocks and sediment using CT-scanning, MRI-scanning, and ground-penetrating radar (GPR) have revolutionized how we view both modern biogenic structures and trace fossils. These instruments and techniques allow the complete reconstruction of modern and ancient traces that would otherwise only be visible as limited two-dimensional features. Previous methods of three-dimensional visualization of trace fossils involved the time-consuming and destructive process of serial sectioning and photographing of hand samples (e.g., Farrow 1975; Knaust 2012a). CT- and MRI-scanning greatly improves the images that may be acquired, reduces the time needed to produce an image, and, perhaps most importantly, does not destroy the original specimen (Gingras et al. 2002b; Knaust 2012a). The production of three dimensional models of trace fossils also improves our ability to directly compare them to equivalently detailed modern analogs (Platt et al. 2010).

Conversely, the ability to image modern biogenic structures in active sediment improves their utility as models of comparison. The application of high-resolution, non-invasive, subsurface imaging technology to view *in situ* biogenic structures is a new area of research, with a growing number of successful GPR applications in soils and sandy media (Buynevich 2011; Kinlaw and Grasmueck 2012; Buynevich et al. 2014). Whereas current GPR units cannot resolve fine-scale burrow morphology, it can indicate the presence of burrows, their lateral extent, orientation, depth, and fill type. This is useful for understanding the lateral and vertical distribution of burrows in different environments.

Assessing the morphology of modern traces typically requires the production of a cast of plaster, resin, or metal. Whereas this technique is almost as old as the science of ichnology, digitization of these casts makes them even more useful models of comparison. Various laser scanners allow for the production of three dimensional digital models of modern biogenic structures from which more detailed measurements can be obtained than those from tape measures or calipers including the mean, median, mode of width, height, length, and circumference of shafts, tunnels, and chambers, as well as properties that cannot normally be properly evaluated such as volume and surface area (e.g., Platt et al. 2010).

The combination of neoichnological studies and the application of new technology will improve our ability to make taxonomic, ecological, and environmental links between modern and ancient traces. In doing so, trace fossils will become an even more vital part of any interpretation of ancient environments and ecosystems.

Acknowledgments I thank Andy Rindsberg and three anonymous reviewers for their helpful comments and suggestions that greatly improved this chapter. I also thank the editors for their invitation to contribute a chapter to this volume.

References

Aber, J. D., & Melillo, J. M. (2001). *Terrestrial ecosystems*. San Diego: Harcourt Academic Press.

Ahlbrandt, T. S., Andrews, S., & Gwynne, D. T. (1978). Bioturbation in eolian deposits. *Journal of Sedimentary Petrology, 48,* 839–848.

Atkinson, R. J. A. (1986). Mud-dwelling megafauna of the Clyde Sea area. *Royal Society of Edinburgh Proceedings, 90B,* 351–361.

Baucon, A., Bordy, E., Brustur, T., Buatois, L. A., Cunningham, T., De, C., et al. (2012). A history of ideas in ichnology. In D. Knaust & R. G. Bromley (Eds.), *Trace fossils as indicators of sedimentary environments* (pp. 3–44). Amsterdam: Elsevier.

Behrensmeyer, A. K., Damuth, J. D., DiMichele, W. A., Potts, R., Sues, H. D., & Wing, S. L. (1992). *Terrestrial ecosystems through time*. Chicago: University of Chicago Press.

Bellosi, E. S., & González, M. G. (2010). Paleosols of the middle Cenozoic Sarmiento Formation, central Patagonia. In R. H. Madden, A. A. Carlini, M. C. Vucetich & R. F. Kay (Eds.), *The paleontology of Gran Barranca. Evolution and environmental change through the middle Cenozoic of Patagonia* (pp. 293–305). Cambridge: Cambridge University Press.

Bellosi, E. S., Laza, J. H., Victoria Sánchez, M., & Genise, J. F. (2010). Ichnofacies analysis of the Sarmiento Formation (middle Eocene-early Miocene) at Gran Barranca, central Patagonia. In R. H. Madden, A. A. Carlini, M. C. Vucetich & R. F. Kay (Eds.), *The paleontology of Gran Barranca. Evolution and environmental change through the Middle Cenozoic of Patagonia* (pp. 306–316). Cambridge: Cambridge University Press.

Bertling, M., Braddy, S. J., Bromley, R. G., Demathiey, G. R., Genise, J., Mikulas, R., et al. (2006). Names for trace fossils: a uniform approach. *Lethaia, 39,* 265–286.

Birkeland, P. W. (1999). *Soils and geomorphology*. New York: Oxford University Press.

Bowen, J., & Hembree, D. (2014). Neoichnology of two spirobolid millipedes: improving the understanding of the burrows of soil detritivores. *Palaeontologia Electronica, 17,* 1–48.

Bown, T. M., & Laza, J. H. (1990). A Miocene termite nest from southern Argentina and its paleoclimatological implications. *Ichnos, 1,* 73–79.

Bromley, R. G. (1996). *Trace fossils: Biology, taphonomy, and applications*. London: Chapman and Hall.

Bromley, R. G., & Asgaard, U. (1975). Sediment structures produced by a spatangoid echinoid: a problem of preservation. *Bulletin of the Geological Society of Denmark, 24,* 261–281.

Bromley, R. G., & Ekdale, A. A. (1986). Composite ichnofabrics and tiering of burrows. *Geological Magazine, 123,* 59–65.

Bromley, R. G., & Frey, R. W. (1974). Redescription of the trace fossil *Gyrolithes* and taxonomic evaluation of *Thalassinoides, Ophiomorpha, Spongeliomorpha. Bulletin of the Geological Society of Denmark, 23,* 311–335.

Buatois, L. A., & Mángano, M. G. (2004). Animal-substrate interactions in freshwater environments: applications of ichnology in facies and sequence stratigraphic analysis of fluvio-lacustrine successions. In D. McIlroy (Ed.), *The applications of ichnology to palaeoenvironmental and stratigraphic analysis* (pp. 311–334). London: Geological Society of London, Special Publication 228.

Buol, S. W., Southard, R. J., Graham, R. C., & McDaniel, P. A. (2003). *Soil genesis and classification*. Ames: Iowa State Press.

Butler, D. R. (1995). *Zoogeomorphology: Animals as geomorphic agents*. Cambridge: Cambridge University Press.

Buynevich, I. V. (2011). Buried tracks: ichnological applications of high-frequency georadar. *Ichnos, 18,* 189–191.

Buynevich, I. V., Curran, H. A., Wiest, L. A., Bentley, A. P. K., Kadurin, S. V., Seminack, C. T., et al. (2014). Near-surface imaging (GPR) of biogenic structures in siliciclastic, carbonate, and gypsum dunes. In D. I. Hembree, B. F. Platt & J. J. Smith (Eds.), *Experimental approaches to understanding fossil organisms: Lessons from the living* (pp. 405–418). Dordrecht: Springer.

Cadée, G. C., & Goldring, R. (2007). The Wadden Sea, cradle of invertebrate ichnology. In W. Miller III (Ed.), *Trace fossils: Concepts, problems, prospects* (pp. 3–13). Amsterdam: Elsevier.

Cambefort, Y., & Hanski, I. (1991). Dung beetle population biology. In I. Hanski & Y. Cambefort (Eds.), *Dung beetle ecology* (pp. 36–50). Princeton: Princeton University Press.

Cantil, L. F., Sánchez, M. V., Sarzetti, L., Molina, A., & Genise, J. F. (2015). Nests and brood balls of *Coprophanaeus (Coprophanaeus) cyanescens* (Olsoufieff, 1924) (Coleoptera: Scarabaeidae: Scarabaeinae). *The Coleopterists Bulletin, 69,* 153–158.

Catena, A., & Hembree, D. I. (2014). Biogenic structures of burrowing skinks: neoichnology of *Mabuya mutifasciata* (Squamata: Scincidae). In D. I. Hembree, B. F. Platt & J. J. Smith (Eds.), *Experimental approaches to understanding fossil organisms: Lessons from the living* (pp. 343–369). Dordrecht: Springer.

Counts, R. R., & Hasiotis, S. T. (2009). Neoichnological experiments with masked chafer beetles (Coleoptera: Scarabaeidae): implications for backfilled continental trace fossils. *PALAIOS, 24,* 74–91.

Dashtgard, S. E., & Gingras, M. K. (2005). Facies architecture and ichnology of recent salt-marsh deposits: Waterside Marsh, New Brunswick, Canada. *Journal of Sedimentary Research, 75,* 596–607.

Gibert, J. M. de, & Saez, A. (2009). Paleohydrological significance of trace fossil distribution in Oligocene fluvial-fan-to-lacustrine systems of the Ebro Basin, Spain. *Palaeogeography, Palaeoclimatology, Palaeoecology, 272,* 162–175.

Dold, P., & Prothero, D. (2003). Magnetic stratigraphy of the lower-middle Miocene Pawnee Creek and Martin Canyon formations, northeastern Colorado. *Journal of Vertebrate Paleontology, 23,* 46.

Doody, J. S., James, H., Ellis, R., Gibson, N., Raven, M., Mahony, S., et al. (2014). Cryptic and complex nesting in the yellow-spotted monitor, *Varanus panoptes*. *Journal of Herpetology, 48,* 363–370.

Doody, J. S., James, H., Colyvas, K., McHenry, C. R., & Clulow, S. (2015). Deep nesting in a lizard, *déjà vu* devil's corkscrew: first helical reptile burrow and deepest vertebrate nest. *Biological Journal of the Linnean Society, 116,* 13–26.

Doonan, T. J., & Stout, I. J. (1994). Effects of gopher tortoise (*Gopherus polyphemus*) body size on burrow structure. *The American Midland Naturalist, 131,* 273–280.

Droser, M. L., & Bottjer, D. J. (1986). A semiquantitative field classification of ichnofabric. *Journal of Sedimentary Petrology, 56,* 558–559.

Dzenowski, N., & Hembree, D. I. (2014). Quantifying vertebrate biogenic structures using modern analogs: the neoichnology of ambystomatid salamanders. In D. I. Hembree, B. F. Platt & J. J. Smith (Eds.), *Experimental approaches to understanding fossil organisms: Lessons from the living* (pp. 305–342). Dordrecht: Springer.

Ekdale, A. A. (1985). Paleoecology of the marine endobenthos. *Palaeogeography, Palaeoclimatology, Palaeoecology, 50,* 63–81.

Ekdale, A. A., & Berger, W. H. (1978). Deep-sea ichnofacies: modern organism traces on land and in pelagic carbonates of the western equatorial Pacific. *Palaeogeography, Palaeoclimatology, Palaeoecology, 23,* 263–278.

Ekdale, A. A., Bromley, R. G., & Pemberton, S. G. (1984). *Ichnology, trace fossils in sedimentology and stratigraphy*. Society of Economic Paleontologists and Mineralogists, Short Course Notes, 15.

Elmes, G. W. (1991). Ant colonies and environmental disturbance. In P. S. Meadows & A. Meadows (Eds.), *The environmental impact of burrowing animals and animal burrows: Symposia of the Zoological Society of London, 63* (pp. 15–32). Oxford: Clarendon Press.

Farrow, G. E. (1975). Techniques for the study of fossil and recent traces. In R. W. Frey (Ed.), *The study of trace fossils: A synthesis of principles, problems, and procedures in ichnology* (pp. 537–554). Berlin: Springer.

Fluegeman, R. H. (2003). Late Eocene-early Oligocene benthic foraminifera in the Gulf Coastal Plain: regional vs. global influences. In D. R. Prothero, L. C. Ivany & E. A. Nesbitt (Eds.), *From greenhouse to icehouse: The marine Eocene-Oligocene transition* (pp. 282–293). New York: Columbia University Press.

Frakes, L. A., Francis, J. E., & Syktus, J. I. (1992). *Climate modes of the Phanerozoic: The history of the Earth's climate over the past 600 million years*. Cambridge: Cambridge University Press.

Frey, R. W. (1968). The Lebensspuren of some common marine invertebrates near Beaufort, North Carolina. 1, Pelecypod burrows. *Journal of Paleontology, 42,* 570–574.

Frey, R. W. (1970). The Lebensspuren of some common marine invertebrates near Beaufort, North Carolina. 2, Anemone burrows. *Journal of Paleontology, 44,* 308–311.

Frey, R. W. (1973). Concepts in the study of biogenic sedimentary structures. *Journal of Sedimentary Petrology, 43,* 6–19.

Frey, R. W. (1975). The realm of ichnology, its strengths and limitations. In R. W. Frey (Ed.), *The study of trace fossils: A synthesis of principles, problems, and procedures in ichnology* (pp. 13–38). Berlin: Springer.

Frey, R. W. (1978). Behavioral and ecological implications of trace fossils. In P. B. Bassan (Ed.), *Trace fossil concepts* (pp. 43–66). Tulsa: Society of Economic Paleontologists and Mineralogists, Short Course Notes, 5.

Frey, R. W., & Pemberton, S. G. (1985). Biogenic structures in outcrops and cores. I. Approaches to ichnology. *Bulletin of Canadian Petroleum Geology, 33,* 72–115.

Frey, R. W., & Pemberton, S. G. (1987). The *Psilonichnus* ichnocoenose and its relationship to adjacent marine and nonmarine ichnocoenoses along the Georgia coast. *Bulletin of Canadian Petroleum Geology, 13,* 333–357.

Frey, R. W., Howard, J. D., & Pryor, W. A. (1978). *Ophiomorpha:* its morphologic, taxonomic, and environmental significance. *Palaeogeography, Palaeoclimatology, Palaeoecology, 23,* 199–229.

Frey, R. W., Curran, H. A., & Pemberton, S. G. (1984). Tracemaking activities of crabs and their environmental significance: the ichnogenus *Psilonichnus*. *Journal of Paleontology, 58,* 333–350.

Galbreath, E. C. (1953). A contribution to the Tertiary geology and paleontology of northeastern Colorado. *University of Kansas Paleontological Contributions, Vertebrata, 4*.

Genise, J. F., & Bown, T. M. (1994). New Miocene scarabeid and hymenopterous nests and early Miocene (Santacrucian) paleoenvironments, Patagonian Argentina. *Ichnos, 3,* 107–111.

Genise, J. F. Bellosi, E. S., & Gonzalez, M. G. (2004). An approach to the description and interpretation of ichnofabrics in palaeosols. In D. McIlroy (Ed.), *The applications of ichnology to palaeoenvironmental and stratigraphic analysis* (pp. 355–382). London: The Geological Society of London, Special Publication, 228.

Genise, J. F., Cantil, L. F., Dinghi, P. A., Sánchez, M. V., & Sarzetti, L. (2013). The aestivation chamber of the giant earthworm *Glossoscolex bergi* (Glossoscolecidae) in the subtropical rainforest of Misiones (Argentina). *Ichnos, 20,* 1161119.

Genise, J. F., Melchor, R. N., Bellosi, E. S., Gonzalez, M. G., & Krause, M. (2007). New insect pupation chambers (pupichnia) from the Upper Cretaceous of Patagonia, Argentina. *Cretaceous Research, 28,* 545–559.

Gingras, M. K., Pickerill, R. K., & Pemberton, S. G. (2002a). Resin casts of modern burrows provide analogs for composite trace fossils. *PALAIOS, 17,* 206–211.

Gingras, M. K., Macmillian, B., Balcom, B. J., Saunders, T., & Pemberton, S. G. (2002b). Using magnetic resonance imaging and petrographic techniques to understand the textural attributes and porosity distribution in *Macaronichnus*-burrowed sandstone. *Journal of Sedimentary Research, 72,* 552–558.

Gingras, M. K., MacEachern, J. A., & Pickerill, R. K. (2004). Modern perspectives on the *Teredolites* ichnofacies: observations from Willapa Bay, Washington. *PALAIOS, 19,* 79–88.

Gingras, M. K., Bann, K. L, MacEachern, J. A., Waldron, J., & Pemberton, S. G. (2007a). A conceptual framework for the application of trace fossils. In J. A. MacEachern, K. L. Bann, M. K. Gingras & S. G. Pemberton (Eds.), *Applied Ichnology* (pp. 1–26). Tulsa: Society for Sedimentary Geology.

Gingras, M. K., Lalond, S. V., Amskold, L., & Konhauser, K. O. (2007b). Wintering chironomids mine oxygen. *PALAIOS, 22,* 433–438.

Glinski, J., & Lipiec, J. (1990). *Soil physical conditions and plant roots.* Boca Raton: CRC Press Inc.

Gobetz, K. E. (2005). Claw impressions in the walls of modern mole (*Scalopus aquaticus*) tunnels as a means to identify fossil burrows and interpret digging movements. *Ichnos, 12,* 227–231.

Halfen, A. F., & Hasiotis, S. T. (2010). Neoichnological study of the traces and burrowing behaviors of the western harvester ant *Pognomyrmex occidentalis* (Insecta: Hymenoptera: Formicidea): paleopedogenic and paleoecologic implications. *PALAIOS, 25,* 703–720.

Halffter, G., & Matthews, E. G. (1966). The natural history of dung beetles of the subfamily Scarabaeninae (Coleoptera, Scarabaenidae). *Folia Entomologica Mexicana, 12/13,* 1–312.

Hasiotis, S. T. (2002). *Continental trace fossils.* Tulsa: Society for Sedimentary Geology.

Hasiotis, S. T. (2003). Complex ichnofossils of solitary and social soil organisms: understanding their evolution and roles in terrestrial paleoecosystems. *Palaeogeography, Palaeoclimatology, Palaeoecology, 192,* 259–320.

Hasiotis, S. T. (2004). Reconnaissance of Upper Jurassic Morrison Formation ichnofossils, Rocky Mountain region, USA: paleoenvironmental, stratigraphic, and paleoclimatic significance of terrestrial and freshwater ichnocoenoses. *Sedimentary Geology, 167,* 177–268.

Hasiotis, S. T. (2007). Continental ichnology: fundamental processes and controls on trace fossil distribution. In W. Miller III (Ed.), *Trace fossils: Concepts, problems, prospects* (pp. 268–284). Amsterdam: Elsevier.

Hasiotis, S. T., & Bown, T. M. (1992). Invertebrate trace fossils: the backbone of continental ichnology. In C. Maples & R. West (Eds.), *Trace fossils: Their paleobiological aspects* (pp. 64–104). Paleontological Society Short Course, Number 5.

Hasiotis, S. T., Kraus, M. J., & Demko, T. M. (2007). Climatic controls on continental trace fossils. In W. Miller III (Ed.), *Trace fossils: Concepts, problems, prospects* (pp. 166–189). Amsterdam: Elsevier.

Hembree, D. I. (2009). Neoichnology of burrowing millipedes: linking modern burrow morphology, organism behavior, and sediment properties to interpret continental ichnofossils. *PALAIOS, 24,* 425–439.

Hembree, D. I. (2013). Neoichnology of the whip scorpion *Mastigoproctus giganteus*: complex burrows of predatory terrestrial arthropods. *PALAIOS, 28,* 141–162.

Hembree, D. I. (2014). Large, complex burrows of terrestrial invertebrates: neoichnology of *Pandinus imperator*. In D. I. Hembree, B. F. Platt & J. J. Smith (Eds.), *Experimental approaches to understanding fossil organisms: Lessons from the living* (pp. 229–263). Dordrecht: Springer.

Hembree, D. I., & Hasiotis, S. T. (2006). The identification and interpretation of reptile ichnofossils in paleosols through modern studies. *Journal of Sedimentary Research, 76,* 575–588.

Hembree, D. I., & Hasiotis, S. T. (2007a). Paleosols and ichnofossils of the White River Formation of Colorado: insight into soil ecosystems of the North American Midcontinent during the Eocene-Oligocene transition. *PALAIOS, 22,* 123–142.

Hembree, D. I., & Hasiotis, S. T. (2007b). Biogenic structures produced by sand-swimming snakes: a modern analog for interpreting continental ichnofossils. *Journal of Sedimentary Research, 77,* 389–397.

Hembree, D. I., & Hasiotis, S. T. (2008). Miocene vertebrate and invertebrate burrows defining compound paleosols in the Pawnee Creek Formation, Colorado, U.S.A. *Palaeogeography, Palaeoclimatology, Palaeoecology, 270,* 349–365.

Hembree, D. I., Johnson, L. M., & Tenwalde, R. W. (2012). Neoichnology of the desert scorpion *Hadrurus arizonensis*: burrows to biogenic cross lamination. *Palaeontologica Electronica, 15,* 1–34.

Hickman, C. S. (2003). Evidence for abrupt Eocene-Oligocene molluscan faunal change in the Pacific Northwest. In D. R. Prothero, L. C. Ivany & E. A. Nesbitt (Eds.), *From greenhouse to icehouse: The marine Eocene-Oligocene transition* (pp. 71–87). New York: Columbia University Press.

Hils, J. M., & Hembree, D. I. (2015). Neoichnology of the burrowing spiders *Gorgyrella inermis* (Mygalomorphae: Idiopidae) and *Hogna lenta* (Araneomorphae: Lycosidae). *Palaeontologica Electronica, 18,* 1–62.

Hole, F. D. (1981). Effects of animals on soil. *Geoderma, 25,* 75–112.

Howard, J. D. (1978). Sedimentology and trace fossils. In P. B. Basan (Ed.), *Trace fossil concepts* (pp. 11–42). Society for Sedimentary Geology Short Course No. 5.

Hunt, A. P., Chin, K., & Lockley, M. G. (1994). The paleobiology of vertebrate coprolites. In S. K. Donovan (Ed.), *The paleobiology of trace fossils* (pp. 221–240). Baltimore: John Hopkins University Press.

Hutchinson, J. H. (1992). Western North American reptile and amphibian record across the Eocene/Oligocene boundary and its climatic implications. In D. R. Prothero & W. A. Berggren (Eds.), *Eocene-Oligocene climatic and biotic evolution* (pp. 451–463). Princeton: Princeton University Press.

Ivany, L. C., Lohmann, K. C., & Patterson, W. P. (2003). Paleogene temperature history of the U.S. Gulf Coast Plain inferred from ∂^{18} O of fossil otoliths. In D. R. Prothero, L. C. Ivany & E. A. Nesbitt (Eds.), *From greenhouse to icehouse: The marine Eocene-Oligocene transition* (pp. 232–251). New York: Columbia University Press.

Jackson, T. P. (2000). Adaptation to living in an open arid environment: lessons from the burrow structure of the two southern African whistling rats, *Parotomys brantsii* and *P. littledalei*. *Journal of Arid Environments, 46,* 345–355.

Johnson, L. M., & Hembree, D. I. (2015). Understanding anuran burrows: neoichnology of the Eastern spadefoot toad, *Scaphiopus holbrooki*. *Palaeontologica Electronica, 18,* 1–29.

Kanazawa, K. (1992). Adaptation of test shape for burrowing and locomotion in spatangoid echinoids. *Palaeontology, 35,* 733–750.

Kinlaw, A. (1999). A review of burrowing by semi-fossorial vertebrates in arid environments. *Journal of Arid Environments, 41,* 127–145.

Kinlaw, A. E., & Grasmueck, M. (2012). Evidence for and geomorphologic consequences of a reptilian ecosystem engineer: the burrowing cascade initiated by the gopher tortoise. *Geomorphology, 157–158,* 108–121.

Knaust, D. (2012a). Methodology and techniques. In D. Knaust & R. G. Bromley (Eds.), *Trace fossils as indicators of sedimentary environments* (pp. 245–271). Amsterdam: Elsevier.

Knaust, D. (2012b). Trace-fossil systematics. In D. Knaust & R. G. Bromley (Eds.), *Trace fossils as indicators of sedimentary environments* (pp. 79–102). Amsterdam: Elsevier.

Krapovickas, V. (2012). Deposits of the Santa Cruz Formation (late early Miocene): paleohydrologic and paleoclimatic significance. In S. F. Vizcaíno, R. F. Kay & M. S. Bargo (Eds.), *Early Miocene paleobiology in Patagonia: High-latitude paleocommunities of the Santa Cruz Formation* (pp. 91–103). Cambridge: Cambridge University Press.

Krapovickas, V., Ciccioli, P. L., Mángano, M. G., Marsicano, C. A., & Limarino, C. O. (2009). Paleobiology and paleoecology of an arid-semiarid Miocene South American ichnofauna in anastomosed fluvial deposits. *Palaeogeography, Palaeoclimatology, Palaeoecology, 284,* 129–152.

Kraus, M. J. (1997). Lower Eocene alluvial paleosols: pedogenic development, stratigraphic relationships, and paleosol/landscape associations. *Palaeogeography, Palaeoclimatology, Palaeoecology, 129,* 387–406.

Kraus, M. J. (1999). Paleosols in clastic sedimentary rocks: their geologic applications. *Earth Science Reviews, 47,* 41–70.

Kraus, M. J., & Hasiotis, S. T. (2006). Significance of different modes of rhizolith preservation to interpreting paleoenvironmental and paleohydrologic settings: examples from Paleogene paleosols, Bighorn Basin, Wyoming. *Journal of Sedimentary Research, 76,* 633–646.

Lavelle, P., & Spain, A. (2001). *Soil ecology*. Dordrecht: Kluwer Academic Publishers.

Lawfield, A. M. W., & Pickerell, R. K. (2006). A novel contemporary fluvial ichnocoenose: unionid bivalves and the *Scoyenia-Mermia* ichnofacies transition. *PALAIOS, 21,* 391–396.

Lee, K. E. (1985). *Earthworms: Their ecology and relationships with soils and land use*. Orlando: Academic Press.

Leopold, E. B., & Denton, M. F. (1987). Comparative age of grassland and steppe east and west of the northern Rocky Mountains. *Annals of the Missouri Botanical Gardens, 74,* 841–867.

Leopold, E. B., Liu, G., & Clay-Poole, S. (1992). Low-biomass vegetation in the Oligocene? In D. R. Prothero & W. A. Berggren (Eds.), *Eocene-Oligocene climatic and biotic evolution* (pp. 399–420). Princeton: Princeton University Press.

MacEachern, J. A., Bann, K. L., Pemberton, S. G., & Gingras, M. K. (2007a). The ichnofacies paradigm: high-resolution paleoenvironmental interpretation of the rock record. In J. A. MacEachern, K. L. Bann, M. K. Gingras & S. G. Pemberton (Eds.), *Applied Ichnology* (pp. 27–64). Tulsa: Society for Sedimentary Geology.

MacEachern, J. A., Pemberton, S. G., Bann, K. L., & Gingras, M. K. (2007b). Departures from the archetypal ichnofacies: effective recognition of physico-chemical stresses in the rock record. In J. A. MacEachern, K. L. Bann, M. K. Gingras & S. G. Pemberton (Eds.), *Applied Ichnology* (pp. 65–93). Tulsa: Society for Sedimentary Geology.

Melchor, R. N., Genise, J. F., & Miquel, S. E. (2002). Ichnology, sedimentology and paleontology of Eocene calcareous paleosols from a palustrine sequence, Argentina. *PALAIOS, 17,* 16–35.

Melchor, R. N., Genise, J. F., Umazano, A. M., & Superina, M. (2012). Pink fairy armadillo meniscate burrows and ichnofabrics from Miocene and Holocene interdune deposits of Argentina: palaeoenvironmental and palaeoecological significance. *Palaeogeography, Palaeoclimatology, Palaeoecology, 350–352,* 149–170.

Mikuś, P., & Uchman, A. (2013). Beetle burrows with a terminal chamber: a contribution to the knowledge of the trace fossil *Macanopsis* in continental sediments. *PALAIOS, 28,* 403–413.

Miller, K. G. (1992). Middle Eocene to Oligocene stable isotopes, climate, and deep-water history: the Terminal Eocene Event? In D. R. Prothero & W. A. Berggren (Eds.), *Eocene-Oligocene climatic and biotic evolution* (pp. 160–177). Princeton: Princeton University Press.

Miller, K. G., Fairbanks, R. G., & Mountain, G. S. (1987). Tertiary oxygen isotope synthesis, sea level history, and continental margin erosion. *Paleoceanography, 2,* 1–19.

Mora, P., Miambi, E., Jiménez, J. J., Decaëns, T., & Rouland, C. (2005). Functional complement of biogenic structures produced by earthworms, termites, and ants in the neotropical savannas. *Soil Biology and Biochemistry, 37,* 1043–1048.

M'rabet, S. M., Henaut, Y., Sepulveda, A., Rojo, R., Calme, S., & Geissen, V. (2007). Soil preference and burrow structure of an endangered tarantula, *Brachypelma vagans* (Mygalomorphae: Theraphosidae). *Journal of Natural History, 41,* 1025–1033.

Mueller, P., & McCann, T. (2014). Ichnology of the Miocene Jaraba Formation, Almazan Basin, NE Spain–morphological characteristics and paleoenvironmental implications of distinct continental trace fossils. *Neues Jahrbuch für Geologie und Paläontologie, 274,* 55–71.

Myers, J. A. (2003). Terrestrial Eocene-Oligocene vegetation and climate in the Pacific Northwest. In D. R. Prothero, L. C. Ivany & E. A. Nesbitt (Eds.), *From greenhouse to icehouse: The marine Eocene-Oligocene transition* (pp. 171–188). New York: Columbia University Press.

Needham, S. J., Worden, R. H., & McIlroy, D. (2005). Experimental production of clay rims by macrobiotic sediment ingestion and excretion processes. *Journal of Sedimentary Research, 75,* 1028–1037.

Neto de Carvalho, C., Baucon, A., & Canilho, S. (2015). 'Meniscate burrow' ichnoguild from the alluvial fan deposits of Sarzedas Basin (upper Miocene, Portugal). In D. McIlroy (Ed.), *Ichnology: papers from ICHNIA III* (pp. 51–61). Geological Association of Canada, Miscellaneous Publications 9.

O'Geen, A. T., & Busacca, A. J. (2001). Faunal burrows as indicators of paleo-vegetation in eastern Washington, USA. *Palaeogeography, Palaeoclimatology, Palaeoecology, 169,* 23–37.

O'Geen, A. T., McDaniel, P. A., & Busacca, A. J. (2002). Cicada burrows as indicators of paleosols in the inland Pacific Northwest. *Soil Science Society of America Journal, 66,* 1584–1586.

Oleinik, A. E., & Marincovich Jr., L. (2003). Biotic response to the Eocene-Oligocene transition. In D. R. Prothero, L. C. Ivany & E. A. Nesbitt (Eds.) *From greenhouse to icehouse: The marine Eocene-Oligocene transition* (pp. 36–56). New York: Columbia University Press.

Osgood, R. G., Jr. (1975). The paleontological significance of trace fossils. In R. W. Frey (Ed.), *The study of trace fossils: A synthesis of principles, problems, and procedures in ichnology* (pp. 87–108). Berlin: Springer.

Pemberton, S. G., & Frey, R. W. (1985). The *Glossifungites* ichnofacies: modern examples from the Georgia coast, U.S.A. In H. A. Curran (Ed.), *Biogenic structures: Their use in interpreting depositional environments* (pp. 273–259). Society of Economic Paleontologists and Mineralogists, Special Publication 35.

Pemberton, S. G., Spila, M., Pulham, A. J., Saunders, T., MacEachern, J. A., Robbins, D., & Sinclair, I. K. (2001). *Ichnology and sedimentology of shallow to marginal marine systems: Ben Nevis and Avalon reservoirs, Jeanne d'Arc basin*. Newfoundland: Geological Association of Canada, Short Course Notes, 15.

Platt, B. F., & Hasiotis, S. T. (2012). Empirical determination of physical controls on megafaunal footprint formation through neoichnological experiments with elephants. *PALAIOS, 27,* 725–737.

Platt, B. F., Hasiotis, S. T., & Hirmas, D. R. (2010). Use of low-cost multistripe laser triangulation (MLT) scanning technology for three-dimensional, quantitative paleoichnological and neoichnological studies. *Journal of Sedimentary Research, 80,* 590–610.

Prothero, D. R., & Berggren, W. A. (1992). Eocene-Oligocene climatic and biotic evolution. Princeton: Princeton University Press.

Prothero, D. R., & Whittlesey, K. E. (1998). Magnetic stratigraphy and biostratigraphy of the Orellan and Whitneyan land-mammal ages in the White River Group. In D. O. Terry Jr., H. E. LaGarry & R. M. Hunt (Eds.), *Depositional environments, lithostratigraphy, and biostratigraphy of the White River and Arikaree Groups (late Eocene to early Miocene North America)* (pp. 39–61). Geological Society of America, Special Paper, 325.

Prothero, D. R., Ivany, L. C., & Nesbitt, E. A. (2003). *From greenhouse to icehouse: The marine Eocene-Oligocene transition*. New York: Columbia University Press.

Ratcliffe, B. C., & Fagerstron, J. A. (1980). Invertebrate lebensspuren of Holocene floodplains: their morphology, origin, and paleoecological significance. *Journal of Paleontology, 54,* 614–630.

Reynolds, T. D., & Wakkinen, W. L. (1987). Characteristics of the burrows of four species of rodents in undisturbed soils in southeastern Idaho. *The American Midland Naturalist, 118,* 245–250.

Retallack, G. J. (1983). *Late Eocene and Oligocene paleosols from Badlands National Park, South Dakota* (Vol. 193). Boulder: Geological Society of America, Special Paper.

Retallack, G. J. (1984). Trace fossils of burrowing beetles and bees in an Oligocene paleosol, Badlands National Park, South Dakota. *Journal of Paleontology, 58,* 571–592.

Retallack, G. J. (1992). Paleosols and changes in climate and vegetation across the Eocene/Oligocene boundary. In D. R. Prothero & W. A. Berggren (Eds.), *Eocene-Oligocene climatic and biotic evolution* (pp. 382–398). Princeton: Princeton University Press.

Retallack, G. J. (2001a). Cenozoic expansion of grasslands and global climatic cooling. *Journal of Geology, 109,* 407–426.

Retallack, G. J. (2001b). *Soils of the past*. Oxford: Blackwell.

Retallack, G. J. (2004). Late Oligocene bunch grassland and early Miocene sod grassland paleosols from central Oregon, U.S.A. *Palaeogeography, Palaeoclimatology, Palaeoecology, 207,* 203–237.

Retallack, G. J., Bestland, E. A., & Fremd, T. J. (2000). *Eocene and Oligocene paleosols of central Oregon* (Vol. 344). Boulder: Geological Society of America, Special Paper.

Retallack, G. J., Orr, W. N., Prothero, D. R., Duncan, R. A., Kester, P. R., & Ambers, C. P. (2004). Eocene-Oligocene extinction and paleoclimatic change near Eugene, Oregon. *Geological Society of America Bulletin, 116,* 817–839.

Rindsberg, A. K. (2012). Ichnotaxonomy: finding patterns in a welter of information. In D. Knaust & R. G. Bromley (Eds.), *Trace fossils as indicators of sedimentary environments* (pp. 45–78). Amsterdam: Elsevier.

Rodríguez-Tovar, F. J. (2007). Substrate firmness controlling nesting behavior of *Bembix oculata* (Hymenoptera, Bembicinae). In R. G. Bromley, L. A. Buatois, M. G. Mángano, J. F. Genise & R. N. Melchor (Eds.), *Sediment-organism interactions: A multifaceted ichnology* (pp. 353–359). Tulsa: Society for Sedimentary Geology.

Sakagami, S. F., & Michener, C. D. (1962). *The nest architecture of the sweat bees (Halictinae): A comparative study of behavior*. Lawrence: University of Kansas Press.

Sánchez, M. V., González, M. G., & Genise, J. F. (2010). Phytolith analysis of *Coprinisphaera*, unlocking dung beetle behaviour, herbivore diets and palaeoenvironments along the middle Eocene-early Miocene of Patagonia. *Palaeogeography, Palaeoclimatology, Palaeoecology, 285,* 224–236.

Sarzetti, L. C., Genise, J. F., & Sanchez, M. V. (2014). Nest architecture of *Oxaea austere* (Andrenidae, Oxaeinae) and its significance for the interpretation of Uruguayan fossil bee cells. *Journal of Hymenoptera Research, 39,* 59–70.

Schäfer, W. (1972). *Ecology and palaeoecology of marine environments*. Edinburgh: Oliver and Boyd.

Schaetzl, R., & Anderson, S. (2009). *Soils: Genesis and geomorphology*. Cambridge: Cambridge University Press.

Seeland, D. (1985). Oligocene paleogeography of the northern Great Plains and adjacent mountains. In R. M. Flores & S. S. Kaplan (Eds.), *Cenozoic paleogeography of the west-central United States* (pp. 187–205). Denver: Rocky Mountain Section, Society of Economic Paleontologists and Mineralogists.

Seilacher, A. (1953). Studien zur Palichnologie. I. Über die Methoden der Palichnologie. *Neues Jahrbuch für Geologie und Paläontologie, Abhandlungen, 96,* 421–452.

Seilacher, A. (1964). Biogenic sedimentary structures. In J. Imbire & N. Newell (Eds.), *Approaches to paleoecology* (pp. 296–316). New York: Wiley.

Seilacher, A. (1967). Bathymetry of trace fossils. *Marine Geology, 5,* 413–428.

Seilacher, A. (1978). Use of trace fossil assemblages for recognizing depositional environments. In P. B. Basan (Ed.), *Trace fossil concepts* (pp. 167–181). Tulsa: Society of Economic Paleontologists and Mineralogists.

Seilacher, A. (1986). Evolution of behavior as expressed in marine trace fossils. In M. H. G. Nitecki & J. A. Kitchell (Eds.), *Evolution of animal behavior: Palaeontological and field approaches* (pp. 62–87). New York: Oxford University Press.

Seilacher, A. (2007). *Trace fossil analysis*. Berlin: Springer.

Seilacher, A., & Seilacher, E. (1994). Bivalvian trace fossils: a lesson from actuo-paleontology. *Courier Forschungsinstitut Senckenberg, 169,* 5–15.

Shorthouse, D. J., & Marples, T. G. (1980). Observations on the burrow and associated behavior of the arid-zone scorpion *Urodacus yaschenkoi* (Birula). *Australian Journal of Zoology, 28,* 581–590.

Smith, A. B., & Crimes, T. P. (1983). Trace fossils formed by heart urchins–a study of *Scolicia* and related traces. *Lethaia, 16,* 79–92.

Smith, J. J., & Hasiotis, S. T. (2008). Traces and burrowing behaviors of the cicada nymph *Cicadetta calliope:* neoichnology and paleoecological significance of extant soil-dwelling insects. *PALAIOS, 23,* 503–513.

Smith, J. J., Hasiotis, S. T., Kraus, M. J., & Woody, D. T. (2008a). Relationship of floodplain ichnocoenoses to paleopedology, paleohydrology, and paleoclimate in the Willwood Formation, Wyoming, during the Paleocene-Eocene Thermal Maximum. *PALAIOS, 23,* 683–699.

Smith, J. J., Hasiotis, S. T., Woody, D. T., & Kraus, M. (2008b). Paleoclimatic implications of crayfish-mediated prismatic structures in paleosols of the Paleogene Willwood Formation, Bighorn Basin, Wyoming, U.S.A. *Journal of Sedimentary Research, 78,* 323–334.

Stucky, R. K. (1992). Mammalian faunas in North America of Bridgerian to Early Arikareean "ages" (Eocene to Oligocene). In D. R. Prothero & W. A. Berggren (Eds.), *Eocene-Oligocene climatic and biotic evolution* (pp. 464–493). New Jersey: Princeton University Press.

Sudd, J. H. (1967). *An introduction to the behavior of ants*. London: Edward Arnold.

Sullivan, R. M., & Holman, J. A. (1996). Squamata. In D. R. Prothero & R. J. Emry (Eds.), *The terrestrial Eocene-Oligocene transition in North America* (pp. 354–372). New York: Cambridge University Press.

Swinehart, J. B., Souders, V. L., Degraw, H. M., & Diffendal, R. F. (1985). Cenozoic paleogeography of western Nebraska. In R. M. Flores & S. S. Kaplan (Eds.), *Cenozoic paleogeography of the west-central United States* (pp. 209–229). Denver: Rocky Mountain Section, Society of Economic Paleontologists and Mineralogists.

Taylor, A. M., & Goldring, R. (1993). Description and analysis of bioturbation and ichnofabric. *Journal of the Geological Society of London, 150,* 141–148.

Taylor, A., Goldring, R., & Gowland, S. (2003). Analysis and application of ichnofabrics. *Earth-Science Reviews, 60,* 227–259.

Tschinkel, W. R. (2003). Subterranean ant nests: trace fossils past and future. *Palaeogeography, Palaeoclimatology, Palaeoecology, 192,* 321–333.

Uchman, A., & Pervesler, P. (2006). Surface lebensspuren produced by amphipods and isopods (crustaceans) from the Isonzo delta tidal flat, Italy. *PALAIOS, 21,* 384–390.

van Donselaar-ten Bokkel Huinink, W. A. E. (1966). *Structure, root systems, and periodicity of savanna plants and vegetation in northern Surinam.* Amsterdam: North-Holland Publishing Company.

Villani, M. G., Allee, L. L., Diaz, A., & Robbins, P. S. (1999). Adaptive strategies of edaphic arthropods. *Annual Review of Entomology, 44,* 233–256.

Weaver, J. E. (1919). *The ecological relations of roots.* Washington: Carnegie Institution of Washington Publication.

Williams, S. C. (1966). Burrowing activities of the scorpion *Anuroctonus phaeodactylus* (Wood) (Scorpionida: Vaejovidae). *Proceedings of the California Academy of Sciences, 34,* 419–428.

Willis, E. R., & Roth, L. M. (1962). Soil and moisture relations of *Scaptocoris divergins* Troeschner (Hemiptera: Cynidae). *Annals of the Entomological Society of America, 55,* 21–32.

Wolfe, J. A. (1992). Climatic, floristic, and vegetational changes across the Eocene/Oligocene boundary in North America. In D. R. Prothero & W. A. Berggren (Eds.), *Eocene-Oligocene climatic and biotic evolution* (pp. 421–436). New Jersey: Princeton University Press.

Yancy, T. E., Elsik, W. C., & Sancay, R. H. (2003). The palynological record of late Eocene climate change, northwest Gulf of Mexico. In D. R. Prothero, L. C. Ivany & E. A. Nesbitt (Eds.), *From greenhouse to icehouse: The marine Eocene-Oligocene transition* (pp. 252–268). New York: Columbia University Press.

Zachos, J. C., Pagani, M., Sloan, L., Thomas, E., & Billups, K. (2001). Trends, rhythms, and aberrations in global climate 65 Ma to present. *Science, 292,* 689–693.

Chapter 11
Fossil Pollen and Spores in Paleoecology

Luke Mander and Surangi W. Punyasena

Abstract The discipline of paleoecology is a multidisciplinary field that uses geological and biological evidence to investigate the past occurrence, distribution and abundance of species and populations on timescales ranging from hundreds to hundreds of millions of years. In this way, paleoecology is broadly concerned with the ecology of the past. In this article, we discuss how paleoecological data are derived from assemblages of fossil pollen and spores, which are dispersed by plants as part of their life cycles, and how this material can be used to reconstruct paleoenvironments. We outline how pollen and spores can be analyzed and classified, and explore the potential strengths and weaknesses of paleoecological data, before considering technological and methodological developments that may play a role in the future development of this discipline.

Keywords Vegetation history • Plants • Morphology Taxonomy • Microscopy • Climate

Background and Overview

The discipline of paleoecology is concerned with the ecology of the past. It uses geological and biological evidence to investigate the past occurrence, distribution and abundance of species and populations on various timescales (Birks and Birks 1980; Seddon 2012). Paleoecology is a multidisciplinary field, and researchers working in this discipline gather data from a variety of geological and biological sources in order to investigate ancient ecology. This includes paleontological data that provide physical evidence of the organisms that were living in the past, sedimentological data that speak to the environments in which these organisms lived and were fossilized, and geochemical data that provide a window into ancient climates and can track the passing of time.

In this review, we are concerned with the paleoecology of ancient plant life as reconstructed using fossil pollen and spores. Pollen and spores are dispersed by plants as part of their life cycle. They are abundant and well preserved in the fossil record, and provide extensive, continuous records of ecological and environmental change. Paleoecologists who undertake this research aim to reconstruct the diversity, composition and abundance of past vegetation. Paleoecologists also use plant macrofossils, including vegetative structures such as leaves and their fragments, and rarer reproductive structures, such as cones. Although these larger fossils are not as abundant in the geologic sediments, when available, they provide detailed taxonomic, structural, and physiological data that may not be available with pollen and spores alone. Macrofossils, therefore, serve a counterpoint to pollen and spore records of Earth's past vegetation. More recent advances in metagenomic analyses of ancient DNA provide similarly detailed taxonomic occurrence data that cannot as of yet be replicated with pollen (Pedersen et al. 2016).

Pollen and spores are extracted primarily from sedimentary deposits. These samples are often unconsolidated sediments that have accumulated at the bottom of lakes, but researchers also examine sediment from peat bogs and sedimentary rock outcrops. Pollen is preserved in a wide range of environments. Non-sedimentary sources of recent pollen material include pollen trapped within moss polsters, packrat middens, ice cores, and within permafrost. In order to extract pollen and spores from these samples, they must be macerated in the laboratory, and this involves using

L. Mander (✉)
Department of Environment, Earth and Ecosystems, The Open University, Milton Keynes, MK7 6AA, UK
e-mail: Luke.Mander@open.ac.uk

S. W. Punyasena
Department of Plant Biology, University of Illinois, Urbana, IL 61801, USA
e-mail: Punyasena@life.illinois.edu

D. A. Croft et al. (eds.), *Methods in Paleoecology: Reconstructing Cenozoic Terrestrial Environments and Ecological Communities*, Vertebrate Paleobiology and Paleoanthropology, https://doi.org/10.1007/978-3-319-94265-0_11

chemicals to dissolve minerals and excess organic material. The taxonomic identification of these pollen and spores involves visual analysis by a human expert using a microscope. Pollen and spores are classified by using specialist terminology to describe the morphological features that characterize each specimen, and each specimen is compared to formal taxonomic descriptions in monographic works, as well as reference material (pollen and spores extracted from the reproductive structures of taxonomically identified plants collected in the field or herbarium).

Paleoecological data can be derived from any period in Earth history that contains fossil pollen and spores. This stretches from the first assemblages of spores produced by land plants in the Ordovician/Darriwillian (Strother et al. 2015) to the present day. Although paleoecological analyses are undertaken on assemblages of pollen and spores throughout Earth history (e.g., Traverse 2007), the discipline of paleoecology has centered on analyses of material from the Quaternary period (~2.6 Ma–present). Similarly, paleoecological analyses have been undertaken on deposits all over the world, from the tropics to the poles, but there has been a historical focus on North America and northwest Europe, and this is reflected to a degree in our review.

Foundational Work and Current Themes

The use of pollen and spores to reconstruct vegetation history, as it is practiced today, dates back to von Post (1916). He showed that pollen in samples of peat could pinpoint the position of modern vegetation zones, such as the southern spruce border in Sweden, and demonstrated that fossil pollen grains could be used to reconstruct their position through time (von Post 1916). The subsequent work of Potonie (1934), Wodehouse (1935) and Erdtman (1952) made attempts to erect a formal system of nomenclature for describing the morphology of pollen and spores, and the legacy of this work echoes in recent syntheses of descriptive terminology that aim to ensure paleoecologists use the same terms to describe the same morphological features (Punt et al. 2007). During this phase of the development of paleoecology, techniques for examining critical features of pollen and spore morphology were developed, such as the Lux-Obscuritas method of analyzing patterns of the organization of pollen and spore walls (Erdtman 1952). The widespread adoption of electron microscopy by paleoecologists in the latter part of the mid-20th century allowed finer morphological details of pollen and spores to be investigated, and this facilitated an increase in the taxonomic precision of pollen and spore records of Earth's vegetation (Fægri et al. 1989). Textbooks outlining paleoecological methods, including instructions on how to sample different environments, how to extract pollen and spores in the laboratory, as well as how to describe them and relate the composition of assemblages of pollen and spores found in sediments to the source vegetation that produced them, began to appear with the publication of Erdtman's *An Introduction to Pollen Analysis* (Erdtman 1943). Today there is a wide selection of such texts available that exhaustively document the discipline of pollen and spore-based paleoecology (e.g., Birks and Birks 1980; Fægri et al. 1989; Traverse 2007).

Many current methods in paleoecology are almost identical to those outlined in early texts. These include many of the laboratory methods used to extract fossil pollen and spores from sediments and reference pollen and spore material from flowers and other reproductive structures. Additionally, the practice of a human analyst classifying pollen and spores into taxonomic groups using morphological features observed through a microscope has remained largely unchanged for the last century. The results of these visual classifications – "pollen counts" – are often summarized in "pollen diagrams" (visual representations of pollen taxon abundances as consecutive vertical line graphs) that are broadly similar to the diagram presented by von Post (1916). Today pollen diagrams are constructed using software such as Tilia (Grimm 1991–2001), C2 (Juggins 2007), and the R-package rioja (Juggins 2015), which allows for the statistical analysis of count data as well as visual interpretation. Over the last forty years, numerical approaches to pollen analysis, advocated by Davis (1963) and Gordon and Birks (1972), have evolved into increasingly sophisticated computational models to quantitatively reconstruct past vegetation cover and paleoenvironments (e.g., Jackson and Williams 2004; Dawson et al. 2016).

Paleoecological data today are used in a wide variety of settings. In addition to the reconstruction of vegetation and climate history, which is a traditional mainstay of paleoecological research, pollen data finds application in comparing the output of Earth system and land surface models to empirical data (e.g., Cowling et al. 2008) and in the identification of ecological baselines that are useful in conservation biology (e.g., Gosling et al. 2009). Pollen data are also used in the provision of data to test phylogenetic hypotheses generated by molecular work such as the use of fossil pollen to constrain the evolutionary origins of plant taxa (e.g., Barreda et al. 2015).

From Plant to Sediment: The Biology, Dispersal and Preservation of Pollen and Spores

The Raw Materials: What Are Pollen and Spores?

Pollen and spores represent a discrete phase in the plant life cycle, which is characterized by a sequence of reproduction known as an alternation of generations (Jarzen and Nichols 1996). This sequence typically alternates between a sporophyte generation carrying a double set of chromosomes (diploid; 2-*n*) and a gametophyte generation carrying a single set of chromosomes (haploid; *n*). There is an evolutionary trajectory towards the elaboration of the sporophyte generation and the reduction of the gametophyte generation (Willis and McElwain 2002). This can be observed in the anatomy of fossil plants, and also in comparative studies of the reproductive characters of primitive plants. In the following sections, we outline the life cycles of major plant groups, and highlight when pollen and spores are produced in these life cycles. This outline of plant life cycles represents a synthesis of information from Playford and Dettmann (1996), Willis and McElwain (2002), Raven et al. (2005), Rudall and Bateman (2007) and Traverse (2007).

Homosporous plants: Extant members of this plant group include non-vascular embryophytes such as liverworts, hornworts and mosses. In these primitive non-vascular plants, the haploid (*n*) gametophyte is generally emphasized while the diploid (2*n*) sporophyte is reduced in size and remains attached to the gametophyte, on which it is dependent for nutrition (Fig. 11.1). More complex extant homosporous plants include most ferns and some lycopods, and in these vascular plants the diploid sporophyte is generally emphasized while the haploid gameophyte is separate and reduced in size (Fig. 11.2).

The generalized life cycle of a homosporous plant is as follows. The haploid gametophyte bears an antheridium and an archegonium. The antheridium produces a male haploid gamete by mitosis and the archegonium produces a female haploid gamete by mitosis. These two gametes fuse when the male gamete swims to the archegonium in a film of water and encounters the female gamete. This results in the production of a diploid zygote. Germination of the zygote results in the production of a diploid sporophyte. Upon reaching maturity, the sporophyte undergoes meiotic division resulting in the production of haploid spores. In primitive non-vascular plants such as liverworts, hornworts and mosses, spores are produced in a capsule (Fig. 11.1). In more advanced homosporous vascular plants, such as certain ferns and some lycopods, spores are produced in sporangia (Fig. 11.2). These spores are released and, upon germination, grow into haploid gametophytes.

Heterosporous plants: Extant members of this plant group include the vascular embryophytes, including some ferns such as *Azolla* and *Marselia*, and in these plants the diploid sporophyte is emphasized. In these plants the diploid sporophyte bears a microsporophyll with a microsporangium, and a megasporophyll with a megasporangium. Meiotic division within the microsporangium results in the production of haploid microspores (Fig. 11.3), which germinate to produce a male microgametophyte with an antheridium. Meiotic division within the megasporangium results in the production of haploid megaspores (Fig. 11.3),

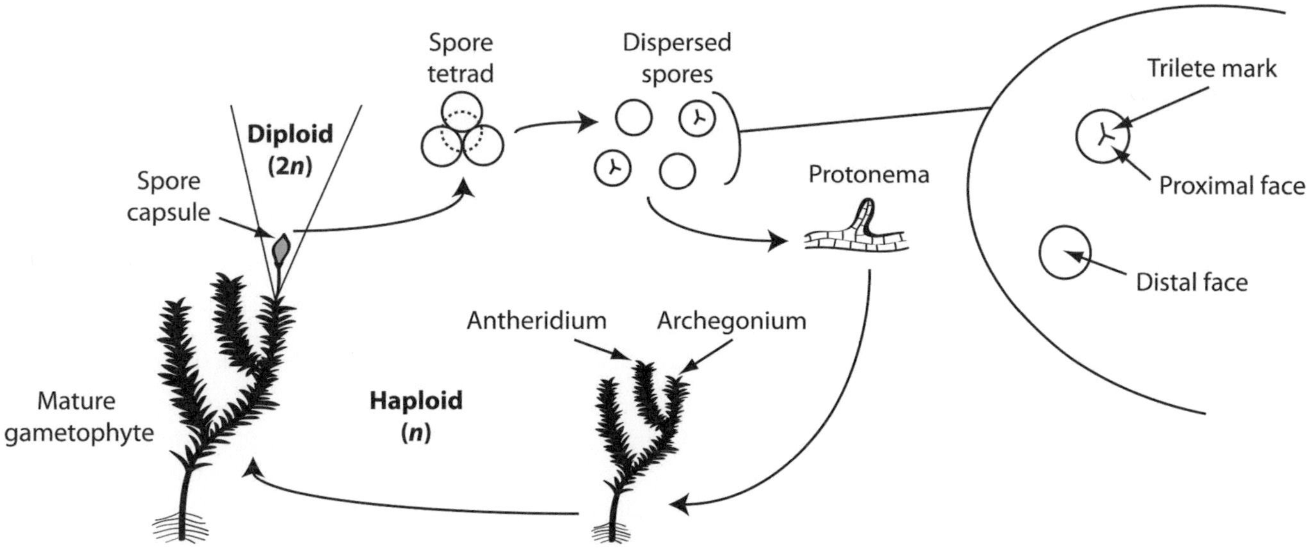

Fig. 11.1 Schematic life cycle of a non-vascular homosporous plant such as a moss in which the haploid gametophyte is emphasized and a diploid spore capsule remains attached to the gametophyte. Not to scale. Based on Playford and Dettmann (1996)

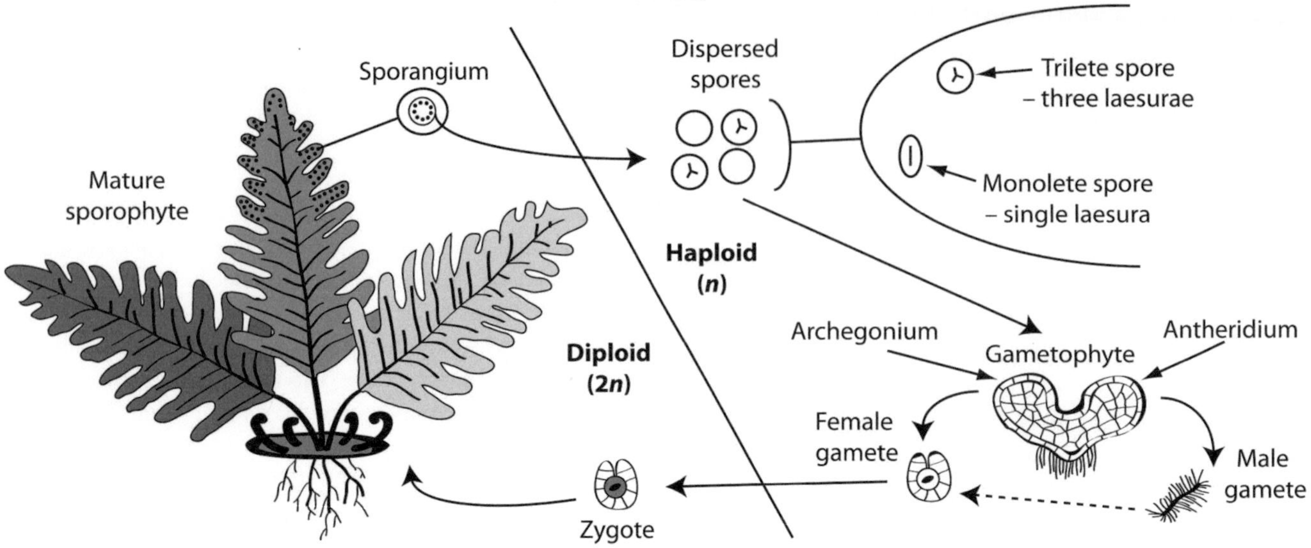

Fig. 11.2 Schematic life cycle of a homosporous fern in which the diploid sporophyte is emphasized and the haploid gametophyte is separate and reduced. Not to scale. Based on Playford and Dettmann (1996)

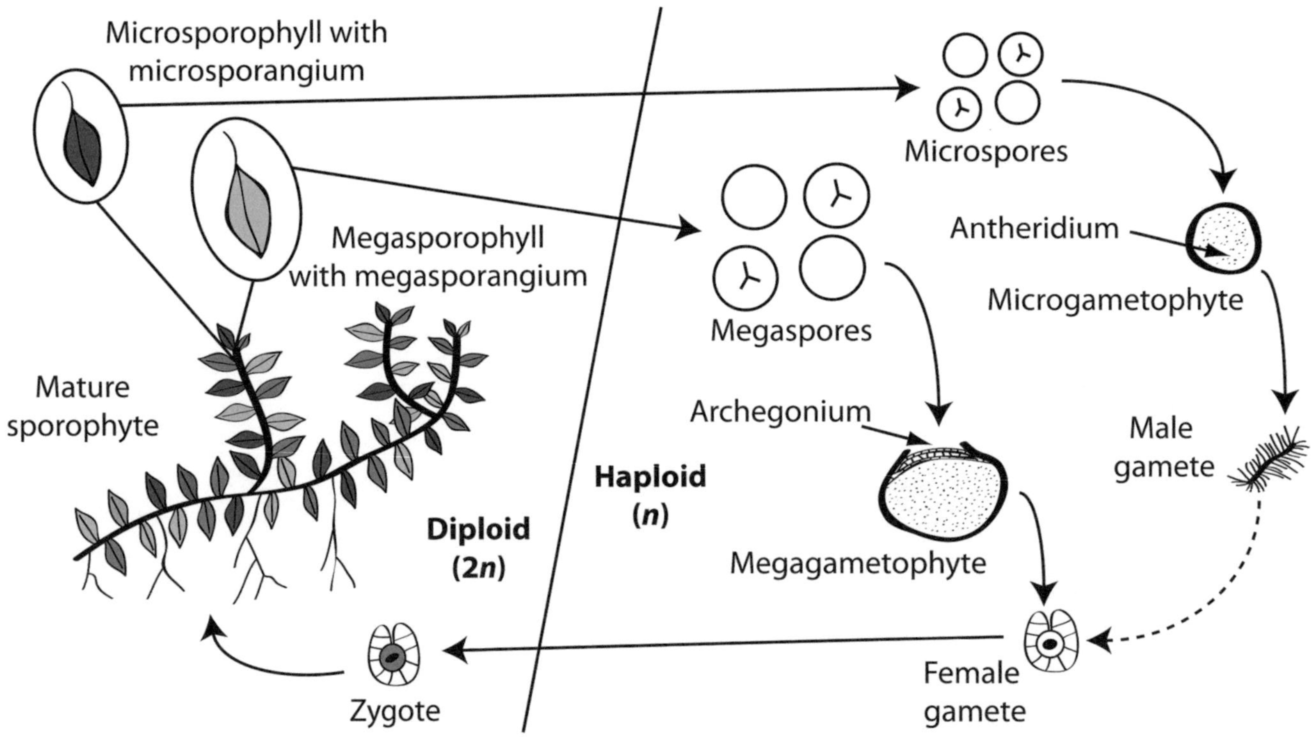

Fig. 11.3 Schematic life cycle of a heterosporous plant, in this case a lycopod, in which the diploid sporophyte is emphasized. Paleoecologists use both microspores and megaspores to reconstruct ancient vegetation. Not to scale. Based on Playford and Dettmann (1996)

which germinate to produce a female megagametophyte with an archegonium. Paleoecologists use both microspores and megaspores to reconstruct vegetation history. The antheridium of the microgametophyte produces a male haploid gamete by mitosis and the archegonium of the megagametophyte produces a female haploid gamete by mitosis. These two gametes fuse when the male gamete swims to the female gamete in a film of water. This results in the production of a diploid embryo. Germination of the embryo results in the production of a diploid sporophyte.

Gymnosperms: Extant seed plants form a monophyletic group with five major clades: cycads, ginkgos, conifers,

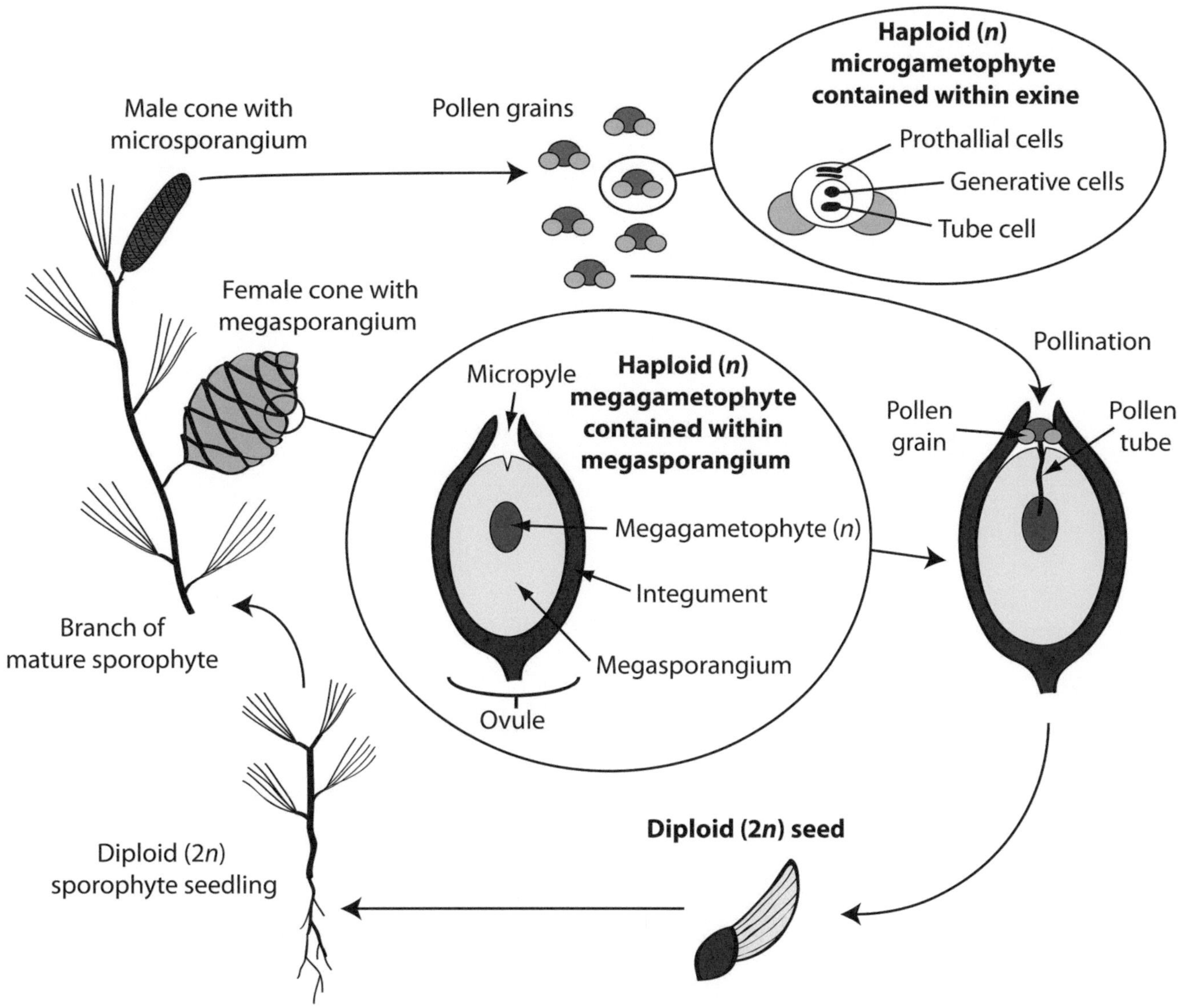

Fig. 11.4 Schematic life cycle of a gymnosperm, in this case a pine tree, in which the diploid sporophyte is emphasized. Not to scale. Based on Raven et al. (2005)

gnetophytes and angiosperms. The first four of these are called gymnosperms and the seeds of these plants are not enclosed in a carpel as they are in angiosperms. In these plants the diploid sporophyte is emphasized, and their generalized life cycle is as follows (Fig. 11.4). The diploid sporophyte bears a male cone with a microsporangium (a pollen sac), and a female cone with a megasporangium. In most gymnosperms male and female cones are borne on the same sporophyte, but in others such as *Ginkgo*, male and female cones are produced by different sporophytes and do not occur on the same sporophyte. Meiotic division within the microsporangium results in the production of haploid microspores, which comprise a microgametophyte contained in a microspore coating, the exine. These microspores are known as pollen grains (Fig. 11.4). The microgametophyte consists of two prothallial cells, a relatively small generative cell with two sperm and in some groups, such as conifers and gnetophytes, there is also a vegetative tube cell that directs the growth of a pollen tube. The megasporangium is contained within an integument and remains attached to the parent sporophyte as an ovule. Meiotic division within the megasporangium results in the production of a haploid megagametophyte with an egg cell.

Upon reaching maturity, the pollen grains are dispersed and alight near the opening of the megasporangium. Pollen is carried through an opening in the integument of the ovule (the micropyle) in a liquid drop and into the vicinity of the megagametophyte. In some groups, such as conifers and gnetophytes, a pollen tube introduces non-motile fertilizing spermatozoids into the vicinity of the egg (siphonogamous

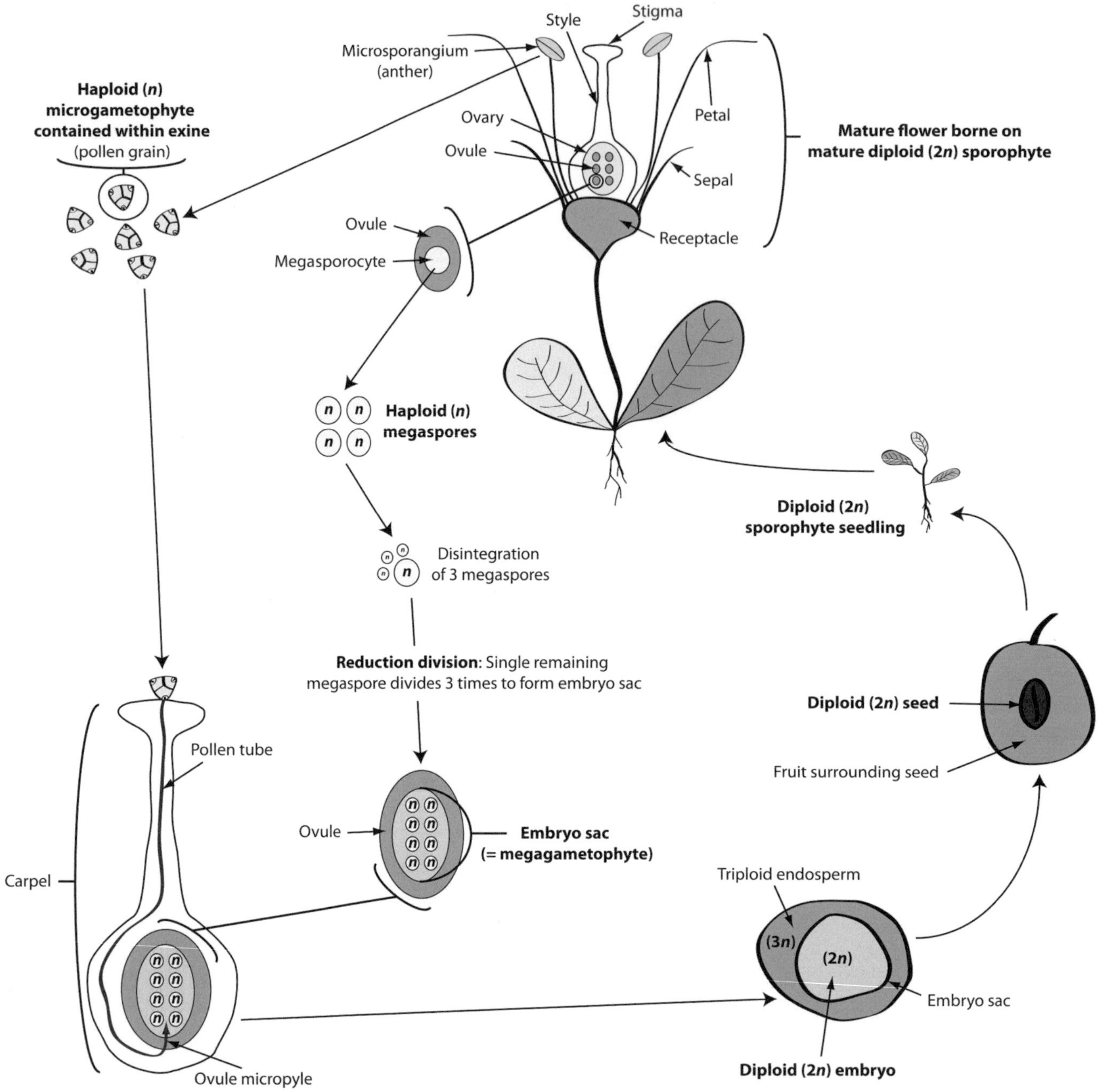

Fig. 11.5 Schematic life cycle of an angiosperm in which the diploid sporophyte is emphasized. Not to scale. Based on Jarzen and Nichols (1996)

fertilization; (Chaloner 1970; Poort et al. 1996)). In other groups, such as cycads and *Ginkgo*, a pollen tube that serves a solely haustorial function develops and absorbs nutrients from the tissues of the megasporangium. After several months of growth, the end of the haustorial pollen tube closest to the pollen grain enlarges and bursts, releasing motile spermatozoids that swim to the egg (zooidogamous fertilization; (Chaloner 1970; Poort et al. 1996)). Fertilization of the egg by the fusion of male and female gametes results in the production of a diploid embryo, which is contained in a coating formed from the ovule integument. This structure is known as a seed. Germination of the seed results in the production of a diploid sporophyte (Fig. 11.4).

Angiosperms: The flowering plants are the fifth major clade of extant seed plants. In these plants the diploid sporophyte is emphasized, and their generalised life cycle is as follows (Fig. 11.5). The diploid sporophyte bears a flower with a microsporangium (an anther) and a megasporangium (an ovule). Meiotic division within the microsporangium results in the production of haploid microspores, which comprise a microgametophyte contained in a microspore coating, the exine. These microspores are known as pollen grains (Fig. 11.5). Each pollen grain contains three nuclei, one of which is the pollen tube nucleus and the other two are non-motile sperm nuclei. The megasporangium is enclosed by the tissues of the carpel. Meiotic division within the megasporangium results in the production of four haploid megaspores. Three of these disintegrate and the remaining functional megaspore undergoes reduction division to form a megagametophyte or embryo sac that has eight nuclei. One of these nuclei is the egg and two of these nuclei are the polar nuclei.

Upon reaching maturity, the pollen grains are dispersed, usually by wind or an animal vector, and alight on the stigma. The pollen germinates on the stigma and a pollen tube is produced. The pollen tube grows down the style (for ~50 cm in the case of *Zea mays*) and enters the ovule via the ovule micropyle. One sperm nucleus fuses with the egg to produce a diploid embryo. The other sperm nucleus fuses with the two polar nuclei to produce a triploid ($3n$) primary endosperm nucleus, which is a source of nutrition for the developing embryo. The diploid embryo develops within the embryo sac. The integuments of the ovule become the seed coat and the tissues of the ovary become the fruit. Germination of the seed results in the production of a diploid sporophyte.

General Relationships Between Pollen and Spores and Vegetation

Pollen and spores are dispersed as part of the life cycles of plants and accumulate in a variety of depositional settings, including lakes, fluvial environments, swamps, and offshore marine sediments (Gastaldo 2001; Traverse 2007). For angiosperm pollen, pollination mechanism (wind-, animal-, or self-pollination) influences the amount of pollen produced and released into the environment, with wind-pollinated species generally producing the most pollen per individual. (Spores and gymnosperm pollen are primarily wind dispersed).

The transport and deposition of pollen and spores in a forested area has a composite nature, with aerial transfer taking place partly through the vegetation (the trunk space), partly above the canopy, and partly through rainfall (Tauber 1967). A primary assumption that is made when using pollen and spores for the purpose of paleoecology is that an interpretable relationship exists between these dispersed pollen and spores and the vegetation that produced them (Birks and Birks 2000). Studies that compare the composition of the source vegetation and assemblages of modern dispersed pollen and spores in surface samples indicate that this assumption is reasonable in many situations, but also highlight that reading patterns of vegetation change from fossil assemblages through time is not straightforward.

Surface sample studies: Mapped patterns of present-day tree pollen percentages in temperate regions of Europe and North America correspond broadly with geographic patterns of relative tree abundance (e.g., Webb 1974; Delcourt et al. 1983, 1984; Huntley and Birks 1983; Prentice et al. 1987). These observations support the idea that, as a first approximation, changes in pollen and spore percentages in samples through time can be interpreted in terms of roughly comparable changes in relative abundances of taxa in the source vegetation (Prentice and Webb 1986). Similarly, although offshore sediments are thought to present a more blurred picture of the source vegetation (Scott et al. 2004), the distribution of pollen and spores in surface samples in marine sediments offshore West Africa broadly corresponds with the spatial distribution of vegetation growing on the land. For example, "the distribution of Poaceae pollen in the marine surface sediments [offshore southern West Africa] shows a good correspondence with the regions of savannah and desert biomes on the adjacent continent" (Dupont et al. 2007, p. 91).

However, the percentage of a given pollen or spore taxon in a sample depends not only on its own abundance, but also on the abundance of one or more others. This leads to relationships between the percentage of pollen and spore taxa in samples and the percentage of their parent plants in the source vegetation that are non-linear (Davis 1963; Prentice and Webb 1986); such non-linearity is known as the "Fagerlind effect" after Fagerlind (1952). Additionally, a general finding from surface sample studies is that although patterns in the relative abundance of different plant taxa are broadly reflected in pollen and spore data, the representation of certain taxa is biased at the expense of others. In the pollen record, certain taxa are commonly overrepresented (e.g., *Pinus*), whereas others are commonly underrepresented (e.g., *Acer*). This is because pollen spectra are biased towards taxa with high pollen productivity and/or slow deposition from the atmosphere (Prentice 1985).

How much space? When assemblages of pollen and spores are examined, the results of surface sample studies allow a researcher to interpret the assemblage in terms of the broad composition of the vegetation. However, these studies provide less information on the source area of the pollen and spores. In particular, it can be extremely difficult to interpret the size of the area from which the pollen and spores under investigation were recruited (Jacobsen and Bradshaw 1981). This is important because if assemblages of pollen and spores were recruited from very small local habitats, then changes in their composition through time could reflect the vagaries of local ecology (Davis 2000). In contrast, if assemblages of pollen and spores were recruited from a much larger area, then changes in their composition may reflect changes in regional climate (Davis 2000).

Empirical surface sample studies have demonstrated that a general relationship exists between the size of the basin into which pollen and spores are deposited and the size of the source area. Most of these studies have focused on the pollen of trees and they show that larger basins collect pollen from a larger area than smaller basins (Janssen 1973; Bradshaw and Webb 1985; Jackson 1990; see Sugita 1994). Pollen grains can be transported thousands of kilometers, but it is thought that the abundance of such long-distance grains in samples is low (Bakker 1985). This general relationship has been confirmed by simulation studies (e.g., Prentice 1985; Sugita 1993), which have highlighted that the pollen loading per unit area of lake surface from local trees declines as lake size increases because the ratio of perimeter to area decreases as lake radius increases (Sugita 1994). Such modelling work has been used to examine the source area of pollen for idealized depositional basins with different radii. For forest hollows of radius 2 m, 30–45% of the pollen deposited is derived from within 50–100 m (Sugita 1994). For small lakes of radius 50 m this source area increases to 300–400 m, and for medium lakes of radius 250 m this source area is 600–800 m (Sugita 1994). In dense tropical forests where the majority of plants are insect pollinated (Bush 1995) and pollen and spore spectra are not always dominated by canopy trees (Fægri 1966), pollen and spores may travel only 10–50 m from the source plant (Kershaw and Strickland 1990; Bush and Rivera 1998).

Plant macrofossils: Macrofossils (leaves, cones, twigs, shoots and other vegetative and reproductive structures) are extremely useful as paleoecological records in themselves (Jackson and Booth 2007; Taylor et al. 2009). For example, paratropical leaf litter deposits have been shown to reflect the heterogeneity of the source vegetation (Burnham 1989), while temperate deciduous forest litter corresponds closely in composition to the source forest and reflects the relative abundance of species comprising more than 10% of the stem-basal area (Burnham et al. 1992). Relative abundances of modern leaf litter of coniferous trees from temperate forests in North America are also significantly correlated with their basal areas in nearby forests (Dunwiddie 1987). In the context of this review, which is focused on the use of pollen and spores in paleoecology, data from macrofossils are complementary to pollen and spore data and highlight certain vegetation-sensing properties of pollen and spores.

In particular, macrofossil data show that some taxa are generally absent or poorly represented in the pollen and spore record. For example, *Abies*, *Pseudotsuga* and *Larix* are typically well represented in macrofossil assemblages, but these taxa are generally absent or poorly represented in assemblages of pollen and spores extracted from the same samples (Dunwiddie 1987; Jackson and Booth 2007). Additionally, they highlight that pollen and spores generally record vegetation at low taxonomic resolution (Jackson and Booth 2007; Mander and Punyasena 2014). For example, macrofossils can demonstrate that the source vegetation at a given site contains several species within a particular genus, or several genera within a particular family, but these multiple taxa may be represented by a considerably smaller number of pollen and spore morphotypes. There are several classic examples of this, including *Picea* (spruce) and *Quercus* (oak) pollen, which are both extremely difficult to classify below the genus level, and Poaceae (grass) pollen, which is very difficult to classify at a taxonomic rank below family (Mander et al. 2013) (see Mander and Punyasena 2014 for a general review of taxonomic resolution).

The preservation of pollen and spores: The outer layer of pollen and spores, the exine, is composed of an organic compound called sporopollenin. This substance is extremely inert and is resistant to the effects of a variety of acid treatments. The tissues that are contained within this tough exine coating, including genetic material, are extremely labile and consequently are rarely found as fossils. Fossilized pollen tubes have been reported as far back in time as the Pennsylvanian (323–299 Ma) (e.g., Rothwell 1972), but pollen and spores that are found as fossils generally represent only the outer exine layer.

However, sporopollenin is susceptible to degradation by strong oxidants in a laboratory setting such as H_2O_2 and CrO_3 (Traverse 2007), and pollen and spores are vulnerable to a variety of damage types on their journey from parent plant to sediment. These include chemical damage, which can involve oxidation as a result of deposition in sediments prone to wet/dry cycles, and microbial attack, and mechanical damage such as tearing, compression and folding (Mander et al. 2012) (Fig. 11.6). Once a pollen grain or spore is sealed in sediment, it is vulnerable to further microbial attack and compression, as well as thermal maturation due to processes such as igneous intrusion (Mander et al. 2012) (Fig. 11.6). Taxa differ in the amount of sporopollenin that is invested in the exine, and thin-walled taxa such as *Populus* may be destroyed by these processes, while thicker-walled taxa may survive. As a result, changes in the relative abundance of thin-walled taxa may reflect taphonomic, rather than ecological factors.

The sensitivity of sporopollenin to oxidation means that certain rock types, such as sediments with a reddish color that is indicative of oxidation, should generally be avoided because these rocks are unlikely to contain pristine pollen and spores. Pollen grains range in size from $\sim$4.5 μm to $\sim$200 μm, and this means that they are likely to be winnowed away during the deposition of relatively coarse-grained rocks such as sandstones. Consequently, such coarse-grained sediments should generally be avoided, and

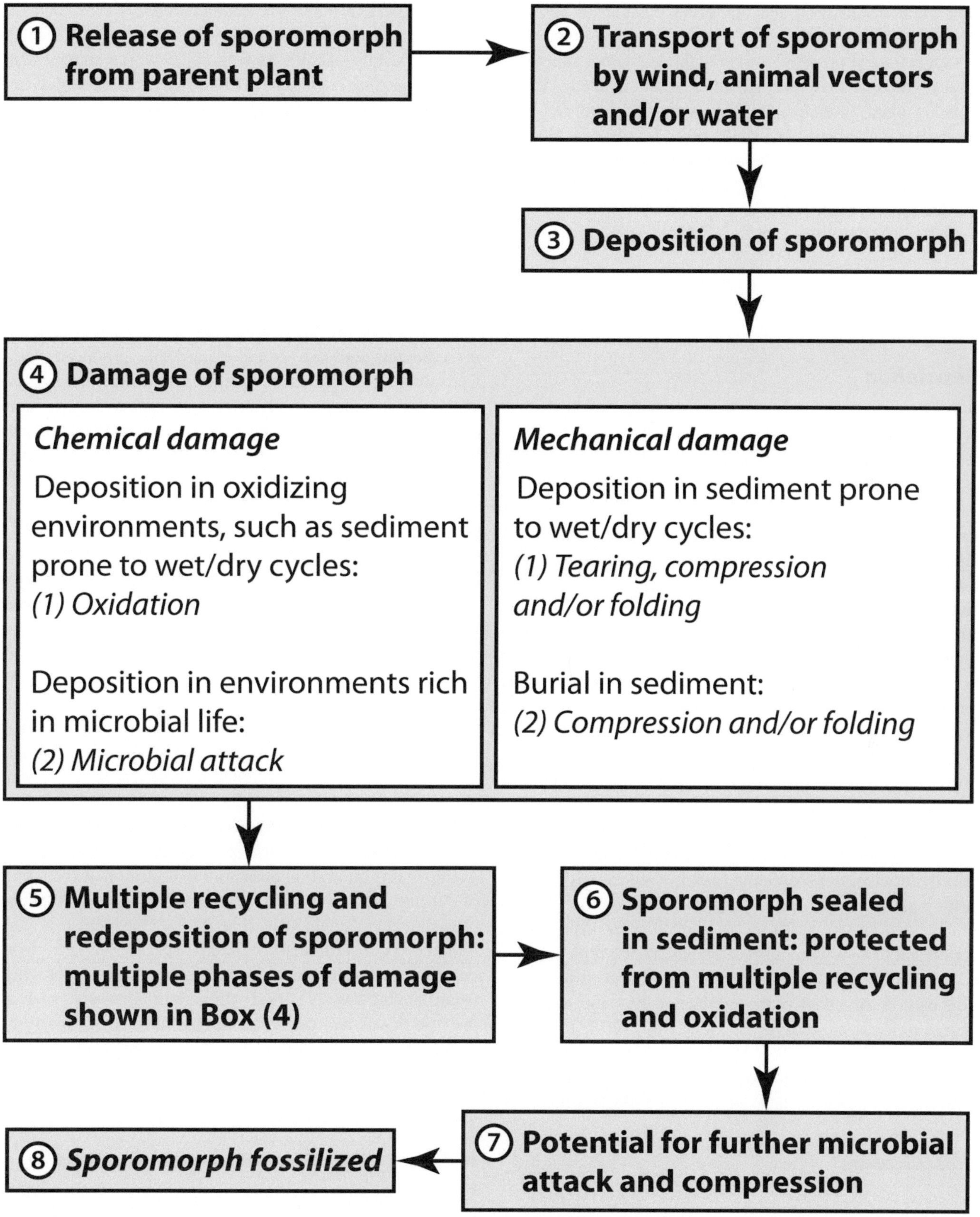

Fig. 11.6 Flow diagram illustrating the types of damage that can affect pollen and spores (referred to in this figure collectively as sporomorphs) in the period between their release and eventual fossilization (from Mander et al. 2012)

finer grained siltstones and mudstones should be targeted instead.

Palynological preparations invariably contain a variety of other fossils that are also tough and resistant to decay. These include plant cuticles, fungal spores, charcoal, and non-pollen palynomorphs. These other fossils can provide additional information on the paleoecology of a sample, and this information is generally complementary to the information derived from pollen and spores.

The Description and Classification of Pollen and Spores

Description

In most paleoecological studies, pollen and spores are classified into taxonomic groups on the basis of their exine morphology. This involves using specialist terminology to describe the morphological features that characterize different specimens. Increasingly, workers are taking a standardized approach to the description of pollen and spores, using published glossaries such as that of Punt et al. (2007), which contain hundreds of specialist terms for the features encountered by the paleoecologist. Morphological features that are typically used to describe pollen and spores include, but are not limited to, their shape, dispersal unit, exine stratification, the nature and arrangement of any apertures, and the nature of any surface ornamentation (Punt et al. 2007).

Some of these morphological features may have functional significance. Pollen apertures, for instance, represent sites where the pollen tube can easily exit the pollen grain, but they also allow for the ingress and egress of water and serve as pathways that guide the folding of pollen grains during desiccation (Wodehouse 1935; Katifori et al. 2010). Additionally, there are suggestions that coarse surface ornamentation may represent an adaptation for insect pollination and that very fine surface ornamentation represents an adaptation for wind pollination, or that surface ornamentation may be involved in the production of static electrical charge to assist the transfer of pollen from vector to stigma (Punt 1986; Chaloner 2013). However, it is not necessary for morphological features that are used to describe a particular pollen grain or spore to have any known biological function.

Classification

Classification of pollen and spores into taxonomic groups involves comparing the morphology of specimens, as recorded by descriptive terminology (Punt et al. 2007), with published identification keys. These keys include broad syntheses such as the *World Pollen and Spore Flora* and *The Northwest European Pollen Flora* (Punt et al. 2003), as well as more locally specific monographic works such as *The Pollen and Spores of Barro Colorado Island* (Roubik and Moreno 1991). These keys do not give complete morphological descriptions, but serve to narrow down the range of possibilities during the classification process. Paleoecologists also make extensive use of reference collections. These are microscope slides of pollen and spores extracted from extant plants, usually from vouchered herbarium specimens, and allow the paleoecologist to establish a robust link between the morphology of a particular pollen grain and its parent plant. It is this link that allows paleoecologists to establish a relationship between the pollen and spores found in sediments and the parent plants that produced them. Consultation of reference material is a crucial step in the classification process, and "any identification must be considered tentative until tested against an actual specimen in the pollen herbarium" (Fægri et al. 1989, p. 237). Reference material that encompasses the full range of natural morphological variation among taxa allows the paleoecologist to compare the morphology of fossil pollen and spores with a wide spectrum of extant taxa (Birks and Birks 1980; Jarzen and Jarzen 2006). Monographs of reference material allow workers to establish pollen and spore types and also provide a basis for work comparing the morphology of pollen and spores from different regions and time periods (Punt et al. 2003).

The classification of pollen and spores into taxonomic groups underpins the discipline of paleoecology, and it produces raw data that allow changes in vegetation diversity and composition to be examined in time and space. Generally speaking, classification at a high taxonomic level, such as the rank of family, is an easier task than classification at a low taxonomic level, such as the rank of species. For example, it is quite straightforward to discriminate the pollen of oak and grass from each other, oak having three colpi and grass having a single pore (Fig. 11.7). However, it is very difficult to discriminate different species of oak from one-another and to discriminate different species of grass from one-another (Fig. 11.7). In certain situations, such as the case of oak and grass pollen, the difficulty of classifying at the species level arises because the pollen of different species is very similar morphologically.

A Prototypical Paleoecological Analysis

In this section of our review, we provide a skeleton of a typical paleoecological analysis. We begin by considering the types of sites that paleoecologists investigate, outline the methods used to release pollen and spores from sediment

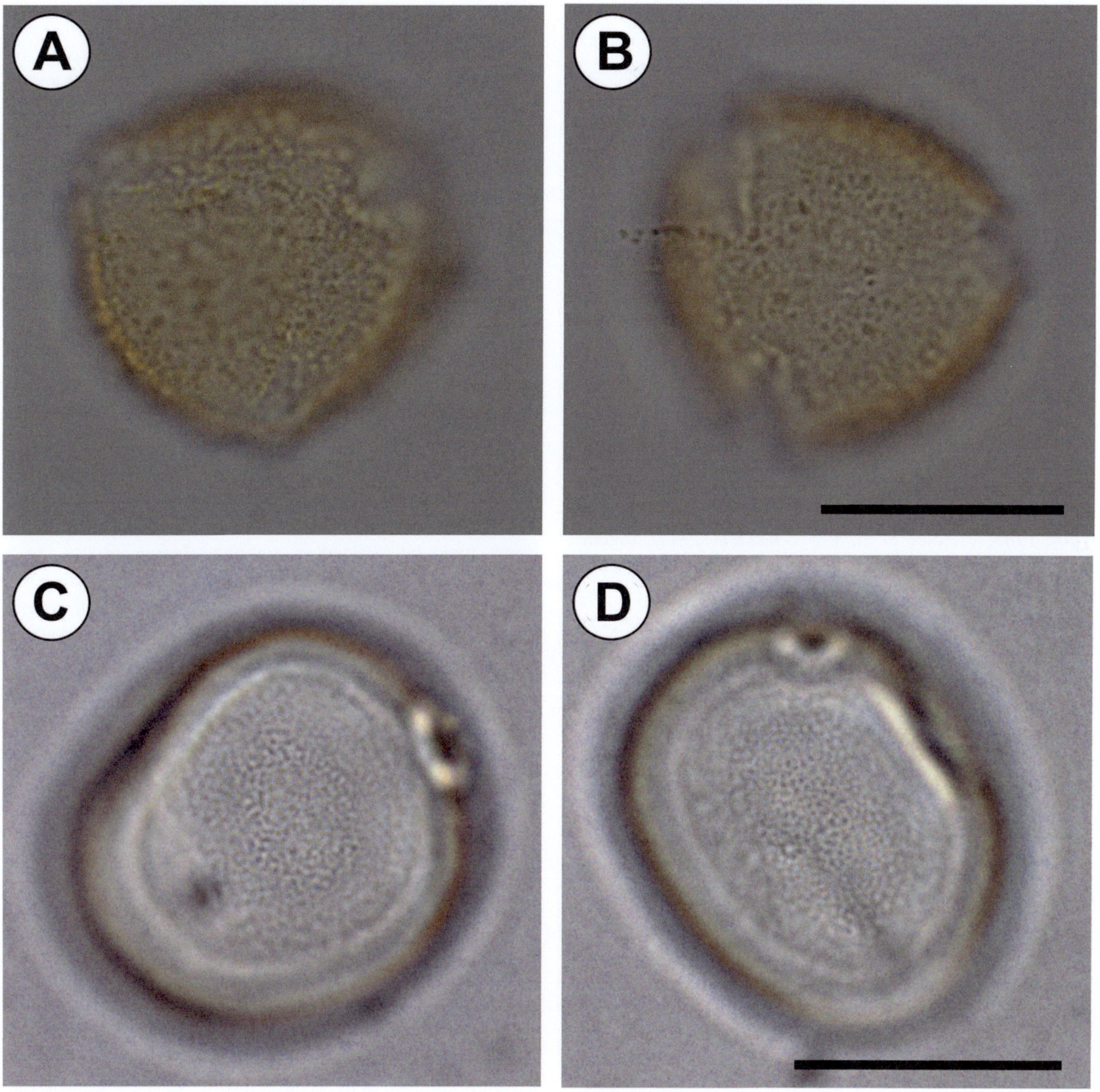

Fig. 11.7 Comparison of pollen produced by two species of oak and two species of grass. Oak pollen: A, *Quercus englemannia*; B, *Quercus virginiana*. Grass pollen: C, *Stipa tenuifolia*; D, *Dactylis glomerata*. Scale bars in B and D represent 20 μm

samples, and highlight the analytical techniques that are routinely used in paleoecology. We also consider the areas of potential strength in paleoecological data, and areas of potential weakness in paleoecological data.

Typical Locations

Almost any location in which pollen and spores accumulate is available for the paleoecologist to study. These include tiny moss polsters within forested areas, peat bogs, sediments in lakes, sediments in abandoned river channels, offshore marine sediments, as well as rocks exposed at outcrop. The choice of location depends on the type of paleoecological study that is intended, and also on the availability of the locations themselves. For example, a classical investigation of the vegetation history of an area over the last few thousand years or so could involve gathering sediments from any of the 304 million natural lakes that cover 4.2 million km^2 of the Earth's surface (Downing et al. 2006). In contrast, a researcher aiming to study pollen and spores preserved in rocks hundreds of millions of years old deep in the geological past is likely to have a more limited set of potential localities from which to gather data.

For classical studies of vegetation history in the Quaternary, a locality that gives a broad integrated picture of the vegetation of the area should be selected. With the results of empirical surface sample studies in mind (Janssen 1973; Webb 1974; Delcourt et al. 1983, 1984; Bradshaw and Webb 1985; Prentice et al. 1987; Jackson 1990), together with guidance from theoretical work (Prentice 1985; Sugita 1993, 1994), the primary sampling site should be large enough to capture pollen and spores from the whole area and not be overwhelmed by pollen and spores produced by local plants (Fægri et al. 1989). Large lakes, such as the 107 ha Rogers Lake in southern Connecticut (Davis 1969), or smaller lakes ($\sim$1 ha) with a simple shape and bathymetry, are the typical localities selected for this purpose. Investigators may choose to study two or more lakes in order to identify patterns that are local versus regional in extent, and so build up a picture of regional vegetation change.

Samples from each of these locations must be taken during a fieldwork campaign. In outcrop situations, samples may be gathered by isolating them from the outcrop using a rock hammer and chisel, and storing them in sample bags. These situations are typical in pre-Quaternary studies where samples are taken from sedimentary successions exposed in cliffs or on the foreshore. Samples gathered in this fashion must be accurately located on a sedimentary log, which provides a graphical representation of the rocks under investigation. Samples gathered from peat may involve a similar process of taking discrete samples, cutting a peat monolith, or may be taken using a coring device. Sampling from a typical lake involves gathering sediment cores from the lake bottom. To do this, researchers construct a transportable raft with scaffolding to support a coring device such as a Livingstone sampler (Wright et al. 1984), and float onto the lake in order to collect sediment cores. A detailed treatment of sampling and coring is provided in Chap. 4 of Fægri et al. (1989). The advantage of this approach is that coring equipment can be taken to remote locations using aircraft or other vehicles, assembled in the field, and then dismantled and returned to the laboratory together with the cores.

Contamination and Reworking

Samples must be cleaned and sealed effectively after sampling to prevent contamination. This contamination can involve transfer of material between different samples and also the inclusion of modern pollen and spores within samples. Contamination with modern pollen can be easy to spot in deep time material; an angiosperm pollen grain in a 350 million-year-old rock sample is unlikely to be anything but contamination, and it can be spotted quickly in countries where there are many exotic plants. However, other forms of contamination can be difficult to detect and has the potential to confound the results of a paleoecological analysis (Scott and Bonnefille 1986). Standard approaches to preventing contamination of more recent geologic samples include working in room free of dust and plant debris. The ideal space is well ventilated but windowless, limiting the introduction of pollen from the atmosphere. Careful sample preparation methods also critical. The exterior layer of sediment must be cleaned or removed before processing to remove modern material that may have come in contact with the sample. Cross-contamination also occurs among samples when sample preparation test tubes and other instruments are not thoroughly cleaned between uses.

Another form of contamination arises from the erosion and redeposition of pollen material. These reworked pollen grains are older than the sedimentary deposits in which they occur. Reworked material can be recognized by the quality of preservation (Mander et al. 2012) and by the identification of morphological types that are biostratigraphic outliers (e.g., Cushing 1964). These outliers should be removed from any paleoecological or evolutionary analysis.

Chronologies

Establishing a well-defined chronology is the basis of paleoecological analysis, as the timing of vegetation changes play an important role in hypothesis testing. In studies of the Late Quaternary, researchers are able to use radiocarbon dating of organic material. Radiocarbon within living organisms is at equilibrium with the atmosphere; the radiocarbon clock begins upon their death. Because the concentration of radiocarbon in the atmosphere varies over millennial timescales, radiocarbon dates must be calibrated in order to provide a chronology in calendar years. These regularly updated calibration curves (IntCal13, SHCal13, and Marine13) are derived from carbon isotope measurements of tree rings, varves, marine foraminifera, uranium-thorium-dated corals and speleothems (Reimer et al. 2013).

Contamination from older or younger sources of carbon is a potential source of error. Best practices include avoiding reliance on small amounts of organic material, which are most easily contaminated by exogenous carbon, and bulk sediments, as algae and aquatic plants may derive carbon from radiocarbon-depleted carbonate rocks (Grimm et al. 2009; Blois et al. 2011). The most accurate age estimates will be derived from the fossil remnants of short-lived herbaceous plants (Grimm et al. 2009).

High-resolution dating can identify periods in time that were characterized by the deposition of large amounts of sediment, and intervals where much time has elapsed but very little sediment has been deposited. Beyond 50,000 year BP, however, the utility of radiocarbon dating diminishes, and it becomes increasingly difficult to gain knowledge of exactly how much time is represented by a given succession of sedimentary rocks. Researchers undertaking paleoecological studies in pre-Quaternary time often report compositional change in vegetation relative to the appearance or disappearance of biostratigraphically sensitive organisms such as ammonites, or relative to some geochemical signature of paleoenvironmental change such as carbon or oxygen isotopes. Absolute dates in such settings can be established using radiometric dating techniques such as U/Pb geochronology (e.g., Schoene et al. 2010).

Laboratory Methods to Release Pollen and Spores from Samples

In order to examine pollen and spores from samples taken from cores, or taken from discrete field samples, it is necessary to macerate the samples in the laboratory, and to prepare the macerated sample residue for viewing under the microscope. These maceration techniques have changed very little over the last 50 years or so, although there is some variation between laboratories, and detailed guidance can be found in standard textbooks (Fægri et al. 1989; Traverse 2007). In this review, we outline standard and widely used methods for releasing pollen and spores from (*i*) soft lake sediments of Quaternary age, and (*ii*) from well-cemented, hard, sedimentary rocks of pre-Quaternary age.

Quaternary lake sediments: The following is adapted from Fægri et al. (1989) we have used these methods with only minor deviations to extract pollen and spores from Quaternary sediments. The choice of mounting medium varies (steps 32–40), and can include silicone oil, glycerine jelly, glycerol or other media. Many Quaternary palynologists prefer the use of viscous liquid media, such as silicone oil, to allow the manual rotation of grains. However, the highest-resolution optical imaging requires solid media with a refraction index of glass (Sivaguru et al. 2018). The following should only be undertaken in a fully equipped laboratory with trained personnel.

(1) Place 0.5–2 cm^3 sample into 50 ml polypropylene tube.
(2) Fill tube with 20 ml 5–10% HCl and place in hot water bath for 20 minutes.
(3) Centrifuge at 3000 rpm for 3 minutes, decant supernatant.
(4) Wash sample with deionized water, centrifuge at 3000 rpm for 3 minutes, decant supernatant.
(5) Fill tube with 10 ml 10% KOH and place in hot water bath for 5 minutes.
(6) Remove samples from water bath and top up with deionized water.
(7) Centrifuge at 3000 rpm for 3 minutes, decant supernatant.
(8) Wash sample with deionized water, centrifuge at 3000 rpm for 3 minutes, decant supernatant.
(9) Fill tube with 20 ml deionized water and sieve sample through 180 μm mesh into second tube, retaining pollen and spores.
(10) Centrifuge at 3000 rpm for 3 minutes, decant supernatant.
(11) Wash sample with deionized water, centrifuge at 3000 rpm for 3 minutes, decant supernatant.
(12) Fill tube with 10 ml 5–10% HCl.
(13) Centrifuge at 3000 rpm for 3 minutes, decant supernatant.
(14) Fill tube with 10 ml 40% HF. Leave for 12 hours in fume hood.
(15) Centrifuge at 3000 rpm for 3 minutes, decant supernatant.
(16) Wash sample with deionized water, centrifuge at 3000 rpm for 3 minutes, decant supernatant.
(17) Fill tube with 20 ml 5–10% HCl and place in hot water bath for 20 minutes.
(18) Centrifuge at 3000 rpm for 3 minutes, decant supernatant.
(19) Wash sample with deionized water, centrifuge at 3000 rpm for 3 minutes, decant supernatant.
(20) Fill tube with 20 ml glacial acetic acid.
(21) Centrifuge at 3000 rpm for 3 minutes, decant supernatant.
(22) Repeat (19) and (20).
(23) In a measuring cylinder, combine 9 parts acetic anhydride with 1 part sulphuric acid (H_2SO_4).
(24) Add 10 ml of this mixture to the sample and place in a hot water bath for 3 minutes.
(25) Remove from water bath and top up with glacial acetic acid.
(26) Centrifuge at 3000 rpm for 3 minutes, decant supernatant.
(27) Fill tube with 20 ml glacial acetic acid.
(28) Centrifuge at 3000 rpm for 3 minutes, decant supernatant.

(29) Wash sample with deionized water, centrifuge at 3000 rpm for 3 minutes, decant supernatant.
(30) Transfer sample to test tube.
(31) Centrifuge sample for 5 minutes at 2000 rpm.
(32) Decant sample.
(33) Add glycerine-water (80% glycerol 20% water) to sample.
(34) Agitate sample using 'lab dancer' test-tube shaker.
(35) Transfer sample to 20 ml test tube.
(36) Centrifuge sample for 5 minutes at 2000 rpm.
(37) Pipette sample residue onto slide.
(38) Add glycerine jelly to residue.
(39) Melt glycerine jelly using hot-plate set at 80°C.
(40) Add cover slip and label; leave to dry overnight.
(41) Clean slide using razor blade, seal edges with clear lacquer, leave to dry overnight.
(42) Inspect pollen and spores using a microscope.

Pre-Quaternary rocks: The following is adapted from Traverse (2007), and one of us (LM) has used this protocol to extract pollen and spores from a variety of Mesozoic and Cenozoic rocks. The choice of mounting medium varies (steps 39–47), and can include silicone oil, glycerine jelly, glycerol or other media. Again, however, the highest-resolution optical imaging requires solid media with a refraction index of glass (Sivaguru et al. 2018). The following should only be undertaken in a fully equipped laboratory with trained personnel.

(1) Crush the rock sample in pestle and mortar.
(2) Dry at 65°C overnight.
(3) Record dry weight and add two *Lycopodium* spore tablets (these tablets contain a known number of spores, and this allows a calculation of pollen concentration).
(4) Transfer sample to polypropylene beaker.
(5) Add ~30 ml 30% cold HCl to sample, wait until reaction has calmed.
(6) Top up beaker with water and leave overnight.
(7) Decant supernatant into waste container.
(8) Add ~30 ml 38% cold HF to sample.
(9) Place sample on orbital shaker, in fume cupboard, and leave (shaking) for 2 hours.
(10) Leave sample to settle overnight.
(11) Decant supernatant into waste container.
(12) Add ~30 ml 30% cold HCl to sample.
(13) Top up beaker with water.
(14) Leave sample to settle overnight.
(15) Decant supernatant into waste container.
(16) Add ~30 ml 38% cold HF to sample.
(17) Place sample on orbital shaker, in fume cupboard, and leave (shaking) for 2 hours.
(18) Leave sample to settle overnight.
(19) Decant supernatant into waste container.
(20) Add ~30 ml 30% cold HCl to sample.
(21) Top up beaker with water.
(22) Leave sample to settle overnight.
(23) Decant supernatant into waste container.
(24) Sieve sample using 250 μm mesh – retain residue.
(25) Sieve sample using 15 μm mesh (10 μm where angiosperms are present).
(26) Transfer residue to small ceramic dish – float in ultrasonic bath to loosen particles.
(27) Sieve sample using 15 μm mesh.
(28) Transfer sample to test tube.
(29) Centrifuge sample for 5 minutes at 2000 rpm.
(30) Decant supernatant.
(31) Fill test tube (two thirds ~10 ml) with $ZnCl_2$ (specific gravity 2.1).
(32) Centrifuge sample for 15 minutes at 800 rpm.
(33) Pipette supernatant sample into 15 μm mesh sieve and wash away any $ZnCl_2$.
(34) Transfer residue to small ceramic dish – float in ultrasonic bath to loosen particles.
(35) Sieve sample using 15 μm mesh.
(36) Transfer sample to test tube.
(37) Centrifuge sample for 5 minutes at 2000 rpm.
(38) Decant sample.
(39) Add glycerine-water (80% glycerol 20% water) to sample.
(40) Agitate sample using 'lab dancer' test-tube shaker.
(41) Transfer sample to 20 ml test tube.
(42) Centrifuge sample for 5 minutes at 2000 rpm.
(43) Pipette sample residue onto slide.
(44) Add glycerine jelly to residue.
(45) Melt glycerine jelly using hot-plate set at 80°C.
(46) Add cover slip and label leave to dry overnight.
(47) Clean slide using razor blade, seal edges with clear lacquer, leave to dry overnight.
(48) Inspect pollen and spores using a microscope.

Data collection and analysis: Routine paleoecological analyses that aim to reconstruct vegetation history involve inspecting pollen and spores using a transmitted light microscope with brightfield illumination. This is a standard workhorse microscopy technique that allows paleoecologists to examine hundreds of specimens quickly. Additionally, the descriptive terminology that is used to describe pollen and spores is mostly composed of terms for morphological features that can be seen using brightfield illumination. This facilitates standardized descriptions, and ensures that researchers use the same terms for the same features. However, there are a variety of other microscopy techniques available to the paleoecologist (Sivaguru et al. 2012). These include the scanning electron microscope, which is useful for examining the surface of pollen and spores at a higher resolution than can be achieved using a light microscope, transmission electron microscopy, which allows high-resolution investigation of internal pollen structures, and fluorescence techniques such as confocal microscopy, which can be useful for viewing some internal structures of pollen and spores, and for reconstructing the three-dimensional shape of specimens. Recent advances in optical super-resolution microscopy additionally allow for unprecedented high-resolution investigations of the internal structure of palynomorphs in three dimensions (Sivaguru et al. 2018).

Researchers will typically count a large number of specimens per sample in order to generate data on the relative abundance of the constituent taxa. The number of specimens counted (known as the "pollen sum") varies. Some workers choose to count a set number of specimens, others count until a certain number of the non-dominant type have been counted, while some count all specimens in an aliquot of residue from a known volume (or weight) of sediment to produce direct measurements of pollen and spore concentrations (see laboratory protocols above; Bennett and Willis 2001). Pollen data can also be reported as pollen influx, which is the pollen concentration normalized by the sediment accumulation rate (grains deposited per year per cm^2 of sediment) (Thompson 1980; Bennett and Willis 2001).

The choice of what is or is not included in the pollen sum and whether percentages, concentrations, or influx is reported is critical, as this will affect paleoecological interpretations and future meta-analyses of these data. Best practice dictates that both the raw numbers and summarization method be documented, alongside the final interpreted pollen spectra. With the proliferation of online databases, such as the North American Pollen Database (http://www.ncdc.noaa.gov/paleo/napd.html), the European Pollen Database (http://www.europeanpollendatabase.net), and the Latin American Pollen Database (http://www.ncdc.noaa.gov/paleo/lapd.html), there are multiple repositories for this critical metadata. To date, the most comprehensive site for data archival is the Neotoma Paleoecology Database (http://www.neotomadb.org), which catalogs both plant and animal fossil abundance data for the Neogene.

Paleoecologists use a variety of techniques to display the results of pollen counts. The use of "pollen diagrams", which display the percentages of taxa through time, date back to von Post (1916), and are still in widespread use today. They are useful for providing an immediate graphical overview of vegetation change through time, and similar diagrams are used to display the accumulation rates of pollen and spores (e.g., Davis 1969). However, over the past 50 years, palynologists have increasingly moved away from subjective interpretations of pollen spectra toward more rigorous, repeatable, quantitative analyses aimed at testing specific ecological or environmental hypotheses. Paleoecologists now use software such as Tilia (Grimm 1991–2001), C2 (Juggins 2007), and Rioja (Juggins 2015) not only to plot pollen diagrams, but also to statistically analyze count data. Multivariate analyses such as detrended correspondence analysis (DCA) and non-metric multidimensional scaling (NMDS) allow the patterns of compositional change in a suite of samples to be visualized in an ordination (Legendre and Birks 2012), while stratigraphically constrained cluster analyses can be used to quantitatively define biozones in a succession (Gordon and Birks 1972; Grimm 1987). These biozones can then be used for the purpose of stratigraphic correlation. Recently developed maximum likelihood methods produce quantitative estimates of the stratigraphic position of isolated samples, and also quantify the level of certainty in those estimates, and this provides a quantitative means of correlating single samples (Punyasena et al. 2012a).

Climatic Interpretation of Pollen Data

Pollen most directly represents the vegetation from which it was derived, with the relative abundance of each taxon influenced not only by the number of mature individuals producing pollen, but the pollen productivity of each individual, and the dispersal mechanism of the species (Fægri et al. 1989). Therefore, there is not a one-to-one representation between pollen abundance and the vegetation from which it was derived (Davis 1963). Despite these ecological and taphonomic filters, however, the vegetation changes captured by pollen are widely accepted as indirectly capturing underlying changes in environmental and climatic conditions.

Reconstruction of paleoclimates from pollen counts can be done in several ways. The simplest approach is through qualitative assessments of analogue biomes. This method requires identification of analogue taxa within a pollen record. Interpretations can be based on the presence/absence of key indicator taxa (e.g., Mayle et al. 2004) or based on the composite pollen spectrum (e.g., von Post 1916).

Most recent applications of an indicator taxon approach derive quantitative climatic values from pollen data by applying knowledge of modern analogues with well-defined biogeographic distributions (Kershaw and Nix 1988; Birks and Birks 1980). A classic example is the study of European holly, ivy, and mistletoe by Iversen (1944). This approach requires substantial knowledge of the autecology of multiple taxa. This data-intensity has become less of a limitation as databases of bioclimatic distributions have accumulated (e.g., Kershaw and Nix 1988) and probabilistic methods have been developed to work with sparse biogeographic data (Bush et al. 2004; Punyasena et al. 2008).

Composite assemblage approaches have similarly moved from qualitative assessments of the similarity of a sample's taxonomic makeup to modern biomes, to quantitative comparisons of pollen spectra between fossil and surface pollen assemblages. Comparison to surface sediments rather than standing vegetation implicitly incorporates many of the taphonomic filters present in fossil pollen assemblages (Wright 1967). Classical ordination methods, such as correspondence analysis (CA) or non-metric multidimensional scaling (NMDS) (Legendre and Birks 2012) allow for fossil assemblages to be matched to their closest modern biome (Williams 2003). Canonical ordinations, such as canonical correspondence analysis (CCA), can be used to directly relate pollen composition to climatic or environmental data (Legendre and Birks 2012). Another widely adopted approach is the use of regression-based transfer functions (Sachs et al. 1977), which use surface samples to relate the abundances of widespread pollen taxa to vegetation abundance. These transfer functions serve as a "correction" of the taphonomic biases of pollen dispersal and preservation, and can then be used to reconstruct paleovegetation. Taken one step further – by mapping these data into climate space and generating pollen-climate response surfaces – regression-based transfer functions can also produce direct estimates of paleoclimate (Bartlein et al. 1986). Other approaches have included the process of "biomization", where pollen spectra are interpreted as plant function types, and these functional types are associated with a specific global biome (Prentice et al. 1996; Prentice and Webb 1998). Biomization allows global comparisons of pollen spectra by focusing on the ecology of the represented plant community rather than its taxonomic composition.

The latest generation of pollen-vegetation models incorporate Bayesian hierarchical models (Paciorek and McLachlan 2009; Garreta et al. 2010; Dawson et al. 2016). Bayesian and related maximum likelihood models are capable of explicitly accounting for and reporting all sources of uncertainty in their environmental reconstructions (Punyasena et al. 2008; Dawson et al. 2016). Although computationally more intensive than traditional regression-based or multivariate approaches, modeling all sources of uncertainty – from the identification of individual grains to climatic distributions of species – should lead to the most transparent and reproducible interpretations of paleoclimate.

Strengths and Weakness of Pollen-Based Paleoecology

Areas of Potential Strength

The consistency of pollen analysis methods over the last century means that pollen data can be aggregated and compared in large meta-analyses. This is particularly true for Late Quaternary samples from well-studied regions like North America and western Europe – the time period and areas for which we have the most existing data. The wide dispersal of pollen and spores, together with the robustness and chemical resistance of the sporopollenin-rich exine (susceptibility to oxidation and the fragility of thin-walled species such as *Populus* notwithstanding), also means that pollen and spores are an ideal fossil group with which to track species in time and space. Additionally, they exhibit a large amount of morphological variation, and although there are a handful of long-standing and difficult classification problems at fine taxonomic levels (Fig. 11.7), pollen and spores in general have sufficient morphological diversity to enable classification.

One of the central strengths of paleoecological data is the provision of information on ecological changes that occur over timescales that are longer than a single human life. This provides information on how life responds to long-term climatic changes that are outside the natural conditions on Earth today. Paleoecological data can also allow scientists to examine past vegetation in pre-industrial times, and these data contribute materially to on-going debate about the biotic effects of anthropogenic climate change. We do not consider the generation of age models in this review, but we note that radiocarbon dates derived from organic matter released from sediment samples can provide accurate dating of sediments, particularly over the last 50,000 years. This provides a means of comparing the temporal patterns of vegetation change from multiple locations and regions, and interpreting these patterns in light of regional and global climate, which allows scientists to examine the rates of long-term ecological processes.

As surface-sample and simulation studies have shown, paleoecological data also have the potential to yield information at multiple spatial scales. At one extreme, paleoecological data gathered from moss polsters and small bogs can allow researchers to reconstruct very local vegetation, while pollen and spores in offshore marine sediments can be used to reconstruct the latitudinal boundaries of entire biomes (Dupont et al. 2007). Such a broad range of spatial scales is amenable to synthesis, and this allows for the reconstruction of vegetation change in relation to climate over entire continents such as North America (Overpeck et al. 1992; Jackson and Williams 2004).

Areas of Potential Weakness

Despite these strengths, there are some inherent weaknesses in paleoecological data. One of these is that the process of classification as it is currently practiced is reliant on the expertise of the analyst. The taxonomic identification of pollen and spores is a highly specialized skill, and the variability in analyst experience means that it is difficult to assess the degree to which different analysts have confidence in the classifications they produce. Similarly, when presented with a pollen diagram that displays fluctuations in the abundance of taxa through time, it is difficult for an independent observer to examine the distribution of any errors made during the classification process, and this undermines, to a degree, the robustness of paleoecological results.

The issue of taxonomic resolution is a substantial challenge for paleoecologists (Mander and Punyasena 2014). The lack of taxonomic precision in the classification of pollen and spores means that most paleoecological data are not reporting patterns at the species level, which is the fundamental unit of evolution. Perhaps more interestingly, the potential taxonomic precision of pollen and spore classifications varies between clades. Some speciose families, such as the Rubiaceae and the Papillionoideae, produce large numbers of pollen morphotypes, some of which are diagnostic to the genus or species level, whereas other large families such as the Poaceae and the Asteraceae contain many species that produce very similar pollen morphotypes. This means that the use of data from pollen and spores to reconstruct plant diversity, even during the Holocene (e.g., Birks et al. 2016) should be interpreted with caution.

Another challenge is the assumption inherent in all paleoenvironmental reconstructions that climatic thresholds for taxa have remained the same over time (i.e., evolution has not occurred). This is very likely not true, as Davis and Shaw (2001) have argued, since a changing climate can potentially serve as a strong selective factor on plant populations. Despite recognizing that autecological preferences may have changed over time, all paleoclimatic interpretations of pollen data largely rely on the assumption that these evolutionary changes are not significant enough to alter our interpretation of pollen data.

There are other problems to paleoenvironmental reconstruction that relate to whether our modern analogues and fossil samples are capturing ecosystems in equilibrium (i.e., climax communities). There is evidence that vegetation and climate have not been in equilibrium for most of the Holocene (Davis 1976). This is further complicated by the recognition that our current landscape has also been significantly altered by anthropogenic influences, so the potential distributions and environmental tolerances of many plant species may be dramatically underestimated. The latter aspect of anthropogenic disturbance is being circumvented by research using historic – rather than current – vegetation records (Dawson et al. 2016). The potential for disequilibrium within post-glacial communities requires that, as is the case for evolution, our interpretations of past environments based on vegetation carry an inherent degree of uncertainty.

Future Developments

In our view, there are two broad areas that are having an immediate impact on the current development of paleoecology. The first is technological advances that are being incorporated into paleoecological research. Microscopy techniques such as super-resolution fluorescence techniques that can gather data on structures as small as ~ 120 nm (e.g., Sivaguru et al. 2012) can increase the amount of morphological information that is recovered from individual specimens, and this has the potential to increase the taxonomic precision of classifications (e.g., Mander and Punyasena 2014). Similar advances in the technology used to analyze data have the potential for similar advances in the speed, accuracy and repeatability of classifications. The use of computational image analysis and machine learning has demonstrable potential for solving classical classification problems in paleoecology, with recent examples including spruce (Punyasena et al. 2012b; Kong et al. 2016) and grass pollen (Mander et al. 2013). However, these approaches require the construction of substantial image libraries and a barrier to progress in this area is the rate of image acquisition. For example, the images in the library of SEM images of grass pollen constructed by Mander et al. (2013) were gathered by a human analyst, and this limited the speed of data acquisition. Nevertheless, the development of scanning electron microscopes that are capable of automated imaged acquisition (e.g., the Zeiss EVO MA microscope) and the development of software to segment and analyze image from pathology scanning microscopes (Tcheng et al. 2016) highlight that technological advances have real potential to speed up the data acquisition process. This is also demonstrated by the integrated system developed by Holt et al. (2011), which uses a robotic microscope to automatically collect images of pollen and software to automatically classify each specimen.

The second is the construction of cyber infrastructure that increases the connectivity between researchers and promotes synthetic data analysis. A key development is the building of databases, including the expansion of the Neotoma database. Unlike the Paleobiology Database (pbdb.org), Neotoma is structured to handle abundance data like pollen counts. However, with the development of advanced programming interfaces (APIs) for both databases, Neotoma and the Paleobiology Database are moving toward data sharing across platforms. As these databases expand their functionality, the hope is that data sharing and public data archival becomes the norm among paleoecologists. These databases provide platforms that allow paleoecologists to synthesize and analyze data from wide geographic areas, and share taxonomic information about groups of organisms. They provide an extended life to paleoecological data, beyond the original paper and analysis. Such databases also increase the temporal scope of paleoecological investigations. Much paleoecological research is still focused on the Quaternary, where pollen and spore data are most abundant and accessible. However, as paleoecological methods are adopted by researchers in deep-time, recent and ancient ecosystems can be compared using comparable data. Such expansion reflects the on-going development of the discipline of paleoecology.

Acknowledgements We are grateful to David Jarzen and the three other reviewers of this chapter for their comments, which helped to clarify the material we have presented here.

References

Bakker, S. V. Z. (1985). Exotic pollen and long-distance wind dispersal at a sub-Antarctic island. *Grana, 24,* 45–54.

Bartlein, P. J., Prentice, I. C., & Webb III, T. (1986). Climatic response surfaces from pollen data for some eastern North American taxa. *Journal of Biogeography, 13,* 35–57

Barreda, V. D., Palazzesi, L., Tellería, M. C., Olivero, E. B., Rine, J. I., & Forest, F. (2015). Early evolution of the angiosperm clade Asteraceae in the Cretaceous of Antarctica. *Proceedings of the National Academy of Sciences, USA, 112,* 10989–10994.

Bennett, K. D., & Willis, K. J. (2001). Pollen. In J. P. Smol, H. J. B. Birks & W. M. Last (Eds.), *Tracking Environmental Change Using Lake Sediments.* Volume 3: *Terrestrial, Algal, and Siliceous Indicators* (pp. 5–32). Dordrecht: Kluwer

Birks, H. H., & Birks, H. J. B. (2000). Future uses of pollen analysis must include plant macrofossils. *Journal of Biography, 27,* 31–35.

Birks, H. J. B., & Birks, H. H. (1980). *Quaternary Paleoecology.* London: Arnold.

Birks, H. J. B., Felde, V. A., & Seddon, A. W. R. (2016). Biodiversity trends within the Holocene. *The Holocene, 26,* 994–1001.

Blois, J. L., Williams, J. W. J., Grimm, E. C., Jackson, S. T., & Graham, R. W. (2011). A methodological framework for assessing and reducing temporal uncertainty in paleovegetation mapping from late-Quaternary pollen records. *Quaternary Science Reviews, 30,* 1926–1939.

Bradshaw, R. H. W., & Webb III, T. (1985). Relationships between contemporary pollen and vegetation data from Wisconsin and Michigan. *Ecology, 66,* 721–737

Burnham, R. J. (1989). Relationships between standing vegetation and leaf litter in a paratropical forest: implications for paleobotany. *Review of Palaeobotany and Palynology, 58,* 5–32.

Burnham, R. J., Wing, S. L., & Parker, G. G. (1992). The reflection of deciduous forest communities in leaf litter: implications for autochthonous litter assemblages from the fossil record. *Paleobiology, 18,* 30–49.

Bush, M. B. (1995). Neotropical plant reproductive strategies and fossil pollen representation. *The American Naturalist, 145,* 594–609.

Bush, M. B., & Rivera, R. (1998). Pollen dispersal and representation in a neotropical rain forest. *Global Ecology and Biogeography Letters, 7,* 379–392.

Bush, M. B., Silman, M. R., & Urrego, D. H. (2004). 48,000 years of climate and forest change in a biodiversity hotspot. *Science, 303,* 827–829.

Chaloner, W. G. (1970). The evolution of miospore polarity. *Geoscience and Man, 1,* 47–57.

Chaloner, W. G. (2013). Three palynological puzzles. *International Journal of Plant Sciences, 174,* 602–607.

Cowling, S. A., Cox, P. M., Jones, C. D., Maslin, M. A., Peros, M., & Spall, S. A. (2008). Simulated glacial and interglacial vegetation across Africa: implications for species phylogenies and trans-Africa migration of plants and animals. *Global Change Biology, 14,* 827–840.

Cushing, E. J. (1964). Redeposited pollen in Late Wisconsin pollen spectra from east-central Minnesota. *American Journal of Science, 262,* 1075–1088.

Davis, M. B. (1963). On the theory of pollen analysis. *American Journal of Science, 261,* 897–912.

Davis, M. B. (1969). Climatic changes in Southern Connecticut recorded by pollen deposition at Rogers Lake. *Ecology, 50,* 409–422.

Davis, M. B. (1976). Pleistocene biogeography of temperate deciduous forests. *Geoscience and Man, 13,* 13–26.

Davis, M. B. (2000). Palynology after Y2K–understanding the source area of pollen in sediments. *Annual Review of Earth and Planetary Sciences, 28,* 1–18.

Davis, M. B., & Shaw, R. G. (2001). Range shifts and adaptive responses to Quaternary climate change. *Science, 292,* 673–679.

Dawson, A., Paciorek, C. J., McLachlan, J. S., Goring, S., Williams, J. W., & Jackson, S. T. (2016). Quantifying pollen-vegetation relationships to reconstruct ancient forests using 19th-century forest composition and pollen data. *Quaternary Science Reviews, 137,* 156–175.

Delcourt, P. A., Delcourt, H. R., & Davidson, J. L. (1983). Modern pollen-vegetation relationships in southeastern United States. *Review of Palaeobotany and Palynology, 39,* 1–45.

Delcourt, P. A., Delcourt, H. R., & Webb III, T. (1984). Atlas of Mapped Distributions of Dominance and Modern Pollen Percentages for Important Tree Taxa of Eastern North America. *American Association of Stratigraphic Palynologists Contribution Series,* No. 14.

Downing, J. A., Prairie, Y. T., Cole, J. J., Duarte, C. M., Tranvik, L. J., Striegl, R. G., et al. (2006). The global abundance and size distribution of lakes, ponds, and impoundments. *Limnology and Oceanography, 51,* 2388–2397.

Dunwiddie, P. W. (1987). Macrofossil and pollen representation of coniferous trees in modern sediments from Washinton. *Ecology, 68,* 1–11.

Dupont, L., Behling, H., Jahns, S., Marret, F., & Kim, J.-H. (2007). Variability in glacial and Holocene marine pollen records offshore from southern Africa. *Vegetation History and Archaeobotany, 16,* 87–100.

Erdtman, G. (1943). *An Introduction to Pollen Analysis.* Waltham: Chronica Botanica.

Erdtman, G. (1952). *Pollen Morphology and Plant Taxonomy.* New York and London: Hafner.

Fægri, K. (1966). Some problems of representativity in pollen analysis. *Palaeobotanist, 15,* 135–140.

Fægri, K., Kaland, P. E., & Krzywinski, K. (1989). *Textbook of pollen analysis by Knut Fægri & Johs Iversen.* Chichester: Wiley.

Fagerlind, F. (1952). The real signification of pollen diagrams. *Botanisker Notiser, 105,* 185–224.

Garreta, V., Miller, P. A., Guiot, J., Hély, C., Brewer, S., Sykes, M. T., et al. (2010). A method for climate and vegetation reconstruction through the inversion of a dynamic vegetation model. *Climate Dynamics, 35,* 371–389.

Gastaldo, R. A. (2001). Plant Taphonomy. In D. E. G. Briggs & P. R. Crowther (Eds.), *Paleobiology II* (pp. 314–317). Oxford: Blackwell Science.

Gordon, A. D., & Birks, H. J. B. (1972). Numerical methods in Quaternary palaeoecology. I. Zonation of pollen diagrams. *New Phytologist, 71,* 961–979

Gosling, W. D., Hanselman, J. A., Knox, C., Valencia, B. G., & Bush, M. B. (2009). Long-term drivers of change in *Polylepis* woodland distribution in the central Andes. *Journal of Vegetation Science, 20,* 1041–1052.

Grimm, E. C. (1987). CONISS: A FORTRAN 77 program from stratigraphically constrained cluster analysis by the method of incremental sum of squares. *Computers and Geosciences, 13,* 13–35.

Grimm, E. C. (1991–2001). *Tillia, TiliaGraph and TGView Software.* Illinois State Museum, Springfield, Illinois, USA.

Grimm, E. C., Maher, L. J., & Nelson, D. M. (2009). The magnitude of error in conventional bulk-sediment radiocarbon dates from central North America. *Quaternary Research, 72,* 301–308.

Holt, K., Allen, G., Hodgson, R., Marsland, S., & Flenley, J. (2011). Progress towards an automated trainable pollen location and classifier system for use in the palynology laboratory. *Review of Palaeobotany Palynology, 167,* 175–183.

Huntley, B., & Birks, H. J. B. (1983). *An Atlas of Past and Present Pollen Maps for Europe 0–13, 000 Years Ago.* Cambridge: Cambridge University Press.

Iversen, J. (1944). *Viscum, Hedera* and *Ilex* as climatic indicators. *Geologiska i Stockholm Förhandlingar, 66,* 463–483.

Jackson, S. T. (1990). Pollen source area and representation in small lakes of the northeastern United States. *Review of Palaeobotany and Palynology, 63,* 53–76.

Jackson, S. T., & Williams, J. W. (2004). Modern analogs in Quarternary paleoecology: here today, gone yesterday, gone tomorrow? *Annual Review of Earth and Planetary Sciences, 32,* 495–537.

Jackson, S. T., & Booth, R. T. (2007). Validation of pollen studies. In S. A. Elias (Ed.), *Encyclopaedia of Quaternary Sciences* (pp. 725–732). Amsterdam: Elsevier Scientific Publishing.

Jacobson, Jr., G. L., & Bradshaw, R. H. W. (1981). The selection of sites for paleovegetational studies. *Quaternary Research, 16,* 80–96.

Janssen, C. R. (1973). Local and regional pollen deposition. In H. J. B. Birks & R. G. West (Eds.), *Quaternary Plant Ecology* (pp. 31–42). Oxford: Blackwell.

Jarzen, D. M., & Nichols, D. J. (1996). Pollen. In J. Jansonius & D. C. McGregor (Eds.), *Palynology: Principles and Applications (American Association of Stratigraphic Palynologists Foundation), vol. 1.* (pp. 261–292).

Jarzen, D. M., & Jarzen, S. A. (2006). Collecting pollen and spore samples from herbaria. *Palynology, 30,* 111–119.

Juggins, S. (2007). *C2 Version 1.5: Software for Ecological and Palaeoecological Data Analysis and Visualisation.* Newcastle upon Tyne: University of Newcastle.

Juggins, S. (2015) rioja: Analysis of Quaternary Science Data, R package version (0.9–9). (http://cran.r-project.org/package=rioja).

Katifori, E., Alben, S., Cerda, E., Nelson, D. R., & Dumais, J. (2010). Foldable structures and the natural design of pollen grains. *Proceedings of the National Academy of Sciences, USA, 107,* 7635–7639.

Kershaw, A. P., & Nix, H. A. (1988). Quantitative palaeoclimatic estimates from pollen data using bioclimatic profiles of extant taxa. *Journal of Biogeography, 15,* 589–602.

Kershaw, A. P., & Strickland, K. M. (1990). A 10 year pollen trapping record from rainforest in northeastern Queensland, Australia. *Review of Palaeobotany and Palynology, 64,* 281–288.

Kong, S., Punyasena, S. W., & Fowlkes, C. C. (2016). Spatially aware dictionary learning and coding for fossil pollen identification. Computer Vision for Microscopy Image Analysis (CVMI) Workshop, In 29th IEEE Conference on Computer Vision and Pattern Recognition (CVPR).

Legendre, P. L., & Birks, H. J. B. (2012). From classical to canonical ordination. In H. J. B. Birks, A. F. Lotter, S. Juggins & J. P. Smol (Eds.), *Tracking Environmental Change Using Lake Sediments, Volume 5: Data Handling and Numerical Techniques* (pp. 201–248). Dordrecht: Springer

Mander, L., & Punyasena, S. W. (2014). On the taxonomic resolution of pollen and spore records of Earth's vegetation. *International Journal of Plant Sciences, 175,* 931–945.

Mander, L., Wesseln, C. J., McElwain, J. C., & Punyasena, S. W. (2012). Tracking taphonomic regimes using chemical and mechanical damage of pollen and spores: an example from the Triassic-Jurassic mass extinction. *PLoS ONE, 7,* e49153.

Mander, L., Li, M., Mio, M., Fowlkes, C. C., & Punyasena, S. W. (2013). Classification of grass pollen through the quantitative analysis of surface ornamentation and texture. *Proceedings of the Royal Society B, 280,* 20131905.

Mayle, F. E., Beerling, D. J., Gosling, W. D., & Bush, M. B. (2004). Responses of Amazonian ecosystems to climatic and atmospheric carbon dioxide changes since the last glacial maximum. *Philosophical Transactions of the Royal Society B, 359,* 499–514.

Overpeck, J. T., Webb, R. S., & Webb, T. III. (1992). Mapping eastern North American vegetation change of the past 18 ka: no-analogues and the future. *Geology, 20,* 1071–1074

Paciorek, C. J., & McLachlan, J. S. (2009). Mapping ancient forests: Bayesian inference for spatio-temporal trends in forest composition using the fossil pollen proxy record. *Journal of the American Statistical Association, 104,* 608–622.

Pedersen, M. W., Ruter, A., Schweger, C., Friebe, H., Staff, R. A., Kjeldsen, K. K., et al. (2016). Postglacial viability and colonization in North America's ice-free corridor. *Nature, 537,* 45–49.

Playford, G., & Dettmann, M. E. (1996). *Spores.* In J. Jansonius & D. C. McGregor (Eds.), *Palynology: Principles and Applications (American Association of Stratigraphic Palynologists Foundation), vol. 1.* (pp. 227–260). Salt Lake City: Publishers Press

Poort, R. J., Visscher, H., & Dilcher, D. L. (1996). Zoidogamy in fossil gymnosperms: the centenary of a concept, with special reference to prepollen of late Paleozoic conifers. *Proceedings of the National Academy of Sciences, USA, 93,* 11713–11717.

Potonie, R. (1934). Zur mikrobotanik der Kohlen und ihrer Verwandten. I. Zur Morphologie der fossilen pollen und sporen. *Arbeiten des Instituts fur Paläobotanik und Petrographie der Brennsteine,* 45–125.

Prentice, I. C. (1985). Pollen representation, source area, and basin size: toward a unified theory of pollen analysis. *Quaternary Research, 23,* 76–86.

Prentice, I. C., & Webb III, T. (1986). Pollen percentages, tree abundances and the Fagerlind effect. *Journal of Quaternary Science, 1,* 35–43.

Prentice, I. C., & Webb III, T. (1998). BIOME 6000: reconstructing global mid-Holocene vegetation patterns from palaeoecological records. *Journal of Biogeography, 25,* 997–1005.

Prentice, I. C., Berglund, B. E., & Olsson, T. (1987). Quantitative forest-composition sensing characteristics of pollen samples from Swedish lakes. *Boreas, 16,* 43–54.

Prentice, C., Guiot, J., Huntley, B., Jolly, D., & Cheddadi, R. (1996). Reconstructing biomes from palaeoecological data: a general method and its application to European pollen data at 0 and 6 ka. *Climate Dynamics, 12,* 185–194.

Punt, W. (1986). Functional factors influencing pollen form. In S. Blackmore & I. K. Ferguson (Eds.), *Pollen and spores: Form and function*. London, UK: Academic Press.

Punt, W., Blackmore, S., Hoen, P. P., & Stafford, P. (2003). Preface to the northwest European pollen flora VIII. *Review of Palaeobotany and Palynology, 123,* vii–ix.

Punt, W., Hoen, P. P., Blackmore, S., Nilsson, S., & Le Thomas, A. (2007). Glossary of pollen and spore terminology. *Review of Palaeobotany and Palynology, 143,* 1–81.

Punyasena, S. W., Mayle, F. E., & McElwain, J. C. (2008). Quantitative estimates of glacial and Holocene temperature and precipitation change in lowland Amazonian Bolivia. *Geology, 36,* 667–670.

Punyasena, S. W., Jaramillo, C., de la Parra, F., & Du, Y. (2012a). Probabilistic correlation of single stratigraphic samples: a generalized approach for biostratigraphic data. *AAPG Bulletin, 96,* 235–244.

Punyasena, S. W., Tcheng, D. K., Wesseln, C., & Mueller, P. G. (2012b). Classifying black and white spruce using layered machine learning. *New Phytologist, 196,* 937–944.

Raven, P. H., Evert, R. F., & Eichhorn, S. E. (2005). *Biology of Plants*. New York: W. H. Freeman and Company Publishers.

Reimer, P. J., Bard, E., Bayliss, A., Beck, J. W., Blackwell, P. G., Ramsey, C. B., et al. (2013). IntCal13 and Marine13 radiocarbon age calibration curves 0–50,000 years cal BP. *Radiocarbon, 55,* 1869–1887.

Rothwell, G. W. (1972). Evidence of pollen tubes in Paleozoic pteridosperms. *Science, 175,* 772–774.

Roubik, D. W., & Moreno, J. E. (1991). *Pollen and spores of Barro Colorado Island*. St. Louis, Missouri: Missouri Botanical Garden.

Rudall, P. J., & Bateman, R. M. (2007). Developmental bases for key innovations in the seed-plant microgametophyte. *Trends in Plant Science, 12,* 318–326.

Sachs, H. M., Webb III, T., & Clark, D. R. (1977). Paleoecological transfer functions. *Annual Review of Earth and Planetary Sciences, 5,* 159–178

Schoene, B., Geux, J., Bartolini, A., Schaltegger, U., & Blackburn, T. J. (2010). Correlating the end-Triassic mass extinction and flood basalt volcanism at the 100 ka level. *Geology, 38,* 387–390.

Scott, L., & Bonnefille, R. (1986). A search for pollen from the hominid deposits of Kromdraai, Sterkfontein and Swartkrans: some problems and preliminary results. *South African Journal of Science, 82,* 380–382.

Scott, L., Marais, E., & Brook, G. A. (2004). Fossil hyrax dung and evidence of Late Pleistocene and Holocene vegetation types in the Namib Desert. *Journal of Quaternary Science, 19,* 829–832.

Seddon, A. W. R. (2012). Paleoecology. In D. Gibson (Ed.), *Oxford bibliographies online: Ecology*. New York: Oxford University Press.

Sivaguru, M., Mander, L., Fried, G., & Punyasena, S. W. (2012). Capturing the surface texture and shape of pollen: a comparison of microscopy techniques. *PLoS ONE, 7,* e39129.a.

Sivaguru, M., Urban, M. A., Fried, G., Wesseln, C. J., Mander, L., & Punyasena, S. W. (2018). Comparative performance of Airyscan and structured illumination superresolution microscopy in the study of the surface texture and 3D shape of pollen. *Microscopy Research and Technique, 81,* 101–114.

Strother, P. K., Traverse, A., & Vecoli, M. (2015). Cryptospores from the Hanadir Shale Member of the Qasim Formation, Ordovician (Darrriwillian) of Saudi Arabia: taxonomy and systematics. *Review of Palaeobotany and Palaeoecology, 212,* 97–110.

Sugita, S. (1993). A model of pollen source area for an entire lake surface. *Quaternary Research, 39,* 239–244.

Sugita, S. (1994). Pollen representation of vegetation in Quaternary sediments: theory and method in patchy vegetation. *Journal of Ecology, 82,* 881–897.

Tauber, H. (1967). Investigations of the mode of transfer of pollen in forested areas. *Review of Palaeobotany and Palynology, 3,* 277–286.

Taylor, T. N., Taylor, E. L., & Krings, M. (2009). *Paleobotany: The biology and evolution of fossil plants* (2nd ed.). Amsterdam: Elsevier.

Tcheng, D. K., Nayak, A. K., Fowlkes, C. C., & Punyasena, S. W. (2016). A visual recognition software application for binary classification and its application to spruce pollen identification. *PLoS ONE, 11,* e0148879.

Thompson, R. (1980). Use of the word "influx" in paleolimnological studies. *Quaternary Research, 14,* 269–270.

Traverse, A. (2007). *Paleopalynology* (2nd ed.). The Netherlands: Springer.

von Post, L. (1967[1916]). Forest tree pollen in Swedish Peat Bog Deposits. *Pollen et Spores, 9,* 378–401. Translation by M. B. Davis & K. Fægri of: von Post, L. 1916. Om skogsträdpollen i sydsvenska torfmosselagerföljder (foredragsreferat). *Geolgiska Föereningen i Stockholm. Föerhandlingar, 38,* 384–390.

Webb III, T. (1974). Corresponding patterns of pollen and vegetation in Lower Michigan: a comparison of quantitative data. *Ecology, 55,* 17–28

Willis, K. J., & McElwain, J. C. (2002). *The Evolution of Plants*. Oxford: Oxford University Press.

Wodehouse, R. P. (1935). *Pollen grains*. New York: McGraw-Hill.

Wright, H. E. (1967). The use of surface samples in Quaternary pollen analysis. *Review of palaeobotany and palynology, 2,* 321–330.

Williams, J. W. (2003). Variations in tree cover in North America since the last glacial maximum. *Global and Planetary Change, 35,* 1–23.

Wright, Jr., H. E., Mann, D. H., & Glaser, P. H. (1984). Piston corers for peat and lake sediments. *Ecology, 65,* 657–659

Chapter 12
Phytoliths in Paleoecology: Analytical Considerations, Current Use, and Future Directions

Caroline A. E. Strömberg, Regan E. Dunn, Camilla Crifò, and Elisha B. Harris

Abstract Phytoliths, microscopic plant silica bodies, are often preserved in modern and fossil soils and sediment, as well as in archaeological contexts. They record unique characteristics of past vegetation and, unlike palynomorphs and macrofossils, are commonly found in direct association with fossil vertebrates, providing vital paleoecological data. Within pre-Quaternary paleoecology, phytolith analysis has so far elucidated Cretaceous-Cenozoic grass evolution, vegetation change including the spread of grasslands, plant-animal co-evolution, and diets of extinct animals. Because deep-time phytolith analysis is a young field, many methodological aspects are not standardized, preventing comparisons among studies and calibration using published modern analogs. On the other hand, the plethora of novel approaches in recent literature points to the potential of phytolith-based paleoecology. Here, we review the present state of phytolith analysis, focusing on pre-Quaternary applications. We discuss the nature and known biases of the phytolith record, current methods, and examples of when phytoliths contributed substantially to knowledge of past ecosystems, and highlight important future avenues of research.

Keywords Phytolith • Paleoecology

Introduction

Plant silica bodies (phytoliths) form inside living plants and are deposited in the soil once the plant dies and decays, where they then form a record of past plant communities (e.g., Piperno 1988). Plant silica bodies were discovered in the 1830s (Struve 1835) and in the last 40 years, phytolith analysis has become a widely used tool in archaeobotany and Quaternary paleoecology (see Rovner 1971; Piperno 1988, 2006). Within these fields, phytoliths are used to, among other things, track the domestication of crop plants (e.g., maize, wheat, rice; Rosen and Weiner 1994; Zhao et al. 1998; Iriarte 2003; Pearsall et al. 2003; Piperno et al. 2009; Perrier et al. 2011), and reconstruct past plant communities (e.g., Kealhofer and Penny 1998; Runge 2001; McMichael et al. 2012a; Piperno et al. 2015), diets (Ciochon et al. 1990; Fox et al. 1996; Henry et al. 2012), and climates (e.g., Bremond et al. 2005b; Lu et al. 2007; Issaharou-Matchi et al. 2016). Although pre-Quaternary phytoliths in sediment or associated with leaf fossils were reported in the literature prior to 2000 (e.g., Swineford and Franks 1959; Baker 1960; Jones 1964; Dumitrička 1973; Mazzoni 1979; Smiley and Huggins 1981), and a few studies used them for taxonomic or even paleoecological purposes (e.g., Thomasson 1978, 1984; Dugas and Retallack 1993; Pinilla and Bustillo 1997), they were not broadly recognized as an important tool for deep-time paleoecological or taxonomic research. However, in the last 15 years, the number of publications that employ phytolith analysis for answering pre-Quaternary paleoecological or evolutionary questions has surged (e.g., Zucol and

C. A. E. Strömberg (✉) · C. Crifò · E. B. Harris
Department of Biology, University of Washington, Seattle, WA 98195, USA
e-mail: caestrom@u.washington.edu

Burke Museum of Natural History & Culture, University of Washington, Seattle, WA 98195, USA

C. Crifò
e-mail: crifoc@uw.edu

E. B. Harris
e-mail: ebhvilhena@gmail.com

R. E. Dunn
Department of Botany, University of Wyoming, Laramie, WY 82071, USA
e-mail: rdunn@fieldmuseum.org

Present Address:
R. E. Dunn
Integrative Research Center, Field Museum of Natural History, Chicago, IL 60605, USA

D. A. Croft et al. (eds.), *Methods in Paleoecology: Reconstructing Cenozoic Terrestrial Environments and Ecological Communities*, Vertebrate Paleobiology and Paleoanthropology, https://doi.org/10.1007/978-3-319-94265-0_12

Brea 2000; Thorn 2001; Strömberg 2002, 2004; Thorn 2004a; Prasad et al. 2005; Strömberg et al. 2007a; Raigemborn et al. 2009; WoldeGabriel et al. 2009; Sánchez et al. 2010; Zucol et al. 2010; Prasad et al. 2011; Strömberg and McInerney 2011; Miller et al. 2012; Strömberg et al. 2013a; Dillhoff et al. 2014; Chen et al. 2015), pointing to increased reliance of this largely untapped line of paleobotanical evidence alongside more traditional methods (palynology, macrofloras). In particular, the occurrence of phytoliths in oxidized sediments that preserve vertebrate faunas but lack pollen or fossil leaves (Fig. 12.1) has made phytoliths a source of information with high relevance for vertebrate paleoecology.

This growing interest in phytolith-based paleoecological inquiry benefits from the rapidly expanding body of work concerning plant silica behavior in living plants and in the environment that provides the necessary background for conducting meaningful phytolith analysis both in submodern and geologic strata. Among the topics under study are silica production and phytolith morphologies within and among plants (e.g., Hodson et al. 2005; Neumann et al. 2017), transport and preservation of phytoliths once they are released from the plant tissues (e.g., Fraysse et al. 2009; Cabanes et al. 2011), the elemental and isotopic composition of phytoliths (Smith and Anderson 2001; Elbaum et al. 2009; Alexandre et al. 2015), and phytolith function(s) (e.g., Sangster et al. 2001; Massey et al. 2006; Strömberg et al. 2016). In addition, numerous publications document progress in methods of extracting phytoliths from sediments (e.g., Parr 2002; Katz et al. 2010), analyzing phytolith assemblages using statistical methods or ordination techniques (e.g., Barboni et al. 2007; Novello et al. 2012; Blinnikov et al. 2013), compensating for biases in the phytolith record (e.g., Blinnikov et al. 2002; Hyland et al. 2013b), and comparing paleoenvironmental inference from phytoliths to other types of data (e.g., Bremond et al. 2005a; Chen et al. 2015). Perhaps the clearest evidence for the potential of plant silica as a paleoecological tool is the recent emergence of numerous, innovative ways in which phytolith assemblage composition, phytolith morphometrics, or phytolith isotopes are used to infer not just plant community composition and vegetation type, but plant physiology (Kelly et al. 1991; Lu et al. 2000; McInerney et al. 2011), habitat openness (Dunn et al. 2015a, 2015b), elevation (An et al. 2015), aridity (Bremond et al. 2005b; Issaharou-Matchi et al. 2016), as well as temperature and other climate variables (Lu et al. 2006; Rossuow 2009; Alexandre et al. 2011b, 2012).

This dramatic increase in applications in phytolith research is not without growing pains. Work is ongoing to provide precise, reliable, and widely applicable analytical protocols to allow for more robust and comparable results (e.g., Madella et al. 1998; Strömberg 2007, 2009a). In particular, a recent effort has been launched to address one of the main challenges in phytolith science, namely the lack of standardization in classification, description, and terminology of phytolith morphotypes, which has long complicated inter-study comparison and communication (ICPN Working Group 2005).

In this paper we will attempt to survey the promising field of pre-Quaternary phytolith research as well as provide a primer for researchers interested in understanding phytoliths as a paleoecological tool. Specifically, we will: (1) review what is currently known about the production of phytoliths in plants and phytolith taphonomy (transport, preservation) and how these factors influence what inferences can be made from the phytolith fossil record; (2) describe the main application of phytoliths in deep-time paleoecology; (3) provide an overview of methods currently in use; and (4) discuss outstanding questions and future directions. We want to stress that we don't intend to be comprehensive with regards to the many methods and approaches that are used in phytolith research in general, but will focus on deep-time (pre-Quaternary) applications.

Definitions and Terminology

Commonly used or key terms are defined in the text but also in Appendix 12.1. Names for phytolith morphotypes are given using the conventions of the International Committee for Phytolith Nomenclature 1.0 (ICPN Working Group 2005) whenever such a name has been given in the literature; however, we do not make up new ICPN 1.0 names. In some cases it is helpful to use the name from the original publication in addition to the ICPN 1.0 name.

What are Phytoliths?

Monosilicic acid (H_4O_4Si) is absorbed by plant roots, transported through the plant body and deposited as solid hydrated silica polymers (SiO_2 * nH_2O; opal-A) inside and around cells when they are still alive (Kumar et al. 2017). This uptake and transport occur both as passive and active

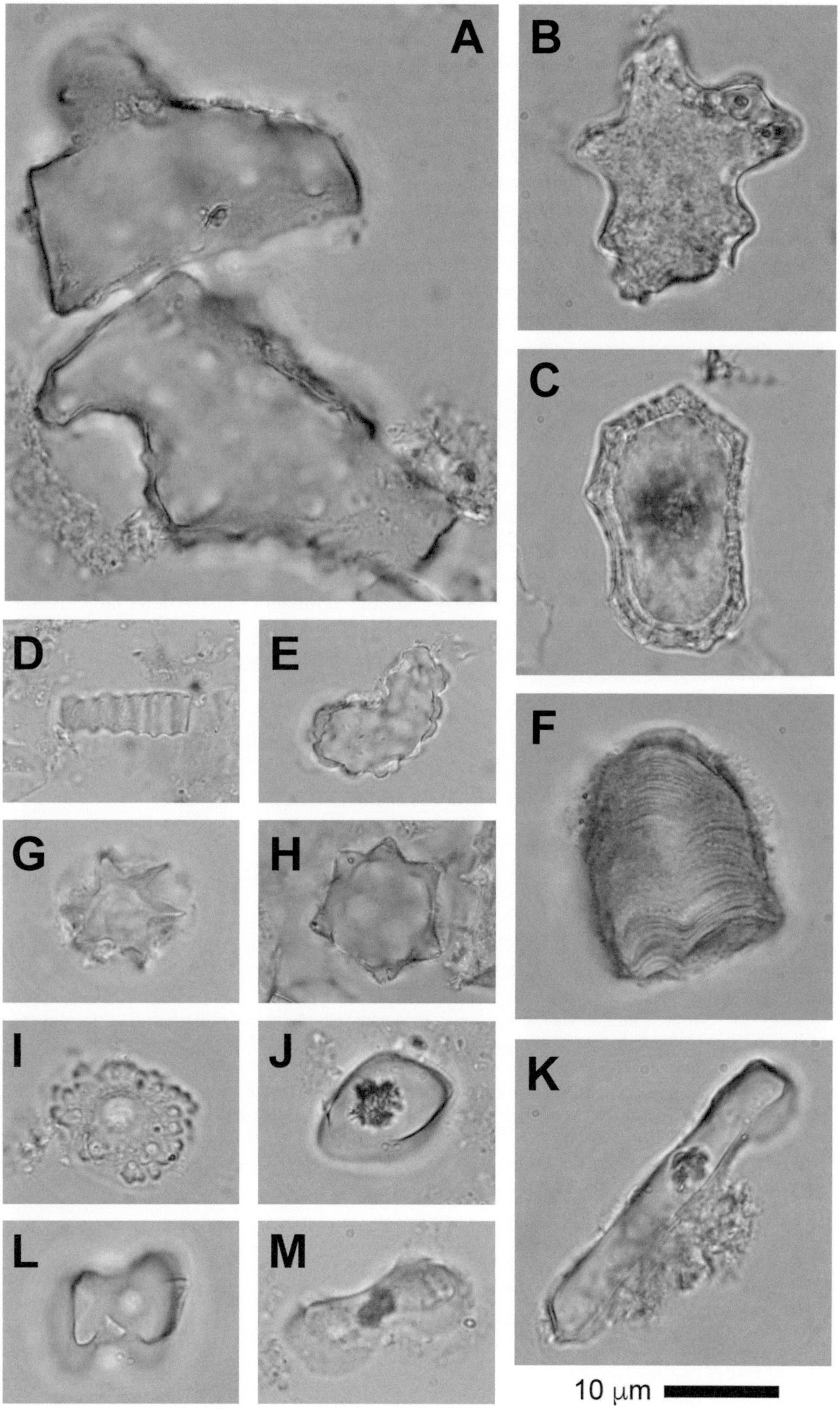

Fig. 12.1 Examples of phytoliths from the Sarmiento Formation. A–G from "forest indicator" taxa; J–M from grasses. A, Oblong tuberculate sclereid. B, Epidermal jigsaw. C, Epidermal polygonal. D, Cylindric sulcate tracheary element. E, Nodular body. F, Vesicular infilling. G, Druse of *Costus*/*Canna* (Zingiberales). H, Globular echinate from palm. I, Conical sedge plate (wetland indicator). J, Rondel GSSCP with truncated top (Poaceae, general). K, Crenate (trapeziform polylobate) GSSCP (open-habitat, Pooideae). L–M, Simple lobate bilobate GSSCPs (open-habitat, PACMAD)

(energy-demanding) processes, depending on the plant and the environmental availability of silica (Ma 2003; Ma and Yamaji 2008). Thus, transpiration appears to be an important factor in pulling water with dissolved silicic acid into the leaves of plants, but cannot fully explain patterns, rates, and timing of silica deposition (Motomura et al. 2002; Faisal et al. 2012; Kumar et al. 2017). Recent work suggests that specific biomolecules may catalyze precipitation of silicic acid as tiny granules of opal-A (Lins et al. 2002; Hodson 2016; Kumar et al. 2017). The silica bodies (phytoliths) that result from aggregation of these granules are microscopic, ranging from infillings of cell walls to cell lumens, partial or full infillings of individual cells, to entire silica skeletons encompassing tens or hundreds of cells (e.g., Piperno 2006; Hodson 2016). In some cases, a silica layer can even form outside of the plant body, between the epidermis and the cuticle (*Equisetum*, Kaufman et al. 1971; Currie and Perry 2007; Law and Exley 2011), and masses of silica are known to be deposited inside the hollow culm cavity of certain bamboos (so-called tabashir; Jones et al. 1966). Plants also deposit minerals in their tissues that are not silica-based, namely calcium oxalates, which are studied within botany and archaebotany (Jones and Bryant 1992; Prychid and Rudall 1999; He et al. 2012). However, because these minerals are typically dissolved along with calcium carbonate cements during silica phytolith extraction, they are rarely visible in fossil samples except as blueish inclusions (in the form of druses, styloids, or raphides) in silicified mesophyll cells or remnants of dissolved lamina in silica cystoliths (Strömberg 2003); for this reason, we will not discuss them further.

A great many functions have been proposed for silica and silicon inside the plant body. Many of these functions relate to silicon as a biologically active element, and include stimulation of photosynthesis, mitigation of stresses, both physical (e.g., radiation damage, water deficit, wind, cold temperatures) and chemical (e.g., nutrient deficiency and excess, low or high soil pH, and metal toxicity), as well as pathogen resistance (Ma et al. 2001; Sangster et al. 2001; Fauteux et al. 2005; Buján 2013). In addition, there is evidence that silica in solid form (as phytoliths) acts as structural support and/or mechanical defense against insect and small mammalian herbivores in extant plants (e.g., Raven 1983; Cooke and Leishman 2011; Hartley and DeGabriel 2016; Strömberg et al. 2016). In contrast, a recent study found little evidence for the longstanding notion that phytoliths in modern grasses deter large mammalian herbivores, and challenged the idea that grass phytoliths evolved for this purpose (Strömberg et al. 2016).

Table 12.1 Characteristics of paleobotanical evidence used for paleoecological reconstruction

		Palynofloras	Phytolith assemblages	Leaf assemblages
Production bias	*Over-represented taxa*	Wind-pollinated taxa (e.g., conifers, grasses), some aquatic taxa	Grasses, some animal-pollinated tropical taxa	Deciduous trees/shrubs
	Under-represented taxa	Animal-pollinated taxa	Some families of dicotyledons, monocotyledons, and non-angiosperm seed plants (e.g., conifers)	Herbs, evergreen trees/shrubs
Preservation bias	*Over-represented taxa*	Taxa with thick-walled pollen/spores (e.g., conifers, elms, ferns)	Taxa with well-silicified phytoliths (e.g., grasses, palms)	Sclerophyllous leaves with thick cuticle
	Under-represented taxa	Taxa with thin-walled pollen/spores (e.g., grasses, aquatic taxa)	Taxa with thin phytoliths (e.g., conifers, some dicotyledons)	Thin leaves with thin cuticle (e.g., aquatic plants, herbs)
	Over-represented environments	Lakes, swamps, lagoons (low pH, negative Eh)	Soils in e.g., floodplains, in particular ash-rich soils, diatomaceous lakes (intermediate pH, positive Eh)	Lakes, swamps, lagoons (low pH, negative Eh); levees (rapid burial)
	Under-represented environments	Soils and deposits on e.g., floodplains, arid environments (high pH, positive Eh)	Swamps, alkaline sediments (high pH)	Soils and deposits on e.g., floodplains, arid environments (high pH, positive Eh)
Taxonomic resolution	*Well resolved*	Woody taxa (trees and shrubs)	Grasses	Many plants (e.g., many angiosperms, conifers, ferns)
	Poorly resolved	Grasses	Woody taxa (trees and shrubs), dicotyledonous herbs	Grasses and (many) other parallel-veined monocotyledons
Temporal resolution		1-many years	$1–10^3$ years	Day(s) – years
Typical spatial resolution		Regional	Local – regional	Local

Characteristics of Phytolith Production

Much as for palynomorphs and leaf fossils, the characteristics of phytoliths as a source of paleoecological information depend in large part on their production patterns among and between plants and the differential preservation potential for different morphotypes, which to some extent follows taxonomic lines (Table 12.1, and below) (e.g., Pearsall 2000; Strömberg 2004; Piperno 2006). Most major groups of land plants contain taxa that deposit substantial numbers of phytoliths in their tissues, but far from all plants are active phytolith accumulators (Fig. 12.2) (Ma and Takahashi 2002). Broadly speaking, the consistently highest silica accumulators fall in the monocotyledons, with grasses and palms as prime examples, ferns, and the sole extant genus in the horsetails, *Equisetum* (Fig. 12.2) (Hodson et al. 2005; Piperno 2006; Trembath-Reichert et al. 2015). Non-angiosperm seed plants, such as conifers, cycads, and Gnetales generally do not accumulate much silica, although exceptions exist (e.g., *Picea*) (Trembath-Reichert et al. 2015; Strömberg et al. 2016). Within non-monocotyledonous angiosperms, several families deposit abundant phytoliths, particularly among tropical trees, vines, and herbs (e.g., Chrysobalanaceae, Cucurbitaceae, Moraceae), but many groups are low producers (e.g., Fabaceae, Solanaceae). In addition, within monocotyledons, several clades exist that do not accumulate much silica (e.g., Juncaceae, Liliaceae) (e.g., Hodson et al. 2005; Katz 2015). Above all, even within clades that contain some of the highest producers, there are subclades or genera that do not produce (e.g., ferns within the Pteridaceae, Sundue 2009). These patterns indicate that even though the genetic framework for active uptake and deposition of silica existed early on in land plants (Trembath-Reichert et al. 2015), active accumulation of silica evolved in parallel or convergently in many clades much later on, perhaps as part of adaptation to particular biotic or abiotic environments (Strömberg et al. 2016). The existing variation means that some low-producing or silica-excluding taxa are underrepresented or "silent" in the phytolith record whereas others, such as grasses, can be overrepresented (Bozarth 1993; Delhon et al. 2003; Strömberg 2004; Piperno 2006; Hyland et al. 2013b). This bias can be compared to the rarity of pollen from insect pollinated plants relative to wind-pollinated conifers in the palynological record, or the relative absence of herbaceous plants in the leaf fossil record (Table 12.1).

Whereas a particular plant produces pollen/spores and leaves with relatively uniform morphology, deposition of silica often occurs at several sites in the plant tissue (e.g., stomata as well as epidermal cells and trichomes). The result is a *multiplicity* of different morphotypes for each plant (Rovner 1971; Piperno 1988). Most phytoliths take on the (relatively conserved) shape of the cell types in which they form in the epidermis, parenchyma, or associated with the vasculature (e.g., stomata, tracheary elements, trichomes). As a consequence, there is much overlap in the shapes of the silica bodies formed inside plants, even between taxa that are not closely related (e.g., epidermal jigsaw in ferns and angiosperms, Dunn et al. 2015b); a phenomenon often referred to as *redundancy* (Rovner 1971; Piperno 1988). Because of this overlap many plants cannot be identified to species, genus or even family based on their phytoliths, without *a priori* knowledge of the taxa present. In Quaternary paleobotany, where the extant flora may provide a meaningful guide to past vegetation, certain taxa can sometimes be identified to genus or species level (Blinnikov et al. 2002; Piperno 2009). In contrast, modern analogs are less commonly available for pre-Quaternary plant communities, meaning that deep-time phytolith analysis has to contend with lower taxonomic resolution, in particular for non-grass plants.

In some taxa, specialized so-called idioblast cells produce shapes that appear largely independent of overall cell shape. Examples include cystoliths (e.g., in Ulmaceae), grass silica short cell phytoliths (GSSCPs), and globular echinate and druses ("stegmata") in palms and Zingiberales (e.g., Prychid et al. 2004; Piperno 2006; Chen and Smith 2013). The phytoliths deposited in these idioblasts are usually highly diagnostic of the group in which they are produced, and there can even be taxonomically diagnostic variation within the group. The prime example is grasses (Poaceae), which display an amazing variation in the shape of their GSSCPs (Rudall et al. 2014), allowing for distinction between grass subclades [e.g., open-habitat grasses in the Pooideae vs. the so-called PACMAD clade (Panicoideae, Arundinoideae, Chloridoideae, Micrairoideae, Aristidoideae, Danthonioideae)], and in some cases even genera (e.g., *Oryza*) (Piperno 2006; Prasad et al. 2011). This is in contrast to the monoporate pollen of grasses, which is fairly homogeneous in morphology across Poaceae and its sister groups (but see Mander et al. 2013), and fossilized grass leaves, which cannot easily be distinguished from other parallel-veined, strap-shaped leaf fossils unless epidermal anatomy of cuticles are present (Thomasson 1987).

However, multiplicity and redundancy is evident within the Poaceae and other taxa with idioblasts as well. Thus, although certain GSSCP forms appear to be taxonomically unique (e.g., so-called vertical bilobates with scooped ends of *Oryza*) (Prasad et al. 2011; Rudall et al. 2014; Neumann et al. 2017), others are produced by many grasses across the phylogeny (e.g., rondels with truncated or keeled tops)

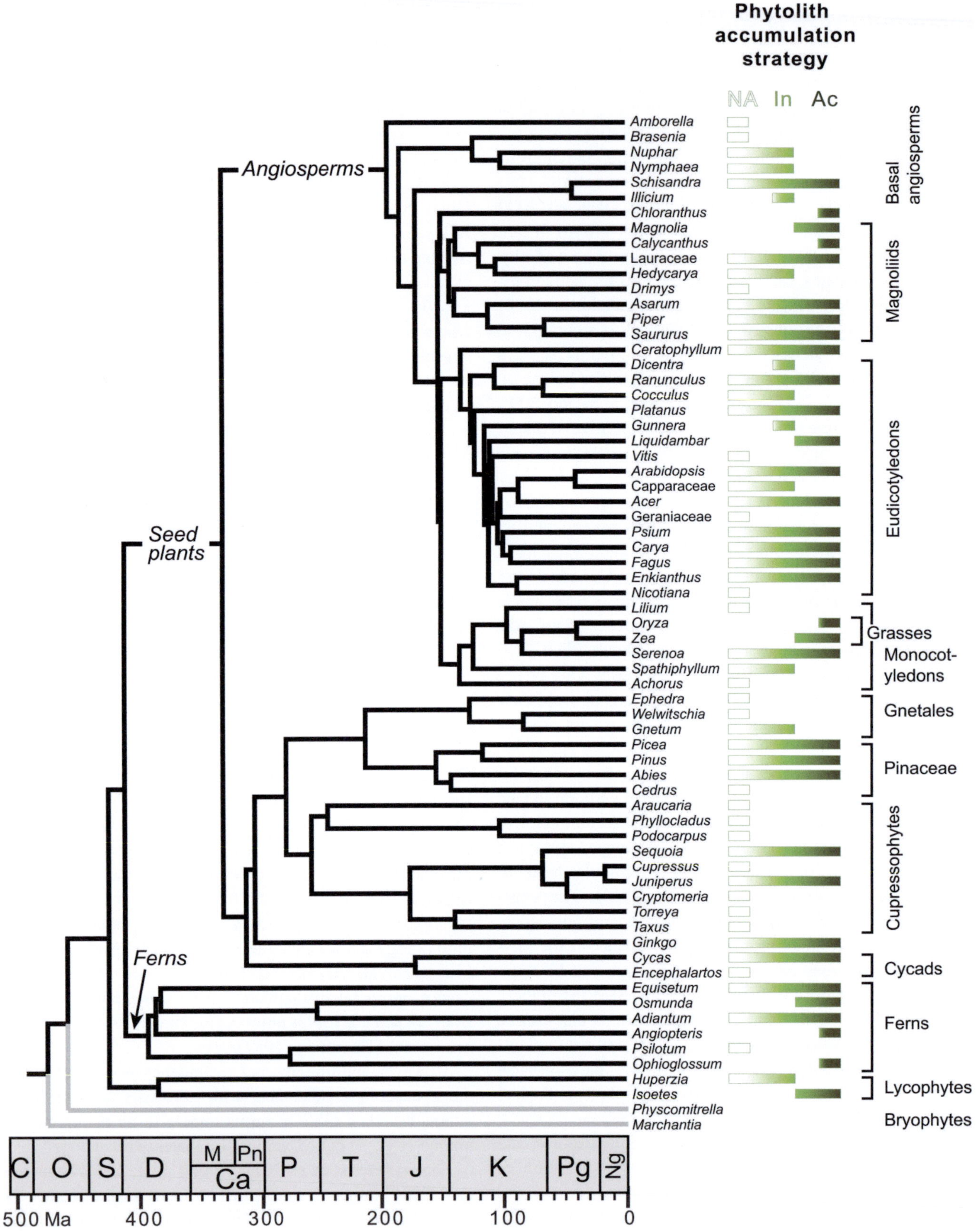

Fig. 12.2 Phytolith/silica production (Phytolith Accumulation Strategy) in vascular plants, modified from Strömberg et al. (2016). Note that currently published silica data are insufficient to determine the production in non-vascular land plants (bryophytes). Phytolith Accumulation Strategy: NA = silica non-accumulators, In = intermediate, Ac = silica accumulators

(Mulholland 1989; Strömberg 2003; Piperno 2006), leading to complexity and sometimes substantial overlap in GSSCP assemblages produced by different grasses (e.g., certain bilobate types are produced in bambusoids, early-diverging grasses, and chloridoids) (Piperno and Pearsall 1998; Strömberg 2003). The problems with overlap in shape among taxa is exacerbated by the way phytoliths have been traditionally described and studied in a 2-D view *in situ* in the grass epidermis (Metcalfe 1960; Watson and Dallwitz 1992; Barboni and Bremond 2009; Rudall et al. 2014; onwards), and the lack of 3-D illustrations of GSSCP morphotypes (but see Fredlund and Tieszen 1994), preventing study of detailed morphology. As a result, GSSCP forms that are different in 3-D are often conflated. One such example are GSSCPs produced in both forest grasses (bambusoid and early-diverging grasses) and in C_4, open-habitat chloridoids that have all been described as "saddle shaped" in 2-D view (Campbell and Kellogg 1987; Watson and Dallwitz 1992), but that look distinctly different when rotated in a liquid medium (Piperno and Pearsall 1998; Strömberg 2004). In addition, the analysis of whether detailed GSSCP shape reflects grass phylogeny has not been updated using more recent phylogenies and classification schemes (e.g., Aliscioni et al. 2012). Recent work has indicated that much better taxonomic resolution is possible when 3-D morphology at high magnification is considered (Piperno 2006), and work is underway to provide a more detailed description of GSSCP in a phylogenetic context using this approach (e.g., Gallaher et al. 2017).

Despite the redundancy among phytolith types formed by many plants, the combination of morphotypes and/or the different relative abundances of morphotypes have the potential to identify a plant at a lower taxonomic level (Miller et al. 2012; Chen and Smith 2013; Strömberg et al. 2013a), or to rule out the presence of a plant. For example, saddle-type GSSCPs combined with collapsed saddles (ICPN 1.0 name: saddle long) may signal the presence of bambusoids or early-diverging grasses rather than chloridoids (Piperno and Pearsall 1998; Strömberg 2003). Similarly, although elongate sinuous long cells are typical of grasses, if they are not accompanied by abundant GSSCPs but instead blocky polyhedrons or "*Pinus* spiny bodies" they were more likely produced by conifers (Klein and Geis 1978; Strömberg 2003). Using entire phytolith "profiles," as opposed to focusing on particular phytolith morphotypes for diagnosing taxa, is a promising future direction for phytolith research (see Gallego et al. 2004).

In summary, the uneven silica accumulation among land plants means that some taxa are not represented in the fossil phytolith record, whereas others tend to be overrepresented (Fig. 12.3). An example of the latter is grass phytoliths, which are ubiquitous and dominant in assemblages from ecosystems where trees and shrubs are poor producers, such as temperate-boreal forests (Table 12.1) (Bozarth 1993; Delhon et al. 2003; Strömberg 2004; Hyland et al. 2013b). Also, because of the pervasive multiplicity and redundancy among phytolith assemblages produced by land plants, phytoliths cannot provide detailed taxonomic information about trees, shrubs, and herbs that lack specialized phytoliths (idioblasts) (Table 12.1). Instead, they can be used to describe relative abundances of plant functional types (e.g., woody plants vs. grasses). On the other hand, certain ecologically important taxa that are often in very low abundance/missing, or not very diagnostic in palynofloras and macrofloras, such as palms, gingers and grasses, tend to be well represented and diagnostic, often to a level below family. Therefore, phytoliths uniquely allow estimation of the diversity of clades represented in grass communities (but only to a coarse level) (Strömberg et al. 2007b). Furthermore, despite these production biases, phytolith assemblages offer accurate data on aspects of vegetation type that are not captured in palynomorph assemblages or macrofloras, as will be described below.

Phytoliths in Soil and Sediment—Preservation and Biases

Phytolith taphonomy is distinct from that of other plant microfossils, such as pollen and spores, for several reasons. First, the chemical composition of opal A renders phytoliths less vulnerable to oxidizing conditions than organic fossils (Retallack 1983a, b); as a result, phytoliths are commonly found in paleosols, whereas pollen/spores are not (Strömberg 2002). Second, compared to pollen/spores, phytoliths have a different size range (2–2,000 micrometers vs. 0.6–100 micrometers for pollen/spores), higher specific gravity (1.5–2.3 vs. 1.45–1.52), and far greater variation in shape (spherical, sheet-like, elongate, irregular vs. generally rounded) (Jones and Beavers 1963; Pearsall 2000; Piperno 2006). In addition, phytoliths are not released from the tops of plants where you typically find pollen sacs or sporangia, but instead from the tops of soils or bottoms of lakes where plants die and decay or discarded plant parts end up. For these reasons, phytoliths may behave more similar to approximately silt-sized soil/sedimentary particles—in particular volcanic glass, which overlap substantially with plant silica in specific gravity, size, and sometimes shapes (George 1924; Strömberg 2003).

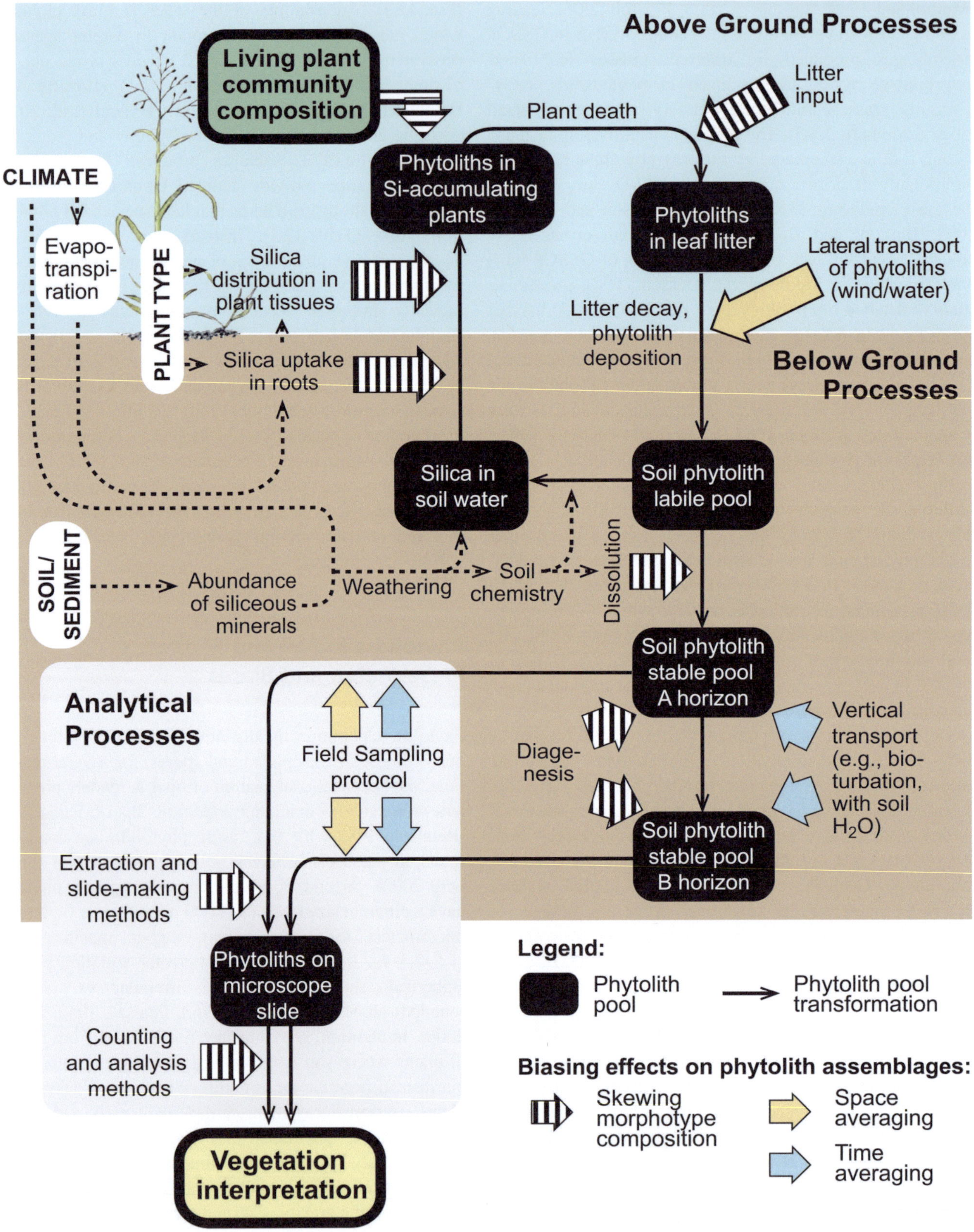

Fig. 12.3 Schematic of phytolith assemblage formation in soils, with influences on phytolith assemblage composition marked with thick arrows. Note that such influences include both the factor that is of interest to the researcher (plant community composition; green bolded box) and factors that obscure this influence and that therefore may bias the vegetation inference (yellow bolded box). The latter factors relate to plant type (e.g., differential phytolith production among plant taxa), climate, soil/sediment type (both influencing e.g., soil chemistry affecting dissolution), and analytical protocol

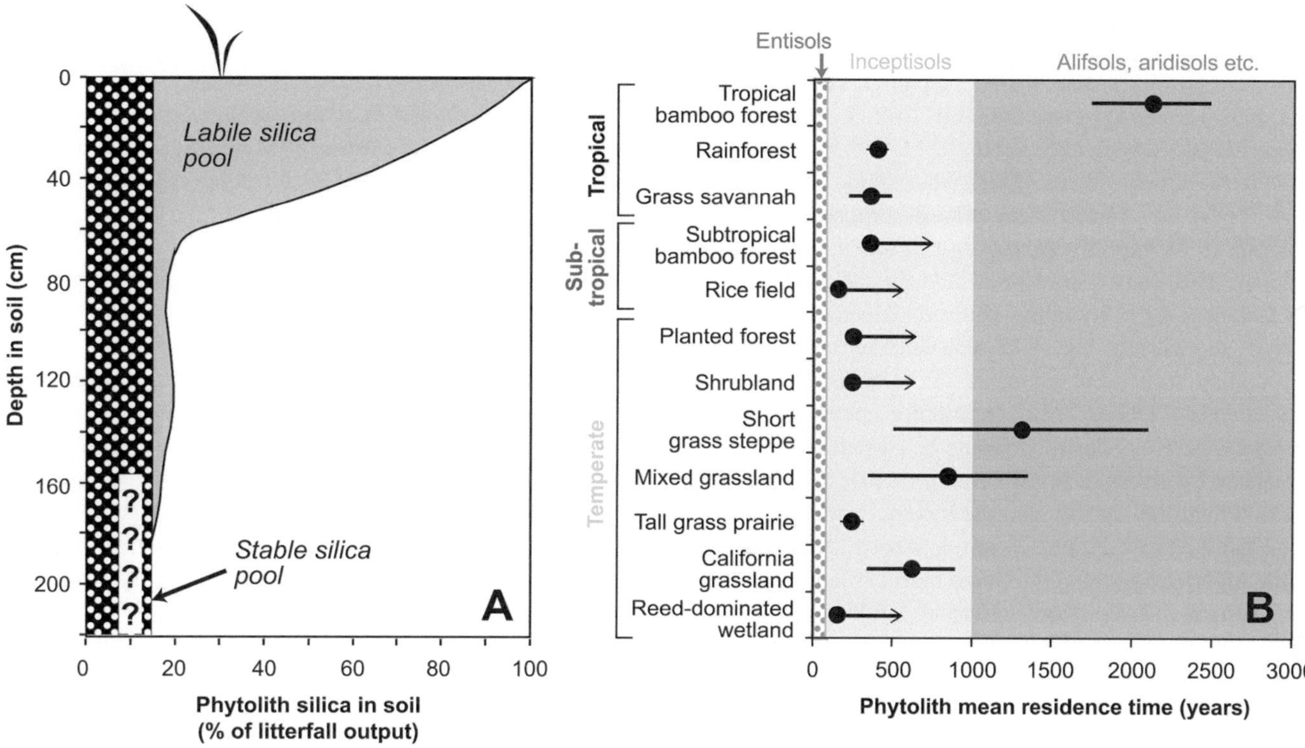

Fig. 12.4 Distribution and turnover of phytoliths in soil. A, Schematic representation of phytolith distribution in soils, showing how the abundance of plant-derived silica decreases with depth in the soil, after Alexandre et al. (1997a). Labile and stable pools differentiated, with question marks showing the potential decrease with depth of the stable phytolith pool (Song et al. 2016). B, Turnover rates of phytoliths in soils under different types of vegetation, based on Table 1 in Song et al. (2016) and references therein. Bars mark the error in turnover rate estimates; arrows indicate minimum estimates

Distribution of Phytoliths in Soils and Sediment

Phytoliths are incorporated into soils or subaerial sediments when leaf litter and plants decay on the surface, or belowground in the case of roots and rhizomes (Fig. 12.3) (Piperno 1988; White et al. 2012). In addition, phytoliths are deposited on the land surface as part of sediment-water suspension (e.g., through overbank flooding) or windblown dust (Fredlund and Tieszen 1994; Alexandre et al. 1997b; Piperno 2001; Osterrieth et al. 2009). Empirical data show that the concentration of biogenic silica is very high near the soil surface, but typically decreases drastically below about 30–80 cm (Fig. 12.4A) (Kelly et al. 1991; Alexandre et al. 1997a; 2011a; Blecker et al. 2006; Gérard et al. 2008; White et al. 2012). Deeper in the soil profile, the phytolith concentrations appear to be stabilized (but see Song et al. 2016). This distributional pattern results from a combination of (bio)chemical dissolution and vertical transport of phytoliths. Specifically, in the top part of the soil, biologically mediated mineralization of organic matter occurs rapidly and the most soluble phytoliths are dissolved; this silica is in large part recycled through (re-)uptake by living plants (Figs. 12.3, 12.4) (Alexandre et al. 1997a, 2011a; White et al. 2012). This "labile" pool of recycled silica ranges from 69% of the total soil silica in short grass prairie to 92% in tropical forest, variation that suggests a strong dependence on climate (Alexandre et al. 1997a; Blecker et al. 2006). The intensity of chemical dissolution is visible as more marked etching or pitting of phytolith morphotypes further down in the soil column (Alexandre et al. 1997a; Meunier et al. 1999; Blecker et al. 2006). Conversely, pedogenic secondary silica (see below), thought to result from dissolved and mobilized amorphous silica from the O and A horizons of the soil, increases with depth (Alexandre et al. 1997a).

The stable pool of phytoliths is subject to vertical transport through bioturbation and illuviation (Fig. 12.3), although the roles of different factors affecting these movement, such as downward water percolation, degree and type of bioturbation, and soil grain size are still debated (Fishkis et al. 2009; 2010a, b; Madella and Lancelotti 2012). Vertical dispersal is thought to typically result in temporally averaged phytolith assemblages across the soil profile. However, several studies involving radiocarbon (C-14) dating of phytoliths in soil profiles (for discussion of the use of phytolith C-14, see Santos et al. 2012; Piperno 2016) have indicated that stratification of the biosilica pool can occur in soils, such that the phytolith record is more temporally resolved than the soil overall (Piperno and Becker 1996; Grave and Kealhofer 1999; McClaran and Umlauf 2000; Kerns et al. 2001; Fishkis et al. 2010a; Morris et al. 2010) (see further *Temporal Resolution And Biases Relating To Vertical Transport*).

Lakes and oceans receive a slightly different biosilica input than soils. Phytoliths can be delivered into lakes through shedding of leaves or reproductive material by the surrounding local vegetation, or after fluvial transport of plant organs (Piperno 1988). In addition, phytoliths eroded from soils near the watercourse can be carried in suspension into lakes by rivers or wind (Piperno 2001; Aleman et al. 2014). Phytoliths and diatoms recovered in deep-sea sediment hundreds of kilometers from Africa, Australia, and North America are thought to have been carried primarily as part of eolian dust from these (upwind) landmasses, although ocean currents may affect distribution closer to the coast (Ehrenberg 1847; Folger et al. 1967; Folger 1970; Locker and Martini 1986). The level of preservation and stratification of phytoliths that are deposited on lake (or ocean) bottoms depends on the degree of bioturbation and the geochemistry of the lake sediments (e.g., pH). For example, lake sediments that reflect hypoxic or anoxic conditions and in which bioturbation is minimal tend to be finely stratified, and can thus preserve extremely well-resolved phytolith records (e.g., Piperno 2001; Strömberg et al. 2007a; Aleman et al. 2014).

Preservation of Phytoliths in Soils and Sediment

The preservation of individual phytoliths in soil depends on several factors, including soil/sediment solution pH, the presence of organic matter, and the potential coating of the phytoliths (Kendrick 2005; Ming Huang et al. 2011; Song et al. 2016). At ambient temperature the solubility of opal-A varies between roughly equivalent to that of silicate minerals to two orders of magnitude higher (e.g., 50–60 mg Si/L vs. 3–7 mg Si/L for quartz; (e.g., Bartoli and Wilding 1980; Fraysse et al. 2010). Dissolution rates for phytoliths are at a minimum at pH 2–3, increasing at higher pH values, especially at > 8.5 (e.g., Fraysse et al. 2006; Ehrlich et al. 2010; Prentice and Webb 2016). Thus, in highly alkaline soils/sediment, phytoliths are often heavily weathered (e.g., Alexandre et al. 1997a; Fredlund and Tieszen 1997b; Madella and Lancelotti 2012), but note that somewhat acidic to neutral soils also produce pitted and etched phytoliths (Borrelli et al. 2008, 2010). Opal-A and similarly easily soluble silica can only remain stable if the interstitial water is oversaturated with silica (Shoji et al. 1994), which is likely why deposits from diatomaceous lakes or those rich in volcanic glass often contain the best-preserved phytolith assemblages (Strömberg 2005; Strömberg et al. 2007b). That said, well-preserved phytoliths can be found in sediment experiencing relatively alkaline conditions as evidenced by abundant $CaCO_3$ cements (Strömberg 2003; Strömberg et al., unpublished data). The higher pH, as well as the catalytic effect of the salt in saltwater suggest that phytolith dissolution is higher in marine environments (Loucaides et al. 2010); nevertheless, pristinely preserved phytoliths have been recovered from marine cores (Locker and Martini 1986; Thorn 2001, 2004b).

The presence of organic molecules enhances solubility and rates of dissolution of opal-A by creating silica-organic complexes even at neutral pH and low temperatures (Drees et al. 1989; Bennett 1991; Kendrick 2005; Fein 2009; Ming Huang et al. 2011). This effect is most pronounced in the root zone, and explains the high turnover of phytoliths in the uppermost A-horizon (see above). This phenomenon may also elucidate the observation that organic rich sediments other than diatomites (which may be saturated with respect to opal-A) are often devoid of phytoliths (Strömberg, unpublished data).

The presence of aluminum and iron may also affect the preservation of phytoliths in soils. Al- and Fe-oxyhydroxides are known to coat mineral grains in soils, preventing dissolution, and could potentially have the same effect on phytoliths (Drees et al. 1989; Cornelis et al. 2011). Notwithstanding, the infrequent yield of biosilica from paleosols with red color indicate that phytoliths are less commonly preserved in such Fe-oxide-rich soils (Strömberg 2003; Strömberg et al., unpublished data). Co-deposition of aluminum and silicon in plant phytoliths has been thought to impede silica dissolution in the soil environment (Bartoli 1985; Carnelli et al. 2002; Cornelis et al. 2011), but some recent experimental work does not support a strong effect of Al on biosilica preservation (Fraysse et al. 2009).

Phytolith shape, size, and degree/type of silicification (e.g., porous vs. dense) further influence rates of dissolution. Laboratory tests indicate that thin, flattened silica bodies with large surface area relative to volume are more readily dissolved compared to more compact shapes, and that porous silica texture is less resistant than the dense silica that typically makes up GSSCPs (Jones and Handreck 1967; Wilding and Drees 1974; Bartoli and Wilding 1980; Cabanes et al. 2011). These results explain why thinly silicified, seemingly fragile phytoliths (e.g., silicified parenchyma) are rare in soil assemblages compared to in living plant material, and why the commonly thin or flattened phytoliths produced by non-grass plants, such as "dicotyledonous" trees [e.g., silicified pavement cells (epidermal polygonal), tracheary elements (cylindric sulcate "tracheid")] are preferentially dissolved in soil (Wilding and Drees 1973, 1974; Bartoli and Wilding 1980; Piperno 1988; Cabanes et al. 2011). Recent experimental work has shown similar dissolution rates for phytoliths from different plant types (conifer, broadleaved tree, tree fern, horsetail) when normalized for surface area (Fraysse et al. 2009), suggesting that textural properties are less important than specific surface area (surface area per unit volume) in determining rates of dissolution (Ehrlich et al. 2010).

In contrast, phytoliths appear to tolerate a range of redox environments (Retallack 1983a), including the highly oxidized floodplain sediments that contain much of the Cenozoic fossil vertebrate record (Strömberg 2002). This redox durability sets phytoliths apart from organically-based paleobotanical remains (palynomorphs, macrofloras) (Table 12.1), and has two major implications for pre-Quaternary paleoecology. First, it has expanded the range of depositional environments preserved in the paleobotanical record to include "upland" ecosystems that previously were not typically represented. Second, it allows for direct comparisons of faunal and floral evidence, providing a basis for robust paleoecological inferences of plant-animal interactions in deep-time (Strömberg 2002, 2011).

In summary, taphonomical processes affecting plant silica tend to favor the preservation of grass phytoliths over non-grass phytoliths. This tendency is in sharp contrast to the lower preservation potential of grass pollen and leaves (Table 12.1). In many temperate ecosystems, the taphonomic bias towards grasses combined with a low production of tree phytoliths results in overrepresentation of grass phytoliths in soil assemblages. In tropical vegetation, where non-grass plants are often abundant phytolith producers, a biasing effect is less apparent.

Diagenesis

Relatively little is known about how diagenesis may bias phytolith assemblages, although it has been proposed that post-burial loss of water would lead to decreased stability of opal-A in general (Cabanes et al. 2011). Opal-A of phytoliths in soil undergoes dissolution and recrystallization, converting into successively more thermodynamically stable forms with quartz as an end member (Williams and Crerar 1985; Williams et al. 1985). Even before visible alterations in texture (see *Preservation Modes of Fossil Phytoliths*), this transformation appears to result in slight optical anisotropy of the phytoliths (from optically isotropic opal-A), as observed under polarizing light (Strömberg 2003). To what extent diagenesis alters the rate and nature of these reactions is currently unclear.

Preservation Modes of Fossil Phytoliths

Fossilized phytoliths, even as old as Cretaceous, can be in pristine condition, with delicate ornamentation or smooth surfaces preserved, and the occluded carbon of GSSCPs and other morphotypes intact (Fig. 12.5J) (e.g., Prasad et al. 2005; Strömberg 2005). In other assemblages, phytolith surfaces appear rough (but not etched) rather than smooth, indicating post-depositional alteration (Fig. 12.5B); such changes appear to affect forest indicator phytoliths more frequently than GSSCPs, making them harder to identify (Strömberg 2003). Other changes in phytolith appearance due to pedogenesis (or diagenesis) include breakage, especially of elongate morphotypes, etching (partial dissolution) where the surface is perforated and eroded (Fig. 12.5C, K, O) ("weathered;" Fredlund and Tieszen 1997b), and material alteration, where the opal-A changes from having glass-like to more crystalline structure, presumably as part of the transformation of opal-A to quartz (Williams and Crerar 1985; Williams et al. 1985). These textural changes also cause shape alterations in the phytoliths, which appear more blocky and crystal-like (Fig. 12.5D, H) (Strömberg 2003).

Fossil phytolith assemblages described to date range from being pristinely preserved (or well preserved but broken), to size sorted (e.g., where only morphotypes up to 20 μms are represented), or containing mostly etched or altered morphotypes (e.g., Strömberg 2003; Strömberg et al. 2013a). Generally speaking, there appears to be a correlation between good preservation of phytoliths and phytolith yield; in other words, low yields often consist of severely etched or altered phytoliths (Madella and Lancelotti 2012; Strömberg et al., unpublished data).

An indicator of biogenic silica dissolution in phytolith preparations is the occurrence of secondarily precipitated silica grains. Secondary silica is pedogenically formed opal-A (most commonly) that results when primary silica from ash or biosilica is dissolved in the O and A horizons and re-precipitated in the subsoil (Kendrick and Graham 2004; Kendrick 2005). It is particularly common in soils in arid, highly seasonal climates as well as in volcaniclastic soils (Retallack 1983b, 1990; Kendrick 2005). In addition to signaling potential dissolution of biogenic silica, which could bias assemblages (see above), pedogenic silica grains can be superficially similar to certain phytolith morphotypes. For example, some types of pedogenic silica resemble globular granulate morphotypes and related forms commonly used to indicate forests (e.g., Barboni et al. 2007); they can be distinguished based on their greater translucency compared to the biogenic silica in fossil assemblages (compare Fig. 12.5M vs. 12.5N) (Strömberg 2004). Likewise, secondarily precipitated silica form laminar structures that can, in some cases, appear superficially similar to vesicular infilling phytoliths (Geis 1973), a common morphotype in woody and herbaceous dicotyledons (compare Fig. 12.5P, 12.5Q vs. 12.5R) (Strömberg 2003, 2004).

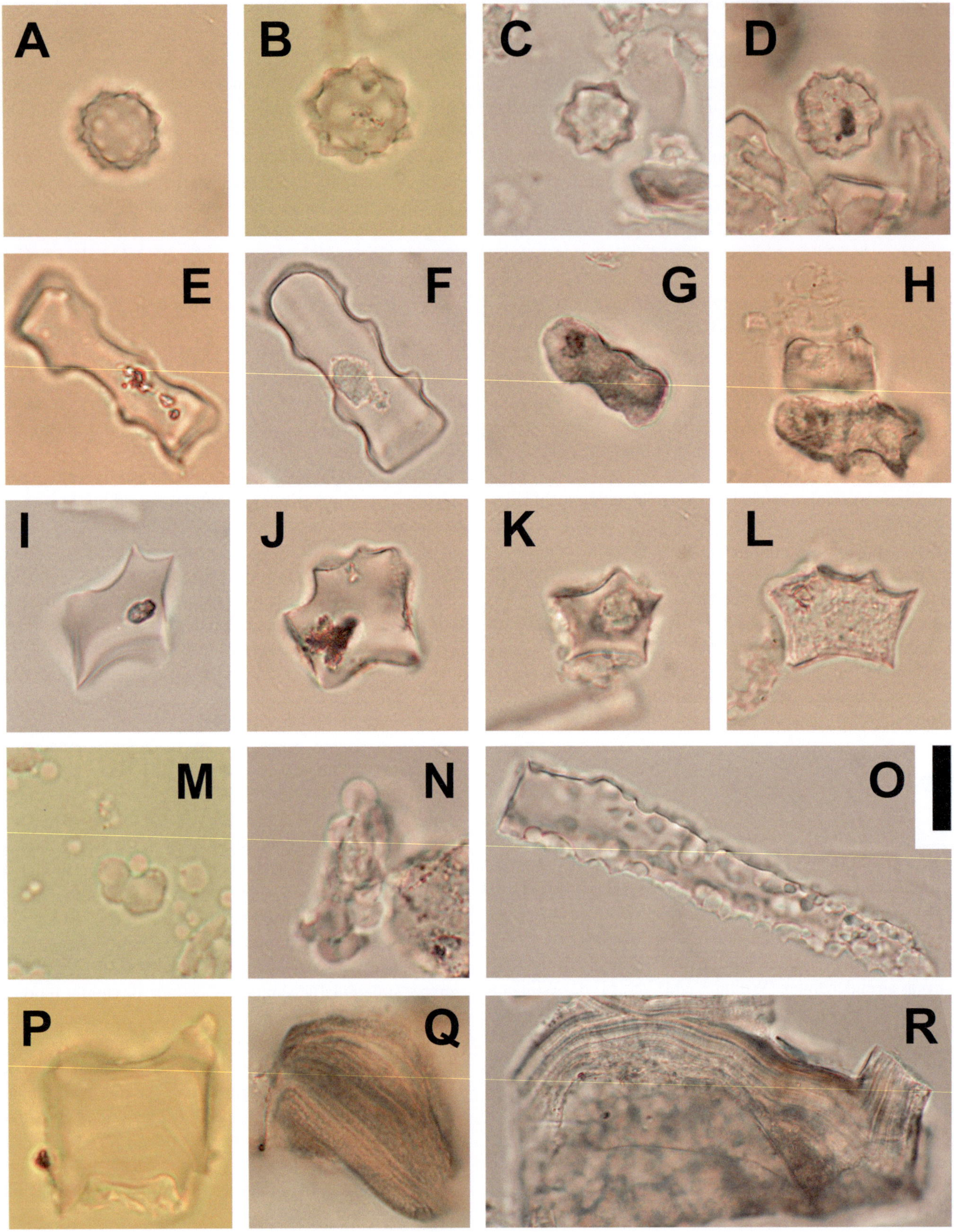
A
B
C
D
E
F
G
H
I
J
K
L
M
N
O
P
Q
R

◄ **Fig. 12.5** Phytolith preservation: examples of modern and fossil phytoliths (Eocene-Miocene of the Great Plains; Strömberg 2003) in different states of preservation. A–D, Globular echinates from palm: A, Modern globular echinate extracted from leaves of *Sabal mexicana* (UWBM PR 6001); B, Slightly etched fossil globular echinate; C, Very etched globular echinate with hollow center; D, Texturally altered globular echinate (note parallel striations in the interior indicating crystal structure). E–H, Crenate GSSCPs (ICPN 1.0 name: trapeziform polylobate) indicative of Pooideae: E, Modern crenate from leaves of *Nasella pulchra* (UWBM PR 5853); F, Fossil crenate, without organic inclusion but otherwise well preserved; G, Texturally altered fossil crenate, but with intact organic inclusion; H, Broken and texturally altered fossil crenate lacking organic inclusion (body at top, note parallel striations in the interior indicating crystal structure). I–L, *Chusquea*-type rondel GSSCP with spiked top (Strömberg 2005), indicative of Bambusoideae: I, Modern *Chusquea*-type rondel extracted from leaves of *Chusquea patens* (UWBM PR 5854); J, Pristinely preserved fossil *Chusquea*-type rondel; K, Etched *Chusquea*-type rondel without the organic inclusion; L, Texturally altered *Chusquea*-type rondel. M, Modern compound sphere phytolith (ICPN 1.0 name possibly: globular granulate, see Strömberg 2003) extracted from leaves of *Ginkgo biloba* (UWBM PR 6100). N, Secondarily precipitated silica in fossil sample. These particles can look very similar to compound sphere phytoliths but can usually be differentiated based on their translucent texture, which is uncommon in fossil phytoliths. O, Heavily etched fossil elongate psilate phytolith, where both the surface and shape are affected by dissolution. P, Modern non-spherical vesicular infilling body extracted from leaves of *Cornus exelsea* (UWBM PR 5868). R, Fossil non-spherical VI body with altered color and texture but otherwise well preserved. S, Secondarily precipitated silica in fossil sample. Such particles can look very similar to non-spherical VI body phytoliths but can usually be differentiated because they are not shaped like plant cells or partial plant cells. Scale bar equals 10 μm. Institutional abbreviation: UWBM = Burke Museum of Natural History and Culture

Spatial and Temporal Resolution of the Phytolith Record

Spatial Resolution and Biases Relating to Lateral Transport

Once freed from plant tissue, phytoliths are usually subject to some degree of transport (Fig. 12.3). Because of the differences in size, shape, specific gravity and mode of release, phytoliths can be expected to, on average, be carried shorter distances than pollen/spores, hence, to preserve a more local signal (Table 12.1). Does this bear out in nature?

The spatial resolution of the pollen/spore and macrofossil record resulting from lateral transport is well-studied (see Birks and Birks 1980; Ellis and Johnson 2013), and suggest that palynomorph assemblages are often regionally mixed (Table 12.1). In contrast, empirical assessment of the localization of the phytolith record in modern ecosystems is limited to a few studies (Folger 1970; Fredlund and Tieszen 1994; Kerns et al. 2001; Bremond et al. 2005a; Aleman et al. 2012, 2014). These use different phytolith classification systems and analytical methods, and are restricted in geographic range; as a result, they are difficult to compare and synthesize into general guidelines for inferring spatial patterns in phytolith analysis.

Fundamentally, the area of vegetation reflected in the phytolith record (i.e., the lateral mixing of phytolith assemblages) depends on how widely phytoliths are transported, which in turn depends on the available dispersal mechanisms. The importance of these mechanisms was traditionally underestimated and soil phytoliths assemblages were considered to reflect local vegetation as a result of *in situ* decay of plant material (Piperno 1988). Current research indicates that phytolith dispersal is controlled by several factors, which vary under different canopy types. The most studied and documented factor is wind. Eolian transport pre-dominates in open habitats, especially after fires, which increase erosion and transport of phytoliths (Piperno 1988; Wallis 2001; Aleman et al. 2014), likely because the phytoliths are released directly into the air from burning plant tissues rather than first being deposited in soils. As a result, phytolith assemblages in these vegetation types may reflect a large spatial scale—indeed, phytoliths are known to travel hundreds of kilometers with wind both on land and over the ocean (e.g., Folger et al. 1967; Folger 1970; Locker and Martini 1986; Piperno 2006; Osterrieth et al. 2009). Wind transport is weaker in forest habitats, where phytolith assemblages reflect more local conditions (Piperno 1988; Bremond et al. 2005a; Piperno 2006; Aleman et al. 2014).

Another factor controlling dispersal is herbivory, whereby animals transport plant material with phytoliths in their guts and deposit them, in the form of dung, away from the growth site (Fredlund and Tieszen 1994). Herbivores can transport phytoliths extra-locally but the distance of transport depends on the animal home range and habits (Fredlund and Tieszen 1994). Thus, within the same habitat, large herbivores (e.g., ungulates), which have large home ranges, are likely to be responsible for more spatial mixing than small herbivores (e.g., rodents) with their typically smaller home ranges (McNab 1963).

Fluvial and colluvial (erosion) factors also affect phytolith dispersal and largely depend on topography and geomorphology (Fredlund and Tieszen 1994). In the case of lake sediments, the spatial scale reflected in phytolith assemblages is shaped by hydrological conditions (Piperno 2001), although eolian input can also be important (Aleman et al. 2014). Thus, in a closed lake (a maar), the phytoliths supplied via water are mainly from the vegetation growing at the lake margin, whereas in a lake with strong fluvial input, phytolith assemblages can be expected to record vegetation from different parts of the watershed (Piperno 1993, 2001; Strömberg et al. 2007a; Aleman et al. 2014).

How may transport influence the composition of the resulting phytolith assemblage? There is experimental evidence that phytolith morphotypes, as a result of their variation in shape and size, exhibit variation in their behavior in fluids such as water, leading to size- or shape sorting (Lentfer et al. 2003; Strömberg 2007). In accordance with this, Folger (1970) found phytolith assemblages carried by wind across the Atlantic to be strongly biased to grains smaller than ~ 21 micrometers. This type of sorting in phytolith assemblages transported by wind or water may result in a bias towards small (<20 micrometer) morphotypes, such as GSSCPs and globular echinates from palms, and against large, multi-cell phytoliths (discussion in Folger 1970; Strömberg et al. 2007a; Selkin et al. 2015).

Because of the potential for spatial mixing and bias of the vegetation signal, knowledge of the depositional environment is critical for interpreting the spatial scales captured in the phytolith record. Nevertheless, most work directly comparing soil phytoliths to standing vegetation shows that local plant communities heavily influence phytolith assemblage composition, and that local vegetation differences are possible to distinguish based on phytoliths—also in open vegetation with important eolian transport (Fredlund and Tieszen 1994; Alexandre et al. 1997b; Barboni et al. 1999; Kerns et al. 2001; Bremond et al. 2005a; Osterrieth et al. 2009; Mercader et al. 2011).

Temporal Resolution and Biases Relating to Vertical Transport

Similarly to pollen/spores and macrofossils, phytolith assemblages preserved in (diatomaceous) lake sediments with distinct lamination are minimally temporally averaged, as each lamina should reflect a time period on the order of seasons-years. These types of deposits can therefore be sampled at a very detailed temporal scale. The same is not true for fossil soils, in which vertical mixing within the soil profile (see above) results in phytolith assemblages that typically represent a long-term average of vegetation composition (Table 12.1) (Fredlund and Tieszen 1994; Aleman et al. 2012). The extent of time averaging is influenced by the sedimentation rate and landscape stability. Paleosols represented in Cenozoic sections sampled for phytoliths (e.g., Entisols, Inceptisols, Alfisols; Soil Survey Staff 2006) typically form over up to 10,000 years (Retallack 2001a; Sheldon and Tabor 2009), implying a potentially very high degree of time averaging.

However, there are several reasons to think that the soil fossil phytolith assemblages in many cases record a much shorter time. First, many of the paleosols from which phytolith assemblages have been described so far in North America, South America, and Eurasia are poorly developed Entisols and Inceptisols with formation times of 10–100 and 100–1,000 years, respectively (Retallack 2001a; Sheldon and Tabor 2009). Second, simply looking at paleosol formation time likely overestimates the formation time of phytolith assemblages. As previously discussed, phytolith assemblages may be stratified in soil profiles (e.g., McClaran and Umlauf 2000; Kerns et al. 2001; Morris et al. 2010), perhaps as a result of continuous aggradation (e.g., in Composite paleosols, Kraus 1999), resulting in a more temporally resolved phytolith pool. In addition, because of continual dissolution of biosilica in actively forming soils, the soil phytolith pool has an average age that is distinct from—and sometimes substantially shorter than—the age of the soil. The tendency for the phytolith pool to turn over can be estimated by the so-called phytolith mean residence time (e.g., Song et al. 2016). Phytolith mean residence time ranges from > 170 years to 2110 ± 372 years, with shorter turnover times in wetter, warmer climates where the intensity of weathering and biological activity in the soil is accelerated (Fig. 12.4B) (e.g., Alexandre et al. 1997a; Blecker et al. 2006; Vivanco and Austin 2006; White et al. 2012; Song et al. 2016). The availability of easily dissolved silica might also play a role, such that residence time is longer on volcanic soils (Song et al. 2016). If a paleosol phytolith assemblage represents a more or less "frozen" snapshot of the soil phytolith pool (i.e., more labile forms are not necessarily removed post-burial), then mean residence time of biosilica (estimated from climate, vegetation, and/or paleosol evidence) could provide a rough guide to how time-averaged it is (Fig. 12.4B). The bottom line is that, even in a soil that has remained stable over several thousands of years (e.g., Alfisols), the bulk of the phytolith pool has turned over several times, and the assemblage that is eventually fossilized should, in most cases, record only the last few hundred years of vegetation.

Size sorting is another potential consequence of vertical translocation of phytoliths. Experimental and field studies have found a decrease of large particles and an increase of small particles with soil depth, indicating preferential downward transport of small particles (Fishkis et al. 2009; Fishkis et al. 2010a, b; Alexandre et al. 2011a). However, in one experiment this sorting appeared to have been counter-acted by earthworm bioturbation (Fishkis et al. 2010b), suggesting that more work is needed before the extent of bias in soil phytoliths assemblages is understood.

Published modern studies thus predict that the richest, least biased, and most well preserved phytolith assemblages derive from the top part of fossilized soil horizons. Curiously, in deep-time work that has explicitly sampled both A- and B- or C-horizons in the same paleosol profiles, phytolith assemblages of comparable preservation and often different morphotype (but not size) composition are reported (Miller et al. 2012; Chen et al. 2015), suggesting that phytolith assemblages may indeed often be stratified within the soil.

Summary: Biases in Phytolith Analysis

Based on the characteristics of phytolith production, transport and preservation described above it can be hypothesized that phytolith assemblages in general should:

1. Be most abundant and best preserved in the A-horizon of fossil soils.
2. Be optimally preserved in well-oxidized (meaning free from organic material), non-basic sediments rich in diatoms or volcanic glass.
3. Be biased against non-grasses in many ecosystems.
4. Reflect local to regional vegetation, depending on vegetation type and depositional environment.
5. Reflect vegetation averaged over several hundreds of years (or less), especially if stratified in paleosols; in laminated lacustrine sediments, phytolith assemblages can reflect up to seasonal variation.

Current Use of Phytoliths In Deep-Time Paleoecology

Phytoliths as Evidence of Plant Evolution, Diversification, and Historical Biogeography

Plant evolution and diversification: Under the assumption that morphology reflects phylogeny, particularly diagnostic phytoliths (e.g., palm stegmata, so-called globular echinates) found in well-dated geologic strata can be used to trace plant lineages through time. Whereas fossil phytoliths of plants outside of the Poaceae have not yet been used for this purpose (Iles et al. 2015), perhaps because of the comparatively low level of taxonomic resolution in many groups (see above), fossil GSSCPs have shown to hold great promise for reconstructing grass diversification. It has long been known that GSSCP shape differs among traditional Poaceae subfamilies (Metcalfe 1960; Twiss et al. 1969; Brown 1984; Mulholland 1989; Mulholland and Rapp 1992; Twiss 1992; Watson and Dallwitz 1992 onwards), and this variation is widely used to identify grass sub-clades in Quaternary samples (e.g., Pearsall 2000; Pearsall et al. 2003; Barboni and Bremond 2009). Based on the work in modern grasses, paleobotanists have relied on GSSCP shape and, if available, phytolith distribution in tissue to infer the presence of particular lineages in the deep-time fossil record (e.g., Retallack et al. 1990; Strömberg 2005; Zucol et al. 2010). For example, the distribution of GSSCPs in the epidermis of grass macrofossils allowed for identification of three modern genera of PACMAD grasses (*Cleistochloa, Distichlis,* and *Stereochlaena*) in the middle Miocene of Kenya, Africa (Retallack et al. 1990; Dugas and Retallack 1993), and GSSCP shape was used to show that open-habitat grasses in the Pooideae and PACMAD had diversified by the Eocene–Oligocene and existed in the Great Plains and Northern Rocky Mountains 7–11 myr before the expansion of grass-dominated habitats there (Strömberg 2004, 2005).

Grass phytolith shape and distribution data have also been mapped onto or incorporated into grass phylogenies for a more rigorous approach to phytolith-based taxonomic inference (Prasad et al. 2005, 2011). The analyses allowed identification of extinct taxa related to rice (in the Oryzoideae) in the Late Cretaceous of Central India, which, when used as calibration points for molecular dating, resulted in substantially older ages (up to 20 million years) for crown-group Poaceae and for the split between the BOP (Bambusoideae, Oryzoideae, Pooideae) and PACMAD clades (Prasad et al. 2011; Christin et al. 2014). These results remain highly controversial, and many researchers question the reliance in these studies on a few key character states (i.e., the vertical bilobates with scooped ends) (see Christin et al. 2014; Kellogg 2015). Putative GSSCPs in monocotyledonous tissues preserved in Late Cretaceous Burmese amber (98.8 Ma; Shi et al. 2012) described by Poinar Jr (2011) might similarly help push back the date for the origin of Poaceae considerably. The debate surrounding these fossils highlights how limited our knowledge is of how GSSCP morphology reflects current understanding of the grass phylogeny (Aliscioni et al. 2012). Nevertheless, some or all of these grass fossils have variously been used to date the Poaceae clade, resulting in new insights with regards to, for example, the timing of the evolution of cold tolerance or C_4 photosynthesis in grasses relative to abiotic change and faunal evolution and ecology (e.g., Bouchenak-Khelladi et al. 2009; Bouchenak-Khelladi et al. 2010; Sandve and Fjellheim 2010; Syme 2012; Bouchenak-Khelladi et al. 2014; Spriggs et al. 2014).

Plant biogeography: Few studies have employed diagnostic phytoliths to infer the biogeographic history of particular clades (but see discussion in Prasad et al. 2005; Strömberg et al. 2007a). We anticipate that such analyses will become more common as the global dataset of Cenozoic and older phytolith assemblages grows and as more refined shape characterization becomes available (Evett and Cuthrell 2016). For example, each taxonomically well-placed Cretaceous grass phytolith will further help narrow down the center of origin and early dispersal paths of Poaceae, a lasting topic of debate among grass systematists (Bremer 2002; Bouchenak-Khelladi et al. 2010) that may also help test biogeographic scenarios relevant to vertebrate paleontology (e.g., Rana and Wilson 2003; Wilson et al. 2007).

Phytoliths as Evidence for Vegetation and Animal Habitats

The most common use of phytolith analysis in the deep-time record is reconstruction of vegetation type, as plant silica assemblages can provide unique information about plant communities and/or habitats for faunas, such as in the Miocene of Africa (Retallack et al. 1990), Greece (Merceron et al. 2016), and Patagonia (Brea et al. 2012). Recently, the use of phytoliths has permitted paleobotanists to address broader questions about vegetation change in response to temporal or spatial changes in the abiotic or biotic environment (e.g., climate, uplift, mass extinction) (e.g., Cotton et al. 2014) or the influence of habitat alteration on animal evolution (e.g., Strömberg 2006). Many of these studies have benefited from integrating (and comparing) phytoliths and other local climatic or environmental proxies. Examples include comparisons among phytoliths, wood and clay mineralogy in the Paleogene of Patagonia (Raigemborn et al. 2009), phytoliths and trace fossils in the Cenozoic of Patagonia (Sánchez et al. 2010), phytoliths, paleosol morphology and isotopes in the Eocene of Montana (Miller et al. 2012), and phytoliths, diatoms, and pollen/spores from the Miocene-Pliocene of Chad (Novello et al. 2015). Below, we highlight five of the main themes in pre-Quaternary phytolith research and the approaches associated with them.

Grassland ecosystem evolution: Because of the rarity of grass macrofossils and pollen in Cenozoic terrestrial deposits, the spread of grass-dominated vegetation was traditionally reconstructed using non-plant proxies, in particular the presence and abundance of large herbivores with high-crowned (hypsodont) cheek teeth (e.g., Osborn 1910; MacFadden 1997), fossilized grassland soils (e.g., Retallack 1997, 2001b), and isotopic evidence of plants with C_4 photosynthesis, assumed to be mainly grasses (e.g., Wang et al. 1994; Cerling et al. 1997). During the last $\sim$ 20 years, well-preserved phytolith assemblages from Eocene–Pliocene deposits from several continents have started to add independent, direct records of vegetation change and grass communities that have allowed evaluation of traditional ideas about the timing and nature of the assembly of grassland ecosystems on different continents (e.g., Strömberg 2005; Strömberg et al. 2007b, 2013a; Zucol et al. 2007; WoldeGabriel et al. 2009; Chen et al. 2015; Novello et al. 2015).

Investigators have differed in their analytical approach to using phytoliths for reconstructing Cenozoic grasslands; here we will outline two of the main methods applied in deep-time (Strömberg 2005; Zucol et al. 2010). Strömberg explicitly discussed the basis for analytical criteria over a series of publications dealing with Cenozoic phytolith assemblages from the Great Plains of North America (Strömberg 2002, 2004, 2005). Initially, Strömberg (2002) adopted the analytical method of Alexandre et al. (1997b), developed for the Holocene–modern of inter-tropical Africa, the so-called *D/P* index, which compares globular granulate forms, produced in abundance by tropical African trees, to all Poaceae morphotypes (see below *Habitat openness*, and Table 12.2). Applying the *D/P* index to the fossil phytoliths suggested that open, grass-dominated habitats had spread in the Central Great Plains by the latest Eocene (Strömberg 2002). Strömberg (2004) later deemed the *D/P* index too narrowly defined for studies outside of East Africa and especially for deep-time studies with no clear modern analogs. Instead, she developed a "general approach" to phytolith analysis for use in pre-Quaternary paleoecology, based on a survey of modern analog work and a quantitative analysis of an extensive modern reference collection (see Strömberg 2003), that attempted to take the known biases of the phytolith fossil record into account. The general approach entails: (1) looking at the whole phytolith assemblage (2–250 micrometers) together, rather than restricting study to a certain size class (e.g., Alexandre et al. 1997b) or studying different size fractions separately (e.g., Runge and Runge 1997; Kealhofer and Penny 1998; Mercader et al. 2000); (2) considering all aspects of a phytolith assemblage, that is, both types and relative amounts of diagnostic phytoliths that make up the GSSCP assemblages (e.g., indicative of open-habitat vs. closed habitat grasses) and the non-grass phytolith component (e.g., diagnostic of palms, Zingiberales, woody and herbaceous dicotyledonous angiosperms) (Fig. 12.1). The latter group of phytoliths was referred to as "forest indicators" because in the Great Plains, where the method was first applied, all non-grass morphotypes appeared to co-vary relative to open-habitat grasses—but Strömberg (2004) stressed that they should be evaluated separately in each new study. Another central aspect of the general approach is the emphasis on (3) looking at *relative* changes in, for example, tree cover through time, not absolute vegetation reconstructions (e.g., 30% forest indicator phytoliths equals savanna woodland) as a way to deal with the potential underrepresentation of non-grass phytoliths (Piperno 1988; Strömberg 2004; Piperno 2006; Hyland et al. 2013b).

The protocol used by Zucol and colleagues (e.g., Raigemborn et al. 2009; Zucol et al. 2010), which remains the most influential approach to deep-time phytolith work in South America, is similar to the general approach in that it considers all morphotypes within the assemblage. It contrasts by analyzing size fractions (e.g., >250 μm, 53–250 μm, 8–53 μm; and <8 μm; Raigemborn et al. 2009) separately, and then combining relative abundances from different size fractions into a total count (A. Zucol, 2015, personal communication). In addition, the methods differ in which morphotypes are used to assess biomass of grasses. Thus, in their grass count, Zucol and colleagues include GSSCPs as well as several other morphotypes that are often and abundantly produced by grasses, such as certain elongate phytoliths (elongate psilate, elongate echinate long cell, and dendritic), trichomes (aciculate and unciform hair cells), and cuneiform and parallipiped bulliform cells (e.g., Zucol et al. 2010). The general approach uses only GSSCPs (Strömberg 2005), excluding other morphotypes because some of them (notably elongates) are found in many other taxa, and their production tends to be linked to environmental factors such as evapotranspiration rather than grass biomass (Piperno 1993; Bremond et al. 2005b; Madella et al. 2009). Additionally, Zucol and colleagues' tally of "dicot" phytoliths encompasses a narrower selection of morphotypes (epidermal jigsaw and multicavate, Zucol et al. 2010) while the "forest indicator" class in the general approach includes a wide range of morphotypes produced by dicots, ferns, and conifers (Strömberg et al. 2007b: Table 2). Because of these differences, it is not possible to directly translate results from one method to the other; however, it seems likely that the method of Zucol and colleagues (2010) will consistently infer a higher relative abundance of grasses than the general approach.

The general approach and similar protocols have been applied to interpret vegetation changes related to the spread of grasslands on several continents (North and South America, Eurasia, Africa). These studies have helped refine or overturn existing ideas about the timing and nature of this vegetation change and the evolution and paleoecology of grazer-type herbivores. Thus, in the Great Plains of North America and in the eastern Mediterranean, phytolith studies showed that pooid (C_3)-dominated, grassy vegetation spread by the early Miocene or even the late Oligocene, several million years before hypsodont ungulate faunas emerged (Strömberg 2004, 2005, 2006, 2011; Strömberg et al. 2007b; see also Merceron et al. 2016). This research has helped refute earlier ideas that the first grasslands were tropical savannas, that is, dominated by C_4 grasses (see discussion in Wang et al. 1994), and allowed documentation of biogeographic variation in the timing of the shift from forest to grasslands (Strömberg 2005; Strömberg and Dunn 2014; Harris et al. 2017). For example, comparison between phytolith assemblages in Turkey and Spain reveals a sharp early-middle Miocene vegetation gradient across the Mediterranean, with forests in the west and grass-dominated plant communities in the east (Pinilla and Bustillo 1997; Strömberg et al. 2007b; Strömberg 2011).

Phytolith assemblages have provided particularly intriguing paleoecological inferences for the Cenozoic of South America. It was long held that "precociously" hypsodont meridiungulate herbivores in the middle Eocene of Patagonia signaled the spread of grassland vegetation (Stebbins 1981; Jacobs et al. 1999; Madden et al. 2010), and the initially studied Eocene phytolith assemblages were interpreted in light of faunas and loess-like sediments to indicate widespread tropical savannas (Fig. 12.6A, B) (Mazzoni 1979; Kay et al. 1999; Zucol et al. 2010). In contrast, a subsequent, high-resolution study of 58 biosilica assemblages from the Sarmiento Formation at Gran Barranca, Patagonia suggested that grasses did not become abundant on the landscape until after the early Miocene, casting doubt on the early grasslands hypothesis for South America and the necessity for grass-dominated vegetation as a context for the evolution of hypsodonty (Fig. 12.6C) (Strömberg et al. 2013a). The Gran Barranca phytolith assemblages, which are dominated by forms typical of (woody) dicotyledons, palms, and include GSSCPs from bambusoid grasses, were initially thought to derive from forest vegetation (Strömberg et al. 2013a), an interpretation at odds with the overall loess-like sedimentology of the Sarmiento Formation (Bellosi 2010; Bellosi and González 2010; Bellosi and Krause 2014), carbon isotopic data indicating dry climates (Kohn et al. 2015), and the common association between hypsodont mammals and open habitats today (e.g., Janis 1988; Williams and Kay 2001; Eronen et al. 2010). However, as will be described below, a new, phytolith-based proxy for habitat openness has allowed a re-interpretation of these data as an open, non-analog palm shrubland (Fig. 12.6D) (Dunn et al. 2015b).

The reason for the widely divergent Eocene vegetation reconstructions between published studies of phytolith assemblages from essentially the same layers in Patagonia (Fig. 12.6B, C) (compare Zucol et al. 2010; Strömberg et al. 2013a), is almost certainly due to the different methods used for extraction and analysis of samples (described above). Work is underway to apply the different approaches to the same samples, with hopes of finding ways to calibrate the variation in vegetation inference and allow direct inter-study comparisons.

Reconstruction of vegetation change beyond grasslands: Phytolith analysis is the most effective method to detect grass-dominated habitats in the fossil record, but it can also be used to reconstruct vegetation change through space and time in non-grassland ecosystems. Although the taxonomic resolution of phytoliths is lower than that of palynofloras and macrofloras (Table 12.1), the relative abundances of various plant functional groups (e.g., different grass

Table 12.2 Phytolith proxies for habitat and climate reconstruction currently in use

Name	Type*	Formula	Definition	Reference
Vegetation Structure Indices				
D/P Index	grass dep.	*D/P*	*D* = circular rugose phytoliths; *P* = dumbell (bilobate) + cross + saddle GSSCPs + point-shaped (trichomes) + fan-shaped (bulliforms) + elongates	Alexandre et al. (1997b)
D/P Index	grass dep.	*D/P*	*D* = rough spherical phytoliths only; *P* = dumbell + cross + saddle + conical + polylobate GSSCPs + point-shaped (trichomes) + fan-shaped (bulliforms)	Bremond et al. (2005a)
D/P Index	grass dep.	Transfer functions relating *D/P* to LAI	*D* = globular granulate; *P* = saddle + cross + dumbell GSSCPs + point-shaped + fan-shaped	Bremond et al. (2005a)
D/P Index	grass dep.	*D/P*°	*D* = globular granulate; *P* = bilobate + cross + saddle + rondel + trapeziform polylobate + trapeziform GSSCPs	Bremond et al. (2008b)
D/P Index	grass dep.	*D/P*	D = sum of globular decorated phytoliths (including nodular), excluding echinate types (palms); *P* = GSSCPs	Neumann et al. (2009)
D/P Index	grass dep.	*D/P*	*D* = rugulose spheres + sclereids; *P* = GSSCPs + elongates + trichomes + bulliforms	Strömberg (2002)
FI/t Ratio	grass dep.	*FI TOT/FI TOT + GSSCP*	*FI TOT* = forest indicator totals; *GSSCP* = grass silica short cell phytolith total	Strömberg et al. (2007b)
D/P Index + Arboreal vs. Non Arboreal Pollen	grass dep.	LAI = 2.216 × ln (*D/P* + 1) + 0.885 × ln (*AP/NAP* + 1)	*D* = rough and smooth spherical; *P* = dumbell + cross + saddle + conical + polylobate GSSCPs; *AP* = Arboreal Pollen; *NAP* = Non-Arboreal Pollen	Aleman et al. (2012)
rLAI (reconstructed LAI)	grass indep.	rLAI = 0.0012 × *PA* + 10.4188 × *PUI* − 13.1621	*PA* = mean anticlinal epidermal phytolith area; *PUI* = mean anticlinal epidermal phytolith undulation	Dunn et al. (2015b)
Climate Indices				
Iph-Aridity Index	grass dep.	chloridoid saddles/(chloridoid saddles + panicoid + dumbell + cross GSSCPs)	Relative proportion of chloridoid grasses of the C_4 grasses	Diester-Haass et al. (1973)
Ic-Climate Index	grass dep.	*Ic*(%) = (rondel + trapeziform polylobate + trapeziform short cell)/(rondel + trapeziform polylobate short cell + trapeziform short cell + saddle + cross + bilobate GSSCPs)	Relative proportion of pooid grasses of all C_3 grasses present	Twiss (1992), Twiss et al. (1987)
Fs-Water stress index	grass dep.	*Fs* = % fan-shaped bulliforms/(sum grass phytoliths – elongates)	Relative proportion of fan-shaped phytoliths of all grass phytolith types excluding elongates	Bremond et al. (2005b)
AET/PET – Priestly– Taylor coefficient	grass dep.	$AET/PET\ (\%) = a_0 Fs + a_1 Iph + a_2(Iph{-}20)^2$	$a_0 = -0.605$; $a_1 = -0.387$; $a_2 = 0.272$; *Fs* = Water Stress Index; *Iph* = Aridity Index	Bremond et al. (2005b)

* Grass dep. = grass dependent; grass indep. = grass independent

subclades, woody and herbaceous dicotyledons, palms, zingibers), and the diversity of morphotypes within these groups have been successfully used to reconstruct changes in non-grassland plant communities through Quaternary time (e.g., Piperno and Becker 1996; Iriarte and Paz 2009; Neumann et al. 2009). Whereas these studies typically rely on comparisons with local, modern ecosystems, a broader set of interpretive guidelines is necessary in the paleontological record (Strömberg 2004; Thorn 2007).

An example of such deep-time work on non-grassland ecosystems is Thorn's (2001) study of late Eocene-early Miocene phytolith assemblages from drill cores from the McMurdo Sound region in the Ross Sea of Antarctica. Although the terrestrial assemblages recovered from these subaerial to marine sediments (Askin and Raine 2000; Raine and Askin 2001) are sparse, they hint at forests initially containing warm- and/or wet-loving palms and panicoid grasses in the? Eocene (or early Oligocene), to forests with rare cool-climate chionochloid grasses in the early Miocene, mirroring the global climatic cooling trend during this time (Thorn 2001). Other studies documenting non-grassy plant communities based on phytoliths include vegetation reconstructions for the Paleogene of Wyoming (Hyland et al. 2013a), late Miocene of Washington State (Dillhoff et al. 2014), and early-middle Miocene of Spain (Pinilla and Bustillo 1997) and China (Strömberg et al. 2007a).

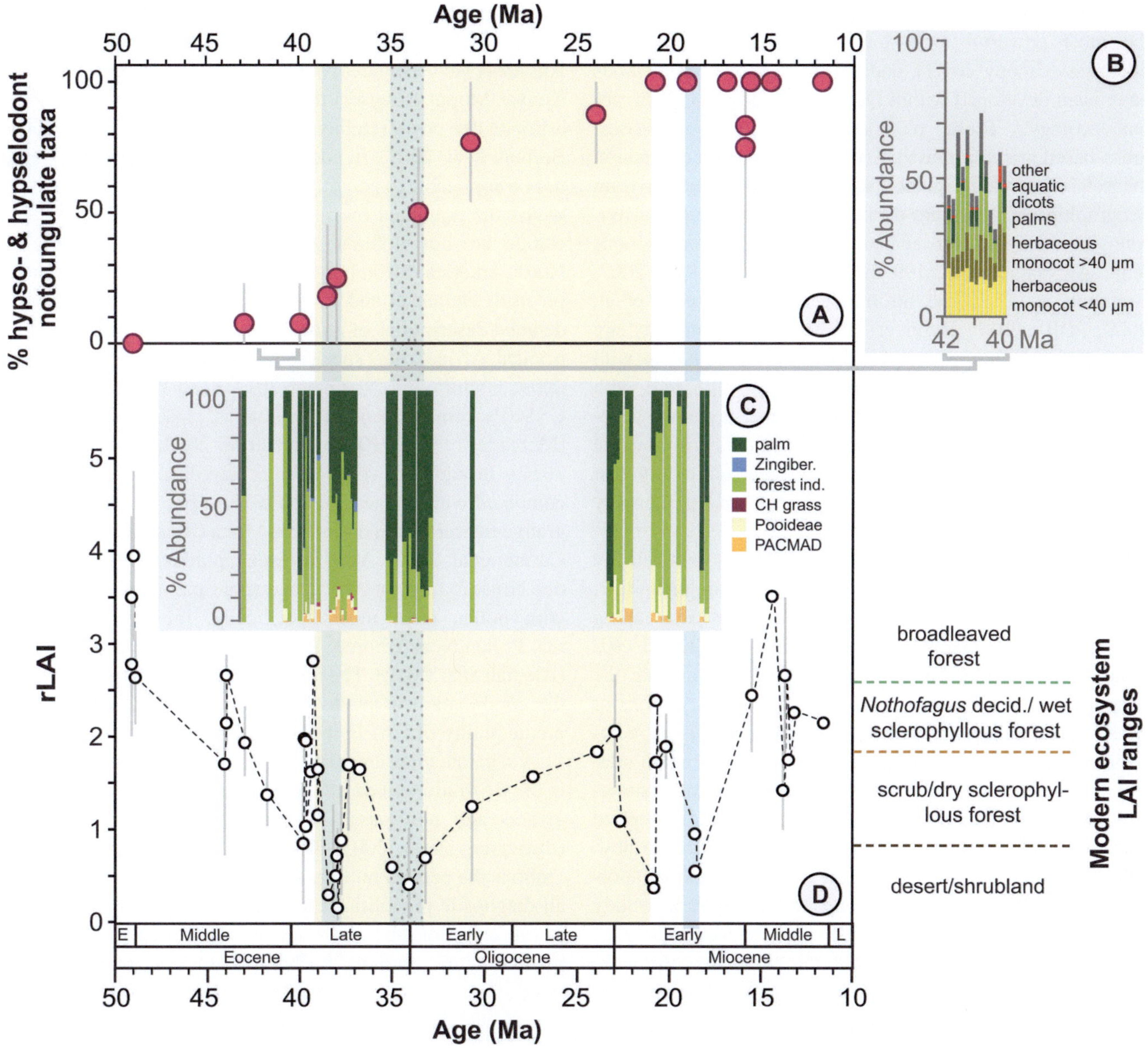

Fig. 12.6 Phytolith studies have been key for solving the riddle of "precocious" hypsodonty in endemic southern South American herbivores. A, The record of faunal hypsodonty in southern South America shows the relative abundance of notoungulate taxa with hypsodont or hypselodont (ever-growing) cheek teeth through time. The dramatic increase in hypsodonty prevalence occurs during the interval when rLAI $\leq$ 1.2 (light yellow field) (Dunn et al. 2015b). B, The relative abundances of selected phytolith functional groups from Gran Barranca Member of the Sarmiento Formation (~ 42–40 Ma), Patagonia, Argentina, indicating abundant grasses and interpreted as a tropical savanna (Zucol et al. 2010). Functional groups are: herbaceous monocots < 40 μm = micro-elongates + brief elongates + GSSCPs, (yellow); herbaceous monocots > 40 μm = various elongates + cuneiform and parallepipedal bulliform cells + acicular hair cells (brown); palms = globular echinate (light green); dicots = epidermal jigsaw + multicavate morphotypes (dark green); aquatic herbs = stellate parenchyma + conical + proteiform + fusiform + selected elongates (red); and other = not defined in text (gray). C, The relative abundances of the phytolith functional groups from Gran Barranca and Cañadón Vaca (~ 43–18.5 Ma), Patagonia, Argentina showing that, overall, open-habitat grasses (orange-PACMAD, yellow-Pooideae) occurred in low abundances (< 25%). Instead, phytolith assemblages were dominated by palms (dark green) and other forest indicators (light green), (Strömberg et al. 2013a). D, The record of rLAI based on epidermal phytolith morphology shows shifts in forest cover throughout the middle Cenozoic (Dunn et al. 2015b). Although there was little grass biomass in these ecosystems, habitats were extremely open during several intervals of time. These low rLAI intervals correspond with episodes of loessic deposition (light blue vertical bars) and abundance of burned phytoliths (light blue with dots) (Selkin et al. 2015). Based on stable isotopes of tooth enamel and bone, local climates during this time have been reconstructed as relatively warm but semi-arid (Kohn et al. 2015). Abbreviations: CH grass = closed-habitat grass (e.g., Bambusoideae), forest ind. = other forest indicators, rLAI = reconstructed Leaf Area Index, Zingiber. = zingibers

Habitat openness: Phytolith analysis constitutes one of the most powerful tools for directly estimating habitat openness (canopy cover), and several quantitative methods have been developed within Quaternary and pre-Quaternary paleoecology. These methods fall into two classes: grass-based and non-grass based. All use modern calibration datasets (either from plants, soils, or both) collected from geographic areas that are deemed similar to the paleoenvironments targeted for ancient vegetation reconstructions (e.g., Alexandre et al. 1997b; Barboni et al. 1999, 2007; Bremond et al. 2005a; Strömberg 2005; Mercader et al. 2009, 2010, 2011; Dunn et al. 2015b), though they are employed in other geographic locations as well. As with most proxy methods, they fundamentally (although with varying degrees of specificity) assume that a certain vegetation type was linked to a specific set of environmental conditions in the past as they are today, and additionally, that phytolith behavior in terms of production and taphonomy has not changed through time.

Today, most open habitats are either dominated by grasses, or at least contain a significant grass component (i.e., grassland, savanna, steppe, shrubland, desert and tundra vegetation types) (Archibold 1995). The grass-based estimates for habitat reconstruction therefore assume that the proportion of grass phytoliths to those from dicotyledons or other "forest indicators" will be an indicator of the degree of habitat openness. Several indices that seek to estimate relative tree cover have been developed based on this assumption. A widely used index is the *D/P* ratio (initially termed "(Dicot./Poa)" = Dicot phytoliths: sum of all Poaceae phytoliths; Table 12.2), which uses the globular granulate morphotype, abundantly produced by certain tropical woody dicotyledons in Africa, to indicate tree cover (Alexandre et al. 1997b). In this scheme, the lower the *D/P* value, the more open the habitat. Refinement of the *D/P* model was made by Bremond et al. (2005a) who derived a transfer function for quantifying habitat openness as Leaf Area Index (LAI, a continuous measure of habitat openness; Asner et al. 2003) from the *D/P* index using a training set of modern soil phytoliths collected in Cameroon. In subsequent papers, the *D/P* ratio has been further modified to exclude certain Poaceae phytolith types that are environmentally controlled and/or not necessarily diagnostic of grasses (e.g., acicular hair cells, bulliform cells, elongates; see Table 12.2) (Bremond et al. 2005b, 2008a). In other studies, *D* has been defined more broadly to include all globular decorated phytoliths including nodular forms (Neumann et al. 2009) as well as globular psilates and blocky types (Mercader et al. 2011), globular echinates from palms (WoldeGabriel et al. 2009), silicified sclereids (Strömberg 2002), or to exclude from the Poaceae sum collapsed saddles typical of closed-habitat grasses (Garnier et al. 2012). In general, all variations of the *D/P* ratio have proven successful in submodern–Holocene vegetation reconstructions in Africa, in particular in the lowlands of tropical Africa (Barboni et al. 2007). However, at higher latitudes (Mediterranean) or higher altitudes in Africa (Mount Kenya, Mount Rungwe) the D/P index is known to underestimate tree cover (Delhon et al. 2003; Bremond et al. 2004; Barboni et al. 2007). In addition, as currently defined, it may not be applicable to geographic regions (and time periods), in which the dominant morphotypes produced by woody vegetation are not globular granulates (e.g., the Cenozoic of North America, Strömberg 2004). Work is ongoing to more precisely define *D* and *P*, building on studies that provide detailed descriptions of phytolith in leaves, wood and bark of tropical African tree species (Runge 1996; Mercader et al. 2009; Collura and Neumann 2017), and that differentiates GSSCPs from African open-habitat vs. closed-habitat grasses (Mercader et al. 2010; Garnier et al. 2012; Neumann et al. 2017). In addition, some authors have used the *D/P* index combined with the general approach to produce a more integrative interpretation of floras in Africa (Neumann et al. 2009; Garnier et al. 2012). Another recent approach for determining tree cover in tropical African savannas merges the *D/P* index with pollen data, more specifically, the ratio of Arboreal Pollen/Non-Arboreal Pollen (AP/NAP) to estimate LAI (Aleman et al. 2012). The transfer function extrapolated from the combination of phytolith and pollen data improves the predictability of LAI in these ecosystems.

A separate grass-based system to quantify habitat structure is the *FI-t* ratio, aimed at studying relative changes between grasses and non-grasses as part of the general approach (Strömberg 2003, 2005; Strömberg et al. 2007b). The *FI-t* ratio is the percent non-grass "forest indicator" phytoliths of all diagnostic phytoliths, except aquatics (Table 12.2); here, forest indicator phytoliths include all forms (e.g., epidermal polygons, various tracheary elements, sclerenchyma, cystoliths, variously decorated globulars) diagnostic of woody and herbaceous dicotyledons, ferns, gymnosperms, and non-grass monocots such as palms (Arecaceae) and gingers (Marantaceae/Costaceae) (Fig. 12.1). Strömberg (2003, 2005) analyzed the GSSCP assemblages separately, in order to take the multiplicity and redundancy among GSSCPs into account; hence, phytoliths from closed habitat grasses (e.g., early-diverging grasses and Bambusoideae) are not included as forest indicators in the *FI-t* ratio. Instead, a full vegetation interpretation relies on study of the GSSCP assemblage composition as well (i.e., comparing the proportions of GSSCPs typically produced by closed-habitat vs. open-habitat grasses to modern grasses using bootstrapping, see below and Strömberg 2005; Strömberg et al. 2007b). As described above (*Grassland ecosystem evolution*), this method has been used to track the emergence of grass-dominated—and presumably open—habitats on several continents (e.g., Strömberg 2005; Strömberg et al. 2007b; Cotton et al. 2012; Chen et al. 2015).

Like other grass-based methods, the general approach/*FI-t* ratio has the potential to underestimate tree cover in ecosystems where the woody component consists of poor silica producers, and in forests with dense, grassy understories; thus, it is better used for tracking *relative* change in vegetation rather than for reconstructing *absolute* tree cover (Strömberg 2004). The general approach also gave misleading results when applied to phytolith assemblages of the middle Eocene-early Miocene of Patagonia (Fig. 12.6). Specifically, the low abundances of open-habitat grass morphotypes, and frequent palms and other forest indicator forms initially led Strömberg and co-authors (2013b) to infer subtropical forests (Fig. 12.6B). Although the 'general approach'/*FI-t* ratio succeeded in showing the scarcity of grasses in these middle Cenozoic ecosystems, it failed to correctly infer habitat openness, as shown by subsequent work using non-grass based approaches (see below) (Dunn et al. 2015b). This case demonstrates that grass based methods such as the 'general approach'/*FI-t* ratio cannot easily resolve vegetation structure when grasses are rare or absent, especially in non-analog systems.

An alternative, taxon-free (non-grass based) approach to estimate vegetation openness is "reconstructed LAI" (*rLAI*; Table 12.2) (Dunn et al. 2015b). This method was developed to circumvent the grass-dependence problem of the *FI-t* and *D/P* approaches. The *rLAI* method uses phenotypic plasticity of pavement cells (preserved as phytoliths) in leaf epidermis, whose shapes and sizes vary in response to different light environments. The approach is based on the observation that leaves of dicotyledons and ferns grown in shade have larger and more undulated epidermal cells than those grown in direct sunlight (e.g., Watson 1942; Kürschner 1997). Dunn et al. (2015b) conducted a modern analog study in Costa Rica to establish an empirical relationship between pavement cell phytolith shape/size and Leaf Area Index; this work resulted in a model for quantifying tree cover (Table 12.2). The method was applied to middle Cenozoic strata of Patagonia and the resulting long term record of LAI is strikingly concordant with other lines of evidence for open, dry habitats (e.g., sedimentology, mammalian hypsodonty, carbon isotopes; Bellosi 2010; Bellosi and González 2010; Madden et al. 2010; Kohn et al. 2015), despite the relative lack of grass phytoliths in the same samples (Fig. 12.6) (Strömberg et al. 2013a; Dunn et al. 2015b). Hence, by using the *rLAI* and *FI-t* methods simultaneously, it is possible to estimate grass abundance and habitat openness independently and differentiate shrublands (that also contain "forest indicator" phytolith morphotypes) from forests.

Late Neogene spread of C_4 grasslands: During the late Miocene–Pliocene, C_4 grasses rose to ecological dominance at low- to mid-latitudes on many continents (e.g., Cerling et al. 1997; Osborne 2008; Edwards et al. 2010). Until recently, this transition was documented using mainly stable carbon isotopes of tooth enamel, paleosol carbonates, carbonized rootlets, and leaf waxes (e.g., Levin et al. 2004; Passey et al. 2009; Cotton et al. 2012; Feakins et al. 2013). An inherent problem with these methods is that they cannot differentiate important aspects of the C_3-C_4 transition, namely (a) whether C_4 grasses replaced C_3 trees/shrubs or C_3 grasses, and (b) which C_4 grasses were important in grass communities (e.g., drought-tolerant species in the Chloridoideae or mesic, fire prone grasses in the Andropogonae); thus, the nature of this ecological shift has remained unclear (see Osborne 2008; Edwards et al. 2010).

Phytolith assemblages have lately helped shed light on this transition owing to their ability to reconstruct grasses vs. woody vegetation and provide information about relative abundances of pooids and different C_3 and C_4 PACMADs in grass communities. To account for the current uncertainty in how informative phytolith morphology is with regard to photosynthetic pathway, Strömberg and McInerney (2011) used two phytolith metrics to establish relative abundance ranges of potential C_4 grasses: a minimum estimate using all GSSCPs typical of chloridoids and panicoids and a maximum estimate using all GSSCPs typical of PACMADs. This approach has permitted a conservative assessment of the importance of C_4 grasses in Neogene ecosystems in both North and South America. Specifically, these studies have shown that the spread of C_4 grasses was geographically complex and influenced by regional and local factors, rather than—or in addition to—global factors (e.g., pCO_2 changes) (Cotton et al. 2012, 2014; Chen et al. 2015) (see also Rossouw et al. 2009; Strömberg 2011). For example, in the central Great Plains, the late Miocene expansion of C_4 grasses occurred at the expense of trees (including palms) and C_3 pooid grasses, indicating that climatic drying may have been an important, regional driver of this change (Strömberg and McInerney 2011; Fox et al. 2012a; but see Fox et al. 2012b). In contrast, in parts of northwestern Argentina, late Miocene-Pliocene phytolith assemblages and stable carbon isotope data suggest that tectonic uplift prevented C_4 grasses from expanding, despite their rise to ecological dominance elsewhere during this time (Cotton et al. 2014).

Many of the C_4 grassland studies involving phytoliths have included stable carbon isotope data from paleosol carbonized rootlets (e.g., Cotton et al. 2012, 2014; Chen et al. 2015) or phytoliths (McInerney et al. 2011), allowing for cross-proxy comparisons. These comparative analyses suggest overall good correspondence between phytoliths and isotopes in terms of broad patterns of C_4 dominance (Smith and White 2004; McInerney et al. 2011; Cotton et al. 2012, 2014; Chen et al. 2015). The discrepancies that do exist point to differences in temporal or spatial averaging among proxies (e.g., Cotton et al. 2012). In addition, these studies have highlighted the need for a more phylogenetically precise phytolith classification to more accurately distinguish PACMADs with C_4 vs. C_3 photosynthesis (discussion in McInerney et al. 2011).

Paleoecological context of hominin evolution: A body of work has used phytolith analysis to provide vital clues to the environmental context of Pliocene-Pleistocene hominin evolution in East and Central Africa (e.g., Albert et al. 2006, 2009; WoldeGabriel et al. 2009; Barboni et al. 2010; Cerling et al. 2010; Rossouw and Scott 2011; Novello et al. 2015; Albert and Esteban 2016). Phytolith analysis in combination with macrofossils and pollen/spores suggests that wooded grasslands, savannas, or grassland-forest mosaics dominated East and Central African lowlands as climates got progressively drier 4–1 Ma, although locally, forests and wetlands existed even in arid phases (Bamford et al. 2006; 2008; WoldeGabriel et al. 2009; Barboni et al. 2010; Albert and Bamford 2012; Barboni 2014; Novello et al. 2015; Habermann et al. 2016). In addition, phytoliths combined with stable carbon isotopic data indicate that the C_4-dominated (wooded) grasslands associated with Pliocene-Pleistocene hominins were preceded by grassy vegetation with mainly (C_3) pooid open-habitat grasses (e.g., Cerling 1992; Edwards et al. 2010; Rossouw and Scott 2011; Barboni 2014). Therefore, although further testing is necessary, phytolith data provide at least tentative support for a broad association between human evolution (of e.g., bipedalism) and the spread of C_4 savannas (the so-called savannah hypothesis, Cerling et al. 2010).

Phytoliths and Animal Diet

Plant silica can provide unique and direct information about plant-animal interactions in deep-time if preserved in a dietary context, such as associated with teeth or in feces.

Phytoliths and teeth: Phytoliths are known to become wedged into the enamel surface of (sub)fossil human and ape teeth, constituting direct, albeit incomplete, paleodietary evidence (e.g., Ciochon et al. 1990; Fox et al. 1996; Gügel et al. 2001). This approach has rarely been applied in studies of older deposits, where the risk for contamination may be prohibitory.

In contrast, a body of work has used phytoliths in fossilized dental calculus to reconstruct the plant aspect of animal diets. Dental calculus (i.e., tartar) is mineralized plaque that adheres to enamel surfaces in many vertebrates (including humans, dogs, cats, equids, and sheep, see Table 12.3) most abundantly near openings to salivary ducts (Jubb et al. 1985). Food particles, including phytoliths, become trapped in the calculus as it forms and subsequently fossilize, forming a time-averaged record of food intake (e.g., Gobetz and Bozarth 2001; Henry et al. 2011). Thirty years of phytolith-based dental calculus research has deduced paleodiets for several mammalian lineages, for example ungulates, proboscideans, and, above all, primates, including humans and various hominids (Table 12.3) (e.g., Fox et al. 1996; Gobetz and Bozarth 2001; Boyadjian et al. 2007; Henry and Piperno 2008; Wesolowski et al. 2010; Henry et al. 2011; Asevedo et al. 2012; Henry 2012). For example, Henry et al.'s (2012) study of phytoliths in dental calculus, stable carbon isotopes, and tooth wear suggested that *Australopithecus sediba* fed on a diversity of C_3 plants (e.g., sedges), likely in a gallery forest in an otherwise C_4-dominated savanna ecosystem. Similarly, phytoliths in calculus have helped map out food consumption in Neanderthals (Henry et al. 2011, 2014). Among the few non-primate applications, Gobetz and Bozarth (2001) used the distribution of types of plant silica bodies extracted from calculus of modern, Central Great Plains ungulates and late Pleistocene mastodons (*Mammut americanum*) to show that these extinct proboscideans ingested more grasses than traditionally thought, and Asevedo et al. (2012) inferred a generalist plant diet for late Pleistocene gomphotheres from Brazil based on phytoliths in fossil plaque.

Despite the successful employment of phytolith analysis in dental calculus in archaeological and Quaternary studies, this method has not been used to explore biotic relationships in deep-time. We sampled dental calculus of middle Miocene-aged (Barstovian; 16.3–13.6 Ma) mammals (equids, oreodonts, camelids, tayassuids; Appendix 12.2) of the Deep River Formation, Montana, and compared phytolith assemblages in calculus to those extracted from associated sedimentary matrix to test the potential for phytoliths-based dietary inference in pre-Quaternary aged strata (see Appendix 12.3 for a full description). We found that phytoliths were preserved in some but not all (7 of 19) dental calculus samples (Appendix 12.2). Among them, two oreodont samples contained phytoliths that were abundant enough to warrant counting. Analysis of these two samples indicated that both oreodonts included grass in their diet, and they showed individual differences in the amount of grass relative to non-grass phytoliths recovered (Fig. 12.7). Whether one oreodont ingested more grass than the other is difficult to say, as at this time, there are no studies explicitly relating calculus phytolith abundance to ingested plant biomass. However, these results are informative as previously the merycoidodontid oreodonts were thought to be primarily browsers or mixed feeders based on their tooth morphology (Lander 1998). The occurrence of grass phytoliths in the oreodont calculus agrees with the overall Eocene–Miocene trend towards more abrasive feeding within this group (Mihlbachler and Solounias 2006).

Table 12.3 Vertebrate species (modern, fossil) studied for their calculus production and phytolith content of calculus

Taxon	Common name (age for fossils)	Calculus production	Phytolith production	Selected reference
Modern				
Bos taurus	Cow	variable	few	Middleton and Rovner (1994)
Ovis aries/Capra hircus	Sheep/Goat	variable	few	Middleton and Rovner (1994)
Sus scrofa	Pig	variable	few	Middleton and Rovner (1994)
Alces alces	American moose	a little	few	Gobetz and Bozarth (2001)
Odocoileus virginianus	White-tailed deer	a little	many	Gobetz and Bozarth (2001)
Tapirus bairdii	Baird's tapir	a little	few	Gobetz and Bozarth (2001)
Fossil/Sub-fossil				
Bos taurus	Cow (Holocene)	a little	a little	Armitage (1975)
Gigantopithecus blacki	Asian ape (Pleistocene)	a little	few	Ciochon et al. (1990)
Homo sapiens	Human (Holocene-submodern)	variable	very few	Fox et al. (1996), Henry and Piperno (2008), Lazzati et al. (2016)
Bison bison	American bison (Holocene)	a lot	many	Gobetz and Bozarth (1998)
Mammuthus columbi	Columbian mammoth (Pleistocene)	(sampled dentine and cementum)	many	Gobetz and Bozarth (2000)
Gigantocamelus frickii	"Giraffe-camel" (Pliocene)	a lot	many	Gobetz and Bozarth (2001)
Mammut americanum	American mastodon (Pleistocene)	variable	many	Gobetz and Bozarth (2001)
Homo neanderthalensis	Neanderthal (Pleistocene)	a lot	few	Henry et al. (2011)
Bovidae, various species	Bovid (Pleistocene)	a little	very few	Brophy et al. (2016)
Merychippus	Horse relative (middle Miocene)	a little	none to few	this study
Tayassuid	Peccary relative (middle Miocene)	a little	none	this study
Leptauchenia	Oreodont (middle Miocene)	a little	none to few	this study
Merycoidodon	Oreodont (middle Miocene)	a little	few to many	this study
Promeryochoerus	Oreodont (middle Miocene)	a little	none to few	this study
Merycoidodontoidea	Unidentified oreodont (middle Miocene)	a little	few to many	this study
Camelidae	Unidentified camelid (middle Miocene)	a little	none	this study

Although our study and previous work are encouraging, there are several caveats that should be kept in mind for applying dental calculus-based phytolith analysis in deep-time. First, the occurrence of dental calculus among animals is not well known beyond humans and certain domesticated species (e.g., sheep, goats, pigs) (Table 12.3). Among the studied taxa, calculus is commonly found in humans, cats, dogs, and pigs, and occasionally in horses, sheep, and goats (Weaver 1964; Jubb et al. 1985). Although other species are known to produce dental calculus (e.g., tapir, moose, and white-tailed deer) (Gobetz and Bozarth 2001), they are comparatively understudied. Second, the appearance and position of calculus vary among animals. For example, on human teeth calculus can be off-white to black in color, be bound tightly to enamel surfaces, and can be found abundantly on lingual surfaces of mandibular incisors and buccal surfaces of first molars (Lieverse 1999). In contrast, in horses, dental calculus is softer, chalky, and easily removed from the cementum surface, but it is usually deposited on the canines, and particularly in male horses (Baker 1979; Jubb et al. 1985; Dixon et al. 1999). The chemistry of calculus also depends on the taxon. For

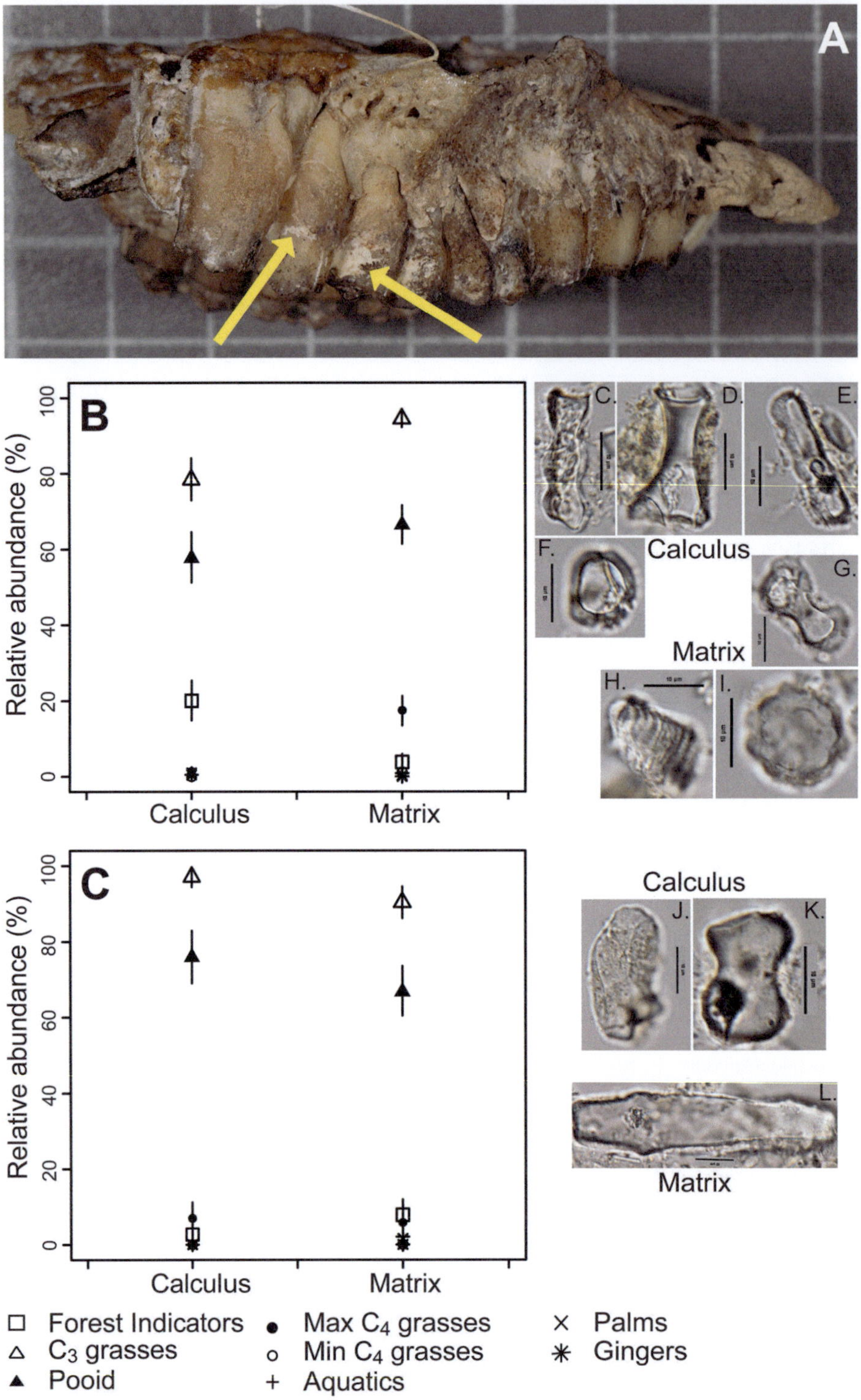

Fig. 12.7 Preliminary study of phytoliths in dental calculus in herbivores from the middle Miocene Deep River Formation, MT. A, Dental calculus on fossil oreodont teeth (marked by arrows). B-C, Relative abundance of phytolith plant functional types (out of the entire diagnostic count) in calculus samples from Miocene oreodonts: B, *Merycoidodon* (UWBM 54624); C, Unidentified oreodont (UWBM 32414). D-M, Examples of phytoliths observed in calculus and matrix samples: D, F, Crenate GSSCP (ICPN 1.0 name: trapeziform polylobate) indicative of Pooideae open-habitat grasses; E, Tall, *Chusquea*-type rondel with spiked top indicative of closed-habitat grass; G, Rondel GSSCP phytolith indicative of Poaceae; H, L, Simple lobate bilobate GSSCP (ICPN 1.0 name: bilobate short cell) indicative of PACMAD open-habitat grasses; I, Cylindric sulcate tracheary element; J, Globular echinate from palm; K, 3D blocky polyhedron; M, Elongate facetate. Scale bars equal 10 μm. Institutional abbreviation: UWBM = Burke Museum of Natural History and Culture

example, dogs produce calculus consisting of calcium phosphate whereas horse plaque mineralizes with calcium carbonate (Jubb et al. 1985). These differences influence the necessary sampling (and chemical extraction) approach for individual taxa, as well as the potential for retrieving enough material for analysis. A broad survey of the production and appearance of calculus, as well as its phytolith content among living herbivores and omnivores is therefore needed to allow for the development of appropriate sampling methods.

Third, because the study of dental calculus in deep-time is in its infancy, the vertebrate paleontology community has not been aware of the potential for finding fossilized plaque on their specimens. Thus, in some of the largest and most well curated collections of Cenozoic mammals, calculus appears to have inadvertently been removed from fossil teeth during preparation (C. Strömberg & E. Harris, personal observation). Aspiring students of paleodiet by means of dental calculus will instead have to rely on unopened casts and new collections.

Phytoliths and fossil feces: Analysis of biosilica assemblages extracted from fossilized dung (coprolites) is relatively commonly used in archaeology and Quaternary paleoecology for inferring ancient diets and environments (Katz et al. 2007; Delhon et al. 2008; Ghosh et al. 2008; Tsartsidou et al. 2008; Bamford et al. 2010; Shahack-Gross 2011; Shahack-Gross et al. 2014; Panadès i Blas et al. 2017). Pre-Quaternary coprolites have not been studied extensively for their phytolith content. In part, this is because herbivore coprolites are less easily recognizable in the field compared to carnivore dung (e.g., Linseele et al. 2013), hence the material available for study has been limited. Another obstacle is that, even when herbivore dung can be identified, it can be difficult, if not impossible to know exactly what animal produced a particular coprolite.

Nevertheless, researchers can sometimes use other information from the dung itself or associated trace fossils, combined with knowledge of the co-occurring fauna, to infer what type of animal was likely responsible. For example, because of the size and plant content of coprolites from the Late Cretaceous of Central India, they could be linked to the only large herbivores in the fauna, titanosaur sauropods (Ghosh et al. 2003; Mohabey and Samant 2003). Phytoliths contained in these coprolites could therefore shed light on dinosaur diets at the end of the Cretaceous (Mohabey and Samant 2003; Prasad et al. 2005, 2011). Similarly, Oligocene fossil scarab dung beetle brood balls from Patagonia, Argentina were tentatively attributed to medium-sized meridiungulate herbivores based on observed relationships between the size of brood balls and dung mass (Halffter and Matthews 1966), and of dung mass and dung producer (e.g., Blueweiss et al. 1978; McCullough 1982), respectively (Strömberg and Stidham 2001). Although no actual dung was preserved in these trace fossils, knowledge of the unique brood ball behavior of the South American scarab beetles (e.g., that they typically pack dung in a layer of clay and that the larvae line the resulting chamber with their feces as they feed on the dung; Halffter and Matthews 1966; Hanski 1983), has permitted interpretation of both the diet and habitat of the dung producers (Strömberg and Stidham 2001; Sánchez et al. 2010). These studies both pointed to diets and habitats with overall very little grass, consistent with later work on paleosol phytoliths from the same sections (Strömberg et al. 2013a).

Climate and Environmental Indicators: Nearest Living Relative Methods

Certain non-grass monocotyledonous plants, such as palms (Arecaceae) and several members of families within the Zingiberales (e.g., spiral gingers, arrowroots, bananas) produce phytoliths that are diagnostic at or below the family level (Piperno 1985, 2006; Chen and Smith 2013). Because they also have relatively narrow environmental tolerances (Dahlgren et al. 1985; Uhl and Dransfield 1987; Greenwood and Wing 1995; Couvreur and Baker 2013; Archibald et al. 2014), they are not only informative for vegetation composition but can also be used to constrain local climate inference (e.g., Strömberg 2004; Chen and Smith 2013; Harris 2016). For example, palms cannot survive prolonged periods of frost; hence they don't typically live in areas with coldest month mean temperatures below 5°C (Greenwood and Wing 1995; Archibald et al. 2014). In addition, this sensitivity is exacerbated under the higher atmospheric CO_2 levels of, for example, the Paleogene (Royer et al. 2002). Members of Zingiberales tend to inhabit frost-free, humid (sub)tropical forests, but the preferred microhabitat differs among families (Chen and Smith 2013). Similarly, Poaceae subclades have highly variable ecologies, including early-diverging and bambusoid forest-dwelling taxa, pooids in cool temperate grasslands, chloridoids in arid grasslands, and sorghum grasses (Andropogonae) that dominate in fire-prone, mesic savannas (Soderstrom and Calderón 1974;

Calderón and Soderstrom 1980; Taub 2001; Edwards and Smith 2010; Quigley and Anderson 2014; Simpson et al. 2016). Many of these groups can be identified by their short cell phytoliths, and reconstruction of grass community composition based on GSSCPs can therefore provide vital clues to local-regional environmental conditions (open vs. closed, dry vs. mesic, hot vs. cold) (Piperno 1988, 2006; Piperno and Pearsall 1998; Strömberg 2004; Strömberg and McInerney 2011).

Based on these strong climatic tolerances/preferences of grasses, several phytolith-based indices have been developed to infer local-regional climate. Among the most commonly employed is the Ic index ("climate index"), which is the percentage of (cool temperate) pooid grasses in the sum of (open-habitat) grasses, and serves as a proxy for regional temperature and climatic zonation along elevation gradients primarily in (sub)tropical areas (e.g., Twiss 1992; Alexandre et al. 1997b; Barboni et al. 2007; Ghosh et al. 2008; Hwang et al. 2012; Calegari et al. 2013). Similarly, the Iph index estimates aridity by looking at the percentage of (commonly arid-adapted) chloridoids within PACMAD grasses (e.g., Diester-Haass et al. 1973; Alexandre et al. 1997b; Bremond et al. 2005b; Barboni et al. 2007; Coe et al. 2014).

Phytoliths of plants that tend to live in or near water or in wetlands, such as sedges, horsetails, and river-weeds have been used to infer proximity to wetlands or bodies of water in the local environment (e.g., Bertoldi de Pomar 1971; Piperno 1988; Strömberg 2004; Guo et al. 2012). The interpretation of water in the local habitat can also be supplemented by information on the relative abundance and composition of other siliceous fossils of diatoms, sponge spicules and chrysophyte cysts (e.g., Piperno 1988; Stevenson et al. 1996; Clarke 2003; Novello et al. 2015).

Climate and Environmental Indicators: Phenotypic Plasticity Methods

Lately, numerous studies have used morphometrics and multivariate statistics to link GSSCP morphology to different aspects of climate, light, and fire. These methods have yet to be applied to deep-time work—and it remains unclear whether they can be applied with precision.

Based on a morphometric study, Lu and Liu (2003a) suggested that xerophytic grasses (Chloridoideae) produce lobate GSSCPs with longer shanks (narrow zone between lobes) than do hygrophytic grasses (e.g., Oryzoideae, Panicoideae). However, these patterns have not been confirmed in experimental studies controlling for water availability, and may be linked to phylogenetic relatedness instead. Work by Wang and Lu (1993) on GSSCP morphology in specific grass species in different locations in China indicated that phytolith shape was constant across an environmental gradient, but that there were slight differences in GSSCP size. Liu et al. (2016) found that phytolith size increases in humid environments with increased mean annual temperature (MAT) in *Phragmites autralis*, but saw the opposite pattern in semi-arid regions where the same grass grows, namely that GSSCPs became smaller as MAT increased.

The formation of certain non-diagnostic phytolith morphotypes (i.e., long cells forming elongate psilate, sinuous, and echinate phytoliths, stomata, and bulliform cell phytoliths) in the leaves of grasses depends in large part on environmental factors, specifically water availability and evapotranspiration (Sangster and Parry 1969; Bremond et al. 2005b; Piperno 2006; Madella et al. 2009). In open habitat grasses, drier climates may result in increased evapotranspiration, leading to enhanced deposition of these non-diagnostic forms (Bremond et al. 2005b). On the other hand, ample access to water in warm climates can lead to comparable transpiration rates and high numbers of bulliform cell phytoliths (Sangster and Parry 1969; Madella et al. 2009); this explains why some bamboos tend to deposit a lot of these morphotypes (Strömberg, unpublished data). Based on the environmental controls on bulliform production, Bremond et al. (2005b) developed an aridity proxy, the so-called fan-shaped (Fs) index, later dubbed the hydrous stress index (Bi) (Messager et al. 2011), defined as the ratio of the bulliform type over the sum of GSSCPs, acicular hair cells, and bulliform cell phytoliths. This index has become widely employed to assess changes in water stress (Messager et al. 2011; Coe et al. 2013, 2014; Parolin et al. 2014; Augustin et al. 2015). Strömberg et al. (2007b) used the ratio of bulliforms to GSSCPs in combination with types of GSSCP phytoliths, whether from mesic or dry adapted grasses, to bolster interpretation of water availability in the habitat (wetland vs. arid).

Light environment has been documented to have significant influence on gross leaf morphology and cell morphology in many taxa (e.g., Givnish 1988; Kürschner 1997; Uhl and Walther 2000). To examine the effects of light environment on grass cell morphology, Dunn et al. (2015a) performed a controlled experiment where they grew five species of grass under four light regimes. They found that in low light conditions, grass long cells (elongate psilate, elongate sinuous) were significantly larger in area and longer than in high light treatments, except in the bamboo

(*Chusquea culeou*), which appeared unaffected by variation in irradiation. The morphology of bilobate, cross, and saddle GSSCP types was unaffected by light, but as for long cells, crenate and polylobate GSSCPs (ICPN 1.0 names: trapeziform polylobate, cylindrical polylobate) were typically longer and larger in shady conditions. Because light environment is a proxy for habitat openness, these results suggest that it may be possible to infer canopy structure from grass phytolith morphology, but the idea has yet to be tested in soil phytolith assemblages across a light gradient (Dunn et al. 2015a).

Fire Regime

Phytoliths that are subjected to heat from fire can become darkened and opaque, have darkened centers, and turn brownish in color; under high fire temperatures they can deform, lose surface ornamentation, and multiple phytoliths can become fused (Parr 2006; Piperno 2006; Albert and Cabanes 2008). The study of such changes in phytolith shape, color, and texture has allowed archaeologists to trace human habitation and use of fire (Kealhofer 1996; Boyd 2002; Albert and Cabanes 2008; McMichael et al. 2012a). In deep-time paleoecological studies, this approach is less commonly employed, but recently burned phytoliths combined with magnetic proxies for fire were used to examine fire frequency across the Eocene–Oligocene Transition in Patagonia, Argentina (Selkin et al. 2015).

Determining whether a phytolith is burned can be problematic in cases where slight color changes or moderate deformation of the phytolith grains has occurred via weathering processes (Evett and Cuthrell 2016), and distinguishing burned from non-burned phytoliths based on color alone is highly subjective. Recent publications have attempted to develop more quantitative, objective ways to detect phytoliths affected by burning. For example, a method by Elbaum et al. (2003) compares the refractive index in burned phytoliths vs. unburned grains, and Pironon et al. (2001) use Raman spectroscopy to measure changes in carbon bonding that occurs upon exposure to high temperatures. However, note that later authors (Evett and Cuthrell 2016) were unsuccessful in repeating this method consistently.

Today, fire is an important contributor in maintaining savannas (Sankaran et al. 2007; Griffith et al. 2014; Lehmann et al. 2014) and in deeper time, it has been implicated in C_4 grassland evolution during the late Miocene (Keeley and Rundel 2005; Hoetzel et al. 2013). To test the C_4 grass vs. fire hypothesis and address other outstanding questions concerning climate-vegetation-fire feedbacks in deep-time, fire histories should be added to other lines of evidence for local paleoecology. Research involving burned phytoliths could potentially contribute substantially to this effort.

Methodology of Phytolith Analysis

Collecting for Phytolith Analysis

What sedimentary facies and type of sediment that should be collected for phytolith analysis depends ultimately on the question being asked (e.g., What was the vegetation around lake X? In what environment did mammal Y live?). However, as a rule of thumb (see *Preservation of Phytoliths in Soils and Sediment*), paleosols developed on sediments rich in volcanic ash and lake sediments with abundant diatoms should produce the best-preserved phytoliths, with the only disadvantage being that phytolith assemblages from diatom- or ash-rich samples can sometimes be "diluted," hence hard to count (Strömberg 2003; Strömberg et al. 2007a). In contrast, on average, very coarse sediments (medium sand and coarser), red beds with abundant clay, and organic rich sediments such as lignites appear to be the least productive (Strömberg et al. 2007b; Dillhoff et al. 2014; Strömberg et al. unpublished data).

Sediment samples collected for phytoliths should be unweathered and void of modern roots; hence, collecting near the modern soil surface should be avoided. Within a paleosol profile, samples for phytolith analysis should preferably be taken in the uppermost part, from the A-horizon (as evidenced by root traces, burrows, and color changes, Retallack 2001a). However, note that paleosols can often be truncated by erosion, in which case it is hard to know how much, or if any, of the uppermost soil horizons are preserved. In such cases the lower A- or the B-horizon, which can sometimes produce relatively well preserved phytoliths serve as substitutes (Strömberg et al. unpublished data). Ongoing work seeks to test how vegetation reconstructions might vary depending on whether phytolith assemblages from the upper vs. the lower A horizon are used (Crifò and Strömberg 2016).

Facies and sediment type determines the temporal and spatial resolution of the phytolith assemblage data (see *Spatial and Temporal Resolution of the Phytolith Record*). In a lake deposit with clear lamina careful sampling can render resolution on the order of years or tens of years (Behrensmeyer and Hook 1992; Liang 2004). Likewise, an ash deposit is likely to have formed over a relatively short amount of time (e.g., Wing et al. 2012). In contrast, paleosol A-horizons have often undergone bioturbation such that time averaging of the silica assemblages may be several hundreds of years (Fig. 12.4). Nevertheless, because there can be temporal stratification of phytolith assemblages within paleosol horizons (see above), sampling within a relatively narrow vertical (e.g., a few centimeter) zone may help minimize time averaging (Fig. 12.3) (e.g., Strömberg 2005; Miller et al. 2012). Similarly, to avoid mixing assemblages of different ages, the lateral extent of the sample should be restricted to 10–15 cm (e.g., Strömberg 2003; Strömberg et al. 2013a). Note that this contrasts sharply with the sampling protocol for modern reference studies, where small subsamples from a larger area (e.g., 30 × 10 m^2; 40 m^2; 1 m^2; or not specified) are combined into a single, mixed sample to represent a larger swath of vegetation (Fredlund and Tieszen 1994; Kerns et al. 2001; Bremond et al. 2005a; Aleman et al. 2012). This methodology uses spatial mixing at the plot level as a way to compensate for the temporal mixing of the vegetation signal that happens in a paleosol. However, no study has so far tested for variability between subsamples, or provided a rationale for the choice of this sampling methodology (but see Crifò and Strömberg 2016).

To reconstruct changes through space, lateral transects can be done following distinct ashes or other horizons that can reliably be inferred to represent the same, relatively narrow time frame (see e.g., WoldeGabriel et al. 2009; Strömberg et al. 2013a; Chen et al. 2015). It is less clear how far apart such samples need to be taken, in particular since spatial averaging differs between vegetation types (Wallis 2001; Piperno 2006). However, Aleman et al. (2012) suggest that a spacing of 20 meters can capture vegetation gradients in an African savanna.

Evaluation of Preservation of Phytolith Assemblages

Assemblages that show a full range of phytolith sizes (<5 micrometers to >80 or >100 micrometers), and have no or few signs of alteration or etching are ideal for quantitative analysis. In contrast, if assemblages appear size sorted (e.g., only morphotypes with a diameter of 10–20 micrometers are represented), caution should be taken when interpreting assemblages (see Selkin et al. 2015). Likewise, if most morphotypes are altered, severely etched, or mixed with a lot of secondary silica (Fig. 12.5), it may signal incomplete preservation, potentially resulting in biased or uninformative counts (e.g., in Hyland et al. 2015). Such assemblages may still provide qualitative information, for example, the relative abundance of short cells vs. palms (similarly-sized morphotypes that are relatively resistant to dissolution and also typically recognizable at high levels of etching, breakage, and alteration) (e.g., Strömberg 2003, unpublished data).

To evaluate the preservational state of a phytolith assemblage Fredlund and Tieszen (1997b) developed a semi-quantitative scheme that classifies the degree of pitting and erosion of elongates and bulliform cells. Strömberg et al. (2007b) recommended studying seven features of the assemblages, namely the presence and abundance of (a) occluded carbon, (b) fine ornamentation, (c) etching, (d) fragmentation of phytoliths, (e) structural/textural alteration of phytoliths, and (f) secondary silicification, to come up with a semi-quantitative assessment of phytolith assemblage preservation; however, this system has so far not been standardized. In the archaeological literature, several quantitative tools to evaluate the likelihood of bias in phytolith assemblages have been proposed, including testing for a correlation between the number of morphotypes identified and phytolith concentration in the acid insoluble fraction (Madella and Lancelotti 2012). The presence of a correlation between these parameters would point to biased phytolith assemblages with lower phytolith yield. Such methods have yet to be adapted for deep-time phytolith analysis.

Phytolith Extraction Methods Currently in Use

In modern (and fossil) soils, phytoliths often constitute <1% of soil total dry weight, and tend to be bound to soil particles (e.g., Jones and Beavers 1964; Blecker et al. 1997). The principle of phytolith extraction is therefore to first "free" the phytoliths by disaggregating the sediment (pretreatment), and then to isolate or concentrate them relative to the remaining sediment (separation). Numerous different approaches to extracting phytoliths from modern and Quaternary soils and sediments exist, and papers continue to be published on this topic (e.g., Rovner 1971; Madella et al. 1998; Piperno 1988; Zhao and Pearsall 1998; Parr 2002; Katz et al. 2010). Most extraction methods so far employed for deep-time studies are variations on standard wet-ashing procedures established by, for example, Piperno (1988); these methods are described in a handful of publications (Zucol and Osterrieth 2002; Strömberg 2004; WoldeGabriel et al. 2009; Zucol et al. 2010).

The basic steps of these extraction protocols are:

(1) *Pretreatment* of the sediment sample, namely dispersion of soil aggregates, which involves:

   (a) removal of cements such as carbonates (using hydrochloric acid) and (less commonly) oxides (using strong acid), which bind the particles together;
   (b) (commonly) deflocculation treatment (using e.g., Na_2H_2EDTA) to prevent the sediment from clumping;
   (c) removal of organics (using an oxidant such as bleach or Schultze's solution and, in very

organic-rich sediments, an additional step using potassium hydroxide to remove humic colloids) and clays (by means of various gravitational settling techniques) that will otherwise obscure the view of phytoliths on the microscope slides;

(2) *Separation and concentration* of biosilica (e.g., phytoliths, diatoms, sponge spicules) from the mineral fraction. This step uses "heavy liquid," a solution with a specific gravity of 2.2–2.4, which allows the lighter biosilica (specific gravity = 1.5–2.3) to rise to the top and the heavier silicate minerals (specific gravity $\geq$ 2.65) to sink;

(3) *Preparation of microscope slides* from the phytolith residue.

Despite the overall similarities in protocols used in paleobotany, vital differences exist. In some studies phytolith yield is divided into different size fractions for separate counting (e.g., Pinilla and Bustillo 1997; Sánchez et al. 2010; Zucol et al. 2010), as is done by many Quaternary workers (e.g., Piperno 1993; Kealhofer and Penny 1998; Mercader et al. 2000), whereas in others, assemblages are studied without fractionation (e.g., Strömberg 2005; WoldeGabriel et al. 2009; Novello et al. 2015). As was discussed previously (*Grassland ecosystem evolution*), counts of separate fractions can not be directly compared, unless a step is added that estimates the number of phytoliths in each fraction. The level of magnification used to study phytoliths (e.g., 500× vs. 1,000×) also varies among studies (compare Strömberg 2005 and WoldeGabriel et al. 2009), which might affect the ability to detect small or faint phytoliths (Strömberg 2003). Finally, there are potential differences in the extent to which phytolith morphotypes are rotated during study (e.g., Strömberg 2002 vs. Strömberg 2005). This is especially important for correctly identifying different GSSCP morphotypes with three-dimensionally complex morphology (Piperno and Pearsall 1998).

Quaternary phytolith research often includes methods to estimate the concentration of phytoliths per soil volume or dry weight by either (1) weighing the phytolith yield and/or counting of a pre-determined aliquot of the yield (Piperno 1993; Albert and Weiner 2001; Esteban et al. 2017), or (2) adding a known number of an exotic agent ("spike", usually in the form of opal beads) to the sample during the extraction process that can then be counted along with the phytoliths on the slide (Powers and Gilbertson 1987; Piperno 1993; Powers-Jones and Padmore 1993). These types of methods have yet to be adapted for deep-time studies. However, preliminary work has shown that the spike method may not be practical for ancient phytolith assemblages because of their often extremely high concentration of phytoliths (Strömberg 2003).

Potential Biases During Phytolith Analysis

Preventing phytolith assemblage bias or damage related to extraction protocols is a major concern for phytolith research and a fair amount of work has been dedicated to curtailing such taphonomical problems (Fig. 12.3) (Fredlund 1986; Piperno 1988; Lentfer and Boyd 1998, 1999, 2000; Madella et al. 1998; Parr 2002; Lentfer et al. 2003; Madella and Lancelotti 2012; Calegari et al. 2013). For the methods used on geological samples, some preliminary testing has been done to ensure that assemblages are not subjected to unnecessary breakage and size-, shape- or density sorting (Strömberg 2003), but more work is needed to verify that protocols used by different workers result in (equally) non-biased biosilica assemblages.

Other steps that can skew phytolith assemblage composition or counts (slide preparation, counting procedure, and count size) have received less attention. The limited work on slide-making protocols conducted so far has indicated that the most prudent approach involves using dry material, to avoid size-sorting in a liquid medium (Strömberg 2007), and that counting should preferably be done using approximately random fields (as opposed to transects) to avoid biases associated with size-sorting across the slide (Dinan and Rowlett 1993; Pearsall 2000; Strömberg 2007).

In terms of assigning the necessary number of phytoliths counted for statistical precision of phytolith inference, only a handful of studies have been done (e.g., Pearsall 2000; Albert and Weiner 2001; Strömberg 2009a, b). Most informative is a study using bootstrapping to investigate the effect on count size on the statistical robustness of phytolith morphotype percentages or ratios (Strömberg 2009a). It found that, in accordance with guidelines established within palynology (e.g., Mosimann 1965; Birks and Birks 1980), the appropriate count size is not a fixed number, but should be determined by the question (e.g., if only the GSSCPs are looked at vs. the whole assemblage) and the composition of individual phytolith assemblages (because statistical errors vary with the value of the percentage or ratio under study). Based on this analysis, Strömberg (2009a) suggested that 200 diagnostic (as opposed to all) phytoliths is a reasonable minimum initial count, and recommended using confidence intervals of percentages and ratios (e.g., using the statistical software R; see Chen et al. 2015) to determine when an appropriate count size has been reached. This count size (200) was also proposed in previous studies, although these did not include measurements of statistical robustness (Piperno 1988; Pearsall 2000; Albert and Weiner 2001).

Analytical Approaches

Traditionally, phytolith research included mainly relative abundance charts like those used in palynology, and most

Quaternary publications feature this type of graph, or some version thereof, as a useful visual presentation of data (Fredlund and Tieszen 1997b; Bremond et al. 2005a; Penny and Kealhofer 2005; Iriarte and Paz 2009; Cordova et al. 2011; Kawano et al. 2012; McMichael et al. 2012b; McCune and Pellatt 2013). However, since the 1990s phytolith workers have started exploring a variety of more sophisticated methods for analyzing count data. Beyond the various vegetation indices (e.g., *D/P*, Iph) discussed earlier (Table 12.2), approaches such as cluster analysis (Blinnikov et al. 2002; Gallego and Distel 2004), Principal Components Analysis (PCA) (Barboni et al. 2007; Mercader et al. 2011; Blinnikov et al. 2013; Dickau et al. 2013), Correspondence Analysis (CA) (Barboni and Bremond 2009; Neumann et al. 2017), ternary plots (Cerling et al. 2010; Esteban et al. 2017), and value-test analysis (Novello et al. 2012) have been applied to modern and Quaternary phytolith assemblages to find temporal or spatial patterns of vegetation change, or to understand similarities and differences among (taxonomic or functional/ecological) plant groups in terms of their phytolith assemblage composition. Discriminant Function Analysis (DFA) has been used to assign phytolith assemblages to *a priori* groupings (e.g., vegetation types, taxa) (Ball et al. 1996, 1999; Bremond et al. 2008a), Non-metric Multi-Dimensional Scaling (NMDS) can help ordinate a combination of phytolith data and environmental variables (Kerns et al. 2001), and regression analysis and CA have been employed to find the relationships between relative abundances of morphotypes, or morphotype shapes, and environmental characters (elevation, temperature etc.) that can then be applied to ancient phytolith assemblages (Lu et al. 2006, 2007; Barboni et al. 2007; Bremond et al. 2008b; Issaharou-Matchi et al. 2016).

Deep-time studies have similarly used pie charts, bar-charts, or tables for representation of data (Pinilla and Bustillo 1997; Thorn 2004a; Strömberg 2005; Strömberg et al. 2013a), as well as various indices or relative abundances of specific morphotypes or morphotype combinations (e.g., *D/P*, *FI-t*, max and min potential C_4) (Strömberg 2002; Strömberg et al. 2007; McInerney et al. 2011; Strömberg and McInerney 2011; Chen et al. 2015). Several publications additionally include cluster analysis (Zucol et al. 2010), regression analysis (Miller et al. 2012), and PCA (Harris et al. 2017) applied in ways comparable to Quaternary publications.

Redundancy in morphology between phytoliths produced by distantly related taxa presents a major obstacle in phytolith studies, in particular when interpreting grass community composition based on GSSCP assemblages. To provide inferences that take redundancy into account, some authors have compared fossil GSSCP assemblages to modern reference material statistically. Examples include statistical hypothesis testing via bootstrapping (Strömberg 2003, 2005; Strömberg et al. 2007b), and multivariate statistical comparisons using the Morisita index to assess similarity (Gallego et al. 2004).

Research attempting to study patterns of phytolith morphology in a phylogenetic context is still in its infancy. Approaches so far involve mainly mapping of character states on phylogenies to infer phylogenetic placement of fossil phytoliths or general patterns of silicification in grasses (Prasad et al. 2005; Rudall et al. 2014). Phytoliths have occasionally been included as characters in total evidence phylogenies (Kellogg and Campbell 1987), and Prasad et al. (2011) used phytolith traits in such an analysis to determine the taxonomic affinities of fossil phytoliths, and to provide a fossil calibration point (Christin et al. 2014), as discussed above.

Modern Phytolith Reference Collections

Use of modern phytolith reference collections (MPRCs): MPRCs from plants and modern soils are the foundation of modern analog studies, and form a vital baseline for deep-time paleoecological studies. Plant-based MPRCs consist of bio-opal extracts from fresh/dried plant tissues of known taxonomic identity to which unknown archaeological or paleontological materials are compared (Pearsall 2016b). Such collections serve importantly in phytolith research because, like herbarium specimens, these vouchers are essential for accurate and consistent identification of phytolith morphotypes. Researchers use the MPRCs to identify diagnostic phytolith types for plant taxa and to understand the variability of phytolith types that are produced in, for example, particular organs (e.g., grass inflorescences, Novello and Barboni 2015), individual genera or species (e.g., *Cucurbita, Zizania*; Piperno et al. 2000; Yost and Blinnikov 2011), larger clades (e.g., Poaceae, Zingerberales; Piperno and Pearsall 1998; Chen and Smith 2013), and/or specific geographic regions (e.g., Ecuador; Pearsall 2016b). Soil-based MPRCs comprise phytolith assemblages extracted from modern soils, which reflect the phytolith production of the local silica accumulating plants in standing (known) and recent vegetation. These assemblages consist mostly of isolated phytoliths, but can include intact tissues that are silicified, such as sheets of silicified leaf epidermis (e.g., Fredlund and Tieszen 1994; Blinnikov 2005; Bremond et al. 2005a; Iriarte and Paz 2009; Mercader et al. 2011; Dunn et al. 2015b).

To date, MPRCs from both soils and plant material have been collected from every continent except Antarctica

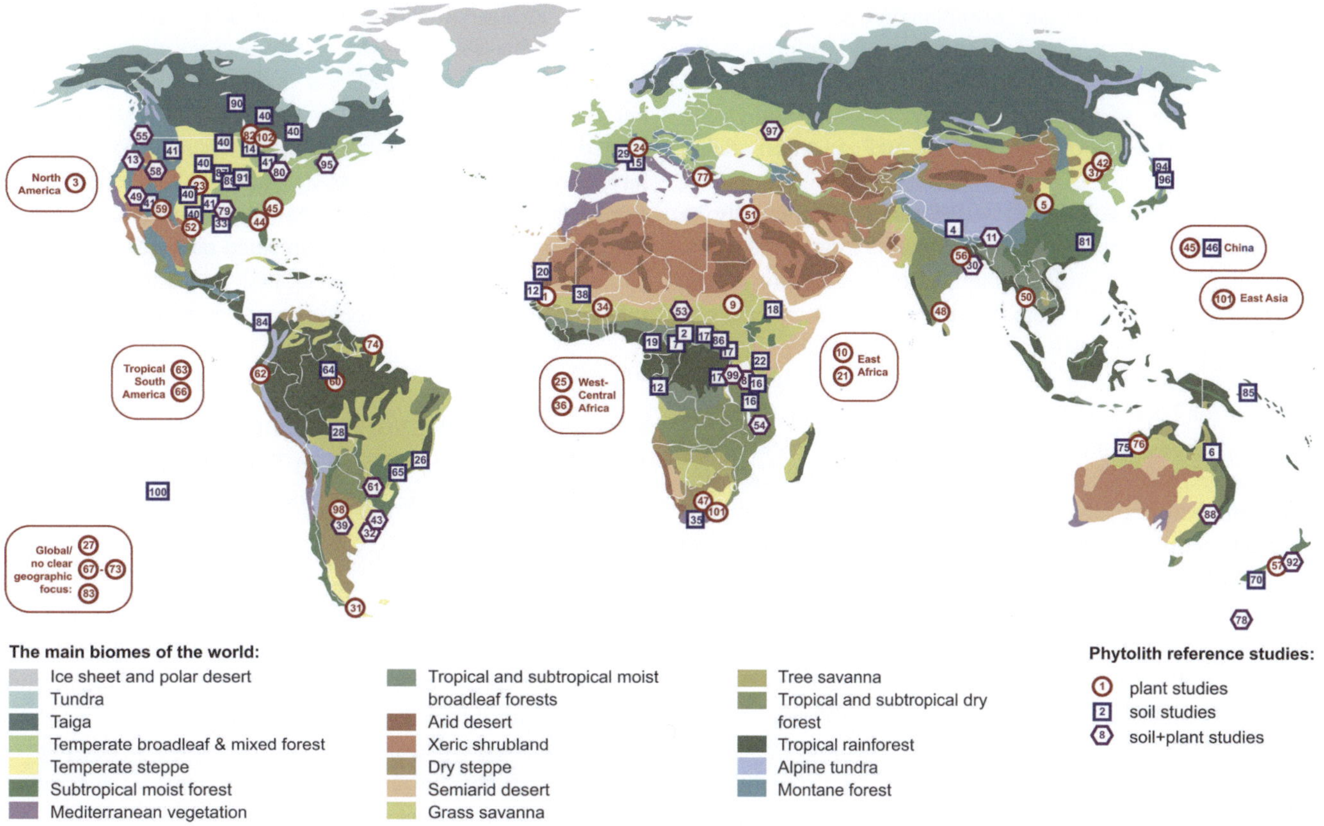

Fig. 12.8 Geographic distribution of plant- and soil-based modern phytolith reference collections/studies (MPRCs). Map showing major biome types after Porse (2008). For references, see Appendix 12.4

(Fig. 12.8, Appendix 12.4). Tropical and temperate regions are fairly evenly sampled, comprising 48% and 52%, respectively, of the total publications, with 56% from plant material and 44% from soil collections. However, as MPRCs have typically been created for specific projects, particular ecosystems and areas have been more intensively sampled, such as grasslands in North America, and grasslands and tropical forests in Africa and South America (Fig. 12.8, Appendix 12.4). In contrast, other regions/vegetation types are more sparsely—or not at all—represented, especially arid regions of North Africa, Eurasia, and Australia, boreal Eurasia, Mexico, the west Pacific tropics, and Andean South America.

Traditionally, publications utilizing MPRCs have focused on figuring unusual (diagnostic) forms of phytoliths per taxon (e.g., Pearsall 1978; Piperno 1985; Bozarth 1987; Piperno 1989; Kealhofer and Piperno 1998; Wallis 2003), meaning there are very few published guides that show the broad range of phytoliths types (and the variation within these types) produced by plant taxa. Most published MPRCs classify phytoliths into morphotypes using shape, texture and ornamentation, symmetry, morphometric data, as well as anatomical origin (ICPN Working Group 2005). However, different researchers have emphasized different aspects of their morphology and, as a result, a number of classification schemes exist (e.g., Twiss et al. 1969; Bertoldi de Pomar 1971; Runge 1996; Bowdery et al. 2001; Strömberg 2003; Zucol et al. 2010). In addition, because workers have chosen different ways to figure their phytolith morphotypes (e.g., SEM micrographs vs. drawings vs. bright-field microscopy images at varying magnification), it is often difficult to make robust comparisons between phytolith morphotypes from different studies.

Therefore, a major task in phytolith research has been developing a consistent classification and nomenclature for phytolith morphotypes. As previously noted, work is underway to standardize phytolith terminology (ICPN Working Group 2005), and there are ongoing efforts to create online databases that serve as outlets and repositories for phytolith images, for example PhytCore (Albert et al. 2016), The Phytoliths of Ecuador (Pearsall 2016a, b), and the Phytolith Gallery (Blinnikov 2016). These advances will greatly improve our chances of creating a globally consistent phytolith classification, allowing for meaningful geographic and taxonomic comparisons among modern phytolith morphotypes and a more useful reference for deep-time paleoecological work. Nevertheless, challenges remain on how to classify and interpret fossil phytolith assemblages from extinct plants and in non-analog ecosystems.

How to create a MPRC: MPRCs should preferably be made from voucher specimens of known taxonomic identity, for instance sampling dried material from herbarium sheets, or using fresh material from botanical gardens or arboreta, part of which is reposited in an herbarium as a voucher. Additionally, fresh or dried plant material can be collected in the field for later phytolith processing in the lab. Different plant organs and tissues (wood, bark, stem, roots, leaves and reproductive tissues, such as seeds, fruits, flowers) are often collected and extracted separately to understand their differential contribution to soil assemblages (e.g., Mulholland 1989; Piperno 1989; Piperno and Pearsall 1998).

To obtain the bio-opal component of the plant material, the organic tissue must be removed (oxidized); several methods to do so exist. Dry ashing involves incineration at high temperatures (450–600°C) in a muffle furnace, whereas wet ashing removes organic materials using strong oxidizing liquids (e.g., Schultze's solution) (Rovner 1972; Piperno 1988). Microwave digestion has also been demonstrated as an expedient method for phytolith isolation (Parr et al. 2001) and most recently, acetolysis has been shown to work effectively (da Costa et al. 2016).

Phytolith extraction from modern soil assemblages is similar to that for fossil samples with an additional recommended step of oxide removal with strong acid prior to heavy liquid floatation (e.g., Zhao and Pearsall 1998).

What types of information should be included in a MPRC?: Given the differential production of phytoliths in plants and the multiplicity and redundancy of phytolith types, MPRCs should seek to include phytolith production tables (e.g., Strömberg 2003; Wallis 2003; Novello and Barboni 2015; Pearsall 2016a; Neumann et al. 2017). Such tables should ideally record whether a plant produces phytoliths and if so, what percent of dry weight material the bio-opal makes up, what types of phytoliths are present, and in what frequencies. MPRC samples can be stored as microscope slides where the sample is mounted in a permanent fixative (e.g., Meltmount, Canada Balsam), or in a viscous liquid that allows for some degree of phytolith rotation, either for a few days as the medium is solidifying (e.g., Permount, Entellan), or permanently (e.g. glycerin jelly, immersion oil). However, note that these latter more flexible mounts are subject to dehydration and are not considered permanent. Phytolith extracts from which slides are made can be stored as a dry powder and can be used for making additional slides when needed. Pearsall (2016a) suggested that some portion of the original plant tissue also be vouchered in the MPRC for additional processing to be performed once the original extract volume has been consumed. Unfortunately, some of the permanent fixatives that have been used in the past are subject to clouding and cracking, rendering microscope slides useless (Pearsall 2016a; Strömberg & Dunn unpublished data). More archival alternatives need to be investigated and implemented to preserve these important reference collections.

Discussion: Unresolved Questions and Future Directions in Phytoliths Applied to Deep-Time Paleoecology

Given its relative youth as a scientific field, it is not surprising that fundamental aspects of phytolith science remain to be better developed, standardized, and more rigorously scrutinized. For phytolith research in general, and for deep-time studies in particular, four major themes stand out that warrant special attention to ensure repeatability of results, meaningful comparisons among studies, and robust use of modern analogs applied to fossil assemblages:

1. **Standardization of morphotypes across studies**

The differences in classification system among different modern and paleontological studies (compare e.g., Bertoldi de Pomar 1971; Runge 1996), and the degree of morphological subdivision ("lumping" vs. "splitting," compare WoldeGabriel et al. 2009; Zucol et al. 2010), limit the usefulness of the modern literature to deep-time work, and also prevent comparison among paleoecological studies from different (or sometimes the same) regions (compare Zucol et al. 2010; Strömberg et al. 2013a). The lack of clear definitions of morphotypes and their morphological range further prevent workers from easily adopting classification schemes used in published works, although exceptions exist (e.g., Strömberg 2003; Zucol and Brea 2005). Future work should therefore seek to establish a formal, standardized classification of phytolith morphotypes based on detailed descriptions and unequivocal diagnoses that can be applied reliably across studies.

2. **Comparability of assemblage counts**

Even though very few phytolith investigators deal with Neogene and older deposits, the differences in analytical approach that exists between studies—in particular whether to analyze fractions separately or not—hinder basic comparisons between results. Even though there is room for different methodologies, work should focus on bridging the analytical gaps between them. In the same vein, evaluation of patterns in different studies depends on the use of standard statistical measures of robustness (e.g., 95% confidence intervals), which in turn is a product of count size (Strömberg 2009a, b). Future work should take care to provide statistical support to ensure the validity of interpretation of patterns (see e.g., Harris et al. 2017).

3. **Constraints on spatial resolution of phytolith assemblages**

To assess spatial heterogeneity of vegetation or determine how representative a particular phytolith assemblage is, it is vital to know how local the vegetation signal is. Currently almost no studies have tested for spatial resolution specifically, and only in certain ecosystems (Fredlund and Tieszen 1994; Kerns et al. 2001; Bremond et al. 2005a; Aleman et al. 2012). Therefore, even though it is likely that there is no straightforward answer to the question of how to determine exactly the areal representation in a sample, the main direction for future studies should be to look at the spatial resolution (of e.g., assemblage composition, phytolith indices, *rLAI*) at different regional and local scales (i.e., within a biome, within an ecosystem, along a transect, and within a plot), and in different ecosystems, focusing on, for example, different forest types varying in canopy openness, and grasslands and savannas varying in grass community composition and tree cover (see Crifò and Strömberg 2016).

4. **Refining nearest living relative and morphometric methods of environmental inference**

To determine how well a phytolith morphotype reflects a certain climate, it is necessary to establish the phylogenetic distribution of the morphotype, and how this distribution correlates with phylogenetic patterns of environmental tolerance or preference among plants. This is rarely done within Quaternary paleoecology, because it is often reasonable to assume that a certain morphotype reflects a local species, and that this species has not changed its ecological niche in the last few million years. However, for deep-time paleoecology such assumptions are tenuous at best (Strömberg 2004). Studies that take a more phylogenetic approach to morphotype circumscription and nearest living relative inference are emerging (Prasad et al. 2005, 2011; Chen and Smith 2013; Strömberg et al. 2013a; Thummel et al. 2017), but substantially more work is needed. In addition, the possibility that plants have changed their tolerance or preference should not be excluded (discussion in Strömberg 2006), and if possible, independent evidence for environmental reconstructions should be incorporated (e.g., from sedimentology, paleosols, stable isotopes, pollen, macrofossils) (see McInerney et al. 2011; Brea et al. 2012; Novello et al. 2015).

Likewise, approaches relying on environmentally determined morphologic variation should seek to verify the extent to which phylogenetic inheritance influences shape. The necessity of accounting for the phylogenetic signal in morphology-based paleoecological proxies was shown recently by Little et al. (2010). These authors found a non-negligible effect of phylogeny on leaf traits used to infer climates (e.g., toothiness). Ongoing work on epidermal cell morphology (cell undulation and cell area) of ~1,000 species of dicotyledonous angiosperms similarly tests to what extent there is phylogenetic signal for these traits within angiosperms, allowing for refinement—and more robust application—of the phytolith-based LAI-proxy (Dunn et al. 2015b; Dunn et al., in prep.).

Acknowledgements We thank Z. Song, D. Conley, and W. Clymans for fruitful discussions about silica in soil. This work was in part supported by the Biology Department, University of Washington, Seattle, and NSF awards EAR-1253713 and EAR-1349530 to C.A.E.S.

Appendix 12.1. Glossary of Common or Key Terms

Term	Explanation	Reference
BOP clade	One of two major lineages within Poaceae containing three subfamilies: Bambusoideae, Oryzoideae and Pooideae. Previously referred to as BEP	Aliscioni et al. (2012) and Soreng et al. (2015)
Bulliform cell	Large inflated cells of the epidermis that occur mostly on the upper side of monocot leaves. These cells are thought to control leaf rolling through water loss and are often a site of silica deposition. The resulting phytoliths are referred to as cuneiform and parallelipipedal bulliform cells (ICPN 1.0)	Piperno (2006)

(continued)

Term	Explanation	Reference
Cuticle	A layer consisting of lipid polymers impregnated with waxes that cover the epidermis of leaves and other above-ground plant organs. The cuticle prevents water loss and protects the plants against contamination	Esau (1965)
Cystolith	Inorganic concretions formed in specialized cells called lithocysts, usually in leaves of plants of certain families (e.g., in the Moraceae and Ulmaceae). Cystoliths often consist of calcium oxalates, silica or both	Esau (1965) and Piperno (2006)
Grass Silica Short Cells (GSSC)	Cells in grass epidermis that are specialized for silicon deposition (idioblasts). Short cells are roughly equidimensional in most taxa and shorter (along the axis of the leaf) than the accompanying long cells within the epidermis. GSSC phytoliths (GSSCP) are the result of complete filling of the GSSCs, in the form of shapes that are more or less diagnostic of different Poaceae subclades (e.g., bilobate, cross, rondel, trapeziform polylobate). GSSCPs are a unique trait of Poaceae	Piperno (2006)
ICPN 1.0	International Code for Phytolith Nomenclature 1.0, a publication that sought to standardize phytolith morphotype naming	ICPN Working Group (2005)
Idioblast cells	Cells which differ markedly from neighboring cells, such as those that synthesize crystals of silica or calcium oxalate in some plant taxa	Esau (1965)
Multiplicity	The fact that many plants produce more than one phytolith morphotype	Piperno (1988) and Rovner (1971)
PACMAD clade	One of two major lineages within Poaceae, the grass family, which contains six subfamilies (Panicoideae, Aristidoideae, Chloridoideae, Micrairoideae, Arundinoideae, and Danthonioideae) for which the clade is named	Aliscioni et al. (2012) and Soreng et al. (2015)
Redundancy	Overlap in morphotype shape among phytoliths produced by more or less phylogenetically distant plants	Piperno (1988) and Rovner (1971)
Stegmata	Specialized silica-containing cells that often surround the vascular bundles in, for example palms, zingibers, and orchids. Stegmata phytoliths are often diagnostic to family or below	Tomlinson (1961)

(continued)

Appendix 12.2. Ungulate Specimens Sampled for Dental Calculus from the Deep River Formation, Montana. Phytolith Assemblages from Taxa in Bold Font Were Analyzed Quantitatively (Fig. 12.7)

Museum Specimen Number	Taxon	Calculus Phytolith Productivity	Matrix Phytolith Productivity
UWBM 40039	*Promerycochoerus*	None	N/A
UWBM 39071	Merycoidodontid	None	N/A
UWBM 59145	*Promerycochoerus*	Some	Some
UWBM 54757	*Leptauchenia*	Some	Many
UWBM 54624	**Merycoidodontid**	**Many**	**Many**
UWBM 53242	*Leptauchenia*	None	N/A
UWBM 53245	Merycoidodontid	None	N/A
UWBM 48113	*Leptauchenia*	Some	Few to many
UWBM 40614	*Merychippus*	None	N/A
UWBM 46831	*Merychippus*	None	N/A
UWBM 48109	*Leptauchenia*	None	N/A
UWBM 48021	*Merychippus*	Some	Many
UWBM 52661	*Merychippus*	None	N/A
UWBM 52796	*Promerycochoerus*	None	N/A
UWBM 54650	*Merychippus*	Some	Few
UWBM 54756	Camelid	None	N/A
UWBM 58038	Tayassuid	None	N/A

(continued)

Museum Specimen Number	Taxon	Calculus Phytolith Productivity	Matrix Phytolith Productivity
UWBM 32414	**Merycoidodontid**	**Many**	**Many**
UWBM 39932	*Leptauchenia*	None	N/A

Appendix 12.3. Analysis of Dental Calculus in Middle Miocene Herbivores (E.B.H.)

Materials and Methods—Fossil teeth used for this study were collected from the Barstovian (16.3–13.6 Ma) Spring Creek 1 locality of the Deep River Formation, Montana (Appendix 12.2). They are held within the vertebrate paleontology collections at the Burke Museum of Natural History and Culture (UWBM), Seattle, Washington.

Fossil teeth specimens were sampled if they preserved a non-matrix film on the dental enamel or cementum surface that resembled dental calculus. The dental films were sampled from the buccal side of the teeth and ranged in color from white to tan (Fig. 12.7). Most dental films were found above the gingival margin. Teeth were first rinsed with DI water and rubbed with Q-tips to ensure the removal of any extraneous phytolith or matrix material. The dental film was gently removed using a dental pick onto a sterile piece of paper and then transferred into glass vials for storage. Samples were lightly crushed (if not already in a powdered form) and treated with hydrochloric acid (HCl) for 24–48 h to remove any calcareous substances. Samples were then transferred to 1.5 ml Eppendorf tubes and spun five to seven times in a mini-centrifuge to remove the HCl. The resulting yields were smeared onto microscope slides with ethanol, mounted with Meltmount, and counted for phytoliths.

In addition to dental calculus, matrix samples associated with each specimen were also collected. Following the method described in Strömberg (2002), and modified from Piperno (1988), matrix samples were processed with acids to remove carbonates, oxidized to remove organic materials, sieved, and gravimetrically separated to yield only the 3–50 μm fraction. Matrix samples were mounted on slides using Meltmount. All phytoliths were identified and categorized following classification used in Strömberg (2005).

Results—Of nineteen calculus samples analyzed, seven preserved phytoliths but only two preserved them in enough abundance to warrant counting (Appendix 12.2). The phytolith productivity of the matrix samples varied from sample to sample but generally contained substantially more phytoliths than the calculus. All calculus and matrix samples from *Merycoidodon* (UWBM 54624) and an unidentified oreodont (UWBM 32414) yielded a wide range of phytolith morphotypes (Table 12.2) of which grass forms—specifically C_3 pooid forms—dominated (Fig. 12.7). The *Merycoidodon* calculus sample contained more forest indicator phytoliths (e.g., closed-habitat grass, dicotyledonous, and palm morphotypes) than the unidentified oreodont specimen. Additionally, there was a statistically significant difference between the calculus and matrix samples for each specimen when comparing the abundance of phytoliths of grasses and woody dicots (Fisher's exact test; *Merycoidodon*: $p \ll 0.05$; unident. oreodont: $p < 0.05$).

Discussion—The fact that the calculus and matrix phytolith assemblages differ lends support to our supposition that we did in fact sample dental calculus. Furthermore, our results suggest that it is reasonable to assume that these calculus phytolith assemblages are representative of the plants these animals ate. Therefore, we hypothesize that these two oreodonts fed mainly on C_3 grasses, but incorporated some browse.

Appendix 12.4. Key to Modern Reference Studies Mapped in Fig. 12.8

No. in Fig. 12.8	Continent	Region	Country/US state	Type of study	Plant types		Vegetation types			Trop./temp. ¥	Reference
					Grass	Other	Grassland	Forest	Other		
1	Africa	West Africa (mainly)	Burkina Faso, Benin, Niger, Nigeria, Ethiopia	Plant	x	–	–	–	–	TR	Neumann et al. (2017)
2	Africa		Central African Republic	Sediment	–	–	x	x	–	TR	Aleman et al. (2012)
3	North America		United States	Plant	x	x	–	–	–	TR-TE	Strömberg (2003)
4	Asia	China-Nepal border		Sediment	–	–	–	x	x	TR-TE	An et al. (2015)
5	Asia	–	China	Plant	–	x	–	–	–	TE	An (2016)
6	Australia	–	Australia	Sediment	–	–	–	x	–	TR	Alexandre et al. (2012)
7	Africa	–	Central African Republic	Sediment			x	x	x	TR	Aleman et al. (2014)
8	Africa	–	Tanzania	Plant, sediment	x	x	x	x	–	TR	Albert et al. (2006, 2015)
9	Africa	Sahelian, Sudanian grassland	Sudan	Plant	x	–	–	–	–	TR	Novello and Barboni (2015)
10	Africa	East Africa (East African lakes)	–	Plant	x	–	–	–	–	TR	Barboni and Bremond (2009)
11	Asia	–	India	Plant, Sediment	x	–	–	x	–	TR-TE	Biswas et al. (2016)
12	Africa	–	Senegal, Congo	Sediment	–	–	x	x	–	TR	Alexandre et al. (1997b)
13	North America	–	United States, WA	Plant, Sediment	x	x	x	x	x	TE	Blinnikov (2005)
14	North America	–	United States, MN	Sediment	–	–	x	–	–	TE	Blinnikov et al. (2013)
15	Europe	–	France	Sediment	–	–	x	x	–	TE	Bremond et al. (2004)
16	Africa	–	Kenya, Tanzania	Sediment	–	–	x	x	–	TR	Bremond et al. (2008b)
17	Africa	Central Africa	Central African Republic, Congo	Sediment	–	–	–	x	–	TR	Runge (1999, 2001)
18	Africa	–	Ethiopia	Sediment	–	–	x	x	x	TR	Barboni et al. (1999, 2007)
19	Africa	–	Cameroon	Sediment	–	–	x	x	x	TR	Bremond et al. (2005a)
20	Africa	–	Senegal, Mauritania	Sediment		–	x	x		TR	Bremond et al. (2005b)
21	Africa	East Africa	Zaire, Rwanda, Kenya	Plant	x	x	–	–	–	TR	Runge (1996)
22	Africa	–	Tanzania	Sediment	–	–	–	–	–	TR	Barboni et al. (2007)
23	North America	Central North American grasslands	United States	Plant	x	–	–	–	–	TE	Brown (1984)

(continued)

24	Europe	Swiss alps	Switzerland	Plant	x	x	–	–	–	TE	Carnelli et al. (2001, 2004)
25	Africa	West/Central Africa	–	Plant	–	x	–	–	–	TR	Collura and Neumann (2017)
26	South America	–	Brazil	Sediment	–	–	x	x	x	TR	Coe et al. (2013, 2014)
27	Global	–	–	Plant	–	x	–	–	–	TR-TE	Chen and Smith (2013)
28	South America	–	Bolivia	Sediment	–		x	x	x	TR	Dickau et al. (2013)
29	Europe	–	France	Sediment	–		x	x		TE	Delhon et al. (2003)
30	Asia	–	India	Plant, Sediment	x	x		x	x	TR	Das et al. (2013a, b)
31	South America	–	Argentina	Plant	x	–	–	–	–	TE	Fernández Pepi et al. (2012)
32	South America	–	Argentina	Plant	x	x	x			TE	Fernandez Honaine et al. (2006), Honaine et al. (2009)
33	North America	–	United States, LA	Sediment	–	–	x	–	x	TR	Fearn (1998)
34	Africa	West Africa (Sahel)	–	Plant	x	–	–	–	–	TR	Fahmy (2008)
35	Africa	–	South Africa	Sediment	–	–	x	x	x	TE	Esteban et al. (2017)
36	Africa	West/Central Africa	–	Plant	–	x		–	–	TR	Eichhorn et al. (2010)
37	Asia	–	China	Plant		x	–	–	–	TE	Ge et al. (2011)
38	Africa	–	Mali	Sediment	–	–	x	x	x	TR	Garnier et al. (2012)
39	South America	–	Argentina	Plant, Sediment	x	–	x		–	TE	Gallego and Distel (2004), Gallego et al. (2004)
40	North America	Great Plains	United States, Canada	Sediment	–		x	–	–	TE	Fredlund and Tieszen (1994) and (1997a)
41	North America	–	United States	Sediment	–	–	x	x	–	TE	Hyland et al. (2013b)
42	Asia	–	China	Plant	x	x	–	–	–	TE	Guo et al. (2012)
43	South America	–	Uruguay	Plant, Sediment	x	x	x	x	x	TR	Iriarte and Paz (2009)
44	North America	–	United States: (GA, LA, FL)	Plant	x	–	–	–	–	TR	Lu and Liu (2003a)
45	North America, Asia	Atlantic and Gulf coasts (United States)	United States, China	Plant	x	–	–	–	–	TR-TE	Lu and Liu (2003b)
46	Asia	–	China	Sediment	–	–	x	x	x	TR-TE	Lu et al. (2006)
47	Africa		South Africa	Plant	x	–	–	–	–	TE	Rossuow (2009)
48	Asia		India	Plant	–	x	–	–	–	TR	Kumari and Kumarasamy (2014)
49	North America		United States, AZ	Plant, sediment	x	x	x	x	–	TE	Kerns (2001)
50	Asia		Thailand	Plant		x			–	TR	Kealhofer and Piperno (1998)
51	Asia		Israel	Plant		x		–	–	TE	Katz et al. (2013)
52	North America		United States, TX	Plant*		x	–	–	–	TE	Jones and Bryant (1992)

(continued)

53	Africa		Chad	Plant, Sediment	x	x	x	–	x	TR	Novello et al. (2012)
54	Africa		Mozambique	Plant, sediment	x	x	x	x	x	TR	Mercader et al. (2010, 2009, 2011)
55	North America		Canada, BC	Plant	x	x	x	x	–	TE	McCune and Pellatt (2013) and McCune et al. (2014)
56	Asia		India	Plant	–	x	–	–	–	TR	Mazumdar and Mukhopadhyay (2009a, 2010a, 2011)
57	Oceania		New Zealand	Plant	x	–	–	–	–	TE	Marx et al. (2004)
58	North America		United States, ID, UT	Plant, sediment	x	x	x		x	TE	Morris et al. (2009a, b, 2010)
59	North America		United States, NM	Plant**	–	x	–	–	–	TE	Morgan-Edel et al. (2015)
60	South America	Amazonas	(Colombia)	Plant	–	x	–	–	–	TR	Morcote-Ríos et al. (2016)
61	South America		Argentina	Plant, Sediment	x	–		x		TR	Montti et al. (2009)
62	South America		Ecuador	Plant	x	x				TR	Pearsall (2016a, b)
63	South America	Tropical South America		Plant	x					TR	Piperno and Pearsall (1998)
64	South America		Brazil	Sediment				x		TR	Piperno and Becker (1996)
65	South America		Brazil	Sediment				x	x	TR	Alexandre et al. (1999)
66	South America, etc.	Tropical		Plant		x				TR	Piperno (1985, 1989)
67	Global			Plant*	x	x				TR-TE	Prychid and Rudall (1999)
68	Global			Plant**		x				TR-TE	Prychid et al. (2003)
69	Global			Plant	x	x				TR-TE	Prychid et al. (2004)
70	Oceania		New Zealand	Sediment			x	x	x	TE	Prebble et al. (2002)
71	Global			Plant		x				TR-TE	Scurfield et al. (1974a, b)
72	Global			Plant	x	(x)				TR-TE	Rudall et al. (2014)
73	Global			Plant		x				TR-TE	Sundue (2009)
74	South America		French Guiana	Plant	x	x				TR	Watling and Iriarte (2013)
75	Australia		Australia, West Australia	Sediment			x	x	x	TR	Wallis (2013)
76	Australia		Australia, West Australia	Plant		x				TR	Wallis (2003)
77	Europe		Greece	Plant	x	x				TE	Tsartsidou et al. (2007)
78	Oceania		New Zealand, Campbell Island	Plant, Sediment	x	x	x		x	TE	Thorn (2004a, 2008)
79	North America		United States, LA	Plant, Sediment	x	x	x	x	x	TR	Tedford (2009)

(continued)

80	North America		United States, OH	Plant, sediment		x		x	–	TE	Wilding and Drees (1971, 1973, 1974)
81	Asia		China	Sediment	–	–		x		TR	Zhao and Piperno (2000)
82	North America		United States, MN	Plant	x	x	–	–	–	TE	Yost and Blinnikov (2011)
83	Global	–	–	Plant, sediment	x	x	x	x	x	TR-TE	Albert et al. (2016)
84	South America		Panama	Sediment				x		TR	Piperno (1993)
85	Asia		New Britain, Papa New Guinea	Sediment			x	x	x	TR	Boyd et al. (1998)
86	Africa		Democratic Republic Congo	Sediment				x	–	TR	Mercader et al. (2000)
87	North America		United States, KS	Sediment	–	–	x	x	–	TE	Kurmann (1985)
88	Australia		Australia, New South Wales	Plant, sediment	x	x			x	TE	Hart (1988)
89	North America	Central Great Plains	United States, KS	Plant	–	x	–	–	–	TE	Bozarth (1985, 1987, 1990, 1992)
90	North America		Canada, Saskatchewan	Sediment	–		x	x	–	TE	Bozarth (1993)
91	North America		United States, MO	Sediment	–		x	x	–	TE	Donohue and Dinan (1993)
92	Oceania		New Zealand	Plant, sediment	x	x	x	x	x	TR-TE	Kondo et al. (1994)
93	Asia	–	Japan	Plant	x	x				TE	Kondo and Pearson (1981), Kondo and Sase (1986)
94	Asia	–	Japan	Sediment	–	–	x	x	–	TE	Inoue and Sase (1996)
95	North America	–	United States, MA	Plant, sediment	x	x	x	x	x	TE	Krauss (1997)
96	Asia	–	Japan	Sediment			x	x	–	TE	Sase and Hosono (2001)
97	Europe	–	Russia, European part	Sediment	x	x	x	x	–	TE	Gol'eva (1997, 2001)
98	South America	–	Argentina	Plant**		x				TE	Borrelli et al. (2011)
99	Africa	–	Uganda/Rwanda	Plant, sediment	x	x	x			TR	Murungi et al. (2016)
100	Oceania (South America)	–	Easter Island, Chile	Sediment	–	–	x?	x?	–	TR	Horrocks et al. (2016)
101	Africa	–	South Africa	Plant	x	x	–	–	–	TE	Cordova (2013), Cordova and Scott (2010)
102	Asia	East Asia	China, East Asia	Plant	x	–	–	–	–	TE-TR	Gu et al. (2013)
103	North America		United States, MN	Plant	x	–	–	–	–	TE	Ollendorf et al. (1988)

* Oxalates; ** Oxalates, silica; ¥ Tropical vs. temperate: TR = tropical, TE = temperate

References

Albert, R. M., & Bamford, M. K. (2012). Vegetation during UMBI and deposition of Tuff IF at Olduvai Gorge, Tanzania (ca. 1.8 Ma) based on phytoliths and plant remains. *Journal of Human Evolution, 63,* 342–350.

Albert, R. M., Bamford, M. K., & Cabanes, D. (2006). Taphonomy of phytoliths and macroplants in different soils from Olduvai Gorge (Tanzania) and the application to Plio-Pleistocene palaeoanthropological samples. *Quaternary International, 148,* 78–94.

Albert, R. M., Bamford, M. K., & Cabanes, D. (2009). Palaeoecological significance of palms at Olduvai Gorge, Tanzania, based on phytolith remains. *Quaternary International, 193,* 41–48.

Albert, R. M., Bamford, M. K., & Esteban, I. (2015). Reconstruction of ancient palm vegetation landscapes using a phytolith approach. *Quaternary International, 369,* 51–66.

Albert, R. M., & Cabanes, D. (2008). Fire in prehistory: an experimental approach to combustion processes and phytolith remains. *Israel Journal of Earth Sciences, 56,* 175–189.

Albert, R. M., Esteban, I. (2016). 9. What early human populations ate: the use of phytoliths for identifying plant remains in the archaeological record at Olduvai. In K. Hardy & L. Kubiak-Martens (Eds.), *Wild Harvest: Plants in the Hominin and Pre-Agrarian Human Worlds* (pp. 171–190). Oxford, UK: Oxbow Books.

Albert, R. M., Ruíz, J. A., & Sans, A. (2016). PhytCore ODB: a new tool to improve efficiency in the management and exchange of information on phytoliths. *Journal of Archaeological Science, 68,* 98–105.

Albert, R. M., & Weiner, S. (2001). Study of phytoliths in prehistoric ash layers from Kebara and Tabun caves using a quantitative approach. In J. D. Meunier & F. Colin (Eds.), *Phytoliths: Applications in Earth sciences and human history* (pp. 251–266). Lisse: A. A. Balkema Publishers.

Aleman, J., Leys, B., Apema, R., Bentaleb, I., Dubois, M. A., Lamba, B., et al. (2012). Reconstructing savanna tree cover from pollen, phytoliths and stable carbon isotopes. *Journal of Vegetation Science, 23,* 187–197.

Aleman, J. C., Canal-Subitani, S., Favier, C., & Bremond, L. (2014). Influence of the local environment on lacustrine sedimentary phytolith records. *Palaeogeography, Palaeoclimatology, Palaeoecology, 414,* 273–283.

Alexandre, A., Basile-Doelsch, I., Delhaye, T., Borshneck, D., Mazur, J., Reyerson, P., et al. (2015). New highlights of phytolith structure and occluded carbon location: 3-D X-ray microscopy and NanoSIMS results. *Biogeosciences, 12,* 863–873.

Alexandre, A., Bouvet, M., & Abbadie, L. (2011a). The role of savannas in the terrestrial Si cycle: a case-study from Lamto, Ivory Coast. *Global and Planetary Change, 78,* 162–169.

Alexandre, A., Crespin, J., Sylvestre, F., Sonzogni, C., & Hilbert, D. W. (2011b). The oxygen isotopic composition of phytoliths from tropical rainforest soils (Queensland, Australia): application of a new paleoenvironmental tool. *Climate of the Past Discussions, 7,* 1693–1735.

Alexandre, A., Crespin, J., Sylvestre, F., Sonzogni, C., & Hilbert, D. W. (2012). The oxygen isotopic composition of phytolith assemblages from tropical rainforest soil tops (Queensland, Australia): validation of a new paleoenvironmental tool. *Climate of the Past, 8,* 307–324.

Alexandre, A., Meunier, J.-D., Colin, F., & Koud, J.-M. (1997a). Plant impact on the biogeochemical cycle of silicon and related weathering processes. *Geochimica et Cosmochimica Acta, 61,* 677–682.

Alexandre, A., Meunier, J.-D., Lezine, A.-M., Vincens, A., & Schwartz, D. (1997b). Phytoliths: indicators of grassland dynamics during the late Holocene in intertropical Africa. *Palaeogeography, Palaeoclimatology, Palaeoecology, 136,* 213–229.

Alexandre, A., Meunier, J.-D., Mariotti, A., & Soubies, F. (1999). Late Holocene phytolith and carbon-isotope record from a Latosol at Salitre, South-Central Brazil. *Quaternary Research, 51,* 187–194.

Aliscioni, S., Bell, H. L., Besnard, G., Christin, P. A., Columbus, J. T., Duvall, M. R., et al. (2012). New grass phylogeny resolves deep evolutionary relationships and discovers C_4 origins. *New Phytologist, 193,* 304–312.

An, X. (2016). Morphological characteristics of phytoliths from representative conifers in China. *Palaeoworld, 25,* 116–127.

An, X., Lu, H., & Chu, G. (2015). Surface soil phytoliths as vegetation and altitude indicators: a study from the southern Himalaya. *Scientific Reports, 5,* 15523.

Archibald, S. B., Morse, G. E., Greenwood, D. R., & Mathewes, R. W. (2014). Fossil palm beetles refine upland winter temperatures in the Early Eocene Climatic Optimum. *Proceedings of the National Academy of Sciences, USA, 111,* 8095–8100.

Archibold, O. W. (1995). *Ecology of world vegetation*. London: Chapman & Hall.

Armitage, P. L. (1975). The extraction and identification of opal phytoliths from the teeth of ungulates. *Journal of Archaeological Science, 2,* 187–197.

Asevedo, L., Winck, G. R., Mothé, D., & Avilla, L. S. (2012). Ancient diet of the Pleistocene gomphothere *Notiomastodon platensis* (Mammalia, Proboscidea, Gomphotheriidae) from lowland mid-latitudes of South America: stereomicrowear and tooth calculus analyses combined. *Quaternary International, 255,* 42–52.

Askin, R., & Raine, J. (2000). Oligocene and Early Miocene terrestrial palynology of the Cape Roberts Drillhole CRP-2/2A, Victoria Land Basin, Antarctica. *Terra Antartica, 7,* 493–501.

Asner, G. P., Scurlock, J. M., & Hicke, J. (2003). Global synthesis of leaf area index observations: implications for ecological and remote sensing studies. *Global Ecology and Biogeography, 12,* 191–205.

Augustin, C. H. R. R., Coe, H. H. G., Chueng, K. F., & Gomes, J. G. (2015). Analysis of geomorphic dynamics in ancient quartzite landscape using phytolith and carbon isotopes, Espinhaço Mountain Range, Minas Gerais, Brazil. *Géomorphologie, 4,* 355–376.

Baker, G. (1960). Fossil opal-phytoliths. *Micropaleontology, 6,* 79–85.

Baker, G. J. (1979). *A study of dental disease in the horse*. Glasgow: University of Glasgow.

Ball, T. B., Gardner, J. S., & Anderson, N. (1999). Identifying inflorescence phytoliths from selected species of wheat (*Triticum monococcum, T. dicoccon, T. dicoccoides*, and *T. aestivum*) and barley (*Hordeum vulgare* and *H. spontaneum*) (Graminae). *American Journal of Botany, 86,* 1615–1623.

Ball, T. B., Gardner, J. S., & Brotherson, J. D. (1996). Identifying phytoliths produced by the inflorescence bracts of three species of wheat (*Triticum monococcum* L., *T. dicoccon* Schrank., and *T. aestivum* L.) using computer-assisted image and statistical analyses. *Journal of Archaeological Science, 23,* 619–632.

Bamford, M., Neumann, F., Pereira, L., Scott, L., Dirks, P., & Berger, L. (2010). Botanical remains from a coprolite from the Pleistocene hominin site of Malapa, Sterkfontein Valley, South Africa. *Palaeontologia Africana, 45,* 23–28.

Bamford, M. K., Albert, R. M., & Cabanes, D. (2006). Plio-Pleistocene macroplant fossil remains and phytoliths from Lowermost Bed II in the eastern palaeolake margin of Olduvai Gorge, Tanzania. *Quaternary International, 148,* 95–112.

Bamford, M. K., Stanistreet, I. G., Stollhofen, H., & Albert, R. M. (2008). Late Pliocene grassland from Olduvai Gorge, Tanzania. *Palaeogeography, Palaeoclimatology, Palaeoecology, 257,* 280–293.

Barboni, D. (2014). Vegetation of Northern Tanzania during the Plio-Pleistocene: a synthesis of the paleobotanical evidences from

Laetoli, Olduvai, and Peninj hominin sites. *Quaternary International, 322,* 264–276.

Barboni, D., Ashley, G. M., Dominguez-Rodrigo, M., Bunn, H. T., Mabulla, A. Z. P., & Baquedano, E. (2010). Phytoliths infer locally dense and heterogeneous paleovegetation at FLK North and surrounding localities during upper Bed I time, Olduvai Gorge, Tanzania. *Quaternary Research, 74,* 344–354.

Barboni, D., Bonnefille, R., Alexandre, A., & Meunier, J.-D. (1999). Phytoliths as paleoenvironmental indicators, West Side Middle Awash Valley, Ethiopia. *Palaeogeography, Palaeoclimatology, Palaeoecology, 152,* 87–100.

Barboni, D., & Bremond, L. (2009). Phytoliths of East African grasses: an assessment of their environmental and taxonomic significance based on floristic data. *Review of Palaeobotany and Palynology, 158,* 29–41.

Barboni, D., Bremond, L., & Bonnefille, R. (2007). Comparative study of modern phytolith assemblages from inter-tropical Africa. *Palaeogeography, Palaeoclimatology, Palaeoecology, 246,* 454–470.

Bartoli, F. (1985). Crystallochemistry and surface properties of biogenic opal. *Journal of Soil Science, 36,* 335–350.

Bartoli, F., & Wilding, L. P. (1980). Dissolution of biogenic opal as a function of its physical and chemical properties. *Proceedings of the Soil Science Society of America, 44,* 873–878.

Behrensmeyer, A. K. & Hook, R. W. (1992). Paleoenvironmental contexts and taphonomic models. In A. K. Behrensmeyer, J. D. Damuth, W. A. DiMichele, R. Potts, H.-D. Sues & S. L. Wing (Eds.), *Terrestrial ecosystems through time. Evolutionary paleoecology of terrestrial plants and animals* (pp. 15–136). Chicago: The University of Chicago Press.

Bellosi, E. (2010). Loessic and fluvial sedimentation in Sarmiento Formation pyroclastics, middle Cenozoic of Central Patagonia. In R. H. Madden, A. A. Carlini, M. G. Vucetich & R. F. Kay (Eds.), *The paleontology of Gran Barranca* (pp. 278–292). Cambridge: Cambridge University Press.

Bellosi, E. & González, M. G. (2010). Paleosols of the middle Cenozoic Sarmiento Formation, Central Patagonia. In R. H. Madden, A. A. Carlini, M. G. Vucetich & R. F. Kay (Eds.), *The paleontology of Gran Barranca* (pp. 293–305). Cambridge: Cambridge University Press

Bellosi, E. S., & Krause, J. M. (2014). Onset of the Middle Eocene global cooling and expansion of open-vegetation habitats in central Patagonia. *Andean Geology, 41,* 29–48.

Bennett, P. C. (1991). Quartz dissolution in organic-rich aqueous systems. *Geochimica et Cosmochimica Acta, 55,* 1781–1797.

Bertoldi de Pomar, H. (1971). Ensayo de clasificación morfológica de los silicofitolitos. *Ameghiniana, 8,* 317–328.

Birks, H. J. B., & Birks, H. H. (1980). *Quaternary palaeoecology*. London: Edward Arnold Ltd.

Biswas, O., Ghosh, R., Paruya, D. K., Mukherjee, B., Thapa, K. K., & Bera, S. (2016). Can grass phytoliths and indices be relied on during vegetation and climate interpretations in the eastern Himalayas? Studies from Darjeeling and Arunachal Pradesh, India. *Quaternary Science Reviews, 134,* 114–132.

Blecker, S. W., McCulley, R. L., Chadwick, O. A., & Kelly, E. F. (2006). Biologic cycling of silica across a grassland bioclimosequence. *Global Biogeochemical Cycles, 20,* GB3023.

Blecker, S. W., Yonker, C. M., Olson, C. G., & Kelly, E. F. (1997). Paleopedologic and geomorphic evidence for Holocene climate variation, Shortgrass Steppe, Colorado, USA. *Geoderma, 76,* 113–130.

Blinnikov, M. S. (2005). Phytoliths in plants and soils of the interior Pacific Northwest, USA. *Review of Palaeobotany and Palynology, 135,* 71–98.

Blinnikov, M. S. (2016). Phytolith gallery. http://web.stcloudstate.edu/msblinnikov/phd/phyt.html.

Blinnikov, M. S., Bagent, C. M., & Reyerson, P. E. (2013). Phytolith assemblages and opal concentrations from modern soils differentiate temperate grasslands of controlled composition on experimental plots at Cedar Creek, Minnesota. *Quaternary International, 287,* 101–113.

Blinnikov, M. S., Busacca, A., & Whitlock, C. (2002). Reconstruction of the Late Pleistocene grassland of the Columbia basin, Washington, USA, based on phytolith records in loess. *Palaeogeography, Palaeoclimatology, Palaeoecology, 177,* 77–101.

Blueweiss, L., Fox, H., Kudzma, V., Nakashima, D., Peters, R., & Sams, S. (1978). Relationships between body size and some life history parameters. *Oecologia, 37,* 257–272.

Borrelli, N., Alvarez, M. F., Osterrieth, M. L., & Marcovecchio, J. E. (2010). Silica content in soil solution and its relation with phytolith weathering and silica biogeochemical cycle in typical Argiudolls of the Pampean Plain, Argentina—A preliminary study. *Journal of Soils and Sediments, 10,* 983–994.

Borrelli, N., Fernández Honaine, M., Altamirano, S. M., & Osterrieth, M. (2011). Calcium and silica biomineralizations in leaves of eleven aquatic species of the Pampean Plain, Argentina. *Aquatic Botany, 94,* 29–36.

Borrelli, N., Osterrieth, M., & Marcovecchio, J. (2008). Interrelations of vegetal cover, silicophytolith content and pedogenesis of typical Argiudolls of the Pampean Plain, Argentina. *Catena, 75,* 146–153.

Bouchenak-Khelladi, Y., Verboom, G. A., Hodkinson, T. R., Salamin, N., Francois, O., Chonghaile, G. N., et al. (2009). The origins and diversification of C_4 grasses and savanna adapted ungulates. *Global Change Biology, 15,* 2397–2417.

Bouchenak-Khelladi, Y., Verboom, G. A., Savolainen, V., & Hodkinson, T. R. (2010). Biogeography of the grasses (Poaceae): a phylogenetic approach to reveal evolutionary history in geographical space and geological time. *Botanical Journal of the Linnean Society, 162,* 543–557.

Bouchenak-Khelladi, Y., Muasya, A. M., & Linder, H. P. (2014). A revised evolutionary history of Poales: origins and diversification. *Botanical Journal of the Linnean Society, 175,* 4–16.

Bowdery, D., Hart, D. M., Lentfer, C. J., & Wallis, L. A. (2001). A universal phytolith key. In J. D. Meunier & F. Colin (Eds.), *Phytoliths: Applications in Earth sciences and human history* (pp. 267–278). Lisse: A. A. Balkema Publishers.

Boyadjian, C. H. C., Eggers, S., & Reinhard, K. (2007). Dental wash: a problematic method for extracting microfossils from teeth. *Journal of Archaeological Science, 34,* 1622–1628.

Boyd, M. (2002). Identification of anthropogenic burning in the paleoecological record of the northern prairies: a new approach. *Annals of the Association of American Geographers, 92,* 471–487.

Boyd, W. E., Lentfer, C. J., & Torrence, R. (1998). Phytolith analysis for a wet tropics environment: methodological issues and implications for the archaeology of Garua Island, West New Britain, Papa New Guinea. *Palynology, 22,* 213–228.

Bozarth, S., & Hofman, J. (1998). Phytolith analysis of bison teeth calculus and impacta from sites in Kansas and Oklahoma. *Current Research in the Pleistocene, 15,* 95–96.

Bozarth, S. R. (1985). Distinctive phytoliths from various dicot species. *Phytolitharien Newsletter, 3,* 7–8.

Bozarth, S. R. (1987). Diagnostic opal morphotypes from rinds of selected *Cucurbita* species. *American Antiquity, 52,* 607–615.

Bozarth, S. R. (1990). Diagnostic opal phytoliths from pods of selected varieties of common beans (*Phaseolus vulgaris*). *American Antiquity, 55,* 98–104.

Bozarth, S. R. (1992). Classification of opal phytoliths formed in selected dicotyledons native to the Great Plains. In G. J. Rapp &

S. C. Mulholland (Eds.), *Phytolith systematics. Emerging issues, vol 1. Advances in Archaeological and Museum Science* (pp. 193–214). New York: Plenum Press.

Bozarth, S. R. (1993). Biosilicate assemblages of boreal forests and aspen parklands. In D. M. Pearsall & D. R. Piperno (Eds.), *Current research in phytolith analysis: Applications in archaeology and paleoecology, vol 10. MASCA Research Papers in Science and Archaeology* (pp. 95–105). Philadelphia: University Museum of Archaeology and Anthropology, University of Pennsylvania.

Brea, M., Zucol, A., & Iglesias, A. (2012). Fossil plant studies from late Early Miocene of the Santa Cruz Formation: paleoecology and paleoclimatology at the passive margin of Patagonia, Argentina. In S. Vizcaíno, R. F. Kay & M. S. Bargo (Eds.), *Early Miocene paleobiology in Patagonia* (pp. 104–128). Cambridge: Cambridge University Press.

Bremer, K. (2002). Gondwanan evolution of the grass alliance of families (Poales). *Evolution, 56,* 1374–1387.

Bremond, L., Alexandre, A., Véla, E., & Guiot, J. (2004). Advantages and disadvantages of phytolith analysis for the reconstruction of Mediterranean vegetation: an assessment based on modern phytolith, pollen and botanical data (Luberon, France). *Review of Palaeobotany and Palynology, 129,* 213–228.

Bremond, L., Alexandre, A., Hély, C., & Guiot, J. (2005a). A phytolith index as a proxy of tree cover density in tropical areas: calibration with Leaf Area Index along a forest–savanna transect in southeastern Cameroon. *Global and Planetary Change, 45,* 277–293.

Bremond, L., Alexandre, A., Peyron, O., & Guiot, J. (2005b). Grass water stress estimated from phytoliths in West Africa. *Journal of Biogeography, 32,* 311–327.

Bremond, L., Alexandre, A., Peyron, O., & Guiot, J. (2008a). Definition of grassland biomes from phytoliths in West Africa. *Journal of Biogeography, 35,* 2039–2048.

Bremond, L., Alexandre, A., Wooller, M. J., Hely, C., Williamson, D., Schafer, P. A., et al. (2008b). Phytolith indices as proxies of grass subfamilies on East African tropical mountains. *Global and Planetary Change, 61,* 209–224.

Brophy, J. K., de Ruiter, D. J., Fortelius, M., Bamford, M., & Berger, L. R. (2016). Pleistocene Bovidae (Mammalia) from Malapa, Gauteng Province, South Africa. *Palaeontologia Electronica, 19,* 1–22.

Brown, D. A. (1984). Prospects and limits of a phytolith key for grasses in the Central United States. *Journal of Archaeological Science, 11,* 345–368.

Buján, E. (2013). Elemental composition of phytoliths in modern plants (Ericaceae). *Quaternary International, 287,* 114–120.

Cabanes, D., Weiner, S., & Shahack-Gross, R. (2011). Stability of phytoliths in the archaeological record: a dissolution study of modern and fossil phytoliths. *Journal of Archaeological Science, 38,* 2480–2490.

Calderón, C. E., & Soderstrom, T. R. (1980). The genera of Bambusoideae (Poaceae) of the American conitinent: keys and comments. *Smithsonian Contributions to Botany, 44,* 1–27.

Calegari, M. R., Madella, M., Vidal-Torrado, P., Pessenda, L. C. R., & Marques, F. A. (2013). Combining phytoliths and $\delta^{13}C$ matter in Holocene palaeoenvironmental studies of tropical soils: an example of an Oxisol in Brazil. *Quaternary International, 287,* 47–55.

Campbell, C. S. & Kellogg, E. A. (1987). Sister group relationships of the Poaceae. In T. R. Soderstrom, K. W. Hilu, C. S. Campbell & M. E. Barkworth (Eds.), *Grass systematics and evolution* (pp. 217–224). Washington, D. C.: Smithsonian Institution Press.

Carnelli, A. L., Madella, M., & Theurillat, J.-P. (2001). Biogenic silica production in selected Alpine plant species and plant communities. *Annals of Botany, 87,* 425–434.

Carnelli, A. L., Madella, M., Theurillat, J.-P., & Ammann, B. (2002). Aluminum in the opal silica reticule of phytoliths: a new tool in palaeoecological studies. *American Journal of Botany, 89,* 346–351.

Carnelli, A. L., Theurillat, J.-P., & Madella, M. (2004). Phytolith types and type-frequencies in subalpine–alpine plant species of the European Alps. *Review of Palaeobotany and Palynology, 129,* 39–65.

Cerling, T. E. (1992). Development of grasslands and savannas in East Africa during the Neogene. *Palaeogeography, Palaeoclimatology, Palaeoecology, 97,* 241–247.

Cerling, T. E., Harris, J. M., MacFadden, B. J., Leakey, M. G., Quade, J., Eisenmann, V., et al. (1997). Global vegetation change through the Miocene/Pliocene boundary. *Nature, 389,* 153–158.

Cerling, T. E., Levin, N. E., Quade, J., Wynn, J. G., Fox, D. L., Kingston, J. D., et al. (2010). Comment on the paleoenvironment of *Ardipithecus ramidus*. *Science, 328,* 1105.

Chen, S. T., & Smith, S. Y. (2013). Phytolith variability in Zingiberales: a tool for the reconstruction of past tropical vegetation. *Palaeogeography, Palaeoclimatology, Palaeoecology, 370,* 1–12.

Chen, S. T., Smith, S. Y., Sheldon, N. D., & Strömberg, C. A. E. (2015). Regional-scale variability in the spread of grasslands in the late Miocene. *Palaeogeography, Palaeoclimatology, Palaeoecology, 437,* 42–52.

Christin, P.-A., Spriggs, E., Osborne, C. P., Strömberg, C. A. E., Salamin, N., & Edwards, E. J. (2014). Molecular dating, evolutionary rates, and the age of the grasses. *Systematic Biology, 63,* 153–165.

Ciochon, R. L., Piperno, D. R., & Thompson, R. G. (1990). Opal phytoliths found on the teeth of the extinct ape *Gigantopithecus blacki*: implications for paleodietary studies. *Proceedings of the National Academy of Sciences, USA, 87,* 8120–8124.

Clarke, J. (2003). The occurrence and significance of biogenic opal in the regolith. *Earth Science Reviews, 60,* 175–194.

Coe, H. H., Macario, K., Gomes, J. G., Chueng, K. F., Oliveira, F., Gomes, P. R., et al. (2014). Understanding Holocene variations in the vegetation of Sao Joao River basin, southeastern coast of Brazil, using phytolith and carbon isotopic analyses. *Palaeogeography, Palaeoclimatology, Palaeoecology, 415,* 59–68.

Coe, H. H. G., Alexandre, A., Carvalho, C. N., Santos, G. M., da Silva, A. S., Sousa, L. O. F., et al. (2013). Changes in Holocene tree cover density in Cabo Frio (Rio de Janeiro, Brazil): evidence from soil phytolith assemblages. *Quaternary International, 287,* 63–72.

Collura, L. V., & Neumann, K. (2017). Wood and bark phytoliths of West African woody plants. *Quaternary International, 434,* 142–159.

Cooke, J., & Leishman, M. R. (2011). Is plant ecology more siliceous than we realise? *Trends in Plant Science, 16,* 61–68.

Cordova, C. E. (2013). C_3 Poaceae and Restionaceae phytoliths as potential proxies for reconstructing winter rainfall in South Africa. *Quaternary International, 287,* 121–140.

Cordova, C. E., Johnson, W. C., Mandel, R. D., & Palmer, M. W. (2011). Late Quaternary environmental change inferred from phytoliths and other soil-related proxies: case studies from the central and southern Great Plains, USA. *Catena, 85,* 87–108.

Cordova, C. E., & Scott, L. (2010). The potential of Poaceae, Cyperaceae, and Restionaceae phytoliths to reflect past environmental conditions in South Africa. In J. Runge (Ed.), *African palaeoenviornments and geomorphic landscape evolution* (pp. 107–133). Boca Raton, FL: CRC Press.

Cornelis, J. T., Delvaux, B., Georg, R. B., Lucas, Y., Ranger, J., & Opfergelt, S. (2011). Tracing the origin of dissolved silicon transferred from various soil-plant systems towards rivers: a review. *Biogeosciences, 8,* 89–112.

Cotton, J. M., Hyland, E. G., & Sheldon, N. D. (2014). Multi-proxy evidence for tectonic control on the expansion of C_4 grasses in northwest Argentina. *Earth and Planetary Science Letters, 395,* 41–50.

Cotton, J. M., Sheldon, N. D., & Strömberg, C. A. E. (2012). High resolution isotopic record of C_4 photosynthesis of a Miocene grassland. *Palaeogeography, Palaeoclimatology, Palaeoecology, 337–338,* 88–98.

Couvreur, T. L. P., & Baker, W. J. (2013). Tropical rain forest evolution: palms as a model group. *BMC Biology, 11,* 48.

Crifò, C., & Strömberg, C. A. E. (2016). Phytoliths in paleoecology: a tool for reconstructing habitat structure, and heterogeneity. *Geologic Society of America Abstracts with Programs, 48.*

Currie, H. A., & Perry, C. C. (2007). Silica in plants: biological, biochemical and chemical studies. *Annals of Botany, 100,* 1383–1389.

da Costa, F. G., Souza, P. C., Klein, D. E., & Bove, C. P. (2016). Application of acetolysis in phytoliths extraction. *Review of Palaeobotany and Palynology, 228,* 93–97.

Dahlgren, R. M. T., Clifford, H. T., & Yeo, P. F. (1985). *The families of the monocotyledons. Structure, evolution, and taxonomy.* New York: Springer.

Das, S., Ghosh, P., Paruya, D. K., Yao, Y.-F., & Li, C.-S. (2013a). Phytolith spectra in respiratory aerial roots of some mangrove plants of the Indian Sunderbans and its efficacy in ancient deltaic environment reconstruction. *Quaternary International, 325,* 179–196.

Das, S., Ghosh, R., & Bera, S. (2013b). Application of non-grass phytoliths in reconstructing deltaic environments: a study from the Indian Sunderbans. *Palaeogeography, Palaeoclimatology, Palaeoecology, 376,* 48–65.

Delhon, C., Alexandre, A., Berger, J.-F., Thiébault, S., Brochier, J.-L., & Meunier, J.-D. (2003). Phytolith assemblages as a promising tool for reconstructing Mediterranean Holocene vegetation. *Quaternary Research, 59,* 48–60.

Delhon, C., Martin, L., Argant, J., & Thiebault, S. (2008). Shepherds and plants in the Alps: multi-proxy archaeobotanical analysis of neolithic dung from "La Grande Rivoire"(Isère, France). *Journal of Archaeological Science, 35,* 2937–2952.

Dickau, R., Whitney, B. S., Iriarte, J., Mayle, F. E., Soto, J. D., Metcalfe, P., et al. (2013). Differentiation of neotropical ecosystems by modern soil phytolith assemblages and its implications for palaeoenvironmental and archaeological reconstructions. *Review of Palaeobotany and Palynology, 193,* 15–37.

Diester-Haass, L., Schrader, H. J., & Thiede, J. (1973). Sedimentological and paleoclimatological investigations of two pelagic ooze cores off Cape Barbas, North-West Africa. *"Meteor" Forschungsergebnisse, C16,* 19–66.

Dillhoff, R. M., Dillhoff, T. A., Jijina, A. P., & Strömberg, C. A. E. (2014). The Vasa Park flora, King County, Washington, USA – a window into the late Miocene of the Pacific Northwest. In W. D. Stevens, O. M. Montiel & P. H. Raven (Eds.), *Paleobotany and biogrography, a festschrift for Alan Graham in his 80th year* (pp. 64–97). St. Louis: Missouri Botanical Garden Press.

Dinan, E. H. & Rowlett, R. M. (1993). Vegetational changes at the Shriver paleo-Indian site, N. W. Missouri: Phytolith analysis as an aid in environmental reconstruction. In D. M. Pearsall & D. R. Piperno (Eds.), *Current research in phytolith analysis: Applications in archaeology and paleoecology, vol 10. MASCA Research Papers in Science and Archaeology* (pp. 73–82). Philadelphia: University Museum of Archaeology and Anthropology, University of Pennsylvania.

Dixon, P., Tremaine, W., Pickles, K., Kuhns, L., Hawe, C., McCann, J., et al. (1999). Equine dental disease Part 1: a long-term study of 400 cases: disorders of incisor, canine and first premolar teeth. *Equine Veterinary Journal, 31,* 369–377.

Donohue, J. A. & Dinan, E. H. (1993). A geoarchaeological analysis of phytolith data from the Bear Creek site, Cedar County, Missouri. In D. M. Pearsall & D. R. Piperno (Eds.), *Current research in phytolith analysis: Applications in archaeology and paleoecology, vol 10. MASCA Research Papers in Science and Archaeology* (pp. 83–94). Philadelphia: University Museum of Archaeology and Anthropology, University of Pennsylvania.

Drees, R. L., Wilding, L. P., Smeck, N. E., & Senkayi, A. L. (1989). Silica in soils: Quartz and disordered silica polymorphs. In J. B. Dixon & S. B. Weed (Eds.), *Minerals in Soil Environments* (pp. 913–974). Madison: Soil Science Society of America.

Dugas, D. P., & Retallack, G. J. (1993). Middle Miocene fossil grasses from Fort Ternan, Kenya. *Journal of Paleontology, 67,* 113–128.

Dumitriča, P. (1973). Phytolitharia. In A. G. Kaneps (Ed.), *Initial reports of the deep sea drilling project, vol XIII. Part 2* (pp. 940–943). Washington, D. C.: U. S. Printing Office.

Dunn, R. E., Strömberg, C. A. E., & Le, T.-Y. (2015a). Light environment and epidermal cell morphology in grasses. *International Journal of Plant Sciences, 176,* 832–847.

Dunn, R. E., Strömberg, C. A. E., Madden, R. H., Kohn, M. J., & Carlini, A. A. (2015b). Linked canopy, climate and faunal evolution in the Cenozoic of Patagonia. *Science, 347,* 258–261.

Edwards, E. J., Osborne, C. P., Strömberg, C. A. E., Smith, S. A., & C_4 Grasses Consortium (2010). The origins of C_4 grasslands: integrating evolutionary and ecosystem science. *Science*, *328*, 587–591

Edwards, E. J., & Smith, S. A. (2010). Phylogenetic analyses reveal the shady history of C_4 grasses. *Proceedings of the National Academy of Sciences, USA, 107,* 2532–2537.

Ehrenberg, C. G. (1847). Passatstaub und Blutregen: Ein grosses organisches unsichtbares Wirken und Leben in der Atmosphäre. *Abhandlungen der Königlichen Akademie der Wissenschaften zu Berlin, 1847,* 269–460.

Ehrlich, H., Demadis, K. D., Pokrovsky, O. S., & Koutsoukos, P. G. (2010). Modern views on desilicification: biosilica and abiotic silica dissolution in natural and artificial environments. *Chemical Review, 110,* 4656–4689.

Eichhorn, B., Neumann, K., & Garnier, A. (2010). Seed phytoliths in West African Commelinaceae and their potential for palaeoecological studies. *Palaeogeography, Palaeoclimatology, Palaeoecology, 298,* 300–310.

Elbaum, R., Melamed-Bessudo, C., Tuross, N., Levy, A. A., & Weiner, S. (2009). New methods to isolate organic materials from silicified phytoliths reveal fragmented glycoproteins but no DNA. *Quaternary International, 193,* 11–19.

Elbaum, R., Weiner, S., Albert, R. M., & Elbaum, M. (2003). Detection of burning of plant materials in the archaeological record by changes in the refractive indices of siliceous phytoliths. *Journal of Archaeological Science, 30,* 217–226.

Ellis, B., & Johnson, K. R. (2013). Comparison of leaf samples from mapped tropical and temperate forests: implications for interpretations of the diversity of fossil assemblages. *PALAIOS*, *28*, 163–177.

Eronen, J. T., Puolamaki, K., Liu, L., Lintulaakso, K., Damuth, J., Janis, C., et al. (2010). Precipitation and large herbivorous mammals I: estimates from present-day communities. *Evolutionary Ecology Research, 12,* 217–233.

Esau, K. (1965). *Plant anatomy* (2nd ed.). New York: Wiley.

Esteban, I., De Vynck, J. C., Singels, E., Vlok, J., Marean, C. W., Cowling, R. M., et al. (2017). Modern soil phytolith assemblages used as proxies for Paleoscape reconstruction on the south coast of South Africa. *Quaternary International, 434,* 160–179.

Evett, R. R., & Cuthrell, R. Q. (2016). A conceptual framework for a computer-assisted, morphometric-based phytolith analysis and classification system. *Journal of Archaeological Science, 68,* 70–78.

Evett, R. R., & Cuthrell, R. Q. (2017). Testing phytolith analysis approaches to estimate the prehistoric anthropogenic burning regime on the central California coast. *Quaternary International, 434,* 78–90.

Fahmy, A. G. (2008). Diversity of lobate phytoliths in grass leaves from the Sahel region, West Tropical Africa: tribe Paniceae. *Plant Systematics and Evolution, 270,* 1–23.

Faisal, S., Callis, K. L., Slot, M., & Kitajima, K. (2012). Transpiration-dependent passive silica accumulation in cucumber (*Cucumis sativus*) under varying soil silicon availability. *Botany, 90,* 1058–1064.

Fauteux, F., Remus-Borel, W., Menzies, J. G., & Belanger, R. R. (2005). Silicon and plant disease resistance against pathogenic fungi. *FEMS Microbiology Letters, 249,* 1–6.

Feakins, S. J., Levin, N. E., Liddy, H. M., Sieracki, A., Eglinton, T. I., & Bonnefille, R. (2013). Northeast African vegetation change over 12 my. *Geology, 41,* 295–298.

Fearn, M. L. (1998). Phytoliths in sediment as indicators of grass pollen source. *Review of Palaeobotany and Palynology, 103,* 75–81.

Fein, J. B. (2009). Experimental and field constraints on the role of silica–organic complexation and silica–microbial interactions during sediment diagenesis. In R. H. Worden & M. Sadoon (Eds.), *Quartz Cementation in Sandstones* (pp. 119–127). Oxford: Blackwell Publishing Ltd.

Fernandez Honaine, M., Zucol, A. F., & Osterrieth, M. L. (2006). Phytolith assemblages and systematic associations in grassland species of the south-eastern Pampean plains, Argentina. *Annals of Botany, 98,* 1155–1165.

Fernández Pepi, M. G., Zucol, A. F., & Arriaga, M. O. (2012). Comparative phytolith analysis of *Festuca* (Pooideae: Poaceae) species native to Tierra del Fuego, Argentina. *Botany, 90,* 1113–1124.

Fishkis, O., Ingwersen, J., Lamers, M., Denysenko, D., & Streck, T. (2010a). Phytolith transport in soil: a field study using fluorescent labelling. *Geoderma, 157,* 27–36.

Fishkis, O., Ingwersen, J., Lamers, M., Denysenko, D., & Streck, T. (2010b). Phytolith transport in soil: a laboratory study on intact soil cores. *European Journal of Soil Science, 61,* 445–455.

Fishkis, O., Ingwersen, J., & Streck, T. (2009). Phytolith transport in sandy sediment: experiments and modeling. *Geoderma, 151,* 168–178.

Folger, D. (1970). Wind transport of land-derived mineral, biogenic, and industrial matter over the North Atlantic. *Deep Sea Research and Oceanographic Abstracts, 2,* 337.

Folger, D., Burckle, L., & Heezen, B. (1967). Opal phytoliths in a North Atlantic dust fall. *Science, 155,* 1243–1244.

Fox, C. L., Juan, J., & Albert, R. M. (1996). Phytolith analysis on dental calculus, enamel surface, and burial soil: information about diet and paleoenvironment. *American Journal of Physical Anthropology, 101,* 101–113.

Fox, D. L., Honey, J. G., Martin, R. A., & Pelaez-Campomanes, P. (2012a). Pedogenic carbonate stable isotope record of environmental change during the Neogene in the southern Great Plains, southwest Kansas, USA: Carbon isotopes and the evolution of C_4-dominated grasslands. *Geological Society of America Bulletin, 124,* 444–462.

Fox, D. L., Honey, J. G., Martin, R. A., & Pelaez-Campomanes, P. (2012b). Pedogenic carbonate stable isotope record of environmental change during the Neogene in the southern Great Plains, southwest Kansas, USA: Oxygen isotopes and paleoclimate during the evolution of C_4-dominated grasslands. *Geological Society of America Bulletin, 124,* 431–443.

Fraysse, F., Pokrovsky, O. S., & Meunier, J.-D. (2010). Experimental study of terrestrial plant litter interaction with aqueous solutions. *Geochimica et Coscmochimica Acta, 74,* 70–84.

Fraysse, F., Pokrovsky, O. S., Schott, J., & Meunier, J.-D. (2006). Surface properties, solubility and dissolution kinetics of bamboo phytoliths. *Geochimica et Coscmochimica Acta, 70,* 1939–1951.

Fraysse, F., Pokrovsky, O. S., Schott, J., & Meunier, J.-D. (2009). Surface chemistry and reactivity of plant phytoliths in aqueous solutions. *Chemical Geology, 258,* 197–206.

Fredlund, G. G. (1986). Problems in the simultaneous extraction of pollen and phytoliths from clastic sediments. In I. Rovner (Ed.), *Plant opal phytolith analysis in archaeology and paleoecology. Proceedings of the 1984 Phytolith Research Workshop, vol 1. Occasional Papers of the Phytolitharien* (pp. 102–111). Raleigh: North Carolina State University.

Fredlund, G. G., & Tieszen, L. L. (1994). Modern phytolith assemblages from the North American Great Plains. *Journal of Biogeography, 21,* 321–335.

Fredlund, G. G., & Tieszen, L. L. (1997a). Calibrating grass phytolith assemblages in climatic terms: application to late Pleistocene assemblages from Kansas and Nebraska. *Palaeogeography, Palaeoclimatology, Palaeoecology, 136,* 199–211.

Fredlund, G. G., & Tieszen, L. L. (1997b). Phytolith and carbon evidence for Late Quaternary vegetation and climate change in the Southern Black Hills, South Dakota. *Quaternary Research, 47,* 206–217.

Gallaher, T. J., Senske, A., Marvet, C., Akbar, S., Wong, S., Hsu, T., Jessett, K., Klahs, P. C., Clark, L. G., & Strömberg, C. A. E. (2017). 3D digitization of grass silica short cell phytoliths (GSSCP's): a digital reference collection to improve their utility for the study of grasses and grassland evolution. *Geological Society of America Abstracts with Programs 49,* 261.

Gallego, L., & Distel, R. A. (2004). Phytolith assemblages in grasses native to Central Argentina. *Annals of Botany, 94,* 865–874.

Gallego, L., Distel, R. A., Camina, R., & Iglesias, R. M. R. (2004). Soil phytoliths as evidence for species replacement in grazed rangelands of central Argentina. *Ecography, 27,* 725–732.

Garnier, A., Neumann, K., Eichhorn, B., & Lespez, L. (2012). Phytolith taphonomy in the middle- to late-Holocene fluvial sediments of Ounjougou (Mali, West Africa). *Holocene, 23,* 416–431.

Ge, Y., Jie, D. M., Sun, Y. L., & Liu, H. M. (2011). Phytoliths in woody plants from the northern slope of the Changbai Mountain (Northeast China), and their implication. *Plant Systematics and Evolution, 292,* 55–62.

Geis, J. W. (1973). Biogenic silica in selected species of deciduous angiosperms. *Soil Science, 116,* 113–130.

George, W. O. (1924). The relation of the physical properties of natural glasses to their chemical composition. *The Journal of Geology, 32,* 353–372.

Gérard, F., Mayer, K. U., Hodson, M. J., & Ranger, J. (2008). Modelling the biogeochemical cycle of silicon in soils: application to a temperate forest ecosystem. *Geochimica et Cosmochimica Acta, 72,* 741–758.

Ghosh, P., Bhattacharya, S. K., Sahni, A., Kar, R. K., Mohabey, D. M., & Ambwani, K. (2003). Dinosaur coprolites from the Late Cretaceous (Maastrichtian) Lameta Formation of India: isotopic and other markers suggesting a C_3 plant diet. *Cretaceous Research, 24,* 743–750.

Ghosh, R., Gupta, S., Bera, S., Jiang, H-e, Li, X., & Li, C.-S. (2008). Ovi-caprid dung as an indicator of paleovegetation and paleoclimate in northwestern China. *Quaternary Research, 70,* 149–157.

Givnish, T. (1988). Adaptation to sun and shade: a whole-plant perspective. *Australian Journal of Plant Physiology, 15,* 63–92.

Gobetz, K., & Bozarth, S. (2000). Diet and paleoecology of Columbian mammoth (*Mammuthus columbi*) determined from phytoliths and diatoms in teeth. *Current Research in the Pleistocene, 17,* 120–122.

Gobetz, K. E., & Bozarth, S. R. (2001). Implications for Late Pleistocene mastodon diet from opal phytoliths in tooth calculus. *Quaternary Research, 55,* 115–122.

Gol'eva, A. A. (1997). Content and distribution of phytoliths in the main types of soils in eastern Europe. In A. Pinilla, J. Juan-Tresserras & M. J. Machado (Eds.), *The state of the art of phytoliths in soils and plants, vol 4. Monografías del Centro de Ciencias Medioambientales* (pp. 15–22). Madrid: CSIC.

Gol'eva, A. A. (2001). *Phytoliths and their informational role in natural and archaeological objects*. Syktuvar Elista, Moscow: Russian Academy of Sciences.

Grave, P., & Kealhofer, L. (1999). Assessing bioturbation in archaeological sediments using soil morphology and phytolith analysis. *Journal of Archaeological Science, 26,* 1239–1248.

Greenwood, D. R., & Wing, S. L. (1995). Eocene continental climates and latitudinal temperature gradients. *Geology, 23,* 1044–1048.

Griffith, D. M., Anderson, T. M., Osborne, C. P., Strömberg, C. A. E., Forrestel, E. J., & Still, C. J. (2014). Biogeographically distinct controls on C_3 and C_4 grass distributions: merging community and physiological ecology. *Global Ecology and Biogeography, 24,* 304–313.

Gu, Y., Zhao, Z., & Pearsall, D. M. (2013). Phytolith morphology research on wild and domesticated rice species in East Asia. *Quaternary International, 287,* 141–148.

Gügel, I. L., Grupe, G., & Kunzelmann, K. H. (2001). Simulation of dental microwear: characteristic traces by opal phytoliths give clues to ancient human dietary behavior. *American Journal of Physical Anthropology, 114,* 124–138.

Guo, M., Jie, D., Liu, H., Luo, S., & Li, N. (2012). Phytolith analysis of selected wetland plants from Changbai mountain region and implications for palaeoenvironment. *Quaternary International, 250,* 119–128.

Habermann, J. M., Stanistreet, I. G., Stollhofen, H., Albert, R. M., Bamford, M. K., Pante, M. C., et al. (2016). In situ 2.0 Ma trees discovered as fossil rooted stumps, lowermost Bed I, Olduvai Gorge, Tanzania. *Journal of Human Evolution*, *9*, 74–87.

Halffter, G., & Matthews, E. G. (1966). The natural history of dung beetles of the subfamily Scarabaeinae (Coleoptera: Scarabaeidae). *Folia Entomologica Mexicana, 12–14,* 1–312.

Hanski, I. (1983). Dung beetles. In H. Lieth & M. J. A. Werger (Eds.), *Tropical rain forest ecosystems, vol 14B. Ecosystems of the World, 1st edn.* (pp. 489–511). Amsterdam: Elsevier.

Harris, E. B. (2016). *Effects of the mid-Miocene climatic optimum on ecosystem structure and plant-animal interactions: A phytolith and stable isotope perspective*. Ph.D. Dissertation, Department of Biology. University of Washington, Seattle, 212 pages.

Harris, E. B., Strömberg, C. A. E., Sheldon, N. D., Smith, S. Y., & Vilhena, D. (2017). Vegetation response during the lead-up to the middle Miocene warming event in the Northern Rocky Mountains. *Palaeogeography, Palaeoclimatology, Palaeoecology, 485*, 401–415.

Hart, D. M. (1988). The plant opal content in the vegetation and sediment of a swamp at Oxford Falls, New South Wales, Australia. *Australian Journal of Botany, 36,* 159–170.

Hartley, S. E., & DeGabriel, J. L. (2016). The ecology of herbivore-induced silicon defences in grasses. *Functional Ecology, 30,* 1311–1322.

He, H., Bleby, T. M., Veneklaas, E. J., Lambers, H., & Kuo, J. (2012). Morphologies and elemental compositions of calcium crystals in phyllodes and branchlets of *Acacia robeorum* (Leguminosae: Mimosoideae). *Annals of Botany, 109,* 887–896.

Henry, A. G. (2012). Recovering dietary information from extant and extinct primates using plant microremains. *International Journal of Primatology, 33,* 702–715.

Henry, A. G., Brooks, A. S., & Piperno, D. R. (2011). Microfossils in calculus demonstrate consumption of plants and cooked foods in Neanderthal diets (Shanidar III, Iraq; Spy I and II, Belgium). *Proceedings of the National Academy of Sciences, USA, 108,* 486–491.

Henry, A. G., Brooks, A. S., & Piperno, D. R. (2014). Plant foods and the dietary ecology of Neanderthals and early modern humans. *Journal of Human Evolution, 69,* 44–54.

Henry, A. G., & Piperno, D. R. (2008). Using plant microfossils from dental calculus to recover human diet: a case study from Tell al-Raqā'i, Syria. *Journal of Archaeological Science, 35,* 1943–1950.

Henry, A. G., Ungar, P. S., Passey, B. H., Sponheimer, M., Rossouw, L., Bamford, M., et al. (2012). The diet of *Australopithecus sediba*. *Nature, 487,* 90–93.

Hodson, M. J. (2016). The development of phytoliths in plants and its influence on their chemistry and isotopic composition. Implications for palaeoecology and archaeology. *Journal of Archaeological Science, 68,* 62–69.

Hodson, M. J., White, P. J., Mead, A., & Broadley, M. R. (2005). Phylogenetic variation in the silicon composition of plants. *Annals of Botany, 96,* 1027–1046.

Hoetzel, S., Dupont, L., Schefuß, E., Rommerskirchen, F., & Wefer, G. (2013). The role of fire in Miocene to Pliocene C_4 grassland and ecosystem evolution. *Nature Geoscience, 6,* 1027–1030.

Honaine, M. F., Osterrieth, M. L., & Zucol, A. F. (2009). Plant communities and soil phytolith assemblages relationship in native grasslands from southeastern Buenos Aires province, Argentina. *Catena, 76,* 89–96.

Horrocks, M., Baisden, T., Flenley, J., Feek, D., Love, C., Haoa-Cardinali, S., et al. (2016). Pollen, phytolith and starch analyses of dryland soils from Easter Island (Rapa Nui) show widespread vegetation clearance and Polynesian-introduced crops. *Palynology, 41,* 339–350.

Hwang, S., Yoon, S.-O., Lee, J.-Y., Kim, H.-S., & Choi, J. (2012). Phytolith analysis and reconstruction of palaeoenvironment at the Nabokri Valley Plain, Buyeo, Korea. *Quaternary International, 254,* 129–137.

Hyland, E., Sheldon, N. D., & Fan, M. (2013a). Terrestrial paleoenvironmental reconstructions indicate transient peak warming during the Early Eocene Climatic Optimum. *Geological Society of America Bulletin, 125,* 1338–1348.

Hyland, E., Sheldon, N. D., & Smith, S. Y. (2013b). Representational bias in phytoliths from modern soils of central North America: implications for paleovegetation reconstructions. *Palaeogeography, Palaeoclimatology, Palaeoecology, 374,* 338–348.

Hyland, E. G., Sheldon, N. D., & Cotton, J. M. (2015). Terrestrial evidence for a two-stage mid-Paleocene biotic event. *Palaeogeography, Palaeoclimatology, Palaeoecology, 417,* 371–378.

ICPN Working Group: Madella, M., Alexandre, A., & Ball, T. (2005). International code for phytolith nomenclature 1.0. *Annals of Botany, 96,* 253–260.

Iles, W. J., Smith, S. Y., Gandolfo, M. A., & Graham, S. W. (2015). Monocot fossils suitable for molecular dating analyses. *Botanical Journal of the Linnean Society, 178,* 346–374.

Inoue, K., & Sase, T. (1996). Paleoenvironmental history of post-Toya ash tephric deposits and paleosols at Iwate Volcano, Japan, using aeolian dust content and phytolith composition. *Quaternary International, 34–36,* 127–137.

Iriarte, J. (2003). Assessing the feasibility of identifying maize through the analysis of cross-shaped size and three-dimensional morphology of phytoliths in the grasslands of southeastern South America. *Journal of Archaeological Science, 30,* 1085–1094.

Iriarte, J., & Paz, E. A. (2009). Phytolith analysis of selected native plants and modern soils from southeastern Uruguay and its implications for paleoenvironmental and archeological reconstruction. *Quaternary International, 193,* 99–123.

Issaharou-Matchi, I., Barboni, D., Meunier, J. D., Saadou, M., Dussouillez, P., Contoux, C., et al. (2016). Intraspecific biogenic silica variations in the grass species *Pennisetum pedicellatum* along an evapotranspiration gradient in South Niger. *Flora – Morphology, Distribution, Functional Ecology of Plants, 220,* 84–93.

Jacobs, B. F., Kingston, J. D., & Jacobs, L. L. (1999). The origin of grass-dominated ecosystems. *Annals of the Missouri Botanical Garden, 86,* 590–643.

Janis, C. M. (1988). An estimation of tooth volume and hypsodonty indices in ungulate mammals, and the correlation of these factors with dietary preferences. In D. E. Russel, J.-P. Santoro & D. Sigogneau-Russel (Eds.), *Teeth revisited: Proceedings of the VIIth International Symposium on Dental Morphology, Paris, France, 1988* (pp. 367–387). Paris: Memoires de Museum National d'Histoire Naturelle, series C.

Jones, J. G., & Bryant, V. M., Jr. (1992). Phytolith taxonomy in selected species of Texas cacti. In G. J. Rapp & S. C. Mulholland (Eds.), *Phytolith systematics. Emerging issues, vol 1. Advances in Archaeological and Museum Science* (pp. 215–238). New York: Plenum Press.

Jones, L. H. P., & Handreck, K. A. (1967). Silica in soils, plants, and animals. *Advances in Agronomy, 19,* 107–149.

Jones, L. H. P., Milne, A. A., & Sanders, J. V. (1966). Tabashir: an opal of plant origin. *Science, 151,* 464–466.

Jones, R. L. (1964). Note on occurrence of opal phytoliths in some Cenozoic sedimentary rocks. *Journal of Paleontology, 38,* 773–775.

Jones, R. L., & Beavers, A. H. (1963). Some mineralogical and chemical properties of plant opal. *Soil Science, 96,* 6.

Jones, R. L., & Beavers, A. H. (1964). Aspects of catenary and depth distribution of opal phytoliths in Illinois soils. *Proceedings of the Soil Science Society of America, 28,* 413–416.

Jubb, K. V. F., Kennedy, P. C., & Palmer, N. (1985). *Pathology of domestic animals* (3rd ed., Vol. 2). Orlando, Florida: Academic Press, Inc.

Katz, O. (2015). Silica phytoliths in angiosperms: phylogeny and early evolutionary history. *New Phytologist, 208,* 642–464.

Katz, O., Cabanes, D., Weiner, S., Maeir, A. M., Boaretto, E., & Shahack-Gross, R. (2010). Rapid phytolith extraction for analysis of phytolith concentrations and assemblages during an excavation: an application at Tell es-Safi/Gath, Israel. *Journal of Archaeological Science, 37,* 1557–1563.

Katz, O., Gilead, I., Bar (Kuthiel), P., & Shahack-Gross, R. (2007). Chalcolithic agricultural life at Grar, northern Negev, Israel: dry farmed cereals and dung-fueled hearths. *Paléorient, 33,* 101–116.

Katz, O., Lev-Yadun, S., & Bar, P. (2013). Plasticity and variability in the patterns of phytolith formation in Asteraceae species along a large rainfall gradient in Israel. *Flora, 208,* 438–444.

Kaufman, P. B., Takeoka, P., Bigelow, W. C., Schmid, R., & Ghosheh, N. S. (1971). Electron microprobe analysis of silica in epidermal cells of *Equisetum*. *American Journal of Botany, 58,* 309–316.

Kawano, T., Sasaki, N., Hayashi, T., & Takahara, H. (2012). Grassland and fire history since the late-glacial in northern part of Aso Caldera, central Kyusyu, Japan, inferred from phytolith and charcoal records. *Quaternary International, 254,* 18–27.

Kay, R. F., Madden, R. H., Vucetich, M. G., Carlini, A. A., Mazzoni, M. M., Re, G. H., et al. (1999). Revised geochronology of the Casamayoran South American Land Mammal Age: climatic and biotic implications. *Proceedings of the National Academy of Sciences, USA, 96,* 13235–13240.

Kealhofer, L. (1996). The human environment during the terminal Pleistocene and Holocene in northeastern Thailand: phytolith evidence from Lake Kumphawapi. *Asian Perspectives, 35,* 229–254.

Kealhofer, L., & Penny, D. (1998). A combined pollen and phytolith record for fourteen thousand years of vegetation change in northeastern Thailand. *Review of Palaeobotany and Palynology, 103,* 83–93.

Kealhofer, L., & Piperno, D. (1998). Opal phytoliths in Southeast Asian flora. *Smithsonian Contributions to Botany, 88,* 1–39.

Keeley, J. E., & Rundel, P. W. (2005). Fire and the Miocene expansion of C_4 grasslands. *Ecology Letters, 8,* 683–690.

Kellogg, E. A. (2015). Fossil record and dates of diversification. In E. A. Kellogg, *Flowering Plants. Monocots* (pp. 103–107). Cham: Springer.

Kellogg, E. A., & Campbell, C. S. (1987). Phylogenetic analyses of the Gramineae. In T. R. Soderstrom, K. W. Hilu, C. S. Campbell, & M. E. Barkworth (Eds.), *Grass systematics and evolution* (pp. 310–322). Washington, D. C.: Smithsonian Institution.

Kelly, E. F., Amundson, R. G., Marino, B. D., & Deniro, M. J. (1991). Stable isotope ratios of carbon in phytoliths as a quantitative method of monitoring vegetation and climate change. *Quaternary Research, 35,* 222–233.

Kendrick, K. J. (2005). Pedogenic silica accumulation. In L. Rattan (Ed.), *Encyclopedia of soil science* (pp. 1251–1253). Taylor & Francis.

Kendrick, K. J., & Graham, R. C. (2004). Pedogenic silica accumulation in chronosequence soils, southern California. *Soil Science Society of America Journal, 68,* 1295–1303.

Kerns, B. K. (2001). Diagnostic phytoliths for a ponderosa pine-bunchgrass community near Flagstaff, Arizona. *The Southwestern Naturalist, 46,* 282–294.

Kerns, B. K., Moore, M. M., & Hart, S. C. (2001). Estimating forest-grassland dynamics using soil phytolith assemblages and $d^{13}C$ of soil organic matter. *Ecoscience, 8,* 478–488.

Klein, R. L., & Geis, J. W. (1978). Biogenic silica in the Pinaceae. *Soil Science, 126,* 145–156.

Kohn, M. J., Strömberg, C. A. E., Madden, R. H., Dunn, R. E., Evans, S., Palacios, A., et al. (2015). Quasi-static Eocene-Oligocene climate in Patagonia promotes slow faunal evolution and mid-Cenozoic global cooling. *Palaeogeography, Palaeoclimatology, Palaeoecology, 435,* 24–37.

Kondo, R., Childs, C., & Atkinson, I. (1994). *Opal phytoliths of New Zealand*. Canterbury, New Zealand: Manaaki Whenua Press.

Kondo, R., & Pearson, T. (1981). Opal phytoliths in tree leaves (Part 2): Opal phytoliths in dicotyledon angiosperm trees leaves. *Research Bulletin of Obihiro University, Series, 1,* 217–230.

Kondo, R., & Sase, T. (1986). Opal phytoliths, their nature and application. *The Quaternary Research, 25,* 31–63.

Kraus, M. J. (1999). Paleosols in clastic sedimentary rocks: their geologic applications. *Earth-Science Reviews, 47,* 41–70.

Krauss, D. A. (1997). *The use of phytolith analysis in paleoenvironmental reconstruction and environmental management on Martha's Vineyard, Massachusetts*. Ph.D. Dissertation, University of Massachusetts, Boston.

Kumar, S., Milstein, Y., Brami, Y., Elbaum, M., & Elbaum, R. (2017). Mechanism of silica deposition in sorghum silica cells. *New Phytologist, 213,* 791–798.

Kumari, I. S., & Kumarasamy, D. (2014). Studies on phytoliths in some marine plants. *International Journal of Plant, Animal and Environmental Sciences, 4,* 1–5.

Kurmann, M. H. (1985). An opal phytolith and palynomorph study of extant and fossil soils in Kansas (U.S.A.). *Palaeogeography, Palaeoclimatology, Palaeoecology, 49,* 217–235.

Kürschner, W. M. (1997). The anatomical diversity of recent and fossil leaves of the durmast oak (*Quercus petraea* Lieblein/*Q. pseudocastanea* Goeppert)—Implications for their use as biosensors of palaeoatmospheric CO_2 levels. *Palaeogeography, Palaeoclimatology, Palaeoecology, 96,* 1–30.

Lander, B. (1998). Oreodontoidea. In C. M. Janis, K. M. Scott, L. L. Jacobs (Eds.), *Evolution of Tertiary mammals in North America. Volume 1: Terrestrial carnivores, ungulates, and ungulatelike mammals* (pp. 402–420). Cambridge: Cambridge University Press.

Law, C., & Exley, C. (2011). New insight into silica deposition in horsetail (*Equisetum arvense*). *BMC Plant Biology, 11,* 112.

Lazzati, A., Levrini, L., Rampazzi, L., Dossi, C., Castelletti, L., Licata, M., et al. (2016). The diet of three Medieval individuals from Caravate (Varese, Italy). Combined results of ICP-MS analysis of trace elements and phytolith analysis conducted on their dental calculus. *International Journal of Osteoarchaeology, 26,* 670–681.

Lehmann, C. E., Anderson, T. M., Sankaran, M., Higgins, S. I., Archibald, S., Hoffmann, W. A., et al. (2014). Savanna vegetation-fire-climate relationships differ among continents. *Science, 343,* 548–552.

Lentfer, C. J., & Boyd, W. E. (1998). A comparison of three methods for the extraction of phytoliths from sediments. *Journal of Archaeological Science, 25,* 1159–1183.

Lentfer, C. J., & Boyd, W. E. (1999). An assessment of techniques for the deflocculation and removal of clays from sediments used in phytolith analysis. *Journal of Archaeological Science, 26,* 31–44.

Lentfer, C. J., & Boyd, W. E. (2000). Simultaneous extraction of phytoliths, pollen and spores from sediments. *Journal of Archaeological Science, 27,* 363–372.

Lentfer, C. J., Cotter, M. M., & Boyd, W. E. (2003). Particle settling times for gravity sedimentation and centrifugation: a practical guide for palynologists. *Journal of Archaeological Science, 30,* 149–168.

Levin, N. E., Quade, J., Simpson, S. W., Semaw, S., & Rogers, M. (2004). Isotopic evidence for Plio-Pleistocene environmental change at Gona, Ethiopia. *Earth and Planetary Science Letters, 219,* 93–110.

Liang, M.-M. (2004). Palynology, palaeocology and palaeoclimate of the Miocene Shanwang Basin, Shangdong Province, China. *Acta Palaeobotanica, Supp. No. 5,* 1–95.

Lieverse, A. R. (1999). Diet and the aetiology of dental calculus. *International Journal of Osteoarchaeology, 9,* 219–232.

Lins, U., Barros, C., Da Cunha, M., & Miguens, F. C. (2002). Structure, morphology, and composition of silicon biocomposites in the palm tree *Syagrus coronata* (Mart.) Becc. *Protoplasma, 220,* 89–96.

Linseele, V., Riemer, H., Baeten, J., De Vos, D., Marinova, E., & Ottoni, C. (2013). Species identification of archaeological dung remains: a critical review of potential methods. *Environmental Archaeology, 18,* 5–17.

Little, S. A., Kembel, S. W., & Wilf, P. (2010). Paleotemperature proxies from leaf fossils reinterpreted in light of evolutionary history. *PLoS ONE, 5,* e15161.

Liu, H., Jie, D., Liu, L., Li, N., Wang, T., & Qiao, Z. (2016). The shape factors of phytoliths in selected plants from the Changbai Mountains and their implications. *Review of Palaeobotany and Palynology, 226,* 44–53.

Locker, S., & Martini, E. (1986). Phytoliths from the southwest Pacific Site 591. *Initial reports of the Deep Sea Drilling Program, 90,* 1079–1084.

Loucaides, S., Behrends, T., & Van Cappellen, P. (2010). Reactivity of biogenic silica: surface versus bulk charge density. *Geochimica et Cosmochimica Acta, 74,* 517–530.

Lu, H.-Y., & Liu, K.-B. (2003a). Morphological variations of lobate phytoliths from grasses in China and the south-eastern United States. *Diversity and Distributions, 9,* 73–87.

Lu, H.-Y., & Liu, K.-B. (2003b). Phytoliths of common grasses in the coastal environments of southeastern USA. *Estuarine Coastal and Shelf Science 58,* 587–600.

Lu, H.-Y., Wu, N.-Q., Liu, K.-B., Jiang, H., & Liu, T.-S. (2007). Phytoliths as quantitative indicators for the reconstruction of past environmental conditions in China II: palaeoenvironmental reconstruction in the Loess Plateau. *Quaternary Science Reviews, 26,* 759–772.

Lu, H.-Y., Wu, N.-Q., Yang, X.-D., Jiang, H., Liu, K.-B., & Liu, T.-S. (2006). Phytoliths as quantitative indicators for the reconstruction of past environmental conditions in China I: Phytolith-based transfer functions. *Quaternary Science Reviews, 25,* 945–959.

Lu, H. Y., Wang, Y. J., Wang, G. A., Yang, H., & Li, Z. (2000). Analysis of carbon isotope in phytoliths from C_3 and C_4 plants and modern soils. *Chinese Science Bulletin, 45,* 1804–1808.

Ma, J. F. (2003). Mechanism of Si uptake in plants. *Fertilizer, 94,* 26–32.

Ma, J. F., Takahashi, E. (2002). *Soil, fertilizer, and plant silicon research in Japan*. Elsevier Science. 294 p.

Ma, J. F., Miyake, Y., & Takahashi, E. (2001). Chapter 2. Silicon as a beneficial element for crop plants. In L. E. Datnoff, G. H. Snyder & G. H. Korndörfer (Eds.), *Silicon in agriculture. Studies in plant science, vol 8* (pp. 17–39). Amsterdam: Elsevier Science B.V.

Ma, J. F., & Yamaji, N. (2008). Functions and transport of silicon in plants. *Cellular and Molecular Life Sciences, 65,* 3049–3057.

MacFadden, B. J. (1997). Origin and evolution of the grazing guild in New World terrestrial mammals. *Trends in Ecology & Evolution, 12,* 182–187.

Madden, R. H., Carlini, A. A., Vucetich, M. G., & Kay, R. F. (Eds.). (2010). *The paleontology of Gran Barranca*. Cambridge: Cambridge University Press.

Madella, M., Jones, M. K., Echlin, P., Powers-Jones, A., & Moore, M. (2009). Plant water availability and analytical microscopy of phytoliths: implications for ancient irrigation in arid zones. *Quaternary International, 193,* 32–40.

Madella, M., & Lancelotti, C. (2012). Taphonomy and phytoliths: a user manual. *Quaternary International, 275,* 76–83.

Madella, M., Powers-Jones, A. H., & Jones, M. K. (1998). A simple method of extraction of opal phytoliths from sediments using a non-toxic heavy liquid. *Journal of Archaeological Science, 25,* 801–803.

Mander, L., Li, M., Mio, W., Fowlkes, C. C., & Punyasena, S. W. (2013). Classification of grass pollen through the quantitative analysis of surface ornamentation and texture. *Proceedings of the Royal Society B, 280,* 20131905.

Marx, R., Lee, D. E., Lloyd, K. M., & Lee, W. G. (2004). Phytolith morphology and biogenic silica concentrations and abundance in leaves of *Chionochloa* (Danthonieae) and *Festuca* (Poeae) in New Zealand. *New Zealand Journal of Botany, 42,* 677–691.

Massey, F. P., Ennos, A. R., & Hartley, S. E. (2006). Silica in grasses as a defence against insect herbivores: contrasting effects on folivores and a phloem feeder. *Journal of Animal Ecology, 75,* 595–603.

Mazumdar, J., & Mukhopadhyay, R. (2009a). Opal phytoliths in three Indian thelypteroid ferns. *Bionature, 29,* 11–15.

Mazumdar, J., & Mukhopadhyay, R. (2009b). Phytoliths of some lycopods. *Indian Fern Journal, 26,* 132–136.

Mazumdar, J., & Mukhopadhyay, R. (2010a). Phytoliths of Ferns II: in some more thelypteroid ferns. *Bionature, 30,* 13–17.

Mazumdar, J., & Mukhopadhyay, R. (2010b). Phytoliths of Ferns III. in Himalayan horsetail and some ferns. *Bionature, 30,* 39–45.

Mazumdar, J., & Mukhopadhyay, R. (2011). Phytoliths of ferns IV: in some aquatic ferns and Chinese brake fern. *Bioresearch Bulletin, 6,* 385–388.

Mazzoni, M. M. (1979). Contribución al conocimiento petrográfico de la Formación Sarmiento, barranca sur del lago Colhue Huapi, provincia de Chubut. *Revista de la Asociación Argentina de Mineralogía, Petrología y Sedimentología, 10,* 33–54.

McClaran, M. P., & Umlauf, M. (2000). Desert grassland dynamics estimated from carbon isotopes in grass phytoliths and soil organic matter. *Journal of Vegetation Science, 11,* 71–76.

McCullough, D. R. (1982). White-tailed deer pellet-group weights. *The Journal of Wildlife Management, 46,* 829–832.

McCune, J. L., & Pellatt, M. G. (2013). Phytoliths of southeastern Vancouver Island, Canada, and their potential use to reconstruct shifting boundaries between Douglas-fir forest and oak savannah. *Palaeogeography, Palaeoclimatology, Palaeoecology, 383,* 59–71.

McCune, J. L., Vellend, M., & Pellatt, M. G. (2014). Combining phytolith analysis with historical ecology to reveal the long-term, local-scale dynamics within a savannah-forest landscape mosaic. *Biodiversity and Conservation, 24,* 609–626.

McInerney, F. A., Strömberg, C. A. E., & White, J. W. C. (2011). The Neogene transition from C_3 to C_4 grasslands in North America: stable carbon isotope ratios of fossil phytoliths. *Paleobiology, 37,* 23–49.

McMichael, C. H., Bush, M. B., Piperno, D. R., Silman, M. R., Zimmerman, A. R., & Anderson, C. (2012a). Spatial and temporal scales of pre-Columbian disturbance associated with western Amazonian lakes. *Holocene, 22,* 131–141.

McMichael, C. H., Piperno, D. R., Bush, M. B., Silman, M. R., Zimmerman, A. R., Raczka, M. F., et al. (2012b). Sparse pre-Columbian human habitation in western Amazonia. *Science, 336,* 1429–1431.

McNab, B. K. (1963). Bioenergetics and the determination of home range size. *The American Naturalist, 97,* 133–140.

Mercader, J., Astudillo, F., Barkworth, M., Bennett, T., Esselmont, C., Kinyanjui, R., et al. (2010). Poaceae phytoliths from the Niassa Rift, Mozambique. *Journal of Archaeological Science, 37,* 1953–1967.

Mercader, J., Bennett, T., Esselmont, C., Simpson, S., & Walde, D. (2009). Phytoliths in woody plants from the Miombo woodlands of Mozambique. *Annals of Botany, 104,* 91–113.

Mercader, J., Bennett, T., Esselmont, C., Simpson, S., & Walde, D. (2011). Soil phytoliths from Miombo woodlands in Mozambique. *Quaternary Research, 75*(138–150), 8.

Mercader, J., Runge, F., Vrydaghs, L., Doutrelepont, H., Ewango, C. E. N., & Juan-Tresseras, J. (2000). Phytoliths from archaeological sites in the tropical forest of Ituri, Democratic Republic of Congo. *Quaternary Research, 54,* 102–112.

Merceron, G., Novello, A., & Scott, R. S. (2016). Paleoenvironments inferred from phytoliths and Dental Microwear Texture Analyses of meso-herbivores. *Geobios, 49,* 135–146.

Messager, E., Lebreton, V., Marquer, L., Russo-Ermolli, E., Orain, R., Renault-Miskovsky, J., et al. (2011). Palaeoenvironments of early hominins in temperate and Mediterranean Eurasia: new palaeobotanical data from Palaeolithic key-sites and synchronous natural sequences. *Quaternary Science Reviews, 30,* 1439–1447.

Metcalfe, C. R. (1960). *Anatomy of the monocotyledons, Volume I, Gramineae*. London: Oxford University Press.

Meunier, J. D., Colin, F., & Alarcon, C. (1999). Biogenic silica storage in soils. *Geology, 27,* 835–838.

Middleton, W. D., & Rovner, I. (1994). Extraction of opal phytoliths from herbivore dental calculus. *Journal of Archaeological Science, 21,* 469–473.

Mihlbachler, M. C., & Solounias, N. (2006). Coevolution of tooth crown height and diet in oreodonts (Merycoidodontidae, Artiodactyla) examined with phylogenetically independent contrasts. *Journal of Mammalian Evolution, 13,* 11–36.

Miller, L. A., Smith, S. Y., Sheldon, N. D., & Strömberg, C. A. E. (2012). Eocene vegetation dynamics in Montana inferred from a high-resolution phytolith record. *Geological Society of America Bulletin, 124,* 1577–1589.

Ming Huang, P., Li, Y., & Sumner, M. E. (Eds.). (2011). *Handbook of soil sciences: Properties and processes* (2nd ed.). Boca Raton, FL: Taylor & Francis Group LLC.

Mohabey, D., & Samant, B. (2003). Floral remains from Late Cretaceous faecal mass of sauropods from central India: implication to their diet and habitat. *Gondwana Geological Magazine, 6,* 225–238.

Montti, L., Honaine, M. F., Osterrieth, M., & Ribeiro, D. G. (2009). Phytolith analysis of *Chusquea ramosissima* Lindm. (Poaceae: Bambusoideae) and associated soils. *Quaternary International, 193,* 80–89.

Morcote-Ríos, G., Bernal, R., & Raz, L. (2016). Phytoliths as a tool for archaeobotanical, palaeobotanical and palaeoecological studies in Amazonian palms. *Botanical Journal of the Linnean Society, 182,* 348–360.

Morgan-Edel, K. D., Boston, P. J., Spilde, M. N., & Reynolds, R. E. (2015). Phytoliths (plant-derived mineral bodies) as geobiological and climatic indicators in arid environments. *New Mexico Geology, 37,* 1–20.

Morris, L. R., Baker, F. A., Morris, C., & Ryel, R. J. (2009a). Phytolith types and type-frequencies in native and introduced species of the sagebrush steppe and pinyon-juniper woodlands of the Great Basin, USA. *Review of Palaeobotany and Palynology, 157,* 339–357.

Morris, L. R., West, N. E., Baker, F. A., Van Miegroet, H., & Ryel, R. J. (2009b). Developing an approach for using the soil phytolith record to infer vegetation and disturbance regime changes over the past 200 years. *Quaternary International, 193,* 90–98.

Morris, L. R., West, N. E., & Ryel, R. J. (2010). Testing soil phytolith analysis as a tool to understand vegetation change in the sagebrush steppe and pinyon-juniper woodlands of the Great Basin Desert, USA. *Holocene, 20,* 697–709.

Mosimann, J. E. (1965). Statistical methods for the pollen analyst: Multinomial and negative multinomial techniques. In B. Kummel & D. Raup (Eds.), *Handbook of paleontological techniques* (pp. 636–673). San Francisco: W. Freeman.

Motomura, H., Mita, N., & Suzuki, M. (2002). Silica accumulation in long-lived leaves of *Sasa veitchii* (Carrière) Rehder (Poaceae–Bambusoideae). *Annals of Botany, 90,* 149–152.

Mulholland, S. C. (1989). Phytolith shape frequencies in North Dakota grasses: a comparison to general patterns. *Journal of Archaeological Science, 16,* 489–511.

Mulholland, S. C. & Rapp, G. J. (1992). A morphological classification of grass silica-bodies. In G. J. Rapp & S. C. Mulholland (Eds.), *Phytolith systematics. Emerging issues, vol 1. Advances in archaeological and museum science* (pp. 65–89). New York: Plenum Press.

Murungi, M. L., McGlynn, G., & Lejju, J. B. (2016). Alpine grassland palaeoecology of the Virunga Volcanoes, East Africa: a new phytolith record from Mt. Muhavura. *Quaternary International, 434,* 102–116.

Neumann, K., Fahmy, A., Lespez, L., Ballouche, A., & Huysecom, E. (2009). The Early Holocene palaeoenvironment of Ounjougou (Mali): Phytoliths in a multiproxy context. *Palaeogeography, Palaeoclimatology, Palaeoecology, 276,* 87–106.

Neumann, K., Fahmy, A. G., Müller-Scheeßel, N., & Schmidt, M. (2017). Taxonomic, ecological and palaeoecological significance of leaf phytoliths in West African grasses. *Quaternary International, 434,* 15–32.

Novello, A., & Barboni, D. (2015). Grass inflorescence phytoliths of useful species and wild cereals from sub-Saharan Africa. *Journal of Archaeological Science, 59,* 10–22.

Novello, A., Barboni, D., Berti-Equille, L., Mazur, J. C., Poilecot, P., & Vignaud, P. (2012). Phytolith signal of aquatic plants and soils in

Chad, Central Africa. *Review of Palaeobotany and Palynology, 178,* 43–58.

Novello, A., Lebatard, A.-E., Moussa, A., Barboni, D., Sylvestre, F., Bourlès, D. L., et al. (2015). Diatom, phytolith, and pollen records from a 10Be/9Be dated lacustrine succession in the Chad basin: insight on the Miocene-Pliocene paleoenvironmental changes in Central Africa. *Palaeogeography, Palaeoclimatology, Palaeoecology, 430,* 85–103.

Ollendorf, A. L., Mulholland, S. C., & Rapp, Jr., G. (1988). Phytolith analysis as a means of plant identification: *Arundo donax* and *Phragmites communis*. *Annals of Botany, 61,* 209–214.

Osborn, H. F. (1910). *The age of mammals in Europe, Asia and North America*. New York: The Macmillan Company.

Osborne, C. P. (2008). Atmosphere, ecology and evolution: what drove the Miocene expansion of C_4 grasslands? *Journal of Ecology, 96,* 35–45.

Osterrieth, M., Madella, M., Zurro, D., & Alvarez, M. F. (2009). Taphonomical aspects of silica phytoliths in the loess sediments of the Argentinean Pampas. *Quaternary International, 193,* 70–79.

Panadès i Blas, X., Bartolomé i Filella, J., Strömberg, C.A.E., Soriano i Tomàs, I., Buckland, P., Serieyssol, K. K., et al. (2017). The utility of livestock dung for reconstructing recent ethnological and environmental histories. *Environmental Archaeology, 22,* 128–146.

Parolin, M., Rasbold, G. G., & Pessenda, L. C. R. (2014). Paleoenvironmental conditions of Campos Gerais, Paraná, since the Late Pleistocene, based on phytoliths and C and N isotopes. In H. H. G. Coe & M. Osterrieth (Eds.), *Synthesis of some phytoliths studies in South America (Brazil and Argentina)* (pp. 149–170). New York: Nova Science Publishers.

Parr, J. F. (2002). A comparison of heavy liquid floatation and microwave digestion techniques for the extraction of fossil phytoliths from sediments. *Review of Palaeobotany and Palynology, 120,* 315–336.

Parr, J. F. (2006). Effect of fire on phytolith coloration. *Geoarchaeology, 21,* 171–185.

Parr, J. F., Dolic, G. L., & Boyd, D. K. (2001). A microwave digestion method for the extraction of phytoliths from herbarium specimens. *Review of Palaeobotany and Palynology, 116,* 203–212.

Passey, B. H., Ayliffe, L. K., Kaakinen, A., Eronen, J. T., Zhang, Z. Q., Cerling, T. E., et al. (2009). Strengthened East Asian summer monsoons during a period of high-latitude warmth? Isotopic evidence from Mio-Pliocene fossil mammals and soil carbonates from northern China. *Earth and Planetary Science Letters, 277,* 443–452.

Pearsall, D. M. (1978). Phytolith analysis of archaeological soils: evidence for maize cultivation in formative Ecuador. *Science, 199,* 177–178.

Pearsall, D. M. (2000). *Paleoethnobotany: A handbook of procedures. 2nd edn.* San Diego: Academic Press.

Pearsall, D. M. (2016a). Phytoliths in the flora of Ecuador: The University of Missouri online phytolith database. http://phytolith.missouri.edu/.

Pearsall, D. M. (2016b). The phytoliths in the flora of Ecuador project: perspectives on phytolith classification, identification, and establishing regional phytolith databases. *Journal of Archaeological Science, 68,* 89–97.

Pearsall, D. M., Chandler-Ezell, K., & Chandler-Ezell, A. (2003). Identifying maize in neotropical sediments and soils using cob phytoliths. *Journal of Archaeological Science, 30,* 611–627.

Penny, D., & Kealhofer, L. (2005). Microfossil evidence of land-use intensification in north Thailand. *Journal of Archaeological Science, 32,* 69–82.

Perrier, X., De Langhe, E., Donohue, M., Lentfer, C., Vrydaghs, L., Bakry, F., Carreel, F., Hippolyte, I., Horry, J. P., Jenny, C., & Lebot, V. (2011). Multidisciplinary perspectives on banana (*Musa* spp.) domestication. *Proceedings of the National Academy of Sciences, USA, 108,* 11311–11318.

Pinilla, A. & Bustillo, A. (1997). Sílicofitolitos en secencies arcillosas con silcretas, Mioceno Medio, Madrid. In A. Pinilla, J. Juan-Tresserras & M. J. Machado (Eds.), *The state of the art of phytoliths in soils and plants, vol 4. Monografias del Centro de Ciencieas Medioambientales* (pp. 255–265). Madrid: CSIC.

Piperno, D. R. (1985). Phytolith analysis and tropical paleo-ecology: production and taxonomic significance of siliceous forms in New World plant domesticates and wild species. *Review of Palaeobotany and Palynology, 45,* 185–228.

Piperno, D. R. (1988). *Phytolith analysis, an archaeological and geological perspective*. San Diego: Academic Press.

Piperno, D. R. (1989). The occurrence of phytoliths in the reproductive structures of selected tropical angiosperms and their significance in tropical paleoecology, paleoethnobotany and systematics. *Review of Palaeobotany and Palynology, 61,* 147–173.

Piperno, D. R. (1993). Phytolith and charcoal records from deep lake cores in the American tropics. In D. M. Pearsall & D. R. Piperno (Eds.), *Current research in phytolith analysis: Applications in archaeology and paleoecology, vol 10. MASCA Research Papers in Science and Archaeology* (pp. 58–71). Philadelphia: University Museum of Archaeology and Anthropology, University of Pennsylvania.

Piperno, D. R. (2001). 11. Phytoliths. In J. P. Smol, H. J. B. Birks & W. M. Last (Eds.), *Terrestrial, algal, and siliceous indicators, vol 3. Tracking environmental change using lake sediments* (pp. 235–251). Dordrecht: Kluwer Academic Publishers.

Piperno, D. R. (2006). *Phytoliths: A comprehensive guide for archaeologists and paleoecologists 1st edn.* New York: AltaMira Press.

Piperno, D. R. (2009). Identifying crop plants with phytoliths (and starch grains) in Central and South America: a review and an update of the evidence. *Quaternary International, 193,* 146–159.

Piperno, D. R. (2016). Phytolith radiocarbon dating in archaeological and paleoecological research: a case study of phytoliths from modern Neotropical plants and a review of the previous dating evidence. *Journal of Archaeological Science, 68,* 54–61.

Piperno, D. R., Andres, T. C., & Stothert, K. E. (2000). Phytoliths in *Cucurbita* and other neotropical Cucurbitaceae and their occurrence in early archaeological sites from the lowland American tropics. *Journal of Archaeological Science, 27,* 193–208.

Piperno, D. R., & Becker, P. (1996). Vegetational history of a site in the central Amazon basin derived from phytolith and charcoal records from natural soils. *Quaternary Research, 45,* 202–209.

Piperno, D. R., McMichael, C., & Bush, M. B. (2015). Amazonia and the Anthropocene: what was the spatial extent and intensity of human landscape modification in the Amazon Basin at the end of prehistory? *The Holocene, 25,* 1588–1597.

Piperno, D. R., & Pearsall, D. M. (1998). The silica bodies of tropical American grasses: morphology, taxonomy, and implications for grass systematics and fossil phytolith identification. *Smithsonian Contributions to Botany, 85,* 1–40.

Piperno, D. R., Ranere, A. J., Holst, I., Iriarte, J., & Dickau, R. (2009). Starch grain and phytolith evidence for early ninth millennium BP maize from the Central Balsas River Valley, Mexico. *Proceedings of the National Academy of Sciences, USA, 106,* 5019–5024.

Pironon, J., Meunier, J. D., Alexandre, A., Mathieu, R., Mansuy, L., Grosjean, A., et al. (2001). Individual characterization of phytoliths: experimental approach and consequences on paleoenvironment understanding. In J.-D. Meunier & F. Colin (Eds.), *Phytoliths: Applications in Earth sciences and human history* (pp. 329–341). Boca Raton, FL: CRC Press.

Poinar, G., Jr. (2011). Silica bodies in the Early Cretaceous *Programinis laminatus* (Angiospermae: Poales). *Palaeodiversity, 4,* 1–6.
Porse, S. (2008). Biomes classified by vegetsert. License: CC-BY 3.0. https://en.wikipedia.org/wiki/Vegetation-/media/File:Vegetation-no-legend.PNG. Database accessed on June 16, 2016.
Powers, A. H., & Gilbertson, D. D. (1987). A simple preparation technique for the study of opal phytoliths from archaeological and Quaternary sediments. *Journal of Archaeological Science, 14,* 529–535.
Powers-Jones, A. H. & Padmore, J. (1993). The use of quantitative methods and statistical analyses in the study of opal phytoliths. In D. M. Pearsall & D. R. Piperno (Eds.), *Current research in phytolith analysis: Applications in archaeology and paleoecology, vol 10. MASCA Research Papers in Science and Archaeology* (pp. 47–56). Philadelphia: University Museum of Archaeology and Anthropology, University of Pennsylvania.
Prasad, V., Strömberg, C. A. E., Alimohammadian, H., & Sahni, A. (2005). Dinosaur coprolites and the early evolution of grasses and grazers. *Science, 310,* 1177–1180.
Prasad, V., Strömberg, C. A. E., Leache, A., Samant, B. R. P., Tang, L., Mohabey, D. M., et al. (2011). Late Cretaceous origin of the rice tribe. *Nature Communications, 2,* 480.
Prebble, M., Schallenberg, M., Carter, J., & Shulmeister, J. (2002). An analysis of phytolith assemblages for the quantitative reconstruction of late Quaternary environments of the Lower Taieri Plain, Otago, South Island, New Zealand I. Modern assemblages and transfer functions. *Journal of Paleolimnology, 27,* 393–413.
Prentice, A. J., & Webb, E. A. (2016). The effect of progressive dissolution on the oxygen and silicon isotope composition of opal-A phytoliths: implications for palaeoenvironmental reconstruction. *Palaeogeography, Palaeoclimatology, Palaeoecology, 453,* 42–51.
Prychid, C. J., Furness, C. A., & Rudall, P. J. (2003). Systematic significance of cell inclusions in Haemodoraceae and allied families: silica bodies and tapetal raphides. *Annals of Botany, 92,* 571–580.
Prychid, C. J., & Rudall, P. J. (1999). Calcium oxalate crystals in monocotyledons: a review of their structure and systematics. *Annals of Botany, 84,* 725–739.
Prychid, C. J., Rudall, P. J., & Gregory, M. (2004). Systematics and biology of silica bodies in monocotyledons. *The Botanical Review, 69,* 377–440.
Quigley, K. M., & Anderson, T. M. (2014). Leaf silica concentration in Serengeti grasses increases with watering but not clipping: insights from a common garden study and literature review. *Frontiers in Plant Science, 5,* 568.
Raigemborn, M., Brea, M., Zucol, A. F., & Matheos, S. (2009). Early Paleogene climate at mid latitude in South America: mineralogical and paleobotanical proxies from continental sequences in Golfo San Jorge basin (Patagonia, Argentina). *Geologica Acta, 7,* 125–145.
Raine, J., & Askin, R. (2001). Terrestrial palynology of Cape Roberts Project Drillhole CRP-3, Victoria Land Basin, Antarctica. *Terra Antartica, 8,* 389–400.
Rana, R. S., & Wilson, G. P. (2003). New Late Cretaceous mammals from the Intertrappean beds of Rangapur, India and paleobiogeographic framework. *Acta Palaeontologica Polonica, 48,* 331–348.
Raven, J. A. (1983). The transport and function of silicon in plants. *Biological Reviews, 58,* 179–207.
Retallack, G. (2001a). *Soils of the past* (2nd ed.). Malden, Massachusetts: Blackwell Science Ltd.
Retallack, G. J. (1983a). Late Eocene and Oligocene paleosols from Badlands National Park, South Dakota. *Geological Society of America Special Paper, 193,* 1–82.
Retallack, G. J. (1983b). A paleopedological approach to the interpretation of terrestrial sedimentary rocks; the mid-Tertiary fossil soils of Badlands National Park, South Dakota. *Geological Society of America Bulletin, 94,* 823–840.
Retallack, G. J. (1990). *Soils of the past: An introduction to paleopedology*. Boston, MA: Hyman Press.
Retallack, G. J. (1997). Neogene expansion of the North American prairie. *PALAIOS, 12,* 380–390.
Retallack, G. J. (2001b). Cenozoic expansion of grasslands and climatic cooling. *Journal of Geology, 109,* 407–426.
Retallack, G. J., Dugas, D. P., & Bestland, E. A. (1990). Fossil soils and grasses of a Middle Miocene East African grassland. *Science, 247,* 1325–1328.
Rosen, A. M., & Weiner, S. (1994). Identifying ancient irrigation: a new method using opaline phytoliths from emmer wheat. *Journal of Archaeological Science, 21,* 125–132.
Rossouw, L., & Scott, L. (2011). Phytoliths and pollen, the microscopic plant remains in Pliocene volcanic sediments around Laetoli, Tanzania. In T. Harrison (Ed.), *Paleontology and geology of Laetoli: Human evolution in context* (pp. 2015–2215). Dordrecht: Springer.
Rossouw, L., Stynder, D. D., & Haarhof, P. (2009). Evidence for opal phytolith preservation in the Langebaanweg 'E' Quarry Varswater Formation and its potential for palaeohabitat reconstruction. *South African Journal of Science, 105,* 223–227.
Rossuow, L. (2009). *The application of fossil grass-phytolith analysis in the reconstruction of late Cainozoic environments in the South African interior.* Ph.D. Dissertation. Bloemfontein, South Africa: University of the Free State.
Rovner, I. (1971). Potential of opal phytoliths for use in paleoecological reconstruction. *Quaternary Research, 1,* 343–359.
Rovner, I. (1972). Note on a safer procedure for opal phytolith extraction. *Quaternary Research, 2,* 591.
Royer, D. L., Osborne, C. P., & Beerling, D. J. (2002). High CO_2 increases the freezing sensitivity of plants: implications for paleoclimatic reconstructions from fossil floras. *Geology, 30,* 963–966.
Rudall, P. J., Prychid, C. J., & Gregory, T. (2014). Epidermal patterning and silica phytoliths in grasses: an evolutionary history. *The Botanical Review, 80,* 59–71.
Runge, F. (1996). Opal Phytolithe in Pflantzen aus dem humiden und semi-ariden Osten Afrikas und ihre Bedeutung fur die Klima- und Vegetationsgeschichte. *Botanische Jahrbüch für Systematik, 118,* 303–363.
Runge, F. (1999). The opal phytolith inventory of soils in central Africa – quantities, shapes, classification, and spectra. *Review of Palaeobotany and Palynology, 107,* 23–53.
Runge, F. (2001). Evidence for land use history by opal phytolith analysis: examples from the central African tropics (eastern Kivu, D. R. Congo). In J. D. Meunier & F. Colin (Eds.), *Phytoliths: Applications in Earth sciences and human history* (pp. 73–85). Lisse: A.A. Balkema Publishers.
Runge, F. & Runge, J. (1997). Opal phytoliths in East African plants and soils. In A. Pinilla, J. Juan-Tresserras & M. J. Machado (Eds.), *The state of the art of phytoliths in soils and plants, vol 4. Monografias del Centro de Ciencieas Medioambientales* (pp. 71–81). Madrid: CSIC.
Sánchez, M. V., Gonzalez, M. G., & Genise, J. F. (2010). Phytolith analysis of *Coprinisphaera*, unlocking dung beetle behaviour, herbivore diets and palaeoenvironments along the Middle Eocene-Early Miocene of Patagonia. *Palaeogeography, Palaeoclimatology, Palaeoecology, 285,* 224–236.
Sandve, S. R., & Fjellheim, S. (2010). Did gene family expansions during the Eocene-Oligocene boundary climate cooling play a role in Pooideae adaptation to cool climates? *Molecular Ecology, 19,* 2075–2088.
Sangster, A. G., Hodson, M. J., & Tubb, H. J. (2001). Silicon deposition in higher plants. In L. E. Datnoff, G. H. Snyder, & G.

H. Korndörfer (Eds.), *Silicon in agriculture* (pp. 85–113). Amsterdam: Elsevier.

Sangster, A. G., & Parry, D. W. (1969). Some factors in relation to bulliform cell silicification in the grass leaf. *Annals of Botany, 33,* 315–323.

Sankaran, M., Ratnam, J., & Hanan, N. (2007). Woody cover in African savannas: the role of resources, fire and herbivory. *Global Ecology and Biogeography, 2007,* 1–10.

Santos, G., Alexandre, A., Southon, J., Treseder, K., Corbineau, R., & Reyerson, P. (2012). Possible source of ancient carbon in phytolith concentrates from harvested grasses. *Biogeosciences, 9,* 1873–1884.

Sase, T. & Hosono, M. (2001). Phytolith record in soils interstratified with late Qauternary tephras overlying the eastern region of Towada volcano. In J.D. Meunier & F. Colin (Eds.), *Phytoliths: Applications in Earth sciences and human history* (pp. 57–71). Lisse: A.A. Balkema Publishers.

Scurfield, G., Anderson, C. A., & Segnit, E. R. (1974a). Silica in woody stems. *Australian Journal of Botany, 22,* 211–229.

Scurfield, G., Segnit, E. R., & Anderson, C. A. (1974b). Silicification of wood. In *Scanning electron microscopy. Part II. Proceedings of the Workshop on Scanning Electron Microscopy and the Plant Sciences*. Chicago: IIT Research Institute.

Selkin, P. A., Strömberg, C. A. E., Dunn, R. E., Kohn, M. J., Carlini, A. A., Davies-Vollum, K. S., et al. (2015). Climate, dust, and fire across the Eocene-Oligocene transition, Patagonia. *Geology, 43,* 567–570.

Shahack-Gross, R. (2011). Herbivorous livestock dung: formation, taphonomy, methods for identification, and archaeological significance. *Journal of Archaeological Science, 38,* 205–218.

Shahack-Gross, R., Boaretto, E., Cabanes, D., Katz, O., & Finkelstein, I. (2014). Subsistence economy in the Negev Highlands: the Iron Age and the Byzantine/Early Islamic period. *Levant, 46,* 98–117.

Sheldon, N. D., & Tabor, N. J. (2009). Quantitative paleoenvironmental and paleoclimatic reconstruction using paleosols. *Earth-Science Reviews, 95,* 1–52.

Shi, G., Grimaldi, D. A., Harlow, G. E., Wang, J., Yang, M., Lei, W., et al. (2012). Age constraint on Burmese amber based on U-Pb dating of zircons. *Cretaceous Research, 37,* 155–163.

Shoji, S., Nanzyo, M., & Dahlgren, R. (1994). *Volcanic ash soils: Genesis, properties and utilization, vol 21.* Amsterdam: Elsevier Science Publishers B.V.

Simpson, K. J., Ripley, B. S., Christin, P. A., Belcher, C. M., Lehmann, C. E., Thomas, G. H., et al. (2016). Determinants of flammability in savanna grass species. *Journal of Ecology, 104,* 138–148.

Smiley, C., & Huggins, L. M. (1981). *Pseudofagus idahoensis*, n. gen. et sp. (Fagaceae) from the Miocene Clarkia Flora from Idaho. *American Journal of Botany, 68,* 741–761.

Smith, F. A., & Anderson, K. B. (2001). Characterization of organic compounds in phytoliths: improving the resolving power of phytoliths $d^{13}C$ as a tool for paleoecological reconstruction of C_3 and C_4 grasses. In J. D. Meunier & F. Colin (Eds.), *Phytoliths: Applications in Earth sciences and human history* (pp. 317–327). Lisse: A.A. Balkema Publishers.

Smith, F. A., & White, J. W. C. (2004). Modern calibration of phytolith carbon isotope signatures for C_3/C_4 paleograssland reconstruction. *Palaeogeography, Palaeoclimatology, Palaeoecology, 207,* 277–304.

Soderstrom, T. R., & Calderón, C. E. (1974). Primitive forest grasses and evolution of the Bambusoideae. *Biotropica, 6,* 141–153.

Soil Survey Staff (2006). *Keys to soil taxonomy* (10 ed.). Washington, D. C.: U. S. Department of Agriculture, Natural Resources Conservation Services.

Song, Z., McGrouther, K., & Wang, H. (2016). Occurrence, turnover and carbon sequestration potential of phytoliths in terrestrial ecosystems. *Earth-Science Reviews, 158,* 19–30.

Soreng, R. J., Peterson, P. M., Romaschenko, K., Davidse, G., Zuloaga, F. O., Judziewicz, E. J., et al. (2015). A worldwide phylogenetic classification of the Poaceae (Gramineae). *Journal of Systematics and Evolution, 53,* 117–137.

Spriggs, E. L., Christin, P. A., & Edwards, E. J. (2014). C_4 photosynthesis promoted species diversification during the Miocene grassland expansion. *PLoS ONE, 9,* e9772210.

Stebbins, G. L. (1981). Coevolution of grasses and herbivores. *Annals of the Missouri Botanical Garden, 68,* 75–86.

Stevenson, R., Bothwell, M., & Loweet, R. (1996). *Algal ecology: Freshwater benthic ecosystems*. San Diego: Academic Press.

Strömberg, C. A. E. (2002). The origin and spread of grass-dominated ecosystems in the Late Tertiary of North America: preliminary results concerning the evolution of hypsodonty. *Palaeogeography, Palaeoclimatology, Palaeoecology, 177,* 59–75.

Strömberg, C. A. E. (2003). *The origin and spread of grass-dominated ecosystems during the Tertiary of North America and how it relates to the evolution of hypsodonty in equids*. Ph.D. Dissertation, University of California, Berkeley.

Strömberg, C. A. E. (2004). Using phytolith assemblages to reconstruct the origin and spread of grass-dominated habitats in the Great Plains during the late Eocene to early Miocene. *Palaeogeography, Palaeoclimatology, Palaeoecology, 207,* 239–275.

Strömberg, C. A. E. (2005). Decoupled taxonomic radiation and ecological expansion of open-habitat grasses in the Cenozoic of North America. *Proceedings of the National Academy of Sciences, USA, 102,* 11980–11984.

Strömberg, C. A. E. (2006). The evolution of hypsodonty in equids: testing a hypothesis of adaptation. *Paleobiology, 32,* 236–258.

Strömberg, C. A. E. (2007). Can slide preparation methods cause size biases in phytolith assemblages?: results from a preliminary study. In M. Madella, D. Zurro, & M. K. Jones (Eds.), *Plants, people and places. Recent studies in phytolith analysis* (pp. 1–12). Oxford: Oxbow Books.

Strömberg, C. A. E. (2009a). Methodological concerns for analysis of phytolith assemblages: does count size matter? *Quaternary International, 193,* 124–140.

Strömberg, C. A. E. (2009b). Reply to comment on "Methodological concerns for analysis of phytolith assemblages: does count size matter?". *Quaternary International, 193,* 143–145.

Strömberg, C. A. E. (2011). Evolution of grasses and grassland ecosystems. *Annual Review of Earth and Planetary Sciences, 39,* 517–544.

Strömberg, C. A. E., Di Stilio, V. S., & Song, Z. (2016). Functions of phytoliths in vascular plants: an evolutionary perspective. *Functional Ecology, 30,* 1286–1297.

Strömberg, C. A. E., & Dunn, R. E. (2014). The vegetational context for rodent evolution in the Pacific Northwest: middle Miocene phytoliths the Mascall Formation of eastern Oregon. *Geological Society of America Abstracts with Program, 46,* 703.

Strömberg, C. A. E., Dunn, R. E., Madden, R. H., Kohn, M. J., & Carlini, A. A. (2013a). Decoupling the spread of grasslands from the evolution of grazer-type herbivores in South America. *Nature Communications, 4,* 1478.

Strömberg, C. A. E., Dunn, R. E., Madden, R. H., Kohn, M. J., & Carlini, A. A. (2013b). Where have all the grasses gone?: new middle Miocene phytolith records reveal that grasslands played a minor role in hypsodonty evolution in southern South America. *Journal of Vertebrate Paleontology Program and Abstracts, 73,* 221.

Strömberg, C. A. E., & McInerney, F. A. (2011). The Neogene transition from C_3 to C_4 grasslands in North America: assemblage analysis of fossil phytoliths. *Paleobiology, 37,* 50–71.

Strömberg, C. A. E. & Stidham, T. A. (2001). Dung beetle brood balls and notoungulate diet. *Journal of Vertebrate Paleontology, 21,* 105A.

Strömberg, C. A. E., Friis, E. M., Liang, M.-M., Werdelin, L., & Zhang, Y.-l., (2007a). Palaeoecology of an Early-Middle Miocene lake in China: preliminary interpretations based on phytoliths from the Shanwang Basin. *Vertebrata PalAsiatica 45,* 145–160.

Strömberg, C. A. E., Werdelin, L., Friis, E. M., & Saraç, G. (2007b). The spread of grass-dominated habitats in Turkey and surrounding areas during the Cenozoic: phytolith evidence. *Palaeogeography, Palaeoclimatology Palaeoecology, 250,* 18–49

Struve, G. A. (1835). *De silica in Plantis Nonnullis*. Ph.D. Dissertation, Humboldt University, Berlin.

Sundue, M. (2009). Silica bodies and their systematic implications in Pteridaceae (Pteridophyta). *Botanical Journal of the Linnean Society, 161,* 422–435.

Swineford, A., Franks, P. C. (1959). Opal in the Ogallala Formation in Kansas. In H. A. Ireland (Ed.), *Silica in sediments* (Vol. 7, pp. 111–120). Tulsa: Society of Economic Paleontologists and Mineralogists.

Syme, A. E. (2012). Diversification rates in the Australasian endemic grass *Austrostipa*: 15 million years of constant evolution. *Plant Systematics and Evolution, 298,* 221–227.

Taub, D. R. (2001). Climate and the U.S. distribution of C_4 grass subfamilies and decarboxylation variants of C_4 photosynthesis. *American Journal of Botany, 87,* 1211–1215.

Tedford, R. A. (2009). *A multi-proxy approach to investigating the latest Holocene (~4,500 Yrs. BP) Vegetational history of Catahoula Lake, Louisiana*. Madison: University of Wisconsin-Madison.

Thomasson, J. R. (1978). Epidermal patterns of the lemma in some fossil and living grasses and their phylogenetic significance. *Science, 199,* 975–977.

Thomasson, J. R. (1984). Miocene grass (Gramineae: Arundinoideae) leaves showing external micromorphological and internal anatomical features. *Botanical Gazette, 145,* 204–209.

Thomasson, J. R. (1987). Fossil grasses: 1820–1986 and beyond. In T. R. Soderstrom, K. W. Hilu, C. S. Campbell & M. E. Barkworth (Eds.), *Grass systematics and evolution* (pp. 159–167). Washington, D. C.: Smithsonian Institution Press.

Thorn, V. C. (2001). Oligocene and early Miocene phytolits from CRP-2/2A and CRP-3, Victoria Land Basin, Antarctica. *Terra Antartica, 8,* 407–422.

Thorn, V. C. (2007). Phytoliths in palaeoecology. *Geology Today, 23,* 153–157.

Thorn, V. C. (2004a). Data report: Phytoliths in drill core sediments from Sites 1165 and 1166, Leg 188, Prydz Bay, East Antarctica. In A. K. Cooper, P. E. O'Brien & C. Richter (Eds.), *Proceedings of the Ocean Drilling Program, Scientific Results, vol 188* (pp. 1–12). College Station: Texas A & M University.

Thorn, V. C. (2004b). Phytoliths from subantarctic Campbell Island: plant production and soil surface spectra. *Review of Palaeobotany and Palynology, 132,* 37–59

Thorn, V. C. (2008). New Zealand sub-Antarctic phytoliths and their potential for past vegetation reconstruction. *Antarctic Science, 20,* 21–32.

Thummel, R. V., Strömberg, C. A. E., & Brightly, W. H. (2017). Evolution of phytolith deposition in modern bryophytes—implications for early land plants. *Geologic Society of America Abstracts with Programs, 49.*

Tomlinson, P. B. (1961). *Anatomy of the monocotyledons, Volume II, Palmae*. Oxford: Oxford University Press.

Trembath-Reichert, E., Wilson, J. P., McGlynn, S. E., & Fischer, W. W. (2015). Four hundred million years of silica biomineralization in land plants. *Proceedings of the National Academy of Sciences, USA, 112,* 5449–5454.

Tsartsidou, G., Lev-Yadun, S., Albert, R.-M., Miller-Rosen, A., Efstratiou, N., & Weiner, S. (2007). The phytolith archaeological record: Strengths and weaknesses evaluated based on a quantitative modern reference collection from Greece. *Journal of Archaeological Science, 34,* 1262–1275.

Tsartsidou, G., Lev-Yadun, S., Efstratiou, N., & Weiner, S. (2008). Ethnoarchaeological study of phytolith assemblages from an agro-pastoral village in Northern Greece (Sarakini): development and application of a Phytolith Difference Index. *Journal of Archaeological Science, 35,* 600–613.

Twiss, P. C. (1992). Predicted world distribution of C_3 and C_4 grass phytoliths. In G. J. Rapp & S. C. Mulholland (Eds.), *Phytolith systematics. Emerging issues, vol. 1. Advances in Archaeological and Museum Science* (pp. 113–128). New York: Plenum Press.

Twiss, P. C., Dort Jr., W., Sorenson, C. J. (1987). Grass-opal phytoliths as climatic indicators of the Great Plains Pleistocene. In W. C. Johnson (Ed.), *Quaternary environments of Kansas, vol 5. Guidebook Series, Kansas Geological Survey* (pp. 179–188). Kansas Geologic Survey.

Twiss, P. C., Suess, E., & Smith, R. M. (1969). Morphological classification of grass phytoliths. *Soil Science Society of America, Proceedings, 33,* 109–115.

Uhl, D., & Walther, H. (2000). Sun leaf or shade leaf? Known facts in the light of new data with implications for palaeobotany. *Feddes Repert, 111,* 165–174.

Uhl, N. W., & Dransfield, J. (1987). *Genera palmarum*. Lawrence, Kansas: Allen Press.

Vivanco, L., & Austin, A. T. (2006). Intrinsic effects of species on leaf litter and root decomposition: a comparison of temperate grasses from North and South America. *Oecologia, 150,* 97–107.

Wallis, L. A. (2001). Environmental history of northwest Australia based on phytolith analysis at Carpenter's Gap 1. *Quaternary International, 83–85,* 103–117.

Wallis, L. A. (2003). An overview of leaf phytolith production patterns in selected northwest Australian flora. *Review of Palaeobotany and Palynology, 125,* 201–248.

Wallis, L. A. (2013). A comparative study of phytolith assemblages in modern sediments from the Kimberley. *Western Australia. Quaternary Australasia, 30,* 6.

Wang, Y., Cerling, T. E., MacFadden, B. J., & Bryant, J. D. (1994). Fossil horses and carbon isotopes: new evidence for Cenozoic dietary, habitat, and ecosystem changes in North America. *Palaeogeography, Palaeoclimatology, Palaeoecology, 107,* 269–280.

Wang, Y. J., & Lu, H. Y. (1993). *The study of phytolith and its application*. Beijing, China: China Ocean Press.

Watling, J., & Iriarte, J. (2013). Phytoliths from the coastal savannas of French Guiana. *Quaternary International, 287,* 162–180.

Watson, L. & Dallwitz, M. J. (1992). Grass genera of the world: descriptions, illustrations, identification, and information retrieval; including synonyms, morphology, anatomy, physiology, phytochemistry, cytology, classification, pathogens, world and local distribution, and references. Version: 23rd April 2010. http://delta-intkey.com/.

Watson, R. (1942). The effect of cuticular hardening on the form of epidermal cells. *New Phytologist, 41,* 223–229.

Weaver, M. (1964). X-ray diffraction study of calculus of the miniature pig. *Archives of Oral Biology, 9,* 75–81.

Wesolowski, V., Ferraz Mendonça de Souza, S. M., Reinhard, K. J., & Ceccantini, G. (2010). Evaluating microfossil content of dental calculus from Brazilian sambaquis. *Journal of Archaeological Science, 37*, 1326–1338.

White, A. F., Vivit, D. V., Schulz, M. S., Bullen, T. D., Evett, R. R., & Aagarwal, J. (2012). Biogenic and pedogenic controls on Si distributions and cycling in grasslands of the Santa Cruz soil chronosequence, California. *Geochimica et Cosmochimica Acta, 94*, 72–94.

Wilding, L. P., & Drees, L. R. (1971). Biogenic opal in Ohio soils. *Soil Science Society of America, Proceedings, 35*, 1004–1010.

Wilding, L. P., & Drees, L. R. (1973). Scanning electron microscopy of opaque opaline forms isolated from forest soils in Ohio. *Soil Science Society of America Proceedings, 37*, 647–650.

Wilding, L. P., & Drees, L. R. (1974). Contributions of forest opal and associated crystalline phases to fine silt and clay fractions of soils. *Clays and Clay Minerals, 22*, 295–306.

Williams, L. A., & Crerar, D. A. (1985). Silica diagenesis; II, General mechanisms. *Journal of Sedimentary Petrology, 55*, 312–321.

Williams, L. A., Parks, G. A., & Crerar, D. A. (1985). Silica diagenesis; I, Solubility controls. *Journal of Sedimentary Petrology, 55*, 301–311.

Williams, S. H., & Kay, R. F. (2001). A comparative test of adaptive explanations for hypsodonty in ungulates and rodents. *Journal of Mammalian Evolution, 8*, 207–229.

Wilson, G. P., Das Sarma, D. C., & Anantharaman, S. (2007). Late Cretaceous sudamericid gondwanatherians from India with paleobiogeographic considerations of Gondwanan mammals. *Journal of Vertebrate Paleontology, 27*, 521–531.

Wing, S. L., Strömberg, C. A. E., Hickey, L. J., Tiver, F., Willis, B., Burnham, R. J., et al. (2012). Floral and environmental gradients on a Late Cretaceous landscape. *Ecological Monographs, 82*, 23–47.

WoldeGabriel, G., Ambrose, S. H., Barboni, D., Bonnefille, R., Bremond, L., Currie, B., et al. (2009). The geological, isotopic, botanical, invertebrate, and lower vertebrate surroundings of *Ardipithecus ramidus*. *Science, 326*, 65–65e5.

Yost, C. L., & Blinnikov, M. S. (2011). Locally diagnostic phytoliths of wild rice (*Zizania palustris* L.) from Minnesota, USA: comparison to other wetland grasses and usefulness for archaeobotany and paleoecological reconstructions. *Journal of Archaeological Science, 38*, 1977–1991.

Zhao, C., & Pearsall, D. M. (1998). Experiments for improving phytolith extractions from soils. *Journal of Archaeological Science, 25*, 587–598.

Zhao, Z., & Piperno, D. R. (2000). Late Pleistocene/Holocene environments in the Middle Yangtze River Valley, China and rice (*Oryza sativa* L.) domestication: the phytolith evidence. *Geoarchaeology: An International Journal, 15*, 203–222.

Zhao, Z. J., Pearsall, D. M., Benfer, R. A., & Piperno, D. R. (1998). Distinguishing rice (*Oryza sativa* Poaceae) from wild *Oryza* species through phytolith analysis, II – Finalized method. *Economic Botany, 52*, 134–145.

Zucol, A. F., & Brea, M. (2000). Análisis fitolítico de la Formación Paraná en la provincia de Entre Ríos. In F. G. Aceñolaza & R. Herbst (Eds.), *El Neógeno de Argentina* (INSUGEO, Serie Correlación Geológica, Vol. 14, pp. 67–76). Tucumán.

Zucol, A. F., & Brea, M. (2005). Sistemática de fitolitos, pautas para un sistema clasificatorio. Un caso en estudio en la Formación Alvear (Pleistoceno inferior). *Ameghiniana, 42*, 685–704.

Zucol, A. F., Brea, M., & Bellosi, E. (2010). Phytolith studies in Gran Barranca (central Patagonia, Argentina): the middle-late Eocene. In R. H. Madden, A. A. Carlini, M. G. Vucetich, & R. F. Kay (Eds.), *The paleontology of Gran Barranca* (pp. 317–340). Cambridge: Cambridge University Press.

Zucol, A. F., Brea, M., Bellosi, E., Carlini, A. A., & Vucetich, M. G. (2007). Preliminary phytolith analysis of Sarmiento Formation in the Gran Barranca (central Patagonia, Argentina). In M. Madella & D. Zurro (Eds.), *Plants, people and places. Recent studies in phytolith analysis* (pp. 189–195). Oxford: Oxbow Books.

Zucol, A. F., & Osterrieth, M. (2002). Técnicas de preparación de muestras sedimentarias para la extracción de fitolitos. *Ameghiniana, 39*, 379–382

Chapter 13
Reconstructing Paleoclimate and Paleoecology Using Fossil Leaves

Daniel J. Peppe, Aly Baumgartner, Andrew Flynn, and Benjamin Blonder

Abstract Plants are strongly influenced by their surrounding environment, which makes them reliable indicators of climate and ecology. The relationship between climate, ecology, plant traits and the geographic distribution of plants based on their climatic tolerances have been used to develop plant-based proxies for reconstructing paleoclimate and paleoecology. These proxies are some of the most accurate and precise methods for reconstructing the climate and ecology of ancient terrestrial ecosystems and have been applied from the Cretaceous to the Quaternary. Despite their utility, the relationships between plant traits and climate that underlie these methods are confounded by other factors such as leaf life-span and phylogenetic history. Work focused on better understanding these confounding factors, incorporating the influence of phylogeny and leaf economic spectrum traits into proxies, expanding modern leaf trait-climate and ecology calibration datasets to additional biogeographic areas and climate regimes, and developing automated computer algorithms for measuring leaf traits are important growing research areas that will help considerably improve plant-based paleoclimate and paleoecological proxies.

D. J. Peppe (✉) · A. Baumgartner · A. Flynn
Terrestrial Paleoclimatology Research Group, Department of Geosciences, Baylor University, Waco, TX 76706, USA
e-mail: Daniel_Peppe@baylor.edu

A. Baumgartner
e-mail: Aly_Baumgartner@baylor.edu

A. Flynn
e-mail: Andrew_Flynn@baylor.edu

B. Blonder
School of Life Sciences, Arizona State University, 427 E Tyler Mall, Tempe, AZ 85287, USA
e-mail: bblonder@gmail.com

Keywords Paleobotany • Leaf physiognomy • Leaf margin analysis • Leaf area analysis • CLAMP • Digital leaf physiognomy • Nearest living relative • Leaf economic spectrum • Leaf vein density • Leaf mass per area

Introduction

Plants are stationary, primary producers whose productivity is based on their ability to photosynthesize and use water efficiently. Terrestrial plants use roots and rhizoids to take in water and minerals from the soil, which are essential for photosynthesis and for maintaining the water content of the plant. Leaves, a plant's primary photosynthetic organ, are used for gas and water exchange between the plant and the atmosphere, and generally have high surface to volume ratios. Plants must therefore balance water loss and uptake, light and carbon dioxide fixation, and heat gain and loss from their environment as these affect their water use efficiency and photosynthetic capacity. Thus for plants to be successful, they must have biochemical and physiological adaptions that are well matched to the conditions they experience in their environment. Given these relationships between plants and their surrounding environment, a plant's traits, such as the size and shape of its leaves, are intrinsically linked to the climate and environment in which it is growing.

A variety of methods have been developed over the last 100 years using the relationships between plant traits, climate, and ecology to develop paleoclimate and paleoecological proxies that can be applied to fossil leaves. In this review, we discuss univariate, multivariate, and taxonomic methods for reconstructing paleoclimate and paleoecology using fossil plants and their advantages and disadvantages. We also discuss new developments and future research directions for leaf-based methods of reconstructing paleoclimate and paleoecology.

D. A. Croft et al. (eds.), *Methods in Paleoecology: Reconstructing Cenozoic Terrestrial Environments and Ecological Communities*, Vertebrate Paleobiology and Paleoanthropology, https://doi.org/10.1007/978-3-319-94265-0_13

Leaf Traits and Climate

The relationships between plant traits, and more specifically leaf traits, and the climate and environment have been noted for over 100 years (e.g., Billings 1905; Bailey and Sinnott 1915, 1916). In particular, the sizes and shapes (physiognomy) of leaves correlate strongly with temperature and moisture, and these relationships have been used to develop proxies for reconstructing paleoclimate (e.g., Bailey and Sinnott 1915, 1916; Webb 1959, 1968; Wolfe 1971, 1978, 1979, 1993, 1995; Dilcher 1973; Dolph and Dilcher 1980a, b; Greenwood 1991, 1992; Wing and Greenwood 1993; Wilf 1997; Wilf et al. 1998; Jacobs 1999; Gregory-Wodzicki 2000; Jacobs 2002; Kowalski and Dilcher 2003; Greenwood et al. 2004; Traiser et al. 2005; Spicer 2007; Adams et al. 2008; Peppe et al. 2011; Teodoridis et al. 2011; Yang et al. 2011, 2015; Spicer 2016). Additionally, the climatic tolerances of plants very strongly influence their geographic distributions (e.g., Von Humboldt and Bonpland 1807; Merriam 1894; Küchler 1964; Holdridge 1967; Whittaker 1975; Larcher and Winter 1981). Thus, a plant community, which is comprised of species with overlapping geographic ranges and environmental tolerances, should broadly reflect the climate in which it is growing, assuming the plant community is in equilibrium with climate (e.g., Svenning and Sandel 2013). The assumption that a plant community is in equilibrium with climate is probably broadly true; however, there are notable situations when plant communities can be out of equilibrium with climate, often referred to as disequilibrium or community disequilibrium, particularly when climate is rapidly changing (e.g., Davis 1986; Svenning and Sandel 2013; Blonder et al. 2015a). The relationship between plant geographic distributions and climate has been used to develop a variety of qualitative and quantitative "nearest living relative" approaches, in which fossil taxa are systematically assigned to a closely related modern plant taxon (i.e., their "nearest living relative") and then the fossil species assemblage is matched to the climatic conditions of its "nearest living relatives" (e.g., Heer 1870, 1878a, b, 1882; Chaney 1938; Axelrod 1948, 1952; MacGinitie 1953; Axelrod and Bailey 1969; Hickey 1977; Kershaw and Nix 1988; Mosbrugger and Utescher 1997).

Univariate and multivariate leaf physiognomic methods and nearest living relative approaches have been applied hundreds, if not thousands, of times to Cretaceous and Cenozoic angiosperm-dominated fossils floras to reconstruct paleoclimate (as an example, see supplement in Little et al. (2010) for a relatively complete list of papers from 1902 to 2010 that have utilized univariate and multivariate leaf physiognomic methods). These climate reconstructions are probably what fossil leaves are best known for among the broader geological and paleoanthropological communities, and there are a number of other excellent detailed reviews on using angiosperm leaves to reconstruct climate (e.g., Chaloner and Creber 1990; Greenwood 2007; Wilf 2008; Jordan 2011; Royer 2012a).

The basic observations underpinning leaf physiognomic paleoclimate methods are the relationships between leaf teeth and mean annual temperature (MAT) and leaf size and mean annual precipitation (MAP). In particular, the average leaf size of all species at a site positively correlate with MAP and the percentage of woody, dicotyledonous (dicot) angiosperm species at a site with toothed leaves and variables related to tooth count and size all negatively correlate with MAT (Figs. 13.1, 13.2; e.g., Bailey and Sinnott 1916; Webb 1968; Dilcher 1973; Wolfe 1979, 1993; Dolph and Dilcher 1980a, b; Givnish 1984; Greenwood 1992; Wilf 1997; Wilf et al. 1998; Jacobs 1999, 2002; Huff et al. 2003; Greenwood et al. 2004; Royer et al. 2005; Peppe et al. 2011; Royer 2012b; Yang et al. 2015). A number of studies have demonstrated that leaf shape in at least some species of angiosperms changes in response to climate both within the lifetime of a single plant (e.g., Royer et al. 2009b; Chitwood et al. 2016a, b) and over evolutionary time scales (e.g., Schmerler et al. 2012). Further, it appears that distinct molecular pathways have evolved in plants to mediate changes in leaf shape in response to changes in photosynthesis, temperature, and light quality (e.g., Chitwood and Sinha 2016), suggesting that plants have evolved mechanisms for changing leaf physiognomy in response to climate and environment.

Leaf size typically scales with local water availability (Fig. 13.2B; e.g., Webb 1968; Givnish 1984; Wilf et al. 1998; Jacobs 1999; Malhado et al. 2009; Wright et al. 2017). Larger leaves tend to be warmer because they have a thicker boundary layer (e.g., Vogel 1970, 2009; Parkhurst and Loucks 1972; Gates 1980; Givnish 1984). Photosynthetic and transpiration rates both increase with increasing leaf temperature up to a maximum of $\sim$35–40°C, at which point plants begin to experience heat stress and photosynthetic and transpiration rates rapidly decrease (e.g., Mooney et al. 1978; Salvuci and Crafts-Bandner 2004; Wahid et al. 2007; Michaletz et al. 2015). As a result, plants tend to have smaller leaves in drier climates to reduce evaporative cooling and larger leaves in wet climates where water loss related to increased transpiration is less important, as a means of balancing heat loss via evaporative cooling and water loss via transpiration (i.e., leaf energy balance) (e.g., Givnish 1984; Montheith and Unsworth 2007). Smaller leaves usually have higher densities of major veins, which makes them less vulnerable to embolisms caused by hydraulic disruptions that occur during drought conditions (Scoffoni et al. 2011), which may also help explain why plants in dry climates have smaller leaves than in wetter climates.

Fig. 13.1 Examples of modern and fossil leaves of woody dicotyledonous angiosperms with entire (A, C) and toothed (B, D) margins. A, Cleared and stained leaf of *Calophyllujm calabea* with an entire margin (NCLC4372). B, Cleared and stained leaf of *Tilia heterophylla* with a toothed margin (NCLC5403). C, Unidentified Paleocene fossil leaf from the Fort Union Formation, Williston Basin, North Dakota, USA with an entire margin (YPM52359). Black arrow indicate location of petiolar attachment. D, *Davidia antiqua*, Paleocene fossil leaf from the Fort Union Formation, Powder River Basin, Montana, USA with a toothed margin (DMNH28307). Large black arrow indicates the leaf's petiole. Small black arrow in upper left indicates one of the leaf's teeth

There are many hypotheses about the functional relationship between leaf teeth in angiosperms and climate. Leaf teeth are sites of increased transpiration (Canny 1990). Thus in cold climates, teeth could be an adaptation for increased carbon uptake via enhanced sap flow early in the growing season, which would allow rapid early season growth ("early season gas-exchange hypothesis"; e.g., Billings 1905; Bailey and Sinnott 1916; Baker-Brosh and Peet 1997; Wing et al. 2000; Royer and Wilf 2006). Leaf teeth also commonly have hydathodes, which are fixed pores in teeth that probably allow for water loss via guttation (Feild et al. 2005). The hydathodes likely also help the plant enhance sap flow, which allows it to release excess root pressure, preventing flooding of intercellular spaces in the leaf and, in colder climates, freeze-thaw embolisms ("root-pressure hypothesis"; Feild et al. 2005). Leaf shape, and in particular dissections of a leaf, are also probably related to bud packing ("bud-packing hypothesis"; Kobayashi et al. 1998; Couturier et al. 2011, 2012; Edwards et al. 2016). Additionally, at least some plants exhibit seasonal heteroblasty, in which there are systematic differences in leaf form associated with position along a branch ("seasonal heteroblasty hypothesis"). More specifically, studies have shown that "preformed" leaves, which are leaves that undergo a period of arrested development within a bud, differ in shape from "neoformed" leaves, which are leaves that develop continuously within a bud (Critchfield 1960, 1971; Edwards et al. 2016).

The early season gas-exchange hypothesis provides a functional explanation for the relationship between leaf teeth and MAT because, based on this hypothesis, species with teeth are able to grow and photosynthesize more rapidly early in the growing season than untoothed species, which is a considerable advantage in colder climates with short growing seasons. The root-pressure hypothesis (e.g., Feild et al. 2005) is not as clearly related to MAT because it is related to water movement and freeze-thaw cycles through the plants, which has at best a weak link to climate. However, this hypothesis might help provide a mechanism to explain how plants maintain enhanced sap flow during rapid leaf expansion early in the growing season. The relationship between leaf teeth, enhanced sap flow and increased water loss also helps explain why toothed species are often more abundant in locally wet environments where the potential negative implications of water loss through leaf teeth are mitigated (e.g., Wolfe 1993; Burnham et al. 2001; Kowalski and Dilcher 2003; Greenwood 2005; Royer et al. 2009a; Peppe et al. 2011). The seasonal heteroblasty hypothesis is also not directly linked to MAT. However, it provides a functional explanation for how temperature may influence leaf development and may help at least partially explain why leaf trait variability appears to be more strongly related to annual climate instead of growing season climate. For example, in *Viburnum*, preformed leaves tend to be rounder, more lobed and toothier than neoformed leaves, which were more elliptical with greatly reduced teeth (Edwards et al. 2016). This seasonal variation in leaf shape could be due to two co-occurring forces: first, the different scenarios of early development (e.g., winter dormancy vs. continual development from bud primordia to leaf flush) and second, the plastic response to different light and temperature regimes experienced in the early versus late season (Edwards et al. 2016). The bud-packing hypothesis is based on the premise that there are constraints on leaf shape that are driven by how the folds and boundaries of leaf primordia are arranged

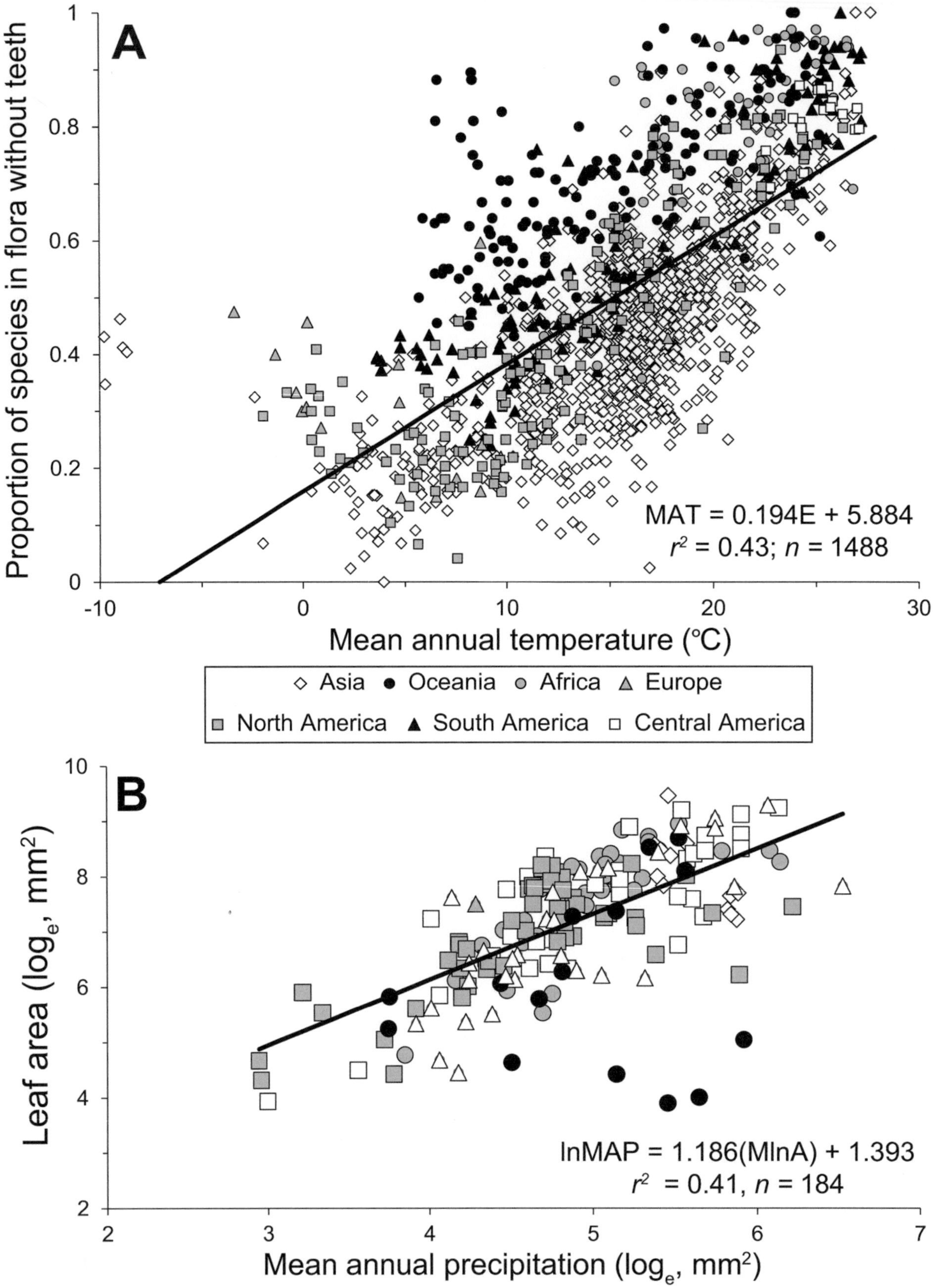

Fig. 13.2 Univariate leaf physiognomic models for estimating mean annual temperature (MAT) and mean annual precipitation (MAP). A, Relationship between the proportion of untoothed species in a flora and MAT for 1488 globally distributed sites. Linear regression fit shown (r^2 = 0.43, F = 1113, $P < 0.001$). Data from Wolfe (1979, 1993), Midgley et al. (1995), Wilf (1997), Kennedy, (1998), Jacobs, (1999, 2002), Burnham et al. (2001), Greenwood et al. (2004), Royer et al. (2005), Hinojosa et al. (2006), Aizen and Ezcurra (2008), Su et al. (2010), Peppe et al. (2011), Chen et al. (2014), Yang et al. (2015). B, Relationship between mean leaf area and MAP for 184 globally distributed sites. Linear regression fit shown (r^2 = 0.41, F = 126, $P < 0.001$). Data from Wilf et al. (1998), Jacobs (1999, 2002), Gregory-Wodzicki (2000), and Peppe et al. (2011)

in a bud, which is unlikely to be driven by climate. However, the bud-packing and seasonal heterblasty hypotheses are very similar in that both argue that climate influences leaf development within the bud, which in turn influences leaf shape, suggesting that climate plays an important role in influencing leaf shape from the beginning of bud development. Although each of these hypotheses help explain some aspects of the relationship between leaf physiognomy and climate, we still lack a physiological or evolutionary mechanism that fully explains how and why the empirical relationships that have been observed between leaf physiognomy and climate in modern floras exists. Work to better understand the mechanism linking climate and leaf physiognomy is ongoing (see for example review in Chitwood and Sinha 2016) and help offer insights that will likely significantly improve leaf physiognomy-climate models in the future.

The occurrence of leaf teeth are likely also influenced by a variety of factors not directly related to climate, such as phylogeny, leaf life-span, and leaf thickness. Phylogenetic history influences the occurrence of leaf teeth (e.g., Little et al. 2010; Burnham and Tonkovich 2011; Hinojosa et al. 2011). This suggests that the leaf teeth-climate relationship observed in modern floras may reflect a combination of plants' responses on both short term (i.e., seasonal) and long term (i.e., evolutionary) time scales and processes of plant community assembly (Schmerler et al. 2012). Plant leaf life-span and leaf thickness also influence the occurrence of leaf teeth as both deciduous species and species with thinner leaves more commonly have teeth than evergreen species and species with thicker leaves (Royer et al. 2012). However, neither plant life-span (i.e., deciduous vs. evergreen) nor leaf thickness correlate strongly with MAT (e.g., Reich et al. 1999; Wright et al. 2004, 2005; Royer et al. 2012), indicating that non-climatic factors also play a role in the occurrence of leaf teeth. The influence of phylogeny, leaf life-span, leaf thickness and other non-climatic factors on leaf physiognomic methods for reconstructing climate is unclear, but may inherently bias paleoclimate estimates (see Royer 2012a for more details). Leaf physiognomic methods that incorporate phylogenetic and leaf ecological information are in development and aim to address the potential influence of phylogeny and other non-climatic drivers of leaf teeth on paleoclimate estimates.

Methods of Paleoclimate Reconstruction

Methods that reconstruct paleoclimate using fossil leaves can be broadly divided into two main types: leaf physiognomic methods and nearest living relative (NLR) methods. Leaf physiognomic methods use the relationship between leaf traits and climate, while NLR approaches use the climatic, and less commonly, the ecological requirements of the taxon that is the nearest living relative to the fossil taxon being studied.

Leaf physiognomic methods – univariate approaches: The relationship between leaf physiognomy and climate was first noted a century ago by Bailey and Sinnott (1915, 1916) and this relationship has become the cornerstone of leaf physiognomic paleoclimate methods (e.g., Wolfe 1971, 1979; Dilcher 1973; Wing and Greenwood 1993; Wolfe 1993; Wilf 1997; Wilf et al. 1998; Jacobs 1999; Gregory-Wodzicki 2000; Greenwood et al. 2004; Spicer 2007, 2016; Peppe et al. 2011). In particular, Bailey and Sinnott (1915, 1916) noted that the proportion of a flora with entire-margined leaves and with large leaves increased with decreasing latitude and suggested that these relationships could be used to estimate ancient climate. Following this pioneering work, paleobotanists used similarities between the relative abundance of species with entire margins and the different size classes of leaves in modern vegetation, as well as Paleogene and Neogene floras to qualitatively infer paleoclimate (e.g., MacGinitie 1953; Wolfe and Hopkins 1967; Hickey 1977; Christophel and Blackburn 1978). Despite the acknowledgement of the relationship between leaf physiognomy and climate and its application to fossil floras and ancient climates, it took more than a half a century before the qualitative relationships noted by Bailey and Sinnott (1915, 1916) were demonstrated quantitatively.

Using floral lists from East Asia, Wolfe (1979) demonstrated a remarkably strong relationship between the percentage of woody dicot species at a site with entire margins, defined as species lacking either marginal teeth or lobes, and MAT. However, although Wolfe (1979) presented a linear regression in his figure that showed the relationship between the percent of species with entire margins and MAT, the statistics and the equation for the regression were not presented. This relationship was then quantified by Wing and Greenwood (1993) using linear regression to develop a predictive equation to estimate MAT based on the percentage of woody dicot species at a site with entire margined species:

$$\mathrm{LMAT} = 30.6P + 1.14 \qquad (1)$$

where LMAT is mean annual temperature estimated using leaf margin analysis and P is defined as:

$$P = \frac{r}{n} \qquad (2)$$

where n is the total number of species sampled, r is the number of species at a site with entire margins, and $(n - r)$ is the number of species with toothed margins. In cases where a species has both entire and non-entire margins, the r score for the species is 0.5.

Table 13.1 Leaf physiognomic models for climate from published literature

Method	Reconstructed climate variable	Equation	Region	r^2	n	SE	Authors
LMA	MAT	(1) MAT = 0.306E + 1.141	East Asia	0.98	34	0.80	Wolfe (1979), Wing and Greenwood (1993)
LMA	MAT	(2) MAT = 0.276E + 1.038	China	0.79	50	1.90	Su et al. (2010)
LMA	MAT	(3) MAT = 0.223E + 6.68	China	0.53	732	3.10	Chen et al. (2014)
LMA	MAT	(4) MAT = 0.2338E + 4.6321	Japan	0.7396	35	1.87	Kennedy et al. (2014)
LMA	MAT	(5) MAT = 0.314E + 0.512	Europe	0.6	1835	1.70	Traiser et al. (2005)
LMA	MAT	(6) MAT = 0.291E − 0.266	North, Central America and Japan	0.76	106	3.40	Wilf (1997)
LMA	MAT	(7) MAT = 0.286E + 2.240	North, Central and South America	0.94	9	2.00	Wilf (1997)
LMA	MAT	(8) MAT = 0.363E + 2.233	North America	0.8	10	3.60	Kowalski and Dilcher (2003)
LMA	MAT	(9) MAT = 0.290E + 1.320	North, Central America	0.91	84	–	Miller et al. (2006)
LMA	MAT	(10) MAT = 0.2471E + 2.3305	North America	0.7919	104	2.63	Kennedy et al. (2014)
LMA	MAT	(11) MAT = 0.250E + 3.42	Temperate Northern Hemisphere	0.87	144	2.20	Wolfe (1993)
LMA	MAT	(12) MAT = 0.275E + 1.36	Temperate Northern Hemisphere	0.768	173	3.30	Gregory-Wodzicki (2000)
LMA	MAT	(13) MAT = 0.249E + 3.53	Temperate Northern Hemisphere (alpine excluded)	0.869	144	2.20	Gregory-Wodzicki (2000)
LMA	MAT	(14) MAT = 0.2475E + 2.9066	Northern Hemisphere	0.81236	149	2.55	Kennedy et al. (2014)
LMA	MAT	(15) MAT = 0.759E − 0.013	New Zealand, the Andes, the Alps	0.571	19	–	Halloy and Mark (1996)
LMA	MAT	(16) MAT = 0.316E − 0.059	South America	0.89	14	1.60	Gregory-Wodzicki (2000)
LMA	MAT	(17) MAT = 0.385E − 10.24	Tropical South America	0.47	30	3.40	Kowalski, (2002)
LMA	MAT	(18) MAT = 0.3825E − 10.9	South America (isotherm)	0.733	16	–	Aizen and Ezcurra (2008)
LMA	MAT	(19) MAT = 0.4236E − 11.37	South America (cell)	0.535	16	–	Aizen and Ezcurra (2008)
LMA	MAT	(20) MAT = 0.2603E + 1.31	South America	0.82	51	2.80	Hinojosa et al. (2011)
LMA	MAT	(21) MAT = 0.4396E − 8.3637	South America	0.9256	14	1.73	Kennedy et al. (2014)
LMA	MAT	(22) MAT = 0.2399E + 3.6916	Southern Africa	0.8567	25	–	Steart et al. (2011)
LMA	MAT	(23) MAT = 0.1094E + 9.5173	Southern Africa	0.3708	14	2.30	Kennedy et al. (2014)
LMA	MAT	(24) MAT = 0.270E − 2.120	Australia	0.63	74	–	Greenwood et al. (2004)
LMA	MAT	(25) MAT = 0.0862E + 17.6547	Pacific Islands	0.3491	6	1.33	Kennedy et al. (2014)
LMA	MAT	(26) MAT = 0.0571E + 16.4785	Australia	0.0559	21	2.34	Kennedy et al. (2014)
LMA	MAT	(27) MAT = 0.0239E + 10.561	New Zealand	0.0174	35	1.69	Kennedy et al. (2014)
LMA	MAT	(28) MAT = 0.257E − 1.8412	Southern Hemisphere	0.40406	90	4.83	Kennedy et al. (2014)
LMA	MAT	(29) MAT = 0.204E + 4.600	North and South America, Japan, and Oceania	0.58	92	4.80	Peppe et al. (2011)
LMA	MAT	(30) MAT = 0.2039E + 3.6562	Oceania, Japan, North America, South America, Southern Africa	0.5733	239	3.97	Kennedy et al. (2014)
LMA	MAT	(31) MAT = 0.194E + 5.884	Global	0.42	1488	4.54	This study

(continued)

Table 13.1 (continued)

Method	Reconstructed climate variable	Equation	Region	r^2	*n*	SE	Authors
LAA	MAT	MAT = 0.115A + 1.192	New Zealand, the Andes, the Alps	0.872	19	–	Halloy and Mark (1996)
DiLP	MAT	MAT = 0.210E + 42.296D − 2.609I − 16.004	North and South America, Japan, and Oceania	0.7	92	4.00	Peppe et al. (2011)
MLR	MAP	(1) MAP = 167.948(apex attenuate) + 377.735 (mesophyll2) + 11.489	Temperate Northern Hemisphere	0.497	74	57.96	Wing and Greenwood (1993)
modified LAA	MAP	(2) MAP = 6.18(mesophyll1 + mesophyll2) + 47.5	Temperate Northern Hemisphere	0.439	74	–	Wilf et al. (1998)
LAA	MAP	(3) lnMAP = 0.484(*M*ln*A*) + 1.78	Temperate Northern Hemisphere	0.612	144	0.47	Gregory-Wodzicki (2000)
LAA	MAP	(4) lnMAP = 0.548(*M*ln*A*) + 0.786	North, Central and South America, Africa	50	50	0.36	Wilf et al. (1998)
LAA	MAP	(5) lnMAP = 0.298(*M*ln*A*) + 2.64	Bolivia	0.52	12	0.22	Gregory-Wodzicki (2000)
LAA	MAP	(6) lnMAP = 1.369(mesophyll1 + mesophyll2) + 3.982	Africa and Bolivia	0.826	42	0.18	Jacobs (2002)
LAA	MAP	(7) lnMAP = 0.309(*M*ln*A*) + 2.566	Tropical Africa and Bolivia	0.734	42	–	Jacobs and Heredeen (2004)
MLR	MAP	(8) lnMAP = 0.6903(mesophyll) + 0.7059(apex acuminate) + 0.884(apex acute) + 0.7542E + 4.9415	Africa	0.794	30	0.19	Jacobs (1999)
MLR	MAP	(9) lnMAP = 1.274(mesophyll1 + mesophyll2) − 1.013(shape elliptic) + 5.198	Africa	0.806	30	0.18	Jacobs (2002)
modified LAA	MAP	(10) lnMAP = 1.321(mesophyll1 + mesophyll2) + 4.018	Africa	0.764	30	0.20	Jacobs (2002)
LAA	MAP	(11) lnMAP = 0.321(*M*ln*A*) + 2.476	Africa	0.661	30	0.24	Jacobs (2002)
LAA	MAP	(12) lnMAP (<260 cm) = 0.354(*M*ln*A*) + 2.167	Tropical Africa, Bolivia, tropical to subtropical Western Hemisphere	0.709	79	–	Jacobs (2002)
LAA	MAP	(13) lnMAP = 0.429(*M*ln*A*) + 1.705	Western Hemisphere, Bolivia, Africa	0.713	92	0.34	Jacobs (2002)
LAA	MAP	(14) MAP = 0.001A + 0.551	New Zealand, the Andes, the Alps	0.713	19	–	Halloy and Mark (1996)
LAA	MAP	(15) lnMAP = 0.283(*M*ln*A*) + 2.92	North and South America, Japan, and Oceania	0.23	92	0.61	Peppe et al. (2011)
LAA	MAP	(16) lnMAP = 0.346(*M*ln*A*) + 2.404	Global	0.41	184	0.93	This study
DiLP	MAP	lnMAP = 0.298A(*M*ln*A*) − 2.717(ln*P*) + 0.279(ln*I*) − 3.033	North and South America, Japan, and Oceania	0.27	92	0.60	Peppe et al. (2011)
MLR	GSP	(1) GSP (<222 cm) = 141.368(mesophyll2) − 136.340(L:W <1) + 93.936(shape elliptic) − 79.774(base round) − 52.386(teeth acute) + 48.050	Temperate Northern Hemisphere	0.804	74	16.00	Gregory and McIntosh (1996)

(continued)

Table 13.1 (continued)

Method	Reconstructed climate variable	Equation	Region	r^2	*n*	SE	Authors
MLR	GSP	(2) GSP = −3.393(leptophyll2) +2.40(apex attenuate) − 2.671(base cordate) + 2.360 (L:W 2-3) + 3.122 (L:W 3-4:1) + 31.6	Temperate Northern Hemisphere (alpine excluded)	0.796	144	–	Wiemann et al. (1998)
MLR	GSP	(3) GSP (<222 cm) = 1.60(apex attenuate) + 2.80 (L:W 2-3) − 45.2	Temperature Northern Hemisphere (alpine excluded)	0.63	144	27.00	Gregory-Wodzicki (2000)
MLR	Wet months precipitation	(1) ln Wet Ppn = 0.8368(mesophyll) + 0.819 (apex acuminate) + 1.1718(apex acute) + 0.8246 (no teeth) + 4.4993	Africa	0.795	30	0.22	Jacobs (1999)
modified LAA	Wet months precipitation	(2) ln Wet Ppn = 1.546(mesophyll1 + mesophyll2) + 6.112	Africa	0.779	30	0.22	Jacobs (2002)
modified LAA	Wet months precipitation	(3) ln Wet Ppn = 1.601(mesophyll1 + mesophyll2) +3.777	Africa and Bolivia	0.833	42	0.21	Jacobs (2002)
modified LAA	Wet months precipitation	(4) ln Wet Ppn = 0.367(*M*ln*A*) + 2.07	Tropical Africa and Bolivia	0.748	42	–	Jacobs and Heredeen (2004)
MLR	Wet months precipitation	3-WET = 110.841(apex attenuate) + 320.457 (L:W 2-3:1) + 179.775(mesophyll2) − 172.859	Temperate Northern Hemisphere	0.583	74	47.23	Wing and Greenwood (1993)
MLR	Dry months precipitation	3-DRY = 45.54(apex attenuate) + 38.186(L:W 2-3:1) − 24.489	Temperature Northern Hemisphere	0.55	74	8.91	Wing and Greenwood (1993)

LMA = leaf margin analysis, LAA = leaf area analysis, MLR = multiple linear regression model, DiLP = digital leaf physiognomy, MAT = mean annual temperature, MAP = mean annual precipitation, GSP = growing season precipitation, wet months precipitation = precipitation during the three wettest months of the year, dry months precipitation = precipitation during the three driest months of the year, E = percent of woody dicots with entire margins, A = leaf area, D = Feret's diameter ratio, I = number of teeth: internal perimeter, P = perimeter ratio, *M*ln*A* = average leaf area of all dicot species

Since Wing and Greenwood's (1993) recalculation of Wolfe's (1979) East Asian leaf margin analysis (LMA) regression, the relationship between leaf margin and MAT has been tested on floras from most major geographic locations and climate biomes on Earth and many global, regional, and site specific LMA regressions have been developed (Fig. 13.1; Table 13.1). Significantly, since Wilf (1997), leaf margin type has been classified as toothed or entire (i.e., non-toothed). Although many factors confound the relationship between margin percentage and MAT, as discussed above and reviewed by others (e.g., Greenwood 2007; Jordan 2011; Peppe et al. 2011; Royer 2012a; Royer et al. 2012), the relationship between leaf margin and MAT is typically statistically significant and similar across geographic regions (Fig. 13.2A). Based on these LMA regressions, LMA has been applied to numerous of Cretaceous and Cenozoic fossil floras to estimate MAT using fossil leaves (see for example studies cited in supplementary materials in Little et al. 2010).

The accuracy and precision of MAT estimates using LMA has been shown to improve when more species are sampled. In his original LMA work, Wolfe (1971) suggested ≥ 30 species were necessary to calculate a reliable MAT estimate, though he suggested that the minimum bar was ≥ 20 species, which has been the practice of most paleobotanists. While the total number of species sampled is important, the proportional recovery of the total flora is more significant. For example, Burnham et al. (2001, 2005) found that in very diverse, extant floras with 55 to >400 species, ≥ 25 of the most common species had to be sampled to make an accurate MAT estimate using LMA. This suggests that LMA should only be applied to well-sampled floras and that MAT estimates made using poorly sampled floras and/or floras with low species richness are unlikely to be accurate.

Sample size also affects the uncertainty of LMA estimates. Based on their reanalysis of the Wolfe (1979) East Asian LMA, Wing and Greenwood (1993) calculated the standard error of LMA estimates of MAT to be ±0.8°C. However, the standard errors of the East Asian LMA regression alone considerably underestimate the uncertainty of leaf margin analysis mean annual temperature (LMAT) estimates. To account for additional uncertainty in LMAT estimates, Wilf (1997) proposed using the binomial sampling error for LMA estimates, which accounts for the standard error of the LMA regression and uncertainty related to sampling the species richness of a fossil flora:

$$\sigma[\mathrm{LMAT}] = \mathrm{c}\sqrt{\frac{\mathrm{P}(1-\mathrm{P})}{n}} \tag{3}$$

where c is the slope of the MAT vs. leaf-margin regression in the dataset used, P is the proportion of species with entire margins, and n is the number of species sampled.

Wilf (1997) suggested the minimum uncertainty of a LMAT estimate was at least ±2.0°C and that the larger of the two errors given by the binomial sampling error and the standard error of the regression if it were >2.0°C was the appropriate uncertainty for LMAT estimates. In Wilf's (1997) study, he noted that the binomial sampling error only accounted for part of the total uncertainty in LMAT estimates. Miller et al. (2006) incorporated the uncertainties associated with the estimation of the proportion of species at a site with teeth (P) into the estimation of uncertainty of LMAT estimates:

$$\sigma[\mathrm{LMAT}] = \mathrm{c}\sqrt{\left[\{1+\varphi(n-1)(P(1-P))\frac{\mathrm{P}(1-\mathrm{P})}{n}\right]} \tag{4}$$

where c is the slope of the MAT vs. leaf-margin regression in the dataset used, P is the proportion of species with entire margins, n is the number of species sampled and φ is the overdispersion factor for P, which was calculated by Miller et al. (2006) to be 0.052.

The Miller et al. (2006) estimation of uncertainty provides the best measurement of the precision of LMAT; however, they noted that there may be additional systematic uncertainties in LMAT based on sampling differing forested environments. Greenwood et al. (2004) documented this effect in a series of sites from Australia that showed a significant correlation between margin percentage and MAT ($r2 = 0.57$, $P < 0.001$), but a larger uncertainty than datasets from East Asia and North and Central America ($\sigma = \pm 3$°C). Based on a large global compilation of sites, Peppe et al. (2011) showed that the uncertainty associated with LMAT estimates using a globally derived LMA is at least ±4°C. Conservatively, given uncertainties associated with sampling (accounted for in Eq. 4) and all of the potential confounding factors that affect the relationship between climate and the proportion of species with teeth uncertainties, which are currently not accounted for in any uncertainty calculation, a minimum uncertainties for LMAT estimates is probably ±5°C (see also Jordan 1997, 2011; Royer 2012a). Despite these uncertainties, LMAT estimates are typically in agreement with other proxies and are generally accurate, and reasonably precise, making the LMA method a useful tool for estimating MAT in the past.

Building on another of Bailey and Sinnott's (1915, 1916) observations that large leaves were more common in warm, mesic environments, several researchers have noted a strong positive relationship between leaf size and mean annual precipitation (MAP) (Webb 1968; Dilcher 1973; Dolph and Dilcher 1980a, b; Hall and Swaine 1981; Givnish 1984). The correlation between leaf size and MAP was first quantified by Givnish (1984) for floras from South America, Central America and Australia. Wilf et al. (1998) expanded on Givnish's (1984) work and used the correlation between the mean of the natural log of species' leaf area (MlnA) and the natural log of MAP (lnMAP) at 50 sites from North America, Central America, South America, and Africa to develop a predictive equation for MAP:

$$\ln(\mathrm{MAP}) = 0.548\mathrm{MlnA} + 0.768 \quad (5)$$

MlnA was calculated for the entire flora by first determining the Raunkier-Webb (Webb 1959) size category for each species and then determining the proportion of species in each of the seven size categories:

$$\mathrm{MlnA} = \sum a_i p_i \quad (6)$$

where a_i is the seven means of the natural log areas of the size categories and p_i is the proportion of species in each size category. This method of estimating MAP using leaf area is known as leaf area analysis (LAA; Table 13.1). The standard error for ln(MAP) for LAA is 0.359 (Wilf et al. 1998).

Following the work of Wilf et al. (1998), the relationship between leaf area and MAP has been tested on a series of floras from Bolivia and predominately Northern Hemisphere temperate regions (Gregory-Wodzicki 2000), floras from equatorial Africa (Jacobs 1999, 2002), and floras from North America, Central America, South America, Asia, and Oceania (Peppe et al. 2011) (Fig. 13.2). These correlations have been used to develop several additional LAA equations to estimate MAP (Table 13.1, Fig. 13.2; Jacobs 1999, 2002; Gregory-Wodzicki 2000; Peppe et al. 2011). Using a dataset of sites from equatorial Africa and Bolivia, Jacobs and Herendeen (2004) also documented a significant relationship between leaf area and precipitation during the three wettest months, which they used to develop a proxy for wet month precipitation (Table 13.1).

Although the relationship between leaf area and mean annual precipitation is significant, LAA estimates of MAP are not accurate and the uncertainties are relatively large. Peppe et al. (2011) used a jackknife-type approach for a dataset of 92 globally distributed sites to gauge the accuracy of LAA, and found that LAA tended to underestimate MAP at sites with relatively high precipitation and overestimate MAP at sites with relatively low precipitation. These results and the original work of Wilf et al. (1998) suggest that LAA is not a precise estimator of MAP and instead is primarily a useful tool for roughly estimating MAP.

Leaf size is also influenced by a variety of other factors including temperature, elevation, soil fertility, and leaf life-span (e.g., Webb 1968; Dolph and Dilcher 1980a, b; Givnish 1984; Wolfe 1995; Halloy and Mark 1996; McDonald et al. 2003; Peppe et al. 2011) confounding the relationship between MAP and leaf size leading to the low accuracy and precision of MAP estimates made using LAA. Additionally, available water appears to be the main control on leaf size (Givnish 1984). In addition to precipitation, plant available water is influenced by many factors such as rates of evapotranspiration, temperature, seasonality, the position of the water table, and hydrologic conductance within the soil. As a result, MAP may not be a good proxy for available water in many situations, which perhaps explains the relatively poor accuracy and precision of LAA.

Although LMA, and to a lesser extent LAA, have been widely adopted by the paleobotanical community and the fossil leaf based MAT and MAP estimates made using the methods are often considered some of the most robust estimates for terrestrial paleoclimates, both methods rely on single characters. Leaf size and shape is remarkablely diverse and many aspects of leaf physiognomy vary with climate (e.g., Wolfe 1993, 1995; Royer et al. 2008, 2009b; Schmerler et al. 2012). Additionally, in at least some cases, multiple leaf physiognomic traits appear to co-vary with climate variables (Dolph and Dilcher 1980a, b; Peppe et al. 2011). Further, although LMA and LAA commonly agree with other climate proxy reconstructions, in many cases, these proxies provide cooler and drier estimates of MAT and MAP than other proxies (e.g., Liang et al. 2003; Fricke and Wing 2004; Wing et al. 2009; Adams et al. 2011). Because LMA and LAA are univariate approaches, the incorporation of additional leaf characters may lead to improvements in paleoclimate estimates.

Leaf physiognomic methods – multivariate approaches: In an attempt to improve on the shortcomings of univariate leaf physiognomic methods, Wolfe (1990, 1993, 1995) developed a multivariate method called Climate-Leaf Analysis Multivariate Program (CLAMP), which currently uses 31 categorical leaf physiognomic traits (see Spicer 2016). In CLAMP the leaf character scores and climate data from a modern training set are analyzed using canonical correspondence analysis (CCA; Wolfe 1995; Kovach and Spicer 1996). The resultant eigenvectors for the climate and leaf physiognomic variables are projected onto the major axes of variation, which roughly correlate with temperature (first axis) and moisture (second axis) (Wolfe 1995). A second-order polynomial fit for a climate variable to the eigenvector is then used to develop a predictive equation for that climate variable, which can then be applied to a fossil flora (Wolfe 1995). In practical terms, this means that the leaf character scores of a fossil flora are projected onto each of the climate variable vectors in the CLAMP calibration dataset and those scores are used to estimate each climate variable. CLAMP estimates MAT, warm month mean temperature, cold month mean temperatures, growing season length, growing season precipitation, mean monthly growing season precipitation, precipitation in the three wettest and three driest months, relative humidity, specific humidity, and enthalpy (Wolfe 1995; Spicer 2016). The standard errors for CLAMP climate estimates are small (Wolfe 1993, 1995;

Spicer et al. 2005); however, the errors are only based on the analytical uncertainty of CCA and do not account for any additional uncertainties related to the leaf character states, character scoring, precision in the climate variables of the calibration data set, or factors that potentially confound leaf character-climate relationships. Thus, the standard errors reported for CLAMP considerably underestimate uncertainty and are artificially precise. For example, at minimum, the uncertainties in MAT estimates made using CLAMP are similar to, if not larger than those for leaf margin analysis (see Fig. 13.1 in Royer 2012a).

CLAMP was originally calibrated with 106 sites with a limited global coverage, and since has been expanded to two primary calibration datasets primarily derived from floras from the Northern Hemisphere: (1) PHYSG3BR, which includes 144 temperature Northern Hemisphere sites, and (2) PHYSG3AR, which includes 173 Northern Hemisphere sites that include sites that experience very cold winter temperatures (e.g., Wolfe 1993, 1995; Wolfe and Spicer 1999; Spicer et al. 2004; Spicer 2016). Since then Spicer and colleagues (see Spicer 2007, 2016) have assessed regional differences in the relationship between climate and CLAMP characters in South America (Gregory-Wodzicki 2000; Kowalski 2002), South Africa (Steart et al. 2011), New Zealand and Australia (Kennedy et al. 2002, 2014) and expanded the calibration datasets to include floras in regions that experience a monsoon climate (Jacques et al. 2011; Khan et al. 2014), floras from the Southern Hemisphere (Kennedy et al. 2014), and floras from temperate and high latitudes in the Northern and Southern Hemispheres (Yang et al. 2015). The CLAMP modern calibration originally used climate data from local climate stations (Wolfe 1993, 1995), but more recent studies follow Spicer et al. (2009) and use climate data extracted from the New et al. (2002) global, gridded climate model. CLAMP analyses were originally performed using a series of spreadsheets and conducting a CCA using the software package CANACO 4.5 (Microcomputer Power, Ithaca, NY, USA). CLAMP analyses can now be performed online (Yang et al. 2011) through the CLAMP website: http://clamp.ibcas.ac.cn/CLAMP_Run_Analysis.html.

Although CLAMP has been widely applied (reviewed in Spicer 2007, 2016), the method has several shortcomings that have been reviewed many times (e.g., Jordan 1997, 2011; Wilf 1997; Wiemann et al. 1998; Wilf et al. 1998; Green 2006; Peppe et al. 2010). The most notable criticisms are that (1) based on the character distributions, CCA may not be the most appropriate technique for predicting climate (Green 2006; Peppe et al. 2010), (2) at least some of the leaves in the original calibration dataset were scored incorrectly (Peppe et al. 2010), (3) many of the CLAMP character states are at least partially ambiguous and the original scoring procedures were poorly defined (Wilf 1997; Wiemann et al. 1998; Wilf et al. 1998, 1999; Green 2006), (4) the CLAMP characters are discrete and categorical despite the fact that leaf traits are continuous (i.e., number of teeth, leaf size) and many of the continuous traits correlate strongly with temperature and/or precipitation (Huff et al. 2003; Royer et al. 2005, 2008; Peppe et al. 2011), and (5) some CLAMP character states only correlate weakly to climate and/or have no physiological explanations for climate causality.

One of the underlying assumptions of CCA is that the variables are unimodally distributed (after Braak 1986). However, Green (2006) and Peppe et al. (2010) showed that many of the CLAMP leaf character distributions were not unimodally distributed. Following this work, Teodoridis et al. (2011) developed transformations of the CLAMP PHYSG3BR and PHYSG3AR calibration datasets that could be used as "correction coefficients" to account for the non-normality of the CLAMP leaf character state data.

Peppe et al. (2010) rescored a subset of Wolfe's (1993, 1995) original calibration sites and demonstrated that the site-mean leaf area of 38 sites in the CLAMP calibration and the leaf area scores of the species in each site were considerably underestimated relative to digitally measured leaf area. When the influence of this underestimation of leaf area was accounted for, it added a considerable amount of uncertainty to all CLAMP climate estimates (Peppe et al. 2010). Spicer and Yang (2010) assessed the influence of leaf size on MAT and enthalpy estimates made using CLAMP and found that when the leaf size character state was removed it increased uncertainty, but caused only relatively small changes in MAT and enthalpy. This result indicates that leaf size contributes relatively little information when predicting at least some climate variables with CLAMP and suggests that some climate predictions could be improved by changing the leaf character states used in CLAMP. Spicer and Yang (2010) only indirectly addressed the main criticism of Peppe et al. (2010), that the leaf area scores in the CLAMP calibration dataset are underestimated, by suggesting that the measured leaf area bias was the result of measuring leaf area directly rather than using the CLAMP leaf area template. This is an important distinction because it means that when performing a CLAMP analysis, leaf area can only be scored using the CLAMP leaf area template and if leaf area is measured directly, even if it is more accurate, it will add considerable uncertainty to the CLAMP estimates.

Several studies have documented issues with the ambiguity of CLAMP character states and original scoring procedures (Wilf 1997; Wiemann et al. 1998; Wilf et al. 1998, 1999; Green 2006; Peppe et al. 2010). Recently, Spicer (2016) updated the character state definitions and scoring protocol in an attempt to eliminate the ambiguities. To test these revisions, Spicer and Yang (2010) had students collect and score leaves from a forest reserve near Beijing, China and found that the resulting CLAMP climate estimates were different, though reasonably similar, suggesting the revised CLAMP scoring protocol and character definitions have helped eliminated some of the problems with the original method.

Although there has been recent work to expand the CLAMP calibration datasets to a wider range of biogeographic regions and climate regimes (e.g., Jacques et al. 2011; Kennedy et al. 2014; Yang et al. 2015), the inherent issues with using categorical variables and leaf characters that are not physiologically linked to climate have not been addressed. Furthermore, many studies have demonstrated that climate estimates made using CLAMP are similar to, or less accurate than, estimates made using LMA and LAA (e.g., Wilf 1997; Wiemann et al. 1998; Gregory-Wodzicki 2000; Kowalski 2002; Kowalski and Dilcher 2003; Hinojosa et al. 2006a, b; Dilcher et al. 2009; Royer 2012a; Smith et al. 2012; West et al. 2015).

Rather than apply CCA, several researchers have used the CLAMP calibration datasets or comparable datasets using the CLAMP characters to develop multiple linear regression models for paleoclimate (e.g., Wing and Greenwood 1993; Gregory 1994; Gregory and McIntosh 1996; Jacobs and Deino 1996; Stranks and England 1997; Wiemann et al. 1998; Gregory-Wodzicki 2000; Kowalski and Dilcher 2003; Lielke et al. 2012). However, these methods have not been widely adopted. Further, because they utilize the CLAMP calibration dataset and leaf character states, they suffer from many of the same problems as CLAMP.

Given the potential problems with CLAMP, and in particular the subjectivity of scoring leaf traits, Huff et al. (2003) proposed digital leaf physiognomy (DiLP), which is a method for measuring leaf traits digitally. The benefits of DiLP are that it increases the reproducibility of leaf character scores and that it measures continuous variables, such as number of teeth and tooth size, that have functional or physiological links to climate. In their preliminary study of three floras, Huff et al. (2003) found that continuous variables related to tooth number and tooth shape varied between floras and with temperature, demonstrating that the incorporation of continuous variables had the potential to improve fossil leaf based paleoclimate models. A subsequent study by Royer et al. (2005) showed that leaves in cold climates are more highly dissected and have more, larger teeth. Using data from 17 sites from the east coast of the United States and Panama, Royer et al. (2005) developed a preliminary multiple linear regression model for predicting MAT that was more accurate and precise than LMA and CLAMP. However, the study was based on sites from a relatively limited range in biogeography and precipitation. Peppe et al. (2011) quantified DiLP leaf physiognomic variables and climate using 92 globally distributed and climatically diverse sites and noted the same patterns between leaf traits and MAT as observed by Huff et al. (2003) and Royer et al. (2005) and also that leaves in wet climates were larger and had fewer, smaller teeth. Using these relationships, Peppe et al. (2011) developed multiple linear regression models for MAT and MAP (Table 13.1):

$$\mathrm{MAT} = 0.210(E) + 42.296(FDR) + 2.609(T:IP) - 16.004 \tag{7}$$

$$\ln(\mathrm{MAP}) = 0.298\ln(A) - 2.71\ln(PR) + 0.279\ln(T:IP) + 3.033 \tag{8}$$

where E is the proportion of species at the site with entire margins, FDR is feret's diameter ratio (diameter of a circle with the same area as the blade area (cm)/longest measurable line across blade area (cm)), $T{:}IP$ is the number of teeth to internal perimeter (number of teeth/perimeter of the leaf after teeth are removed (cm)), A is leaf area, and PR is perimeter ratio (perimeter of the leaf blade/internal perimeter (cm)/perimeter of the leaf after teeth are removed (cm)). The leaf character states used in these models are defined in Peppe et al. (2011). The standard error for the MAT model is ±4.0°C and for the MAP model is 0.60.

DiLP estimates of MAT and MAP are significantly more accurate than MAT estimates made using LMA and marginally better than MAP estimates using LAA (Peppe et al. 2011). DiLP MAT estimates made for several fossil floras are also in better agreement with other proxy reconstructions of MAT than LMAT estimates (Peppe et al. 2011). This suggests that the DiLP method potentially offers improvements over other existing leaf physiognomic methods. DiLP also offers three additional advantages over CLAMP, LMA, and LAA: (1) it minimizes ambiguity in leaf character scoring because computer algorithms process most of the measurements; (2) the DiLP models use mostly continuous variables, such as feret's diameter ratio and number of teeth: internal perimeter; and (3) the leaf characters used in the DiLP models have a functional connection to climate and include traits that display phenotypic plasticity between growing seasons (Royer et al. 2009b) indicating the traits can respond quickly to climate change.

Despite its advantages, a major limitation of DiLP is that some of the characters used in the DiLP MAT and MAP models (feret's diameter ratio and leaf area) require relatively complete preservation of leaves. Unfortunately, fossil leaves are typically fragmented (e.g., Greenwood 1992, 2005, 2007; Spicer et al. 2005), which means that it may not be possible to score all DiLP characters for all fossil taxa. In a sensitivity experiment using modern leaves, Royer et al. (2005) tested the effect of fragmentary leaves on LMA and DiLP and found that in all cases fragmentary leaves led to less precise climate estimates, but when leaves were 50% intact, LMA estimates of MAT had larger errors than

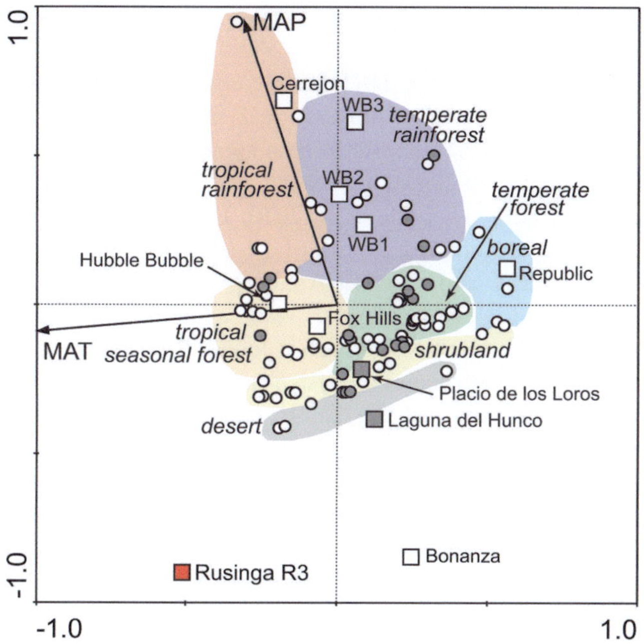

Fig. 13.3 Canonical correspondence analysis plot of the digital leaf physiognomy (DiLP) calibration dataset and fossil sites based on leaf physiognomy from Peppe et al. (2011) with modern biomes mapped onto plot. Extant sites indicated by circles and fossil sites indicated by squares. Site from the Northern Hemisphere are indicated by white symbols and sites from the Southern Hemisphere by gray symbols. The Rusinga R3 fossil site (Michel et al. 2014), which is indicated by a red square, falls well outside the leaf physiognomic space of the extant floras in the current DiLP calibration demonstrating a limitation of the applicability of the method

estimates made with DiLP. Additionally, Royer et al. (2005) found that while multiple specimens should be scored when possible, only one specimen per species needed to be digitally scored to provide an accurate MAT estimate using DiLP. These results suggest that although there is some information lost when leaves are fragmentary, DiLP may still be able to provide reasonable estimates of paleoclimate when fragmentary leaves are scored. In an attempt to address the fact that fossil leaves are commonly incomplete, Peppe et al. (2011) developed a detailed protocol for processing incomplete leaves. In practice DiLP characters can be scored in most fragmentary leaves, provided that at least 25% of the leaf is preserved and the fragment preserves both the margin and the mid-vein. Additionally, recent work documented a strong relationship between leaf vein density and leaf size (e.g., Sack et al. 2012), and was used to develop a "vein scaling" transfer function for estimating leaf area from fragmentary fossils (Merhofer et al. 2015). The incorporation of leaf area estimates made using vein scaling also has the potential to increase the number of fragmentary fossil specimens that can be used to reconstruct MAT and MAP.

DiLP has been successfully applied to several fossil floras (Peppe et al. 2011; Flynn et al. 2016), which suggests that the method has great potential to be more widely applied. However, there are some examples in which the climate estimates made using DiLP were unreasonable based on other lines of evidence (Peppe et al. 2011). For example, estimates for the Laguna del Hunco flora from Argentina, were cooler and drier than expected (Peppe et al. 2011), based on the habitat preference of some of the species present in the flora (Wilf et al. 2009). This discrepancy may be because the physiognomic-climate space of South America is not fully characterized in the current DiLP calibration.

There are also some floras to which DiLP cannot be applied. Peppe et al. (2011) recommend using CCA as an initial check to confirm that a fossil floras physiognomic space is captured by the calibration data. Peppe et al. (2011) and Peppe (unpublished data) have analyzed two floras that plotted outside the calibrated physiognomic space: Bonanza, an Eocene flora from North America, and Rusinga R3, a Miocene flora from Kenya (Michel et al. 2014; Fig. 13.3). Peppe et al. (2011) argued that the Bonanza flora was likely on outlier because the fossil flora was a mixture of two habitats resulting in a unique site mean physiognomic space. There are no modern floras from Africa in the current DiLP calibration, and thus the Rusinga R3 flora from Kenya likely samples a currently uncalibrated physiognomic space (Fig. 13.3). Future work expanding the calibration dataset to more sites from the Southern Hemisphere, and particularly from Africa, may help improve DiLP so that the method can be applied to a wider range of floras.

Shortcomings and limitations of physiognomic methods: Although leaf physiognomic methods of reconstructing paleoclimate have been widely applied and generate reasonably accurate and precise estimates, they are not without their inherent disadvantages. There are a variety of factors that affect and confound the relationships between leaf physiognomy and climate. In this review, we will briefly discuss three of the important factors that limit leaf physiognomic methods: life history, phylogeny, and plant ecology. For a review of some of the other confounding factors, such as taphonomy, local water availability, position in the canopy, soil fertility, and nutrient availability, see the recent reviews by Greenwood (2007), Jordan (2011), and Royer (2012a).

The original leaf margin analysis calibration data and most subsequent calibrations, as well as most of the calibration data that underlies CLAMP and DiLP, are primarily derived from Northern Hemisphere temperate floras (e.g., Wolfe 1979, 1993, 1995; Wilf 1997; Kowalski and Dilcher 2003; Su et al. 2010; Peppe et al. 2011). However, the relationship between leaf physiognomy and climate, and particularly between the presence of leaf teeth and MAT, is distinctly different in the Southern Hemisphere where floras typically have a higher percentage of untoothed species than their Northern Hemisphere counterparts (e.g., Wolfe 1979;

Upchurch and Wolfe 1987; Gregory-Wodzicki 2000; Kowalski 2002; Greenwood et al. 2004; Hinojosa et al. 2006a; Aizen and Ezcurra 2008; Hinojosa et al. 2011; Steart et al. 2011; Kennedy et al. 2014). Further, in Africa, Australia, and New Zealand the correlation between leaf size and MAT is as strong, if not stronger, than between leaf margin and MAT (Webb 1968; Greenwood 1992, 1994; Jacobs 1999, 2002).

These hemispheric differences could be due to regional differences in environment, such as thermal seasonality or soil fertility, phylogenetic differences, leaf life-span (i.e., lack of temperate deciduous forests in the Southern Hemisphere), or differences in plant ecology (e.g., Upchurch and Wolfe 1987; Jordan 1997; Greenwood et al. 2004; Jordan 2011; Peppe et al. 2011). Researchers have also found that there are considerable differences between physiognomy and climate when comparing the CLAMP PHYSG3BR and PHYSG3AR to regional floras in South America, Australia, and New Zealand (e.g., Gregory-Wodzicki 2000; Kennedy et al. 2002, 2014; Kowalski 2002; Hinojosa et al. 2006a; Steart et al. 2011). There are also notable differences in the relationship between leaf physiognomy and climate in some of the DiLP characters (e.g., number of teeth vs. temperature and leaf area vs. temperature) between sites from the different hemisphere, and particularly from Australia and New Zealand (Peppe et al. 2011). Finally, although much research has focused on the regional differences in the relationships between climate and physiognomy, most of the regional datasets have not been added to the larger calibration dataset or used to develop new models (e.g., Steart et al. 2011 from South Africa). As a result, most LMA regressions and the most commonly used CLAMP calibration datasets PHYSG3BR and PHYSG3AR have few to no floras from Africa, South America, and Oceania. The net effect is the most leaf physiognomic models do a relatively poor job reconstructing the paleoclimate of Southern Hemisphere floras (Kowalsk 2002; Greenwood et al. 2004; Hinojosa et al. 2006a; Peppe et al. 2011).

Two of the hypothesized explanations for the differences between Northern and Southern Hemisphere floras, phylogeny and plant ecology, likely play a role in their differences and also strongly influence the relationships globally. Although there is evidence that leaves can phenotypically respond to climate change rapidly (Royer et al. 2009b; Chitwood et al. 2016a, b), there is clearly a genetic control on leaf shape and tooth characters (reviewed in Chitwood and Sinha 2016). Additionally, there are many families of plants that are overwhelmingly toothed (e.g., Betulaceae, Juglandaceae) or untoothed (Lauraceae, Myrtaceae). However, in *Viburnum* there are evolutionary correlated shifts in leaf shape and leaf tooth morphology that are coincident with climate (Schmerler et al. 2012), suggesting that in at least some taxa, leaf physiognomy responds to climate on evolutionary timescales.

More generally, the phylogenetic composition of a flora strongly can strongly influence leaf physiognomy (Little et al. 2010). There appears to be a significant phylogenetic signal in the relationship between leaf margin and MAT (Jones et al. 2009; Little et al. 2010; Hinojosa et al. 2011) and, using a restricted dataset of 17 sites from the eastern United States and Central America, in the relationship between DiLP characters and MAT as well (Little et al. 2010). Little et al. (2010) also found that phylogenetic generalized least squares analysis that included phylogenetic information improved the ability to predict margin state (toothed vs. entire) from MAT, but with much larger uncertainty and a flatter regression slope. Based on these results, Little et al. (2010) argue that LMA is so strongly affected by phylogenetic bias that it should not be used. However, although there is phylogenetic signal in leaf traits, there is a global relationship between leaf traits and climate (Peppe et al. 2011; Yang et al. 2015) suggesting there is likely geographic and evolutionary sorting of taxa and preferred plant morphology relative to climate. Further, recent work by Lawing et al. (2016) suggests that the work of Little et al. (2010) may be affected by phylogenetic autocorrelation and that their conclusions may not be valid unless the geographic distribution of tree species also has strong phylogenetic inertia, which seems unlikely given the rapid movement of plants in response to changing climate in the Quaternary and today (e.g., Comes and Kadereit 1998; Davis and Shaw 2001; Kelly and Goulden 2008). However, if plant communities and/or species are in climate disequilibrium (e.g., Svenning and Sandel 2013; Rokataorinivo et al. 2013; Blonder et al. 2015a), there is the potential for paleoclimate to have influenced modern plant distributions, and thus modern leaf trait-climate relationships, which would considerably confound leaf physiognomic based paleoclimate estimates. Regardless, phylogenetic history likely plays an important role in influencing climate-leaf physiognomy relationships, and future work should focus on incorporating phylogeny into future leaf physiognomic methods.

Plant growth habit and leaf life-span also influences leaf physiognomy. In particular, deciduous taxa are more likely to be toothed than evergreen taxa (Peppe et al. 2011; Royer et al. 2012) and leaf life-span (deciduous vs. evergreen) also influences tooth size and number (Peppe et al. 2011). Thus, the growth habits of species within a plant community likely have strong influence on leaf physiognomy. As an example, floras from the Southern Hemisphere tend to have more species that are evergreen relative to Northern Hemisphere floras; therefore, it is possible that leaf life-span plays a significant role in driving the higher percentage of untoothed species in the Southern Hemisphere (e.g., Upchurch and

Wolfe 1987; Greenwood et al. 2004; Peppe et al. 2011). Using a data set of species, instead of sites for which all of the leaf physiognomic calibrations are optimized, Royer et al. (2012) found that the inclusion of leaf thickness and deciduousness with other DiLP variables and a multiple linear regression model of leaf thickness, deciduousness, and leaf margin reduced the standard errors of both models predictions of MAT. Leaf thickness and deciduousness are not measureable in fossils, but leaf mass per area, which is a proxy for leaf life-span (Royer et al. 2007), can be estimated. The incorporation of leaf mass per area into a model with other DiLP characters also reduces the standard error of the models prediction of MAT (Royer et al. 2012). These preliminary results suggest that although plant ecology clearly has an effect on leaf physiognomy, its incorporation into paleoclimate models might improve the accuracy and precision of the paleoclimate models.

Nearest living relative methods: The geographic distributions of plants are strongly influenced by their environmental and climatic tolerances (e.g., Merriam 1894; Küchler 1964; Holdridge 1967; Whittaker 1975). Under the assumption that the climatic tolerances of fossil taxa and fossil plant assemblages are similar to their modern "nearest living relatives" (NLR), for more than 100 years paleobotanists have used this information to estimate ancient climate based on fossil plant assemblages (e.g., Heer 1870, 1878a, b, 1882; Chaney 1938; Axelrod 1948, 1952; MacGinitie 1953; Axelrod and Bailey 1969; Hickey 1977; Kershaw and Nix 1988; Mosbrugger and Utescher 1997). Probably beginning with the work of Heer (e.g., 1870, 1878a, b, 1882), numerous NLR qualitative and quantitative methods have been developed and applied primarily to Cenozoic floras. Both qualitative and quantitative NLR methods use the climatic and ecological requirements of extant plant taxa, which means that these methods rely upon accurate taxonomic identification of plant fossil material. Additionally, NLR methods are not limited to woody dicot angiosperm leaves alone, and any reliably identified plant organ can be utilized.

Qualitative NLR methods rely on individual indicator taxa to derive paleoclimate and paleoenvironment from a fossil flora. For example, the occurrence of palms in a fossil flora would be used to infer warm and wet conditions with cold months above freezing because the palm family Arecaceae is found predominately in tropical to sub-tropical regions with limited to no winter freezing (e.g., Larcher and Winter 1981; Wing and Greenwood 1993, 1995). However, a considerable limitation of these qualitative NLR approaches is that they do not provide quantitative estimates of paleoclimate (i.e., a MAT or MAP prediction) and the accuracy of the paleoclimate interpretation can be strongly influenced by taphonomic removal of important indicator taxa as well as the level of taxonomic identification of the indicator taxa.

Quantitative NLR methods build on the indicator taxa approach to quantitatively estimate paleoclimate. Originally quantitative NLR methods used a few selected fossil taxa and assessed their overlapping climate preferences, which resulted in broad "coexistence intervals" that were used to estimate paleoclimate (e.g., Grichuk 1969; Hickey 1977; Hickey et al. 1988; Zagwijn and Hager 1987). This approach was modified independently first as bioclimatic analysis (Kershaw and Nix 1988; Greenwood et al. 2003, 2005), and then as the coexistence approach (CA) of Mosbrugger and Utescher (1997). Other NLR methods also use the same fundamental premise, but assess the climatic ranges and/or distributions differently (e.g., Thompson et al. 2012; Harbert and Nixon 2015). All NLR methods have four primary assumptions: (1) there are modern taxa that are systematically closely related to identified fossil taxa (i.e., their "nearest living relative"); (2) the climate requirements of the fossil taxon are similar to those of the nearest living relatives; (3) the climate tolerances of the nearest living relatives can be derived from its modern distribution; and (4) the climate data for the nearest living relatives is good quality (Mosbrugger and Utescher 1997; Utescher et al. 2014).

In CA all identified fossil taxa are conferred to nearest living relatives. The climatic tolerances of each fossil's nearest living relative are then compared to find an interval where all nearest living relatives could coexist (Fig. 13.4). The climatic ranges of many vascular plant taxa have been compiled in the Palaeoflora Database (PFDB; www.palaeoflora.de), which is the dataset used for CA. Individual taxa that do not overlap with the rest of the taxa from a site are considered outliers and are removed from the analysis. This subjective removal of outliers has come under criticism (Grimm and Denk 2012; Grimm et al. 2016; Grimm and Potts 2016). CA has been used to reconstruct seven different aspects of temperature and precipitation: MAT, warm month mean temperature, cold month mean temperature, MAP, monthly minimum precipitation, monthly maximum precipitation, and warm month mean precipitation and has a reported precision of 0.1°C and 1 mm precipitation (Mosbrugger and Utescher 1997). This uncertainty is only the analytical uncertainty of the CA calculation for the coexisting climatic range, and thus is artificially precise and considerably underestimates the total uncertainty of the climate estimate.

Another quantitative NLR method is Bioclimatic Analysis (Kershaw and Nix 1988; Kershaw 1997). Just as in CA, in Bioclimatic Analysis fossil taxa are conferred the nearest living relatives, which are compared to determine the climate interval where all of the nearest living relatives could coexist originally using the PFDB and more recently using climatic ranges of taxa using the Global Biodiversity Information

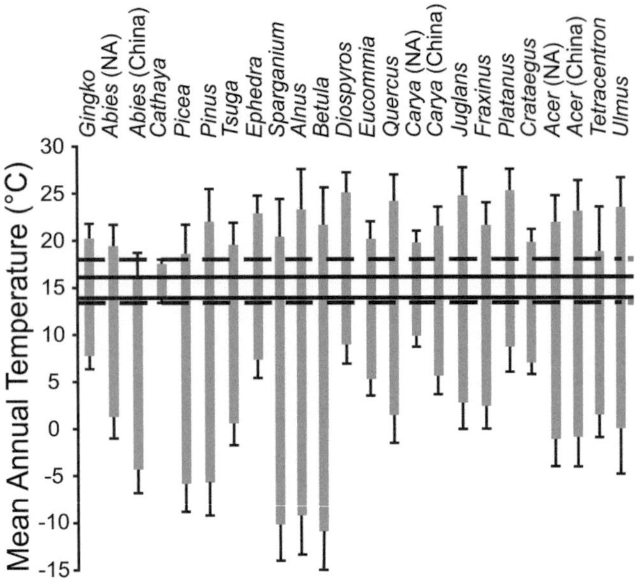

Fig. 13.4 Coexistence approach analysis of the latest Eocene Florissant Fossil Beds flora (Baumgartner and Meyer 2014). The dashed line indicates the estimated coexistence interval using the full range of climate data, while the solid line indicates the estimated coexistence interval between the 10–90th percentiles. Both coexistence intervals are limited by *Abies* and *Cathaya*; these MAT intervals are estimated as 13.4–18.0°C (dashed line) and 13.9–16.2°C (solid line). The age of the flora precludes assigning nearest living relatives at the species level and some genera (i.e. *Betula* and *Alnus*) have relatively large climatic ranges for MAT. There are no outliers in this example

Facility (GBIF, http://www.gbif.org/) and the WorldClim global climate models (Hijmans et al. 2005). However, unlike CA, which uses the entire climatic range of all taxa to calculate a climatic variable, Bioclimatic Analysis initially limited coexistence intervals to the 10th (lower limit) to 90th (upper limit) or 5th to 95th percentile values of the total climatic range for all of the nearest living relative taxa (Kershaw and Nix 1988; Kershaw 1997; Greenwood et al. 2003, 2005). Subsequent iterations of bioclimatic analysis, however, calculated the estimate as the mean value of the 10th and 90th percentile of the maximum and minimum values across all taxa recorded in a sample, with the total range between these plotted as an "error bar" (e.g., Greenwood et al. 2005; Eldrett et al. 2009, 2014; Pross et al. 2012; Smith et al. 2012; Reichgelt et al. 2013).

Climate Reconstruction Analysis using Coexistence Likelihood Estimation (CRACLE) is a new quantitative NLR method that uses global scale specimen collection data to infer species climate tolerance by calculating the maximum joint likelihood of species coexistence (Harbert and Nixon 2015). CRACLE uses species distribution data from specimen collections obtained from the GBIF and the WorldClim global climate models (Hijmans et al. 2005) to determine species-climate tolerance profiles. When compared to other quantitative NLR methods, MAT estimates made using CRACLE are more accurate and precise suggesting this method may offer considerable improvements over other NLR approaches. However, unless modified, CRACLE is only applicable to the Holocene and late Pleistocene because it requires modern geographic distributions of fossil taxa to make climatic estimates. Additionally, a major limitation of CRACLE is that for it to be applied to the fossil record, it assumes realized niche stasis, which has been demonstrated to be invalid (e.g., Jackson and Overpeck 2000; Veloz et al. 2012), and that plant communities are in equilibrium with climate, which also may not always be a valid assumption (e.g., Svenning and Sandel 2013; Rokataorinivo et al. 2013; Blonder et al. 2015b).

The mutual climatic range (MCR) method estimates paleoclimate from the present climatic tolerance ranges of the nearest living relative of fossil taxa and originally was methodologically nearly identical to CA (Sinka and Atkinson 1999; Sharpe 2002). More recently, Thompson et al. (2012) modified the method and currently MCR uses factor-analysis based climate space to plot coldest month temperature against coldest month precipitation and warmest month temperature against warmest month precipitation and the estimated temperature and precipitation for a given time of year (i.e., coldest month, warmest month) is made based on where the climate space overlaps. The MCR method can only be applied to the Quaternary and has only been applied to a few fossil assemblages (Thompson et al. 2012). Applications to modern coleopteran assemblages demonstrates the estimates made using MCR are conservative when compared to meteorological data, but that the method has systematic biases (winter estimates too warm, summer estimates to cool in Britain) (Thompson et al. 2012). Additionally, if neither the absence nor proportion of taxa in the flora can be used to delimit a smaller climatic range, MCR estimates an overly large climatic range (Thompson et al. 2012).

Shortcomings and limitations of nearest living relative methods: All NLR methods rely heavily on accurate climate data for the distribution of the living relative, and precise taxonomic identification. However, resolution of climate estimates decreases at higher taxonomic levels: species-level analyses tend to produce narrow climate intervals, while family-level analyses tend to produce wide climate intervals. Additionally, species turnover within a genus that leads to niche shifts and changes in realized niches through time (e.g., Veloz et al. 2012), are not captured be this approach. This is a problem when trying to apply NLR methods to the fossil record because while younger fossil floras, such as those from the Quaternary, often have a NLR identified at the genus or even species level, older fossil floras, such as those from the Paleogene and Neogene, are often associated with a NLR at the genus or family level where climate intervals are larger and assumptions about the relative stasis of modern climatic distributions of a genus or family may not be appropriate. As a result, NLR methods are typically restricted to Cenozoic floras, though the methods have sometimes been used on assemblages from the late Cretaceous (e.g., Mosbrugger and Utescher 1997; Utescher et al. 2014; Fletcher et al. 2014).

Although NLR methods, and in particular CA, have been applied to a large number of Cenozoic floras (e.g., Utescher et al. 2014), they have recently come under major criticism based on their assumptions about species distributions, the lack of statistical framework, and how climate variables are calculated for past plant communities based on modern plant distributions (Grimm and Potts 2016). Additionally, there are significant issues with the PFDB, which is used in CA. Studies by Grimm and Denk (2012), Grimm et al. (2016) and Grimm and Potts (2016) have shown that the MAT ranges in the PFDB are incorrect when compared to climate stations or WorldClim (Hijmans et al. 2005) and even when using corrected climate data the precision of MAT ranges was between 5–10°C (Grimm and Denk 2012). These criticisms raise major questions about the ability of quantitative NLR methods to accurately and precisely reconstruct the paleoclimate of ancient plant communities.

Leaf Traits and Ecology

In addition to a relationship between leaf traits and climate, there are a suite of leaf structural and functional traits that are correlated to each other and to environmental gradients. This covariation of leaf traits is known as the leaf economic spectrum (LES; Wright et al. 2004, 2005; Reich 2014). The LES is a framework for organizing leaf functional ecology, which reflects a series of trade-offs among leaf traits that are related to carbon balance. Plants on the "fast-return" end of the spectrum, characterized by rapid resource acquisition, typically have a short leaf life-span (less than 12 months), high photosynthetic and respiration rates, low leaf mass per area (M_A), high nutrient concentrations (principally nitrogen and phosphorus), and fast growth rates, while the inverse is true for plants with slow resource acquisition on the "slow-return" end of the spectrum (Reich et al. 1997; Westoby et al. 2002; Diaz et al. 2004; Wright et al. 2004, 2005; Reich 2014). The interrelationships between these traits are largely independent of phylogeny in seed plants (Ackerly and Reich 1999), have been observed in all vascular plant groups (Wright et al. 2004; Karst and Lechowicz 2007; Peppe et al. 2014), and probably reflect evolutionary trade-offs in leaf design to maximize resource acquisition vs. performance for the environment in which a plant is living (Shipley et al. 2006).

Recently, researchers have suggested that venation networks are linked to the LES either as the underlying determinant of leaf mass per area, leaf life-span, carbon assimilation rate, and nitrogen concentration ("vein origin" hypothesis; Blonder et al. 2011, 2013; Blonder et al. 2014a) or that venation networks are only one axis of many that determine hydraulic conductance and carbon economics, and in turn the plant's relative growth rate ("flux trait network" hypothesis; Sack and Scoffoni 2013; Sack et al. 2013, 2014). Regardless of which hypothesis is correct, both highlight the importance of vein networks in the LES.

While in modern leaves these LES variables are easily measured, they cannot be directly measured in fossil leaves. Instead a series of different proxies based on features that are measureable in fossil leaves, such as leaf area, petiole width, and vein density, have been developed to estimate leaf mass per area M_a, leaf life-span (LLS), carbon assimilation rate, and respiration rate (e.g., Brodribb et al. 2007; Boyce et al. 2009; Brodribb and Feild 2010; Royer et al. 2007, 2010, 2012; Blonder et al. 2011; Feild et al. 2011b; Sack and Scoffoni 2013; Sack et al. 2013, 2014; Blonder and Enquist 2014; Peppe et al. 2014). These reconstructed LES variables can then be used to reconstruct aspects of the paleoecology of the fossil plant (e.g., leaf life-span, hydraulic conductance), as well as other features of the ecosystem and environment in which the plant lived. Since the majority of these methods are developed based on relationships within the LES, they are applicable across a variety of plant groups and can be used throughout the geologic record (e.g., Royer et al. 2007, 2010, 2012; Boyce et al. 2009; Brodribb and Feild 2010; Blonder et al. 2011, 2014b; Feild et al. 2011b; Crifo et al. 2014; Peppe et al. 2014), though some aspects of the LES, such as carbon assimilation and respiration rates are probably heavily dependent on past levels of atmospheric CO_2 and O_2 (e.g., De Boer et al. 2012; McElwain et al. 2016).

Methods of Paleoecological Reconstruction

Most leaf economic spectrum traits cannot be directly measured in fossils. However, a proxy for M_a, which is a core LES character and also correlates strongly with leaf thickness and leaf life-span (Wright et al. 2004; Royer et al. 2012), has been developed that is applicable to fossil plants (Royer et al. 2007, 2010; Peppe et al. 2014). The method uses the scaling relationship between petiole width (PW), leaf area (A), and M_a. The relationship between PW and A to M_a is based on the biomechanics of the leaves since heavier leaves need more support, which necessitates larger, thicker petioles (Royer et al. 2007, 2010; Peppe et al. 2014). Leaves with high M_a are more commonly evergreen and thicker,

Table 13.2 Leaf mass per area models for woody dicot angiosperms, herbaceous angiosperms, broadleaf gymnosperms, and ferns

Taxonomic Group	Model	a	b	r^2	Authors
Woody dicot angiosperms	$\log(M_A) = a + b\log(PW^2/A)$	3.07	0.382	0.55	Royer et al. (2007)
Herbaceous angiosperms	$\log(M_A) = a + b\log(PW^2/A)$	3.015	0.3076	0.44	Royer et al. (2010)
Broadleaf gymnosperms	$\log(M_A) = a + b\log(PW^2/A)$	2.245	0.2204	0.32	Royer et al. (2010)
Ferns	$\log(M_A) = a + b\log(PW^4/A)$	4.207	0.252	0.44	Peppe et al. (2014)

which in addition to providing structural support also allows long-term nutrient transport (Wright et al. 2004).

Proxies for leaf mass per area have been developed for all woody dicot angiosperms (Royer et al. 2007), herbaceous angiosperms, broadleaf gymnosperms (Royer et al. 2010), and ferns (Peppe et al. 2014) (Table 13.2). Each of these groups has significantly different scaling relationships between PW and A to M_a and the relationships in angiosperms and broadleaf gymnosperms are explained by a different biomechanical model than ferns. The relationships between area normalized PW and M_a in angiosperms and broadleaf gymnosperms are best explained by modeling a leaf as a pole supporting a singular mass and assumes that the cross-sectional area of the petiole is proportional to leaf mass (Royer et al. 2007, 2010). The relationships between PW and A to M_a in ferns are best explained by a cantilever beam model that assumes the petiole width and length are independent, that petiole shape is relatively invariant, and that the leaf mass is proportional to the flexural rigidity of the entire petiole (Peppe et al. 2014).

The M_a of fossil leaves can be used to estimate a number of paleoecological variables. First, greater M_a is correlated with longer leaf life-span and can be used to reconstruct the leaf life-span of a fossil flora (Royer et al. 2007, 2010; Peppe et al. 2014). Fossil leaves with a $M_a < \sim 87$ g m^{-2} can be broadly categorized as deciduous and those with a $M_a > \sim 129$ g m^{-2} can be broadly categorized as evergreen, with the transition between predominately deciduous to predominately evergreen occurring at ~ 111 g m^{-2} (Royer et al. 2007). The variance in M_a and leaf life-span can also be used to infer seasonality in paleoecosystems with greater variance in M_a in plant populations related to greater seasonality (Wright et al. 2005). Additionally, the distribution of M_a of a flora also appears to be environmentally specific (Royer et al. 2010), suggesting that the distribution of M_a in a fossil flora can be compared to modern distributions and used as a proxy for environment. Finally, M_a correlates with leaf nitrogen content (Royer et al. 2005), and could potentially be used as a proxy for percent nitrogen, which could be used to infer nutrient cycling in fossil forest ecosystems.

The interaction of plants and their herbivores has a long geologic record and can be used to infer a variety of ecological variables (Labandeira and Currano 2013). Reconstructions of M_a combined with measurements of insect-plant interactions via their feeding traces can be used to reconstruct aspects of plant ecology such as nutrient content and ecosystem disturbance (Royer et al. 2007; Currano et al. 2008, 2011). Leaves with greater M_a have lower diversity, lower density, and lower area removed by insect herbivores when compared to leaves with low M_a (Coley 1983; Royer et al. 2007). This relationship between insect damage and M_a is probably because leaves with high M_a are typically thicker and/or tougher, possess a greater amount and concentration of chemical deterrents, and have lower leaf nitrogen content, which makes them less nutritious and less palatable for insect herbivores (Coley 1983; Royer et al. 2007). Additionally, changes in the diversity and density of insect feeding traces can be used to infer large-scale changes in ecology following ecological disturbances, including changes in climate and recovery after mass extinction events (e.g., Wilf et al. 2006; Currano et al. 2008, 2011; Donovan et al. 2016).

Recent studies have suggested a link between vein networks and the LES (Blonder et al. 2011, 2013, 2014); however, this remains controversial (e.g., Sack and Scoffoni 2013; Sack et al. 2013, 2014; Li et al. 2015). If there is a relationship between vein networks and the leaf economic spectrum, measurements of vein networks in fossil plants could be used to reconstruct other LES variables, such as leaf life-span and M_a, which in turn would allow for paleoecological reconstruction. These proxies are in development, but offer considerable promise. In addition to its relationship to the LES, vein networks, and specifically vein density, is a proxy for water use efficiency and gas exchange, making it a powerful tool for assessing gas and water exchange capacity of plants (Sack et al. 2005; Sack and Frole 2006; Brodribb et al. 2007; Boyce et al. 2009; Brodribb and Feild 2010). For example, angiosperms have much higher vein densities than plants from all other groups, living or extinct, which allows angiosperms to have much higher rates of transpiration (Boyce et al. 2009). This connection between high vein density and high rates of transpiration in angiosperms has been used to infer that the evolution of angiosperms and their increase in maximum leaf vein density through the Cretaceous enabled the evolution and expansion of tropical rain forests, because transpiration rates in other plant groups are too low to maintain the precipitation abundance and evenness necessary for tropical rainforest ecosystems (Boyce et al. 2009, 2010; Feild et al. 2011a). The distribution of vein densities within a flora differs significantly between canopy and understory plants and between sites with different forest structures (Crifo et al. 2014). Thus, it is possible to infer ancient forest structure, and its constituent understory and canopy taxa, from the range of vein densities within a fossil flora (Crifo et al. 2014).

Shortcomings and Limitations of Paleoecological Methods

Methods of paleoecological reconstruction are very powerful tools for assessing aspects of plant ecology (e.g., leaf life-span, hydraulic conductance) and a ecosystem's environment (e.g., ecosystem disturbance, forest structure); however, these methods are relatively new and have not been widely applied. The primary reason for this is that they require excellent preservation of venation (vein density proxies), and the entire leaf area and at least part of the leaf petiole (M_A proxies). Due to taphonomic constraints, these proxies cannot be applied to many fossil floras, or in some cases, can only be applied to the best preserved specimens within a flora such that the reconstructed ecological variables may not be representative of the entire flora. As an example, Blonder et al. (2014b) were only able to analyze <20% of more than 6,000 collected fossil specimens, suggesting the potential for significant taphonomic biases in leaf vein density analyses. Additionally, though there appears to be a relationship between leaf vein density and forest structure (Crifo et al. 2014) and M_A and environment (Royer et al. 2010), these relationships are based on relatively limited datasets with small sample sizes. Thus, it's unclear if these relationships are representative globally and to all ecosystems.

Application of Paleoclimate and Paleoecological Proxies

Leaf physiognomic methods for reconstructing paleoclimate have been applied hundreds, if not thousands of times to fossil floras from the Cretaceous through the Quaternary, and leaf trait methods for reconstructing paleoecology have begun to be applied to fossil plants throughout Earth's history. The following is a brief example of an application of these methods for reconstructing paleoclimate and paleoecology by Michel et al. (2014) that made it possible to reconstruct the climate and to determine the climatic, environmental, and ecological biome of the fossil site. The study of Michel et al. (2014) documented a diverse ssemblage of fossil vertebrates, leaves, and tree stump and root casts and used the fossil data to reconstruct the paleoclimate and paleoenvironment of the Early Miocene R3 fossil site on Rusinga Island, western Kenya. The R3 fossil site is of particular interest because of the co-occurrence of multiple species of Early Miocene catarrhines, including *Dendropithecus macinnesi* and *Ekembo heseloni*, as well as several vertebrate fossil taxa that suggest a more closed environment. Prior to the work of Michel et al. (2014), the paleoenvironment of the site was relatively poorly constrained. Michel et al. (2014) mapped a series of fossil tree stump casts of varying diameter preserved within the same stratigraphic layer across the R3 site. Based on the density of the tree stump casts across the landscape, they reconstructed the ecosystem as a dense, closed-canopy, multi-storied forest. Abundant fossil leaves were found in a thin mudstone layer stratigraphically directly above the fossil tree stump casts. The flora was made up of 29 morphotypes, which can be considered a rough proxy for biological species (see Ash et al. 1999 and Peppe et al. 2008 for a review of the morphotype method), representing 27 dicot and 2 monocotyledonous angiosperms. The dicots were used to reconstruct the paleoclimate and paleoecology of the site.

The mean annual temperature and mean annual precipitation of the site was reconstructed using LMA and LAA, respectively (Fig. 13.5). Mean annual temperature reconstructed using the LMA regressions of Wolfe (1979), Miller et al. (2006), Kowalski and Dilcher (2003), and Peppe et al. (2011) ranged between 22.6°C and 34.5°C (Fig. 13.5). Mean annual precipitation reconstructed using the LAA

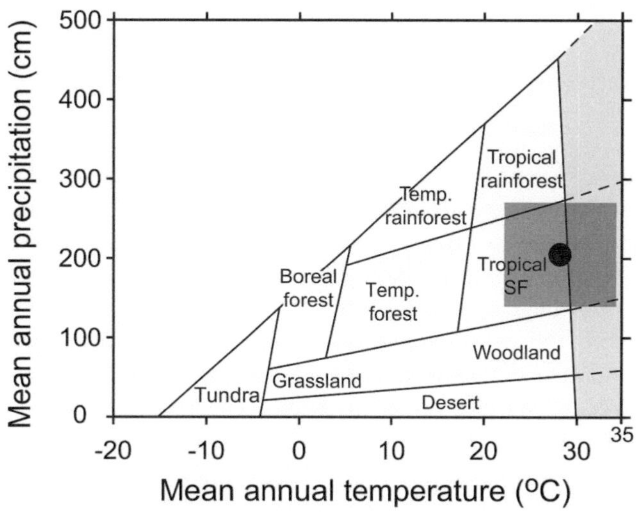

Fig. 13.5 Climatic reconstruction for the Rusinga R3 fossil site from Michel et al. (2014). Black circle is the average is of all mean annual temperature and mean annual precipitation estimates. Shaded square is the range of all climate estimates. Biomes follow Whittaker (1975). Area shaded in light gray indicates mean annual temperatures higher than typical conditions on Earth today. The dashed lines represent projections of biomes outside of modern climate space

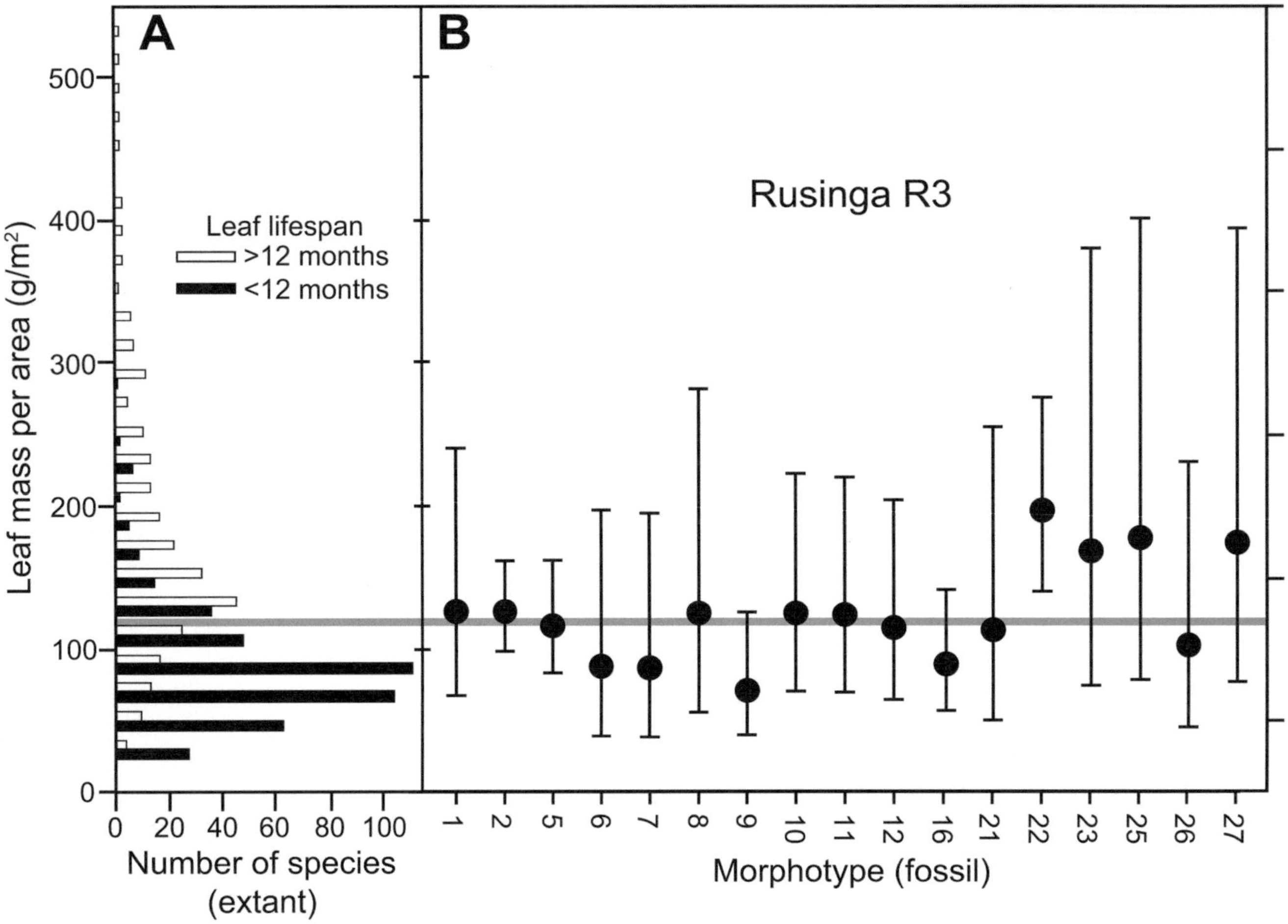

Fig. 13.6 Estimated leaf mass per area (M_a) for Rusinga R3 flora and comparison to extant vegetation. A, M_a and leaf lifespan for 678 species of angiosperms, ferns, and gymnosperms modified from Royer et al. (2010). M_a and leaf lifespan data from Wright et al. (2004). M_a bin is 20 g/m^2. Gray line at 129 g/m^2 indicates the upper bound of the transition between species whose leaves are predominately deciduous (leaf life-span <12 month) versus those that are predominately evergreen (leaf lifespan >12 months). B, Estimated M_a for 17 morphotypes from the Rusinga R3 flora showing that the majority of morphotypes were likely evergreen (10 of 17 reconstructed). Errors are 95% prediction intervals. M_a data and morphotypes from Michel et al. (2014)

regression of Wilf et al. (1998), Gregory-Wodzicki (2000), Jacobs and Herendeen (2004) and Peppe et al. (2011) ranged from 1,394 to 2,618 mm (Fig. 13.5). Wet month precipitation estimates using the method of Jacobs and Herendeen (2004) indicate that there was relatively little precipitation during the dry season. This evidence for seasonality is also supported by analyses of the paleosols at the R3 site. The paleoclimate reconstructions for the site indicate a tropical seasonal forest biome (Fig. 13.5; Whittaker 1975). Interestingly, the forest density evidence for the R3 site are very similar to modern tropical seasonal forests that support medium and large bodied primate communities, such as Barro Colorado Island (BCI), Panama and Ituri Forest, Democratic Republic of Congo. Only univariate methods were utilized on the R3 flora because, as discussed above, the leaf physiognomic space of the fossil flora is not captured by the modern DiLP dataset (Fig. 13.3). This demonstrates one of the limitations of the existing multivariate methods, and the difficulty of reconstructing the paleoclimate of African fossil floras.

Leaf mass per area was reconstructed for seventeen of the twenty seven dicot morphotypes from R3 (Figs. 13.6, 13.7). Leaf mass per area reconstructions for the flora indicate that the species were predominantly evergreen, but that there was an important component of deciduous taxa (Fig. 13.6). The distribution of M_a at the site is most similar to non-riparian sites (Fig. 13.7), which is unsurprising given the sedimentological and paleosol evidence from the site. The paleoclimate reconstructions for the flora are most similar to BCI site (MAT = 26.5°C, MAP = 2620 mm), and the distribution of M_a at the sites is similar, though BCI has a higher proportion of species low M_a than the R3 flora (Fig. 13.7D). The R3 flora is also somewhat similar to the non-riparian, wet tropical rainforest of Buena Vista, Puerto Rico, though

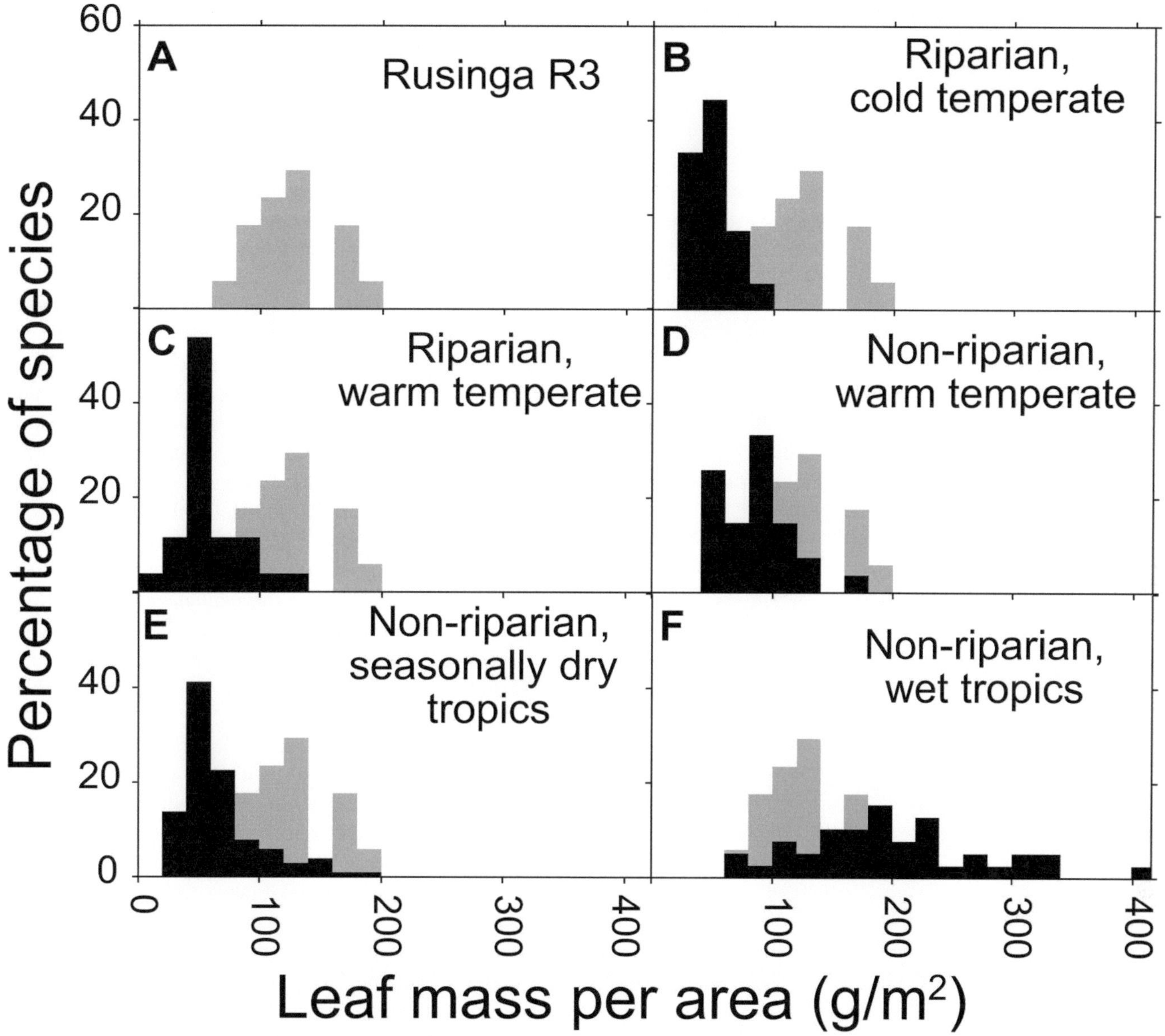

Fig. 13.7 Comparison of the reconstructed M_a distribution of the Rusinga R3 flora with extant sites from varying climate biomes and environments. M_a and climate data for extant sites from Peppe et al. (2011). The fossil and extant sites are comprised of exclusively wood dicot angiosperms. The M_a bin size is 20 g/m^2. A, Rusinga R3 (replicated in gray in all panels). B, Connecticut River, near Middletown, Connecticut, USA. C, Big Hammock Wildlife Refuge Area, Georgia, USA. D, Pee Dee State Park, South Carolina, USA; E, Barro Colorado Island, Panama. F, Buena Vista, Puerto Rico

the wet tropics site has a larger proportion of species with much higher M_a than reconstructed for the R3 flora (Fig. 13.7E). Similar to the paleoclimate estimates, the M_a reconstructions for R3 also suggest a non-riparian tropical seasonal forest or tropical rainforest biome.

The paleoclimate and paleoecological reconstructions for the R3 fossil site using fossil leaves indicate a tropical seasonal forest biome and a warm and seasonally wet climate (Figs. 13.5, 13.6, 13.7). These results are also in agreement with reconstructions of the forest density and paleonvironmental reconstructions based on the paleosols at the site and the composition of the vertebrate fauna. Together, the multiproxy evidence presented by Michel et al. (2014) indicates that the R3 fossil locality sampled a widespread, dense, multistoried, closed-canopy tropical seasonal forest set in a warm and seasonally wet climate. When combined with the primate fossil record from the R3 site, the paleoclimate and paleoecological results of Michel et al. (2014) demonstrate the importance of forested environments in the evolution of early apes in Africa. The Michel et al. (2014) study is an example of the power and utility of combining fossil leaf based estimates of paleoclimate and paleoecology to reconstruct a terrestrial ecosystem, and is a good model for fossil leaf based research focused on terrestrial ecosystem and climate reconstruction.

Future Research Directions

Fossil plants, and particularly fossil leaves, are powerful tools for reconstructing paleoclimate and paleoecology. However, despite the power of the paleoclimate and paleoecology methods that can be applied to fossils, they have important limitations.

One of the most important potential shortcomings of leaf physiognomic methods is the possible influence of phylogeny and life history on leaf trait-climate relationships. Recently, the vertebrate paleontology community has begun to develop a series of trait-based approaches that use relationships between functional traits and environment to reconstruct climate and assess the responses of faunal communities to environmental change called ecometrics (e.g., Eronen et al. 2010a; Eronen et al. 2010b; Polly et al. 2011; Lawing et al. 2012, 2016; Polly and Sarwar 2014). These methods have addressed potential issues related to phylogenetic influences on functional traits by resampling techniques and by applying phylogenetic corrections to the datasets (Eronen et al. 2010b; Lawing et al. 2012; Polly and Sarwar 2014; Lawing et al. 2016). The utilization of these types of methods when revising and developing new leaf physiognomic methods for reconstructing paleoclimate has considerable potential and is an important future research direction. Additionally, as noted above, the leaf physiognomic-climate relationships are different between the Northern and Southern Hemispheres. The development of new calibration datasets that incorporate more floras from the Southern Hemisphere, and particularly Africa and South America, and from tropical regions will add phylogenetic diversity to the calibration datasets and may make it possible to ascertain the driver behind the hemispheric differences, especially if phylogenetic corrections are applied to the datasets.

Leaf life-span also has an important effect on leaf trait-climate relationships. Royer et al. (2012) found that the incorporation of leaf thickness, deciduousness, and M_A into species-specific leaf physiognomic models reduced the standard errors of the models. However, this result was based on a species-based dataset, instead of a site-based calibration, on which all leaf physiognomic models are based. Thus, it's uncertain if these results are broadly applicable to existing leaf physiognomic models. Using a dataset of taxa from 17 floras from a range of MATs, Blonder and Enquist (2014) demonstrated a relationship between vein density and MAT, which they used to develop a predictive model for MAT. However, the relationships between vein density and MAT were different in temperate and tropical floras, such that they developed one model with different parameter values for temperature floras and tropical floras. Taken together, the results of Royer et al. (2012) and Blonder and Enquist (2014) suggest that the incorporation of functional traits, such as vein density and M_A, into leaf physiognomic models could improve their precision and accuracy and are an interesting research direction to pursue.

In addition to potential shortcomings with the methods, making the measurements necessary to apply paleoclimate and paleoecological proxies can be time consuming and trait scoring can be ambiguous. Recently, there has been a push to develop computer-based algorithms that are automated to measure and score leaf traits and to identify plants using leaves (e.g., Cope et al. 2012; Corney et al. 2012a, b; Price et al. 2012; Green et al. 2014; MacLeod and Steart 2015; Wilf et al. 2016; Liao et al. 2017). Automated measurements of leaf size and shape would eliminate ambiguities in leaf trait scoring, increase scoring throughout, make it easier to measure continuous variables, and potentially help aid in the identification of fossil plants. The potential of automated or semi-automated leaf trait scoring and identification is an exciting new development that has the potential to revolutionize the way leaf traits are scored and help to significantly improve leaf trait-climate and ecological proxies.

One of the fundamental shortcomings of leaf-based climate reconstruction is that we still do not understand the physiological and/or evolutionary mechanisms that drive the empirical relationships observed between leaf physiognomy and climate. As a result, leaf physiognomic proxies are empirically derived and traits used for extrapolation may not have a direct functional link to climate or ecology and/or may be strongly driven by phylogeny and life history. This might lead to biased paleoclimate and paleoecological estimates, particularly in deep time. Research focused on the evolution of leaf form and on the genetic underpinnings of leaf shape and development is making significant strides towards addressing this shortcoming (e.g., Pérez-Pérez et al. 2010; Schmerler et al. 2012; Edwards and Donoghue 2013; Chitwood et al. 2014, 2016a, b; Dkhar and Pareek 2014; Rodriguez et al. 2014; Vaughan 2015; Chitwood and Sinha 2016; Coneva et al. 2017). A more complete understanding of the genetic controls on leaf size and shape and the drivers of leaf shape evolution will make it possible for us to better understand how environment and climate influence leaf physiognomy, as well as provide a basis for understanding plant species' distributions on climatic gradients. This will allow for the development of phylogenetically and physiologically informed plant-based paleoclimatic and paleoecological proxies, which has the potential to revolutionize the field and our ability to reconstruct ancient climate and ecology.

Acknowledgements We would like to thank Denise Su, Scott Simpson, and Darin Croft for inviting us to contribute to the *Latest Methods in Reconstructing Cenozoic Terrestrial Environments and Ecological Communities* workshop and to this special volume. We thank Dana Royer for many helpful discussions about these topics and review of this manuscript and David Greenwood for his helpful and constructive review. This work was supported by the National Science Foundation (grant EAR-1325552).

References

Ackerly, D. D., & Reich, P. B. (1999). Convergence and correlations among leaf size and function in seed plants: a comparative test using independent contrasts. *American Journal of Botany, 86,* 1272–1281.

Adams, J. M., Green, W. A., & Zhang, Y. (2008). Leaf margins and temperature in the North American flora: recalibrating the paleoclimatic thermometer. *Global and Planetary Change, 60,* 523–534.

Adams, J. S., Kraus, M. J., & Wing, S. L. (2011). Evaluating the use of weathering indices for determining mean annual precipitation in the ancient stratigraphic record. *Palaeogeography, Palaeoclimatology, Palaeoecology, 309,* 358–366.

Aizen, M. A., & Ezcurra, C. (2008). Do leaf margins of the temperate forest flora of southern South America reflect a warmer past? *Global Ecology and Biogeography, 17,* 164–174.

Ash, A., Ellis, B., Hickey, L. J., Johnson, K. R., Wilf, P., & Wing, S. (1999). Manual of leaf architecture – morphological description and categorization of dicotyledonous and net-veined monocotyledonous angiosperms (65 p.). Washington D. C.: The Leaf Architecture Working Group, Smithsonian Institution.

Axelrod, D. I. (1948). Climate and evolution in western North America during middle Pliocene time. *Evolution*, 127–144.

Axelrod, D. I. (1952). A theory of angiosperm evolution. *Evolution,* 29–60.

Axelrod, D. I., & Bailey, H. P. (1969). Paleotemperature analysis of tertiary floras. *Palaeogeography, Palaeoclimatology, Palaeoecology, 6,* 163–195.

Bailey, I. W., & Sinnott, E. W. (1915). A botanical index of Cretaceous and Tertiary climates. *Science, 41,* 831–834.

Bailey, I. W., & Sinnott, E. W. (1916). The climatic distribution of certain types of angiosperm leaves. *American Journal of Botany, 3,* 24–39.

Baker-Brosh, K. F., & Peet, R. K. (1997). The ecological significance of lobed and toothed leaves in temperate forest trees. *Ecology, 78,* 1250–1255.

Baumgartner, K. A., & Meyer, H. W. (2014). Coexistance climate analysis of the late Eocene Florissant flora, Colorado. *Geological Society of America – Abstract with Programs, 46,* 490.

Billings, F. H. (1905). Precursory leaf serrations of *Ulmus*. *Botanical Gazette*, 224–225.

Blonder, B., & Enquist, B. J. (2014). Inferring climate from angiosperm leaf venation networks. *New Phytologist, 204,* 116–126.

Blonder, B., Nogués-Bravo, D., Borregaard, M. K., Donoghue, I., John, C., Jørgensen, P. M., et al. (2015a). Linking environmental filtering and disequilibrium to biogeography with a community climate framework. *Ecology, 96,* 972–985.

Blonder, B., Nogues-Bravo, D., Borregaard, M., Lessard, J.P., Violle, C., Svenning, J.C., et al. (2015b). Linking environmental filtering and disequilibrium to biogeography with a community climate. *Ecology, 96*, 972–985.

Blonder, B., Violle, C., Bentley, L. P., & Enquist, B. J. (2011). Venation networks and the origin of the leaf economics spectrum. *Ecology Letters, 14,* 91–100.

Blonder, B., Violle, C., Bentley, L. P., & Enquist, B. J. (2014a). Inclusion of vein traits improves predictive power for the leaf economic spectrum: a response to Sack et al. (2013). *Journal of Experimental Botany, 65*, 5109–5114.

Blonder, B., Royer, D. L., Johnson, K. R., Miller, I., & Enquist, B. J. (2014b). Plant ecological strategies shift across the Cretaceous-Paleogene boundary. *PLoS Biology, 12,* e1001949.

Blonder, B., Violle, C., & Enquist, B. J. (2013). Assessing the causes and scales of the leaf economics spectrum using venation networks in *Populus tremuloides*. *Journal of Ecology, 101,* 981–989.

Borregaard, M., Lessard, J. P., Violle, C., Svenning, J. C., Rahbek, C., & Enquist, B. (2015). Linking environmental filtering and disequilibrium to biogeography with a community climate. *Ecology, 96,* 972–985.

Boyce, C. K., Brodribb, T. J., Feild, T. S., & Zwieniecki, M. A. (2009). Angiosperm leaf vein evolution was physiologically and environmentally transformative. *Proceedings of the Royal Society B: Biological Sciences, 276,* 1771–1776.

Boyce, C. K., Lee, J.-E., Feild, T. S., Brodribb, T. J., & Zwieniecki, M. A. (2010). Angiosperms helped put the rain in the rainforests: the impact of plant physiological evolution on tropical biodiversity 1. *Annals of the Missouri Botanical Garden, 97,* 527–540.

Brodribb, T. J., & Feild, T. S. (2010). Leaf hydraulic evolution led a surge in leaf photosynthetic capacity during early angiosperm diversification. *Ecology Letters, 13,* 175–183.

Brodribb, T. J., Feild, T. S., & Jordan, G. J. (2007). Leaf maximum photosynthetic rate and venation are linked by hydraulics. *Plant Physiology, 144,* 1890–1898.

Burnham, R. J., Ellis, B., & Johnson, K. R. (2005). Modern tropical forest taphonomy: does high biodiversity affect paleoclimatic interpretations? *PALAIOS, 20,* 439–451.

Burnham, R. J., Pitman, N. C. A., Johnson, K. R., & Wilf, P. (2001). Habitat-related error in estimating temperatures from leaf margins in a humid tropical forest. *American Journal of Botany, 88,* 1096–1102.

Burnham, R. J., & Tonkovich, G. S. (2011). Climate, leaves, and the legacy of two giants. *New Phytologist, 190,* 514–517.

Canny, M. (1990). Tansley review no. 22 what becomes of the transpiration stream? *New Phytologist, 114,* 341–368.

Chaloner, W. G., & Creber, G. T. (1990). Do fossil plants give a climatic signal? *Journal of the Geological Society, 147,* 343–350.

Chaney, R. W. (1938). Paleoecological interpretations of Cenozoic plants in western North America. *The Botanical Review, 4,* 371–396.

Chen, W.-Y., Su, T., Adams, J. M., Jacques, F. M., Ferguson, D. K., & Zhou, Z.-K. (2014). Large-scale dataset from China gives new insights into leaf margin–temperature relationships. *Palaeogeography, Palaeoclimatology, Palaeoecology, 402,* 73–80.

Chitwood, D. H., Ranjan, A., Martinez, C. C., Headland, L. R., Thiem, T., Kumar, R., et al. (2014). A modern ampelography: a genetic basis for leaf shape and venation patterning in grape. *Plant Physiology, 164,* 259–272.

Chitwood, D. H., Klein, L. L., O'Hanlon, R., Chacko, S., Greg, M., Kitchen, C., et al. (2016a). Latent developmental and evolutionary shapes embedded within the grapevine leaf. *New Phytologist, 210,* 343–355.

Chitwood, D. H., Rundell, S. M., Li, D. Y., Woodford, Q. L., Tommy, T. Y., Lopez, J. R., et al. (2016b). Climate and developmental plasticity: interannual variability in grapevine leaf morphology. *Plant Physiology, 170,* 1480–1491.

Chitwood, D. H., & Sinha, N. R. (2016). Evolutionary and environmental forces sculpting leaf development. *Current Biology, 26,* R297–R306.

Coley, P. D. (1983). Herbivory and defensive characteristics of tree species in a lowland tropical forest. *Ecological monographs, 53,* 209–234.
Comes, H. P., & Kadereit, J. W. (1998). The effect of Quaternary climatic changes on plant distribution and evolution. *Trends in Plant Science, 3,* 432–438.
Coneva, V., Frank, M. H., de Luis Balaguer, M. A., Li, M., Sozzani, R., & Chitwood, D. H. (2017). Genetic architecture and molecular networks underlying leaf thickness in desert-adapted tomato *Solanum pennellii*. *Plant Physiology, 175,* 376–391.
Cope, J. S., Corney, D., Clark, J. Y., Remagnino, P., & Wilkin, P. (2012). Plant species identification using digital morphometrics, A review. *Expert Systems with Applications, 39,* 7562–7573.
Corney, D., Clark, J. Y., Tang, H. L., & Wilkin, P. (2012a). Automatic extraction of leaf characters from herbarium specimens. *Taxon, 61,* 231–244.
Corney, D. P., Tang, H. L., Clark, J. Y., Hu, Y., & Jin, J. (2012b). Automating digital leaf measurement: the tooth, the whole tooth, and nothing but the tooth. *PLoS ONE, 7,* e42112.
Couturier, E., Du Pont, S. C., & Douady, S. (2011). The filling law: a general framework for leaf folding and its consequences on leaf shape diversity. *Journal of Theoretical Biology, 289,* 47–64.
Couturier, E., Brunel, N., Douady, S., & Nakayama, N. (2012). Abaxial growth and steric constraints guide leaf folding and shape in Acer pseudoplatanus. *American Journal of Botany, 99,* 1289–1299.
Crifo, C., Currano, E. D., Baresch, A., & Jaramillo, C. (2014). Variations in angiosperm leaf vein density have implications for interpreting life form in the fossil record. *Geology, 42,* 919–922.
Christophel, D. C., & Blackburn, D. T. (1978). Tertiary megafossil flora of Maslin Bay, South Australia: a preliminary report. *Alcheringa, 2,* 311–319.
Critchfield, W. B. (1960). Leaf dimorphism in *Populus trichocarpa*. *American Journal of Botany, 47,* 699–711.
Critchfield, W. B. (1971). Shoot growth and heterophylly in *Acer*. *Journal of the Arnold Arboretum, 52,* 240–266.
Currano, E. D., Jacobs, B. F., Pan, A. D., & Tabor, N. J. (2011). Inferring ecological disturbance in the fossil record: a case study from the Late Oligocene of Ethiopia. *Palaeogeography, Palaeoclimatology, Palaeoecology, 309,* 242–252.
Currano, E. D., Wilf, P., Wing, S. L., Labandeira, C. C., Lovelock, E. C., & Royer, D. L. (2008). Sharply increased insect herbivory during the Paleocene-Eocene Thermal Maximum. *Proceedings of the National Academy of Sciences, USA, 105,* 1960–1964.
Davis, M. B. (1986). Climatic instability, time, lags, and community disequilibrium. In J. Diamond & T. J. Case (Eds.), *Community ecology* (pp. 269–284). New York, NY: Harper & Row.
Davis, M. B., & Shaw, R. G. (2001). Range shifts and adaptive responses to Quaternary climate change. *Science, 292,* 673–679.
De Boer, H. J., Eppinga, M. B., Wassen, M. J., & Dekker, S. C. (2012). A critical transition in leaf evolution facilitated the Cretaceous angiosperm revolution. *Nature Communications, 3,* 1221.
Diaz, S., Hodgson, J. G., Thompson, K., Cabido, M., Cornelissen, J. H. C., Jalili, A., et al. (2004). The plant traits that drive ecosystems: evidence from three continents. *Journal of Vegetation Science, 15,* 295–304.
Dilcher, D. (1973). A paleoclimatic interpretation of the Eocene floras of southeastern North America. In A. Graham (Ed.), *Vegetation and vegetational history of northern Latin America* (pp. 39–59). Amsterdam: Elsevier.
Dilcher, D. L., Kowalski, E. A., Wiemann, M. C., Hinojosa, L. F., & Lott, T. A. (2009). A climatic and taxonomic comparison between leaf litter and standing vegetation from a Florida swamp woodland. *American Journal of Botany, 96,* 1108–1115.
Dkhar, J., & Pareek, A. (2014). What determines a leaf's shape? *EvoDevo, 5,* 47.
Dolph, G. E., & Dilcher, D. (1980a). Variation in leaf size with respect to climate in Costa Rica. *Biotropica, 12,* 91–99.
Dolph, G. E., & Dilcher, D. (1980b). Variation in leaf size with respect to climate in the tropics of the Western Hemisphere. *Bulletin of the Torrey Botanical Club, 107,* 154–162.
Donovan, M. P., Iglesias, A., Wilf, P., Labandeira, C. C., & Cuneo, N. R. (2016). Rapid recovery of Patagonian plant-insect associations after the end-Cretaceous extinction. *Nature Ecology and Evolution, 1,* 12.
Edwards, E. J., & Donoghue, M. J. (2013). Is it easy to move and easy to evolve? Evolutionary accessibility and adaptation. *Journal of Experimental Botany, 64,* 4047–4052.
Edwards, E. J., Spriggs, E. L., Chatelet, D. S., & Donoghue, M. J. (2016). Unpacking a century-old mystery: winter buds and the latitudinal gradient in leaf form. *American Journal of Botany, 103,* 975–978.
Eldrett, J. S., Greenwood, D. R., Harding, I. C., & Huber, M. (2009). Increased seasonality through the Eocene to Oligocene transition in northern high latitudes. *Nature, 459,* 969–974.
Eldrett, J. S., Greenwood, D. R., Polling, M., Brinkhuis, H., & Sluijs, A. (2014). A seasonality trigger for carbon injection at the Paleocene-Eocene Thermal Maximum. *Climate of the Past, 10,* 759–769.
Eronen, J., Puolamäki, K., Liu, L., Lintulaakso, K., Damuth, J., Janis, C., et al. (2010a). Precipitation and large herbivorous mammals i: estimates from present-day communities. *Evolutionary Ecology Research, 12,* 217–233.
Eronen, J. T., Polly, P. D., Fred, M., Damuth, J., Frank, D. C., Mosbrugger, V., et al. (2010b). Ecometrics: the traits that bind the past and present together. *Integrative Zoology, 5,* 88–101.
Feild, T. S., Brodribb, T. J., Iglesias, A., Chatelet, D. S., Baresch, A., Upchurch, G. R., et al. (2011a). Fossil evidence for Cretaceous escalation in angiosperm leaf vein evolution. *Proceedings of the National Academy of Sciences, USA, 108,* 8363–8366.
Feild, T. S., Sage, T. L., Czerniak, C., & Iles, W. J. D. (2005). Hydathodal leaf teeth of *Chloranthus japonicus* (Chloranthaceae) prevent guttation-induced flooding of the mesophyll. *Plant, Cell & Environment, 28,* 1179–1190.
Feild, T. S., Upchurch Jr, G. R., Chatelet, D. S., Brodribb, T. J., Grubbs, K. C., Samain, M.-S., et al. (2011b). Fossil evidence for low gas exchange capacities for early Cretaceous angiosperm leaves. *Paleobiology, 37,* 195–213.
Fletcher, T. L., Greenwood, D. R., Moss, P. T., & Salisbury, S. W. (2014). Paleoclimate of the Late Cretaceous (Cenomanian–Turonian) portion of the Winton Formation, central-western Queensland, Australia: new observations based on CLAMP and bioclimatic analysis. *PALAIOS, 29,* 121–128.
Flynn, A. G., Peppe D. J., Abbuhl, B., & Williamson, T. E. (2016). Paleoclimate and Paleoecology of Diverse Early Paleocene Fossil Flora from the San Juan Basin, New Mexico, USA. *XIV International Palynologival Congress – X International Organisation of Paleobotany Conference*.
Fricke, H. C., & Wing, S. L. (2004). Oxygen isotope and paleobotanical estimates of temperature and δ^{18}O-latitude gradients over North America during the Early Eocene. *American Journal of Science, 304,* 612–635.
Gates, D. M. (1980). *Biophysical ecology*. New York: Springer-Verlag.
Givnish, T. J. (1984). Leaf and canopy adaptations in tropical forests. In E. Medina, H. A. Mooney & C. Vasquez-Yanes (Eds.), *Physiological ecology of plants of the wet tropics* (pp. 51–84). The Hague: Junk.

Green, W. A. (2006). Loosening the CLAMP: an exploratory graphical approach to the Climate Leaf Analysis Multivariate Program. *Palaeontologia Electronica, 9,* 1–17.

Green, W. A., Little, S. A., Price, C. A., Wing, S. L., Smith, S. Y., Kotrc, B., et al. (2014). Reading the leaves: a comparison of leaf rank and automated areole measurement for quantifying aspects of leaf venation. *Applications in Plant Sciences, 2,* 1400006.

Greenwood, D. R. (1991). The Taphonomy of plant macrofossils. Ch. 7. In S. K. Donovan (Ed.), *The processes of fossilization* (pp. 141–169). London: Belhaven Press.

Greenwood, D. R. (1992). Taphonomic constraints on foliar physiognomic interpretations of late Cretaceous and Tertiary paleoclimates. *Review of Palaeobotany and Palynology, 71,* 149–190.

Greenwood, D. R. (1994). Palaeobotanical evidence for Australian Tertiary climates. In R. S. Hill (Ed.), *History of the Australian vegetation: Cretaceous to recent* (pp. 44–59). Cambridge: Cambridge University Press.

Greenwood, D. R. (2005). Leaf margin analysis: taphonomic constraints. *PALAIOS, 20,* 498–505.

Greenwood, D. R. (2007). Fossil angiosperm leaves and climate: from Wolfe and Dilcher to Burnham and Wilf. *Advances in Angiosperm Paleobotany and Paleoclimatic Reconstruction, 258,* 95–108.

Greenwood, D. R., Moss, P. T., Rowett, A. I., Vadala, A. J., & Keefe, R. L. (2003). Plant communities and climate change in southeastern Australia during the early Paleogene. *Special Papers-Geological Society of America,* 365–380.

Greenwood, D. R., Archibald, S. B., Mathewes, R. W., & Moss, P. T. (2005). Fossil biotas from the Okanagan Highlands, southern British Columbia and northeastern Washington state: climates and ecosystems across an Eocene landscape. *Canadian Journal of Earth Sciences, 42,* 167–185.

Greenwood, D. R., Wilf, P., Wing, S., & Christophel, D. C. (2004). Paleotemperature estimation using leaf-margin analysis: is Australia different? *PALAIOS, 19,* 129–142.

Greenwood, D. R., & Wing, S. L. (1995). Eocene continental climates and latitudinal temperature gradients. *Geology, 23,* 1044–1048.

Gregory-Wodzicki, K. M. (2000). Relationship between leaf morphology and climate, Bolivia: implications for estimating paleoclimate from fossil floras. *Paleobiology, 26,* 668–688.

Gregory, K. (1994). Palaeoclimate and palaeoelevation of the 35 Ma Florissant flora, Front Range, Colorado. *Palaeoclimates, 1,* 23–57.

Gregory, K. M., & McIntosh, W. C. (1996). Paleoclimate and paleoelevation of the Oligocene Pitch-Pinnacle flora, Sawatch Range, Colorado. *Geological Society of America Bulletin, 108,* 546–561.

Grichuk, V. (1969). *An attempt to reconstruct certain elements of the climate of the Northern Hemisphere in the Atlantic Period of the Holocene*. In M. I. Neishtadt (Ed.), *Golosten* (pp. 41–57). Moscow: 8th INQUA Congress, Izd-vo Nauka.

Grimm, G. W., & Denk, T. (2012). Reliability and resolution of the Coexistence Approach—a revalidation using modern-day data. *Review of Palaeobotany and Palynology, 172,* 33–47.

Grimm, G. W., & Potts, A. J. (2016). Fallacies and fantasies: the theoretical underpinnings of the Coexistence Approach for palaeoclimate reconstruction. *Climate of the Past, 12,* 611–622.

Grimm, G. W., Bouchal, J. M., Denk, T., & Potts, A. (2016). Fables and foibles: a critical analysis of the Palaeoflora database and the Coexistence approach for palaeoclimate reconstruction. *Review of Palaeobotany and Palynology, 233,* 216–235.

Hall, J. B., & Swaine, M. D. (1981). *Distribution and ecology of vascular plants in a tropical rain forest: Forest vegetation in Ghana.* Dordrecht: Springer Science+Business Media.

Halloy, S. R. P., & Mark, A. F. (1996). Comparative leaf morphology spectra of plant communities in New Zealand, the Andes and the European Alps. *Journal of the Royal Society of New Zealand, 26,* 41–78.

Harbert, R. S., & Nixon, K. C. (2015). Climate reconstruction analysis using coexistence likelihood estimation (CRACLE): a method for the estimation of climate using vegetation. *American Journal of Botany, 102,* 1277–1289.

Heer, O. (1870). Die Miocene flora und fauna Spitzbergens. *Flora Fossilis Artica, 1–3,* 1–98.

Heer, O. (1878a). *Florae fossilis Sachalinensis. Flora Fossilis Artica, 4–5,* 1–61.

Heer, O. (1878b). Fossilen flora Sibiriens und des Amurlandes. *Flora Fossilis Artica, 4–5,* 1–58.

Heer, O. (1882). Die fossile flora Gronlands, Erster Theil. *Flora Fossilis Artica, 6,* 1–147.

Hickey, L. J. (1977). Stratigraphy and paleobotany of the Golden Valley Formation (Early Tertiary) of western North Dakota. *Geological Society of America Memoir, 150,* 1–183.

Hickey, L. J., Johnson, K. R., & Dawson, M. R. (1988). The stratigraphy, sedimentology, and fossils of the Haughton Formation: a post-impact crater-fill, Devon Island, NWT, Canada. *Meteoritics, 23,* 221–231.

Hijmans, R. J., Cameron, S. E., Parra, J. L., Jones, P. G., & Jarvis, A. (2005). Very high resolution interpolated climate surfaces for global land areas. *International Journal of Climatology, 25,* 1965–1978.

Hinojosa, L. F., Armesto, J. J., & Villagran, C. (2006a). Are Chilean coastal forests pre-Pleistocene relicts? Evidence from foliar physiognomy, palaeoclimate, and phytogeography. *Journal of Biogeography, 33,* 331–341.

Hinojosa, L. F., Pesce, O., Yabe, A., Uemura, K., & Nishida, H. (2006b). Physiognomical analysis and paleoclimate of the Ligorio Márquez fossil flora, Ligorio Márquez Formation, 46°45′s, Chile. In H. Nishida (Ed.), *Post-Cretaceous floristic changes in southern Patagonia, Chile* (pp. 45–55). Tokyo: Chuo University Press.

Hinojosa, L. F., Pérez, F., Gaxiola, A., & Sandoval, I. (2011). Historical and phylogenetic constraints on the incidence of entire leaf margins: insights from a new South American model. *Global Ecology and Biogeography, 20,* 380–390.

Holdridge, L. R. (1967). *Life zone ecology*. San Jose, Costa Rica: Tropical Science Center.

Huff, P. M., Wilf, P., & Azumah, E. J. (2003). Digital future for paleoclimate estimation from fossil leaves? Preliminary results. *PALAIOS, 18,* 266–274.

Jacobs, B. F. (1999). Estimation of rainfall variables from leaf characters in tropical Africa. *Palaeogeography, Palaeoclimatology, Palaeoecology, 145,* 231–250.

Jacobs, B. F. (2002). Estimation of low-latitude paleoclimates using fossil angiosperm leaves: examples from the Miocene Tugen Hills, Kenya. *Paleobiology, 28,* 399–421.

Jacobs, B. F., & Deino, A. L. (1996). Test of climate-leaf physiognomy regression models, their application to two Miocene floras from Kenya, and $^{40}Ar/^{39}Ar$ dating of the Late Miocene Kapturo site. *Palaeogeography, Palaeoclimatology, Palaeoecology, 123,* 259–271.

Jacobs, B. F., & Herendeen, P. S. (2004). Eocene dry climate and woodland vegetation in tropical Africa reconstructed from fossil leaves from northern Tanzania. *Palaeogeography, Palaeoclimatology, Palaeoecology, 213,* 115–123.

Jackson, S. T., & Overpeck, J. T. (2000). Responses of plant populations and communities to environmental changes of the late Quaternary. *Paleobiology, 26,* 194–220.

Jacques, F. M., Su, T., Spicer, R. A., Xing, Y., Huang, Y., Wang, W., et al. (2011). Leaf physiognomy and climate: are monsoon systems different? *Global and Planetary Change, 76,* 56–62.

Jones, C. S., Bakker, F. T., Schlichting, C. D., & Nicotra, A. B. (2009). Leaf shape evolution in the South African genus *Pelargonium* l'her. (Geraniaceae). *Evolution, 63,* 479–497.

Jordan, G. J. (1997). Uncertainty in palaeoclimatic reconstructions based on leaf physiognomy. *Australian Journal of Botany, 45,* 527–547.

Jordan, G. J. (2011). A critical framework for the assessment of biological palaeoproxies: predicting past climate and levels of atmospheric CO_2 from fossil leaves. *New Phytologist, 192,* 29–44.

Karst, A. L., & Lechowicz, M. J. (2007). Are correlations among foliar traits in ferns consistent with those in the seed plants? *New Phytologist, 173,* 306–312.

Kelly, A. E., & Goulden, M. L. (2008). Rapid shifts in plant distribution with recent climate change. *Proceedings of the National Academy of Sciences, USA, 105,* 11823–11826.

Kennedy, E. M. (1998). *Cretaceous and Tertiary megafloras from New Zealand and their climate signals*. Ph.D. Dissertation, Open University.

Kennedy, E. M., Arens, N. C., Reichgelt, T., Spicer, R. A., Spicer, T. E., Stranks, L., et al. (2014). Deriving temperature estimates from Southern Hemisphere leaves. *Palaeogeography, Palaeoclimatology, Palaeoecology, 412,* 80–90.

Kennedy, E. M., Spicer, R. A., & Rees, P. M. (2002). Quantitative palaeoclimate estimates from the late Cretaceous and Paleocene leaf floras in the northwest of the South Island, New Zealand. *Palaeogeography, Palaeoclimatology, Palaeoecology, 184,* 321–345.

Kershaw, A. (1997). A bioclimatic analysis of early to middle Miocene Brown Coal floras, Latrobe Valley, south-eastern Australia. *Australian Journal of Botany, 45,* 373–387.

Kershaw, A., & Nix, H. (1988). Quantitative palaeoclimatic estimates from pollen data using bioclimatic profiles of extant taxa. *Journal of Biogeography*, 589–602.

Khan, M. A., Spicer, R. A., Bera, S., Ghosh, R., Yang, J., Spicer, T. E. V., et al. (2014). Miocene to Pleistocene floras and climate of the eastern Himalayan Siwaliks, and new palaeoelevation estimates for the Namling-Oiyug Basin. *Tibet. Global and Planetary Change, 113,* 1–10.

Kobayashi, H., Kresling, B., & Vincent, J. F. (1998). The geometry of unfolding tree leaves. *Proceedings of the Royal Society of London B: Biological Sciences, 265*, 147–154.

Kovach, W. L., & Spicer, R. (1996). Canonical correspondence analysis of leaf physiognomy: a contribution to the development of a new palaeoclimatological tool. *Paleoclimates, 2,* 125–138.

Kowalski, E. A. (2002). Mean annual temperature estimation based on leaf morphology: a test from tropical South America. *Palaeogeography, Palaeoclimatology, Palaeoecology, 188,* 141–165.

Kowalski, E. A., & Dilcher, D. L. (2003). Warmer paleotemperatures for terrestrial ecosystems. *Proceedings of the National Academy of Sciences, USA, 100,* 167–170.

Küchler, A. W. (1964). *Potential natural vegetation of the conterminous United States*. New York: American Geographical Society.

Labandeira, C. C., & Currano, E. D. (2013). The fossil record of plant-insect dynamics. *Annual Review of Earth and Planetary Sciences, 41,* 287–311.

Larcher, W., & Winter, A. (1981). Frost susceptibility of palms: Experimental data and their interpretations. *Principes, 25,* 143–152.

Lawing, A. M., Eronen, J. T., Blois, J. L., Graham, C. H., & Polly, P. D. (2016). Community functional trait composition at the continental scale: The effects of non-ecological processes. *Ecography, 13,* 1–13.

Lawing, A. M., Head, J. J., & Polly, P. D. (2012). The ecology of morphology: the ecometrics of locomotion and macroenvironment in north american snakes. In J. Louys (Ed.), *Paleontology in ecology and conservation* (pp. 117–146). Heidelberg: Springer.

Li, L., McCormack, M. L., Ma, C., Kong, D., Zhang, Q., Chen, X., et al. (2015). Leaf economics and hydraulic traits are decoupled in five species-rich tropical-subtropical forests. *Ecology letters, 18,* 899–906.

Liang, M.-M., Bruch, A., Collinson, M., Mosbrugger, V., Li, C.-S., Sun, Q.-G., et al. (2003). Testing the climatic estimates from different palaeobotanical methods: an example from the middle Miocene Shanwang flora of China. *Palaeogeography, Palaeoclimatology, Palaeoecology, 198,* 279–301.

Liao, F., Peng, J., & Chen, R. (2017). LeafletAnalyzer, an automated software for quantifying, comparing and classifying blade and serration features of compound leaves during development, and among induced mutants and natural variants in the legume *Medicago truncatula*. *Frontiers in plant science, 8*, 915.

Lielke, K., Manchester, S., & Meyer, H. (2012). Reconstructing the environment of the northern Rocky Mountains during the Eocene/Oligocene transition: Constraints from the palaeobotany and geology of south-western Montana, USA. *Acta Palaeobotanica, 52,* 317–358.

Little, S. A., Kembel, S. W., & Wilf, P. (2010). Paleotemperature proxies from leaf fossils reinterpreted in light of evolutionary history. *PLoS ONE, 5,* e15161.

MacGinitie, H. D. (1953). *Fossil plants of the Florissant Beds*. Colorado: Carnegie Institution of Washington.

MacLeod, N., & Steart, D. (2015). Automated leaf physiognomic character identification from digital images. *Paleobiology, 41,* 528–553.

Malhado, A., Malhi, Y., Whittaker, R., Ladle, R., Ter Steege, H., Phillips, O., et al. (2009). Spatial trends in leaf size of Amazonian rainforest trees. *Biogeosciences, 6,* 1563–1576.

McDonald, P. G., Fonseca, C. R., Overton, J. M., & Westoby, M. (2003). Leaf-size divergence along rainfall and soil-nutrient gradients: Is the method of size reduction common among clades? *Functional Ecology, 17,* 50–57.

McElwain, J. C., Montañez, I., White, J. D., Wilson, J. P., & Yiotis, C. (2016). Was atmospheric CO_2 capped at 1000 ppm over the past 300 million years? *Palaeogeography, Palaeoclimatology, Palaeoecology, 441,* 653–658.

Merkhofer, L., Wilf, P., Haas, M. T., Kooyman, R. M., Sack, L., Scoffoni, C., et al. (2015). Resolving Australian analogs for an Eocene Patagonian paleorainforest using leaf size and floristics. *American Journal of Botany, 102,* 1160–1173.

Merriam, C. H. (1894). Laws of temperature control of the geographic distribution of terrestrial animals and plants. *National Geographic Magazine, 6,* 229–238.

Michaletz, S. T., Weiser, M. D., Zhou, J., Kaspari, M., Helliker, B. R., & Enquist, B. J. (2015). Plant thermoregulation: energetics, trait–environment interactions, and carbon economics. *Trends in Ecology & Evolution, 30,* 714–724.

Michel, L. A., Peppe, D. J., Lutz, J. A., Driese, S. G., Dunsworth, H. M., Harcourt-Smith, W. E., et al. (2014). Remnants of an ancient forest provide ecological context for early Miocene fossil apes. *Nature Communications, 5,* 3236.

Miller, I. M., Brandon, M. T., & Hickey, L. J. (2006). Using leaf margin analysis to estimate mid-Cretaceous (Albian) paleolatitude of the Baja BC block. *Earth and Planetary Science Letters, 245,* 95–114.

Mooney, H. A., Björkman, O., & Collatz, G. J. (1978). Photosynthetic acclimation to temperature in the desert shrub, *Larrea divaricata* I. Carbon dioxide exchange characteristics of intact leaves. *Plant Physiology, 61,* 406–410.

Monteith, J., & Unsworth, M. (2007). Principles of environmental physics, Academic Press.

Mosbrugger, V., & Utescher, T. (1997). The coexistence approach – a method for quantitative reconstructions of tertiary terrestrial palaeoclimate data using plant fossils. *Palaeogeography, Palaeoclimatology, Palaeoecology, 134,* 61–86.

New, M., Lister, D., Hulme, M., & Makin, I. (2002). A high-resolution data set of surface climate over global land areas. *Climate research, 21,* 1–25.

Parkhurst, D. F., & Loucks, O. (1972). Optimal leaf size in relation to environment. *The Journal of Ecology, 60,* 505–537.

Peppe, D. J., Hickey, L. J., Miller, I. M., & Green, W. A. (2008). A morphotype catalogue, floristic analysis, and stratigraphic description of the Aspen Shale flora (Cretaceous-Albian) of southwestern Wyoming. *Bulletin of the Peabody Museum of Natural History, 49,* 181–208.

Peppe, D. J., Lemons, C. R., Royer, D. L., Wing, S. L., Wright, I. J., Lusk, C. H., et al. (2014). Biomechanical and leaf-climate relationships: a comparison of ferns and seed plants. *American Journal of Botany, 101,* 338–347.

Peppe, D. J., Royer, D. L., Cariglino, B., Oliver, S. Y., Newman, S., Leight, E., et al. (2011). Sensitivity of leaf size and shape to climate: global patterns and paleoclimatic applications. *New Phytologist, 190,* 724–739.

Peppe, D. J., Royer, D. L., Wilf, P., & Kowalski, E. A. (2010). Quantification of large uncertainties in fossil leaf paleoaltimetry. *Tectonics, 29,* TC3015.

Pérez-Pérez, J. M., Esteve-Bruna, D., & Micol, J. L. (2010). QTL analysis of leaf architecture. *Journal of Plant Research, 123,* 15–23.

Polly, P. D., Eronen, J. T., Fred, M., Dietl, G. P., Mosbrugger, V., Scheidegger, C., et al. (2011). History matters: Ecometrics and integrative climate change biology. *Proceedings of the Royal Society of London B, Biological Sciences, 278,* 1131–1140.

Polly, P. D., & Sarwar, S. (2014). Extinction, extirpation, and exotics: Effects on the correlation between traits and environment at the continental level. *Annales Zoologici Fennici, 51,* 209–226.

Price, C. A., Wing, S., & Weitz, J. S. (2012). Scaling and structure of dicotyledonous leaf venation networks. *Ecology letters, 15,* 87–95.

Pross, J., Contreras, L., Bijl, P. K., Greenwood, D. R., Bohaty, S. M., Schouten, S., et al. (2012). Persistent near-tropical warmth on the Antarctic continent during the early Eocene epoch. *Nature, 488,* 73–77.

Rakotoarinivo, M., Blach-Overgaard, A., Baker, W. J., Dransfield, J., Moat, J., & Svenning, J.-C. (2013). Palaeo-precipitation is a major determinant of palm species richness patterns across Madagascar: a tropical biodiversity hotspot. *Proceedings of the Royal Society of London B, Biological Sciences, 280,* 20123048.

Reich, P. B. (2014). The world-wide 'fast–slow'plant economics spectrum: A traits manifesto. *Journal of Ecology, 102,* 275–301.

Reich, P. B., Ellsworth, D. S., Walters, M. B., Vose, J. M., Gresham, C., Volin, J. C., et al. (1999). Generality of leaf trait relationships: a test across six biomes. *Ecology, 80,* 1955–1969.

Reich, P. B., Walters, M. B., & Ellsworth, D. S. (1997). From tropics to tundra: global convergence in plant functioning. *Proceedings of the National Academy of Sciences, USA, 94,* 13730–13734.

Reichgelt, T., Kennedy, E. M., Mildenhall, D. C., Conran, J. G., Greenwood, D. R., & Lee, D. E. (2013). Quantitative palaeoclimate estimates for Early Miocene southern New Zealand: evidence from Foulden Maar. *Palaeogeography, Palaeoclimatology, Palaeoecology, 378,* 36–44.

Rodriguez, R. E., Debernardi, J. M., & Palatnik, J. F. (2014). Morphogenesis of simple leaves: regulation of leaf size and shape. *Wiley Interdisciplinary Reviews: Developmental Biology, 3,* 41–57.

Royer, D. L. (2012a). Climate reconstruction from leaf size and shape: New developments and challenges. *The Paleontological Society Papers, 18,* 195–212.

Royer, D. L. (2012b). Leaf shape responds to temperature but not CO_2 in *Acer rubrum. PLoS ONE, 7,* e49559.

Royer, D. L., Kooyman, R. M., Little, S. A., & Wilf, P. (2009a). Ecology of leaf teeth: a multi-site analysis from an Australian subtropical rainforest. *American Journal of Botany, 96,* 738–750.

Royer, D. L., McElwain, J. C., Adams, J. M., & Wilf, P. (2008). Sensitivity of leaf size and shape to climate within *Acer rubrum* and *Quercus kelloggii. New Phytologist, 179,* 808–817.

Royer, D. L., Meyerson, L. A., Robertson, K. M., & Adams, J. M. (2009b). Phenotypic plasticity of leaf shape along a temperature gradient in *Acer rubrum. PLoS ONE, 4,* e7653.

Royer, D. L., Miller, I. M., Peppe, D. J., & Hickey, L. J. (2010). Leaf economic traits from fossils support a weedy habit for early angiosperms. *American Journal of Botany, 97,* 438–445.

Royer, D. L., Peppe, D. J., Wheeler, E. A., & Niinemets, U. (2012). Roles of climate and functional traits in controlling toothed vs. untoothed leaf margins. *American Journal of Botany, 99,* 915–922.

Royer, D. L., Sack, L., Wilf, P., Lusk, C. H., Jordan, G. J., Niinemets, U., et al. (2007). Fossil leaf economics quantified: calibration, Eocene case study, and implications. *Paleobiology, 33,* 574–589.

Royer, D. L., & Wilf, P. (2006). Why do toothed leaves correlate with cold climates? Gas exchange at leaf margins provides new insights into a classic paleotemperature proxy. *International Journal of Plant Sciences, 167,* 11–18.

Royer, D. L., Wilf, P., Janesko, D. A., Kowalski, E. A., & Dilcher, D. L. (2005). Correlations of climate and plant ecology to leaf size and shape: Potential proxies for the fossil record. *American Journal of Botany, 92,* 1141–1151.

Sack, L., & Frole, K. (2006). Leaf structural diversity is related to hydraulic capacity in tropical rain forest trees. *Ecology, 87,* 483–491.

Sack, L., & Scoffoni, C. (2013). Leaf venation: structure, function, development, evolution, ecology and applications in the past, present and future. *New Phytologist, 198,* 983–1000.

Sack, L., Scoffoni, C., John, G. P., Poorter, H., Mason, C. M., Mendez-Alonzo, R., et al. (2013). How do leaf veins influence the worldwide leaf economic spectrum? Review and synthesis. *Journal of Experimental Botany, 64,* 4053–4080.

Sack, L., Scoffoni, C., John, G. P., Poorter, H., Mason, C. M., Mendez-Alonzo, R., et al. (2014). Leaf mass per area is independent of vein length per area: avoiding pitfalls when modelling phenotypic integration (reply to). *Journal of Experimental Botany, 65,* 5115–5123.

Sack, L., Scoffoni, C., McKown, A. D., Frole, K., Rawls, M., Havran, J. C., et al. (2012). Developmentally based scaling of leaf venation architecture explains global ecological patterns. *Nature Communications, 3,* 837.

Sack, L., Tyree, M. T., & Holbrook, N. M. (2005). Leaf hydraulic architecture correlates with regeneration irradiance in tropical rainforest trees. *New Phytologist, 167,* 403–413.

Salvucci, M. E., & Crafts-Brandner, S. J. (2004). Relationship between the heat tolerance of photosynthesis and the thermal stability of Rubisco activase in plants from contrasting thermal environments. *Plant Physiology, 134,* 1460–1470.

Schmerler, S. B., Clement, W. L., Beaulieu, J. M., Chatelet, D. S., Sack, L., Donoghue, M. J., et al. (2012). Evolution of leaf form correlates with tropical–temperate transitions in *Viburnum* (Adoxaceae). *Proceedings of the Royal Society B, Biological Sciences, 279,* 3905–3913.

Scoffoni, C., Rawls, M., McKown, A., Cochard, H., & Sack, L. (2011). Decline of leaf hydraulic conductance with dehydration: relationship to leaf size and venation architecture. *Plant Physiology, 156,* 832–843.

Sharpe, S. E. (2002). Constructing seasonal climograph overlap envelopes from Holocene packrat midden contents, Dinosaur National Monument, Colorado. *Quaternary Research, 57,* 306–313.

Shipley, B., Lechowicz, M. J., Wright, I., & Reich, P. B. (2006). Fundamental trade-offs generating the worldwide leaf economics spectrum. *Ecology, 87,* 535–541.

Sinka, K., & Atkinson, T. (1999). A mutual climatic range method for reconstructing palaeoclimate from plant remains. *Journal of the Geological Society, 156,* 381–396.

Smith, R. Y., Basinger, J. F., & Greenwood, D. R. (2012). Early Eocene plant diversity and dynamics in the Falkland flora, Okanagan Highlands, British Columbia, Canada. *Palaeobiodiversity and Palaeoenvironments, 92,* 309–328.

Spicer, R. A. (2007). Recent and future developments of CLAMP: building on the legacy of Jack A. Wolfe. *Courier Forschungsinstitut Senckenberg, 258,* 109–118.

Spicer, R. A. (2016). *CLAMP, climate leaf analysis multivariate program*, The Open University.
Spicer, R. A., Herman, A. B., & Kennedy, E. M. (2004). Foliar physiognomic record of climatic conditions during dormancy: climate leaf analysis multivariate program (CLAMP) and the cold month mean temperature. *The Journal of Geology, 112,* 685–702.
Spicer, R. A., Herman, A. B., & Kennedy, E. M. (2005). The sensitivity of CLAMP to taphonomic loss of foliar physiognomic characters. *PALAIOS, 20,* 429–438.
Spicer, R. A., Valdes, P. J., Spicer, T. E. V., Craggs, H. J., Srivastava, G., Mehrotra, R. C., et al. (2009). New developments in CLAMP: calibration using global gridded meteorological data. *Palaeogeography, Palaeoclimatology, Palaeoecology, 283,* 91–98.
Spicer, R. A., & Yang, J. (2010). Quantification of uncertainties in fossil leaf paleoaltimetry: Does leaf size matter? *Tectonics, 29*, TC6001.
Steart, D. C., Spicer, R. A., & Bamford, M. K. (2011). Is southern Africa different? An investigation of the relationship between leaf physiognomy and climate in southern African mesic vegetation. *Review of Palaeobotany and Palynology, 162,* 607–620.
Stranks, L., & England, P. (1997). The use of a resemblance function in the measurement of climatic parameters from the physiognomy of woody dicotyledons. *Palaeogeography, Palaeoclimatology, Palaeoecology, 131,* 15–28.
Su, T., Xing, Y. W., Liu, Y. S., Jacques, F. M. B., Chen, W. Y., Huang, Y. S., et al. (2010). Leaf margin analysis: a new equation from humid to mesic forests in China. *PALAIOS, 25,* 234–238.
Svenning, J.-C., & Sandel, B. (2013). Disequilibrium vegetation dynamics under future climate change. *American Journal of Botany, 100,* 1266–1286.
Teodoridis, V., Mazouch, P., Spicer, R. A., & Uhl, D. (2011). Refining CLAMP—investigations towards improving the Climate Leaf Analysis Multivariate Program. *Palaeogeography, Palaeoclimatology, Palaeoecology, 299,* 39–48.
ter Braak, C. J. F. (1986). Canonical correspondence analysis: A new eigenvector technique for multivariate direct gradient analysis. *Ecology, 67,* 1167–1179.
Thompson, R., Anderson, K., Pelltier, R., Strickland, L., Bartlein, P., & Shafer, S. L. (2012). Quantitative estimation of climatic parameters from vegetation data in North America by the mutual climatic range technique. *Quaternary Science Reviews, 51,* 18–39.
Traiser, C., Klotz, S., Uhl, D., & Mosbrugger, V. (2005). Environmental signals from leaves–a physiognomic analysis of European vegetation. *New Phytologist, 166,* 465–484.
Upchurch, G., & Wolfe, J. (1987). Mid-Cretaceous to Early Tertiary vegetation and climate: Evidence from fossil leaves and woods. In E. M. Friis, W. Chaloner & P. R. Crane (Eds.), *Origins of angiosperms and their biological consequences* (pp. 75–106). Cambridge: Cambridge University Press.
Utescher, T., Bruch, A. A., Erdei, B., François, L., Ivanov, D., Jacques, F. M. B., et al. (2014). The coexistence approach—theoretical background and practical considerations of using plant fossils for climate quantification. *Palaeogeography, Palaeoclimatology, Palaeoecology, 410,* 58–73.
Vaughan, J. (2015). Investigating the genetic basis of natural leaf shape variation in *Arabidopsis thaliana*, Ph.D. Dissertation, University of York.
Veloz, S. D., Williams, J. W., Blois, J. L., He, F., Otto-Bliesner, B., & Liu, Z. (2012). No-analog climates and shifting realized niches during the late Quaternary: implications for 21st-Century predictions by species distribution models. *Global Change Biology, 18,* 1698–1713.
Vogel, S. (1970). Convective cooling at low airspeeds and the shapes of broad leaves. *Journal of Experimental Botany, 21,* 91–101.
Vogel, S. (2009). Leaves in the lowest and highest winds: temperature, force and shape. *New Phytologist, 183,* 13–26.
Von Humboldt, A., & Bonpland, A. (1807). *Essai sur la géographie des plantes : accompagné d'un tableau physique des régions équinoxiales, fondé sur des mesures exécutées, depuis le dixième degré de latitude boréale jusqu'au dixième degré de latitude australe, pendant les années 1799, 1800, 1801, 1802 et 1803*. Paris: Chez Levrault, Schoell et compagnie. Translated by Sylvie Romanowski as Essay on the geography of plants. In S. T. Jackson (Ed.). (2009). *Essay on the geography of plants* (pp. 57–143). Chicago: University of Chicago Press.
Wahid, A., Gelani, S., Ashraf, M., & Foolad, M. R. (2007). Heat tolerance in plants: an overview. *Environmental and Experimental Botany, 61*(3): 199–223.
Webb, L. J. (1959). A physiognomic classification of Australian rain forests. *Journal of Ecology, 47,* 551–570.
Webb, L. J. (1968). Environmental relationships of structural types Australian rain forest vegetation. *Ecology, 49,* 296–311.
West, C. K., Greenwood, D. R., & Basinger, J. F. (2015). Was the Arctic Eocene 'rainforest' monsoonal? Estimates of seasonal precipitation from early Eocene megafloras from Ellesmere Island, Nunavut. *Earth & Planetary Science Letters, 427,* 18–30.
Westoby, M., Falster, D. S., Moles, A. T., Vesk, P. A., & Wright, I. J. (2002). Plant ecological strategies: some leading dimensions of variation between species. *Annual Review of Ecology and Systematics, 33,* 125–159.
Whittaker, R. H. (1975). *Communities and ecosystems*. New York: Macmillan.
Wiemann, M. C., Manchester, S. R., Dilcher, D. L., Hinojosa, L. F., & Wheeler, E. A. (1998). Estimation of temperature and precipitation from morphological characters of dicotyledonous leaves. *American Journal of Botany, 85,* 1796–1802.
Wilf, P. (1997). When are leaves good thermometers? A new case for leaf margin analysis. *Paleobiology, 23,* 373–390.
Wilf, P. (2008). Fossil angiosperm leaves: paleobotany's difficult children prove themselves. *Paleontological Society Papers, 14,* 319–333.
Wilf, P., Wing, S. L., Greenwood, D. R., & Greenwood, C. L. (1998). Using fossil leaves as paleoprecipitation indicators: an Eocene example. *Geology, 26,* 203–206.
Wilf, P., Wing, S. L., Greenwood, D. R., & Greenwood, C. L. (1999). Using fossil leaves as paleoprecipitation indicators: an Eocene example: Reply. *Geology, 27,* 91–92.
Wilf, P., Labandeira, C. C., Johnson, K. R., & Ellis, B. (2006). Decoupled plant and insect diversity after the end-Cretaceous extinction event. *Science, 313,* 1112–1115.
Wilf, P., Little, S. A., Iglesias, A., del Carmen Zamaloa, M., Gandolfo, M. A., Cuneo, N. R., & Johnson, K. R. (2009). Papuacedrus (Cupressaceae) in Eocene Patagonia: a new fossil link to Australasian rainforests. *American Journal of Botany, 96*(11): 2031–2047.
Wilf, P., Zhang, S., Chikkerur, S., Little, S. A., Wing, S. L., & Serre, T. (2016). Computer vision cracks the leaf code. *Proceedings of the National Academy of Sciences, USA, 113,* 3305–3310.
Wing, S., Bao, H., & Koch, P. L. (2000). An Eocene cool period? Evidence for continental cooling during the warmest part of the Cenozoic. In B. T. Huber, K. G. MacCleod & S. Wing (Eds.), *Warm climates in Earth history* (pp. 197–237). Cambridge, UK: Cambridge University Press.
Wing, S., & Greenwood, D. R. (1993). Fossils and fossil climate: the case for equable continental interiors in the Eocene. *Philosophical Transactions of the Royal Society of London Series B, 341,* 243–252.
Wing, S. L., Herrera, F., Jaramillo, C. A., Gomez-Navarro, C., Wilf, P., & Labandeira, C. C. (2009). Late Paleocene fossils from the Cerrejon Formation, Colombia, are the earliest record of neotropical rainforest. *Proceedings of the National Academy of Sciences, USA, 106,* 18627–18632.

Wolfe, J., & Hopkins, D. (1967). Climatic changes recorded by Tertiary land floras in northwestern North America. In K. Hatai (Ed.), *Tertiary correlations and climatic changes in the Pacific* (pp. 67–76). Tokyo, Japan: Eleventh Pacific Science Congress.

Wolfe, J. A. (1971). Tertiary climatic fluctuations and methods of analysis of Tertiary floras. *Palaeogeography, Palaeoclimatology, Palaeoecology, 9,* 27–57.

Wolfe, J. A. (1978). A paleobotanical interpretation of Tertiary climates in the Northern Hemisphere: data from fossil plants make it possible to reconstruct Tertiary climatic changes, which may be correlated with changes in the inclination of the Earth's rotational axis. *American Scientist, 66,* 694–703.

Wolfe, J. A. (1979). Temperature parameters of humid to mesic forests of eastern Asia and relation to forests of other regions in the Northern Hemisphere and Australasia. *United States Geological Survey Professional Paper, 1106,* 1–37.

Wolfe, J. A. (1990). Palaeobotanical evidence for a marked temperature increase following the Cretaceous/Tertiary boundary. *Nature, 343,* 153–156.

Wolfe, J. A. (1993). A method of obtaining climatic parameters from leaf assemblages. *United States Geological Survey Bulletin, 2040,* 1–71.

Wolfe, J. A. (1995). Paleoclimatic estimates from Tertiary leaf assemblages. *Annual Review of Earth and Planetary Sciences, 23,* 119–142.

Wolfe, J. A., & Spicer, R. A. (1999). Fossil leaf character states; multivariate analyses. In T. P. Jones & N. P. Rowe (Eds.), *Fossil plants and spores: Modern techniques* (pp. 233–239). London, United Kingdon: Geological Society.

Wright, I. J., Reich, P. B., Cornelissen, J. H. C., Falster, D. S., Groom, P. K., Hikosaka, K., et al. (2005). Modulation of leaf economic traits and trait relationships by climate. *Global Ecology and Biogeography, 14,* 411–421.

Wright, I. J., Reich, P. B., Westoby, M., Ackerly, D. D., Baruch, Z., Bongers, F., et al. (2004). The worldwide leaf economics spectrum. *Nature, 428,* 821–827.

Wright, I. J., Dong, N., Maire, V., Prentice, I. C., Westoby, M., Díaz, S., et al. (2017). Global climatic drivers of leaf size. *Science, 357,* 917–921.

Yang, J., Spicer, R. A., Spicer, T. E., Arens, N. C., Jacques, F., Su, T., et al. (2015). Leaf form–climate relationships on the global stage: an ensemble of characters. *Global Ecology and Biogeography, 24,* 1113–1125.

Yang, J., Spicer, R. A., Spicer, T. E., & Li, C.-S. (2011). 'CLAMP online': a new web-based palaeoclimate tool and its application to the terrestrial Paleogene and Neogene of North America. *Palaeobiodiversity and Palaeoenvironments, 91,* 163–183.

Zagwijn, W., & Hager, H. (1987). Correlations of continental and marine Neogene deposits in the south-eastern Netherlands and the lower Rhine District. *Mededelingen van de Werkgroep voor Tertiaire en Kwartaire Geologie, 24,* 59–78.

Chapter 14
Three-Dimensional Geometric Morphometrics in Paleoecology

Sabrina C. Curran

Abstract Quantification and analysis of shape is an important component of many paleoecological studies. Geometric morphometrics is a powerful shape analysis tool that allows its user to compare entire regions of morphology, visualize shape differences between groups, and create visualizations based on real data. This method is rapidly becoming the standard for data collection and analysis in many fields such as anthropology, biology, ecology, forensics, paleontology, and zoology. Here, the basic procedures of geometric morphometrics are reviewed and a case study on the ecomorphology of the cervid calcaneus is provided to illustrate how geometric morphometrics can be used in paleoecological studies.

Keywords Shape analysis • Surface • Outline • Landmark • Ecomorphology • Multivariate analysis • Artiodactyla • Cervidae

Introduction

Identifying and interpreting differences in morphology is a central theme to many paleoecological studies, including palynology (Mander and Punyasena 2018), phytolith studies (Strömberg et al. 2018), macrobotanical analysis (Peppe et al. 2018), dental mesowear and microwear (Green and Croft 2018), functional morphology (Dunn 2018; Evans and Pineda-Munoz 2018), and ecomorphology (Barr 2018). This chapter reviews three-dimensional (3D) geometric morphometrics methods currently being employed in paleoecological contexts.

Geometric morphometrics (GM) is rapidly becoming the standard method in biological and paleontological studies of morphology. It is widely used in ecological studies of masticatory behavior and locomotor behavior (e.g., Aguilar-Medrano 2017; Fernández et al. 2016; Ritzman et al. 2016; Taylor et al. 2016; van Heteran et al. 2016). Operating with a robust definition of shape and resting on a firm statistical foundation, GM methods are appropriate for a wide variety of shape-related questions, many of which had existed as qualitative descriptions previously. GM allows researchers to provide a high-resolution description of shape, analyze the entirety of an anatomical feature as a unit, take into account the 3D structure of shape, visualize shape differences, and even create hypothetical shapes based on actual data. Although not appropriate for every paleoecological study, GM is a powerful tool for shape-related research, especially for contexts wherein the researcher is interested in complex morphology.

It should be noted that there are many more shape analysis methods than are covered by this chapter, such as Elliptical Fourier Analysis (Slater et al. 2010; Panagiotopoulou 2009), Eigenshape Analysis (MacLeod 1999; Sheets et al. 2006), joint-surface angular interfaces (Tocheri et al. 2008, 2011), dental topographic analysis (Evans and Fortelius 2008; Eronen et al. 2010; Smits and Evans 2012), and superquadrics (Sommer et al. 2006). The method covered here (Geometric Morphometrics) was selected due to its ubiquity and flexibility. Although this method can be applied to the morphology of almost any set of specimens that have homologous features, I will focus on biological specimens and specifically skeletal elements (generally referred to as "objects" here) and applications in ecological functional morphology (ecomorphology). A case study on Pleistocene Eurasian Cervidae (deer and relatives) ecomorphology will be used to illustrate how one can use geometric morphometrics in paleoecological research.

S. C. Curran (✉)
Department of Sociology and Anthropology, Ohio University, 124 Bentley Annex, Athens, OH, USA
e-mail: currans@ohio.edu

D. A. Croft et al. (eds.), *Methods in Paleoecology: Reconstructing Cenozoic Terrestrial Environments and Ecological Communities*, Vertebrate Paleobiology and Paleoanthropology, https://doi.org/10.1007/978-3-319-94265-0_14

Historical and Theorectical Background

What we still refer to as solid geometry was mostly eloquently articulated by Euclid of Alexandria in the third century BCE in the *Elements*, which are considered some of the most influential mathematical texts of all time (Boyer 1991). Euclidean geometry defines the space in which most of the techniques of quantitative shape analysis work and form the basis of how we think about shape to this day. Quantification of Euclidean dimensions (length, width, depth, etc.) dominate studies of functional morphology. Such measures (referred to as traditional morphometrics) are useful in describing shape variation; however, they are difficult to apply to very complex shapes such as joint surfaces. Analyses using traditional morphometrics are further complicated by the effects of body size differences, which can be great in biological studies. Geometric morphometrics, which is described in detail below, is well-suited for addressing these issues in the analysis of complex shapes.

Although the origins of GM can be traced back to 1977 with David Kendall's work on stone circles (or even conceptually back to D'Arcy Thompson's deformation grids in 1917), the foundational principles set forth by Bookstein (1978) were worked out in a series of workshops and conferences in the 1980s and 1990s. These meetings resulted in what are referred to as the "color books", for each published proceedings or text has a dominant cover color (Red: Bookstein et al. 1985; Blue: Rohlf and Bookstein 1990; Orange: Bookstein 1991; Black: Marcus et al. 1993; White: Marcus et al. 1996; and later, Green: Zelditch et al. 2004; Yellow: Cardini and Loy 2013). Over the past twenty-eight years, GM methodology has evolved from a "revolution" (Rohlf and Marcus 1993), to a "synthesis" (Bookstein 1996), to a "paradigm" (Adams et al. 2013).

Geometric morphometrics offers a different way of evaluating the shape of an object than is afforded by traditional morphometric studies (Mitteroecker and Gunz 2009; Baab et al. 2012; Adams et al. 2013; Cooke and Terhune 2015; McNulty and Vinyard 2015). In GM, data are collected as two-dimensional or three-dimensional coordinates across a region of interest that contain all of the information that one would describe qualitatively and/or measure with traditional methods. For example, say we are analyzing the shape of hominin crania when viewed from posterior. We could describe the cranium of *Australopithecus* as being somewhat bell-shaped, *Homo neanderthalensis* as 'en bombe' or oval-shaped, and *Homo sapiens* as being shaped like the end of a loaf of bread. To quantify such shapes with traditional morphometrics, we could take a series of measurements from the posterior cranium, such as cranial height (the distance between osteometric points basion and bregma) or cranial breadth (the distance between left and right euryon) and create ratios from them. However, we would still be missing information about the regions between those measurements including the overall curvature of the cranial vault, which can provide a more detailed understanding of cranial morphology and facilitate comparisons between taxa. With GM, we could analyze the outline of the cranium (from a two-dimensional photograph) or the entire surface of the cranium (from three-dimensional scans). Not only would we maximize retention of information about the overall shape of the cranium throughout the analysis, but we would also be able to visualize the differences found in the collected data since we can apply vector transformations to the original data, which is much more difficult to do with traditional morphometrics. Thus, GM allows morphologists to quantify features that are traditionally described qualitatively. Further, we can even create hypothetical specimens based on the digitized data. For these reasons and others, geometric morphometrics is a powerful tool for the morphologist. In paleoecology, shape analysis is important for studies of both ecological functional morphology (ecomorphology) and taxonomy (used in many ecological methods).

In GM, the shape of an object is digitized as a series of 2D or 3D landmark coordinates that are superimposed in a single coordinate system (described in detail below) for analysis. The resulting coordinates (called Procrustes coordinates) are the raw data that can be analyzed using standard multivariate methods used in the study of shape. However, unlike traditional morphometric data, GM analyses maintain all the interrelationships of the original coordinates are maintained, including interlandmark distance and angles (Slice 2007). Thus, the variation in shape can easily be visualized and hypothetical specimens can be created. For example, in a study of variation in cervid femoral head shape as it relates to locomotor behavior in different habitats (Curran et al. 2016), the average femoral head shape for cervids in different habitats or the shape changes along principal components axes can be visualized (Fig. 14.1). Furthermore, one can explore hypothetical femoral head shapes in multivariate space based on the specimens in the space (such as the average specimen in the center of Fig. 14.1) allowing researchers to ask questions about their data that may not have been possible before.

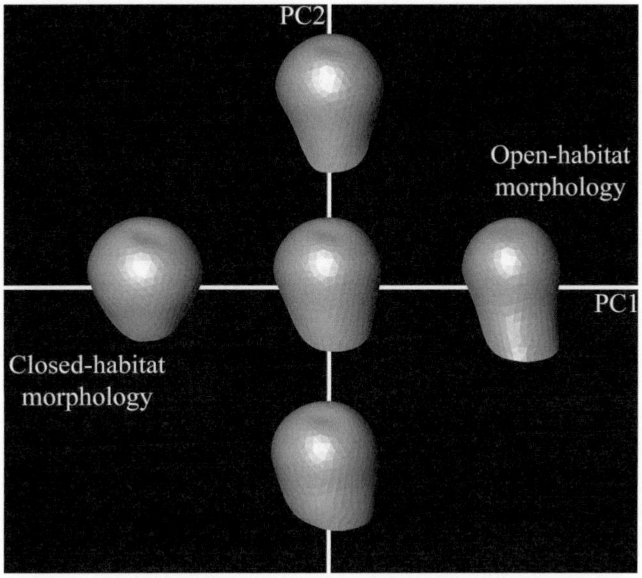

Fig. 14.1 Visualizations of cervid femoral head shape (in superior view) differences along PC1 (68% of variation) and PC2 (26% of variation), with the average specimen at the origin. Open-adapted cervids are distributed along the positive sides of PCs 1 and 2, while closed-adapted cervids are along the negative sides of the axes

Methodology

Geometric morphometrics (GM) analysis begins with the selection and digitization of landmarks, which are points that are homologous, or at least operationally homologous, between specimens. For semilandmark analysis, curves or surfaces of a region are considered homologous when anchored by landmarks (Bookstein 1991, 1996/1997; Gunz et al. 2005) and are digitized as a series of coordinates across a region. Selection of landmarks is important because they must be present in all specimens in the analysis and provide sufficient coverage of the anatomical region in question. It should be noted that these landmarks are homologous in a geometrical rather than biological sense (that is, the landmark location, itself, is likely not the unit of selection, but rather a part of the biologically homologous region; Gunz et al. 2005). Further, not all landmarks are considered to be equal. Ideally, landmarks are placed at discrete anatomical loci, such as the intersection of cranial sutures or foramina (Bookstein 1991). Such landmarks are called Type I and carry the highest repeatability. For example, a Type I landmark on the human skull, the craniometric point 'nasion' is defined as the intersection of the frontal bone and the two nasal bones. Barring pathology or damage, this landmark is easy to locate and present on all human skulls. Type II landmarks are those placed on anatomical features that include tips of protuberances, bases of pits, and maxima of curved features, such as the tip of the (unworn) canine (Bookstein 1991). Type III landmarks are commonly used as the end points of conventional (caliper) measurements, such as lengths and widths. The craniometric point 'euryon', for example, is found on each side of the cranium to record its greatest breadth. Type III landmarks are defined in reference to other features are the least repeatable and perhaps the least likely to be equivalent among individuals. In landmark-based GM analyses, it is best to use as many Type I and II landmarks as possible.

If the region of interest is devoid of landmarks, as is the case for joint surfaces and other smooth features such as much of the human cranial vault, then semilandmark analyses may be used (Bookstein 1996/1997). Although the semilandmark coordinates are not necessarily homologous, the region of interest from which the data are collected (curve or surface) is considered homologous (Perez et al. 2006). Semilandmarks can be collected in a variety of ways. For a 2D analysis, points are usually digitized from an image using software (e.g., the tps suite by Rohlf (2015)). In 3D analyses, points are digitized either directly from a specimen using a MicroScribe (in Excel or visualized in real time using CAD software such as Rhinoceros®), or extracted from models created with a 3D scanning device (laser scanners, structured-light scanners, CT-scanners, etc.) or photogrammetry. With many of the data collection tools, coordinates will need to be extracted from digitized models using software such as Landmark Editor (Wiley et al. 2005) or several R packages (see below). Landmarks are generally selected as a starting and end points (minimally) along the feature and then semilandmarks are digitized from a curve or surface in between those landmarks. As with a landmark-based analysis, the number of landmarks on each specimen must be equal and thus the user either must specify a number of semilandmarks across the region of interest or resample (Curran 2012; Gunz et al. 2005; Harcourt-Smith et al. 2008) a surface or outline created from the digitized coordinates. Next, a sliding procedure is conducted on the collected coordinates wherein semilandmarks 'slide' along tangents to the curve or surface from which they are derived until all semilandmarks are in roughly homologous locations (Gunz and Mitteroecker 2013). This process starts with a reference or template specimen and is iteratively repeated based on a mean configuration. The resulting semilandmarks then are projected back down on to the original curve or surface of the target specimen (Gunz et al. 2005; Gunz and Mitteroecker 2013).

Once all landmarks or semilandmarks have been digitized for all specimens, their coordinates are submitted to a Generalized Procrustes Analysis (GPA). Specimens are typically digitized in slightly different orientations and positions, which will introduce differences in the coordinates that are irrelevant to the shape of the specimen. GPA superimposition removes this extraneous information about location, isometric size, and orientation.

To remove information about a specimen's digitized/scanned location, the centroid of each specimen configuration is moved in space to a common origin. A specimen's centroid is found by calculating the mean *x*, mean *y*, (and mean *z* for 3D analyses) coordinate of every point in its configuration (the "center of gravity" for all the landmarks; McNulty 2003, p. 61). Subtracting the specimen's mean *x* from all of its *x*-coordinates, the mean *y* from its *y*-coordinates, and the mean *z* from its *z*-coordinates translates the specimen to the origin; when applied to all specimens, this eliminates any differences between objects due to their location.

Especially important to the study of biological variation is that isometric size, which is often one of the greatest differences between specimens in traditional morphometric analyses (Falsetti et al. 1993; Jungers et al. 1995), can be removed as a confounding factor in the initial stages of the GPA, however, allometric shape variation still remains in the data (Klingenberg 2016). Landmark configurations are scaled isometrically by removing size differences among specimens by dividing each coordinate of a specimen by that specimen's centroid size, a measure of size that has no correlation to shape, by setting all configurations to a centroid size of one (Zelditch et al. 2004). Centroid size is calculated as the square root of the sum of squared distances of a specimen's landmarks to its centroid (Bookstein 1991; Dryden and Mardia 1998; Mitteroeker and Gunz 2009). It should be noted that with certain research questions it may be undesirable to remove size from an analysis (Klingenburg 2016) since size, itself, is an ecological variable. In fact, GM methods allow the researcher to parse out size and shape and examine them individually or in concert.

To remove information about a specimen's digitized/scanned orientation, coordinate configurations are rotated to minimize the distance between each specific landmark across all specimens. This is an iterative least-squares process in which sum-of-squared distances between each landmark are minimized (Slice 2005). In the first iteration, a mean configuration (reference configuration) of translated and scaled specimens is produced on to which all specimen configurations (target configurations) are then superimposed. A new mean configuration is then calculated and the process is repeated until further superimpositions produce negligible differences in distance minimization.

There are several software packages available to conduct a GPA (and other analyses), such as Morphologika (O'Higgins and Jones 2006), Morpheus (Slice 2002), MorphoJ (Klingenberg 2011), EVAN toolbox (EVAN Society, http://www.evan-society.org/), PAST (Hammer et al. 2001), and through R packages such as geomorph (Adams and Otarola-Castrillo 2013) and Morpho (Schlager 2017). Many of these are freely available through the Stony Brook morphometric website (http://life.bio.sunysb.edu/morph/), which is a useful starting point for GM studies in general. Consult Appendix 14.1 for other software packages.

Following a GPA, the objects are aligned to share a common coordinate system, which can be organized in a matrix or 3D array, and can be used as shape variables (Slice 2005), called Procrustes coordinates. These coordinates are the data that are submitted to multivariate analyses. For a 3D analysis, each *x, y, z* value for each landmark is input as a variable in future analyses (Cardini and Elton 2007; Cardini et al. 2015), thus (for example) a 3D configuration with 20 landmarks will have 60 variables for multivariate analyses. Here, a word of caution must be extended. Geometric morphometrics is a powerful method for analysis of shape differences; however, there are specific criteria that must be met for its use. In most studies of shape variance, including those based in traditional morphometrics, large sample sizes are desired. In the case of GM, large sample sizes are often necessary due to the large number of variables inherent to its methods This is particularly important if one is considering doing any multivariate analysis that involves matrix inversion, such as canonical variates analysis (described below). Further, the user should have a robust understanding of the assumptions in both geometric morphometrics (beyond the basics covered here-the reader is referred to Slice (2007) as a good starting point and to the "color" books listed above) and any multivariate analyses (Neff and Marcus 1980) they might use in order to avoid violating those assumptions and obtaining false results (the 'garbage in, garbage out' principle).

The following is an overview of some of the multivariate methods available, and it not meant to be exhaustive. GM is a rapidly growing field and these methods continue to be improved upon and new methods appear frequently. If you are interested in exploring the distribution of morphological variation in your sample, then principal components analysis (PCA) is a typical starting point, however, there is recent criticism of the use of this method in GM (Bookstein 2016, 2017b). Currently, PCA is used in two main applications in GM studies-data reduction and exploration. PCA recombines the original variables by rigidly rotating the data and creating new axes (the principal components, or PCs) that are mathematically uncorrelated (orthogonal) and which summarize sample variance in lower dimensionality. In doing so, the user now has fewer variables (the PCs, along which each specimen is scored), with the first PC summarizing the most variation in the sample, and each subsequent PC summarizing less and less. Typically, the majority of sample variance can be summarized in far fewer PCs than the number of original variables. The PCs are usually plotted against one another and the user then explores the distribution of specimens in the PC morphospace.

There are several methods available to explore the relationship between shape (e.g., Procrustes coordinates) and other variables such as sex, size, time, phylogeny, and ecological variables (e.g., temperature, precipitation, diet, trophic level, etc.). Multivariate regression and multiple multivariate regression can be used to test for correlations between one or more independent variables (such as those listed above) with several dependent variables (shape data; Frost et al. 2003; Klingenberg 2011; Cáceres et al. 2014). Partial Least-Squares (PLS; sometimes called 2-Block partial least-squares, 2B-PLS; Rohlf and Corti 2000) is commonly used to investigate covariance of two sets of data (for example, shape and environmental variables or two different shapes) without the assumption of one data set as a predictor variable (Monteiro et al. 2003; Colangelo et al. 2010; Piras et al. 2012; Cáceres et al. 2014; Meloro et al. 2014).

When the research goal is to classify unknown specimens into preselected (*a priori*) groups (sex, diet type, habitat, etc.) or to evaluate distinctness of *a priori* groups, linear discriminant analysis (LDA; two groups) or canonical variates analysis (CVA; multiple groups) is frequently used. Like PCA, CVA recombines the variables to create new variables (canonical variates, or CVs) to maximize variability between groups relative to variability within groups (Neff and Marcus 1980). However, unlike PCA, CVA is not a rigid rotation of the original data and distorts the original distribution of specimens by scaling distances between groups by within-group variance (Cooke and Terhune 2015). To proceed, the number of cases (specimens) should be greater than the number of variables (Kovarovic et al. 2011). In analyses with very high numbers of variables, the CVA will always show group separation, even if there is no biological reality to such a distribution (Mitteroecker and Bookstein 2011). However, group separation can be evaluated by the results of a cross-validation analysis and premutation tests (McNulty et al. 2006; Adams and Collyer 2015). Cross-validation removes one specimen or subsets of specimens, re-computes the CVA, and then classified the removed specimen(s) into a group. A likelihood of group membership (posterior probability) is calculated for each reclassified specimen and average correct reclassifications rates can be calculated for the training set (based on specimens of known group membership), which indicates how well the analysis will perform with unknown cases. One can also test the significance of group separation by using a permutation test based on known Mahalanobis squared distance (D^2; the squared distance between centroids, or the multivariate mean for each group) compared to randomly resampled (permuted) data between groups (Neff and Marcus 1980; McNulty et al. 2006; Curran 2009).

To overcome some of the short-comings of LDA and CVA, one can use a between-groups PCA (bgPCA), wherein group means, rather than individual specimens are used to calculate the PC axes and then all specimens are scored (and plotted) along those axes (Mitteroecker and Bookstein 2011; Gunz et al. 2012; Renaud et al. 2015; Cooke and Terhune 2015). Like CVA, a cross-validation can then be used to evaluate the correct reclassification rate for the bgPCA (Mitteroecker and Bookstein 2011).

Beyond providing a new way to collect and analyze shape data, GM also allows for visualizations of shape differences in a dataset, which is much more difficult to do with traditional morphometric data. One can visualize difference in landmark positions between specimens using thin-plate splines (TPS; Bookstein 1989; Zelditch et al. 2004), wire-frame models, and warped surfaces (Wiley et al. 2005; Klingenberg 2013). TPS visualizes differences between specimens (or averages of groups of specimens) using deformations on a grid (originally envisioned by Thompson in 1917). These deformations demonstrate regions where a uniform grid with plotted landmarks of a reference coordinate configuration would have to be deformed in order to keep the landmarks of a target configuration in the in the same locations on the grid (Bookstein 1989; Klingenberg 2013).

One could also visualize specific points in a morphospace that are not sampled by actual specimens, such as a group mean shape, or shape differences along a principal component axis. In the case study below, wire-frame visualizations were created for the extremes of a morphological feature under investigation. Using principal component scores from a PCA on the specimen's Procrustes coordinates, a single discriminant function can be produced to separate two predefined groups, such as comparing the calcanei between open habitat adapted cervids versus closed habitat adapted. The scores of each specimen along the discriminant function are regressed against the Procrustes coordinates for each specimen, which produces a vector of coefficients correlated with a grouping variable, such as habitat type (McNulty et al. 2006). When this vector is added to and subtracted from the consensus configuration (for the entire data set), it produces two new configurations that demonstrate the differences between the two extremes (such as "Open" and "Closed") along the vector (Frost et al. 2003). The vector can be multiplied in order to amplify shape differences, which allows for more obvious differentiation between the extremes by increasing the magnitude (but not direction) of the shape change. These changes can be visualized on wire-frame models for point data (the case for the study here) or across a surface for 3D scanned data. It should be noted that the only real data are the landmarks and semi-landmarks; all transformations between them are interpolated (Klingenberg 2013).

Geometric morphometrics offers users a method for shape analysis that operates under a specific and clear definition of shape (the geometrical information that remains when the variation due to location, orientation, and scale are removed from an object (Slice 2007)), which enhances objectivity and replicability. Its basic theoretical principles (as described above) are well-established and the software (if not the hardware) to conduct GM analyses are free and becoming increasingly user-friendly. As such, GM methods are used ubiquitously throughout studies of anthropology, biology, and paleontology. Quantification and analysis of entire regions of shape and the maintenance of the interrelationships within the landmark configuration make geometric morphometrics a very appealing type of shape analysis, however, potential users should always ask if it is the appropriate method for their research question. While 2D analyses require only digital images, 3D analysis requires some type of digitizer (MicroScribe, laser or white-light scanner, CT-scanner, etc.), which can be expensive and processing of collected data can be time intensive. The benefit to the latter, of course, is that many biological features are three-dimensional and analysis in two-dimensions may result in loss of important information. However, with the limitations imposed by the necessity of large sample sizes and expensive equipment, in some cases a study using traditional morphometrics may be more appropriate, depending on one's question. In short, if your question can be answered with a few measurements, then do so. However, if you are interested in complex morphology with differences that can be quite subtle, geometric morphometrics may be a good fit for your research.

Case Study: Cervid Ecomorphology

The following case study is intended to give the reader a sense of the entire process of GM from data collection to results. This example is in no way meant to suggest the only manner in which to use this method, but instead to illustrate a specific question for which it is appropriate. Other examples of geometric morphometrics in ecomorphology and paleoecology include studies on reptiles and amphibians (Adams 2011; review in Kaliontzopoulou 2011), primates (White 2009; Cooke 2011), felids (Christiansen 2008; Walmsley et al. 2012), canids (Raia 2004), ursids (Figueirido et al. 2005; Figueirido and Soibelzon 2010; van Heteren et al. 2016), mustelids (Fabre et al. 2015; Bottom-Divet et al. 2017), voles (McGuire 2010), rodents (Cano et al. 2013; Morgan and Alvarez 2012), bovids (Forrest et al. 2018), and deer (Curran 2009, 2012, 2015), the latter of which will be illustrated below.

The deer family Cervidae, contains approximately 40 different species in five tribes (Gilbert et al. 2006). They live in a wide variety of habitats, including non-sand deserts (e.g., *Odocoileus hemionus*, mule deer), grasslands (e.g., *Ozotoceros bezoarticus*, pampas deer), tundra (e.g., *Rangifer tarandus*, reindeer/caribou), woodlands (e.g., *Caproelus caproelus*, roe deer), temperate forests (e.g., *Alces alces*, moose), and tropical forests (e.g., *Pudu puda*, pudu). In these different habitats, they must be able to escape predators and each habitat has different levels of vegetative complexity, topographic relief, and substrate firmness, which in turn require different forms of locomotion and predator-escape behaviors. For example, the reindeer/caribou inhabits open tundra and uses speed in flight and herding behavior to escape predators. Morphologically, having a stable joint surface with low risk of hip joint dislocation (at the cost of joint mobility) is of primary importance. Alternatively, the pudu occupies tropical forests and must move through complex vegetation and topographical relief when escaping predators. Joint stability is sacrificed for joint mobility in order to change direction quickly while fleeing. Adaptations for different locomotor strategies can be evaluated in joint surface morphology, whole bone shape, and in specific regions of skeletal elements (for further studies using GM on post-crania, see also Arias-Martorell et al. 2015; Bottom-Divet et al. 2017; Muñoz et al. 2017).

Ecomorphological studies have been conducted on the closely related family of Bovidae (antelope and relatives) using traditional morphometric methods. Notably, work by Kappelman (1988, 1991) demonstrated that several morphological aspects of the femur, especially femoral head shape and shape of the medial patellar margin correlate to different locomotor (habitat) adaptations. Subsequently, there were several ecomorphological studies of bovid post-crania (Plummer and Bishop 1994; Scott et al. 1999; DeGusta and Vrba 2003, 2005; Weinand 2007; Kovarovic and Andrews 2007; Plummer et al. 2008, 2015), although few started with explicit functional morphological hypotheses, with the recent exception of Barr's (2014, 2015, 2018) studies of the astragalus. Cervids have received much less attention than bovids in post-cranial ecomorphology studies, although several species were included by Kovarovic and Andrews (2007).

To fill this void, I collected three-dimensional landmark and semi-landmark data from across rear-limb skeletal elements in 33 extant and (minimally) five extinct species of deer

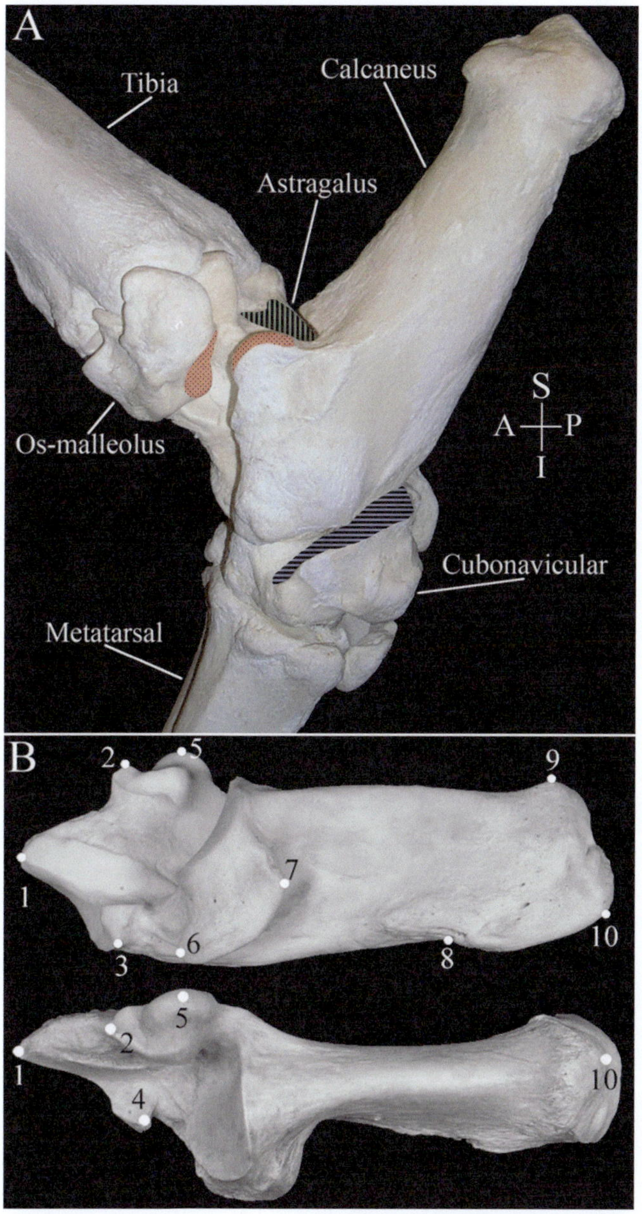

Fig. 14.2 A, Lateral view of left-side hock joint in articulation with skeletal elements labeled and joint surfaces highlighted: astragalus-calcaneus joint (vertical green and black stripes); os malleolus-calcaneus joint (pink and black stippling); cubonavicular-calcaneus joint (blue and black horizontal stripes). Orientation abbreviations: A, anterior; I, inferior; P, posterior; S, superior. B, Ten landmarks used in this study (numbered) on a right-side calcaneus in medial (top) and supero-anterior (bottom) views; both views are oriented with anterior to left

(many specimens were assigned to only Cervidae (gen. + sp. unknown) and exhibited a wide variety of sizes) to test the hypothesis that rear limb morphology reflects cervid locomotion in different habitats and substrates. Each extant species was assigned to a habitat type: Open (deserts, open grasslands, tundra), Intermediate Open (ecotones, tall grasslands, swamps), Intermediate Closed (woodlands/shrublands), and Closed (forests) (Curran 2009, 2012, 2015). Data were collected for 37 morphological features, but I will limit discussion here to the landmark-based geometric morphometric study of the calcaneus (n = 123 for extant specimens and n = 47 for fossil specimens) for illustrative purposes (Fig. 14.2).

In cervids, the calcaneus acts as a Class 1 lever, with the calcaneal-astragalar joint surface (vertically-striped area in Fig. 14.2A) as the fulcrum, the calcaneal body as the lever arm and the distal limb as the load arm. An increase in calcaneal length (approximated by landmarks 1 and 10 in Fig. 14.2B) increases the lever arm and thus increases the mechanical advantage of the contracting muscle that attaches onto the calcaneal tuber but reduces the speed of movement at the ankle. Conversely, a shorter calcaneus provides less power but a quicker movement. Cervids that utilize saltatorial locomotion (bounding leaps) to escape predators, which is an effective adaptive strategy common in more densely vegetated ecotones and closed habitats have longer calcanei than cervids that are cursorial in open habitats (Hildebrand 1985) with cervids that stott (leaping with all four feet contacting the substrate at the same time and mostly found in open mountainous regions) having the longest calcanei. The anterior process of the calcaneus articulates with three bones: anteriorly with the os malleolus (stippled region in Fig. 14.2A and landmark 5 in Fig. 14.2B), inferiorly with the cubonavicular (horizontally-striped region in Fig. 14.2A and between landmarks 1 and 3 in Fig. 14.2B), and at several places with the astragalus (vertically-striped region in Fig. 14.2A), but most especially on the sustentaculum tali (medial border demarked by landmarks 6 and 7 in Fig. 14.2B). Orientation of these joint surfaces affects the rest position of the calcaneus prior to lower limb extension, which, at the hock joint, is almost completely restricted to the parasagittal plane in cervids. The rest position of the calcaneus is hypothesized to determine how far the calcaneus moves along the astragalus when the gastrocnemius muscle is contracted (Fig. 14.3). If the calcaneus sits more vertically on the articulations with the cubonavicular and astragalus, then it moves a shorter distance, which allows for a quick but less powerful force to be generated. This morphology is associated with open-adapted cursors that increase speed by increasing the number of times the same foot strikes the ground per second (the pace cycle). A more horizontal rest position means that the calcaneus travels farther with the contraction of the gastrocnemius, which generates more power, but is slower. This morphology is associated with saltatorial runners that flee predators in a series of leaps that allow for maneuverability in more closed habitats.

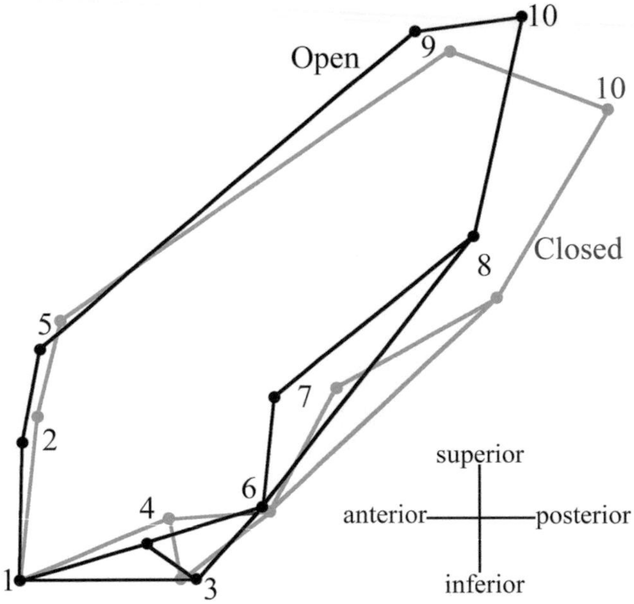

Fig. 14.3 Comparative rest position of the calcaneus of open-adapted (black wire-frame model) and closed-adapted (gray wire-frame model) cervids with landmarks numbered

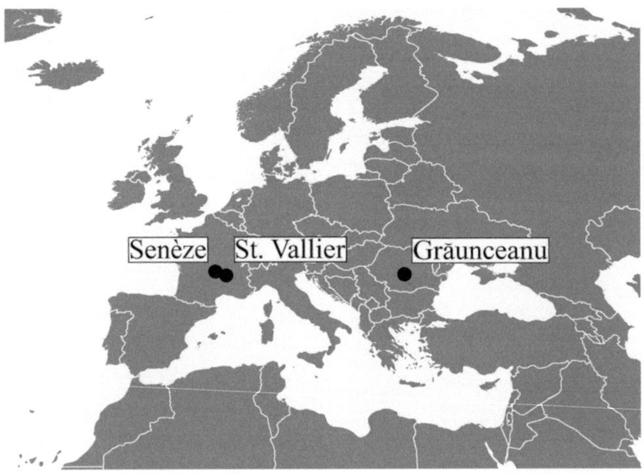

Fig. 14.4 Map of Pleistocene European fossil sites in this study

Once the comparative sample of extant cervid morphology is established, it can be used as a training set to classify cervid specimens of unknown locomotor behaviors (i.e., fossils). Here, this will be applied to cervids from three early Pleistocene sites (St. Vallier and Senèze in France and Grăunceanu in Romania; see Fig. 14.4), which contain several very large deer taxa, including *Cervalces* and *Eucladoceros* (Table 14.1). The three sites are slightly older (Nomade et al. 2014) than the first appearance of *Homo erectus* in Eurasia at Dmanisi, Georgia (maximally 1.85 Ma; Ferring et al. 2011), and contain some of the same cervid genera, however, specimens from Dmanisi were not available for this GM analysis. Using multiple habitat proxies (including macromammals (Gabunia et al. 2000; Palmqvist 2002; Lordkipanidze et al. 2007; Tappen et al. 2007; Hemmer et al. 2011), micromammals (Furió et al. 2010), herpetofauna (Blain et al. 2014), pollen (Gabunia et al. 2000; Messager et al. 2010a), phytoliths (Messager et al. 2010a, b), Dmanisi has been reconstructed as a variable but mostly arid, heterogeneous environment dominated by grasses, with more closed habitats such as woodlands and forests available nearby. The cervid taxa from Dmanisi include *Cervus* cf. *nestii* and *Eucladoceros* aff.

Table 14.1 Summary of fossil specimens in this study

Country	Site	Date	Cervids Present	*n*
Romania	Grăunceanu	Early Pleistocene	*Eucladoceros* sp.	17
France	St. Vallier	Early Pleistocene (2.4–2.1 Ma)	*Croizetoceros ramosus medius*	2
			Eucladoceros ctenoides vireti	3
			Cervidae (sp. + gen. unknown)	17
France	Senèze	Early Pleistocene (2.2–2.0 Ma)	*Alces* sp.	1
			Eucladoceros ctenoides senezensis	2
			Cervus sp.	1
			Cervidae (sp. + gen. unknown)	4
			Total	47

Table 14.2 Calcaneus landmarks (Cal LM) used in this study. Note: black dots in Figure 14.2B are landmarks that were not homologous and thus eliminated from the analysis

Landmark	Location description
1	Anterior point of cubonavicular articular surface
2	Anteroinferior point of lateral astragular articular surface
3	Anteroposterior point of cubonavicular articular surface
4	Superomedial point of anterior astragular articular surface
5	Anterosuperior point of os malleous articular surface
6	Inferomedial point of sustentacular tali
7	Superomedial point of sustentacular tali
8	Posteroinferior point of gactronemius enthesis
9	Anterosuperior point of calceneal tuber
10	Posterosuperior point of caleanal tuber

tegulensis, both of which are associated with "dense to open forests and shrub landscapes" (Lordkipanidze et al. 2007, Supplementary Table S2). To compare the reconstructed paleohabitat of Dmanisi to other European sites that do not contain hominins, GM ecomorphological analyses were conducted on cervid post-crania from St. Vallier, Senèze, and Grăunceanu.

Using a MicroScribe® MLX 3D point digitizer (http://www.3d-microscribe.com), *x, y,* and *z* coordinates for ten landmarks (Fig. 14.2B, Table 14.2) were directly uploaded into a spreadsheet (in this case, Excel). Once all data were collected on extant and extinct specimens, they were submitted to Morphologika (O'Higgins and Jones 2006) for Generalized Procrustes Analysis to translate them to the same origin, rotate them into the same orientation, and scale them to the same isometric size (as described above) (Fig. 14.3). The resulting Procrustes coordinates were then used in the following analysis.

The main goal of this study was to provide a new paleoecological proxy using cervid remains, which are often the most frequent taxa in Eurasian fossil assemblages. However, there are two main confounding factors (phylogenetic relatedness and body size) that need to be assessed before proceeding with the ecomorphological proxy analysis. Since cervids belong to the same family, they may share similar morphology due to common ancestry and this may bias the ecomorphological signal. To test for this, a phylogenetic signal was calculated in MorphoJ (Klingenberg 2011) by comparing the sum of branch lengths of a phylogenetic tree (=tree length) based on known phylogenetic relationships of the extant taxa (from 10kTrees; Arnold et al. 2012) and Procrustes coordinates (means for each species) to trees generated with randomly permuted tip taxa. The null hypothesis of no phylogenetic signal in the morphological data is rejected if 5% or more of the permuted tree lengths are shorter than the original (known lengths) tree. A significant p-value ($p \leq 0.05$) indicates a relationship between phylogeny and morphology. In this study, the *p*-value is 0.7105, a non-significant result.

To assess the relationship between allometric size differences (since isometric size has been removed from the data during the GPA) and calcaneal morphology, a multivariate linear regression (Monteiro 1999) was performed on body size of extant taxa and the Procrustes coordinates in the geomorph package for R (Adams and Otárola-Castillo

2013). Cervids vary in body size by two orders of magnitude (4.65 kg in *Pudu mephistophiles* to 513 kg in *Alces alces*; Novak 1999) and the metric data have to be log-transformed before any comparison can be made. To test for a relationship between size and shape before proceeding with the ecomorphological analysis, the natural log of centroid size (lnCS) was regressed against the natural log of body weight (lnBW) for extant cervids, resulting in $R^2 = 0.87$ and $p < 0.0001$. The lnCS of the Procrustes configurations was used as a proxy for body size in this analysis. Here, the correlation between lnCS and shape (Procrustes coordinates) was non-significant ($p = 0.3057$).

Once phylogeny and allometry are shown to not have a large impact on morphology (or adjustments are made for them if they do have an impact), Procrustes coordinates of the cervid calcaneus can be used as a paleohabitat proxy. Extant cervid specimens with a known habitat preference were used as a training set for canonical variates analysis (CVA) to categorize unknown (fossil) specimens into a habitat type. The habitat types used here are similar to those used in bovid ecomorphological studies, in that the continuum of habitats is broken down into the four categories described above (for details refer to Curran 2012). Since geometric morphometric analyses usually have a high number of variables (three for each Procrustes coordinate-the *x, y,* and *z* values in a 3D analysis) and since CVA requires more cases (specimens) than variables, data (including extant and extinct specimens) were first reduced using PCA. To avoid overfitting (making the analysis too specific to the extant training set and possibly reducing the ability to accurately assign fossil specimens to a habitat category; Kovarovic et al. 2011), the number of PCs used in further analyses was determined by the lowest number of PCs that returned the highest reclassification rate but retained at least 90% of the original sample variance (Sheets et al. 2006; Curran 2015). Here, 17 PCs, summarizing 94.56% of the total sample variance, were submitted to CVA.

Specimens in the CVA extant training sample were reclassified into their known habitat type with a mean accuracy rate of 61.7% using the cross-validation method and 79.1% with the resubstitution method (Table 14.3), results that are 2.47 and 3.16 times (respectively) better than by chance alone. To test if the habitat types were significantly different from each other, pair-wise non-parametric permutation tests were conducted on the centroids (group mean Procrustes coordinates) for each habitat type (after McNulty et al. 2006; see Curran 2015). All habitat types were significantly different at (minimally) $p = 0.0031$ (Table 14.3).

Using the linear combinations calculated for the CVA, fossil deer specimens from three Eurasian paleontological sites were categorized into one of the four habitat types.

The calcaneal data overwhelmingly indicate open and intermediate open habitats were available in both France and Romania during the Early Pleistocene (Fig. 14.5, Table 14.4, Appendix 14.2). When combined, cervids with open-adapted calcanei make up 57.4% of the assemblages and intermediate open-adapted calcanei comprise a further 34%, for a pooled total of 91.4%. These results are supported by reconstructions from these sites using different proxies (as reviewed in Curran 2015).

There was only one specimen that was classified into the closed habitat category, belonging to *Cervalces gallicus* from

Table 14.3 Reclassification results (by %) and permutation tests for the extant specimens (training set). InterClosed= Intermediate Closed, InterOpen = Intermediate Open

	Closed	InterClosed	InterOpen	Open
Resubstitution				
Closed	92.86	0.0	0.0	7.14
InterClosed	0.0	81.82	9.09	9.09
InterOpen	5.56	11.11	66.67	16.67
Open	0.0	9.38	15.63	75.0
Cross-validation				
Closed	71.43	7.14	14.29	7.14
InterClosed	27.27	45.45	9.09	18.18
InterOpen	5.56	14.81	61.11	18.52
Open	0.0	12.5	18.75	68.75
	InterOpen	InterClosed	Closed	
Permutation tests				
Open	<0.0001	0.0003	<0.0001	
InterOpen	–	0.0014	<0.0001	
InterClosed	–	–	<0.0001	

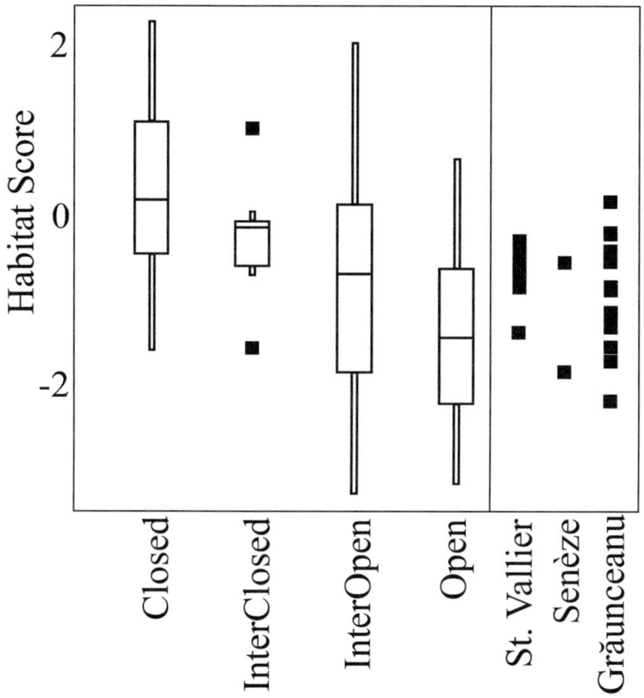

Fig. 14.5 Habitat Score analysis box-and-whiskers diagram. Extant comparative specimens are plotted on the left by habitat type, and specimens from the three fossil sites are plotted on the right. Black horizontal line represents the median, the box encloses the 25th and 75th percentiles, and thin vertical bars (whiskers) include specimens in the range of up to 1.5 times the distance between the quartiles

the French site of Senèze. This species was hypothesized by Breda (2008) to have occupied similar habitats to the modern moose, *Alces alces*, although this was considered contentious as *Cervalces gallicus* had extremely large antlers. The results from this study show that this species had a calcaneus adapted to closed habitats, indicating that there were some forested regions available near the site of Senèze.

A large number of the fossil calcanei in this analysis have not been identified to taxon. While this study is not intended to resolve this, there is an intriguing pattern in the data. When specimen scores for the first canonical variate were regressed against the natural log of centroid size (lnCS), there was a distinct patterning to the distribution of the fossil specimens in this morphospace. Although the R^2 was low (0.092), the p-value indicated significance ($p < 0.0001$). In Figure 14.6A, it can be seen that many of the fossil specimens plot to the negative side on the first canonical variate, indicating that their morphology is not closed-adapted. When the fossil specimen distribution is examined without the extant training set, two or three groups of specimens are evident. In Figure 14.6B, the asterisks are fossil specimens that are assigned only to Cervidae. In this distribution, there is a clear *Eucladoceros* group and a *Croizetoceros* group. Near each are several specimens of unknown taxon and then a group of specimens that falls in between the two groups (larger asterisks). It is possible that the larger specimens belong to *Eucladoceros*, that the smallest specimens belong to *Croizetoceros*, and possibly that the medium/small-sized group belongs to *Cervus philisi*, which is known from known from both St. Vallier and Senèze, France (Valli 2004a, b; Delson et al. 2006). This hypothesis is supported by the habitat assignments for these specimens (see Appendix 14.2): the large specimens were mainly assigned to intermediate open habitat category, which is mainly comprised of *Eucladoceros*, while the medium/small- and small-sized specimens without taxonomic assignments were only assigned to the open habitat category, as were the *Croizetoceros* specimens.

It should be noted that the Dmanisi faunal assemblage contains members of the genus *Eucladoceros*, which were assigned to a more closed habitat (Lordkipanidze et al. 2007, Supplementary Information), however, the analysis here and elsewhere (see especially Curran 2015) indicate a more open-adapted signal for their post-cranial morphology. Although it is likely that several habitat types were available at Dmanisi, there is a substantial grassland signal from palynological records at the time of hominin occupation (Messager et al. 2010a). Coupled with the reconstructions of relatively

Table 14.4 Fossil specimen habitat assignments

Site		Open	InterOpen	InterClosed	Closed
Grăunceanu	*n*	9	5	3	0
	(%)	52.9	29.4	17.6	0
St. Vallier	*n*	14	8	0	0
	(%)	63.6	36.4	0	0
Senèze	*n*	4	3	0	1
	(%)	50	37.5	0	12.5
Totals	*n*	27	16	3	1
	(%)	57.4	34	6.4	2.1

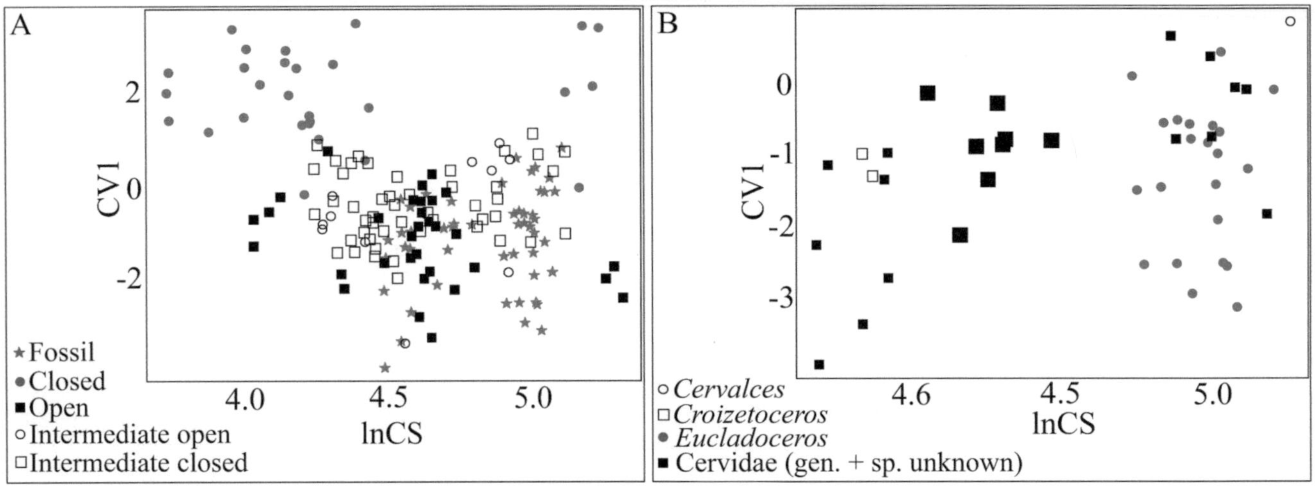

Fig. 14.6 A, Specimens plotted in a morphospace of canonical variate one (CV1) versus the natural log of centroid size (lnCS). B, Only the fossil specimens in the analysis

open habitats and similarity of taxa between Dmanisi and the three European sites examined here, open to intermediate open paleohabitats were likely available across Eurasia at the time of *Homo erectus*'s first arrival there. It is yet to be seen if hominins with similar dates to Dmanisi will also be found in Europe, for there does not seem to have been an ecological or geographic barrier to European dispersal (Moncel 2010).

Future Prospects

In 1993, Rohlf and Marcus declared geometric morphometrics "a revolution in morphometrics" (p. 129). Ten years later, Adams et al. (2004, p. 15) wrote, "We anticipate that over the next few years the geometric morphometric methods that were developed to address shortcomings of what was then the traditional approach to shape analysis will themselves evolve into a standard research protocol … perhaps even becoming the new "traditional" morphometrics". At this point, thirteen years later, in many fields (anthropology, biology, forensics, paleontology, mammalogy, zoology, etc.), geometric morphometrics is nearing the standard method for shape analysis. This is evidenced in the dramatic growth of publications using GM over the last fifteen years (Fig. 14.7, an update to Adams et al. 2004), a

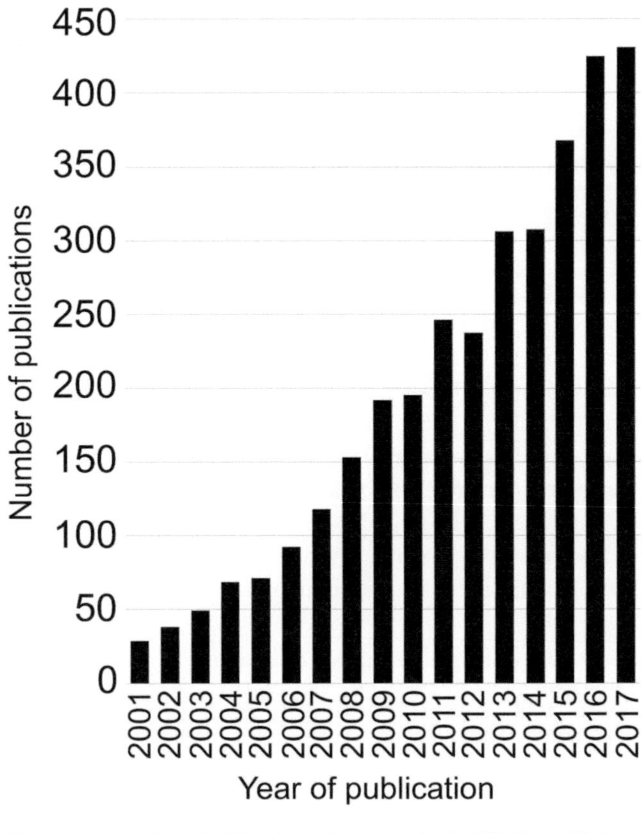

Fig. 14.7 Number of publications that appear in the ISI: Web of Science database per year with a "geometric morphometrics" keyword search

large increase in the number of courses and workshops being offered at universities, professional meetings, and as stand-alone entities (see the Stony Brook morphometrics page at http://life.bio.sunysb.edu/morph/notices.html for a listing of some of these), and the availability of on-line databases such as MorphoSource (Copes et al. 2016), MorphoBrowser (Evans et al. 2007), and others (see Davies et al. 2017). These databases are of special importance, as they make data widely available, storable, and in some cases, reduce issues of inter-observer error (von Cramon-Taubadel et al. 2007; Robinson and Terhune 2017; Shearer et al. 2017).

Greater accessibility to GM data and instruction coupled with open access software, the availability of photogrammetry (Evin et al. 2016), and the decreasing cost of 3D digitizers is partially responsible for the proliferation of GM studies. Equally responsible for the rise in GM's usage are the practitioners of the methods itself. Although GM can be considered a "mature" (Slice 2007, p. 273) method, its base assumptions and applications are constantly being revised (Adams et al. 2013; Bookstein 2017a, b). Recently, methods have been presented for the integration of GM with functional morphology (Bookstein 2015), finite elements analysis (Parr et al. 2012; O'Higgins and Milne 2013; Smith et al. 2015), and genetics (Sherwood and McNulty 2011; Bo et al. 2014; Geiger et al. 2016; Martinez-Abadias et al. 2016). Great strides are also being made in the virtual reconstruction and corrections to deformation of fossil specimens (Gunz et al. 2009; Tallman et al. 2014; Bauer and Harvati 2015; Millella et al. 2015), visualization of shape differences (Klingenberg 2013), and the automation of data collection (Boyer et al. 2011, 2015; Gonzalez et al. 2016). There is no doubt that GM methods will continue to be improved and be integrated further into studies of shape.

In the end, it can be said that GM has great intuitive appeal to researchers interested in shape analysis. There is a certain romantic quality to its mathematics and analysis technique. Thin-plate splines and wire-frame models elegantly display differences in shapes that can be described qualitatively, yet GM data allow for statistical analyses that are not possible with qualitative descriptors. With GM, we can not only quantify shape differences that hitherto may have been described only qualitatively, we can go beyond that and virtually create shapes in a manner that was not previously possible. It is no wonder then that GM has become a standard in the tool kit of morphologists.

Acknowledgements Special thanks to Darin Croft, Scott Simpson, and Denise Su for organizing the excellent "Latest Methods in Cenozoic Paleoecology" symposium and to the Cleveland Museum of Natural History for hosting the event. Funding for this research was provided by National Science Foundation Doctoral Dissertation Improvement Grant BCS-0824607, University of Minnesota Doctoral Dissertation Fellowship, University of Minnesota Herb E. Wright Fellowship in Paleoecology, and University of Minnesota, Department of Anthropology Block Grants. For access to and assistance with collections I thank Alexandru Petculescu (Institute of Speleology, Bucharest, Romania), Aurelian Popescu (Museum of Oltenia, Craiova, Romania), Abel Prieur (Collections de Geologie, Universite C. Bernard Lyon 1, France), Didier Berthet (Collection du Musee des Confluences, Lyon, France), Michi Schulenberg (Chicago Field Museum, USA), Eileen Westwig (American Museum of Natural History, USA), Linda Gordon (National Museum of Natural History, USA), and Chris Conroy (UC Berkeley Museum of Vertebrate Zoology, USA). Thanks to Adam Sylvester for the femoral head visualization in Fig. 14.1 and to the editors and reviewers whose insights greatly improved this manuscript.

Appendix 14.1

Software Packages

Edgewarp: http://brainmap.stat.washington.edu/edgewarp//
EVAN Toolbox: http://evan-society.org/node/42
geomorph (R package): https://cran.r-project.org/web/packages/geomorph/index.html
Landmark editor: http://graphics.idav.ucdavis.edu/research/EvoMorph
MorphoJ: http://www.flywings.org.uk/MorphoJ_page.htm
Morpho (R package): https://cran.r-project.org/web/packages/Morpho/index.html
Morphologika2: https://sites.google.com/site/hymsfme/downloadmorphologica
PAST: http://folk.uio.no/ohammer/past/
R: http://www.r-project.org/

On-line Resources

MorphoBrowser: http://pantodon.science.helsinki.fi/mor phobrowser/
MorphoSource: http://morphosource.org/
Stony Brook Morphometrics page: http://life.bio.sunysb.edu/morph/

Appendix 14.2

ID	Species	Site	Assigned Habitat	Closed	InterClosed	InterOpen	Open
FR684a	Cervidae (m)	Senèze	Open	0.0096	0.0792	0.0839	0.8273
FR684b	Cervidae (m)	Senèze	Open	0.0091	0.1811	0.0838	0.726
FR684c	Cervidae (m)	Senèze	Open	0.0439	0.1076	0.1281	0.7204
FR683a	Cervidae (m)	Senèze	Open	0.0022	0.1576	0.2168	0.6235
FR662a	Cervus (l)	Senèze	InterOpen	0.0094	0.0248	0.5111	0.4547
FR608a	Eucladoceros senezensis	Senèze	InterOpen	0.0938	0.036	0.6379	0.2323
FR574a	Eucladoceros senezensis	Senèze	InterOpen	0.0037	0.001	0.9376	0.0577
FR537a	Alces (Libralces)	Senèze	Closed	0.5834	0.0318	0.1544	0.2305
FR2817	Cervidae (s)	St. Vallier	Open	0	0.0015	0.0013	0.9972
FR2821	Cervidae (s)	St. Vallier	Open	0	0.0021	0.0025	0.9954
FR2816	Cervidae (s)	St. Vallier	Open	0	0.0063	0.0112	0.9825
FR2819	Cervidae (s)	St. Vallier	Open	0.001	0.002	0.0212	0.9758
FR2818	Cervidae (s)	St. Vallier	Open	0.0001	0.0254	0.0189	0.9556
FR2791	Cervidae (l)	St. Vallier	Open	0.0004	0.0157	0.1087	0.8752
FR2808	Cervidae (m)	St. Vallier	Open	0.0067	0.03	0.0906	0.8727
FR2810	Cervidae (s)	St. Vallier	Open	0.0052	0.0199	0.1021	0.8727
FR2805	Cervidae (m)	St. Vallier	Open	0.0002	0.0995	0.1293	0.771
FR2813	Cervidae (s)	St. Vallier	Open	0.0036	0.0231	0.2835	0.6898
FR497716	Croizetoceros ramosus	St. Vallier	Open	0.0025	0.0437	0.2825	0.6713
FR2823	Cervidae (m)	St. Vallier	Open	0.0085	0.0337	0.3265	0.6314
FR2809	Cervidae (m)	St. Vallier	Open	0.0627	0.2448	0.137	0.5555
FR497661	Croizetoceros ramosus	St. Vallier	Open	0.005	0.0068	0.4675	0.5207
FR2130	Eucladoceros senezensis	St. Vallier	InterOpen	0.0096	0.0123	0.696	0.2821
FR2120	Eucladoceros senezensis	St. Vallier	InterOpen	0.0021	0.0123	0.7241	0.2616
FR2792	Cervidae (l)	St. Vallier	InterOpen	0.0053	0.0051	0.8284	0.1612
FR2796	Cervidae (l)	St. Vallier	InterOpen	0.1394	0.0171	0.7031	0.1405
FR2131	Eucladoceros senezensis	St. Vallier	InterOpen	0.1063	0.0043	0.8121	0.0772
FR2795	Cervidae (l)	St. Vallier	InterOpen	0.0288	0.0193	0.878	0.0739
FR2797	Cervidae (l)	St. Vallier	InterOpen	0.0221	0.0124	0.9081	0.0575
FR2798	Cervidae (l)	St. Vallier	InterOpen	0.0267	0.0004	0.9689	0.004
ROIV-V.1	Eucladoceros sp.	Grăunceanu	Open	0	0.017	0.0187	0.9642

References

Adams, D. C. (2011). Quantitative genetics and evolution of head shape in *Plethodon* salamanders. *Evolutionary Biology, 38,* 278–286.

Adams, D. C., Rohlf, F. J., & Slice, D. E. (2004). Geometric morphometrics: ten years of progress following the 'revolution'. *Italian Journal of Morphology, 71,* 5–16.

Adams, D. C., & Otárola-Castillo, E. (2013). Package 'geomorph': geometric morphometric analysis of 2d/3d landmark data. R package version 1.0.

Adams, D. C., Rohlf, F. J., & Slice, D. E. (2013). A field comes of age: geometric morphometrics in the 21st century. *Hystrix: The Italian Journal of Mammology, 24,* 7–14.

Adams, D. C., & Collyer, M. L. (2015). Permutation tests for phylogenetic comparative analyses of high-dimensional shape data: what you shuffle matters. *Evolution, 69,* 823–829.

Aguilar-Medrano, R. (2017). Ecomorphology and evolution of the pharyngeal apparatus of benethic damselfish (Pomacentridae, subfamily Stegastinae). *Marine Biology, 164,* 21.

Arias-Martorell, J., Tallman, M., Potau, J. M., Bello-Hellegouarch, G., & Pérez- Pérez, A. (2015). Shape analysis of the proximal humerus in orthograde and semi-orthograde primates: correlates of suspensory behavior. *American Journal of Primatology, 77,* 1–19.

Arnold, C., Matthews, L. J., & Nunn, C. L. (2012). The 10k trees website: a new online resource for primate phylogeny. *Evolutionary Anthropology, 19,* 114–118.

Baab, K. L., McNulty, K. P., & Rohlf, F. J. (2012). The shape of human evolution: a geometric morphometrics perspective. *Evolutionary Anthropology, 21,* 151–165.

Barr, W. A. (2014). Functional morphology of the bovid astragalus in relation to habitat: controlling phylogenetic signal in ecomorphology. *Journal of Morphology, 275,* 1201–1216.

Barr, W. A. (2015). Paleoenvironments of the Shungura Formation (Plio-Pleistocene: Ethiopia) based on ecomorphology of the bovid astragalus. *Journal of Human Evolution, 88,* 97–107.

Barr, W. A. (2018). Ecomorphology. In D. A. Croft, D. F. Su & S. W. Simpson (Eds.), *Methods in paleoecology: Reconstructing Cenozoic terrestrial environments and ecological communities* (pp. 337–347). Cham: Springer.

Bauer, C. C., & Harvati, K. (2015). A virtual reconstruction and comparative analysis of the KNM-ER 42700 cranium. *Anthropologischer Anzeiger, 72,* 129–140.

Blain, H.-A., Agustí, J., Lordkipanidze, D., Rook, L., & Delfino, M. (2014). Paleoclimatic and paleoenvironmental context of the Early Pleistocene hominins from Dmanisi (Georgia, Lesser Caucasus) inferred from the herpetofaunal assemblage. *Quaternary Science Reviews, 105,* 136–150.

Bo, W., Wang, Z., Xu, F., Fu, G., Sui, Y., Wu, W., et al. (2014). Shape mapping: Genetic mapping meets geometric morphometrics. *Briefings in Bioinformatics, 15,* 571–581.

Bookstein, F. L. (1978). *The measurement of biological shape and shape change* (Lecture Notes in Biomathematics, Vol. 24). New York: Springer-Verlag.

Bookstein, F. L. (1989). Principal warps: thin-plate splines and the decomposition of deformations. *Institute of Electrical and Electronics Engineers, Transactions on Pattern Analysis and Machine Intelligence, 11,* 567–585.

Bookstein, F. L. (1991). *Morphometric tools for landmark data: Geometry and biology*. Cambridge: Cambridge University Press.

Bookstein, F. L. (1996a). Biometrics, biomathematics and the morphometric synthesis. *Bulletin of Mathematical Biology, 58,* 313–365.

Bookstein, F. L. (1996/1997). Landmark methods for forms without landmarks: morphometrics of group differences in outline shape. *Medical Imaging Analysis, 1,* 225–243.

Bookstein, F. L. (2015). The relation between geometric morphometrics and functional morphology, as explored by Procrustes interpretation of individual shape measures pertinent to function. *The Anatomical Record, 298,* 314–327.

Bookstein, F. L. (2016). The inappropriate symmetries of multivariate statistical analysis in geometric morphometrics. *Evolutionary Biology, 43,* 277–313.

Bookstein, F. L. (2017a). A newly noticed formula enforces fundamental limits on geometric morphometric analyses. *Evolutionary Biology, 44,* 522–541.

Bookstein, F. L. (2017b). A method of factor analysis for shape coordinates. *American Journal of Physical Anthropology, 164,* 221–245.

Bookstein, F. L., Chernoff, B., Elder, R., Humphries, J., Smith, G., & Strauss, R. (1985). *Morphometrics in evolutionary biology*. (Special Publication No. 15). Philadelphia: Academy of Natural Sciences.

Bottom-Divet, L., Cornette, R., Houssaye, A., Fabre, A.-C., & Herrel, A. (2017). Swimming and running: a study of the convergence in long bone morphology among semiaquatic mustelids (Carnivora: Mustelidae). *Biological Journal of the Linnean Society, 121,* 38–49.

Boyer, C. B. (1991). *A history of mathematics* (2nd ed.). Hoboken: Wiley.

Boyer, D., Lipman, Y., Clair, E. St., Puente, J., Funkhouser, T., Patel, B., et al. (2011). Algorithms to automatically quantify the geometric similarity of anatomical surfaces. *Proceedings of the National Academy of Sciences, USA, 108,* 18221–18226.

Boyer, D. M., Puente, J., Gladman, J. T., Glynn, C., Mukherjee, S., Yapuncich, G. S., et al. (2015). A new fully automated approach for aligning and comparing shapes. *The Anatomical Record, 298,* 249–276.

Breda, M. (2008). Palaeoecology and palaeoethology of the Plio-Pleistocene genus *Cervalces* (Cervidae, Mammalia) in Eurasia. *Journal of Vertebrate Paleontology, 28,* 886–899.

Cáceres, N., Meloro, C., Carotenuto, F., Passaro, F., Sponchiado, J., Melo, G. L., et al. (2014). Ecogeographical variation in skull shape of capuchin monkeys. *Journal of Biogeography, 41,* 501–512.

Cano, A. R. G., Fernández, M. H., & Álvarez-Sierra, M. Á. (2013). Dietary ecology of Murinae (Muridae, Rodentia): a geometric morphometric approach. *PLoS ONE, 11,* e79080.

Cardini, A., & Elton, S. (2007). Sample size and sampling error in geometric morphometric studies of size and shape. *Zoomorphology, 126,* 121–134.

Cardini, A., & Loy, A. (Eds.). (2013). Virtual morphology and evolutionary morphometrics in the new millennium. *Hystrix: The Italian Journal of Mammology, 24(1).*

Cardini, A., Seetah, K., & Barker, G. (2015). How many specimens do I need? Sampling error in geometric morphometrics: testing the sensitivity of means and variances in simple randomized selection experiments. *Zoomorphology, 134,* 149–163.

Christensen, P. (2008). Evolution of skull and mandible shape in cats (Carnivora: Felidae). *PLoS ONE, 3,* e2807.

Colangelo, P., Castiglia, R., Franchini, P., & Solano, E. (2010). Patterns of shape variation in the eastern African gerbils of the genus *Gerbilliscus*, USA (Rodentia, Muridae): environmental correlations and implications for taxonomy and systematics. *Mammalian Biology, 75,* 302–310.

Cooke, S. B. (2011). Paleodiet of extinct platyrrhines with emphasis on the Caribbean forms: three-dimensional geometric morphometrics of mandibular second molars. *The Anatomical Record, 294,* 2073–2091.

Cooke, S. B., & Terhune, C. E. (2015). Form, function, and geometric morphometrics. *The Anatomical Record, 298,* 5–28.

Copes, L. E., Lucas, L. M., Thostenson, J. O., Hoekstra, H. E., & Boyer, D. M. (2016). A collection of non-human primate computed tomography scans housed in MorphoSource, a repository for 3D data. *Scientific Data, 3,* 160001.

Curran, S. (2009). Hominin paleoecology and cervid ecomorphology. Ph.D. Dissertation, University of Minnesota.

Curran, S. (2012). Expanding ecomorphological methods: geometric morphometric analysis of Cervidae post-crania. *Journal of Archaeological Science, 39,* 1172–1182.

Curran, S. (2015). Examining *Eucladoceros* ecomorphology using geometric morphometrics. *The Anatomical Record, 298,* 291–313.

Curran, S., Terhune, C., Sylvester, A., Gogol, S., & Hubbard, J. (2016). New ecomorphological proxies for paleohabitat reconstructions: Geometric morphometric analyses of cervid joint surface morphology. *PaleoAnthropology Society* meetings, 12 April (2016).

Davies, T. G., Rahman, I. A., Lautenschlager, S., Cunningham, J. A., Asher, R. J., Barrett, P. M., et al. (2017). Open data and digital morphology. *Proceedings of the Royal Society B: Biological Sciences, 284,* 20170194.

DeGusta, D., & Vrba, E. (2003). A method for inferring paleohabitats from the functional morphology of bovid astragali. *Journal of Archaeological Science, 30,* 1009–1022.

DeGusta, D., & Vrba, E. (2005). Methods for inferring paleohabitats from the functional morphology of bovid phalanges. *Journal of Archaeological Science, 32,* 1099–1113.

Delson, E., Faure, M., Guérin, C., Aprile, A., Argant, J., Blackwell, B., et al. (2006). Franco-American renewed research at the Late Villafranchian locality of Senèze (Haute-Loire, France). *Courier Forschunginstitut Senckenberg, 256,* 275–290.

Dunn, R. H. (2018). Functional morphology of the postcranial skeleton. In D. A. Croft, D. F. Su & S. W. Simpson (Eds.), *Methods in paleoecology: Reconstructing Cenozoic terrestrial environments and ecological communities* (pp. 23–36). Cham: Springer.

Dryden, I. L., & Mardia, K. V. (1998). *Statistical shape analysis.* New York: Wiley.

Eronen, J. T., Evans, A. R., Jernvall, J., & Fortelius, M. (2010). The impacts of regional climate on the evolution of mammals: a case study using fossil horses. *Evolution, 64,* 398–408.

Evans, A. R., & Fortelius, M. (2008). Three-dimensional reconstruction of tooth relationships during carnivoran chewing. *Palaeontologia Electronica, 11*(2), 10A.

Evans, A. R., & Pineda-Munoz, S. (2018). Inferring mammal dietary ecology from dental morphology. In D. A. Croft, D. F. Su & S. W. Simpson (Eds.), *Methods in paleoecology: Reconstructing Cenozoic terrestrial environments and ecological communities* (pp. 37–51). Cham: Springer.

Evans, A. R., Wilson, G. P., Fortelius, M., & Jernvall, J. (2007). High-level similarity of dentitions in carnivorans and rodents. *Nature, 445,* 78–81.

Evin, A., Souter, T., Hulme-Beaman, Ameen, C., Allen, R., Viacava, P., et al. (2016). Use of close-range photogrammetry in zooarchaeology: creating accurate 3D models of wolf crania to study dog domestication. *Journal of Archaeological Science: Reports, 9,* 87–93.

Fabre, A.-C., Cornette, R., Goswami, A., & Peigné, S. (2015). Do constraints associated with locomotor habitat drive the evolution of forelimb shape? *Journal of Anatomy, 226,* 596–610.

Falsetti, A. B., Jungers, W. L., & Cole III, T. M. (1993). Morphometrics of the callitrichid forelimb. *International Journal of Primatology, 14,* 551–572.

Fernández, P. J., Holowka, N. B., Demes, B., & Jungers, W. L. (2016). Form and function of the human and chimpanzee forefoot: implications for early hominin bipedalism. *Scientific Reports, 6,* 30532.

Ferring, R., Oms, O., Agusti, J., Berna, F., Niotadze, M., Sheila, T., et al. (2011). Earliest human occupation at Dmanisi (Georgian Caucasus) dated to 1.85–1.78 Ma. *Proceedings of the National Academy of Sciences, USA, 108,* 10432–10436.

Figueirido, B., Palmqvist, P., & Pérez-Claros, J. A. (2005). Ecomorphological correlates of craniodental variation in bears and paleobiological implications for extinct taxa: an approach based on geometric morphometrics. *Journal of Zoology, 277,* 70–80.

Figueirido, B., & Soibelzon, L. H. (2010). Inferring palaeoecology in extinct tremarctine bears (Carnivora, Ursidae) using geometric morphometrics. *Lethaia, 43,* 209–222.

Forrest, F. L., Plummer, T. W., & Raaum, R. L. (2018). Ecomorphological analysis of bovid mandibles from Laetoli Tanzania using 3D geometric morphometrics: implications for hominin paleoenvironmental reconstruction. *Journal of Human Evolution, 114,* 20–34.

Frost, S. R., Marcus, L. F., Bookstein, F. L., Reddy, D. P., & Delson, E. (2003). Cranial allometry, phylogeography, and systematics of large-bodied papionins (Primates: Cercopithecinae) inferred from geometric morphometric analysis of landmark data. *The Anatomical Record, 275A,* 1048–1072.

Furió, M., Agustí, J., Mouskhelishvili, A., Sanisidro, Ó., & Santos-Cubedo, A. (2010). The paleobiology of the extinct venomous shrew *Beremendia* (Sorcidae, Insectivora, Mammalia) in relation to the geology and paleoenvironment of Dmanisi (Early Pleistocene, Georgia). *Journal of Vertebrate Paleontology, 30,* 928–942.

Gabunia, L., Vekua, A., & Lordkipanidze, D. (2000). The environmental contexts of early occupation of Georgia (Transcaucasia). *Journal of Human Evolution, 38,* 785–802.

Gilbert, C., Ropiquet, A., & Hassanin, A. (2006). Mitochondrial and nuclear phylogenies of Cervidae (Mammalia, Ruminantia): systematics, morphology, and biogeography. *Molecular Phylogenetics and Evolution, 40,* 101–117.

Geiger, M. F., Schreiner, C., Delmastro, G. B., & Herder, F. (2016). Combining geometric morphometrics with molecular genetics to investigate a putative hybrid complex: a case study with barbels *Barbus* spp. (Teleostei: Cyprinidae). *Journal of Fish Biology, 88,* 1038–1055.

Gonzalez, P. N., Barbeito-Andres, J., D'Addona, L. A., Bernail, V., & Perez, S. I. (2016). Technical note: Performance of semi and fully automated approaches for registration of 3D surface coordinates in geometric morphometric studies. *American Journal of Physical Anthropology, 160,* 169–178.

Green, J. L., & Croft, D. A. (2018). Using dental mesowear and microwear for dietary inference: a review of current techniques and applications. In D. A. Croft, D. F. Su & S. W. Simpson (Eds.), *Methods in paleoecology: Reconstructing Cenozoic terrestrial environments and ecological communities* (pp. 53–73). Cham: Springer.

Gunz, P., Mitteroecker, P., & Bookstein, F. L. (2005). Semilandmarks in three dimensions. In D. E. Slice (Ed.), *Modern morphometrics in physical anthropology* (pp. 73–98). New York: Kluwer Academic/Plenum Publishers.

Gunz, P., & Mitteroecker, P. (2013). Semilandmarks: a method for quantifying curves and surfaces. *Hystrix: The Italian Journal of Mammology, 24,* 103–109.

Gunz, P., Mitteroecker, P., Neubauer, S., Weber, G. W., & Bookstein, F. L. (2009). Principles for the virtual reconstruction of hominin crania. *Journal of Human Evolution, 57,* 48–62.

Gunz, P., Ramsier, M., Kuhrig, M., Hublin, J.-J., & Spoor, F. (2012). The mammalian bony labyrinth reconsidered, introducing a comprehensive geometric morphometric approach. *Journal of Anatomy, 220,* 529–543.

Hammer, Ø., Harper, D. A. T., & Ryan, P. D. (2001). PAST: Paleontological statistics software package for education and data analysis. *Palaeontologia Electronica, 4*(1:4), 9 pp.

Harcourt-Smith, W. E., Tallman, M., Frost, S. R., Wiley, D. F., Rohlf, F. J., & Delson, E. (2008). Analysis of selected hominoid joint surfaces using laser scanning and geometric morphometrics: a preliminary report. In E. J. Sargis & M. Dagosto (Eds.), *Mammalian evolutionary morphology: A tribute to Frederick S. Szalay* (pp. 373–383). Berlin: Springer Science + Business Media.

Hemmer, H., Kahlke, R.-D., & Vekua, A. (2011). A cheetah *Acinonyx pardinesis* (Croizet et Jobert, 1828) s.l. at the hominin site of Dmanisi (Georgia) – A potential prime meat supplier in Early Pleistocene ecosystems. *Quaternary Science Reviews, 30,* 2703–2714.

Hildebrand, M. (1985). Walking and running. In M. Hildebrand, D. M. Bramble, K. F. Liem, & D. B. Wake (Eds.), *Functional vertebrate morphology* (pp. 38–57). Cambridge, MA, USA: Belknap Press.

Jungers, W. L., Falsetti, A. B., & Wall, C. E. (1995). Shape, relative size, and size-adjustments in morphometrics. *Yearbook of Physical Anthropology, 38,* 137–161.

Kappelman, J. (1988). Morphology and locomotor adaptations of the bovid femur in relation to habitat. *Journal of Morphology, 198,* 119–130.

Kappelman, J. (1991). The paleoenvironment of *Kenyapithecus* at Fort Ternan. *Journal of Human Evolution, 20,* 95–129.

Kaliontzopoulou, A. (2011). Geometric morphometrics in herpetology: modern tools for enhancing the study of morphological variation in amphibians and reptiles. *Basic and Applied Herpetology, 25,* 5–32.

Kendall, D. G. (1977). The diffusion of shape. *Advances in Applied Probability, 9,* 428–430.

Klingenberg, C. P. (2011). MorphoJ: an integrated software package for geometric morphometrics. *Molecular Ecology Resources, 11,* 353–357.

Klingenberg, C. P. (2013). Visualizations in geometric morphometrics: how to read and how to make graphs showing shape changes. *Hystrix, the Italian Journal of Mammalogy, 24,* 15–24.

Klingenberg, C. P. (2016). Size, shape, and form: concepts of allometry in geometric morphometrics. *Development, Genes, and Evolution, 226,* 113–137.

Kovarovic, K., & Andrews, P. (2007). Bovid post-cranial ecomorphological survey of the Laetoli paleoenvironment. *Journal of Human Evolution, 52,* 663–680.

Kovarovic, K., Aiella, L. C., Cardini, A., & Lockwood, C. A. (2011). Discriminant function analyses in archaeology: are classification rates too good to be true? *Journal of Archaeological Science, 38,* 3006–3018.

Lordkipanidze, D., Jashashvili, T., Vekua, A., De León, M. S. P., Zollikofer, C. P., Rightmire, G. P., et al. (2007). Postcranial evidence from early *Homo* from Dmanisi, Georgia. *Nature, 449,* 305–310.

MacLeod, N. (1999). Generalizing and extending the eigenshape method of shape space visualization and analysis. *Paleobiology, 25,* 107–138.

Mander, L., & Punyasena, S. W. (2018). Fossil pollen and spores in paleoecology. In D. A. Croft, D. F. Su & S. W. Simpson (Eds.), *Methods in paleoecology: Reconstructing Cenozoic terrestrial environments and ecological communities* (pp. 213–232). Cham: Springer.

Marcus, L. F., Bello, E., & García-Valdecasas, A. (Eds.). (1993). *Contribution to morphometrics*. Madrid: Monografias del Museo Nacional de Ciencias Naturales.

Marcus, L. F., Corti, M., Loy, A., Naylor, G. J. P., & Slice, D. E. (Eds.). (1996). *Advances in morphometrics*. (NATO ASI Series A: Life Sciences, Vol. 284). New York: Plenum.

Martinez-Abadias, N., Mateu, R., Niksic, M., Russo, L., & Sharpe, J. (2016). Geometric morphometrics on gene expression patterns within phenotypes: a case study on limb development. *Systematic Biology, 65,* 194–211.

McGuire, J. L. (2010). Geometric morphometrics of vole (*Microtus californicus*) dentition as a new paleoclimate proxy: shape change along geographic and climatic clines. *Quaternary International, 212,* 198–205.

McNulty, K. P. (2003). *Geometric morphometric analyses of extant and fossil hominoid craniofacial morphology*. Ph.D. Dissertation, City University of New York.

McNulty, K. P., Frost, S. R., & Strait, D. S. (2006). Examining affinities of the Taung child by developmental simulation. *Journal of Human Evolution, 51,* 274–296.

McNulty, K. P., & Vinyard, C. J. (2015). Morphometry, geometry, function and the future. *The Anatomical Record, 298,* 328–333.

Meloro, C., Cáceres, N., Carotenuto, F., Passaro, F., Sponchiado, J., Melo, G. L., et al. (2014). Ecogeographical variation in skull morphometry of howler monkeys (Primates: Atelidae). *Zoologischer Anzeiger, 253,* 345–359.

Messager, E., Lordkipanidze, D., Kvavadze, E., Ferring, C. R., & Voinchet, P. (2010a). Palaeoenvironmental reconstruction of Dmanisi site (Georgia) based on palaeobotanical data. *Quaternary International, 223–224,* 20–27.

Messager, E., Lordkipanidze, D., Delhon, C., & Ferring, C. R. (2010b). Palaeoecological implications of the Lower Pleistocene phytolith record from the Dmanisi Site (Georgia). *Palaeogeography, Palaeoclimatology, Palaeoecology, 288,* 1–13.

Millella, M., Zollikofer, C. P. E., & Ponce de León, M. S. (2015). Virtual reconstruction and geometric morphometrics as tools for paleopathology: a new approach to study rare developmental disorders of the skeleton. *The Anatomical Record, 298,* 335–345.

Mitteroecker, P., & Bookstein, F. (2011). Linear discrimination, ordination, and the visualization of selection gradients in modern morphometrics. *Evolutionary Biology, 38,* 100–114.

Mitteroecker, P., & Gunz, P. (2009). Advances in geometric morphometrics. *Evolutionary Biology, 36,* 235–247.

Moncel, M.-H. (2010). Oldest human expansions in Eurasia: favoring and limiting factors. *Quaternary International, 223–224,* 1–9.

Monteiro, L. R. (1999). Multivariate regression models and geometric morphometrics: the search for causal factors in the analysis of shape. *Systematic Biology, 48,* 192–199.

Monteiro, L. R., Duarte, L. C., & dos Reis, S. F. (2003). Environmental correlates of geographical variation in skull and mandible shape of the punar´e rat *Thrichomys apereoides* (Rodentia: Echimyidae). *Journal of Zoology, London, 261,* 47–57.

Morgan, C. C., & Álvarez, A. (2012). The humerus of South American caviomorph rodents: shape, function and size in a phylogenetic context. *Journal of Zoology, 290,* 107–116.

Muñoz, N. A., Cassini, G. H., Candela, A. M., & Vizcaíno, S. F. (2017). Ulnar articular surface 3D landmarks and ecomorphology of small mammals: a case study of two early Miocene typotheres (Notoungulata) from Patagonia. *Earth and Environmental Science Transactions of the Royal Society of Edinburgh, 106,* 315–323.

Neff, N. A., & Marcus, L. F. (1980). *A survey of multivariate methods for systematics*. New York: American Museum of Natural History.

Nowak, R. (1999). *Walker's mammals of the world,* (Vol. II, pp. 1091–1133). Baltimore: John Hopkins University Press.

Nomade, S., Pastre, J. F., Guillou, H., Faure, M., Guérin, C., Delson, E., et al. (2014). $^{40}Ar/^{39}Ar$ constraints on some French landmark Late Pliocene to Early Pleistocene large mammalian paleofaunas: paleoenvironmental and paleoecological implications. *Quaternary Geochronology, 21,* 2–15.

O'Higgins, P., & Jones, N. (2006). Tools for statistical shape analysis. Hull York Medical School. http://hyms.fme.googlepages.com/resources.

O'Higgins, P., & Milne, N. (2013). Applying geometric morphometrics to compare changes in size and shape arising from finite element analyses. *Hystrix, the Italian Journal of Morphology, 24,* 126–132.

Palmqvist, P. (2002). On the community structure of the large mammals assemblage from Dmanisi (East Georgia, Caucasus). In M. D. Renzi, M. V. P. Alonso, M. Belinchón, E. Penalver, P. Montoya & A. Márquez-Aliaga (Eds.), *Current topics in taphonomy and fossilization* (pp. 361–383). Valencia: Ayuntamentio de Valencia.

Panagiotopoulou, O. (2009). Finite element analysis (FEA): applying an engineering method to functional morphology in anthropology and human biology. *Annals of Human Biology, 36,* 609–623.

Parr, W. C. H., Wroe, S., Chamoli, U., Richards, H. S., McCurry, M. R., Clausen, P. D., et al. (2012). Toward integration of geometric morphometrics and computational biomechanics: new methods for 3D virtual reconstruction and quantitative analysis of Finite Element Models. *Journal of Theoretical Biology, 301,* 1–14.

Peppe, D. J., Baumgartner, A., Flynn, A., & Blonder, B. (2018). Reconstructing paleoclimate and paleoecology using fossil leaves. In D. A. Croft, D. F. Su & S. W. Simpson (Eds.), *Methods in paleoecology: Reconstructing Cenozoic terrestrial environments and ecological communities* (pp. 287–315). Cham: Springer.

Perez, S. I., Bernal, V., & Gonzalez, P. N. (2006). Differences between sliding semi-landmark methods in geometric morphometrics, with an application to human craniofacial and dental variation. *Journal of Anatomy, 208,* 769–784.

Piras, P., Marcolini, F., Claude, J., Venture, J., Kotsakis, T., & Cubo, J. (2012). Ecological and functional correlates of molar shape variation in European populations of *Arvicola* (Arvicolinae, Rodentia). *Zoologischer Anzeiger, 251,* 335–343.

Plummer, T. W., & Bishop, L. C. (1994). Hominid paleoecology at Olduvai Gorge, Tanzania as indicated by antelope remains. *Journal of Human Evolution, 27,* 47–75.

Plummer, T. W., Bishop, L. C., & Hertel, F. (2008). Habitat preference of extant African bovids based on astragalus morphology: operationalizing ecomorphology for palaeoenvironmental reconstruction. *Journal of Archaeological Science, 35,* 3016–3027.

Plummer, T. W., Ferraro, J. V., Louys, J., Hertel, F., Alemseged, Z., Bobe, R., et al. (2015). Bovid ecomorphology and hominin paleoenvironments of the Shungura Formation, lower Omo River Valley, Ethiopia. *Journal of Human Evolution, 88,* 108–126.

Raia, P. (2004). Morphological correlates of tough food consumption in large land carnivores. *Italian Journal of Zoology, 71,* 45–50.

Renaud, S., Dufour, A.-B., Hardouin, E. A., Ledevine, R., & Auffray, J.-C. (2015). Once upon multivariate analyses: when they tell several stories about biological evolution. *PLoS ONE, 10,* e0132801.

Ritzman, T. B., Terhune, C. E., Gunz, P., & Robinson, C. A. (2016). Mandibular ramus shape of *Australopithecus sediba* suggests a single variable species. *Journal of Human Evolution, 10,* 54–64.

Robinson, C., & Terhune, C. E. (2017). Error in geometric morphometric data collection: combining data from multiple sources. *American Journal of Physical Anthropology, 164,* 62–75.

Rohlf, F. J. (2015). The tps series of software. *Hystrix, the Italian Journal of Mammalogy, 26,* 9–12.

Rohlf, F. J., & Bookstein, F. L. (Eds.). (1990). *Proceedings of the Michigan Morphometrics Workshop.* University of Michigan Museum of Zoology Special Publication 2.

Rohlf, F. J., & Marcus, L. F. (1993). A revolution in morphometrics. *Trends in Ecology & Evolution, 8,* 129–132.

Rohlf, F. J., & Corti, M. (2000). Use of two-block partial least-squares to study covariation in shape. *Systematic Biology, 49,* 740–753.

Schlager, S. (2017). Morpho and Rvcg – Shape analysis in R. In G. Zheng, S. Li & G. Szekely (Eds.), *Statistical shape and deformation analysis* (pp. 217–256). Cambridge: Academic Press.

Scott, R., Kappelman, J., & Kelley, J. (1999). The paleoenvironment of *Sivapithecus parvada. Journal of Human Evolution, 36,* 245–274.

Shearer, B. M., Tallman, M., Cooke, S. B., Halenar, L. B., Reber, S. L., Plummer, J., et al. (2017). Evaluating causes of error in landmark-based data collection using scanners. *PLoS ONE, 12,* e0187452.

Sheets, H. D., Covino, K. M., Panasiewicz, J. M., & Morris, S. R. (2006). Comparison of geometric morphometric outline methods in the discrimination of age-related differences in feather shape. *Frontiers in Zoology, 3,* 15.

Sherwood, R. J., & McNulty, K. P. (2011). Dissecting the genetic architecture of craniofacial shape. In P. E. Lestrel (Ed.), *Biological shape analysis: Proceedings of the 1st international symposium* (pp. 145–171). Singapore: World Scientific.

Slater, G. J., Figuerirido, B., Louis, L., Yang, P., & Van Valkenburgh, B. (2010). Biomechanical consequences of rapid evolution in the polar bear linage. *PLoS ONE, 5,* e13870.

Slice, D. E. (2002). Morpheus et al. Department of Ecology and Evolution. State University of New York, Stony Brook.

Slice, D. E. (2005). Modern morphometrics. In D. E. Slice (Ed.), *Modern morphometrics in physical anthropology* (pp. 1–45). New York: Kluwer Academic/Plenum Publishers.

Slice, D. E. (2007). Geometric morphometrics. *Annual Review of Anthropology, 36,* 261–281.

Smith, A. L., Benazzi, S., Ledogar, J. A., Tamvada, K., Smith, L. C. P., Weber, G. W., et al. (2015). Biomechanical implications of intraspecific shape variation in chimpanzee crania: moving toward an integration of geometric morphometrics and finite element analysis. *The Anatomical Record, 298,* 122–144.

Smits, P. D., & Evans, A. R. (2012). Functional constraints on tooth morphology in carnivorous mammals. *BMC Evolutionary Biology, 12,* 146.

Sommer III, H. J., Eckhardt, R. B., & Shiang, T. Y. (2006). Superquadratic modeling of cranial and cerebral shape and asymmetry. *American Journal of Physical Anthropology, 129,* 189–195.

Strömberg, C. A. E., Dunn, R. E., Crifò, C., & Harris, E. B. (2018). Phytoliths in paleoecology: analytical considerations, current use, and future directions. In D. A. Croft, D. F. Su & S. W. Simpson (Eds.), *Methods in paleoecology: Reconstructing Cenozoic terrestrial environments and ecological communities* (pp. 233–285). Cham: Springer.

Tallman, M., Amenta, A., Delson, E., Frost, S. R., Ghosh, D., Klukkert, Z. S., et al. (2014). Evaluation of a new method of fossil retrodeformation by algorithmic symmetrization: crania of papionins (Primates: Cercopithecidae) as a test case. *PLoS ONE, 9,* e100833.

Tappen, M. J., Lordkipanidze, D., Bukshianidze, M., Vekua, A., & Ferring, R. (2007). Are you in or out (of Africa)? In T. R. Pickering, K. Schick, & N. Toth (Eds.), *Breathing life into fossils: Taphonomic studies in honor of C. K. Brain* (pp. 119–135). Bloomington: Stone Age Institute Press.

Taylor, A. C., Lautenschlager, S., Qi, Z., & Rayfield, E. J. (2016). Biomechanical evaluation of different musculoskeletal arrangements in *Psittacosaurus* and implications for cranial function. *The Anatomical Record, 300,* 49–61.

Thompson, D. W. (1917). *On growth and form.* London: Cambridge University Press.

Tocheri, M. W., Solhan, C. R., Orr, C. M., Femiani, J., Frohlich, B., Groves, C. P., et al. (2011). Ecological divergence and medial cuneiform morphology in gorillas. *Journal of Human Evolution, 60,* 171–184.

Tocheri, M. W., Orr, C. M., Jacofsky, M. C., & Marzke, M. W. (2008). The evolutionary history of the hominin hand since the last

common ancestor of *Pan* and *Homo*. *Journal of Anatomy, 212,* 544–562.
Valli, A. M. F. (2004a). Cervidae from the late Pleistocene deposit (Mid-Villafranchian) of Saint-Vallier, (Drôme, France). *Geobios, 37,* S191–S232.
Valli, A. M. F. (2004b). Taphonomy of Saint-Vallier (Drôme, France), the reference locality for biozone MN17 (Upper Pliocene). *Lethaia, 37,* 337–350.
van Heteren, A. H., MacLarnon, A., Soligo, C., & Rae, T. C. (2016). Functional morphology of the cave bear (*Ursus spelaeus*) mandible: a 3D geometric morphometric analysis. *Organisms, Diversity & Analysis, 16,* 299–314.
von Cramon-Taubadel, N., Frazier, B. C., & Lahr, M. M. (2007). The problem of assessing landmark error in geometric morphometrics: theory, methods, and modifications. *American Journal of Physical Anthropology, 134,* 24–35.
Walmsley, A., Elton, S., Louys, J., Bishop, L. C., & Meloro, C. (2012). Humeral epiphyseal shape in the Felidae: the influence of phylogeny, allometry, and locomotion. *The Journal of Morphology, 273,* 1424–1438.
Weinand, D. C. (2007). A study of parametric versus non-parametric methods for predicting paleohabitat from Southeast Asia bovid astragali. *Journal of Archaeological Science, 34,* 1774–1783.
White, J. (2009). Geometric morphometric investigation of molar shape diversity in modern lemurs and lorises. *The Anatomical Record, 292,* 701–719.
Wiley, D. F., Amenta, N., Alcantara, D. A., Ghosh, D., Kil, Y. J., Delson, E., Harcourt-Smith, W., Rohlf, F. J., St. John, K., & Hamann, B. (2005). Evolutionary morphing. *IEEE Visualization, Proceedings*, 431–438.
Zelditch, M. L., Swiderski, D. L., Sheets, H. D., & Fink, W. L. (2004). *Geometric morphometrics for biologists: A primer*. London: Elsevier Academic Press.

Chapter 15
Ecomorphology

W. Andrew Barr

Abstract The goal of ecomorphology is to identify morphological variation that is related to ecology (e.g., dietary preference or locomotor habits), with the aim of inferring ecological traits from morphological traits. This chapter reviews the basic principles of ecomorphology and provides many examples of ecomorphic studies in a range of taxa, with an emphasis on bovids (antelope and relatives). The focus of this chapter is on "applied" ecomorphology, which refers to studies that use ecomorphology as a tool to reconstruct environments. The chapter summarizes some of the strengths and weaknesses of the applied ecomorphic approach, and discusses future directions for studies using this methodology.

Keywords Functional morphology • Environmental reconstruction • Functional traits • Bovidae • Phylogenetic signal

Background

The aim of ecomorphology is to link morphological form with function in a particular ecological context (Bock and von Wahlert 1965; Bock 1989). Ecomorphology can be an end unto itself for understanding the autecology (sensu Wing et al. 1992) of extinct organisms, or it can be a means towards the end of characterizing entire extinct communities and the habitats supporting them. The distinction between ecomorphology and traditional functional morphology is not always clear-cut. The term "ecomorphology" is broader than "functional morphology" as the latter term generally refers explicitly to biomechanics, such as the ratio of force and resistance arms in skeletal lever systems and the resulting functional outcomes (Dunn 2018). Ecomorphology builds upon functional morphology but encompasses a broader view of how biomechanical functions are put to use in ecological contexts in ways that are adaptive for organisms (Bock 1994).

"Applied" ecomorphology is a term that is useful for describing studies employing ecomorphic techniques to reconstruct the proportions of organisms in a community with different adaptations, usually in an effort to understand some aspect of the sampled habitat (e.g., vegetation structure). Applied ecomorphology is primarily concerned with inferences about habitat structure rather than the adaptations of individuals or species, although inferences about individuals and species form the foundation of applied studies. Because the focus of this volume is on methods for reconstructing past environments, the examples used in this chapter are heavily biased towards applied ecomorphology studies rather than more classic examples of ecomorphic analysis (see Wainwright and Reilly 1994). Applied ecomorphic studies based on bovids (antelopes and relatives) have been particularly common (Kappelman 1991; Plummer and Bishop 1994; Scott et al. 1999; DeGusta and Vrba 2003; Kovarovic and Andrews 2007; Plummer et al. 2008; White et al. 2009; Barr 2014) and have served as important sources of evidence for the environmental context of human evolution.

Types of Ecomorphic Analyses

Dietary Ecomorphology

Many different ecomorphic methods have been used to reconstruct dietary preferences of extinct animals. I begin by discussing the most commonly used characteristic –

W. Andrew Barr (✉)
Department of Anthropology, Center for the Advanced Study of Human Paleobiology, The George Washington University, Washington, DC, USA
e-mail: wabarr@gmail.com

D. A. Croft et al. (eds.), *Methods in Paleoecology: Reconstructing Cenozoic Terrestrial Environments and Ecological Communities*, Vertebrate Paleobiology and Paleoanthropology, https://doi.org/10.1007/978-3-319-94265-0_15

hypsodonty – and then address other characteristics useful for relating craniodental form with diet. Examples from a variety of taxonomic groups are discussed, but the primary focus is on ungulates (hoofed mammals), particularly bovid artiodactyls.

Ungulates are often categorized either as grazers, which mostly subsist on monocot grasses, or browsers, which preferentially select dicot leaves and stems from bushes and trees (Janis 2008; Damuth and Janis 2011; Cerling et al. 2015). Mixed-feeders form a third group and typically consume significant amounts of vegetation from both categories (Hofmann and Stewart 1972). These gross categories tend to conflate the type of food acquired (grasses vs. dicots) with the method of food acquisition (i.e., bulk consumption vs. selective feeding; see Spencer 1995). Furthermore, additional factors, especially feeding height relative to the ground, often strongly influences dental form due to variation in the amount of exogenous grit in the diet, regardless of the type of food consumed (Damuth and Janis 2011). Nonetheless, the browser/grazer dichotomy persists and is a useful generalization that is applicable to characterizing the diets of fossil and modern ungulates (Gordon and Prins 2008; Cerling et al. 2015).

One of the most salient features related to dietary function is a tooth's mechanical resistance to wear. A variety of factors influence wear resistance (Van Valen 1960), but by far the most important factor is the quantity of enamel present in a tooth. While some organisms, like hominoids (Martin 1985), tend to increase wear resistance by increasing enamel thickness, ungulates tend to increase wear resistance by increasing tooth height (Fortelius 1985). The term "hypsodont" is used to describe teeth that have high crowns, allowing an organism's dentition to remain functional for a long period of time in the face of extreme wear (see Fig. 15.1). In contrast to low-crowned (brachydont) teeth, hypsodont teeth often have tall enamel pillars that are buttressed by cementum. Unworn teeth in hypsodont ungulates with complex crown morphology are capable of very little functional shearing but quickly wear to expose enamel ridges. Hypsodont teeth continue to erupt from the jaw during much of the lifespan of the animal, exposing fresh enamel ridges that preserve tooth function even as an abrasive diet quickly wears down teeth (Janis and Fortelius 1988; Damuth and Janis 2011). Traditionally, hypsodonty was often discussed as an adaptation to grazing (Simpson 1951; Stebbins 1981; Jacobs et al. 1999), and phytoliths (literally "plant rocks," structural silica incorporated into the tissues of many plants, especially grasses; see Strömberg et al. 2018) were thought to be the dominant source of abrasive particles (Baker et al. 1959). However, recent studies suggest that exogenous environmental grit may be a major source of abrasive particles contributing to tooth wear (Hoffman et al. 2015 and references therein; but see Spradley et al. 2016). Therefore, it is critical to emphasize that hypsodonty is adaptive for resisting dental wear, not for eating grass per se (Damuth and Janis 2014).

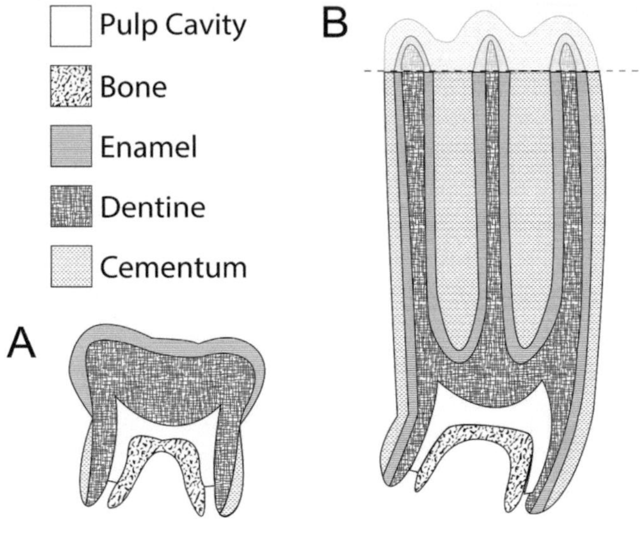

Fig. 15.1 Idealized coronal sections through: A, brachydont human molar; and B, hypsodont horse molar. Dotted line on hypsodont molar indicates wear level. A hypsodont molar retains functional enamel ridges even as abrasive wear reduces the height of the tooth crown during the life of the organism. Redrawn after Janis and Fortelius (1988, their Figure 8)

Many distantly-related mammalian groups have evolved hypsodonty to cope with high rates of wear. Janis and Fortelius (1988: Table 2) detail a wide range of groups with hypsodont species including; suids (pigs and relatives), rhinocerotids, equids (horses and kin), proboscideans (elephant relatives), notoungulates (extinct South American ungulates), macropodoid marsupials (wallabies and kangaroos), xenarthrans (sloths, armadillos, and relatives), lagomorphs (rabbits and pikas), and many rodent groups. Hypsodonty is a reliable ecomorphic indicator of an abrasive diet that has evolved convergently in many mammalian lineages.

The relationship between tooth crown height and diet is complicated by the occurrence of hypselodonty: teeth that grow and erupt continuously throughout an animal's lifespan. Among the groups noted above, hypselodonty has evolved in notoungulates, xenarthrans, lagomorphs, and rodents. In effect, ever-growing teeth are infinitely tall, because the tooth crown continues to form throughout life as teeth are worn, even if the crown is not necessarily tall at any given point in an organism's lifespan. Therefore, linking the height of the tooth crown with diet or habitat can be difficult in hypselodont species (Bargo et al. 2006a).

For ungulates, Janis (1988) showed that a simple hypsodonty index (HI), calculated as M3 height divided by M3 buccolingual width, effectively differentiates between extant browsers and grazers. Exclusive browsers generally have HI values < 2, and exclusive grazers typically show HI values > 4.5 (Damuth and Janis 2011). HI is commonly used to infer the diet of newly described species (e.g., Faith et al. 2012) and for broader paloeocological studies to predict diet in a taxonomically diverse array of species (e.g., Reed 1997; Janis et al. 2000; Flynn et al. 2003; Croft et al. 2008). In taxa with highly modified dentitions, such as rodents, other hypsodonty metrics may be more appropriate. Williams and Kay (2001) used a hypsodonty metric of M1 height scaled to M1 crown area because M1 is the largest tooth in many rodent lineages and rodent molars vary in shape to a greater degree than those of ungulates.

Efforts have been made to link hypsodonty to measureable aspects of climate at broad geographic scales and through geological time (Fortelius et al. 2002). So-called "ecometric" studies (see Vermillion et al. 2018) have demonstrated that the geographic distribution of mean hypsodonty values for large mammals in a community correlates with mean annual precipitation (MAP) (Eronen et al. 2010a; Polly et al. 2011). This approach seeks to characterize the distribution of an ecomorphic trait (hypsodonty) in a community and correlate this community value with underlying climate variables. As such, this type of approach is a promising way forward for "applied" ecomorphology because it provides a means of connecting the morphology of individuals with habitat variables that may be understood with reference to modern analogue habitats. The analysis of dental traits other than hypsodonty is discussed elsewhere in this volume (Evans and Pineda-Munoz 2018).

Craniodental Dietary Ecomorphology

Craniodental morphology has been linked to dietary ecology in a wide array of animals and has been particularly well-studied in ungulates (Mendoza et al. 2002; Clauss et al. 2008; Raia et al. 2010). In ungulates, one of the best-understood craniodental differences between grazing and browsing forms is the overall shape of the muzzle (Janis and Ehrhardt 1988; Solounias and Moelleken 1993; Pérez–Barbería and Gordon 2001; Tennant and MacLeod 2014). Modern grazing taxa such as equids (horses), warthogs (*Phacochoerus* spp.), the white rhinoceros (*Ceratotherium simum*), and alcelaphin bovids (wildebeest and relatives) tend to have wide (blunt or U-shaped) muzzles, reflecting a non-selective feeding strategy of cropping grasses. Browsers, such as black rhinoceros (*Diceros bicornis*) and cephalophin bovids (duikers), have narrower (pointed or V-shaped) muzzles that reflect their more selective browsing habits (see Fig. 15.2). Spencer (1995, 1997) identified several additional craniodental features related to dietary and habitat preferences in bovids. Grazers are characterized by flexed braincases, deep and wide mandibles, and generally long faces, whereas browsing taxa tend to have the converse (Spencer 1995). Spencer (1995) also showed that premolar row length differentiates taxa that inhabit secondary grasslands, which have relatively short premolar rows, from those dwelling in edaphic grassland, which have longer premolar rows. These established relationships between craniodental morphology and diet/habitat preference make it possible to confidently characterize the autecology of fossil bovids when adequate craniodental evidence is available.

The correlations between craniodental morphology and diet described above are general relationships that are not specific to ungulates. For instance, grazing macropodoid marsupials are convergent with grazing ungulates in having broad incisors, deep and wide mandibular angles, and acute basicranial angles (Janis 1990). Similarly, extinct South American ground sloths appear to exhibit the same relationship between muzzle shape and dietary selectivity as ungulates, with wide-muzzled sloths inferred to be

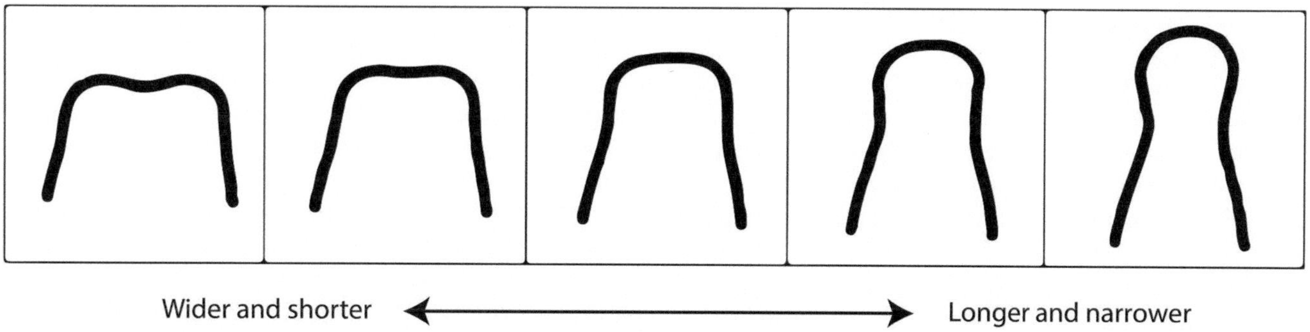

Fig. 15.2 Diagrams illustrating the first principal component of shape variation in the snouts of extant ruminants, from Tennant and MacLeod (2014). The left side of the figure shows snouts more typical of grazers, while the right side shows snouts more typical of browsers. These diagrams illustrate shape change across a dietary spectrum and none corresponds with a particular taxon

non-selective bulk feeders and narrow-muzzled forms inferred to be more selective (Bargo et al. 2006b; Bargo and Vizcaíno 2008; Shockey and Anaya 2011). Similar dietary trends in muzzle and mandible shape have been shown for extinct equids (Raia et al. 2010; Bernardes et al. 2013) and have also been inferred for extinct South American relatives of armadillos known as glyptodonts (Vizcaíno et al. 2011).

Various aspects of craniodental morphology have been linked to broad dietary categories in rodents (Samuels 2009). For example, carnivorous rodents have elongate incisors and reduced cheek tooth areas, insectivorous rodents have elongate snouts and degenerate dentitions, and herbivorous rodents have elongate molar tooth rows and a wider skull and rostrum (Samuels 2009). These characterizations apply to rodents from diverse phylogenetic groups, indicating convergence in these morphological features as dietary adaptations. Furthermore, some characteristics such as wide rostra in herbivorous rodents parallel differences in rostrum shape seen among ungulate dietary groups (discussed above) and thus point to general biomechanical principles driving these shape differences.

Craniodental ecomorphology has been well studied in carnivorans (members of order Carnivora, such as cats, dogs, and bears) with respect to feeding style and prey capture and how these traits have changed through evolutionary time (e.g., Radinsky 1981; Van Valkenburgh 1988, 1989; Lewis 1997; Wesley-Hunt 2005; Werdelin and Wesley-Hunt 2010, 2014; Figueirido et al. 2011; Meloro 2011a, b; Meloro and O'Higgins 2011; Meloro et al. 2015). Carnivorans can be classified into broad dietary categories ranging from hypercarnivores that specialize on flesh consumption to meat/bone eaters to omnivores that have broad diets (Van Valkenburgh 1989). As might be expected, species in these categories possess distinctive craniodental adaptations related to their diet. Hypercarnivores (e.g., lions and tigers) tend to have extremely well-developed carnassial teeth (specialized for shearing flesh) and little to no tooth area devoted to grinding (Van Valkenburgh 1989). Meat/bone eaters (e.g., hyenas) have somewhat less well-developed carnassials and are characterized by extremely large and robust premolars for crushing bones (Van Valkenburgh 1989). Omnivorous carnivorans (e.g., bears and raccoons) have weakly developed carnassial teeth and a relatively large proportion of tooth area devoted to grinding (Van Valkenburgh 1989).

Canine tooth morphology is related to killing behaviors in predatory carnivorans. Felids (cats and relatives) have long, strong canines that are used to kill prey with deep bites that often contact bone (Van Valkenburgh and Ruff 1987). Canids (dogs and relatives) kill prey with shallower bites that do not penetrate deeply and less frequently result in canine-bone contact (Van Valkenburgh and Ruff 1987). Thus, canids and other carnivorans that kill with shallow slashing wounds exhibit mediolaterally compressed canines characterized by lower bending strengths than felids (Van Valkenburgh and Ruff 1987).

Locomotor Ecomorphology

A large proportion of the fossil record consists of isolated postcranial fossils, which often cannot be precisely identified taxonomically (e.g., to genus or species). However, such specimens can be analyzed using "taxon-free" (Andrews and Hixson 2014) ecomorphological methods. In general, locomotor ecomorphology aims to link morphological traits with habitat-specific locomotor performance. This link may be based on habitual substrate use, as in studies of cercopithecoid primate (Old World monkey) ecomorphology, which have focused on distinguishing terrestrial, semi-terrestrial, and arboreal forms (Gebo and Sargis 1994; Elton 2001, 2002; Gosselin-Ildari 2013; Elton et al. 2016). Studies of carnivoran postcranial ecomorphology have focused both on substrate use (fossorial, terrestrial, arboreal) and locomotor speed (Van Valkenburgh 1987; Polly 2010; Walmsley et al. 2012; Meloro et al. 2013). A recent study of rodent postcranial ecomorphology focused on distinguishing locomotor categories based on substrate use (Samuels and Van Valkenburgh 2008).

For ungulates, the link between morphology and locomotion has been made with regard to habitat-specific modes of predator avoidance. Ungulate taxa living in open, grassland-type habitats face intense pressure from cursorial (fast-running) carnivorans. Many open-country taxa rely on running speed to out-pace these predators (Jarman 1974) and thus have adaptations for rapid locomotion over flat substrates. Cursorial adaptations include elongate distal limb elements (Scott 1985) and joint adaptations that restrict joint motion to the parasagittal plane (Kappelman 1988). However, it is important to note that not all cursorial adaptations are the result of selection for increased running speed. In camelids, cursorial adaptations may have arisen for efficient long-distance dispersal rather than to outpace cursorial carnivorans (Janis et al. 2002).

Ungulate taxa that occupy dense forests with closed canopies are faced with complicated three-dimensional substrates that include undergrowth, fallen trees, and other obstacles. Cursoriality is a less effective predator-avoidance strategy in these environments, and forest-dwelling taxa tend to rely more on crypsis (hiding) to avoid predation. Thus, cursorial adaptations are less pronounced in forest-adapted taxa, and adaptations towards joint mobility are emphasized. In bovids, locomotor ecomorphology has been used to predict habitat preferences based on the morphology of femora (Kappelman 1988, 1991; Kappelman et al. 1997; Janis et al. 2012), metapodials (Plummer and Bishop 1994; Scott et al. 1999), astragali (DeGusta and Vrba 2003; Plummer et al. 2008, 2015; Barr 2014, 2015), phalanges (DeGusta and Vrba 2005; Louys et al. 2013), and a variety of other skeletal elements (Kovarovic and Andrews 2007). Similar methods linking postcranial morphology to predator avoidance have also been applied to suids (pigs; Bishop et al. 1999, 2006) and cervids (deer; Curran 2012, 2015).

Steps in an Applied Ecomorphic Analysis

This section details the steps in a typical applied ecomorphic analysis. As a practical example, this will be based on my own ecomorphic study of bovid astragali.

First, traits to be measured or observed are identified, ideally with reference to specific functional hypotheses linking each trait to the ecological variable of interest (e.g., dietary preference or locomotor category). These traits may be typical linear measurements or composite variables derived from an ordination of geometric morphometric data (e.g., Curran 2012, 2018). In my analysis of bovid astragali, I mined the literature of previous ecomorphic work on astragali (DeGusta and Vrba 2003; Weinand 2007; Plummer et al. 2008) as well as the functional morphology literature on artiodactyl ankle joints (Schaeffer 1947, 1948; Alexander and Bennett 1987) in order to obtain a candidate list of traits.

It is advisable to restrict the number of traits used in an ecomorphic analysis, as using too many traits can lead to "over-fitting," which produces artificially high classification accuracies (Kovarovic et al. 2011). One approach to deal with this issue is to use phylogenetic comparative methods in order to identify the traits that are most closely related to the ecological category of interest (Barr and Scott 2014). Traits that are less clearly related to ecology can then be removed from further analyses (Barr 2014). Other approaches exist for dealing with this problem (Louys et al. 2013; Elton et al. 2016), and this is also discussed further below.

The next step is to use relevant comparative species to determine how well the measured variables reflect the ecological categories of interest. Comparative species are often chosen among closely-related taxa, but this may not always be possible, especially when studying the ecomorphology of extinct species with no modern relatives (Chen and Wilson 2015). Comparative species must be classified into one of several ecological categories of interest. For instance, in bovid locomotor studies (e.g., Kappelman 1991; Plummer and Bishop 1994; Kappelman et al. 1997; Scott et al. 1999; DeGusta and Vrba 2003; Kovarovic and Andrews 2007), species are typically classified as occupying one of several habitat types such as open grassland, closed forest, or montane habitat. For studies relating to diet, the classification scheme reflects dietary categories, such as hypercarnivore, meat/bone, or omnivore in the case of carnivorans (Van Valkenburgh 1989).

Once variables are measured and extant species are classified into categories, Discriminant Function Analysis (DFA) (Manly 2004) is frequently used to quantify the relationship between the traits and the ecological categories of interest. Wilk's lambda (Manly 2004: 46) and its associated p-value can be used to determine whether a DFA is statistically significant. If a DFA is shown to be statistically significant, a variety of metrics can be used to judge the accuracy of its classifications. The simplest metric is termed "resubstitution" (DeGusta and Vrba 2003; Plummer et al. 2008). Resubstitution involves making categorical predictions for each specimen with DFA equations derived from the original dataset and computing the percent accuracy of these predictions. However, making "predictions" for specimens using equations derived from those specimens is problematic and results in overestimation of predictive success (Kovarovic et al. 2011). Cross-validation gets around this problem by withholding each specimen of the dataset one-by-one and recalculating the DFA each time. Classifications for each specimen are thus made using equations that do not include that particular specimen. Cross-validation offers a more realistic picture of DFA prediction success (Kovarovic et al. 2011).

Finally, the same measurements can be taken from fossils and added to the DFA equations derived from the extant comparative study in order to predict their most likely ecological category. These categorical predictions can then be used to infer distribution of the ecological trait of interest in the community. In my study of the bovids from the Shungura Formation in Ethiopia (Barr 2015), I measured 234 fossil astragali using the same measurement scheme I had previously validated in extant species (Barr 2014). I used the DFA predictions generated from these fossils to demonstrate that the Shungura Formation bovid community was dominated by individuals adapted to closed environments at 3.4 Ma, while the community at 2.4 Ma included more open-adapted individuals. These kinds of results can be used to infer changes in ecological communities and environmental conditions through geological and evolutionary time.

Strengths of Ecomorphology

A widely-cited strength of ecomorphology is that the method is "taxon-free" (Andrews and Hixson 2014), meaning that ecomorphic inferences do not rely on taxonomic identifications; rather, they are based on functional morphological traits. The taxon-free nature of ecomorphology is a strength in two ways. First, because it is taxon-free, ecomorphology does not require the assumption of taxonomic uniformitarianism; i.e., that a fossil species must have had the same ecological characteristics as modern representatives of the same taxonomic group (e.g., tribe or subfamily). This assumption often underlies paleoenvironmental interpretations of faunal lists, for instance. To take a hypothetical example: tragelaphin bovids (bushbuck and relatives) may be identified at a site based on horn core evidence, and this may be taken to indicate the presence of woodland habitats because modern tragelaphins are browsing taxa, and several modern species occupy dense woodland and forest environments. This assumption may be problematic if there is within-lineage evolution in ecological characteristics because fossil species may differ from their modern relatives. In contrast to this uniformitarian approach, the ecomorphic approach infers ecological information for each specimen based directly on functional links between morphology and ecology. In an applied ecomorphic study, a site would only be interpreted as reflecting a woodland environment if fossil remains showed functional traits that could be functionally linked to a browsing diet or locomotion adapted for woodland habitats.

Secondly, the taxon-free nature of ecomorphology means that much larger samples are available for ecological characterizations. In many cases, only relatively complete craniodental specimens, which are rare in the fossil record, can be identified to species and analyzed via taxonomy-based methods. Fossil limb bones and other less-complete material that can only be identified to a higher taxonomic level (e.g., to family) can still be used for ecomorphic reconstructions. Moreover, as alluded to above, taxonomic identification in bovids relies heavily on horn cores, which in some taxa are only found in males. By contrast, both sexes have limb bones and teeth, common targets of ecomorphic studies. Furthermore, molar teeth and some postcranial elements (e.g., astragali) are extremely durable and tend to be well represented in all paleontological collections. Thus, studying postcranial elements and teeth may provide a clearer picture of the relative abundances of bovids with different ecological characteristics as compared to taxonomy-based analyses that rely preferentially on horn cores and cranial material.

Shortcomings of Ecomorphology

One of the major shortcomings of ecomorphology is related to the categorization schemes used in ecomorphic studies (e.g., browser vs. grazer for diet, or forest vs. grassland for habitat). Some species are easily assigned unambiguously to categories on the extremes of ecological continua (e.g., committed grazers or dedicated browsers), but many others are intermediate with respect to these continua (e.g., mixed-feeding species). Species assigned to such intermediate ecological categories can show significant variation. For instance, some mixed-feeding species might prefer grass while others prefer browse even though all mixed feeders consume both vegetation types. This heterogeneity within ecological categories is the result of parsing continuous ecological variation into discrete analytical units. Assigning taxa to a particular category is almost always an oversimplification, but more precise methods of ecological characterization are possible. For example, GIS analysis of species distributions with respect to ecological gradients can be used to achieve a more quantitative picture of a species' habitat preferences (Meloro et al. 2013).

Although the "taxon-free" nature of ecomorphology has been discussed as a strength of the method, there are several ways in which taxonomy (and phylogeny) can influence the results of an ecomorphic study. First, ecomorphology is not completely taxon-free because specimens must generally be identified at least to family, as many ecomorphic predictive equations apply within families (Andrews and Hixson 2014). Secondly, the taxonomic composition of the comparative sample can influence the outcome of a DFA (Klein et al. 2010). For instance, if a modern comparative sample is heavily dominated by alcelaphin bovids (committed grazers) and cephalophin bovids (committed browsers), the DFA will produce very high classification accuracies and will appear to succeed very well in predicting diet. However, the same DFA may not perform nearly as well for mixed-feeding taxa that were not well represented in the extant sample. This problem can be mitigated to some extent by careful selection of the extant comparative sample.

Even if all modern species were to be included in a comparative sample, the phylogenetic relationships of extant taxa would need to be carefully considered. This is because phylogenetic signal can cause a DFA to produce high classification accuracies that are easy to over-interpret in terms of the functional link between morphological traits and ecological categories (Barr 2014). This problem stems from the fact that species are related in a hierarchical phylogeny, with some species in the extant sample more closely related to one another than to other species in the sample. Since closely-related species tend to resemble each other in morphology as well as ecology, each species in the data set may not represent a separate instance of adaptation for a particular diet or locomotor strategy. This can raise issues of what is known as phylogenetic pseudo-replication; closely related species are not statistically independent, and it is inappropriate to treat them as such (Felsenstein 1985).

The issue is illustrated in Fig. 15.3, which shows two hypothetical phylogenetic relationships for woodland and grassland-adapted bovids. In Fig. 15.3A, woodland and grassland taxa are scattered across the two clades, and there is a perfect correspondence between the morphological trait and the habitat category. This implies a very strong functional relationship between the trait represented by the green circles and woodland habitats, because species in different clades have adapted convergently to the same habitat several times. Therefore, extrapolating outside this sample (i.e., predicting habitat preference based on trait values in fossils) could be done with relative confidence; there are multiple instances of convergent morphological adaptation to habitat supporting this ecomorphic linkage.

In Fig. 15.3B, the woodland taxa comprise a phylogenetic group to the exclusion of the grassland-adapted taxa. There is still a perfect correspondence between morphological traits and habitat preference, but there are no instances of convergence (independent evolution of the same trait). Adaptation to habitat occurred in ancestral taxa (green and yellow triangles), and descendant taxa within each clade share habitat preferences and morphological traits because they inherited them from their common ancestor. Traits that have a distribution that tracks phylogenetic topology in this way are said to have high phylogenetic signal (Münkemüller et al. 2012). Clearly, the morphological traits in the extant taxa in Fig. 15.3B must be functionally compatible with the taxon's habitat preference; otherwise, these taxa would either go extinct or adapt morphologically. Thus, traits with high phylogenetic signal are not necessarily uninformative for ecomorphology. Nevertheless, when traits have high phylogenetic signal, extrapolating ecomorphic predictions to taxa outside of the comparative sample may be more difficult, because the functional relationship between traits is confounded by phylogenetic relatedness.

Figure 15.3 presents extreme cases for purposes of illustrating the issue. In practice, morphological traits tend to have more subtle and complex relationships to both phylogenetic and ecological groupings. Traits that exhibit some level of phylogenetic signal often also exhibit a functional signal above and beyond that predicted based on phylogenetic relationships (Scott and Barr 2014). Thus, the presence of phylogenetic signal by no means precludes the use of a trait in an ecomorphic study, but it does argue for careful consideration about what factors drive variation in that trait. One approach to dealing with the issue of phylogenetic signal is to screen traits for a significant functional relationship using phylogenetic comparative methods and to exclude traits that do not have strong functional signals exceeding that expected based on phylogenetic relationships (Barr 2014). A different approach is to perform DFA first and subsequently search for examples of morphological convergence among distantly-related species in the same ecological category as well as morphological divergence among closely-related species in different ecological categories (Louys et al. 2013). The benefits and limitations of these and other methods for dealing with phylogeny in applied ecomorphic studies continues to be discussed (Elton et al. 2016).

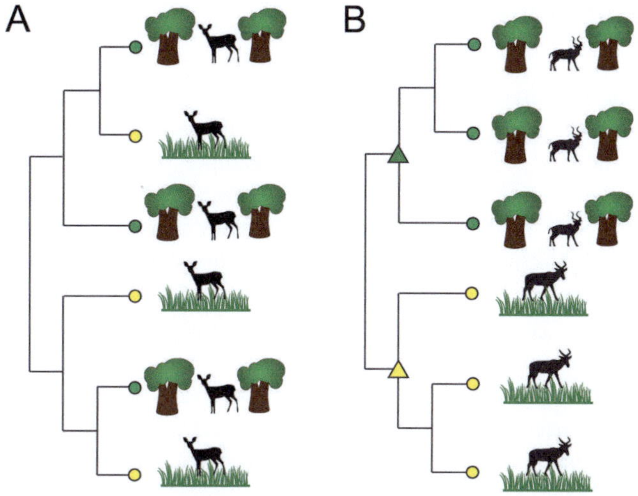

Fig. 15.3 Two hypothetical scenarios showing different relationships between morphological traits (colored circles) and habitat preferences (grassland or woodland). Colored triangles represent morphological traits in extinct ancestral taxa. In scenario A, there is a perfect correspondence between morphology and habitat, and there are several instances of convergent evolution of the same morphological trait in a particular habitat. In scenario B, there is still a perfect correspondence between habitat and morphology, but there are no examples of convergent evolution of morphological traits. Rather, each species inherited morphological traits and habitat preferences from a common ancestor. Scenario A offers the strongest possible evidence for a functional link between morphology and habitat, while the evidence for a functional link in scenario B is not as well supported because it lacks evidence of convergent evolution

Future Prospects

The biggest challenge facing applied ecomorphic studies is understanding what results mean in terms of interpretable ecological variables. Take, for example, a bovid fossil assemblage for which postcranial ecomorphology indicates that 85% of the bovids are open-country adapted and 15% are light-cover adapted. These proportions provide critical primary data regarding the adaptations of the bovid community. However, these proportions are not straightforward to interpret in terms of what they reveal about inferred habitats in the past, even if we assume perfect fidelity (i.e., 100% accuracy) in habitat predictions for each fossil.

What is the likelihood that the hypothetical sample above could be drawn from the bovid community in an open-arid environment such as the Etosha National Park in Namibia, a habitat dominated by grassland and low-density mopane savanna (du Plessis 1999)? What is the likelihood of drawing the same sample from a grassland community in the Serengeti ecosystem? Both systems have extensive grasslands, but the Serengeti receives more precipitation than Etosha, and the two areas likely present distinct ecological challenges and affordances. The methods currently used in ecomorphology do not offer a formal way of excluding or including either of these possible modern analogs based on the relative proportions of bovid adaptations in the fossil sample. One clear path forward is to examine the proportions of bovids with different adaptations in modern ecosystems to link these proportions to potential modern analogues. This type of analysis has been done only on a limited basis with bovid locomotor categories (Plummer and Bishop 1994). Taxonomically broader community structure analysis has been more successful in linking the distribution of ecological traits with environmental variables and/or modern analogues (Andrews et al. 1979; Reed 1997, 1998; Flynn et al. 2003; Kovarovic et al. 2018). In the future, a major focus in applying ecomorphic methods should be in establishing rigorous ways of interpreting the results of these methods. So-called "ecometric" methods, which link community values of ecomorphic traits with climatic correlates (Fortelius et al. 2002; Eronen et al. 2010a, b; Polly et al. 2011; Vermillion et al. 2018) may offer a way forward.

References

Alexander, R. M., & Bennett, M. B. (1987). Some principles of ligament function, with examples from the tarsal joints of the sheep (*Ovis aries*). *Journal of Zoology, 211,* 487–504.

Andrews, P., & Hixson, S. (2014). Taxon-free methods of palaeoecology. *Annales Zoologici Fennici, 51,* 269–284.

Andrews, P., Lord, J. M., & Nesbit-Evans, E. M. (1979). Patterns of ecological diversity in fossil and modern mammalian faunas. *Biological Journal of the Linnean Society, 11,* 177–205.

Baker, G., Jones, L. H. P., & Wardrop, I. D. (1959). Cause of wear in sheeps' teeth. *Nature, 184,* 1583–1584.

Bargo, M. S., & Vizcaíno, S. F. (2008). Paleobiology of Pleistocene ground sloths (Xenarthra, Tardigrada): biomechanics, morphogeometry and ecomorphology applied to the masticatory apparatus. *Ameghiniana, 45,* 175–196.

Bargo, M. S., De Iuliis, G., & Vizcaíno, S. F. (2006a). Hypsodonty in Pleistocene ground sloths. *Acta Palaeontologica Polonica, 51,* 53.

Bargo, M. S., Toledo, N., & Vizcaíno, S. F. (2006b). Muzzle of South American Pleistocene ground sloths (Xenarthra, Tardigrada). *Journal of Morphology, 267,* 248–263.

Barr, W. A. (2014). Functional morphology of the bovid astragalus in relation to habitat: controlling phylogenetic signal in ecomorphology. *Journal of Morphology, 275,* 1201–1216.

Barr, W. A. (2015). Paleoenvironments of the Shungura Formation (Plio-Pleistocene: Ethiopia) based on ecomorphology of the bovid astragalus. *Journal of Human Evolution, 88,* 97–107.

Barr, W. A., & Scott, R. (2014). Phylogenetic comparative methods complement discriminant function analysis in ecomorphology. *American Journal of Physical Anthropology, 153,* 663–674.

Bernardes, C., Sicuro, F. L., Avilla, L. S., & Pinheiro, A. E. P. (2013). Rostral reconstruction of South American hippidiform equids: new anatomical and ecomorphological inferences. *Acta Palaeontologica Polonica, 58,* 669–678.

Bishop, L., Hill, A., & Kingston, J. (1999). Paleoecology of Suidae from the Tugen Hills, Baringo, Kenya. In P. Andrews & P. Banham (Eds.), *Late Cenozoic environments and hominid evolution: A tribute to Bill Bishop* (pp. 99–111). London: Geological Society, London.

Bishop, L. C., King, T., Hill, A., & Wood, B. (2006). Palaeoecology of *Kolpochoerus heseloni* (=*K. limnetes*): a multiproxy approach. *Transactions of the Royal Society of South Africa, 61,* 81–88.

Bock, W. J. (1989). From biologische Anatomie to ecomorphology. *Netherlands Journal of Zoology, 40,* 254–277.

Bock, W. J. (1994). Concepts and methods in ecomorphology. *Journal of Biosciences, 19,* 403–413.

Bock, W. J., & von Wahlert, G. (1965). Adaptation and the form-function complex. *Evolution, 19,* 269–299.

Cerling, T. E., Andanje, S. A., Blumenthal, S. A., Brown, F. H., Chritz, K. L., Harris, J. M., et al. (2015). Dietary changes of large herbivores in the Turkana Basin, Kenya from 4 to 1 Ma. *Proceedings of the National Academy of Sciences, USA, 112,* 11467–11472.

Chen, M., & Wilson, G. P. (2015). A multivariate approach to infer locomotor modes in Mesozoic mammals. *Paleobiology, 41,* 280–312.

Clauss, M., Kaiser, T., & Hummel, J. (2008). The morphophysiological adaptations of browsing and grazing mammals. In I. J. Gordon & H. H. T. Prins (Eds.), *The ecology of browsing and grazing. Ecological Studies, 195,* 47–88.

Croft, D. A., Flynn, J. J., & Wyss, A. R. (2008). The Tinguiririca fauna of Chile and the early stages of "modernization" of South American mammal faunas. *Arquivos do Museu Nacional, 66,* 191–211.

Curran, S. C. (2012). Expanding ecomorphological methods: geometric morphometric analysis of Cervidae post-crania. *Journal of Archaeological Science, 39,* 1172–1182.

Curran, S. C. (2015). Exploring *Eucladoceros* ecomorphology using geometric morphometrics. *The Anatomical Record, 298,* 291–313.

Curran, S. C. (2018). Three-dimensional geometric morphometrics in paleoecology. In D. A. Croft, D. F. Su & S. W. Simpson (Eds.), *Methods in paleoecology: Reconstructing Cenozoic terrestrial environments and ecological communities* (pp. 317–335). Cham: Springer.

Damuth, J., & Janis, C. M. (2011). On the relationship between hypsodonty and feeding ecology in ungulate mammals, and its utility in palaeoecology. *Biological Reviews, 86,* 733–758.

Damuth, J., & Janis, C. M. (2014). A comparison of observed molar wear rates in extant herbivorous mammals. *Annales Zoologici Fennici, 51,* 188–200.

DeGusta, D., & Vrba, E. (2003). A method for inferring paleohabitats from the functional morphology of bovid astragali. *Journal of Archaeological Science, 30,* 1009–1022.

DeGusta, D., & Vrba, E. (2005). Methods for inferring paleohabitats from the functional morphology of bovid phalanges. *Journal of Archaeological Science, 32,* 1099–1113.

Dunn, R. H. (2018). Functional morphology of the postcranial skeleton. In D. A. Croft, D. F. Su & S. W. Simpson (Eds.), *Methods in paleoecology: Reconstructing Cenozoic terrestrial environments and ecological communities* (pp. 23–36). Cham: Springer.

du Plessis, W. P. (1999). Linear regression relationships between NDVI, vegetation and rainfall in Etosha National Park, Namibia. *Journal of Arid Environments, 42,* 235–260.

Elton, S. (2001). Locomotor and habitat classification of cercopithecoid postcranial material from Sterkfontein Member 4, Bolt's Farm and Swartkrans Members 1 and 2, South Africa. *Palaeontologia Africana, 37,* 115–126.

Elton, S. (2002). A reappraisal of the locomotion and habitat preference of *Theropithecus oswaldi. Folia Primatologica, 73,* 252–280.

Elton, S., Jansson, A.-U., Meloro, C., Louys, J., Plummer, T. W., & Bishop, L. C. (2016). Exploring morphological generality in the Old World monkey postcranium using an ecomorphological framework. *Journal of Anatomy, 228,* 534–560.

Eronen, J. T., Polly, P. D., Fred, M., Damuth, J., Frank, D. C., Mosbrugger, V., et al. (2010a). Ecometrics: the traits that bind the past and present together. *Integrative Zoology, 5,* 88–101.

Eronen, J. T., Puolamäki, K., Liu, L., Lintulaakso, K., Damuth, J., Janis, C. M., et al. (2010b). Precipitation and large herbivorous mammals I: estimates from present-day communities. *Evolutionary Ecology Research, 12,* 217–233.

Evans, A. R., & Pineda-Munoz, S. (2018). Inferring mammal dietary ecology from dental morphology. In D. A. Croft, D. F. Su & S. W. Simpson (Eds.), *Methods in paleoecology: Reconstructing Cenozoic terrestrial environments and ecological communities* (pp. 37–51). Cham: Springer.

Faith, J. T., Potts, R., Plummer, T. W., Bishop, L. C., Marean, C. W., & Tryon, C. A. (2012). New perspectives on middle Pleistocene change in the large mammal faunas of East Africa: *Damaliscus hypsodon* sp. nov. (Mammalia, Artiodactyla) from Lainyamok, Kenya. *Palaeogeography, Palaeoclimatology, Palaeoecology, 361–362,* 84–93.

Felsenstein, J. (1985). Phylogenies and the comparative method. *American Naturalist, 125,* 1–15.

Figueirido, B., MacLeod, N., Krieger, J., De Renzi, M., Pérez-Claros, J. A., & Palmqvist, P. (2011). Constraint and adaptation in the evolution of carnivoran skull shape. *Paleobiology, 37,* 490–518.

Flynn, J. J., Wyss, A. R., Croft, D. A., & Charrier, R. (2003). The Tinguiririca Fauna, Chile: biochronology, paleoecology, biogeography, and a new earliest Oligocene South American Land Mammal "Age". *Palaeogeography, Palaeoclimatology, Palaeoecology, 195,* 229–259.

Fortelius, M. (1985). Ungulate cheek teeth: developmental, functional, and evolutionary interrelations. *Acta Zoologica Fennica, 180,* 1–76.

Fortelius, M., Eronen, J., Jernvall, J., Liu, L., Pushkina, D., Rinne, J., et al. (2002). Fossil mammals resolve regional patterns of Eurasian climate change over 20 million years. *Evolutionary Ecology Research, 4,* 1005–1016.

Gebo, D. L., & Sargis, E. J. (1994). Terrestrial adaptations in the postcranial skeletons of guenons. *American Journal of Physical Anthropology, 93,* 341–371.

Gordon, I. J., & Prins, H. H. T. (Eds.). (2008). *The ecology of browsing and grazing. Ecological Studies* (Vol. 195). Berlin: Springer.

Gosselin-Ildari, A. (2013). *The evolution of cercopithecoid locomotion: A morphometric, phylogenetic, and character mapping approach.* Ph.D. Dissertation, Stony Brook University.

Hoffman, J. M., Fraser, D., & Clementz, M. T. (2015). Controlled feeding trials with ungulates: a new application of in vivo dental molding to assess the abrasive factors of microwear. *Journal of Experimental Biology, 218,* 1538–1547.

Hofmann, R. R., & Stewart, D. R. M. (1972). Grazer or browser: a classification based on the stomach-structure and feeding habits of East African ruminants. *Mammalia, 36,* 226–240.

Jacobs, B. F., Kingston, J. D., & Jacobs, L. L. (1999). The origin of grass-dominated ecosystems. *Annals of the Missouri Botanical Garden, 86,* 590–643.

Janis, C. M. (1988). An estimation of tooth volume and hypsodonty indices in ungulate mammals, and the correlation of these factors with dietary preference. In D. E. Russell, J. P. Santoro & D. Sigogneau-Russell (Eds.), *Teeth revisited: Proceedings of the VII international symposium on dental morphology, Mémoirs de Museé d'Histoire Naturelle, Paris, Serie C, Volume 53* (pp. 371–391). Paris: Editions du Muséum.

Janis, C. M. (1990). Correlation of cranial and dental variables with body size in ungulates and macropodids. *Memoirs of The Queensland Museum, 28,* 349–366.

Janis, C. M. (2008). An evolutionary history of browsing and grazing ungulates. In I. J. Gordon & H. H. T. Prins (Eds.), *The ecology of browsing and grazing. Ecological Studies, 195,* 21–45.

Janis, C. M., & Ehrhardt, D. (1988). Correlation of relative muzzle width and relative incisor width with dietary preference in ungulates. *Zoological Journal of the Linnean Society, 92,* 267–284.

Janis, C. M., & Fortelius, M. (1988). On the means whereby mammals achieve increased functional durability of their dentitions, with special reference to limiting factors. *Biological Reviews of the Cambridge Philosophical Society, 63,* 197.

Janis, C. M., Damuth, J., & Theodor, J. M. (2000). Miocene ungulates and terrestrial primary productivity: where have all the browsers gone? *Proceedings of the National Academy of Sciences, USA, 97,* 7899–7904.

Janis, C. M., Theodor, J. M., & Boisvert, B. (2002). Locomotor evolution in camels revisited: a quantitative analysis of pedal anatomy and the acquisition of the pacing gait. *Journal of Vertebrate Paleontology, 22,* 110–121.

Janis, C. M., Shoshitaishvili, B., Kambic, R., & Figueirido, B. (2012). On their knees: distal femur asymmetry in ungulates and its relationship to body size and locomotion. *Journal of Vertebrate Paleontology, 32,* 433–445.

Jarman, P. J. (1974). The social organisation of antelope in relation to their ecology. *Behaviour, 48,* 215–267.

Kappelman, J. (1988). Morphology and locomotor adaptations of the bovid femur in relation to habitat. *Journal of Morphology, 198,* 119–130.

Kappelman, J. (1991). The paleoenvironment of *Kenyapithecus* at Fort Ternan. *Journal of Human Evolution, 20,* 95–129.

Kappelman, J., Plummer, T. W., Bishop, L., Duncan, A., & Appleton, S. (1997). Bovids as indicators of Plio-Pleistocene paleoenvironments in East Africa. *Journal of Human Evolution, 32,* 229–256.

Klein, R. G., Franciscus, R. G., & Steele, T. E. (2010). Morphometric identification of bovid metapodials to genus and implications for taxon-free habitat reconstruction. *Journal of Archaeological Science, 37,* 389–401.

Kovarovic, K., & Andrews, P. (2007). Bovid postcranial ecomorphological survey of the Laetoli paleoenvironment. *Journal of Human Evolution, 52,* 663–680.

Kovarovic, K., Aiello, L. C., Cardini, A., & Lockwood, C. A. (2011). Discriminant function analyses in archaeology: are classification rates too good to be true? *Journal of Archaeological Science, 38,* 3006–3018.

Kovarovic, K., Su, D. F. & Lintulaakso, K. (2018). Mammal community structure analysis. In D. A. Croft, D. F. Su, & S. W. Simpson (Eds.), *Methods in paleoecology: Reconstructing*

Cenozoic terrestrial environments and ecological communities (pp. 349–370). Cham: Springer.
Lewis, M. E. (1997). Carnivoran paleoguilds of Africa: implications for hominid food procurement strategies. *Journal of Human Evolution, 32,* 257–288.
Louys, J., Montanari, S., Plummer, T. W., Hertel, F., & Bishop, L. C. (2013). Evolutionary divergence and convergence in shape and size within African antelope proximal phalanges. *Journal of Mammalian Evolution, 20,* 239–248.
Manly, B. F. J. (2004). *Multivariate statistical methods: A primer* (3rd ed.). New York: Chapman & Hall/CRC.
Martin, L. (1985). Significance of enamel thickness in hominoid evolution. *Nature, 314,* 260–263.
Meloro, C. (2011a). Feeding habits of Plio-Pleistocene large carnivores as revealed by the mandibular geometry. *Journal of Vertebrate Paleontology, 31,* 428–446.
Meloro, C. (2011b). Locomotor adaptations in Plio-Pleistocene large carnivores from the Italian Peninsula: palaeoecological implications. *Current Zoology, 57,* 269–283.
Meloro, C., & O'Higgins, P. (2011). Ecological adaptations of mandibular form in fissiped Carnivora. *Journal of Mammalian Evolution, 18,* 185–200.
Meloro, C., Elton, S., Louys, J., Bishop, L. C., & Ditchfield, P. (2013). Cats in the forest: predicting habitat adaptations from humerus morphometry in extant and fossil Felidae (Carnivora). *Paleobiology, 39,* 323–344.
Meloro, C., Clauss, M., & Raia, P. (2015). Ecomorphology of Carnivora challenges convergent evolution. *Organisms Diversity & Evolution, 15,* 711–720.
Mendoza, M., Janis, C. M., & Palmqvist, P. (2002). Characterizing complex craniodental patterns related to feeding behaviour in ungulates: a multivariate approach. *Journal of Zoology, 258,* 223–246.
Münkemüller, T., Lavergne, S., Bzeznik, B., Dray, S., Jombart, T., Schiffers, K., et al. (2012). How to measure and test phylogenetic signal. *Methods in Ecology and Evolution, 3,* 743–756.
Pérez-Barbería, F. J., & Gordon, I. J. (2001). Relationships between oral morphology and feeding style in the Ungulata: a phylogenetically controlled evaluation. *Proceedings of the Royal Society of London. Series B: Biological Sciences, 268,* 1023.
Plummer, T. W., & Bishop, L. C. (1994). Hominid paleoecology at Olduvai Gorge, Tanzania as indicated by antelope remains. *Journal of Human Evolution, 27,* 47–75.
Plummer, T. W., Bishop, L. C., & Hertel, F. (2008). Habitat preference of extant African bovids based on astragalus morphology: operationalizing ecomorphology for palaeoenvironmental reconstruction. *Journal of Archaeological Science, 35,* 3016–3027.
Plummer, T. W., Ferraro, J. V., Louys, J., Hertel, F., Alemseged, Z., Bobe, R., et al. (2015). Bovid ecomorphology and hominin paleoenvironments of the Shungura Formation, lower Omo River Valley, Ethiopia. *Journal of Human Evolution, 88,* 108–126.
Polly, P. (2010). Tiptoeing through the trophics: geographic variation in carnivoran locomotor ecomorphology in relation to environment. In A. Goswami & A. Friscia (Eds.), *Carnivoran evolution: New views on phylogeny, form, and function* (pp. 347–410). Cambridge: Cambridge University Press.
Polly, P. D., Eronen, J. T., Fred, M., Dietl, G. P., Mosbrugger, V., Scheidegger, C., et al. (2011). History matters: ecometrics and integrative climate change biology. *Proceedings of the Royal Society of London B: Biological Sciences, 278,* 1131–1140.
Radinsky, L. B. (1981). Evolution of skull shape in carnivores: 1. Representative modern carnivores. *Biological Journal of the Linnean Society, 15,* 369–388.
Raia, P., Carotenuto, F., Meloro, C., Piras, P., & Pushkina, D. (2010). The shape of contention: adaptation, history, and contingency in ungulate mandibles. *Evolution, 64,* 1489–1503.
Reed, K. E. (1997). Early hominid evolution and ecological change through the African Plio-Pleistocene. *Journal of Human Evolution, 32,* 289–322.
Reed, K. E. (1998). Using large mammal communities to examine ecological and taxonomic structure and predict vegetation in extant and extinct assemblages. *Paleobiology, 24,* 384–408.
Samuels, J. X. (2009). Cranial morphology and dietary habits of rodents. *Zoological Journal of the Linnean Society, 156,* 864–888.
Samuels, J. X., & Van Valkenburgh, B. (2008). Skeletal indicators of locomotor adaptations in living and extinct rodents. *Journal of Morphology, 269,* 1387–1411.
Schaeffer, B. (1947). Notes on the origin and function of the artiodactyl tarsus. *American Museum Novitates, 1356,* 1–24.
Schaeffer, B. (1948). The origin of a mammalian ordinal character. *Evolution, 2,* 164–175.
Scott, K. (1985). Allometric trends and locomotor adaptations in the Bovidae. *Bulletin of the American Museum of Natural History, 179,* 197–288.
Scott, R. S., & Barr, W. A. (2014). Ecomorphology and phylogenetic risk: implications for habitat reconstruction using fossil bovids. *Journal of Human Evolution, 73,* 47–57.
Scott, R. S., Kappelman, J., & Kelley, J. (1999). The paleoenvironment of *Sivapithecus parvada*. *Journal of Human Evolution, 36,* 245–274.
Shockey, B. J., & Anaya, F. (2011). Grazing in a new late Oligocene mylodontid sloth and a mylodontid radiation as a component of the Eocene-Oligocene faunal turnover and the early spread of grasslands/savannas in South America. *Journal of Mammalian Evolution, 18,* 101–115.
Simpson, G. G. (1951). *Horses: The story of the horse family in the modern world and through sixty million years of history*. Oxford: Oxford University Press.
Solounias, N., & Moelleken, S. M. C. (1993). Dietary adaptation of some extinct ruminants determined by premaxillary shape. *Journal of Mammalogy, 74,* 1059–1071.
Spencer, L. M. (1995). Morphological correlates of dietary resource partitioning in the African Bovidae. *Journal of Mammalogy, 76,* 448–471.
Spencer, L. M. (1997). Dietary adaptations of Plio-Pleistocene Bovidae: implications for hominid habitat use. *Journal of Human Evolution, 32,* 201–228.
Spradley, J. P., Glander, K. E., & Kay, R. F. (2016). Dust in the wind: how climate variables and volcanic dust affect rates of tooth wear in central american howling monkeys. *American Journal of Physical Anthropology, 159,* 210–222.
Stebbins, G. L. (1981). Coevolution of grasses and herbivores. *Annals of the Missouri Botanical Garden, 68,* 75–86.
Strömberg, C. A. E., Dunn, R. E., Crifò, C., & Harris, E. B. (2018). Phytoliths in paleoecology: analytical considerations, current use, and future directions. In D. A. Croft, D. F. Su & S. W. Simpson (Eds.), *Methods in paleoecology: Reconstructing Cenozoic terrestrial environments and ecological communities* (pp. 233–285). Cham: Springer.
Tennant, J. P., & MacLeod, N. (2014). Snout shape in extant ruminants. *PLoS ONE, 9,* e112035.
Van Valen, L. (1960). A functional index of hypsodonty. *Evolution, 14,* 531–532.
Van Valkenburgh, B. (1987). Skeletal indicators of locomotor behavior in living and extant carnivores. *Journal of Vertebrate Paleontology, 7,* 162–182.

Van Valkenburgh, B. (1988). Trophic diversity in past and present guilds of large predatory mammals. *Paleobiology, 14,* 155–173.

Van Valkenburgh, B. (1989). Carnivore dental adaptations and diet: a study of trophic diversity within guilds. In J. L. Gittleman (Ed.), *Carnivore behavior, ecology, and evolution* (pp. 410–436). Ithaca: Cornell University Press.

Van Valkenburgh, B., & Ruff, C. B. (1987). Canine tooth strength and killing behaviour in large carnivores. *Journal of Zoology, 212,* 379–397.

Vermillion, W. A., Polly, P. D., Head, J. J., Eronen, J. T., & Lawing, A. M. (2018). Ecometrics: a trait-based approach to paleoclimate and paleoenvironmental reconstruction. In D. A. Croft, D. F. Su & S. W. Simpson (Eds.), *Methods in paleoecology: Reconstructing Cenozoic terrestrial environments and ecological communities* (pp. 371–392). Cham: Springer.

Vizcaíno, S. F., Cassini, G. H., Fernicola, J. C., & Bargo, M. S. (2011). Evaluating habitats and feeding habits through ecomorphological features in glyptodonts (Mammalia, Xenarthra). *Ameghiniana, 48,* 305–319.

Wainwright, P. C., & Reilly, S. M. (1994). *Ecological morphology: Integrative organismal biology*. Chicago: University of Chicago Press.

Walmsley, A., Elton, S., Louys, J., Bishop, L. C., & Meloro, C. (2012). Humeral epiphyseal shape in the Felidae: the influence of phylogeny, allometry, and locomotion. *Journal of Morphology, 273,* 1424–1438.

Weinand, D. C. (2007). A study of parametric versus non-parametric methods for predicting paleohabitat from Southeast Asian Bovid astragali. *Journal of Archaeological Science, 34,* 1774–1783.

Werdelin, L., & Wesley-Hunt, G. D. (2010). The biogeography of carnivore ecomorphology. In A. Goswami & A. Friscia (Eds.), *Carnivoran evolution: New views on phylogeny, form and function* (pp. 225–245). Cambridge: Cambridge University Press.

Werdelin, L., & Wesley-Hunt, G. D. (2014). Carnivoran ecomorphology: patterns below the family level. *Annales Zoologici Fennici, 51,* 259–268.

Wesley-Hunt, G. D. (2005). The morphological diversification of carnivores in North America. *Paleobiology, 31,* 35–55.

White, T. D., Ambrose, S. H., Suwa, G., Su, D. F., DeGusta, D., Bernor, R. L., et al. (2009). Macrovertebrate paleontology and the Pliocene habitat of *Ardipithecus ramidus*. *Science, 326,* 67–93.

Williams, S. H., & Kay, R. F. (2001). A comparative test of adaptive explanations for hypsodonty in ungulates and rodents. *Journal of Mammalian Evolution, 8,* 207–229.

Wing, S., Sues, H., Potts, R., DiMichele, W., & Behrensmeyer, A. (1992). Evolutionary paleoecology. In A. K. Behrensmeyer, J. D. Damuth, W. A. DiMichele, R. Potts, H.-D. Sues & S. L. Wing (Eds.), *Terrestrial ecosystems through time: Evolutionary paleoecology of terrestrial plants and animals* (pp. 1–14). Chicago: University of Chicago Press.

Chapter 16
Mammal Community Structure Analysis

Kris Kovarovic, Denise F. Su, and Kari Lintulaakso

Abstract Fundamentally rooted in Odum's niche concept, mammal community studies are based on the understanding that each resident species reveals information about its environment through its adaptations to specific resources and landscape features. Ecologists view the community's profile of strategies for exploiting particular spatial and dietary niches; a quantitative summary of these strategies when compared across locales from a variety of habitat types demonstrates striking similarities in the communities that live in similar habitats regardless of their location. Recognizing that communities can be compared across space, paleoecologists implemented community studies across time in an effort to reconstruct past environments. This synecological approach to paleoenvironmental reconstruction may be thought of as holistic, since it is not restricted to a single mammal family. However, thorough explorations of how fossil and extant communities differ have revealed significant dissimilarities brought about by the taphonomic history of paleontological assemblages. Techniques have been developed for addressing differences between the modern comparative community sample and the paleontological sample to which it is compared, but recent research conducted by both neo- and paleoecologists has suggested that there are unappreciated differences between modern habitats, as well.

Keywords Community analysis • Ecological diversity Ecomorphology • Niche • Taxon-free • Synecology

Introduction

Mammalian remains are often abundant at paleontological sites around the world and represent a critical source of information about the Earth's biological past. Individual species and entire assemblages can reveal the unique characteristics of the mammalian community in a given time and place, contributing to our understanding of how lifeforms evolved in different settings over geological time.

Many classic post-Darwin paleontological studies focused largely on morphological descriptions of mammalian remains for the sake of understanding species and, by extension, their reconstructed evolutionary relationships. Influential mid-20th century developments in modern ecological theory, however, spurred something of a revolution in mammalian paleo-sciences. A step-change in thinking arose with the extension of Odum's (1953) classic niche concept into the work of paleontologists who were interested in general interactions between the environment and mammalian evolution and reconstructing ancient environments. Paleontologists came to appreciate more fully that mammals could provide information about their habitats, and studies utilizing their remains in this manner added to a growing corpus of paleoecological literature.

There are numerous ways that mammals form an integral part of paleoecological research, from isotopic work (see Higgins 2018) to ecomorphology (see Barr 2018), but the focus here is on the analysis of entire communities, referred to variously as community structure analysis or ecological diversity analysis among other monikers. We review the

K. Kovarovic (✉)
Department of Anthropology, Durham University, South Road, Durham, DH1 3LE, United Kingdom
e-mail: kris.kovarovic@durham.ac.uk

D. F. Su
Department of Paleobotany and Paleoecology, Cleveland Museum of Natural History, 1 Wade Oval Drive, Cleveland, OH, United States of America
e-mail: dsu@cmnh.org

K. Lintulaakso
Department of Geosciences and Geography, University of Helsinki, Gustaf Hällströmin katu 2a, 00014 Helsinki, Finland
e-mail: kari.lintulaakso@gmail.com

D. A. Croft et al. (eds.), *Methods in Paleoecology: Reconstructing Cenozoic Terrestrial Environments and Ecological Communities*, Vertebrate Paleobiology and Paleoanthropology, https://doi.org/10.1007/978-3-319-94265-0_16

theoretical underpinning for this synecological approach and its scientific development, consider its pros and cons compared to other available methods, and look ahead to further refinements that will allow us to elucidate ever more nuanced information about ancient environments.

Ecological Theory: The Underpinning of Community Structure Analysis

The Niche Concept

Eugene P. Odum is often credited with being one of, if not *the,* founder of the modern science of ecology. He was certainly not the only person in pre- or post-Darwin times to discuss connections between mammals and their environments (e.g., Hutchinson 1957, 1959) or to overtly outline the nature of relationships between the two (such as predator-prey interactions). However, Odum was a prime mover in an influential mid-20th century shift in ecological thinking. The general principles he championed are still persuasive and, despite some having been shown to be flawed in some respects, his work in the 1950s can be described almost as "contagious" in many scientific circles, particularly in debates surrounding the utility of evolutionary versus ecosystem ecology. One area where his intellectual input contributed greatly to the development of paleoecology is in the ecological niche concept, earlier defined by Elton (1927). Odum described this in the 2nd edition (1959) of his popular 1953 text *Fundamentals of Ecology* as "the position or status of an organism within its community and ecosystem resulting from the organism's structural adaptations, physiological responses, and specific behavior (inherited and/or learned)" (Odum 1959, p. 27). This he contrasted to a habitat, the physical space occupied by a species, employing an analogy that has endured for decades: the habitat is the species' "address", but the niche is its "profession".

The very language used here conveys something that is dynamic rather than static. It suggests that a niche is defined by relationships between individual species and between species and their physical environments and, furthermore, that these relationships can change over time. Indeed, by emphasizing that *structural adaptations* are a part of the very definition of a niche, an evolutionary time scale is invoked. Bodily structures are adaptations and, as such, they can tell us something about the niche(s) to which an organism was adapted (Harrison 1962). This isn't a circular concept so much as an empirical approach that enables us to work "backwards" – as paleontologists often do – by comparing what can be observed in the present with remains from the past; in this case, the hard tissue adaptive structures of the skeleton and dentition.

Although ecologically meaningful mammalian structural adaptations (also known as "ecomorphologies", a concatenation of the term "ecological functional morphologies") may be explicitly studied to infer past environmental conditions (see Barr 2018), community analysis uses information derived from these adaptations, as well as other proxies, to classify individual species into categories that broadly describe their preferred niches. The concept of a niche as expressed by Odum is also particularly pertinent here, as it makes clear that individual species have a place – their "position or status" – within a community, as well as within the ecosystem as a whole. The implicit idea is that the entire community can yield a greater amount of information about its environment than individual taxa, which occupy a limited range of niches, and therefore cannot provide a composite picture. Simply put, the whole is greater than the sum of its parts.

Mammalian Adaptations to the Niche

The exploitable niches that are available in a given habitat are conditioned by a combination of both biotic and abiotic factors, such as the different kinds of vegetation (trees, grass, herbs, shrubs, etc.) and the particular qualities of the soils in which they grow (drainage, nutrients, texture, inorganic composition, etc.) (Lomolino et al. 2006). These characteristics dictate what types of mammals can live in a given place and how they behave; a locale with nutrient-rich, moist but well-drained soil supporting heavy vegetation cover will have a very different mammalian community than a locale with dry, sandy soils and sparse vegetation cover. For example, tropical and subtropical rain forests or similar biomes with high annual average temperature and precipitation, where fruit resources are available year round, support more purely frugivorous mammals than areas with seasonal precipitation, which have higher numbers of herbivorous (folivorous) taxa (Kay and Madden 1997; Hanya et al. 2011; Pineda-Munoz et al. 2016).

A habitat can be divided into two distinct niches that are relevant for paleoecology: a spatial niche, which relates to the physical space in which a species moves during its life cycle, and the trophic niche, which relates to the diet on which a species relies to meet its nutritional and energy requirements. All mammals need a place to live and food to eat; thus, they are adapted to (or "fill") a particular spatial and trophic niche.

Soft-tissue structures are every bit as adaptive as hard tissue structures but are of limited use to paleontologists given their resistance to fossilization. However, the preservation of skeletal and dental remains, although impacted by numerous taphonomic factors, yields a wealth of material from the past that can be related back to spatial and trophic niche exploitation. Mammals negotiate and utilize space with different locomotor repertoires as evidenced by the morphology of their postcranial skeleton, particularly elements of the fore- and hind limb, which vary according to the substrate to which the species is adapted (e.g., Kappelman 1988).

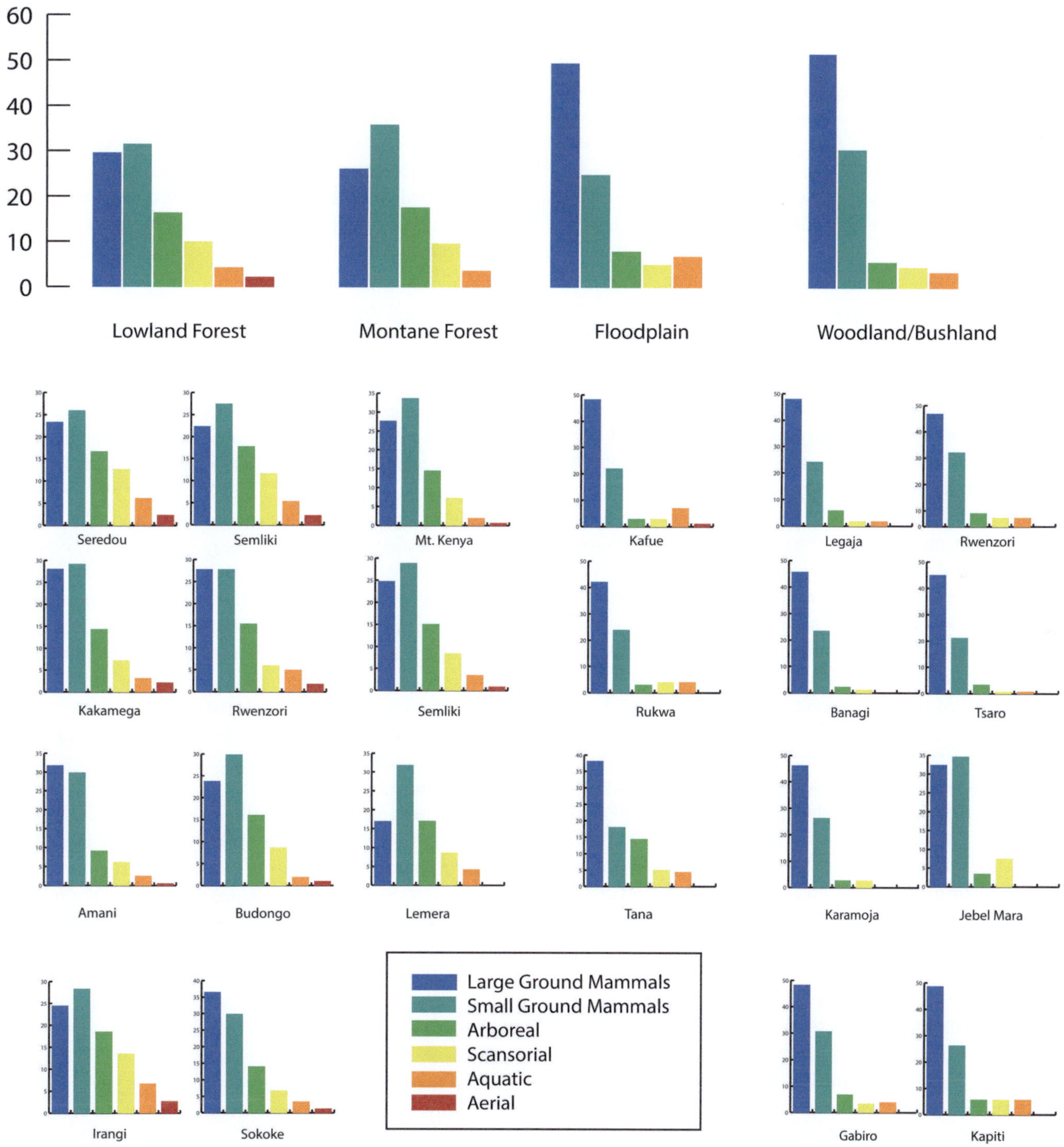

Fig. 16.1 Locomotor niche exploitation profiles from four habitat types in Africa, based on data from 22 modern mammalian communities. Data from Andrews et al. (1979)

Teeth, in comparison to postcranial elements, provide evidence regarding the trophic niche to which a species is adapted. Animals have specific dietary requirements and, thus, dental morphologies that are indicative of a carnivorous, insectivorous, frugivorous, or herbivorous diet. Combination niche exploitation strategies are particularly common in herbivores, which may be further defined as browsers (preferring dicots) or grazers (preferring grass) and, quite frequently, a mix of the two, which often relates to the seasonal availability of food sources. It is also common for herbivores on the browsing end of the spectrum to consume fruits and for insectivores to consume small mammals and/or fruit. True omnivores are rare, and the morphology of their dentition tends to point to a specific combination of dietary strategies rather than an equal reliance on all food resources (Pimm and Lawton 1978; Yodzis 1984; Pineda-Munoz and Alroy 2014).

Community Structure

The "structure" in a community structure analysis is essentially a summary of niche exploitation profiles. The primary interest is how the mammalian species fill trophic and spatial niches with their dietary and locomotor adaptations, although in many cases, body size is also considered as ecologically relevant and given the same consideration as the two primary niche profiles (Andrews et al. 1979). Summaries can be helpfully visualized in histograms, which demonstrate quite effectively a key concept of this analytical approach to environmental reconstruction: habitats with a similar array of available niches have similar community structure profiles, regardless of their geographical location. Figure 16.1, originally published by Andrews et al. (1979), expresses this in a series of histograms organized according to five broad habitat types. Profiles for the relative proportion of species in taxonomic orders, average body mass categories, and locomotor and trophic niche categories are presented. By comparing these profiles across the habitat categories, it is easy to see where differences between them occur in their mammalian faunas. For example, compare the lowland and montane forest locomotor profiles with those from non-forest habitats. Species in the large ground mammal category, which are entirely terrestrial, are found in much greater numbers in non-forests (approximately 50% of the fauna), whereas forests have a higher proportion of species classified as small ground mammals, which practice some form of arboreality in addition to exploiting the terrestrial niche. These data are from 22 modern localities in Africa, but the patterns can also be viewed on a broader geographic scale in datasets that extend to multiple regions and continents (Andrews 1996).

The phenomenon noted above is evident if one compares the profiles of the Central American tropical forest and Chinese paratropical forest in Fig. 16.2 (data from Andrews and Humphrey 1999). Although separated by some 14,000 km and millions of years of evolutionary history, these profiles are visibly similar, with approximately equal numbers of species in each category. The differences appear relatively minor; Chinese paratropical locales have a small number of aerial species that are lacking in the sampled Central American communities, which themselves have a slightly higher proportion of terrestrial taxa. But, overall, their general patterns are the same. This structural similarity in their faunal communities is indicative of the fact that all forests present a similar array of exploitable niches for resident species to fill.

Significance for Understanding Paleoecology

Paleoenvironmental community studies are fundamentally rooted in the present, based on observations of mammalian behavior in habitats where they are known to live. Demonstrating the existence of similarities in patterns of community structure in broad habitat categories was originally largely based on extant communities (e.g., Karr 1971; Fleming 1973), but an obvious next step for paleoecologists was to extend this principle to the fossil record. While direct observations of past mammalian dietary and locomotor habits cannot be made, their postcranial and dental remains provide evidence for their adaptive niches, from which community structure profiles can be constructed.

Andrews and colleagues (1979) developed the community structure technique for interpreting paleoecology based on early studies of species diversity in fossil and modern avian and mammalian communities (Kurtén 1952; Harrison 1962; Simpson 1964; Karr 1971). They noted that these early studies (Simpson 1964; Karr 1971) showed changes in the diversity of animal and plant species following a north-south gradient on different continents: as latitude decreased, the number of species increased. Additionally, the structure of the communities, i.e., distribution of body sizes and dietary and locomotor adaptations, also changed with latitude (Fleming 1973). They showed that community structure varies significantly with habitat type within a single latitudinal zone but that when both latitude and habitat type are similar, community structure is constant, regardless of differences in continent or species number and composition. They

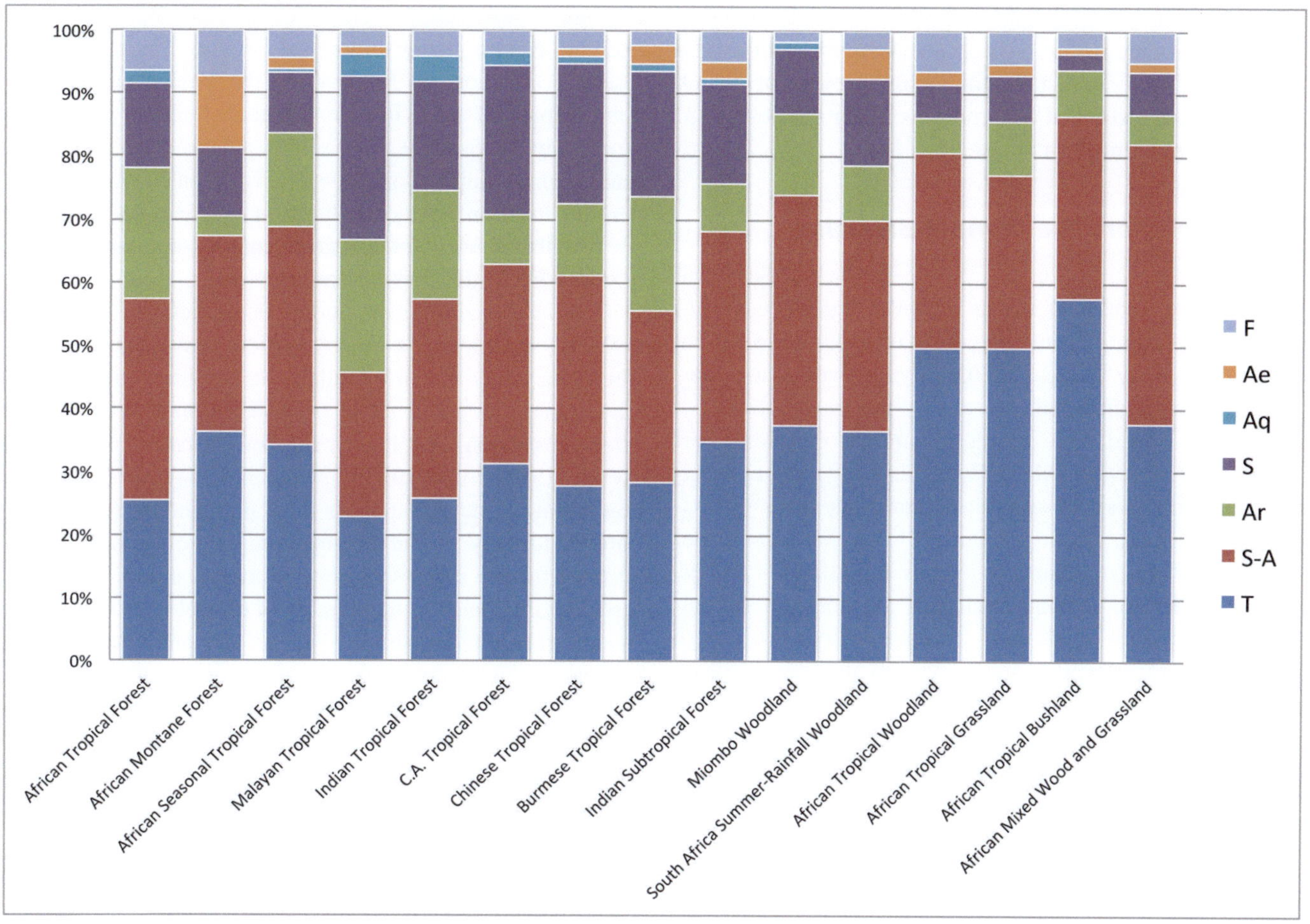

Fig. 16.2 Locomotor niche exploitation profiles for 15 habitat types from across the world, based on data from 44 modern mammalian communities. Data from Andrews and Humphrey (1999, p. 287). F = fossorial, Ae = aerial, Aq = aquatic, S = scansorial, Ar = arboreal, S-A = scansorial-arboreal, T = terrestrial

proposed that the habitat of a faunal community can be determined based on its community structure, irrespective of species composition, and that this predictive quality has direct application for paleoecology. By establishing patterns of community structure for a series of modern faunal communities and comparing them to those calculated for fossil communities, it is possible to make direct comparisons and inferences regarding the habitats from which the fossil assemblages may have come. This methodology was quickly integrated by many researchers into studies of the paleoecology of fossil hominoid-bearing faunal communities (e.g., Andrews and Van Couvering 1979; Van Covering 1980; Nesbit Evans et al. 1981) and is regularly incorporated into paleoecological studies of mammalian fossil sites today (e.g., Reed 1997; Kovarovic et al. 2002; Flynn et al. 2003; Su and Harrison 2007; Croft et al. 2008; Su et al. 2009; Meloro and Kovarovic 2013).

Paleoenvironmental Reconstruction

Bearing in mind the theoretical considerations of community structure analysis, there are also practical aspects to consider when implementing this method as a form of paleoenvironmental reconstruction. Decisions must be made regarding the selection of extant localities for the comparative dataset and how mammalian community niche exploitation profiles are compiled and categorized. The issue of categorization is particularly critical as neither mammalian niche exploitation behavior nor habitats are discrete entities, existing instead on a spectrum of related variables. Nevertheless, the method requires that species are classified according to their preferred niches and localities to their broad habitat or biome, which reduces the representation of their natural biological diversity. A balance must be struck between using a classification scheme that is nuanced enough to provide useful environmental interpretations and one that is also applicable to fossil species for which behavior is reconstructed from proxies rather than observed directly as in the modern world.

Extant Comparative Sample Selection

Assembling a dataset of extant localities against which to compare the target paleontological locality requires a careful and often lengthy process of identifying natural localities (i.e., ones not significantly altered by human habitation or activity) associated with complete mammalian species lists (Andrews et al. 1979). Many faunal lists for modern localities were compiled for specific research questions or conservation projects, and it is often the case that they record only the small or large fauna or species in particular families. Only localities for which faunal lists are considered complete are viable for community structure analyses, although establishing completeness itself may be difficult. Furthermore, the conflation of gamma (=regional) with alpha (=locality) species richness is also problematic, particularly when the faunal list has been assembled based on distribution maps, which may incorporate "expected occurrences" (i.e., a species is absent but marked as being present because the habitat is similar to where it is known to occur; for example, see Happold 1987). Needless to say, faunal lists based on alpha species richness are more appropriate for community structure analyses and should be used in favor of those based on the regional occurrence of taxa. Both of these factors can pose a challenge in selecting the appropriate comparative modern community. Fortunately, with the advent of electronic databases and data sharing requirements, there are a number of resources that can be useful in this regard; examples include species occurrence datasets, such as those published by the National Center for Ecological Analysis and Synthesis (NCEAS, https://www.nceas.ucsb.edu/), Ecological Archives (http://esapubs.org/archive/default.htm), DryadLab (http://datadryad.org/), and Biological Inventories of the World's Protected Areas (http://www.ice.ucdavis.edu/bioinventory/bioinventory.html), as well as specific peer-reviewed journals like Check List (http://biotaxa.org/cl/index) that focus on data related to species geographic distributions. A good starting point for finding information to assist with standardizing these occurrence data (which quite often are from national parks and similar areas) is the World Database on Protected Areas (WDPA, https://www.protectedplanet.net/).

In addition to being suitable from a sampling standpoint, localities should generally encompass the entire spectrum of ecological possibilities. However, it may be possible to exclude certain major biomes depending on the specific research question or location of the paleontological site being subjected to environmental reconstruction. There may be reasonable grounds for excluding temperate and Arctic tundra communities, for example, if the paleocommunity is derived from an equatorial location where other indicators suggest that it was not physiognomically similar to tundra (e.g., desert in the tropical realm).

Gross vegetation physiognomy is, in fact, one of the best defining characteristics for any habitat classification system. There are many systems available in the literature, some of which span the entire range of global biomes, but they are often based on specific vegetation species associations (White 1983; Eiten 1992; Lawesson 1994). But, just as a community structure analysis does not account for the taxonomic identity of the mammalian species, it also cannot account for the species of vegetation in the habitats. Structural variables such as the number of canopies, crown height, overall tree, bush, or shrub cover and the presence of grassy or herbaceous ground cover are more relevant. Of course, such systems can be rather coarse, lumping several habitat types into broad groups that obscure ecological nuances that exist between them. The system used by Andrews et al. (1979) in Fig. 16.1 considers only four categories, including one for all woodlands and bushlands; although it is rather reductive, it also presents a series of rules that make it reasonably easy to place each locality into a habitat group. The more nuanced the habitat classification system, the more information that can be extracted from a paleocommunity, but the harder it becomes to categorize the habitat groups in the first place. Since they exist on a continuous spectrum of vegetation cover, the difficulty is in knowing where to draw the line to create discrete, analyzable categories. For this reason, paleoecologists have drawn on current research and descriptions of extant vegetation communities and habitats (particularly White 1983 and Olson et al. 1983, 1985) to devise clearly-defined classification systems that can be applied to the paleontological record (Andrews et al. 1979; Reed 1997; Andrews and Humphrey 1999). Table 16.1 (originally from Kovarovic et al. 2002) summarizes the habitat categories of 22 modern localities used in a community structure analysis of the Upper Ndolanya Beds of Laetoli, a Pliocene hominin-bearing site. The habitat type column represents the broad classification system employed, while a more detailed description of each locality provides the basis for the categorization and aids in subsequent interpretations.

It is important to note that vegetation structure and diversity are strongly correlated with annual rainfall (Gentry 1988); consequently, species richness and community

Table 16.1 Twenty-two modern African localities classified by habitat. Modified from Kovarovic et al. (2002), which is based on the habitat classification system first outlined by Andrews et al. (1979)

Locality	Country	Habitat type	Description
Irangi	Zaire	Tropical forest	Lowland evergreen forest; primary and secondary-riverine cultivation types
Seredou	Guinea	Tropical forest	Lowland evergreen forest
Semliki a	Uganda	Tropical forest	Lowland semi-deciduous forest
Budongo	Uganda	Tropical forest	Lowland semi-deciduous forest; secondary and monotypic types
Semliki b	Uganda	Montane forest	Montane evergreen forest continuous with lowland
Mt. Kenya	Kenya	Montane forest	Montane evergreen forest; primary and secondary
Lemera	Zaire	Montane forest	Montane evergreen forest in river valley
Amani	Tanzania	Seasonal tropical forest	Intermediate semi-deciduous forest
Kakamega	Kenya	Seasonal tropical forest	Intermediate semi-deciduous forest, secondary with glades, hills with bush
Rwenzori Forest	Uganda	Seasonal tropical forest	Lowland semi-deciduous forest
Sokoke-Gedi	Kenya	Seasonal tropical forest	Lowland deciduous forest and woodland
Tana River	Kenya	Mixed wood and grassland	Floodplain grassland with patches of deciduous forest
Gabiro	Rwanda	Mixed wood and grassland	Wooded grassland next to lake flats
Legaja	Tanzania	Mixed wood and grassland	Lake in short grassland with woodland on the fringes
Rukwa	Tanzania	Mixed wood and grassland	Floodplain grassland with bordering woodland and swamps
Rwenzori	Uganda	Tropical woodland	Short grass plains-bush-woodland mosaic
Banagi	Tanzania	Tropical woodland	Deciduous woodland and bushland
Kafue	Zambia	Tropical grassland	Floodplain grassland with bordering woodland
Serengeti	Tanzania	Tropical grassland	Short grass plains with small patches of bush
Karamoja	Uganda	Tropical arid bushland	Arid bushland
Kapiti Plains	Kenya	Tropical arid bushland	Grassland-bushland mosaic
Tsavo	Kenya	Tropical arid bushland	Bushland and arid bush

structure will also vary in relation to annual rainfall (Kay and Madden 1997). It has been shown that in tropical South American faunal communities, there is significant positive correlation between rainfall and species richness and abundances of frugivores, arboreal mammals, and herbivorous mammals, among other attributes (Kay and Madden 1997). In fact, it is possible to use these relationships to predict annual rainfall in past environments (Kay and Madden 1997). Thus, it is important to take into account the diversity of annual rainfall sampled by the comparative modern communities when reconstructing the paleoenvironments of past communities.

Fossil Assemblage Suitability

Not all paleontological localities can be subjected to a community structure analysis. The assemblage must preserve a reasonably high proportion of the mammalian fauna that existed. Although most sites will suffer some amount of taphonomic loss depending on a variety of ecological and depositional conditions, an assemblage that is noticeably lacking in overall species number, or richness, is not a good candidate for this methodology (Andrews 1996). It is difficult to judge the completeness of a community in the fossil record, but some basic protocols can be applied to determine whether an assemblage is suitable.

One point to consider is the number of taxa that are required purely from the standpoint of multivariate statistical robusticity. From this perspective, even an assemblage with a low number of species, as few as five or six (Louys 2007; Louys et al. 2009), can be used. However, communities with such low species richness are obviously depauperate when compared to the richness observed in natural modern communities. Theoretically speaking, habitats that are structurally and topographically complex, such as forests and complex woodlands, provide a greater number of niches to exploit and therefore have greater richness (Reed 2008). In the tropical realm, it is uncommon to find communities with fewer than 40 taxa across all body sizes; this value is approached only by communities inhabiting quite open environments (e.g., Serengeti Plains; Andrews and Nesbit Evans 1979; Andrews et al. 1979). While this might be a useful threshold to consider given that the number is reflected by extant faunas, it is an unreasonable expectation of many fossil communities, which tend to average lower than 40 taxa (see, for example, Andrews and Nesbit Evans 1979) for reasons pertaining to taphonomy (Badgley et al. 1995), the intensity of scientific sampling (it is known that bigger samples tend to have more species; Croft 2013; Su 2016) and our ability to correctly identify all individual fossil species from fragmentary remains. Furthermore, the proportion of each taxon in the sample and the distributional pattern of the taxa that make up the faunal community are all factors that affect the accuracy of predicting the correct habitat association for the fauna (Mares and Willig 1994). Even when these factors are known, as in modern communities, a large proportion of the fauna is often required to be confident of the habitat association of the community (Mares and Willig 1994). In an extensive exploration of the species richness required to effectively distinguish between modern habitats, Louys et al. (2009) found that this varied depending on broad habitat type as defined by vegetation structure (closed, mixed, or open). They provide empirical support for what makes intuitive sense; richness is generally correlated with broad habitat type, so it is expected that a greater number of species might be required to distinguish between those that are taxonomically richer. A minimum of 32 species was identified as the threshold for distinguishing between communities adapted for closed and mixed habitats, whereas only 12 were required to distinguish between open and mixed, and eight for open and closed (Louys et al. 2009).

Ecovariable Categories

Community structure analysis considers three variables that reflect the niche exploitation of the species, also termed ecovariables: body size, locomotor adaptation, and feeding preference. Their expression provides information on how the species of a mammalian community fills the spatial and trophic niches of the ecosystem.

Body size (mass): Body size is one of the most fundamental biological characteristics of mammals. It has been shown to correlate significantly with other aspects of biology and behavior and is an important factor in space utilization for an organism. Various studies have shown a correlation between body size distributions of mammalian communities and their environments (e.g., Legendre 1986, 1989; Gingerich 1989; Croft 2001; Flynn et al. 2003; but also see Rodríguez 1999 for a dissenting opinion), although it appears to be the least discriminating of the three ecovariables (Andrews 1996; Reed 1997; Lintulaakso and Kovarovic 2016). Body size classification in extant mammals is relatively straightforward, done by calculating the mean of the species' weight range. Reconstructing body size in fossil mammals is more complicated, and is based on the consistent isometric and allometric relationships between various skeletal dimensions and body mass (e.g., Gould 1975; Alexander 1977; Fortelius 1990; Gingerich et al. 1982; Scott 1990; Legendre 1986; references in Damuth and MacFadden 1990; see Hopkins 2018 for detailed discussion and additional references). For the most part, because community structure analysis relies on the mammalian species list, and species identification is dependent on craniodental material, the methodology uses dental dimensions as body size proxies. It is an added advantage that teeth are the most identifiable and preserved element in the mammalian fossil record and that various dimensions are known to correlate with body size, among them molar surface area and tooth row length (Gould 1975; Gingerich et al. 1982; Fortelius 1990; Hopkins 2008, 2018). It is worth noting that although there are many factors that can influence the robustness of body size estimation (see Hopkins 2018 for detailed discussion), precision is not as crucial in community structure analysis because species are categorized in ranges of body masses that compensate for any inexactness in the predication. Body size categories vary from study to study and can have as few as three categories (<1 kg, 1–45 kg, >45 kg; Andrews et al. 1979) to as many as 12 categories (see Table 2 in Townsend et al. 2010). The division of body size categories is arbitrary to a certain extent, but it is important to consider the body size distribution of the target paleocommunity so that body size categories are not overly or under-divided.

Locomotor adaptation (substrate categories): Locomotor adaptations are generally considered in terms of the substrate or spatial niche for which the organisms are adapted (Harrison 1962; Eisenberg 1981). Assignment of a locomotor class should be based on the morphological adaptations of the limb bones that allow the species to move within its niche (see Dunn 2018). However, there are instances, particularly within a fossil assemblage, where it is often not possible to associate isolated postcranial elements to taxonomically identified craniodental materials; in such cases, it may be necessary to categorize a species based on specimens of from other sites or, as a last resort, from taxonomic analogy.

Locomotor categories are relatively broad and emphasize ecological niche structure; the animals in a community are categorized to correspond to the physical layers in the environment (Andrews et al. 1979). The categories proposed by Andrews et al. (1979) and commonly used in community structure analyses are: (1) aerial, the space above the vegetation to which aerial mammals, such as bats, are restricted; (2) arboreal, mammals that are found solely in the upper canopy or small branch zone of trees; (3) scansorial, mammals that are found in the middle and lower canopy and often climb up and down trees; (4) terrestrial-scansorial, terrestrial mammals that are not restricted to the ground and often utilize the lower branches of trees; (5) terrestrial, mammals that are restricted to the ground; (6) fossorial, mammals that burrow in the ground; (7) aquatic, mammals that inhabit bodies of water. Combination niche categories are also possible, e.g., terrestrial-arboreal or terrestrial-aquatic, as some species equally exploit the tree canopy and the ground, or underground dens and the land surface, albeit often for different sorts of activities like eating, sleeping and rearing young. There are also specialist niche exploitation categories, such as cursorial, which indicates a species that is well-adapted for running long distances over relatively open ground (see also Barr 2018), although they are generally considered part of the terrestrial category. Later studies have further combined the arboreal and scansorial categories and re-named terrestrial-scansorial to terrestrial-arboreal and included all mammals that use both substrates (e.g., Reed 1997; Su and Harrison 2007), while others have further refined the categories to accommodate the suite of specializations that allow the animal to exploit more than one ecological zone (e.g., Kovarovic et al. 2002). The less refined the categories, the easier it will be to assign fossil taxa to the appropriate category; however, it may also result in a decreased ability to distinguish between faunal communities. Most studies exclude the aerial category in their analysis due to the paucity of bats in the fossil record and the rarity of gliding mammals in both modern and fossil communities (Andrews et al. 1979; Reed 1997; Kovarovic et al. 2002; Su and Harrison 2007; among others).

Dietary adaptations (trophic categories): Trophic categories are assigned on the basis of the primary dietary adaptation of the animal, usually inferred from tooth morphology (specialization of molar cusps, degree of hypsodonty, canine morphology or reduction; see Evans and Pineda-Munoz 2018), wear patterns (mesowear and microwear; see Green and Croft 2018), and/or stable carbon isotope composition (see Higgins 2018). The use of multiple proxies to infer the diet of fossil mammals can sometimes result in contradictory reconstructions where the lines of evidence do not align; for example, stable carbon isotope data may suggest that the species consumed both browse and graze even though its teeth are hypsodont (e.g., alcelaphin species from Laetoli; see Kingston and Harrison 2007). This can reflect the difference between fundamental niche and realized niche (Hutchinson 1957) and, where these data are obtainable, categorization should be based on what a species is actually eating at that particular site rather than its morphological adaptation. Furthermore, in some cases, so little of a species is preserved or its ecological behavior is so little understood that one must default to the behavior of its relatives (extinct or extant) or "average" what is known about its genus or family.

The trophic or dietary categories to be used could include: (1) carnivore, animals that consume meat as a majority of their diet irrespective of taxonomic classification; can be further subdivided into animals that consume only meat, animals that consume meat and bone, and animals that consume meat and invertebrates; (2) insectivores, animals that eat mostly insects irrespective of taxonomic classification; (3) grazers, animals that mostly eat grass or specific types of grass from edaphic grasslands (i.e., where drainage is impeded and there is often seasonal flooding); (4) browsers, animals that mostly eat products of dicot bushes and trees; (5) mixed feeders, animals that eat both dicot and monocot leaves; (6) frugivores, animals that eat mostly fruits; can be further refined to animals that eat fruits and leaves and animals that eat fruits and insects (but often combined because it is difficult to distinguish these two categories based on ecomorphology); (7) omnivores, animals whose diet is wide-ranging and consists of food from more than two trophic categories; (8) root and tuber eaters, animals that specialize in roots and tubers. The use of categories varies across studies; the difference is primarily in the degree of refinement in category definitions and the use of combination categories (e.g., carnivore vs. carnivore-meat, carnivore-bone, carnivore-invertebrate; see Reed 1997 vs. Kovarovic et al. 2002). Additionally, at sites that predate the middle to late Miocene spread of true C_4 grasslands, herbivorous animals are more correctly categorized as folivorous "open habitat feeders" and "closed habitat feeders" (see Dunn et al. 2015). As is the case with choosing locomotor and substrate categories, it is easier to assign a fossil taxon to a broader category, particularly if the ecology of the fossil species is not well known; however, this may also result in greater overlap in the community structure of faunal communities and make it more difficult to distinguish among them.

Multivariate Statistics for a Multivariate Problem

Since its inception, analyses of community structure have included univariate, bivariate, and multivariate comparisons. Profiles of community structure in histograms (e.g., Andrews et al. 1979a; Nesbit Evans et al. 1981; Kovarovic et al. 2002) and bivariate plots of ecovariables (e.g., Reed 1997, 2008; Kovarovic et al. 2002) are useful interpretative tools and can demonstrate ecological relationships; for example, the proportion of frugivorous mammals is inversely related to that of grazing mammals, with forest habitats having more frugivorous species and open habitats having more grazing species (see Fig. 16.3). Univariate statistical tests are also appropriate when testing the null hypothesis that mean proportions of ecovariables are the same between habitat categories. For example, Reed (2008) used a single factor ANOVA to test the null hypothesis followed by the "Tukey honestly significant difference for uneven sample sizes" post-hoc test for differences between pairs of habitats for each ecovariable.

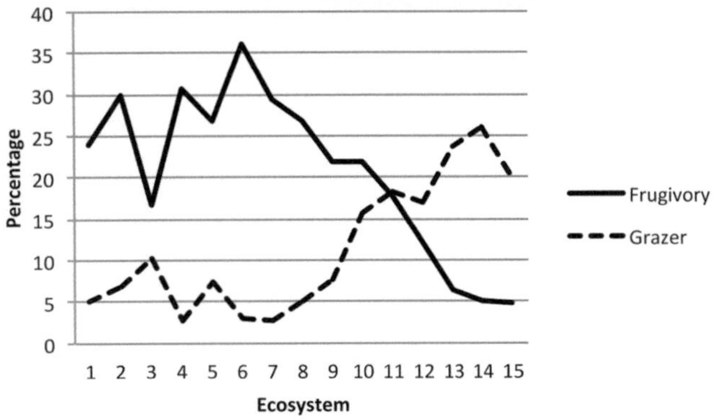

Fig. 16.3 Percentage of frugivorous and grazing species in modern ecosystems. More open habitats (ecosystems 10–15) have higher proportions of grazing species and lower proportions of frugivorous species. More wooded habitats (ecosystems 1–9) have the opposite pattern, with higher proportions of frugivorous species and lower proportions of grazing species. Ecosystems: 1 = African tropical forest, 2 = African montane forest, 3 = African seasonal tropical forest, 4 = Malayan tropical forest, 5 = Indian tropical forest, 6 = Central American tropical forest, 7 = Chinese paratropical forest, 8 = Burmese subtropical forest, 9 = Indian subtropical forest, 10 = Miombo Woodland, 11 = South African summer-rainfall woodland, 12 = African mixed wood and grassland, 13 = African tropical woodland, 14 = African tropical grassland, 15 = African tropical arid bushland

Although other approaches can be used, community structure is fundamentally a multivariate problem and requires robust multivariate statistical techniques. It is important to note that multivariate analysis does not actually provide environmental data; rather, it organizes the data by ordination or classification in such way as to allow for visualization of the data and environmental interpretations (Andrews 1996). Different multivariate analyses have been used in studies of community structure and are summarized as follows.

(1) **Hierarchical clustering analysis**: A type of cluster analysis in which a hierarchy is built using a measure of distance between pairs of observations, typically shown as a dendrogram (for examples of application, see Andrews 1996; Kovarovic et al. 2002; Su 2011).
(2) **Principal components analysis (PCA)**: An ordination technique in which the data are transformed into uncorrelated variables called principal components. The first principal component contains as much of the variability of the data as possible, and each remaining component contains as much of the remaining variability as possible (Zuur et al. 2007). The results are usually presented as a bivariate plot of its first and second principal components, which should contain most of the variance of the data (for examples of application, see Kovarovic et al. 2002; Flynn et al. 2003; Su and Harrison 2007).
(3) **Principal coordinates analysis**: An ordination technique similar to PCA that uses a distance matrix instead of data transformation (Zuur et al. 2007); it is sometimes preferred to PCA because correlations within the ecological data matrix are generally weak (Andrews 1996; this reference also provides an example of its application). Principal coordinates analysis can be followed by a canonical analysis (called a discriminant analysis when the interest is in group membership and a correlation analysis when the interest is in relationships with environmental variables) that allows for constrained ordination on the basis of any distance or dissimilarity measure (Anderson and Willis 2003; see Fig. 16.4 for an example of its application).
(4) **Non-metric multidimensional scaling**: A non-metric ordination method designed to circumvent the assumption of linearity in principal coordinates analysis. Instead, it uses rank order information in a dissimilarity matrix to carry out iterative multidimensional scaling that produces differences in distribution and clustering that is analogous to the components in principal components analysis (Andrews 1996; this reference also provides an example of its application).
(5) **Correspondence analysis**: A geometric technique used to visualize the associations between the levels of a two-way contingency table; rows and columns of the table are displayed as points in a low-dimensional space that is consistent with their associations in the table (Greenacre 1984). For examples, see Greenacre and Vrba (1984), Reed (2008), Kovarovic et al. (2013), and Su and Haile-Selassie (in press).
(6) **Discriminant function analysis**: A statistical technique used to predict a categorical dependent variable (i.e., group membership) by one or more continuous or binary independent predictor variables. See Croft et al. (2008) and Su and Haile-Selassie (in press) for examples of the application of this technique and Kovarovic et al. (2011) for a review of some important considerations of this method.

The choice of which multivariate analysis to use depends on the questions being asked by the researcher as well as the data themselves, as analyses often have different assumptions (see citations above for computational details). Different analyses are not mutually exclusive and are often used in conjunction with one another. With analyses that deal with proportional data, as is the case with community structure data, transformation of the raw data is often necessary. Sokal and Rohlf (1995) proposed using the arcsine square root transformation for these types of data, and this has been widely adopted since then by ecologists and paleoecologists alike (Warton and Hui 2011). There are other types of transformation that can be used (e.g., logit transformation; Warton and Hui 2011); the choice of transformation and, importantly, its necessity, is determined by the research question and the assumptions made about the data (Clarke and Warwick 1994; Kindt and Coe 2005).

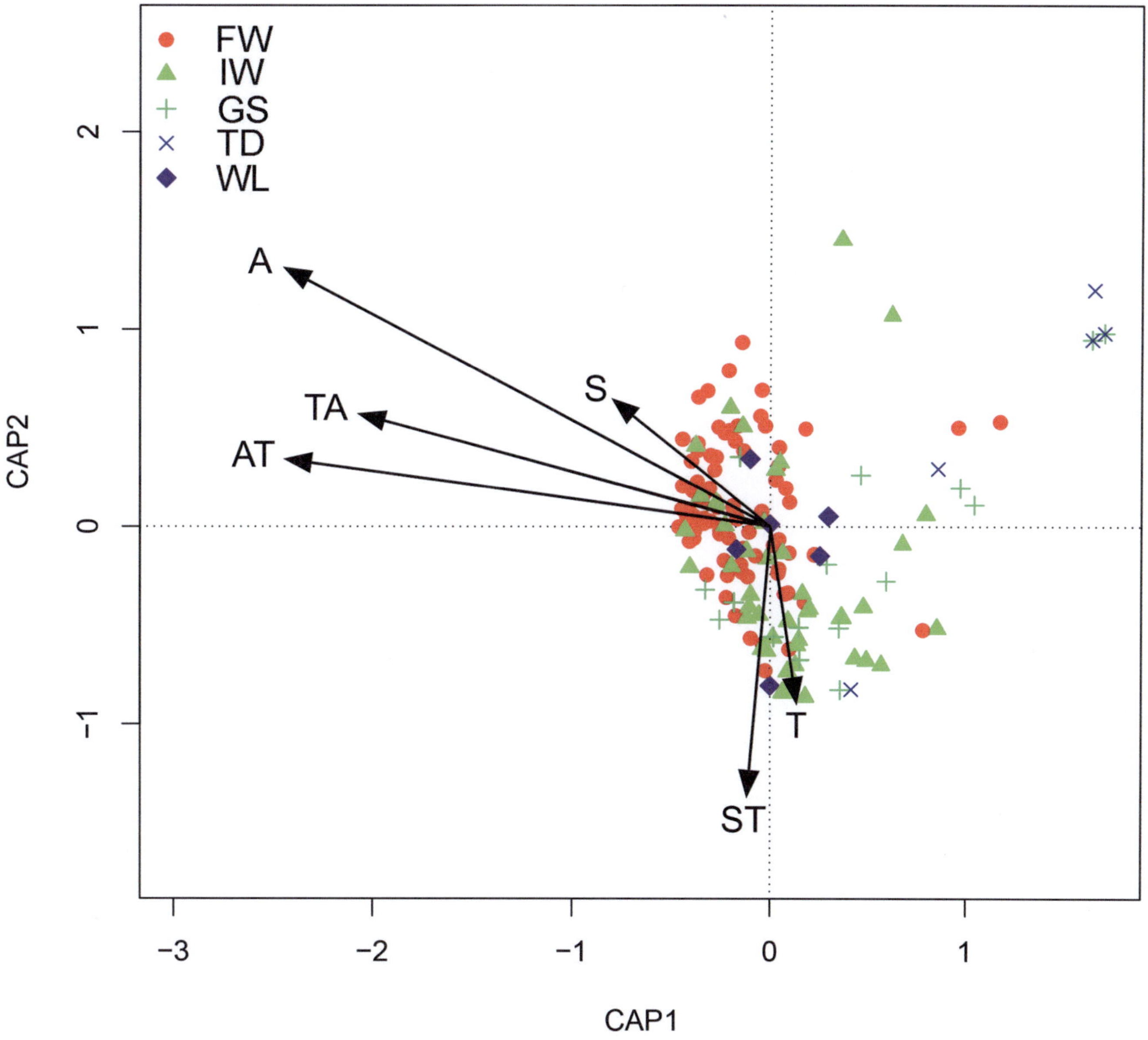

Fig. 16.4 Canonical analysis of principal coordinates (CAP; Anderson and Willis 2003) for 169 tropical mammal communities (see Lintulaakso and Kovarovic 2016). The communities are analyzed by the number of mammalian species within locomotion groups. Communities are grouped by Olson et al. (1983) major ecosystems. Two main principal coordinates are displayed. CAP1 is associated with a low number of arboreal species. Therefore, forested, more closed environments (FW) can be identified by a high negative correlation of low numbers of arboreal (A) or arboreal–terrestrial (AT, TA) species. CAP2 is associated with a low number of subterranean–terrestrial (ST) or terrestrial (T) species. Here, the more open environments (IW, GS) can be identified by a high negative correlation of low numbers of subterranean–terrestrial or terrestrial species. Tundra and desert (TD) has a relatively high positive correlation with both CAP1 and CAP2, having a low number of any species within the locomotion groups. Wetland (WL) communities are scattered all over the plot, having no real association with any of the CAP axes. This plot also shows some outliers that could be investigated further as a potential source of error. For example, the outliers may pertain to a locality that was misclassified a priori into an ecosystem group, the species list for the locality may not be complete, or the species list may have been collected from a number of ecosystems within a single locality. Major ecosystems (Olson et al. 1983) are: major forest and woodland (FW), interrupted woods (IW), grass and shrub complexes (GS), tundra and desert (TD), and major wetlands (WL)

Strengths of Community Analysis

The utility of the community structure approach for reconstructing past environments is multi-faceted: it makes use of what is often ample and easily accessible data (fossil remains); it is grounded in a solid theoretical context, the main principles of which (i.e., adaptation, niche exploitation) can be clearly observed in the modern natural world, so comparative data are plentiful and reliable; and it is based on a complete mammalian community as defined by the ecological relationships that exist between species and their environments. Some challenges do exist to utilizing the approach, and these are detailed later in this chapter, but here we outline some of the more nuanced advantages to selecting this method over others, illustrated with examples.

An All-Inclusive Approach

Community structure analysis is holistic in that it can include both micro- and macromammals. This is desirable because they relate to different aspects of their environment, including scale, and therefore provide complementary information on niche exploitation. Smaller taxa tend to more regularly exploit fruits and seeds and are able to more efficiently exploit the arboreal and underground substrates in comparison to larger, generally more terrestrial species. Small species are also prey animals for larger carnivores reliant on meat consumption, while smaller species with carnivorous diets tend to eat a combination of small animal protein and insects, or insects in combination with other resources. These differences suggest that eliminating smaller body mass categories from an analysis would also eliminate important ecological information provided by the many frugivorous, insectivorous, fossorial, and arboreal taxa present. These niches, in essence, would be under-represented, and discrimination of habitat types would be based on a limited array of niches, thus blurring the boundaries between them.

However, many paleontological sites present a taphonomic bias against the preservation of smaller-bodied species that can make their inclusion analytically challenging (e.g., Kovarovic et al. 2002; see "Analytical Challenges" below). Additionally, some taxonomic units are traditionally left out of this type of analysis for other practical and ecological reasons. Volant mammals (i.e., Chiroptera) are excluded because of their rarity in the fossil record (Andrews et al. 1979; Reed 1998; Kovarovic et al. 2002; Mendoza et al. 2004; Louys et al. 2011). Although this group displays diverse ecological niche adaptations (bats vary considerably in body mass and diet across a wide spectrum) and therefore should be useful for discriminating among modern habitats, they are of limited utility in the fossil record. Mammals restricted entirely to aquatic environments, such as Cetacea (whales and dolphins) and Sirenia (manatees and dugongs), are also often left out of such analyses because it is the terrestrial habitat that is generally of interest at paleontological sites (although marine mammal community analyses can be conducted in their own right; e.g., Schick et al. 2011). Including an aquatic or semi-aquatic category is often done to allow for categorization of mammals that are partly adapted for an aquatic lifestyle but that still exploit the terrestrial environment in some fashion, such as hippopotami, otters, beavers, and similar species. This allows some element of water availability to be considered implicitly without direct reference to the geological evidence.

"Taxon-Free": Ecology Versus Species Diversity

When taken as a sum of available niches and the proportion of species that exploit them, as in the case of a community structure analysis, a paleohabitat can be viewed without specific reference to the identity of the resident species; in a sense, this method may be thought of as "taxon-free". This is perhaps one of the most important aspects of the method, since paleospecies can be difficult to classify taxonomically (see discussion in "Analytical Challenges" below). Taxonomic considerations in a paleocommunity analysis are thus restricted to the determination of how many species are present (which, admittedly, is not always an easy task) rather than their exact identities or phylogenetic relationships, a somewhat separate area of study often under revision. What matters for a community structure analysis is what the individual species were doing as evidenced by their adaptations to particular niches. Diversity of ecological relationships and niche exploitation strategies employed by a community are far more relevant than number of species alone.

Being unencumbered by taxonomy removes the reliance on "indicator species" in paleoecological reconstructions. This approach, which is implicitly imbued with the principle of taxonomic uniformitarianism, was a favored analytical tool in earlier paleoecological work (e.g., Coe 1980; Vrba 1980; Greenacre and Vrba 1984). Much rigorous research went into determining which species are ecologically consistent in the modern world based on the idea that they were likely to have behaved similarly in the past; one frequently cited example of this approach is the Antilopine-Alcelaphine Criterion (AAC) developed by Vrba and colleagues (Vrba 1980; Greenacre and Vrba 1984), which was applied to many sites in the East and South African paleoanthropological record (Kappelman 1984; Potts 1988; Shipman and Harris 1988). The AAC method was based on observations of extant bovid tribes and their tendency towards the consistent exploitation of specific trophic and spatial niches. Some tribes, such as the antilopines and alcelaphines, are known to prefer the resources of open, arid habitats where they comprise 60% or more of the bovid fauna. When compared to the proportion of species in other tribal pairs such as Bovini and Reduncini, representing wetter, closed habitats, and Tragelaphini plus Aepycerotini (the impala, *Aepyceros melampus*), indicating dry and closed habitats, this criterion was understood to be a powerful predictor of paleoenvironment, particularly gross vegetation physiognomy. However, despite strong evidence that ecological similarities in closely-related bovid taxa may be the result of evolutionary constraint and canalization of traits, not all tribes have had such stable ecological strategies over the same time scales (Sponheimer and Lee-Thorp 2003; Faith et al. 2014; Behrensmeyer 2015; Cerling et al. 2015) or are known to include species that are evolutionarily sensitive to ecological change, such as the alcelaphine hartebeest, *Alcelaphus buselaphus* (e.g., Flagstad et al. 2001) or the impala and its extinct relatives in the genus *Aepyceros* (Faith et al. 2014). For reasons such as these, indicator species are approached with some caution. The AAC method has not been abandoned, but it is used more conservatively and almost always in conjunction with other proxies, such as non-mammalian indictors, stable isotopes, and/or depositional conditions (e.g., Schrenk et al. 1995; Sponheimer and Lee-Thorp 2003; Campisano and Feibel 2007; Bedaso et al. 2013). Community structure analyses are also often discussed in light of the taxonomic composition of the target community with particular reference to specific indicator taxa, but this is used to inform or support the main interpretations derived from the community analysis itself (e.g., Su 2016).

Time and Space

Mammalian communities, being sensitive to ecological conditions, are especially useful when reconstructing habitats across a broad stretch of time or space (e.g., Fortelius et al. 2002). This is borne out by research in extant communities, particularly in small mammal communities, which experience turnover and re-assembly rapidly enough that they can be observed over historical time periods or after a known environmental event that shifted habitat distributions, as well as on spatial scales where observations of presence/absence and ecological behavior are easily made (e.g., Williams et al. 2002; Thibault et al. 2004; Thibault and Brown 2008). Macromammal evidence is less well established but also suggests that communities change quickly in response to shifts in climate and that these changes are recorded in assemblages of skeletal and dental remains on modern landscapes (e.g., Western and Behrensmeyer 2009).

In paleontological locations where the fauna is known to have changed taxonomically over time, a full community analysis can bring greater insight to the ecological conditions that supported these different communities. Good examples of this approach are from the paleoanthropological sites of Laetoli and Olduvai in the southern Rift Valley in northern Tanzania, where a significant amount of paleoecological work has been conducted in an effort to understand the environmental context of three hominin genera throughout the Pliocene and Pleistocene: *Australopithecus, Paranthropus*, and *Homo*.

At Laetoli, *Australopithecus afarensis* is found in the 3.8–3.6 Ma Upper Laetoli Beds and *Paranthropus aethiopicus* in the 2.66 Ma Upper Ndolanya Beds (Leakey and Harris 1987; Harrison 2011). Community structure analyses indicated that the older beds were more wooded (Andrews 1989; Reed 1997) than early interpretations based on indicator species or single-family studies that emphasized the presence of terrestrial and grazing taxa (Leakey and Harris 1987). The composite mammalian community instead suggests that a range of habitats and resources were present, including grass cover that supported the grazing guild; however, the proportion of frugivores indicates that fruiting vegetation was present, as well (Su and Harrison 2007, 2015). Both the overall locomotor and trophic niche profiles are most similar to modern Serengeti woodland habitats (Andrews 1989). In the same location, approximately one million years later, the community structure looks somewhat different, notably in the greater preponderance of terrestrial and grazing species at the expense of browsers and arboreal taxa (Kovarovic et al. 2002). Even when the greater taphonomic loss of small-bodied species in this stratum is taken into account, the large mammal community most closely resembles a semi-arid bushland habitat (Kovarovic et al. 2002). This apparent trend towards increased aridity is also seen elsewhere in the region. Thirty-six kilometers north of Laetoli, where the extensive deposits of Olduvai Gorge record the evolutionary history of hominins in great detail, the 2.0–1.8 Ma Bed I sediments (Deino 2012) indicate a change in climatic conditions towards greater aridity, and the mammal community suggests a shift in the vegetation towards more open habitats at the end of the sequence (Fernandez-Jalvo et al. 1998). However, later in time, the mammalian community structure suggests that even in light of continued increasing aridity (Cerling and Hay 1986), the woody vegetation cover and the mammals it supported had stabilized. Although studies had long acknowledged differences in the Bed II fauna above and below the 1.74 Ma Lemuta Member (Leakey 1971; Hay 1976, 1990; Gentry and Gentry 1978), community structure analyses comparing Lower Bed II, where *Paranthropus* and *Homo habilis* remains have been recorded, to Upper Bed II deposits in which *Paranthropus* persists and *Homo erectus* emerges, demonstrated that the faunal differences are most likely explained by taphonomy; both of the Bed II mammal communities were woodland-adapted (Kovarovic et al. 2013).

Work at Olduvai reminds us that mammal communities vary not only over time, but also over space. A long program of research at the site, particularly in Bed I and lowermost Bed II, has focused on assessing landscape use of resident hominins (Blumeschine and Peters 1998; Blumeschine et al. 2003, 2012a, 2012b), which necessarily considers both habitat and geology. Recursions of the paleo-lake Olduvai and periods where springs and river-fed wetlands were the main water sources shaped the landscape and modified habitat availability intermittently (Ashley et al. 2009; Stanistreet 2012). These hydrological phases would have supported different vegetation complexes across the landscape. A full-scale spatial community structure analysis has yet to be undertaken, but faunal composition does differ spatially (Cushing 2002), and it can be hypothesized that areas of higher relief, where drainage was good and woody growth was possible, supported different communities from those located in areas of poorer drainage with marshy or otherwise more open habitats.

One difficulty in assessing community structure on a landscape scale is that in addition to the fact that mammals are not static users of their habitats and move around considerably, taphonomic conditions may not allow full communities to have been preserved at each place of interest, forcing researchers to "average" the fauna across a large area (see discussion in "Analytical Challenges", below). However, some sites such as Aramis, Ethiopia, where *Ardipithecus ramidus* remains are known at 4.4 Ma, demonstrate that it may be possible to provide a more refined interpretation of community structure variation across a landscape (White et al. 2009). Here, the mammals identified along a 7-km transect situated on a southeasterly gradient vary in their functional adaptions to both the trophic and spatial niche. The entire mammalian community's niche exploitation profile was not considered for both niches, but the selected macromammals studied suggest full community differences between open habitats in the southeast and woodlands, with which *Ardipithecus* is associated, elsewhere.

Analytical Challenges

Locality Size and Habitat Homogeneity

One issue to consider when conducting community structure analysis concerns the size and scale of the areas from which modern faunal communities derive. Very often, they are entire national parks or even entire regions of a country, which are likely to encompass multiple habitat types. This results in faunal lists with species that are not actually found in the same area or habitat in nature but artificially appear on the same faunal list due to the large scale of the faunal survey in question (e.g., Wilson 1975; Happold 1987). This makes it difficult to directly compare fossil and modern faunal communities because fossil sites are often much more constrained in size than the area from which modern comparative faunal communities are derived (Su and Harrison 2007; Louys et al. 2009) and also decreases the resolution of the interpretations possible. Ideally, the geographic scale of the comparative modern communities should match that of the fossil community of interest (Croft 2013).

It is also important to choose modern localities carefully to minimize latitudinal variation and to represent uniform topographical zones and vegetation types (Andrews et al. 1979) although, in theory, the size of the area should not be a significant consideration as long as the habitat is homogeneous across the area covered. The most straightforward way to deal with this issue is to simply avoid faunal communities suspected of representing several major habitat types or to divide them as much as possible according to specific habitats that can be well-defined within the overall area. Alternatively, a statistical approach can be used to identify species lists whose compositions are significantly distinct and under- or over-represent particular species groups. Louys et al. (2009) suggest a taxonomic distinctness analysis based on a measure developed by Warwick and Clarke (1995; Clarke and Warwick 2001) that can determine communities that have a higher than average taxonomic distinctness, potentially the result of sampling multiple habitats across a large area. These can then be excluded from further community analysis since they do represent a natural, habitat-specific fauna.

Defining and Assessing Paleospecies

Although it is not necessary to taxonomically identify individual fossil specimens in a community, discrete taxonomic units still need to be identified. Often material is too fragmentary to allow this, and it is common to see that fossil assemblages have fewer species overall when compared to modern communities (see "Fossil Assemblage Suitability", above); this may relate not only to under-sampling of the fauna generally, but also to an inability to identify all unique paleotaxa. It may be difficult to distinguish between multiple species in a particularly diverse family or tribe with numerous closely-related taxa, with the result that faunal lists may be composed of generic or tribal groups (e.g., "Alcelaphini A" and "Alcelaphini B"). These distinctions may be based on differences in body mass because the morphology is so similar. In such cases, as long as they can be recognized as distinct taxonomic units and ecovariable categories assigned to them, they should be included in the analysis.

Taphonomy and Sampling

There are many taphonomic factors that impact the composition of the final fossil assemblage in such a way that it no longer reflects the composition of the original living community. Most of them involve differential destruction and transportation of bones on the landscape due to fluvial factors, carnivore and rodent activities, weathering, and trampling (e.g., Voorhies 1969a, b; Behrensmeyer 1975; Behrensmeyer and Hill 1980; Blumenschine 1988). Furthermore, collection strategies and/or sampling, regardless of how thorough, are probably not sampling the entirety of what was once the living community. For example, at Laetoli, all specimens identifiable by taxon or anatomical elements are collected (Harrison 2011). This is an intensive collecting strategy that ensures a representative fossil assemblage, but even then, smaller taxa are significantly under-represented. Size-abundance analysis of herbivores >1 kg indicates that species <10 kg are under-represented in the Laetoli fossil assemblage and that the assemblage is taphonomically biased by surface processes (Su 2005). It is important to note that because community structure analysis relies on faunal lists, many of these taphonomic biases are ameliorated and such biases may not need to be corrected prior to conducting an analysis (Andrews et al. 1979). It is, however, useful to identify the taphonomic biases present in the paleontological assemblage of interest for interpreting the results.

Another important taphonomic consideration is the general under-representation of small mammals in fossil assemblages. Although they are useful habitat indicators in their own right and can be subjected to their own community structure analysis to aid paleoecological interpretations (e.g., Reed 2007; Avery 2007; Stoetzel et al. 2011), they present unique challenges to paleoenvironmental inferences based on community structure. They are often accumulated under different taphonomic regimes from the large mammals in fossil assemblages (Andrews 1990a, b; Reed 1997, 2007) or are so poorly preserved that they are under-represented in the fossil community (Andrews et al. 1979; Reed 1997; Su and Harrison 2007). Based on these factors, many researchers make the decision to exclude them from the analysis to reduce the need to consider this taphonomic bias in interpreting results (cutoffs of 500 g, 1 kg, and 4 kg are commonly used; e.g., Reed 1997, 2008; Su and Harrison 2007; Louys et al. 2009; Su 2011, 2016). Where small species are clearly biased in an assemblage, a statistical solution may be to employ a sampling technique that modifies the comparative modern data such that the body mass profiles match that of the target paleontological community (Kovarovic et al. 2013). However, eliminating smaller body mass categories would also eliminate important ecological information, and the analysis would thus be based on a limited array of niches that may not discriminate between different habitats equally well (Mares and Willig 1994).

An isotaphonomic approach can also minimize taphonomic effects (Behrensmeyer et al. 1992). Clyde and Gingerich (1998) only included Paleocene localities in their study of the Bighorn Basin, Wyoming that satisfied several taphonomic criteria (i.e., that sampled the same paleosol facies), to investigate the impact of the latest Paleocene thermal maximum and the earliest Wasatchian immigration event on the local mammalian community structure. Even though this approach may result in fewer localities, it provides some guarantee that any observed faunal changes are not due to taphonomic differences. Indeed, Clyde and Gingerich (1998) were able to demonstrate that short-term climatic change and its associated immigration event had a sudden and long-term effect on the Bighorn community structure.

An important consideration of sampling is stratigraphic control in fossil assemblages. Time-averaging can inflate species richness when taxa that may actually be separated by tens or hundreds of thousands of years are considered to be part of the same community. This creates a problem when comparing with modern assemblages, which are essentially sampling a single point in time. In order to fully appreciate faunal and environmental change through time, it is necessary to have tight stratigraphic control and to sample as small of a temporal unit as possible when conducting community analyses. One of the best examples of this is the examination of faunal and environmental change through the Siwalik Formation of northern Pakistan by Barry et al. (2002). By dividing the five-million-year interval (10.7–5.7 Ma) into 100-Kyr units, they were able to demonstrate a high level of background turnover in the Siwalik record along with brief and irregularly-spaced turnover pulses likely due to climatic and environmental factors. Without this fine-grained control of temporal resolution, it would not have been possible to demonstrate so conclusively the lack of environmental and faunal stasis in the Siwalk record.

Further Considerations and Future Directions

A Dynamic Biome

In community structure analysis and other faunal-based methods of paleoenvironmental reconstruction, we are using ecological information derived from animals as a proxy for the vegetation structure that was present. We are, in effect, assuming that mammalian community structures in similar biomes are convergent regardless of their location in time and space can be distinguished based on the fauna (and their adaptations). However, biomes are dynamic entities, changed by temporal, geographic and scale-dependent events that also change the structure of the associated mammalian communities.

Temporal factors: Temporal changes in the climate and physical environment occur over ecological (days to millennia) and geological (orbital to tectonic) time frames; floral and faunal communities have to adjust to these changes (Overpeck et al. 2003). The circadian response of vegetation to daily differences in the amount of solar energy can be observed as the movement of leaves or the opening and closing of flowers, for example, but it is change that occurs on a longer time scale that can impact on paleoenvironmental reconstruction.

Within decadal and centennial time frames, primary or secondary succession of vegetation alters plant species composition of the biome and, as a result, mammalian community structure. For example, after a disturbance to a forested area, whether by storm, wildfire, or some other event, pioneering species invade the new area quite rapidly and begin a new succession of plants and animals. The animals are often r-selected species early on, but more K-selected animals eventually inhabit the forest as the succession continues (Pianka 1970). Thus, within a relatively short time span, the faunal community structure can change in the same geographic location, and this may impact paleoenvironmental reconstruction.

Climatic events, like ice ages, can cause millennial changes in vegetation (Overpeck et al. 2003; Hessler et al. 2010). Pollen records from both marine and terrestrial sediment cores demonstrate that during the last glacial period, as a result of global temperature change, cold- and warm-adapted plant species altered their distributions (Overpeck et al. 2003; Hessler et al. 2010). As plant composition and the biomes gradually change, mammalian communities respond to these changes by regional- or continental-scale migration, extinction, or by subspecies-level selection (Overpeck et al. 2003), often resulting in a shift in community structure.

Larger changes in biomes occur over geological time scales, both at the orbital and tectonic levels. The orbital level, which spans 10,000 to 100,000 years, includes glacial-interglacial cycles and relates to the orbital forces described by Milankovitch (Ruddiman 2007). These cycles have repeatedly formed and broken up biomes, causing extinction and speciation events in both plant and mammalian communities. Some biomes in the past have no present-day equivalent. Biomes are also affected by plate tectonics and continental drift. These changes take millions of years and cause major changes in biomes by creating new mountain ranges, opening and closing water courses, and altering global climatic conditions. At this scale, mammalian communities can change drastically in species composition and community structure as they are shaped by major extinction and speciation events (Janis 1993).

Geographic factors: Geography is significant for the evolution of biomes and their associated mammalian fauna. Both latitudinal and altitudinal change or the presence of a continental or island environment alter climatic variability and the rate of dispersion and speciation. The amount of solar energy received is directly related to latitude; low and mid-latitudes have high solar radiation and higher latitudes have low solar radiation (Ruddiman 2007). Similar phenomena can be observed with patterns of species richness, which is generally higher in the tropics and lower towards the poles. Several hypotheses have been proposed to explain this pattern, including: (1) warm climates have faster rates of microevolution (Gillman et al. 2009); and (2) the tropics have more energy available in usable form, which results in more organisms and, hence, more species (=species richness-energy hypothesis; Brown 1981). While these have not yet been sufficiently tested, it is clear that latitude and the amount of solar energy have important implications for climate, vegetation and mammalian communities. A similar pattern exists with altitude. Atmospheric temperature decreases by 6.5°C/km as altitude increases (Coe 2009), and this has a significant impact on the vegetation and the associated mammalian community (e.g., Geise et al. 2004).

Other geographical factors such as the size, shape, and type of the observed area (continent or smaller island) also impact vegetation type and structure of the mammalian community. Studies of island biogeography have noted that islands closer to continents have higher species richness than those further away, as they are easier to reach by migrating populations (MacArthur and Wilson 1967). In addition, larger islands have more resources than smaller ones; thus, there is a higher probability of immigration of continental animals (MacArthur and Wilson 1967).

Geographic scale: Changes in vegetation or mammalian communities can also be analyzed at different geographical scales, from the micro- to macro-level. At the macro-level, the analyzed scale is at the continental level where local changes are insignificant from an analytical perspective. For example, as the climate became increasingly arid during the Miocene, grass-dominated open habitats spread while closed forested areas decreased. This had a direct impact on mammalian herbivore tooth structure, which evolved from low-crowned (brachydont) to medium (mesodont) or high-crowned (hypsodont) teeth in response to their changing trophic resources (Eronen et al. 2010a; Strömberg 2011). At the meso-level, changes in vegetation and mammalian communities are observed within continents. A change in climate may cause the disappearance of one biome and the appearance or increase in size of another, as was the case during the late Pleistocene and Holocene, when the same general climatic events affected both South America and Africa. African mammalian communities remained almost unchanged while those of South America changed dramatically (also due in part to human activity; Cione et al. 2003; Barnosky and Lindsey 2010; Barnosky et al. 2016), leading to present-day faunal dissimilarities between the two continents (De Vivo and Carmignotto 2004). At the micro-level, changes in vegetation and mammalian communities are observed in local areas such as a specific island, mountain, valley, or basin. For example, high levels of endemism exist not only in the Philippine islands as a whole, but within individual islands, as well. Repeated rising and falling sea levels during the Pleistocene joined and separated the Philippine islands numerous times, forming the so-called Pleistocene island groups with unique sets of species (Catibog-Singa et al. 2006).

Implications for community structure analysis: Most of the above-mentioned temporal, geographic and scale-dependant events relate to historical contingency, or the effect of the order and timing of past events on community assembly. Abiotic events like wildfires, floods, storms, and earthquakes change species arrival history of a location or area. These biotic events have different priority effects on the order in which species arrive to the primary or secondary succession area. Although quite often the succession may follow an expected order, the consequences of species arrival history can cause differences in the structure of communities, as effects are amplified over time and space via population growth and interactions (Fukami 2015). The results of this are that even in relatively similar environments, faunal composition can be different due to their different histories. This suggests that historical contingency may be an important factor to consider when interpreting the results of community structure analyses (see "Moving Forward" for more discussion and Croft 2006 for a documented example).

Moving Forward

The more nuanced our understanding of the relationship between a biome and its associated mammalian community (and its ecological structure), the better we can infer past environmental and climatic conditions based on proxies. There is much that can be done to refine community structure analyses, thus guaranteeing ever more robust and detailed environmental reconstructions. We suggest that development in the following three areas would be most beneficial to improving the methodology:

(1) **Historical contingency**: As discussed previously, paleoecological studies rely on the assumption of community convergence and, indeed, there are general similarities in community structure between localities of similar latitudes on different continents in tropical and northern (Palearctic and Nearctic) zones. Nevertheless, this apparently only applies to broad locomotor and dietary variables in the tropics (Lintulaakso and Kovarovic 2016) and between extreme environments that are similar, such as desert and steppe communities with high levels of water stress and low complexity in vegetation structure (Rodríguez at al. 2006). Historical effects are likely to be responsible for demonstrated differences between communities in similar habitats (Losos 1996; Rodríguez et al. 2006; Graham and Fine 2008; Fukami 2015; Lintulaasko and Kovarovic 2016), but it is not yet clear how historical contingency impacts the application of community structure analysis to paleoenvironmental reconstruction or how frequently we may simply be analyzing environments for which there are no modern analogs (Soligo and Andrews 2005; Semken et al. 2010). Thus, a better understanding of the effects of historical contingency should be considered as one of the next steps in improving the method.

(2) **Heterogeneous paleoenvironments**: Community structure analysis is an excellent way of reconstructing the spatial and trophic niches that were available to past communities. But, it can be a coarse approximation that forces habitats into discrete and broadly-defined categories, glossing over the nuanced differences among them (regardless of the mechanisms that are responsible for them). Steps are being taken to move away from categorical habitats and environments, with greater emphasis placed on the multidimensional spectrum of many environments. Louys et al. (2015) outlined techniques for describing habitats and environments using a number of biotic and abiotic quantitative variables. They calculated arboreal heterogeneity of a particular locality in terms of the proportions of heavy, moderate, light, and absent tree canopy cover to reconstruct the paleoenvironment (Louys et al. 2015). The incorporation of a continuous multidimensional spectrum of environmental conditions rather than a simplistic partitioning of habitats into categories will be essential for-providing a more nuanced and accurate reconstruction of

paleoenvironment and a better understanding of the relationship between the biome and their associated faunal communities.

(3) **Paleoclimate reconstructions**: The method as it is generally applied also lacks the ability to clearly identify paleoclimatic factors, focusing largely on a reconstruction of the physical environment rather than climate. However, the use of averages of ecometric traits (species traits that are easily measurable, whose structure is closely related to their function, and whose function interacts directly with local environment) in communities has been shown to be connected to precise measures of climate variables, such as the amount of precipitation or temperature (Eronen et al. 2010b; Fortelius et al. 2016; Žliobaitė et al. 2016). These types of data allow us to characterize physical habitats at the regional scale, linking them specifically to primary production (Fortelius et al. 2016).

Community structure analysis is rooted in the niche concept and the idea that faunal communities and the adaptations of their members can provide information about their environment. It is a powerful tool for reconstructing paleoenvironments. While it does not rely on a single mammalian taxon, it does incorporate ecometrics and taxon-specific analyses (dietary and locomotor reconstructions for specific taxa) to arrive at a holistic and community-level perspective on paleoenvironmental reconstruction. With further development and refinement, this methodology can help move us away from categorical descriptions of paleoenvironments and emphasize the continuous nature of ecological variables. This will allow for more nuanced reconstructions of paleoenvironments and a better understanding of the relationship between biotic communities and their environments.

Acknowledgements KK and KL gratefully acknowledge Denise Su, Darin Croft and Scott Simpson for the invitation to participate in the "Latest Methods in Reconstructing Cenozoic Terrestrial Environments and Ecological Communities" symposium in Cleveland 2015. KK sincerely thanks the organizers for support and assistance with childcare during the symposium and for useful discussions throughout. We also thank Peter Andrews for his continued advice and insight.

References

Alexander, R. M. (1977). Allometry of the limbs of antelopes (Bovidae). *Journal of Zoology, London, 183,* 125–146.

Anderson, M. J., & Willis, T. J. (2003). Canonical analysis of principal coordinates: a useful method of constrained ordination for ecology. *Ecology, 84,* 511–525.

Andrews, P. (1989). Paleoecology of Laetoli. *Journal of Human Evolution, 18,* 173–181.

Andrews, P. (1990a). *Owls, caves, and fossils*. Chicago: The University of Chicago Press.

Andrews, P. (1990b). Small mammal taphonomy. In E. H. Lindsey, V. Fahlbusch & P. Mein (Eds.), *European Neogene mammal chronology* (pp. 487–494). New York: Plenum Press.

Andrews, P. (1996). Paleoecology and hominoid palaeoenvironments. *Biological Reviews, 71,* 257–300.

Andrews, P., & Humphrey, L. (1999). African Miocene environments and the transition to early hominines. In T. G. Bromage & F. Schrenk (Eds.), *African biogeography. Climate change and human evolution* (pp. 282–300). Oxford: Oxford University Press.

Andrews, P., & Nesbit Evans, E. (1979). The environment of *Ramapithecus* in Africa. *Paleobiology, 5,* 22–30.

Andrews, P., & Van Couvering, J. H. (1975). Paleoenvironments in the East African Miocene. In F. Szalay (Ed.), *Approaches to primate paleobiology* (pp. 62–103). Basel: Karger.

Andrews, P., Lord, J. M., & Nesbit Evans, E. M. (1979). Patterns of ecological diversity in fossil and modern mammalian faunas. *Biological Journal of the Linnean Society, 11,* 177–205.

Ashley, G. M., Tactikos, J. C., & Owen, R. B. (2009). Hominin use of springs and wetlands: paleoclimate and archaeological records from Olduvai Gorge (1.79–1.74 Ma). *Palaeogeography, Palaeoclimatology, Palaeoecology, 272,* 1–16.

Avery, D. M. (2007). Micromammals as palaeoenvironmental indicators of the southern African Quaternary. *Transactions of the Royal Society of South Africa, 62,* 17–23.

Badgley, C., Bartels, W. S., Morgan, M. E., Behrensmeyer, A. K., & Raza, S. M. (1995). Taphonomy of vertebrate assemblages from the Paleogene of northwestern Wyoming and the Neogene of northern Pakistan. *Palaeogeography, Palaeoclimatology, Palaeoecology, 115,* 157–180.

Barnosky, A. D., & Lindsey, E. L. (2010). Timing of Quaternary megafaunal extinction in South America in relation to human arrival and climate change. *Quaternary International, 217,* 10–29.

Barnosky, A. D., Lindsey, E. L., Villavicencio, N. A., Bostelmann, E., Hadly, E. A., Wanket, J., et al. (2016). Variable impact of late-Quaternary megafaunal extinction in causing ecological state shifts in North and South America. *Proceedings of the National Academy of Sciences, USA, 113,* 856–861.

Barr, W. A. (2018). Ecomorphology. In D. A. Croft, D. F. Su & S.W. Simpson (Eds.), *Methods in paleoecology: Reconstructing Cenozoic terrestrial environments and ecological communities* (pp. 337–347). Cham: Springer.

Barry, J. C., Morgan, M. E., Flynn, L. J., Pilbeam, D., Behrensmeyer, A. K., Raza, S. M., Khan, I. A., Badgley, C, Hicks, J., Kelley, J. (2002). Faunal and environmental change in the late Miocene Siwaliks of northern Pakistan. *Paleobiology, 28,* 1–71.

Bedaso, Z. K., Wynn, J. G., Alemseged, Z., & Geraads, D. (2013). Dietary and paleoenvironmental reconstruction using stable isotopes of herbivore tooth enamel from middle Pliocene Dikika, Ethiopia: implication for *Australopithecus afarensis* habitat and food resources. *Journal of Human Evolution, 64,* 21–38.

Behrensmeyer, A. K. (1975). Taphonomy and paleoecology in the hominid fossil record. *Yearbook of Physical Anthropology, 19,* 36–50.

Behrensmeyer, A. K. (2015). Four million years of African herbivory. *Proceedings of the National Academy of Sciences, USA, 112,* 11428–11429.

Behrensmeyer, A. K., & Hill, A. P. (1980). *Fossils in the making: Vertebrate taphonomy and paleoecology*. Chicago: The University of Chicago Press.

Behrensmeyer, A. K., Damuth, J. D., DiMichele, W. A., Potts, R., Sues, H.-D., & Wing, S. L. (1992). *Terrestrial ecosystems through time: Evolutionary paleoecology of terrestrial plants and animals*. Chicago: The University of Chicago Press.

Blumenschine, R. J. (1988). An experimental model of the timing of hominid and carnivore influence on archaeological bone assemblages. *Journal of Archaeological Science, 15,* 483–502.

Blumenschine, R. J., & Peters, C. R. (1998). Archaeological predictions for hominid land use in the paleo-Olduvai Basin, Tanzania, during lowermost Bed II times. *Journal of Human Evolution, 34,* 565–607.

Blumenschine, R. J., Peters, C. R., Masao, F. T., Clarke, R. L., Deino, A. L., Hay, R. L., et al. (2003). Late Pliocene *Homo* and hominid land use from western Olduvai Gorge, Tanzania. *Science, 299,* 1217–1221.

Blumenschine, R. J., Masao, F. T., Stollhofen, H., Stanistreet, I. G., Bamford, M. K., Albert, R. M., et al. (2012a). Landscape distribution of Oldowan stone artifact assemblages across the fault compartments of the eastern Olduvai Lake Basin during early lowermost Bed II times. *Journal of Human Evolution, 63,* 384–394.

Blumenschine, R. J., Stanistreet, I. G., & Masao, F. T. (2012b). Olduvai Gorge and the Olduvai landscape paleoanthroplogy project. *Journal of Human Evolution, 63,* 247–250.

Brown, J. H. (1981). Two decades of homage to Santa Rosalia: toward a general theory of diversity. *American Zoologist, 21,* 877–888.

Campisano, C. J., & Feibel, C. S. (2007). Connecting local environmental sequences to global climate patterns: evidence from the hominin-bearing Hadar Formation, Ethiopia. *Journal of Human Evolution, 53,* 515–527.

Catibog-Singa, C., Catibog, C. S., & Heaney, L. R. (2006). *Philippine biodiversity: Principles and practice*. Haribon Foundation.

Cerling, T. E., & Hay, R. L. (1986). An isotopic study of paleosol carbonates from Olduvai Gorge. *Quaternary Research, 25,* 63–78.

Cerling, T. E., Andanje, S. A., Blumenthal, S. A., Brown, F. H., Chritz, K. L., Harris, J. M., et al. (2015). Dietary changes of large herbivores in the Turkana Basin, Kenya from 4 to 1 Ma. *Proceedings of the National Academy of Sciences, USA, 112,* 11467–11472.

Cione, A. L., Soibelzon, L., & Tonni, E. P. (2003). The broken zig-zag; Late Cenozoic large mammal and tortoise extinction in South America. *Revista del Museo Argentino de Ciencias Naturales, Nueva Serie, 5,* 1–19.

Clarke, K. R., & Warwick, R. M. (1994). *Change in marine communities: An approach to statistical analysis and interpretation*. Primer-E Ltd: Plymouth, UK.

Clarke, K. R., & Warwick, R. M. (2001). A further biodiversity index applicable to species lists: variation in taxonomic distinctness. *Marine Ecology Progress Series, 216,* 265–278.

Clyde, W. C., & Gingerich, P. D. (1998). Mammalian community response to the latest Paleocene thermal maximum: an isotaphonomic study in the northern Bighorn Basin, Wyoming. *Geology, 26,* 1011–1014.

Coe, H. (2009). Atmospheric energy and the structure of the atmosphere. In C. N. Hewitt & A. V. Jackson (Eds.), *Atmospheric science for environmental scientists* (pp. 54–82). West Sussex: Wiley.

Coe, M. (1980). The role of modern ecological studies in the reconstruction of palaeoenvironments in sub-Saharan Africa. In A. K. Behrensmeyer & A. P. Hill (Eds.), *Fossils in the making: Vertebrate taphonomy and paleoecology* (pp. 55–67). Chicago: The University of Chicago Press.

Croft, D. A. (2001). Cenozoic environmental change in South American as indicated by mammalian body size distributions (cenograms). *Diversity and Distributions, 7,* 271–287.

Croft, D. A. (2006). Do marsupials make good predators? Insights from predator-prey diversity ratios. *Evolutionary Ecology Research, 8,* 1193–1214.

Croft, D. A. (2013). What constitutes a fossil mammal community in the early Miocene Santa Cruz Formation? *Journal of Vertebrate Paleontology, 33,* 401–409.

Croft, D. A., Flynn, J. J., & Wyss, A. R. (2008). The Tinguiririca Fauna of Chile and the early stages of "modernization" of South American mammal faunas. *Arquivos do Museu Nacional, 66,* 191–211.

Cushing, A. E. (2002). The landscape zooarchaeology and paleontology of Plio-Pleistocene Olduvai, Tanzania and their implications for early hominid ecology. Ph.D. Dissertation, Rutgers, the State University of New Jersey.

Damuth, J., & MacFadden, B. J. (1990). *Body size in mammalian paleobiology: Estimation and biological implications*. Cambridge: Cambridge University Press.

De Vivo, M., & Carmignotto, A. P. (2004). Holocene vegetation change and the mammal faunas of South America and Africa. *Journal of Biogeography, 31,* 943–957.

Deino, A. L. (2012). $^{40}Ar/^{39}Ar$ dating of Bed I, Olduvai Gorge, Tanzania, and the chronology of early Pleistocene climate change. *Journal of Human Evolution, 63,* 251–273.

Dunn, R. E., Strömberg, C. A. E., Madden, R. H., Kohn, M. J., & Carlini, A. A. (2015). Linked canopy, climate, and faunal change in the Cenozoic of Patagonia. *Science, 347,* 258–261.

Dunn, R. H. (2018). Functional morphology of the postcranial skeleton. In D. A. Croft, D. F. Su & S. W. Simpson (Eds.), *Methods in paleoecology: Reconstructing Cenozoic terrestrial environments and ecological communities* (pp. 23–36). Cham: Springer.

Eisenberg, J. F. (1981). *The mammalian radiations: An analysis of trends in evolution, adaptation, and behavior.* Chicago: The University of Chicago Press.

Eiten, G. (1992). How names are used for vegetation. *Journal of Vegetation Science, 3,* 419–424.

Elton, C. S. (1927). *Animal ecology*. New York: MacMillan Co.

Eronen, J. T., Puolamäki, K., Liu, L., Lintulaakso, K., Damuth, J., Janis, C., et al. (2010a). Precipitation and large herbivorous mammals II: application to fossil data. *Evolutionary Ecology Research, 12,* 235–248.

Eronen, J. T., Polly, P. D., Fred, M., Damuth, J., Frank, D. C., Mosbrugger, V., et al. (2010b). Ecometrics: the traits that bind the past and present together. *Integrative Zoology, 5,* 88–101.

Evans, A. R., & Pineda-Munoz, S. (2018). Inferring mammal dietary ecology from dental morphology. In D. A. Croft, D. F. Su & S. W. Simpson (Eds.), *Methods in paleoecology: Reconstructing Cenozoic terrestrial environments and ecological communities* (pp. 37–51). Cham: Springer.

Faith, J. T., Tryon, C. A., Peppe, D. J., Beverly, E. J., & Blegen, N. (2014). Biogeographic and evolutionary implications of an extinct late Pleistocene impala from the Lake Victoria Basin, Kenya. *Journal of Mammalian Evolution, 21,* 213–222.

Fernandez-Jalvo, Y., Denys, C., Andrews, P., Williams, T., Dauphin, Y., & Humphrey, L. (1998). Taphonomy and paleoecology of Olduvai Bed-I (Pleistocene, Tanzania). *Journal of Human Evolution, 34,* 137–172.

Flagstad, Ø., Syversten, P. O., Stenseth, N. C., & Jakobsen, K. S. (2001). Environmental change and rates of evolution: the phylogeographic pattern within the hartebeest complex as related to climatic variation. *Proceedings of the Royal Society B, 268,* 667–677.

Fleming, T. H. (1973). Numbers of mammal species in North and Central American forest communities. *Ecology, 54,* 555–563.

Flynn, J. J., Wyss, A. R., Croft, D. A., & Charrier, R. (2003). The Tinguiririca Fauna, Chile: biochronology, paleoecology, biogeography, and a new earliest Oligocene South American Land Mammal 'Age'. *Palaeogeography, Palaeoclimatology Palaeoecology, 195,* 229–259.

Fortelius, M. (1990). Problems using fossil teeth to estimate body sizes of extinct mammals. In J. Damuth & B. J. MacFadden (Eds.), *Body size in mammalian paleobiology: Estimation and biological implications* (pp. 207–288). New York: Cambridge University Press.

Fortelius, M., Eronen, J., Jernvall, J., Liu, L., Pushkina, D., Rinne, J., et al. (2002). Fossil mammals resolve regional patterns of Eurasian climate change over 20 million years. *Evolutionary Ecology Research, 4,* 1005–1016.

Fortelius, M., Žliobaitė, I., Kaya, F., Bibi, F., Bobe, R., Leakey, L., et al. (2016). An ecometric analysis of the fossil mammal record of the Turkana Basin. *Philosophical Transactions of the Royal Society B, 371,* 20150232.

Fukami, T. (2015). Historical contingency in community assembly: integrating niches, species pools, and priority effects. *Annual Review of Ecology, Evolution, and Systematics, 46,* 1–23.

Geise, L., Pereira, L. G., Bossi, D. E. P., & Bergallo, H. G. (2004). Pattern of elevational distribution and richness of non volant mammals in Itatiaia National Park and its surroundings, in southeastern Brazil. *Brazilian Journal of Biology, 64,* 599–612.

Gentry, A. H. (1988). Changes in plant community diversity and floristic composition on environmental and geographical gradients? *Annals of the Missouri Botanical Garden, 75,* 1–34.

Gentry, A. W., & Gentry, A. (1978). Fossil Bovidae (Mammalia) of Olduvai Gorge, Tanzania, Part II. *Bulletin of the British Museum of Natural History (Geology), 30,* 1–83.

Gillman, L. N., Keeling, D. J., Ross, H. A., & Wright, S. D. (2009). Latitude, elevation and the tempo of molecular evolution in mammals. *Proceedings of the Royal Society of London B: Biological Sciences, 276,* 3353–3359.

Gingerich, P. D. (1989). New earliest Wasatchian mammalian fauna from the Eocene of northwestern Wyoming: composition and diversity in a rarely sampled high-floodplain assemblage. *University of Michigan Papers on Paleontology, 28,* 1–97.

Gingerich, P. D., Smith, B. H., & Rosenberg, K. (1982). Allometric scaling in the dentition of primates and prediction of body weight from tooth size in fossils. *American Journal of Physical Anthropology, 58,* 81–100.

Gould, S. J. (1975). On the scaling of tooth size in mammals. *American Zoologist, 15,* 351–362.

Graham, C. H., & Fine, P. V. (2008). Phylogenetic beta diversity: linking ecological and evolutionary processes across space in time. *Ecology letters, 11,* 1265–1277.

Green, J. L., & Croft, D. A. (2018). Using dental mesowear and microwear for dietary inference: a review of current techniques and applications. In D. A. Croft, D. F. Su & S. W. Simpson (Eds.), *Methods in paleoecology: Reconstructing Cenozoic terrestrial environments and ecological communities* (pp. 53–73). Cham: Springer.

Greenacre, M. J. (1984). *Theory and applications of correspondence analysis*. New York: Academic Press.

Greenacre, M. J., & Vrba, E. S. (1984). Graphical display and interpretation of antelope census data in African wildlife areas, using correspondence analysis. *Ecology, 65,* 984–997.

Hanya, G., Stevenson, P., van Noordwijk, M., Te Wong, S., Kanamori, T., Kuze, N., et al. (2011). Seasonality in fruit availability affects frugivorous primate biomass and species richness. *Ecography, 34,* 1009–1017.

Happold, D. C. D. (1987). *The mammals of Nigeria*. Oxford: Clarendon Press.

Harrison, J. L. (1962). The distribution of feeding habits among animals in a tropical rain forest. *Journal of Animal Ecology, 31,* 53–63.

Harrison, T. (2011). Laetoli revisited: renewed paleontological and geological investigations at localities on the Eyasi Plateau in northern Tanzania. In T. Harrison (Ed.), *Paleontology and geology of Laetoli: Human evolution in context. Volume 1: Geology, geochronology, paleoecology and paleoenvironment* (pp. 1–15). Dordrecht: Springer Press.

Hay, R. L. (1976). *The geology of Olduvai Gorge: A study of sedimentation in a semiarid basin.* Berkeley: University of California Press.

Hay, R. L. (1990). Olduvai Gorge: a case history in the interpretation of hominid paleoenvironments in East Africa. In L. F. Laporte (Ed.), *Establishment of a geological framework for paleoanthropology* (pp. 23–37). Geological Society of America Special Paper 242.

Hessler, I., Dupont, L., Bonnefille, R., Behling, H., González, C., Helmens, K. F., et al. (2010). Millennial-scale changes in vegetation records from tropical Africa and South America during the last glacial. *Quaternary Science Reviews, 29,* 2882–2899.

Higgins, P. (2018). Isotope ecology from biominerals. In D. A. Croft, D. F. Su & S. W. Simpson (Eds.), *Methods in paleoecology: Reconstructing Cenozoic terrestrial environments and ecological communities* (pp. 99–120). Cham: Springer.

Hopkins, S. S. B. (2008). Reassessing the mass of exceptionally large rodents using toothrow length and area as proxies for body mass. *Journal of Mammalogy, 89,* 232–243.

Hopkins, S. S. B. (2018). Estimation of body size in fossil mammals. In D. A. Croft, D. F. Su & S. W. Simpson (Eds.), *Methods in paleoecology: Reconstructing Cenozoic terrestrial environments and ecological communities* (pp. 7–22). Cham: Springer.

Hutchinson, G. E. (1957). The multivariate niche. *Cold Spring Harbor Symposium of Quantitative Biology, 22,* 415–421.

Hutchinson, G. E. (1959). Homage to Santa Rosalia or why are there so many kinds of animals? *The American Naturalist, 93,* 145–159.

Janis, C. M. (1993). Tertiary mammal evolution in the context of changing climates, vegetation, and tectonic events. *Annual Review of Ecology and Systematics, 24,* 467–500.

Kappelman, J. (1984). Plio-Pleistocene environments of Bed I and lower Bed II, Olduvai Gorge, Tanzania. *Palaeogeography, Palaeoclimatology, Palaeoecology, 48,* 171–196.

Kappelman, J. (1988). Morphology and locomotor adaptations of the bovid femur in relation to habitat. *Journal of Morphology, 198,* 119–130.

Karr, J. R. (1971). Structure of avian communities in selected Panama and Illinois habitats. *Ecological Monographs, 41,* 207–233.

Kay, R. F., & Madden, R. H. (1997). Mammals and rainfall: paleoecology of the middle Miocene at La Venta (Colombia, South America). *Journal of Human Evolution, 32,* 161–199.

Kindt, R, & Coe, R. (2005). *Three diversity analysis. A manual and software for common statistical methods for ecological and biodiversity studies*. Nairobi: World Agroforestry Centre (ICRAF).

Kingston, J. D., & Harrison, T. (2007). Isotopic dietary reconstructions of Pliocene herbivores at Laetoli: implications for early hominin paleoecology. *Palaeogeography, Palaeoclimatology, Palaeoecology, 243,* 272–306.

Kovarovic, K., Andrews, P., & Aiello, L. C. (2002). The palaeoecology of the Upper Ndolanya Beds at Laetoli, Tanzania. *Journal of Human Evolution, 43,* 395–418.

Kovarovic, K., Aiello, L. C., Cardini, A., & Lockwood, C. A. (2011). Discriminant function analyses in archaeology: are classifications rates too good to be true? *Journal of Archaeological Science, 38,* 3006–3018.

Kovarovic, K., Slepkov, R., & McNulty, K. P. (2013). Ecological continuity between Lower and Upper Bed II, Olduvai Gorge, Tanzania. *Journal of Human Evolution, 64,* 538–555.

Kurtén, B. (1952). The Chinese Hipparion fauna. *Commentationes Biological, Societas Scientiarum Fennica, 13,* 1–82.

Lawesson, J. E. (1994). Some comments on the classification of African vegetation. *Journal of Vegetation Science, 5,* 441–444.

Leakey, M. D. (1971). *Olduvai Gorge: Excavations in Beds I and II 1960–1963* (Vol. 3). Cambridge: Cambridge University Press.

Leakey, M. D., & Harris, J. M. (Eds.). (1987). *Laetoli: A Pliocene site in northern Tanzania*. Oxford: Clarendon Press.

Legendre, S. (1986). Analysis of mammalian communities from the late Eocene and Oligocene of Southern France. *Paleovertebrata, 16,* 191–212.

Legendre, S. (1989). Les communautés de mammifères du Paléogène (Eocène supérieur et Oligocène) d'Europe occidentale: structures, milieux et evolution. *Münchner Geowissenschafliche Abhandlungen A, 16,* 1–110.

Lintulaakso, K., & Kovarovic, K. (2016). Diet and locomotion, but not body size, differentiate mammal communities in worldwide tropical ecosystems. *Palaeogeography, Palaeoclimatology, Palaeoecology, 454,* 20–29.

Lomolino, M. V., Riddle, B. R., Brown, J. H., & Brown, J. H. (2006). *Biogeography*. Sunderland: Sinauer Associates.

Losos, J. B. (1996). Phylogenetic perspectives on community ecology. *Ecology, 77,* 1344–1354.

Louys, J. (2007). *Ecology and extinction of southeast Asia's megafauna*. Ph.D. Dissertation, University of New South Wales.

Louys, J., Travouillon, K. J., Bassarova, M., & Tong, H. (2009). The use of protected natural areas in palaeoecological analyses: assumptions, limitations and application. *Journal of Archaeological Science, 36,* 2274–2288.

Louys, J., Meloro, C., Elton, S., Ditchfield, P., & Bishop, L. C. (2011). Mammal community structure correlates with arboreal heterogeneity in faunally and geographically diverse habitats: implications for community convergence. *Global Ecology and Biogeography, 20,* 717–729.

Louys, J., Meloro, C., Elton, S., Ditchfield, P., & Bishop, L. C. (2015). Analytical framework for reconstructing heterogeneous environmental variables from mammal community structure. *Journal of Human Evolution, 78,* 1–11.

MacArthur, R. H., & Wilson, E. O. (1967). *Theory of island biogeography*. Princeton: Princeton University Press.

Mares, M. A., & Willig, M. R. (1994). Inferring biome associations of Recent mammals from samples of temperate and tropical faunas: paleoecological considerations. *Historical Biology, 8,* 31–48.

Meloro, C., & Kovarovic, K. (2013). Spatial and ecometric analyses of the Plio-Pleistocene large mammal communities of the Italian peninsula. *Journal of Biogeography, 40,* 1451–1462.

Mendoza, M., Goodwin, B., & Criado, C. (2004). Emergence of community structure in terrestrial mammal-dominated ecosystems. *Journal of Theoretical Biology, 230,* 203–214.

Nesbit Evans, E. M., Van Couvering, J. A. H., & Andrews, P. (1981). Paleoecology of Miocene sites in western Kenya. *Journal of Human Evolution, 10,* 99–116.

Odum, E. P. (1953). *Fundamentals of ecology*. Philadelphia: W.B. Saunders Company.

Odum, E. P. (1959). *Fundamentals of ecology* (2nd ed.). Philadelphia: W.B. Saunders Company.

Olson, J. S., Watts, J. A., & Allison, L. J. (1983). *Carbon in live vegetation of major world ecosystems*. Oak Ridge: Oak Ridge National Laboratory.

Olson, J. S., Watts, J. A., & Allison, L. J. (1985). *Major world ecosystem complexes ranked by carbon in live vegetation: A database*. Oak Ridge: Carbon Dioxide Information Analysis Center, U.S. Department of Energy, Oak Ridge National Laboratory.

Overpeck, J., Whitlock, C., & Huntley, B. (2003). Terrestrial biosphere dynamics in the climate system: past and future. In K. D. Alverson, R. Bradley & T. F. Peterson (Eds.), *Paleoclimate, global change and the future* (pp. 81–103). Berlin: Springer.

Pianka, E. R. (1970). On r-and K-selection. *The American Naturalist, 104,* 592–597.

Pimm, S. L., & Lawton, J. H. (1978). On feeding on more than one trophic level. *Nature, 275,* 542–544.

Pineda-Munoz, S., & Alroy, J. (2014). Dietary characterization of terrestrial mammals. *Proceedings of the Royal Society B: Biological Sciences, 281,* 20141173.

Pineda-Munoz, S., Evans, A. R., & Alroy, J. (2016). The relationship between diet and body mass in terrestrial mammals. *Paleobiology, 42,* 659–669.

Potts, R. (1988). *Early hominid activities at Olduvai Gorge*. New York: Aldine de Gruyter.

Reed, D. N. (2007). Serengeti micromammals and their implications for Olduvai paleoenvironments. In R. Bobe, Z. Alemseged & A. K. Behrensmeyer (Eds.), *Hominin environments in the East African Pliocene: An assessment of the faunal evidence* (pp. 217–255). Dordrecht: Springer.

Reed, K. E. (1997). Early hominid evolution and ecological change through the African Plio-Pleistocene. *Journal of Human Evolution, 32,* 289–322.

Reed, K. E. (1998). Using large mammal communities to examine ecological and taxonomic structure and predict vegetation in extant and extinct assemblages. *Paleobiology, 24,* 384–408.

Reed, K. E. (2008). Paleoecological patterns at the Hadar hominin site, Afar Regional State, Ethiopia. *Journal of Human Evolution, 54,* 743–768.

Rodríguez, J. (1999). Use of cenograms in mammalian palaeoecology. A critical review. *Lethaia, 32,* 331–347.

Rodríguez, J., Hortal, J., & Nieto, M. (2006). An evaluation of the influence of environment and biogeography on community structure: the case of Holarctic mammals. *Journal of Biogeography, 33,* 291–303.

Ruddiman, W. F. (2007). *Earth's climate: Past and future*. New York: Macmillan.

Schick, R. S., Halpin, P. N., Read, A. J., Urban, D. L., Best, B. D., Good, C. P., et al. (2011). Community structure in pelagic marine mammals at large spatial scales. *Marine Ecology Progress Series, 434,* 165–181.

Schrenk, F., Bromage, T. G., Gorthner, A., & Sandrock, O. (1995). Paleoecology of the Malawi Rift: vertebrate and invertebrate faunal contexts of the Chiwondo Beds, northern Malawi. *Journal of Human Evolution, 28,* 59–70.

Scott, K. M. (1990). Postcranial dimensions of ungulates as predictors of body mass. In J. Damuth & B. J. MacFadden (Eds.), *Body size in mammalian paleobiology: Estimation and biological implications* (pp. 301–355). New York: Cambridge University Press.

Semken, H. A., Jr., Graham, R. W., & Stafford, T. W., Jr. (2010). AMS 14C analysis of Late Pleistocene non-analog faunal components from 21 cave deposits in southeastern North America. *Quaternary International, 217,* 240–255.

Shipman, P., & Harris, J. (1988). Habitat preference and paleoecology of *Australopithecus boisei* in Eastern Africa. In F. E. Grine (Ed.), *Evolutionary history of the "robust" australopithecines* (pp. 343–381). New York: Aldine de Gruyter.

Simpson, G. G. (1964). Species density of North American Recent mammals. *Systematic Zoology, 13,* 57–73.

Sokal, R. R., & Rohlf, F. J. (1995). *Biometry: The principles and practice of statistics in biological research*. New York: W. H. Freeman and Company.

Soligo, C., & Andrews, P. (2005). Taphonomic bias, taxonomic bias and historical non-equivalence of faunal structure in early hominin localities. *Journal of Human Evolution, 49,* 206–229.

Sponheimer, M., & Lee-Thorp, J. A. (2003). Using carbon isotope data of fossil bovid communities for paleoenvironmental reconstruction. *South African Journal of Science, 99,* 273–275.

Stanistreet, I. G. (2012). Fine resolution of early hominin time, Beds I and II, Olduvai Gorge, Tanzania. *Journal of Human Evolution, 63,* 300–308.

Stoetzel, E., Marion, L., Nespoulet, R., El Hajraoui, M. A., & Denys, C. (2011). Taphonomy and paleoecology of the late Pleistocene to middle Holocene small mammal succession of El Harhoura 2 cave (Rabat-Témara, Morocco). *Journal of Human Evolution, 60,* 1–33.

Strömberg, C. A. (2011). Evolution of grasses and grassland ecosystems. *Annual Review of Earth and Planetary Sciences, 39,* 517–544.

Su, D. F. (2005). *The paleoecology of Laetoli, Tanzania: Evidence from the mammalian fauna of the Upper Laetolil Beds.* Ph.D. Dissertation, New York University.

Su, D. F. (2011). Large mammal evidence for the paleoenvironment of the Upper Laetoli and Upper Ndolanya beds of Laetoli, Tanzania. In T. Harrison (Ed.), *Paleontology and geology of Laetoli: Human evolution in context. Volume 1: Geology, geochronology, paleoecology and paleoenvironment* (pp. 381–392). Dordrecht: Springer.

Su, D. F. (2016). The taphonomy and paleoecology of Korsi Dora Vertebrate Locality 1, Woranso-Mille Study Area, Ethiopia. In Y. Haile-Selassie & D. F. Su (Eds.), *The postcranial anatomy of* Australopithecus afarensis*: New insights from KSD-VP-1/1* (pp. 25–37). New York: Springer.

Su, D. F., & Haile-Selassie, Y. (in press). Mosaic habitats at Woranso-Mille (Ethiopia) during the Pliocene and implications for *Australopithecus* paleoecology. In S. Reynolds & R. Bobe (Eds.), *African paleoecology and human evolution.* Cambridge: Cambridge University Press.

Su, D. F., & Harrison, T. (2007). The paleoecology of the Upper Laetoli Beds at Laetoli. In R. Bobe, Z. Alemseged & A. K. Behrensmeyer (Eds.), *Hominin environments in the East African Pliocene: An assessment of the faunal evidence* (pp. 279–313). Dordrecht: Springer.

Su, D. F, & Harrison, T. (2015). The paleoecology of the upper Laetolil Beds, Laetoli, Tanzania: A review and synthesis. *Journal of African Earth Sciences 101,* 405–419.

Su, D. F., Ambrose, S. H., Degusta, D., & Haile-Selassie, Y. (2009). Paleoenvironment. In Y. Haile-Selassie & G. WoldeGabriel (Eds.), Ardipithecus kadabba: *Late Miocene evidence from the Middle Awash, Ethiopia* (pp. 521–547). Berkeley: University of California Press.

Thibault, K. M., & Brown, J. H. (2008). Impact of an extreme climatic event on community assembly. *Proceedings of the National Academy of Sciences, USA, 105,* 3410–3415.

Thibault, K. M., White, E. P., & Morgan Ernest, S. K. (2004). Temporal dynamics in the structure and composition of a desert rodent community. *Ecology, 85,* 2649–2655.

Townsend, K. E. B., Rasmussen, D. T., Murphey, P. C., & Evanoff, E. (2010). Middle Eocene habitat shifts in the North American western interior: A case study. *Palaeogeography, Palaeoclimatology, Palaeoecology, 297,* 144–158.

Van Couvering, J. A. H. (1980). Community evolution in East Africa during the late Cenozoic. In A. K. Behrensmeyer & A. P. Hill (Eds.), *Fossils in the making: Vertebrate taphonomy and paleoecology* (pp. 272–298). Chicago: The University of Chicago Press.

Voorhies, M. R. (1969a). Sampling difficulties in reconstructing late Tertiary mammalian communities. *Proceedings of the North American Paleontological Convention, Part E,* pp. 454–468.

Voorhies, M. R. (1969b). *Taphonomy and population dynamics of an early Pliocene vertebrate fauna, Knox County, Nebraska.* Laramie: University of Wyoming.

Warton, D. I., & Hui, F. K. (2011). The arcsine is asinine: the analysis of proportions in ecology. *Ecology, 92,* 3–10.

Warwick, R. M., & Clarke, K. R. (1995). New 'biodiversity' measures reveal a decrease in taxonomic distinctness with increasing stress. *Marine Ecology Progress Series, 129,* 301–305.

Western, D., & Behrensmeyer, A. K. (2009). Bone assemblages track animal community structure over 40 years in an African savannah ecosystem. *Science, 324,* 1061–1064.

White, F. (1983). *The vegetation of Africa: A descriptive memoir to accompany the Unesco/AETFAT/UNSO vegetation map of Africa.* Paris: Unesco.

White, T. D., Ambrose, S. H., Suwa, G., Su, D. F., DeGusta, D., Bernor, R. L., et al. (2009). Macrovertebrate paleontology and the Pliocene habitat of *Ardipithecus ramidus. Science, 326,* 87–93.

Williams, S. E., Marsh, H., & Winter, J. (2002). Spatial scale, species diversity, and habitat structure: small mammals in Australian tropical rain forest. *Ecology, 83,* 1317–1329.

Wilson, V. J. (1975). Mammals of the Wankie National Park. *Museum Memoirs of the National Museums and Monuments of Rhodesia, 4,* 1–147.

Vrba, E. S. (1980). The significance of bovid remains as indicators of environment and predation patterns. In A. K. Behrensmeyer & A. P. Hill (Eds.), *Fossils in the making: Vertebrate taphonomy and paleoecology* (pp. 247–271). Chicago: The University of Chicago Press.

Yodzis, P. (1984). How rare is omnivory? *Ecology, 65,* 321–323.

Žliobaitė, I., Rinne, J., Tóth, A. B., Mechenich, M., Liu, L., Behrensmeyer, A. K., et al. (2016). Herbivore teeth predict climatic limits in Kenyan ecosystems. *Proceedings of the National Academy of Sciences, USA, 113,* 12751–12756.

Zuur, A. F., Leno, E. N., & Smith, G. M. (2007). *Analysing ecological data.* New York: Springer.

Chapter 17
Ecometrics: A Trait-Based Approach to Paleoclimate and Paleoenvironmental Reconstruction

Wesley A. Vermillion, P. David Polly, Jason J. Head, Jussi T. Eronen, and A. Michelle Lawing

Abstract Ecometrics is a trait-based approach to study ecosystem variability through time. An ecometric value is derived from describing the distribution of functional traits at the community level, which may arise by environmental filtering, extinction, or convergence. An ecometric relationship describes the correspondence between spatially explicit ecometric values and corresponding environmental variation. Transfer functions and maximum likelihood approaches have been developed with modern trait-environment relationships to reconstruct paleotemperature, paleoprecipitation, and paleovegetation cover given the distribution of functional traits within a community. Because the focus for this approach is on the traits and not on species, it is transferable through space and time and can be used to compare novel communities. In this paper we review the concepts and history of ecometric analysis and then describe practical methods for implementing an ecometric study.

Keywords Functional morphology • Paleoecology
Maximum likelihood • Geographic distribution • Taxon-free

W. A. Vermillion · A. M. Lawing (✉)
Department of Ecosystem Science and Management, Texas A&M University, College Station, TX, USA
e-mail: alawing@tamu.edu

P. D. Polly
Departments of Earth & Atmospheric Sciences, Biology, and Anthropology, Indiana University, Bloomington, IN, USA

J. J. Head
Department of Zoology and University Museum of Zoology, University of Cambridge, Cambridge, CB23EJ, UK

J. T. Eronen
Ecosystems and Environment Research Programme and Helsinki Institute of Sustainability Science (HELSUS), University of Helsinki, Helsinki, Finland

Summary

One of the factors that sort species geographically into communities is the filtering of species by their functional morphologies along climate and environmental gradients (e.g., mammalian herbivores with higher molars eat more gritty and abrasive vegetation and therefore more species with higher molars live in places with more gritty and abrasive vegetation), thus, the morphology has a direct relationship with its environmental condition. In this sorting process, the distribution of functional traits within a community moves toward an optimal environmental condition. These functional traits may be described in terms of their distributions within a community and descriptors of the distribution of community level functional traits have been defined as ecometrics (Eronen et al. 2010a; Polly et al. 2011). Thus, ecometrics are traits described at the community level that have a relationship with environmental condition. Ecometrics are spatially variable, because the composition of species and traits in communities vary through geography. If species are perfectly sorted, the trait distributions in communities will parallel variation in the environmental condition to which the traits are functionally linked (i.e., the ecometric correlation equals one). If the ecometric correlation is strong, one can estimate the most likely environmental condition given a specific ecometric value using a transfer function or likelihood estimation. If the ecometric correlation is not strong, one may still be able to estimate the most likely environmental condition, but the likelihood surface would be more flat, which would produce a more ambiguous reconstruction. The power of this approach is that functional trait measurements are transferable through space and through time, regardless of the specific species that make up the community. For this reason, ecometrics is a useful approach to reconstruct paleoclimate and paleoenvironment (Damuth et al. 1992).

D. A. Croft et al. (eds.), *Methods in Paleoecology: Reconstructing Cenozoic Terrestrial Environments and Ecological Communities*, Vertebrate Paleobiology and Paleoanthropology, https://doi.org/10.1007/978-3-319-94265-0_17

Terms

Functional trait: a behavioral, biochemical, morphological, phenological, physiological, or structural characteristic that has a physical or chemical interaction with a specific environmental factor. This characteristic is expressed as a phenotype of individual organisms and it is considered relevant to the response of organisms to the surrounding abiotic conditions. A functional trait is something that can be observed and measured.

Geographic sorting: the process by which trait distributions are arranged in geographic space along gradients of environmental conditions.

Taxon free: not depending on taxonomic identity (i.e., it is a descriptor based on trait values rather than on presence or absence of particular taxa). For example, using palms as indicators of frost-free conditions is a taxon-based approach, whereas using counts of stomatal density as indicators of carbon dioxide concentration is a taxon-free approach (Beerling et al. 2011).

Ecometric value: a statistic summarizing the distribution of a functional trait within a community at a single geographic location. Means, variances, proportions, and extremes have been used as descriptors in previous studies; however, any parameter that describes a distribution of functional traits may be used.

Ecometric patterning: geography, or spatial variation, of ecometric values across communities. Perfect patterning represents optimal distributions of populations and species. The lack of patterning could represent a lack of useable covariation, a mismatch between the trait and the environment, or it could result from the inappropriate choice of environmental correlates.

Ecometric correlation: product-moment correlation between ecometric variation and variation in the environmental condition. High correlations evince a close match between the functional requirements of local communities and their abiotic condition.

Ecometric likelihood: the probability of finding a particular environmental condition given the ecometric value of a community at any single geographic location.

Ecometric load: the amount of mismatch between the distribution of functional traits and the selective optimum for the local environment. High loads may indicate risk to changing environmental conditions. Note that high ecometric loads can be present even with high ecometric correlations if traits are weakly but consistently sorted.

Ecometric optimum: the optimal distribution of an ecometric trait in a particular environment. The optimum is more likely to be a distribution rather than a single trait value, especially in locally heterogeneous environments (e.g., a savannah habitat favors both large and small body sizes).

Transfer function: a function that represents the relationship between the ecometric trait values and a relevant environmental variable.

Theoretical and Historical Background

Functional traits are features that organisms use to interact with their environment and with other organisms (Diaz and Cabido 2001; Violle et al. 2007). Functional traits of species within a community ultimately mediate the relationship between organisms' morphology and their ecology and link organisms to biotic and abiotic conditions present at particular geographic locations at a particular point in time (Ricklefs and Travis 1980; Chapin 1993; Poff 1997). A certain combination of biotic and abiotic conditions will thus favor traits that maximize an organism's performance, either through natural selection or geographic sorting. Natural selection is an evolutionary process that acts on the variation in a population, where individuals that thrive in their environment produce the most offspring and disproportionately share their traits with the next generation (Darwin and Wallace 1858). Species that move geographically in response to environmental change may follow an environment for which they are already adapted and form new assemblages as a result (Jackson and Overpeck 2000). We are not concerned with the trait values of particular species, but with the distribution of traits within communities (e.g., mean, standard deviation, and range; Damuth et al. 1992). Even weak sorting processes can produce trait gradients at the community level, as has been demonstrated in climate gradients in plant leaf traits (Wolfe 1979; Reich et al. 1997; Wilf 1997; Wright et al. 2004; Royer et al. 2005; Wing et al. 2005; Cornwell and Ackerly 2009; Peppe et al. 2011), large mammalian herbivores dental traits (Janis and Fortelius 1988; Damuth et al. 1992; Fortelius et al. 2002; Mendoza et al. 2005; Damuth and Janis 2011; Eronen et al. 2010b, 2012), carnivoran locomotor traits (Polly 2010), snake locomotor traits (Lawing et al. 2012), and mammal life history, physiological, and range traits (Lawing et al. 2017). The study of functional traits that have a relationship with climate or environmental gradients at the community level has been termed ecometrics to distinguish it from ecomorphology of particular species or clades (Eronen et al. 2010a; Polly et al. 2011). Figure 17.1 shows a few examples of the morphology of ecometric traits.

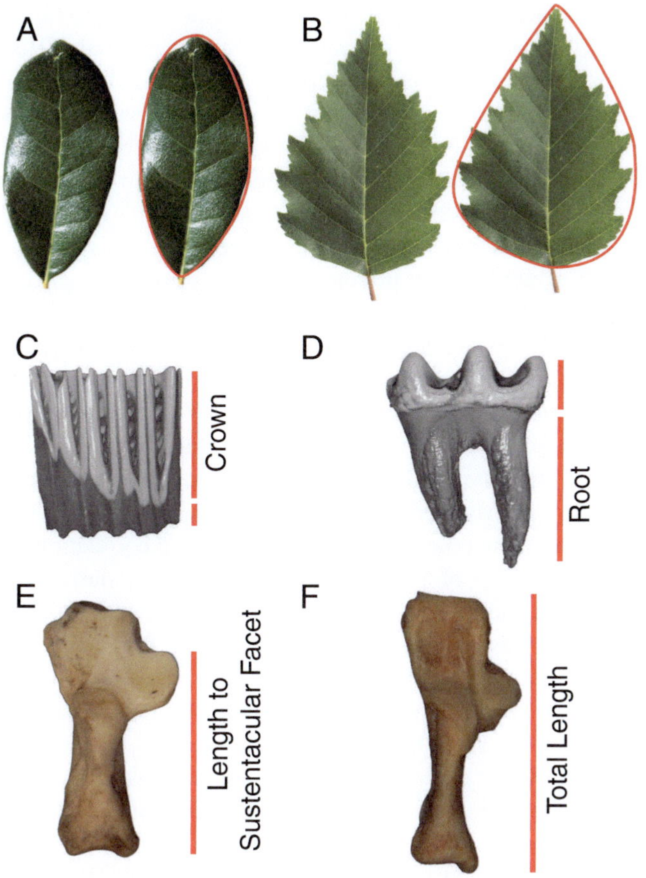

Fig. 17.1 A, Example of an entire leaf margin from a live oak leaf modified from an image by Jeremy Stovall. The leaf margin is entire when its margin is not punctuated. An easy way to determine if the margin is entire or to quantify the entirety of the margin is to overlay the image of the leaf with an oval, as is depicted in the figure with the red outline. Entire leaf margins are characteristic of hotter environments. B, Example of a non-entire leaf margin from a dwarf birch modified from an image by Fox Valley. Non-entire leaf margins are characteristic of cooler environments. C, Example of hypsodonty in a rodent modified from Tapaltsyan et al. (2015). Hypsodonty is measured by the crown to root ratio. Hypsodont rodents are grazers and are characteristic of arid environments with more gritty vegetation. D, Example of brachydonty in a rodent modified from Tapaltsyan et al. (2015). Brachydont rodents are browsers and are found in more wet environments with mixed vegetation. E, Example of a calcaneum from a raccoon modified from an image on boneid.net. Gear ratio is measured on the calcaneum and is the ratio of the length of the sustentacular facet to the total length of the calcaneum. The gear ratio for the raccoon is typical of a plantigrade animal and is characteristic for animals living in high vegetation cover (e.g., dense woodlands). F, Example of a calcaneum from a panther modified from an image on boneid.net. The gear ratio in the panther is typical of a digitigrade animal and is characteristic of animals that are adapted to run through open habitats

The relationship between the distribution of functional traits in a community and the corresponding environmental factors in the local environment is likely to be complex. Some trait-environment relationships favor a single ecometric optimum in each local environment such that all species tend to have the same trait value. Stomatal density on plant leaves might be expected to have a single optimal value across all members of a community because carbon dioxide concentrations and evapotranspiration conditions are the same for all. However, locally heterogeneous environments are likely to favor a heterogeneous ecometric optimum. For example, savannah environments tend to favor a combination of large body masses (for animals that can travel long distances to procure food in a comparatively resource poor habitat and defend themselves in the open) and small body masses (for species that can subsist on the resources found in a limited area and can find cover in grasses or tree tops). The nature of the ecometric optimum will dictate how it should be measured. An optimum that favors a single trait value can always be adequately represented with the community mean, but an optimum that favors a complex distribution may be better characterized by variance, skewness, or other measure. Temporal variation in environmental conditions may also shape the distribution of the ecometric optimum. For example, Žliobaitė et al. (2016) recently showed that recurring extreme conditions (like drought) were better predictors of dental traits in Kenyan herbivores than average conditions, indicating that recurring limiting conditions may be important drivers of community functional trait distributions.

Historical Background

Bailey and Sinnott (1915) conducted the first study of community-level functional traits. They showed plant communities that have more leaves with ridges, or non-entire margins, occur in moist temperate climates that experience warm summers and cold winters. Conversely, plant communities that have leaves with no ridges, or entire leaf margins, occur in dry climates including tropic, alpine and arctic regions, moors, steppes, deserts, saline situations, and other physiologically dry climates. They used the proportion of species within a community with a specified phenotype as their metric to describe the community, what we refer to as an ecometric trait value.

Since Bailey and Sinnott (1915) demonstrated the variation in community leaf shape across a temperature gradient, researchers have used leaf shape and other traits to reconstruct paleotemperature. Most notably, Wolfe (1979, 1993) quantified the relationship between leaf margins and temperature, which led to many quantitative paleoclimate reconstructions from leaf assemblages, called leaf-margin analysis (e.g., Wilf 1997; Greenwood et al. 2004; Yang et al. 2007). Contemporary to and since Wolfe's work, researchers have been interested in identifying the function of leaf teeth (see Royer and Wilf (2006) for a history and discussion).

The ecometric approach has grown out of the formalization by Damuth et al. (1992) of "taxon-free" characterizations of animal communities. These authors were interested in understanding long-term evolutionary and ecological processes. They argued that distributions of traits such as body size, dietary type, locomotor specialization, and predator-to-prey ratios serve as descriptive indices that can be used to compare important aspects of community function in the Paleozoic, Mesozoic, and Cenozoic or between faunal and floral realms in a single time period, because they capture information about productivity, resource availability, community physiognomy, and resource partitioning, comparisons that cannot be made if community function is characterized by the species that dominate it (such as grassland communities). They noted that what makes an approach "taxon-free" is the focus on distributions of traits for inferring paleoenvironment (i.e., the ecometric distributions), not the inferential pathways that lead to the understanding of the trait-environment relationship, which may be founded on taxon-based observations. The fact that many traits have a limited phylogenetic distribution (e.g., teeth are limited to vertebrates, leaf stomata are limited to plants) means that taxon-free approaches, including ecometrics, are not completely free of time, place, and taxon, but comparisons between communities that have no species in common are nevertheless possible. Andrews and Hixson (2014) recently reviewed and critiqued taxon-free approaches.

Ecometric Examples

One of the best-understood ecometrics is the molar crown height of herbivorous large mammals. Different diets vary in the amount of wear they produce: species that eat abrasive or tough foods, or foods of poor nutritive quality, usually have hypsodont teeth, which have a high crown relative to the root. Hypsodonty is a morphology adaptated to high rates of tooth wear (Janis and Fortelius 1988). Proximal factors including abrasiveness of plant material, such as phytolith content and ambient grit and dust, are therefore likely to play a role in selection for tooth crown height (Janis and Fortelius 1988; Strömberg et al. 2013; Fortelius et al. 2014). Tooth crown height correlates with precipitation in the modern world and geologic past (Fortelius et al. 2002; Eronen et al. 2010a, b; Fortelius et al. 2014). Eronen et al. (2010b) used regression trees to show that there is a strong correlation (65.8%) between crown tooth height and regional precipitation, between diet and precipitation (66.5%), and when diet and tooth crown height are combined (74.2%). Community hypsodonty levels, therefore, have a strong correlation with annual precipitation, which affects both plant tissue properties and the amount of local grit (Fortelius et al. 2014) and may be used as a paleoprecipitation proxy that constrain regional details about vegetation patterns and climate models (Fortelius et al. 2002).

Together with other dental traits, hypsodonty can additionally be used to estimate more indirect properties like net primary productivity (Liu et al. 2012) or more proximate ones like production of volcanic ash (Strömberg et al. 2013). Such complex combinations of proximal and distal environmental factors are likely to be associated with all ecometric traits (Polly and Head 2015). Žliobaitė et al. (2016) showed that dental traits are closely linked to vegetation greenness (normalized difference vegetation index [NDVI]) in addition to precipitation and temperature. This is reassuring as net primary productivity (NPP) estimates used in Liu et al. (2012) are computed using precipitation as an input and are highly dependent on it. NDVI is a direct observation of vegetation greenness and is independent of precipitation and temperature measurements. The NDVI depends on climatic conditions and reflects NPP and thus the availability and quality of herbivores' food. Using data from present-day seasonal environments in Africa, Žliobaitė et al. (2016) demonstrate that the dental traits show strong correlations to non-availability of preferred plant foods (e.g., during dry seasons or longer dry periods), rather than the properties of average foods consumed. This is the main functional link between climate and herbivore teeth and closely follows the suggestion by Owen-Smith (2002) (see also discussion in Liu et al. 2012 and Fortelius et al. 2014).

Other dental characteristics (such as dental wear patterns, dental structure, and tooth crown complexity) have also been used as estimates of diet in mammals (Fortelius and Solounias 2000; Evans et al. 2007; Eronen et al. 2010a; Wilson et al. 2012; Evans 2013; Saarinen 2014; Saarinen and Karme 2017). Wilson et al. (2012) used the dental shape descriptor of orientation patch count (OPC), which is a measure of dental occlusal complexity that correlates well with feeding ecology in extant mammals, to determine the paleodiet of the extinct mammalian clade Multituberculata. They were able to identify a shift from a more carnivorous or omnivorous diet among early multituberculates into a more herbivorous diet around 85 million years ago. This shift coincided with the rise of angiosperms. Since vegetation patterns are driven by climatic conditions, these other dental characteristics could potentially be used as climatic proxies. Evans (2013) reviewed dental shape descriptors, which can be used in ecometric studies and will be useful for describing environmental conditions in the past.

Another well-studied trait in both endothermic and ectothermic vertebrates is body size, initiated by the work of Bergmann (1847). Mean body size in non-flying mammals has been shown to increase as temperature decreases, and therefore mean body size increases geographically toward the poles (Rodriquez et al. 2008; Eronen et al. 2010a). It has also been documented in ectothermic animals that body size is positively correlated with mean ambient temperature (Makarieva et al. 2005). This has allowed for the use of ectothermic animals, to be used to estimate paleotemperatures (Head et al. 2009; Polly et al. 2011; Head et al. 2013).

Table 17.1 Examples of ecometrics in plants, birds, mammals, and snakes. Environmental conditions are mean annual temperature (MAT), annual precipitation (AP), atmospheric carbon dioxide (pCO_2), net primary productivity (NPP), trophic position (TP), dietary classification (DC), and vegetation cover (VC)

Group	Functional Trait	Environmental Condition	Citation
Plants	Leaf margin	MAT	Wolfe (1979)
Plants	Leaf shape	MAT	Wolfe (1990), Royer et al. (2005), Peppe et al. (2011)
Plants	Stomata counts	pCO_2	Beerling et al. (2002)
Birds	Body mass	MAT	Meiri and Dayan (2003)
Mammals	Body mass	MAT	Damuth et al. (1992)
Mammals	Tooth morphology	AP; TP; DC; NPP	Evans (2013)
Ungulates	Hypsodonty	AP	Fortelius et al. (2002)
Carnivorans	Limb proportion	VC; MAT; AP	Polly (2010)
Snakes	Body length	MAT	Head et al. (2009)
Snakes	Tail proportion	VC	Lawing et al. (2012)

Examples of ecometrics in plants, birds, mammals, and snakes are detailed in Table 17.1, showing the functional trait and the environmental condition for which it is related.

Ecometric Patterning

Geographic sorting of traits along environmental gradients is one of three processes by which ecometric patterning can emerge, along with evolutionary adaptation and extinction (Polly and Head 2015). Environmental gradients are shown to filter species geographically (Belmaker and Jetz 2012). The filtering effects of environmental gradients may sort species geographically by their traits, resulting in local communities with similar trait values that are correlated with local environmental conditions. The ecometric approach was formalized as a way of synthesizing data from the ecological present and geological past to measure responses of ecosystems to climate change in order to better predict the risks posed by global change over the next millennia (Eronen et al. 2010a; Polly et al. 2011). It has been used to investigate the impacts and risk of extinction (Wilson et al. 2012; Polly and Sarwar 2014; Polly and Head 2015) and to understand non-ecological processes (Lawing et al. 2017). The relative roles of geographic sorting, local adaptation, and extinction have been investigated by process-based simulation (Polly and Head 2015). Ecometrics and its paleontological precursors have been used to estimate paleobiotic conditions using traits (Damuth et al. 1992). For example, ectotherm body size (Head et al. 2009, 2013) and leaf shape (Wolfe 1993; Little et al. 2010) have been used as paleothermometers and tooth crown height, hypsodonty, has been used to estimate paleoaridity (Fortelius et al. 2002).

The ecometric approach may be used to forecast community responses to anthropogenic climate change (Barnosky et al. 2017). The concept of ecometric load, which is the mismatch between community-level functional trait values and optimal performance, potentially allows us to estimate the vulnerability of a community to climate or environmental change when the optimal performance of the traits along an environmental gradient is known (Polly and Head 2015). If changing conditions increase the ecometric load, communities are more likely to become vulnerable.

Methods for using ecometrics to characterize paleoclimate and paleoenvironments have been expanded extensively. For example, Head et al. (2009) made a significant push forward on the construction of ecometric models and the application of the transfer function. Instead of solely using correlative models for inference, they used a mechanistic model from the physiology of poikilothermic metabolism along with modern observations of boid snakes and the temperatures they live in to predict paleotemperature at 58–60 Ma (Paleocene) in the neotropics. They found a minimum mean annual temperature of 30–34°C was needed in the Cerrejon Formation in Colombia during this time in order for the large boid, *Titanoboa*, to survive. This is several degrees warmer than the mean annual temperature of 26–27°C found in that area of the world.

The effects of faunal sampling, extinction, and extirpation on ecometric patterns were investigated by Polly and Sarwar (2014) using resampling and rarefaction methods with calcaneum gear ratios. Gear ratios are related to the locomotor style of an animal and, in turn, related to the animal's habitat (see also Curran 2018). Polly and Sarwar (2014) showed that extinction, extirpation, and range change have minimal effects on ecometric correlations when they affect less than a quarter of the species in North American carnivoran communities. If the correlation between gear ratio and an environmental variable, such as vegetation cover, is either high or low, then extinction of more than 25% of species will alter the correlations, but the strongly and weakly correlated environmental variables will still be distinguished. Local extinctions, extirpation, caused a decline in ecometric correlations; however, up to 75% of species could be lost this way before ecometric patterns were completely lost. This result suggested that even in fossil faunal assemblages with a small proportion of the total fauna, the ecometric patterns may still be recoverable if enough independent assemblages are considered. These authors also showed that global extinction and large-scale extirpation have a less predictable effect on ecometric correlations than do local extinctions. How the ecometric pattern is affected by global extinction depends on which organisms go extinct and what their ecometric correlation was. So, if a species, such as a polar bear, is a large contributor to local ecometric means, and if its range is correlated to an extreme of the environmental variable, such as zero vegetation cover, then its extinction could drive correlation up or down dependent on the circumstance.

Ecometric Modeling

Polly et al. (2016a) used stochastic modeling to assess the role of population level processes in the formation of ecometric patterns. They systematically varied parameters like local selection intensity, probability of dispersal, probability of extirpation, gene flow, and ancestral trait value to understand the interaction between geographic sorting processes, trait selection, and clade dynamics in producing ecometric patterns. These authors introduced the concept of ecometric load, an analogy with genetic or mutational load, as the mismatch between the distribution of functional traits in a community (or a group of communities) and the selective optimum for those traits in the local environment (Polly et al. 2016a). In cases where ecometric load can be measured, it may help predict whether environmental change is likely to make a community more vulnerable to reorganization or extinction. Interestingly, their models frequently produced parallel trait evolution in different clades, rather than the kinds of phylogenetically correlated trait distributions that are observed in many clades. For example, calcaneum gear ratio, which has strong ecometric sorting in the real world, also has a strong phylogenetic correlation with about 60% of its variance being explained by phylogeny (Polly et al. 2017). The lack of phylogenetic correlation in the models suggested that empirically observed examples of trait-based phylogenetic sorting of species into communities in the modern world results from large-scale turnover in environments.

Ecometrics and Ecogeographical Rules

Ecogeographical rules often describe patterns that we would classify as ecometric. Ecogeographic rules are hypotheses about how morphological variation changes along environmental or geographic gradients. They may be rules regarding within-species variation or between-species variation. Across species variation has been the most important component of variation to consider for the ecometric approach.

The most notable ecogeographic rule is Bergmann's Rule (Bergmann 1847; Blackburn et al. 1999). It states that body mass correlates with temperature and this has been demonstrated both intraspecifically (Ashton 2002; Freckleton et al. 2003) and interspecifically (Blackburn and Hawkins 2004; Diniz-Filho et al. 2007). Other ecogeographical rules include Allen's rule (length of appendages in endotherms positively correlate with temperature; Allen 1877), Gloger's rule (pigmentation is correlated with humidity within endothermic species; Gloger 1833), Fox's rule (among small Australian mammals during community assembly, it is more likely that species entering a community will represent different functional groups; Fox 1987) and Jordan's rule (there is an inverse relationship between meristic characters and water temperature; Lincoln and Clark 1982). See Gaston et al. (2008) for a review of ecogeographic rules.

Implementing and Ecometric Analysis

In order to document the existence of an ecometric pattern, three types of data are required: geographic ranges of species, abiotic conditions, and functional trait measurement of species. The spatial resolution of the data and the density of sampling of traits and environmental variables within local communities will depend on the scale of the functional relationship and on the question being addressed. Here we focus on ecometric patterns that emerge at regional, continental, or global scales, but the same principles could be applied to patterns that emerge on landscape scales. To simplify large-scale analysis, we recommend measuring trait distributions by sampling the trait once for each species making up the local community and making the assumption that each species has the same trait value everywhere it occurs. Finer scale analyses might benefit from sampling the frequencies of traits among the individuals in a local community to account for variation in abundance of the species making up the community, or from measuring the local values of traits within each community. Because of our choice of scale, our analysis requires information about the geographic range where species are known to live or where they have the potential to live. For our purposes, local community composition includes all the species whose ranges overlap at a sampling point, an assumption that is reasonable for coarse scale analysis, but which might be unreasonable for a finer-scale landscape analysis.

Geographic Range

Geographic range data are often available from field guides and are becoming more readily available online as spatial shapefiles. For example, NatureServe (www.natureserve.org) and International Union for Conservation of Nature host a website that has an option to download spatial data that depict the ranges of species as polygons (www.iucnredlist.org). The ranges are typically drawn by experts and represent the general areas of occupation of a species. Species occurrences may be used to draw range maps, if no range maps are available, so long as the occurrence data are relatively randomly distributed across the entire environmental range of the species. Once the geographic ranges are obtained, a sampling scheme for measuring the trait-environment relationship needs to be determined. Sampling sites may be systematically or randomly scattered across the extent of the study area. A list of species that occur at each sample site should then be compiled. Some sources of data for geographic ranges include the NatureServe and IUCN for mammals, birds, amphibians, coral and some fishes (www.natureserve.org, www.iucnredlist.org). Plant distribution data for North America and Europe are available from USDA PLANTS database (plants.usda.gov) and from the European Environmental Agency (eea.europa.eu). The Global Biodiversity Information Facility contains species occurrences (longitude and latitude) of both plants and animals (www.gbif.org). A geographic information system (GIS) is useful to work with geographic ranges and other geographic information. Open source systems are available, such as QGIS, DIVA, or even R. Proprietary software is also available; the most popular products are developed through ESRI. Researchers should check with their institutions for GIS services and licensing.

Environmental Data

Environmental factors can be measured directly in the field at sample sites or collected from maps representing the geographic variability of the environmental condition. For every sample site, the environmental variables that are functionally linked to the trait of interest should be sampled. PRISM Climate Group and WorldClim databases provide relatively high-resolution datasets of climate for the globe (www.prism.oregonstate.edu and www.worldclim.org). Global potential vegetation is available from UW SAGE (www.washington.edu/research/tools/sage/); global historical vegetation coverage is available from Oak Ridge National Laboratories (https://doi.org/10.3334/ornldaac/419); global digital elevation models and annual productivity are available from NASA (earthobservatory.nasa.gov/) and USGS (nationalmap.gov/elevation.html); global soil distributions are available from International Institute for Applied Systems Analysis (www.iiasa.ac.at/); global land cover is available from the European Space Agency (www.esa.int/); global nitrogen deposition is available from Oak Ridge National Laboratory (www.ornl.gov); global freeze and thaw status is available from the National Snow and Ice Data Center (nsidc.org/); global terrestrial ecosystems are available from World Wildlife Fund (www.worldwildlife.org). In addition to all of these data sources, there are numerous others to find online and in the published literature. One may use any reliable map as a reference for the environmental condition at sample localities.

Functional Traits

Functional traits, such as body size, locomotor strategy, dentition, or shape, should be measured or categorized for each species that has the functional trait of interest within each community. Ideally, one would measure the traits from all the species that occur at each sampling location; however, a species value may be used for all of the locations in which that species occurs. Species values for traits may be collected from measuring specimens directly or from obtaining measurements from the published literature. Typically only one value for each species is necessary; however, if there is strong geographic variation, the analysis can include species values that are geographically explicit. If the functional trait is variable within species to the extent that it causes performance differences, then individuals should be sorted along the environmental gradient, but if that variation does not cause performance differences with respect to the environmental gradient, then it is reasonable to ignore the within species variation. The functional traits may be continuous (e.g., size, shape, or ratios of morphological elements) or categorical (e.g., locomotor strategy, substrate use, or dentition type) in nature.

The equipment necessary for collecting the measurements of functional traits greatly varies depending on the trait of interest. One may collect trait data from published literature, and thus, no equipment is necessary. If measurements are taken directly from specimens, the researcher might want to photograph specimens or use calipers to take direct measurements from specimens. If photographs are taken to investigate specimens, the orientation of the camera and the orientation of the specimen should be consistent for all photographs and photos should include a scale. For classifications or counts, no equipment is necessary, unless the researcher requires magnification. For geometric morphometrics, a camera and scale bar, a 3D scanner, or other digitizing equipment is necessary.

Data Analysis

For each sample site, a list of the relevant species that co-occur there must be assembled. The distribution of functional traits in the community members can then be measured. For the remaining examples, we will use the trait mean to summarize that distribution, but other distributional statistics such as variation, standard deviation, range, or median may be appropriate depending on the expected relationship between the trait and its associated environmental parameter. It is useful to assemble data from all sampling sites into a table with the mean trait values and the associated environmental values in columns and sites in the rows. If there is more than one environmental variable associated with the trait of interest, then in this table each variable requires its own column. Ecometric correlation can easily be calculated from this table as the Pearson's correlation coefficient (r), or a nonparametric equivalent, of the two columns as a measure of correspondence between the environmental parameter and the ecometric trait. If the performance relationship between trait and environment is known, the ecometric load can be calculated as the mean absolute difference between the observed trait means and the expected values for all rows in the table (i.e., the residual between the observed and expected values). Note that ecometric load is largely independent of the ecometric correlation, except insofar that a weak correlation requires at least some mismatch between performance optimum and realized trait mean and thus cannot have a load of zero (Polly et al. 2016a). Ecometric patterns with strong correlations can also have strong loads. Establish a transfer function to determine the specific relationship between the two variables (i.e., it estimates function coefficients). If there are fossil localities to estimate past abiotic conditions, then apply the transfer function to the mean of the functional traits measured from the fossils at a single fossil site. Calculate confidence limits for the estimated abiotic condition.

Ecometric estimates of paleoenvironment can be made from a regression-based transfer function or from a likelihood distribution of environmental values given an observed ecometric value. If regression-based methods are used, reduced major axis (RMA) regression may be the most appropriate if there is uncertainty in the independent variable. Transfer functions are prediction equations derived from regressing the environmental variable of interest onto the predictive trait value. Different transfer functions are used depending on the type of trait variable and the type of environmental variable. Other considerations include the type of predicted relationship between the trait and environment and the statistical fitting procedures.

Transfer functions have been typically calibrated with modern distributions of species and their functional traits. For optimal use of the transfer function, for both reconstructing abiotic condition and tracking trait distributions through time, the functional trait should be either directly fossilizable or able to be estimated from fossil specimens. For example, hypsodonty may be estimated directly from fossil specimens (Fortelius et al. 2002) and body size may be accurately estimated from fragmentary skeletal and dental remains (Alexander 1989; MacFadden 1990).

Strengths of Approach

Ecometrics allows us to quantify the dynamic relationships between organisms and their environment. Arguably, the primary strength is that ecometric traits are predictable and transferable through space and time. Since ecometrics uses taxon-free descriptors of community characteristics, it allows for the comparison of community changes in the modern world to those in deep time (Eronen et al. 2010a; Polly et al. 2011). The species that make up the community in the modern world do not need to be the same as those that make up the paleocommunity; the only thing the two communities need to share is the functional trait under study. Because ecometrics focuses on traits that are directly related to environmental conditions, it can be applied to questions across geographic and temporal scales (Wolfe 1994; Weiher et al. 1998; Thompson et al. 2001; Fortelius et al. 2002). Some examples of questions include how have environmental conditions changed in the past, what are the past rates of change in environmental conditions, how were specific environmental conditions distributed across a landscape, how do communities assemble and why are there consistent trait distributions across communities of very different taxonomic composition, what drives diversity gradients and do they change through time, and how does extinction impact traits distributions. The quality of crossing spatial and temporal scales allows ecometrics to be applied from local to global community levels and for the comparison of ecometric patterning across scales. In addition, many ecometric traits are easily measured on both animal and plant fossils (Alexander 1989; MacFadden 1990; Fortelius et al. 2002; Royer et al. 2005; Head et al. 2009), so we can begin to integrate across different depositional environmental and data types.

Biases and Shortcomings

The incomplete nature of the fossil record may influence the quality of data in ecometric studies. However, Polly and Sarwar (2014) showed that even if only 25% of the species of a community are found in a fossil locality, the ecometric patterns still appear. Taphonomic issues with the fossil preservation should also be taken into account, because the functional traits used in ecometric studies need to be measurable or inferred from the fossil remains. It has often not been the case where this is an issue (Alexander 1989; MacFadden 1990; Fortelius et al. 2002; Royer et al. 2005; Head et al. 2009; Peppe et al. 2011; Wilson et al. 2012) and there are various ways to remove the taphonomic influence from the fossil, thus returning it, or a model of it, to its original state (i.e., retrodeformation methods; Webster and Hughes 1999; Angielczyk and Sheets 2007).

Several important assumptions are made with the ecometric approach, including, that the ecometric relationship doesn't change through time and that the full range of morphologies and environmental conditions are represented (i.e., they are not truncated – there are no biased extinctions and only analogous climate conditions). If these are reasonable assumptions for the ecometric trait in question, one may use an ecometric relationship to reconstruct paleoclimate and paleoenvironment, given a couple more assumptions about the fossil localities. The community of fossils within the fossil localities being studied should represent the actual community of species that co-occurred with the depositional environment. In addition, the functional traits should be measurable in the preserved fossils. More work is needed to quantify coefficients describing the relationship between functional traits and environmental conditions to understand if and how they change through time and space.

More research needs to be done concerning intraspecific variation. Intraspecific variation, at least in regards to carnivoran calcaneum gear ratios, is high, and it does not follow community level ecometric patterning. Models seem to support the idea that this is possible due to the high level of gene flow within populations, but the overall signal at the community level overshadows the intraspecific variation of the populations.

Ecometrics Example

In this section we provide an example of how mean annual temperature can be estimated from the distribution of body masses in North American mammals using a likelihood approach. In addition to presenting the method, we also discuss the theory for why ecometric distributions are related to local environment and how our methodology is related to that theory.

When the relationship between trait performance and environment is known, the likelihood function can, in theory, be determined *a priori* (Arnold 1983). Performance-environment relationships can sometimes be estimated for traits recovered in fossils using tools such as finite element analysis, hydrodynamic or mechanical principles, or functional morphological experimentation (e.g., Baumiller et al. 1991; Rayfield 2007; Wroe 2008; Shino and Suzuki 2011; Stayton 2011; Polly et al. 2016b). However, in many cases the performance-environment relationship is unknown, even when the trait-function relationship is well understood. For example, the functional role of hind-limb gear ratio of mammals is understood (it is related through physical lever principles to the efficiency of hind limb extension, and thus to the mechanics of walking, springing, climbing, and running), but its distribution in a particular environment is difficult to predict *a priori* because it is likely to be influenced by a combination of factors such as vegetation physiognomy, terrain ruggedness, substrate, snow cover, and predator-prey interactions (Klein et al. 1987; Polly 2008, 2010; Crête and Larivìere 2003). In such cases, likelihoods can be estimated directly from empirical data by making the assumption that the distribution of traits in modern communities is close enough to equilibrium to be representative. Polly and Sarwar (2014) found this to be a reasonable assumption for carnivore limb ratios, but care should be taken before generalizing that result to other traits.

Likelihoods for paleoenvironmental parameters can be estimated empirically by projecting environmental variables into an ecometric space, which is a mathematical space whose axes are statistical descriptors of the distribution of traits in local communities (Fig. 17.2A). Ecometric space is in some ways the conceptual opposite of Hutchinson's (1957) niche space, in which a species' niche is the volume it occupies in a multidimensional space whose axes are environmental variables. Translated into ecometric terms, the niche of a species is circumscribed by the performance of the functional traits that relate it to the environmental factors. An ecometric space inverts that relationship by mapping the distribution of environmental parameters on trait-based axes, except that those axes describe the distribution of traits in community assemblages rather than the traits of a single species. Figure 17.2A shows mean annual temperature mapped into an ecometric space whose axes are defined by the distribution of log body mass (kg) of mammals.

To create this ecometric space we sampled mammal faunas at 50 km intervals across the entirety of North America. At each sampling point we calculated the mean and standard deviation of the body masses (kg) of the local mammals from the data compiled by Smith et al. (2003) and we recorded the local mean annual temperature (MAT) from Hijmans et al. (2005).

Estimating the likelihood of MAT from community trait characteristics is then a three-step process. First, the trait data should be binned so that a frequency distribution of the environmental variable can be calculated. The number of bins is arbitrary, but it should be coarse enough to encompass a reasonable number of communities in the densest areas of the trait space, yet fine enough to be biologically

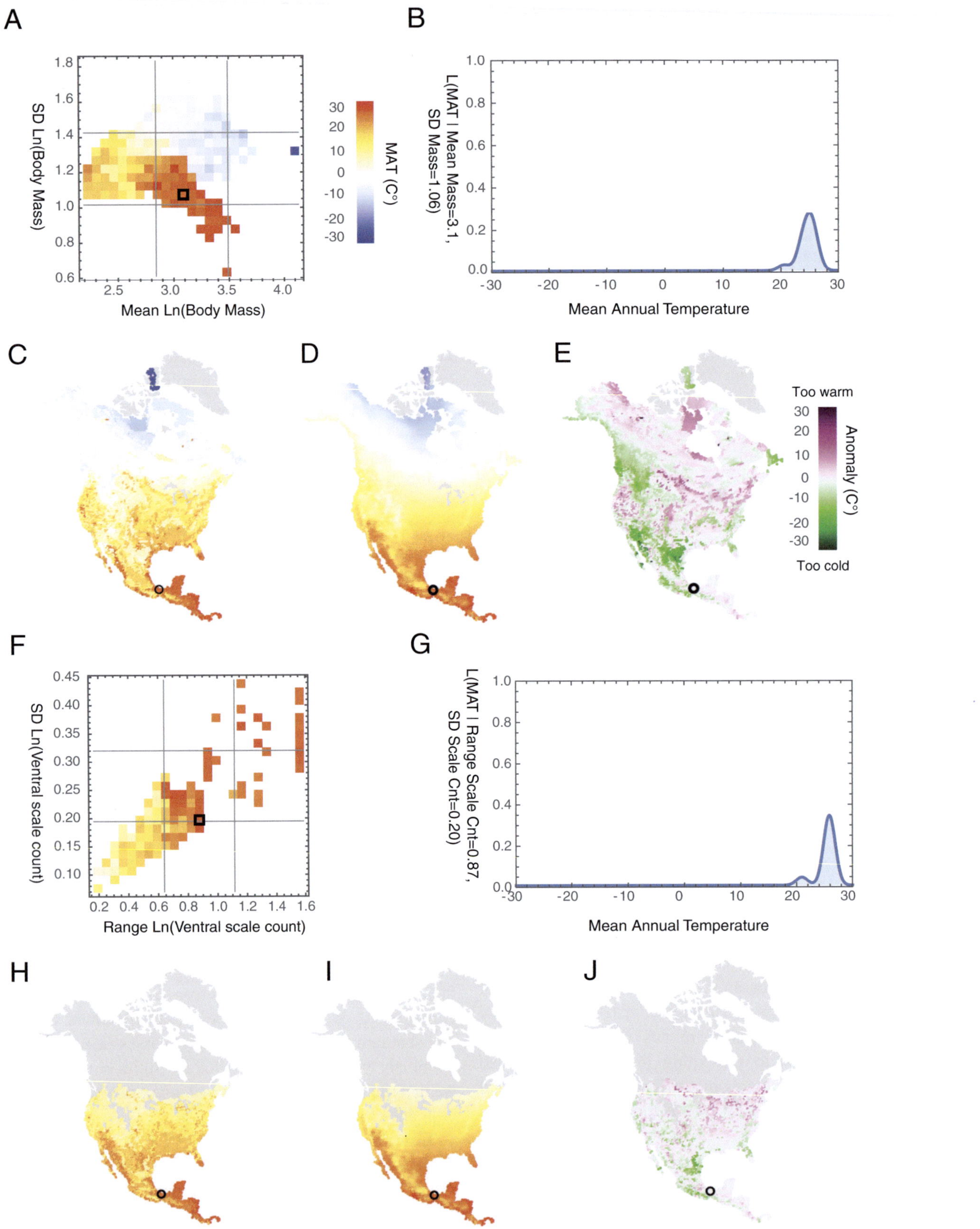

Fig. 17.2 A, Ecometric space for community-level mammalian body mass showing the maximum likelihood estimation of mean annual temperature (MAT) for each bin. B, Likelihood function for MAT based on mammal body mass at the point in Central America highlighted by the black circles in C–E. C, MAT estimated from mammalian body mass. D, Observed MAT. E, anomaly between estimated and observed MAT. F, Ecometric space for snake body size showing the maximum likelihood estimate for MAT in each bin. G, Likelihood function for MAT at the same point in Central America based on snake size. H, MAT estimated from combined mammalian body mass and snake size. I, Observed MAT. J, Anomaly between combined estimate and observed MAT

meaningful. The ideal bin size would be about as wide as the standard errors of the means and standard deviations in the bin. We divided our ecometric space into 625 bins (25 × 25).

Second, the MAT observations in each trait bin were themselves binned to tabulate their frequency distribution. The number of temperature bins is also arbitrary (we used a bin size of 1°C). Once binned, a discrete probability density function (PDF) can be estimated by dividing the bin counts by the total number of observations. A continuous PDF can also be fit to the frequency data. Figure 17.2B shows a continuous PDF for temperature in communities with a mean ln body mass between 3.05 (21.11 kg) and 3.13 (22.87 kg) and a standard deviation between 1.05 and 1.10 (black box in Fig. 17.2A) based on a Gaussian kernel density estimate with bandwidth of 1°C.

With these pieces in place, a maximum likelihood estimate of the environmental variable (MAT) can be made by measuring the trait (body mass) mean and standard deviation at a site of interest and then finding the environmental value that maximizes the PDF. The maximum likelihood estimates of MAT for all the sampling locations in North America are shown in Fig. 17.2C, compared with the real MAT values in Fig. 17.2D. This estimate (MAT at sampling locations) is reasonable, but tends to be too low in high elevation areas and too warm at lower elevations (see anomaly map in Fig. 17.2E). On average, the discrepancy between the real MAT and the estimate based on mammal body masses was 3.9°C, which serves as an approximate standard error (but one that is probably underestimated because the likelihood functions were estimated from the same data on which they were tested).

One of the primary advantages of using likelihood and ecometric spaces is that the likelihoods of alternative reconstructions can be compared. For example, the maximum likelihood estimate of MAT for the Central American location highlighted by the black circles in Fig. 17.2C–E is 24.95°C with $L(24.95) = 0.274$ derived from the PDF for that location (Fig. 17.2B). The real MAT is 25.3°C, which has $L(25.3) = 0.265$, which is clearly much more plausible with a likelihood ratio of 0.967 than an MAT of 20°C, where $L(20.0) = 0.024$ produces a likelihood ratio of only 0.088. This approach can be developed into a formal likelihood ratio test of alternative hypotheses about paleoenvironment, and it can be combined with prior probabilities from independent paleoenvironmental proxies into a Bayesian framework (Polly and Head 2015).

The second advantage of the likelihood approach is that incommensurable traits can be combined. Mammalian body mass distributions were fairly good predictors of MAT, as might be expected by extrapolating Bergmann's rule, which postulates that subspecies in colder climates will have larger body mass than their conspecifics elsewhere (Bergmann 1847; Scholander 1955; Mayr 1963), to the level of species and communities. However, homeothermic mammals are well-insulated from climate and their body size may not be as good predictors of MAT as that of ectotherms (Head et al. 2009). Because large ectothermic animals require warm ambient temperatures to sustain their growth rate, the size range of species in hot climates is expected to be greater than in cold climates (Makarieva et al. 2005). Using a dataset of ventral scale count as a proxy for size assembled by Lawing et al. (2012), we repeated the likelihood estimation for MAT in North America using the range of log scale count and its standard deviation as the dimensions of the ecometric space (Fig. 17.2F). These two parameters are both measures of dispersion and are therefore correlated, yet pick out different aspects of the distribution of body size in local communities. These data yield a similar likelihood function for the site in Central America, but with a narrower peak of predicted MAT (Fig. 17.2G). Because the likelihood functions are probabilities, they can be combined by multiplying them and renormalizing so that the area under them is 1.0. The resulting function gives the likelihood of MAT given the distribution of body masses in mammals and the distribution of ventral scale counts in snakes. This combined likelihood function provides a better estimate of temperature than either trait does alone (Fig. 17.2H–J). The mean anomaly between estimated and real temperature based on the combined estimate is only 2.4°C, much better than with mammals alone. Sacrificed is the ability to estimate MAT in the northern part of the continent where the climate is too cold for snakes to live.

A fully worked ecometrics example and R code is documented in Appendix I.

Future Prospects

Several aspects of ecometrics need to be addressed with future research. These areas include—intraspecific variation (but see Polly et al. 2017 for a recent example), abundance, phylogenetics, combining multiple ecometrics, conservation, and ecosystem services. Polly et al. (2016a) took first steps to identify and define a theory of ecogeography, moving from pattern to process with ecometric modeling. They used modeling to simulate the effects of changing microevolutionary processes on the ecometric patterning seen at the community level. As this area of interest progresses, eventually we hope to be able to observe ecometric patterns and determine which microevolutionary and macroevolutionary processes caused the patterns we see.

Another future development in the field of ecometrics is the prospect of marine ecometrics (Wallin 1991; Yasuhara et al. 2015). Due to the depositional environments found in marine systems, marine fossils are the majority of fossils found

(Sepkoski 1978). Marine fossils are often used in isotopic studies to reconstruct paleoecology (Bowen 1964; Anderson and Arthur 1983; Krantz et al. 1987; Wefer and Berger 1991; Rodrigues et al. 2014; Huyghe et al. 2015; Reich et al. 2015). The changes in isotopic fractionation may follow geographical gradients similar to an ecometric (Marchais et al. 2013; Lynch-Stieglitz et al. 2015; Mackenzie et al. 2014; Beard et al. 2015). However, the isotopic fractionation process is not a morphological trait, it is a metabolic functional trait, and as such, follows different biological, physical, and chemical rules.

New advances in remote sensing capabilities could alter the accuracy of our estimates of both species geographic ranges and the environmental condition in which they occur. Remote sensing appears like it will be quite useful when studying changes in plant community structures over large geographic regions. It allows for a level of detail in which single plants are recognizable across regional scales (Jones and Vaughan 2010).

Geometric morphometrics will also allow for the study of ecometrics when a particular functional trait changes size across age, but doesn't change shape (or vice versa). New advances will also allow for 3D modeling of functional traits. This may allow for the identification of aspects of the morphology that are better ecometrics than those being used. It can also allow for better ecometrics since we will no longer be using linear or volumetric measurements, but instead use the full 3D shape. Recent work by Polly (2010) and Wilson et al. (2012) has shown that this is a viable avenue of future inquiry.

Acknowledgements This contribution is the result of collaborative work supported by the Integrative Climate Change Biology program (iCCB) of the International Union of Biological Sciences (IUBS). We thank Rachel Short, Darin Croft, and two anonymous reviewers for providing valuable feedback on the manuscript. PDP and JJH were supported by NSF grants EAR 1338298 and 1338028.

Appendix 17.1. Ecometrics Workflow and R Code

This section demonstrates an ecometric modeling workflow using the R Statistical Programing Language. To demonstrate these models in R, we will work with spatial data. There are special functions in two packages, *raster* and *sp*, that allow for relatively quick processing of spatial information (Bivand et al. 2013; Hijmans 2015; Pebesma and Bivand 2005). We will use climate data from the worldclim database (Hijmans et al. 2005) and we will use trait data body mass and hypsodonty from the PanTHERIA database (Jones et al. 2009) and from Eronen et al. (2010b). The code below can be typed directly into an R console or can be entered into an R script file. A bold word indicates that the word is a function. To start the analysis, load the two required libraries. If they are not installed yet on your computer, install them with the function install.packages().

```
library(raster)
## Loading required package: sp
library(sp)
```

Load the sampling locations and look at the first six rows of data with the functions read.csv() and head(). The first function read.csv() is a wrapper for another function called read.table(), which can be used in place of read.csv(), if the data are in tab delimited format. Use the help() function to see the documentation associated with each function.

```
points <- read.csv("data/SamplingPoints.csv")
head(points)
##  GLOBALID Longitude Latitude
## 1  103148 -42.1727 83.26264
## 2  103149 -38.3442 83.26264
## 3  103150 -34.5156 83.26264
## 4  103151 -30.6871 83.26264
## 5  103152 -26.8586 83.26264
## 6  103235 -79.4690 82.81348
```

Plot the sampling locations with the function plot() to visualize the geographic distribution of the sampling locations. In this example, we use 50 km equidistant points sampled across North America (Fig. 17.A1). These are the same points used in Polly (2010).

```
plot(points[,2:3], col = "gray", pch = 16)
```

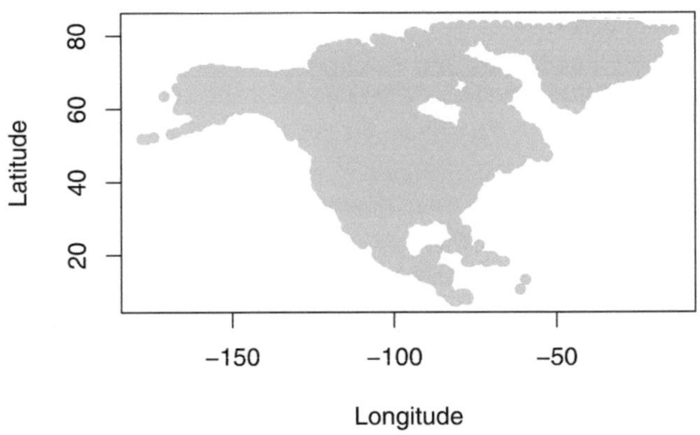

Fig. 17.A1 An example of output in the R Statistical Programing Language from calling the plot function for plotting the latitude and longitude of 50 km equidistant points sampled from across North America

Download raster climate data from the worldclim database using the getData() function from the package raster that we loaded with the library() function (Hijmans et al. 2005; Hijmans 2015). In this example, we download the 10-minute resolution, but if you would like to try a higher resolution data set, then change the argument named res to 2.5 or 0.5. Extract the temperature and precipitation for each sampling location using the extract() function.

```
bioclim <- getData("worldclim", down-
load = T, path = "data", var = "bio", res = 10)
```

Extract the temperature for each sampling location.

```
temperature <- extract(bioclim[[1]], points[,2:3])
```

Calculate the temperature range for all the sampling localities to make a plot of the temperature. We add one to the range to make the range equal to index values that we can use to subset the color function. The R language starts the subset of data at an index value of 1.

Calculate the color value associated with each temperature value and the temperature values associated with even breaks to assign legend values. Figure 17.A2 is a map of the mean annual temperatures.

```
temp_range <- 1 + max(temperature, na.rm = T) -
min(temperature, na.rm = T)
colfunc_temp <- colorRampPalette(c
("darkblue", "blue", "gray", "yellow",
"red"))(temp_range)[1 + temperature - min
(temperature, na.rm = T)]
h <- hist(temperature, breaks = 5)
plot(points[,2:3], col = colfunc_temp,
 pch = 16, main = "Mean Annual Temperature
(C)")
legend("bottomright", legend = h
$breaks/10, pch = 16, col =
colorRampPalette(c("darkblue", "blue", "gray",
"yellow", "red"))(length(h$breaks)))
```

Extract the precipitation for each sampling locality.

```
precipitation <- extract(bioclim[[12]], points
[,2:3])
```

Calculate the precipitation range for all the sampling localities to make a plot of the precipitation. Also, calculate color value associated with each precipitation value and the precipitation values associated with even breaks to assign legend values. Figure 17.A3 is a map of the precipitation values.

```
precip_range <- 1 + max(log(precipitation),
na.rm = T) - min(log(precipitation), na.rm = T)
colfunc_pr <- colorRampPalette(c("brown", "green"))
(precip_range)[1 + log(precipitation) -
min(log(precipitation), na.rm = T)]
h <- hist(log(precipitation), breaks = 5)
plot(points[,2:3], col = col-
func_pr, pch = 16, main = "Precipitation (mm)")
legend(-36.25, 60.5, legend = round(exp(h$breaks)),
pch = 16, col =
colorRampPalette(c("brown", "green"))(length
(h$breaks)))
```

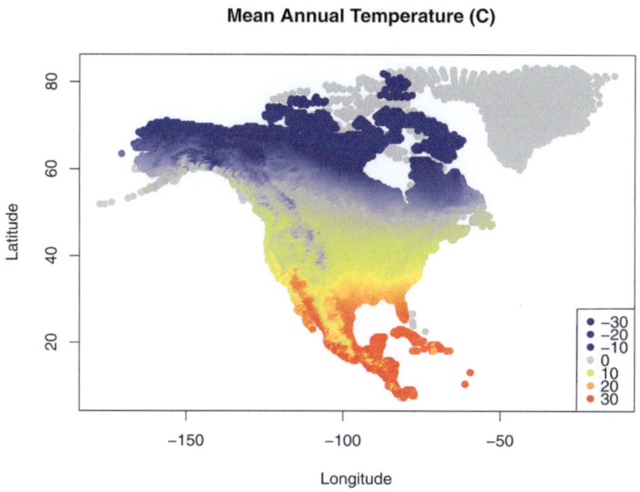

Fig. 17.A2 A heat map of the mean annual temperature (°C), where the hotter colors represent warmer places and cooler colors represent colder places

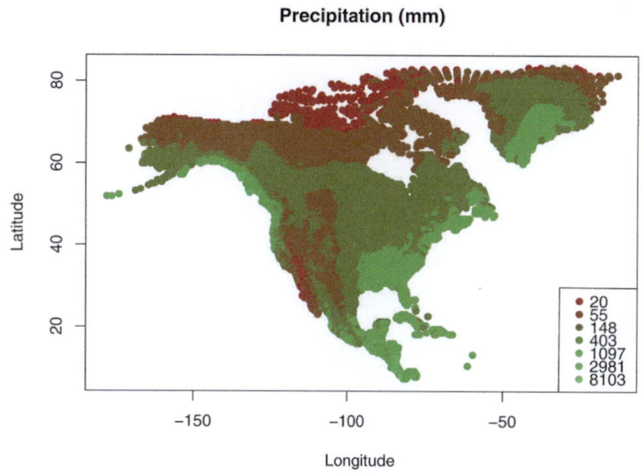

Fig. 17.A3 A heat map of the annual precipitation (mm), where greener colors represent wetter places and browner colors represent drier places

Compile the climate variables into a new data.frame called climate. Remove the variables that are taking up memory with the rm() function if your memory is getting sluggish.

```
climate <- cbind(points, temperature, precipitation)
#rm(bioclim, temperature, precipitation, points)
```

Visually check the climate variables for normality and if they are not mostly normally distributed, transform them for normality (Fig. 17.A4).

```
head(climate)
##  GLOBALID Longitude Latitude temperature precipi-
tation
## 1  103148 -42.1727 83.26264     -169        139
## 2  103149 -38.3442 83.26264     -170        141
## 3  103150 -34.5156 83.26264     -175        149
## 4  103151 -30.6871 83.26264     -185        166
## 5  103152 -26.8586 83.26264     -180        139
## 6  103235 -79.4690 82.81348     -207         90

hist(climate[,4], main = "", xlab = "Mean Annual
Temperature", col = "gray")
```

Temperature appears to be reasonably normally distributed, so now we check precipitation (Fig. 17.A4).

```
hist(climate[,5], main = "", xlab = "Annual Precipita-
tion", col = "gray")
```

Precipitation appears to be log distributed (Fig. 17.A5). We log transform this variable to get it closer to normality (Fig. 17.A6).

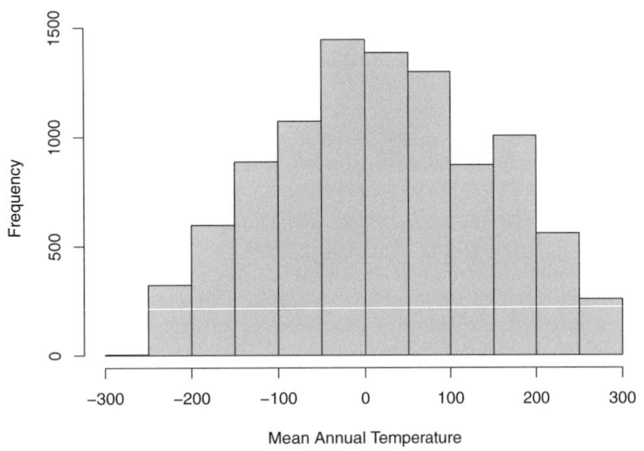

Fig. 17.A4 A histogram of the mean annual temperature (°C × 10) of all of the sampling locations

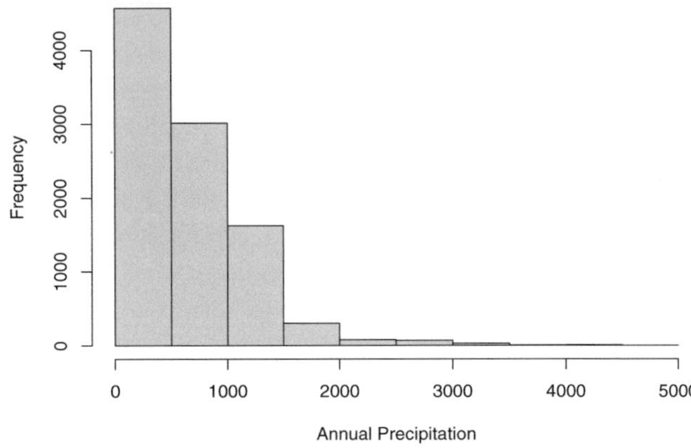

Fig. 17.A5 A histogram of annual precipitation (mm) of all the sampling locations

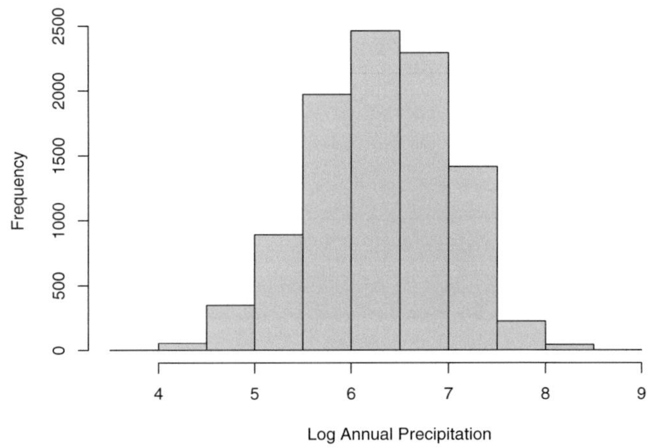

Fig. 17.A6 A histogram of log annual precipitation (mm) of all sampling locations

```
climate[,5] <- log(climate[,5])
hist(climate[,5], main = "", xlab = "Log Annual Pre-
cipitation", col = "gray")
```

Next, we read in the trait data from a folder called data. We assign the row names of the new data frame to the names of the taxon within the dataset. We look at the first six rows of the trait data frame with the head() function. The two traits that we use in this example are body mass and hypsodonty index. Body mass is reported in grams and is the mass of any adult reported in the PanTHERIA database (Jones et al. 2009) from live or freshly-killed specimens. These include captive, wild, provisioned, or unspecified populations and include male, female, and sex unspecified individuals. The mean for each species is reported for each species. The second trait that we use is an index for hypsodonty from Eronen et al. (2010b).

```
traits <- read.csv("data/NAmammalTraits.csv")
rownames(traits) <- traits$TaxonName
head(traits)
##        TaxonName BodyMass hypsodonty_index
## Didelphis virginiana Didelphis
virginiana 3.387760              1
## Aplodontia rufa     Aplodontia rufa 2.906448       3
## Sciurus carolinensis  Sciurus carolinensis
2.736715         1
## Sciurus griseus    Sciurus griseus 2.847480        1
## Sciurus niger            Sciurus niger    NA       1
## Tamiasciurus douglasii Tamiasciurus
douglasii 2.352183           1
```

Now we read in shapefiles containing polygons that represent the geographic ranges for all of the species of interest. These specific shape files were obtained from IUCN Redlist using their spatial data download option (www.iucnredlist.org). If you are dealing with large shapefiles, then this step will take a reasonable amount of processing time.

```
geography <-
shapefile
("data/TERRESTRIAL_MAMMALS/TERRESTRIAL_MAMMALS.
shp")
```

Next we create a list of species at each sampling locality by first turning the sampling points into spatial points with the function SpatialPoints(). We assign the coordinate reference system of our spatial points to a proj4string to match the coordinate reference system of the spatial polygons representing the geographic ranges. We then create a list with the function over(). If you are dealing with large shapefiles, keep in mind that the over() function will take a reasonable amount of time to process.

```
sp <- SpatialPoints(climate[,2:3], proj4string = CRS
(proj4string(geography)))
o <- over(sp, geography, returnList = T)
```

The sample size at each site is calculated by determining the length of the vector returned for each site. The ecometric for body mass and hypsodonty index are summarized for the community level distribution. Here, we summarize with the mean.

```
richness <- unlist(lapply(o, function(x) length(-
traits[x$binomial,"hypsodonty_index"])))
ecometric_bodymass <- unlist(lapply(o, function(x)
mean(traits[x$binomial,"BodyMass"],
na.rm = T)))
ecometric_hypsodonty <- unlist(lapply(o, function(x)
mean(traits[x$binomial,"hypsodonty_index"], na.rm =
T)))
```

First Approximation with Transfer Function

Now we create a model describing the relationship between traits and climate. First, we consider the relationship between hypsodonty and precipitation. We build a simple linear model to describe the variation in precipitation due to the variation in hypsodonty using the function lm(). We only use sites that we have data for more than five species. We look at a summary of the model using the function summary (). Both the intercept and the coefficient (here the coefficient represents the slope of the linear relationship) are not zero ($p < 0.001$). The amount of explained variation (R^2) is 30%. We then make a scatterplot of those variables to look at the general spread of data and add the linear model with the function abline() (Fig. 17.A7).

```
model_mass <- lm(climate[richness > 5,4] ~
ecometric_bodymass[richness > 5])
summary(model_mass)
## Call:
## lm(formula = climate[richness > 5, 4] ~
ecometric_bodymass[richness > 5])
## Residuals:
##    Min    1Q Median    3Q   Max
## -170.72 -65.22 -24.73  43.71 375.19
## Coefficients:
##             Estimate Std. Error t value Pr(>|t|)
## (Intercept)       471.512    6.571  71.76  <2e-16 ***
## ecometric_bodymass[richness > 5] -153.559    2.304
-66.64  <2e-16 ***
## ---
## Signif. codes:  0 "***" 0.001 "**" 0.01 "*" 0.05 "."
0.1 " " 1
## Residual standard error: 92.6 on 8651 degrees of
freedom
```

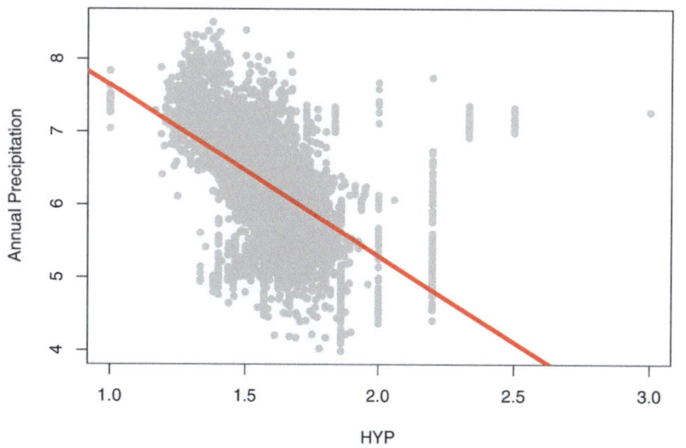

Fig. 17.A7 A scatterplot of hypsodonty and log annual precipitation with a trend line

```
## (15 observations deleted due to missing data)
## Multiple R-squared: 0.3392, Adjusted R-squared: 0.3391
## F-statistic: 4440 on 1 and 8651 DF, p-value: < 2.2e-16
plot(ecometric_bodymass[richness > 5], climate[richness > 5,4], ylab = "MAT", xlab = "Body Mass", pch = 16, col = "gray")
curve(model_mass$coefficients[1] + model_mass$coefficients[2] * x, col = "red", lwd = 4, add = T)
```

From this model, we can see there is some predictive power in this transfer function, but the linear model does not capture the relationship well. In the next section we will show how to estimate annual precipitation from hypsodonty with a maximum likelihood approach that better captures the relationship between annual precipitation and hypsodonty.

Now we create a model describing the relationship between the body mass and temperature. We build a linear model to describe the variation in body mass due to the variation in temperature using the function lm(). We look at a summary of the model using the function summary(). Both the intercept and all the coefficients are significantly different from zero ($p < 0.001$). The amount of explained variation (R^2) is approximately 34%. We then make a scatterplot of those variables to look at the general spread of data and add the model with the function curve() (Fig. 17.A8).

```
model_mass <- lm(climate[richness > 5,4] ~
ecometric_bodymass[richness > 5])
summary(model_mass)
## Call:
## lm(formula = climate[richness > 5, 4] ~
ecometric_bodymass[richness > 5])
## Residuals:
##     Min     1Q Median     3Q    Max
## -170.72 -65.22 -24.73  43.71 375.19
## Coefficients:
##          Estimate Std. Error t value Pr(>|t|)
## (Intercept)      471.512    6.571 71.76 <2e-16 ***
## ecometric_bodymass[richness > 5] -153.559   2.304 -66.64 <2e-16 ***
## ---
## Signif. codes: 0 '***' 0.001 '**' 0.01 '*' 0.05 '.' 0.1 ' ' 1
## Residual standard error: 92.6 on 8651 degrees of freedom
## (15 observations deleted due to missing data)
## Multiple R-squared: 0.3392, Adjusted R-squared: 0.3391
## F-statistic: 4440 on 1 and 8651 DF, p-value: < 2.2e-16
plot(ecometric_bodymass[richness > 5],
climate[richness > 5,4], ylab = "MAT", xlab
= "Body Mass", pch = 16, col = "gray")
curve(model_mass$coefficients[1] +
model_mass$coefficients[2] * x, col = "red", lwd
= 4, add = T)
```

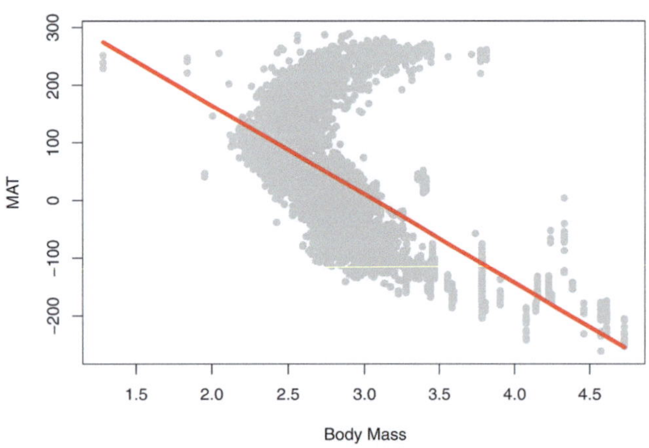

Fig. 17.A8 A scatterplot of body mass and mean annual temperature with a trend line

From this model, we can see that, again, there is some predictive power in this transfer function, but the linear model does not capture the relationship well. In the next section we will show how to estimate mean annual temperature from body mass with a maximum likelihood approach that better captures the relationship between the two.

The coefficients that were estimated in both of these models can be used to estimate paleotemperature and precipitation. Confidence limits can also be calculated given the input dataset. It is important to note that the size of the confidence limits will vary with climate. For example, between 5 C and 28 C, there is a stronger relationship with body size than below or above those temperatures. Hypsodonty has high variability throughout the precipitation range present in North America; however, there is a central tendency about the average relationship between precipitation and hypsodonty index that is useful in reconstructing paleoprecipitation with confidence limits.

Maximum Likelihood Estimation

Although transfer functions, while easy to apply and adequate for first approximations, assume a fairly simple one-to-one relationship between environment and trait means. Combining different traits that have functional relationships with the same environmental factor is also awkward with conventional regression-based transfer functions, especially if the traits are fundamentally different in kind or scale (e.g., body mass measured in kg and humerus shape measured in Procrustes units).

An alternative strategy is to estimate the likelihood of environmental parameters given the distribution of traits in a community (Lawing et al. 2012; Polly and Head 2015). This approach, like many likelihood or Bayesian methods, requires far fewer assumptions about the statistical distributions of variables and it allows otherwise incommensurable data to be combined into the same estimate.

To begin, we need to create another variable at the community level, namely the standard deviation, to use in the maximum likelihood estimate of temperature.

```
sd_ecometric_bodymass <- unlist(lapply(o,
function(x)
 sd(traits[x$binomial,"BodyMass"], na.rm = T)))
```

We create bins using the body mass variable and extract the break points for each bin.

```
#bin the community level trait distribution into 25X25
#first take the range of each
mtemp <- range(ecometric_bodymass, na.rm = T)
sdtemp <- range(sd_ecometric_bodymass, na.rm = T)
#get the break points for the mean and sd
mbrks <- seq(mtemp[1], mtemp[2], diff(mtemp)/25)
sdbrks <- seq(sdtemp[1], sdtemp[2], diff(sdtemp)/25)
#assign bin codes for each
mbc <- .bincode(ecometric_bodymass, breaks = mbrks)
sdbc <- .bincode(sd_ecometric_bodymass, breaks =
sdbrks)
```

We calculate the temperature for each bin.

```
#calculate the data for the raster
obj <- array(NA,dim = c(25,25))
for(i in 1:25){
 for(j in 1:25){
  dat <- round(temperature[which(mbc==i & sdbc==j)]/
10)
  obj[26 - j,i] <- mean(dat, na.rm = T)
 }
}
```

Next, we create a raster to store the body mass and temperature data for bins.

```
#make a raster
r <- raster(extent(0,25,0,25), resolution = 1)
#set the values to the obj
r <- setValues(r,obj)
```

Plot the raster and highlight the bin that we will use to extract data to show an example of that maximum likelihood estimate (Fig. 17.A9).

```
#make an empty plot
plot(1:25, 1:25, type = "n", xlim = c(1,25), ylim = c
(1,25),
   xaxs = "i", yaxs = "i", asp = 1, axes = F, xlab ="",
ylab = "")
#add the rectangle/box
rect(0, 1, 25, 25, lwd = 3)
#add the raster data
plot(r, col = colorRampPalette(c("darkblue", "blue",
  "grey","yellow", "red"))(round(maxValue(r) -
  minValue(r))), add = T)
#this is mean = 3.1, 12, and sd = 1.08, 10
rect(11, 9, 12, 10, lwd = 4)
```

The colors in this raster plot show the Mean Annual Temperature (MAT) maximum likelihood estimate given the associated mean and standard deviation of each bin.

We extract the data for the highlighted bin and plot the kernel density with a Gaussian kernel (Fig. 17.A10). This shows the distribution of the likelihood surface.

```
#grab all the data for that box
dat <- round(temperature[which(mbc==12 & sdbc==10)]/
```

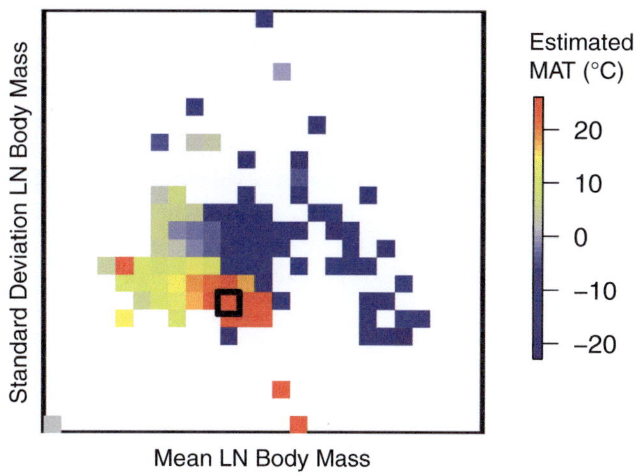

Fig. 17.A9 Ecometric space for the trait body mass with mean body mass on the x-axis and standard deviation of body mass on the y-axis. The colors represent the estimate of mean annual temperature for each mean and standard deviation combination. The hotter colors represent higher mean annual temperature and the cooler colors represent lower mean annual temperatures

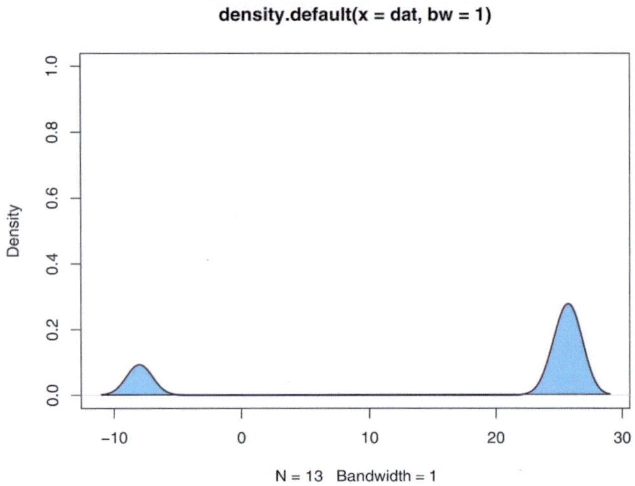

Fig. 17.A10 Kernel density estimation from one combination of mean and standard deviation of body mass

```
10)
#plot the kernel density with gaussian kernel,
bandwidth = 1
mod <- density(dat, bw = 1)
plot(mod, ylim = c(0,1), col = "darkblue", lwd = 2)
polygon(mod$x, mod$y, col = "skyblue")
```

This likelihood surface shows a bimodal distribution of the most likely temperature. Although it is bimodal, it is much more likely that the temperature falls on the warm end of the spectrum, as opposed to the cold end.

Next, we calculate the maximum likelihood for all bins.

```
modmax <- array(NA, dim = length(points[,1]))
mod <- list()
for(i in 1:length(points[,1])){
 if(!(is.na(mbc[i]) | is.na(sdbc[i]))){
     dat <- round(temperature[which(mbc==mbc[i] &
sdbc==sdbc[i])]/10)
   mod[[i]] <- density(dat, bw = 1)
   modmax[i] <- mod[[i]]$x[which.max(mod[[i]]$y)]
}}
modmax <- round(modmax*10)
```

We only use bins with more than the number of species specified as the cutoff. Here we use seven. This means that there needs to be at least seven species recorded at each location to be included in the estimate.

```
cutoff <- 7
```

To plot the maximum likelihood temperature estimate from the ecometric values, we create a color palette for the temperature estimates. In addition, we save the histogram with five break points to a variable to use in plotting (refer back to Fig. 17.A4).

```
colfunc_eco    <-    colorRampPalette(c("darkblue",
"blue",  "gray",  "yellow",  "red"))(temp_range)[1  +
modmax - min(modmax, na.rm = T)]
h <- hist(temperature, main = "", xlab = "Mean
Annual Temperature", col = "gray", breaks = 5)
```

We map the maximum likelihood temperature estimate from body mass (Fig. 17.A11).

```
plot(points[,2:3], col = "gray", pch = 16)
points(points[richness > cutoff, 2:3], col =
colfunc_eco[richness > cutoff], pch = 16)
legend(-31.5, 61, legend = h$breaks/10, pch = 16,
col = colorRampPalette(c("darkblue",
"blue", "gray", "yellow", "red"))(length(h$breaks)))
```

Next we plot the actual temperature to compare with the estimated temperature (refer back to Fig. 17.A2).

```
plot(points[,2:3], col = "gray", pch = 16, main = "Mean
Annual Temperature (C)")
points(points[richness > cutoff,2:3], col =
colfunc_temp[richness > cutoff], pch = 16)
legend(-31.5, 61, legend = h$breaks/10, pch = 16,
col = colorRampPalette(c("darkblue",
"blue", "gray", "yellow", "red"))(length(h$breaks)))
```

We plot the anomaly to visualize the difference between the estimated and actual Mean Annual Temperature (Fig. 17.A12).

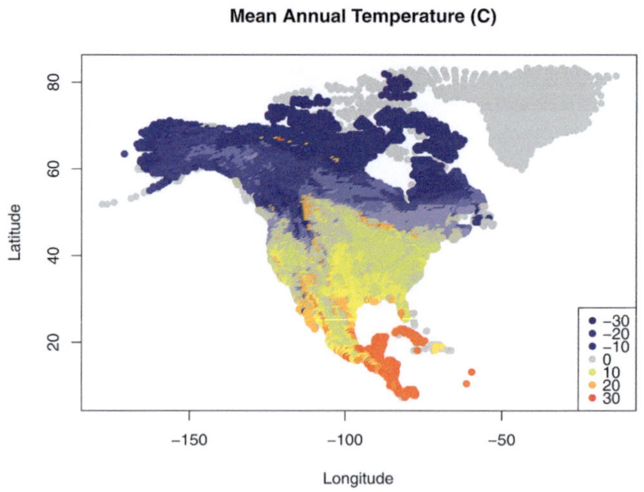

Fig. 17.A11 A heat map of the estimate of mean annual temperature (°C) from the mean and standard deviation of body mass. The hotter colors represent higher temperature estimates and the cooler colors represent lower temperature estimates

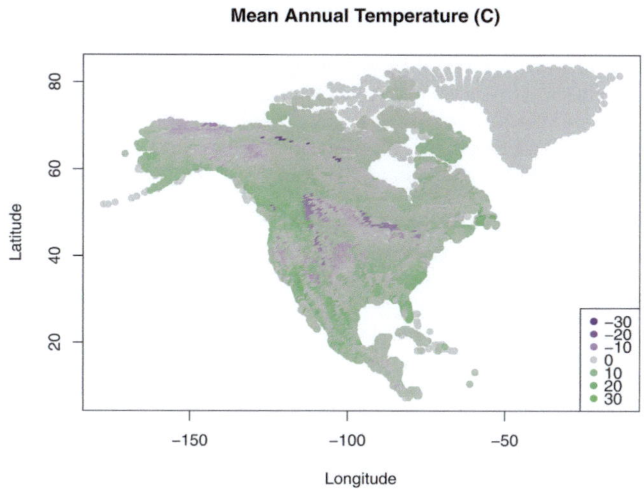

Fig. 17.A12 Anomaly between the observed mean annual temperature and the estimated mean annual temperature. The purple colors represent negative anomalies, up to −30°C. The green colors represent positive anomalies, up to 30°C

```
plot(points[,2:3], col = "gray", pch = 16)
anom <- temperature - modmax
colfunc_anom <- colorRampPalette(c("purple", "grey",
"green"))(max(anom, na.rm = T) -
min(anom, na.rm = T))[1 + anom - min(anom, na.rm = T)]
points(points[richness > cutoff, 2:3], col =
colfunc_anom[richness > cutoff], pch = 16)
legend(-31.5, 61, legend = h$breaks/10, pch = 16,
col = colorRampPalette(c("purple", "grey", "green"))
(length(h$breaks)))
```

The anomaly between the estimated Mean Annual Temperature and the actual Mean Annual Temperature shows that most of the temperature estimates are less that 1°C divergent from the actual Mean Annual Temperature.

References

Alexander, R. M. (1989). Optimization and gaits in the locomotion of vertebrates. *Physiological Reviews, 69,* 1199–1227.

Allen, J. A. (1877). The influence of physical conditions in the genesis of species. *Radical Review, 1,* 108–140.

Anderson, T. F., & Arthur, M. A. (1983). Stable isotopes of oxygen and carbon and their application to sedimentologic and paleoenvironmental problems. In M. A. Arthur, T. F. Anderson, I. R. Kaplan, J. Veizer & L. S. Land (Eds.), *Stable isotopes in sedimentary geology* (pp. 1–151). Tulsa, OK: SEPM Short Course.

Andrews, P., & Hixson, S. (2014). Taxon-free methods of palaeoecology. *Annales Zoologici Fennici, 51,* 269–284.

Angielczyk, K. D., & Sheets, H. D. (2007). Investigation of simulated tectonic deformation in fossils using geometric morphometrics. *Paleobiology, 33,* 125–148.

Arnold, S. J. (1983). Morphology, performance and fitness. *American Zoologist, 23,* 347–361.

Ashton, K. G. (2002). Do amphibians follow Bergmann's rule? *Canadian Journal of Zoology, 80,* 708–716.

Bailey, I. W., & Sinnott, E. W. (1915). A botanical index of Cretaceous and Tertiary climates. *Science, 41,* 831–834.

Barnosky, A. D., Hadly, E. A., Gonzalez, P., Head, J. J., Polly, P. D., Lawing, A. M., et al. (2017). Merging paleobiology with conservation biology to guide the future of terrestrial ecosystems. *Science, 355,* eaah4787.

Baumiller, T. K., LaBarbera, M., & Woodley, J. D. (1991). Ecology and functional morphology of the isocrinid *Cenocrinus asterius* (Linnaeus) (Echinodermata: Crinoidea): in situ and laboratory experiments and observations. *Bulletin of Marine Science, 48,* 731–748.

Beard, J. A., Ivany, L. C., & Runnegar, B. (2015). Gradients in seasonality and seawater oxygen isotopic composition along the early Permian Gondwanan coast, SE Australia. *Earth and Planetary Science Letters, 425,* 219–231.

Beerling, D., Lomax, B., Royer, D., Upchurch, G., & Kump, L. (2002). An atmospheric pCO_2 reconstruction across the Cretaceous-Tertiary boundary from leaf megafossils. *Proceedings of the National Academy of Sciences, USA, 99,* 7836–7840.

Beerling, D. J., Fox, A., Stevenson, D. S., & Valdes, P. J. (2011). Enhanced chemistry-climate feedbacks in past greenhouse worlds. *Proceedings of the National Academy of Sciences, USA, 108,* 9770–9775.

Belmaker, J., & Jetz, W. (2012). Regional pools and environmental controls of vertebrate assemblages. *American Naturalist, 179,* 512–523.

Bergmann, C. (1847). Ueber die Verhältnisse der Wärmeökonomie der Thiere zu ihrer Grösse. *Gottinger studien, 3,* 595–708.

Bivand, R. S., Pebesma, E., & Gomez-Rubio, V. (2013). *Applied spatial data analysis with R* (2nd ed.). New York, NY: Springer.

Blackburn, T. M., Gaston, K. J., & Loder, N. (1999). Geographic gradients in body size: a clarification of Bergmann's rule. *Diversity and Distributions, 5,* 165–174.

Blackburn, T. M., & Hawkins, B. A. (2004). Bergmann's rule and the mammal fauna of northern North America. *Ecography, 27,* 715–724.

Bowen, R. (1964). Oxygen isotope paleotemperature measurements on Mesozoic Belemnoidea and their importance in paleoclimatic studies. In U. Colombo & G. D. Hobson (Eds.), *Advances in organic geochemistry* (pp. 271–283). New York, NY: Pergamon.

Chapin III, F. S. (1993). Functional role of growth forms in ecosystem and global processes. In J. R. Ehleringer & C. B. Field (Eds.), *Scaling physiological processes: Leaf to globe* (pp. 287–312). San Diego, CA: Academic Press.

Cornwell, W. K., & Ackerly, D. D. (2009). Community assembly and shifts in plant trait distributions across an environmental gradient in coastal California. *Ecological Monographs, 79,* 109–126.

Crête, M., & Lariviere, S. (2003). Estimating the costs of locomotion in snow for coyotes. *Canadian Journal of Zoology, 81,* 1808–1814.

Curran, S. C. (2018). Three-dimensional geometric morphometrics in paleoecology. In D. A. Croft, D. F. Su & S. W. Simpson (Eds.), *Methods in paleoecology: Reconstructing Cenozoic terrestrial environments and ecological communities* (pp. 317–335). Cham: Springer.

Damuth, J. D., Jablonski, D., Harris, R. M., Potts, R., Stucky, R. K., Sues, H. D. & Weishampel, D. B. (1992). Taxon-free characterization of animal communities. In A. K. Beherensmeyer, J. D. Damuth, W. A. diMichele, R. Potts, H. D. Sues & S. L. Wing (Eds.) *Terrestrial ecosystems through time: Evolutionary paleoecology of terrestrial plants and animals* (pp. 183–203). Chicago: University of Chicago Press.

Damuth, J. D., & Janis, C. M. (2011). On the relationship between hypsodonty and feeding ecology in ungulate mammals and its utility in paleaeoecology. *Biological Reviews, 86,* 733–758.

Darwin, C. R., & Wallace, A. R. (1858). On the tendency of species to form varieties; and on the perpetuation of varieties and species by natural means of selection. *Journal of the Proceedings of the Linnean Society of London. Zoology, 3,* 45–50.

Díaz, S., & Cabido, M. (2001). Vive la différence: plant functional diversity matters to ecosystem processes. *Trends in Ecology & Evolution, 16,* 646–655.

Diniz-Filho, J. A. F., Bini, L. M., Rodriguez, M. A., Rangel, T. F. L., & Hawkins, B. A. (2007). Seeing the forest for the trees: partitioning ecological and phylogenetic components of Bergmann's rule in European Carnivora. *Ecography, 30,* 598–608.

Eronen, J. T., Polly, P. D., Fred, M., Damuth, J., Frank, D. C., Mosbrugger, V., et al. (2010a). Ecometrics: the traits that bind the past and present together. *Integrative Zoology, 5,* 88–101.

Eronen, J., Puolamaki, K., Liu, L., Lintulaakso, K., Damuth, J., Janis, C., et al. (2010b). Precipitation and large herbivorous mammals I: estimates from present-day communities. *Evolutionary Ecology Research, 12,* 217–233.

Eronen, J. T., Fortelius, M., Micheels, A., Portmann, F. T., Puolamäki, K., & Janis, C. M. (2012). Neogene aridification of the Northern Hemisphere. *Geology, 40,* 823–826.

Evans, A. R. (2013). Shape descriptors as ecometrics in dental ecology. *Hystrix, the Italian Journal of Mammalogy, 24,* 133–140.

Evans, A. R., Wilson, G. P., Fortelius, M., & Jernvall, J. (2007). High-level similarity of dentitions in carnivorans and rodents. *Nature, 445,* 78–81.

Fortelius, M., & Solounius, N. (2000). Functional characterization of ungulate molars using the abrasion-attrition wear gradient: a new method for reconstructing paleodiets. *American Museum Novitates, 3301,* 1–36.

Fortelius, M., Eronen, J. T., Jernvall, J., Liu, L. P., Pushkina, D., Rinne, J., et al. (2002). Fossil mammals resolve regional patterns of Eurasian climate change over 20 million years. *Evolutionary Ecology Research, 4,* 1005–1016.

Fortelius, M., Eronen, J. T., Kaya, F., Tang, H., Raia, P., & Puolamäki, K. (2014). Evolution of Neogene mammals in Eurasia: environmental forcing and biotic interactions. *Annual Review of Earth and Planetary Sciences, 42,* 579–604.

Fox, B. J. (1987). Species assembly and the evolution of community structure. *Evolutionary Ecology, 1,* 201–213.

Freckleton, R. P., Harvey, P. H., & Pagel, M. (2003). Bergmann's rule and body size in mammals. *The American Naturalist, 161,* 821–825.

Gaston, K. J., Chown, S. L., & Evans, K. L. (2008). Ecogeographical rules: elements of a synthesis. *Journal of Biogeography, 35,* 483–500.

Gloger, C. L. (1833). *Das Abwandern der Vogel durch Einfluss des Klimas*. Breslau, Prussia: A. Schulz.

Greenwood, D. R., Wilf, P., Wing, S. L., & Christophel, D. C. (2004). Paleotemperature estimation using leaf-margin analysis: is Australia different? *PALAIOS, 19,* 129–142.

Head, J. J., Bloch, J. I., Hastings, A. K., Bourque, J. R., Cadena, E. A., Herrera, F. A., et al. (2009). Giant boid snake from the Palaeocene neotropics reveals hotter past equatorial temperatures. *Nature, 457,* 715–717.

Head, J. J., Gunnell, G. F., Holroyd, P. A., Hutchison, J. H., & Ciochon, R. L. (2013). Giant lizards occupied herbivorous mammalian ecospace during the Paleogene greenhouse in Southeast Asia. *Proceedings of the Royal Society of London, Series B, 280,* 20130665.

Hijmans, R. J. (2015). raster: Geographic data analysis and modeling. *R package version 2.3-40.*

Hijmans, R. J., Cameron, J. L., Parra, P. G., Jones, P. G., & Jarvis, A. (2005). Very high resolution interpolated climate surfaces for global land areas. *International Journal of Climatology, 25,* 1965–1978.

Hutchinson, G. E. (1957). Concluding remarks. *Cold Spring Harbor Symposiums on Quantitative Biology, 22,* 415–427.

Huyghe, D., Lartaud, F., Emmanuel, L., Merle, D., & Renard, M. (2015). Palaeogene climate evolution in the Paris Basin from oxygen stable isotope ($\delta^{18}O$) compositions of marine molluscs. *Journal of the Geological Society, 172,* 576–587.

Jackson, S. T., & Overpeck, J. T. (2000). Responses of plant populations and communities to environmental changes of the late Quaternary. *Paleobiology, 26,* 194–220.

Janis, C. M., & Fortelius, M. (1988). On the means whereby mammals achieve increased functional durability of their dentitions, with special reference to limiting factors. *Biological Reviews of the Cambridge Philosophical Society, 63,* 197–230.

Jones, H. G., & Vaughan, R. A. (2010). *Remote sensing of vegetation: Principles, techniques, and applications*. New York, NY: Oxford University Press.

Jones, K. E., Bielby, J., Cardillo, M., Fritz, S. A., O'Dell, J., Orme, C. D. L., et al. (2009). PanTHERIA: a species-level database of life history, ecology, and geography of extant and recently extinct mammals. *Ecology, 90,* 2648–2648.

Klein, D. R., Meldgaard, M., & Fancy, S. G. (1987). Factors determining leg length in *Rangifer tarandus*. *Journal of Mammalogy, 68,* 642–655.

Krantz, D. E., Williams, D. F., & Jones, D. S. (1987). Ecological and paleoenvironmental information using stable isotope profiles from living and fossil molluscs. *Palaeogeography, Palaeoclimatology, Palaeoecology, 58,* 249–266.

Lawing, A. M., Eronen, J. T., Blois, J. L., Graham, C. H., & Polly, P. D. (2017). Community functional trait composition at the continental scale: the effects of non-ecological processes. *Ecography, 40,* 651–663.

Lawing, A. M., Head, J. J., & Polly, P. D. (2012). The ecology of morphology: the ecometrics of locomotion and macroenvironment in North American snakes. In J. Louys (Ed.), *Paleontology in ecology and conservation* (pp. 117–146). Berlin: Springer, Berlin Heidelberg.

Lincoln, R. J. B., & Clark, G. A. (1982). *A dictionary of ecology, evolution and systematics*. Cambridge, UK: Cambridge University Press.

Little, S. A., Kembel, S. W., & Wilf, P. (2010). Paleotemperature proxies from leaf fossils reinterpreted in light of evolutionary history. *PLoS ONE, 5,* e15161.

Liu, L., Puolamäki, K., Eronen, J. T., Ataabadi, M. M., Hernesniemi, E. & Fortelius, M. (2012). Dental functional traits of mammals resolve productivity in terrestrial ecosystems past and present. *Proceedings of the Royal Society of London B: Biological Sciences*, 20120211.

Lynch-Stieglitz, J., Polissar, P. J., Jacobel, A. W., Hovan, S. A., Pockalny, R. A., Lyle, M., et al. (2015). Glacial-interglacial changes in central tropical Pacific surface seawater property gradients. *Paleoceanography, 30,* 423–438.

MacFadden, B. J. (1990). *Body size in mammalian paleobiology: Estimation and biological implications*. Cambridge, UK: Cambridge University Press.

Makarieva, A. M., Gorshkov, V. G., & Li, B. L. (2005). Temperature-associated upper limits to body size in terrestrial poikilotherms. *Oikos, 111,* 425–436.

MacKenzie, K. M., Longmore, C., Preece, C., Lucas, C. H., & Trueman, C. N. (2014). Testing the long-term stability of marine isoscapes in shelf seas using jellyfish tissues. *Biogeochemistry, 121,* 441–454.

Marchais, V., Schaal, G., Grall, J., Lorrain, A., Nerot, C., Richard, P., et al. (2013). Spatial variability of stable isotope ratios in oysters (*Crassostrea gigas*) and primary producers along an estuarine gradient (Bay of Brest, France). *Estuaries and Coasts, 36,* 808–819.

Mayr, E. (1963). *Animal species and evolution.* Cambridge, Massachusetts: Belknap Press of Harvard University Press.

Mendoza, M., Janis, C. M., & Palmqvist, P. (2005). Ecological patterns in the trophic-size structure of large mammal communities: a 'taxon-free' characterization. *Evolutionary Ecology Research, 7,* 505–530.

Owen-Smith, R. N. (2002). *Adaptive herbivore ecology: From resources to populations in variable environments.* Cambridge University Press.

Pebesma, E. J., & Bivand, R. S. (2005). Classes and methods for spatial data in R. *R News, 5,* 2.

Peppe, D. J., Royer, D. L., Cariglino, B., Oliver, S. Y., Newman, S., et al. (2011). Sensitivity of leaf size and shape to climate: global patterns and paleoclimatic applications. *New Phytologist, 190,* 724–739.

Poff, N. L. (1997). Landscape filters and species traits: towards mechanistic understanding and prediction in stream ecology. *Journal of the North American Benthological Society, 16,* 391–409.

Polly, P. D. (2008). Adaptive zones and the pinniped ankle: a 3D quantitative analysis of carnivoran tarsal evolution. In E. Sargis & M. Dagosto (Eds.), *Mammalian evolutionary morphology: A tribute to Frederick S. Szalay* (pp. 165–194). Dordrecht: Springer.

Polly, P. D. (2010). Tiptoeing through the trophics: geographic variation in carnivoran locomotor ecomorphology in relation to environment. In A. Goswami & A. Friscia (Eds.), *Carnivoran evolution: New views on phylogeny, form, and function* (pp. 347–410). Cambridge, UK: Cambridge University Press.

Polly, P. D., Eronen, J. T., Fred, M., Dietl, G. P., Mosbrugger, V., Scheidegger, C., et al. (2011). History matters: ecometrics and integrative climate change biology. *Proceedings of the Royal Society of London B: Biological Sciences*, rspb20102233.

Polly, P. D., & Head, J. J. (2015). Measuring Earth-life transitions: ecometric analysis of functional traits from late Cenozoic vertebrates. In P. D. Polly, J. J. Head & D. L. Fox (Eds.), *Earth-life transitions: Paleobiology in the context of Earth system evolution* (pp. 21–46). The Paleontological Society Papers 21. New Haven, CT: Yale Press.

Polly, P. D., & Sarwar, S. (2014). Extinction, extirpation, and exotics: effects on the correlation between traits and environment at the continental level. *Annales Zoologici Fennici, 51,* 209–226.

Polly, P. D., Lawing, A. M., Eronen, J. T., & Schnitzler, J. (2016a). Processes of ecometric patterning: modeling functional traits, environments, and clade dynamics in deep time. *Biological Journal of the Linnaean Society, 118,* 39–63.

Polly, P. D., Stayton, C. T., Dumont, E. R., Pierce, S. E., Rayfield, E. J. & Angielczyk, K. (2016b). Combining geometric morphometrics and finite element analysis with evolutionary modeling: towards a synthesis. *Journal of Vertebrate Paleontology, 36,* e1111225.

Polly, P. D., Fuentes-Gonzales, J., Lawing, A. M., Bormet, A. K., & Dundas, R. G. (2017). Clade sorting has a greater effect than local adaptation on ecometric patterns in Carnivora. *Evolutionary Ecology Research, 18,* 187–200.

Rayfield, E. J. (2007). Finite element analysis and understanding the biomechanics and evolution of living and fossil organisms. *Annual Review of Earth and Planetary Sciences, 35,* 541–576.

Reich, P. B., Walters, M. B., & Ellsworth, D. S. (1997). From tropics to tundra: global convergence in plant functioning. *Proceedings of the National Academy of Sciences, USA, 94,* 13730–13734.

Reich, S., Warter, V., Wesselingh, F. P., Zwaan, J. C., Lourens, L., & Renema, W. (2015). Paleoecological significance of stable isotope ratios in Miocene tropical shallow marine habitats (Indonesia). *PALAIOS, 30,* 53–65.

Ricklefs, R. E., & Travis, J. (1980). A morphological approach to the study of Avian community organization. *The Auk, 97,* 321–338.

Rodrigues, G. B., Fauth, G., Santos, R. V., Koutsoukos, E. A. M., & Colin, J.-P. (2014). Tracking paleoecological and isotopic changes through the K-Pg boundary from marine ostracodes: the Poty quarry section, northeastern Brazil. *Cretaceous Research, 47,* 105–116.

Rodríguez, M. Á., Olalla-Tárraga, M. Á., & Hawkins, B. A. (2008). Bergmann's rule and the geography of mammal body size in the Western Hemisphere. *Global Ecology and Biogeography, 17,* 274–283.

Royer, D. L., & Wilf, P. (2006). Why do toothed leaves correlate with cold climates? Gas exchange at leaf margins provides new insights into a classic paleotemperature proxy. *International Journal of Plant Science, 167,* 11–18.

Royer, D. L., Wilf, P., Janesko, D. A., Kowalski, E. A., & Dilcher, D. L. (2005). Correlations of climate and plant ecology to leaf size and shape: potential proxies for the fossil record. *American Journal of Botany, 92,* 1141–1151.

Saarinen, J. (2014). Ecometrics of large herbivorous land mammals in relation to climatic and environmental changes during the Pleistocene. Ph.D. Dissertation, Uppsala University.

Saarinen, J., & Karme, A. (2017). Tooth wear and diets of extant and fossil xenarthrans (Mammalia, Xenarthra)–Applying a new mesowear approach. *Palaeogeography, Palaeoclimatology, Palaeoecology, 476,* 42–54.

Scholander, P. F. (1955). Evolution of climatic adaptation in homeotherms. *Evolution, 9,* 15–26.

Sepkoski, J. J. (1978). A kinetic model of Phanerozoic taxonomic diversity I. Analysis of marine orders. *Paleobiology, 4,* 223–251.

Shino, Y., & Suzuki, Y. (2011). The ideal hydrodynamic form of the concavo-convex productide brachiopod shell. *Lethaia, 44,* 329–343.

Smith, F. A., Lyons, S. K., Ernest, S. K. M., Jones, K. E., Kauffman, D. M., Dayan, T., et al. (2003). Body mass of late Quaternary mammals. *Ecology, 84,* 3403.

Stayton, C. T. (2011). Biomechanics on the half shell: functional performance influences patterns of morphological variation in the emydid turtle carapace. *Zoology, 114,* 213–223.

Strömberg, C. A. E., Dunn, R. E., Madden, R. H., Kohn, M. J., & Carlini, A. A. (2013). Decoupling the spread of grasslands from the evolution of grazer-type herbivores in South America. *Nature Communications, 4,* 1478.

Tapaltsyan, V., Eronen, J. T., Lawing, A. M., Sharir, A., Janis, C., Jernvall, J., et al. (2015). Continuously growing rodent molars result from a predictable quantitative evolutionary change over 50 million years. *Cell Reports, 11,* 673–680.

Thompson, J. N., Reichman, O. J., Morin, P. J., Polis, G. A., Power, M. E., Sterner, R. W., Couch, C. A., Gough, L., Holt, R., Hooper, D. U., Keesing, F., Lovell, C. R., Milne, B. T., Molles, M. C., Roberts, D. W., & Strauss, S. Y. (2001). Frontiers of ecology: as ecological research enters a new era of collaboration, integration, and technological sophistication, four frontiers seem paramount for understanding how biological and physical processes interact over multiple spatial and temporal scales to shape the Earth's biodiversity. *BioScience, 51,* 15–24.

Violle, C., Navas, M. L., Vile, D., Kazakou, E., Fortunel, C., Hummel, I., et al. (2007). Let the concept of trait be functional! *Oikos, 116,* 882–892.

Wallin, M. (1991). Ecometric analysis of factors regulating eutrophication effects in coastal waters: a case study of marine fish farms. Ph.D. Dissertation, Uppsala University.

Webster, M., & Hughes, N. C. (1999). Compaction-related deformation in Cambrian olenelloid trilobites and its implications for fossil morphometry. *Journal of Paleontology, 73,* 355–371.

Wefer, G., & Berger, W. H. (1991). Isotope paleontology: growth and composition of extant calcareous specie. *Marine Geology, 100,* 207–248.

Weiher, E., Clarke, G. D. P., & Keddy, P. A. (1998). Community assembly rules, morphological dispersion, and the coexistence of plant species. *Oikos, 81*, 309–322.

Wilf, P. (1997). When are leaves good thermometers? A new case for leaf margin analysis. *Paleobiology, 23,* 373–390.

Wilson, G. P., Evans, A. R., Corfe, I. J., Smits, P. D., Fortelius, M., & Jernvall, J. (2012). Adaptive radiation of multituberculate mammals before the extinction of dinosaurs. *Nature, 483,* 457–460.

Wing, S. L., Harrington, G. L., Smith, F. A., Bloch, J. I., Boyer, D. M., & Freeman, K. H. (2005). Transient floral change and rapid global warming at the Paleoeocene-Eocene boundary. *Science, 310,* 993–996.

Wolfe, J. A. (1979). Temperature parameters of humid to mesic forests of eastern Asia and relation to forests of other regions of the northern hemisphere and Australasia. *Geological Survey Professional Paper*, 1106.

Wolfe, J. A. (1990). Palaeobotanical evidence for a marked temperature increase following the Cretaceous/Tertiary boundary. *Nature, 343*, 153–156.

Wolfe, J. A. (1993). A method of obtaining climatic parameters from leaf assemblages. *Bulletin of the US Geological Survey*, 2040.

Wolfe, J. A. (1994). Tertiary climatic changes at middle latitudes of western North America. *Palaeogeography, Palaeoclimatology, Palaeoecology, 108,* 195–205.

Wright, I. J., Reich, P. B., Westoby, M., Ackerly, D. D., Baruch, Z., Bongers, F., et al. (2004). The worldwide leaf economics spectrum. *Nature, 428,* 821–827.

Wroe, S. (2008). Cranial mechanics compared in extinct marsupial and extant African lions using a finite-element approach. *Journal of Zoology, 274,* 332–339.

Yang, J., Wang, Y. F., Spicer, R. A., Mosbrugger, V., Li, C. S., & Sun, Q. G. (2007). Climatic reconstruction at the Miocene Shanwang basin, China, using leaf margin analysis, CLAMP, coexistence approach, and overlapping distribution analysis. *American Journal of Botany, 94,* 599–608.

Yasuhara, M., Tittensor, D. P., Hillebrand, H., & Worm, B. (2015). Combining marine macroecology and palaeoecology in understanding biodiversity: microfossils as a model. *Biological Reviews, 92,* 195–215.

Žliobaitė, I., Rinne, J., Tóth, A. B., Mechenich, M., Liu, L., Behrensmeyer, A. K., et al. (2016). Herbivore teeth predict climatic limits in Kenyan ecosystems. *Proceedings of the National Academy of Sciences, USA, 113,* 12751–12756.

Chapter 18
Making Sense of the Evidence: Synthesizing Paleoecological Data

Denise F. Su and Darin A. Croft

Abstract Paleontologists are increasingly concerned with understanding the biology of extinct species and their environments. This has resulted in a proliferation of new techniques and methodologies that provide a wealth of new data for understanding the paleobiology of extinct species and paleoecological relationships between them and their environment. However, combining such analyses to reconstruct an ancient terrestrial ecosystem can present many challenges, particularly when different lines of evidence appear to provide contradictory information. Here, we discuss some of the ways that taphonomy can bias primary data and how awareness of such issues can increase the accuracy of paleoecological studies. We discuss two geographically and temporally disparate fossil sites as examples of paleoecological reconstruction (the Pliocene of East Africa and the early Oligocene of central Chile) and possible ways of reconciling apparently contradictory data and analyses. We conclude that reconstructing ancient ecosystems requires: (1) accurate data and knowledge of potential biases; (2) a thorough understanding of the analytical techniques being applied and their *a priori* assumptions; and (3) a willingness to recognize non-analog species, habitats, and communities when necessary.

Keywords Taphonomy • Non-analog • Scale • Paleoecological reconstruction

D. F. Su (✉)
Department of Paleobotany and Paleoecology, Cleveland Museum of Natural History, Cleveland, OH 44106, USA
e-mail: dsu@cmnh.org

D.A. Croft
Department of Anatomy, Case Western Reserve University, 10900 Euclid Ave., Cleveland, OH 44106, USA
e-mail: dcroft@case.edu

Introduction

The field of paleontology has changed dramatically and rapidly over the past several decades. Although identifying species in the fossil record and determining their evolutionary relationships (systematics) is still a primary focus of paleontological studies, an increasing number of investigations now focus on understanding the biology of extinct species (paleobiology) and the biotic and abiotic contexts in which they lived (paleoecology). With this greater emphasis in paleobiological and paleoecological studies has come a proliferation of techniques (as detailed in this volume) for understanding how extinct species lived and moved, the habitats in which they lived, and the climatic regimes that affected them and their habitats. These advances have revolutionized our understanding of the paleobiology and paleoecology of past organisms (e.g., Hopley et al. 2007; Goillot et al. 2009; DeMiguel et al. 2010; papers in Harrison 2011a, b; Biasetti et al. 2012; Merceron et al. 2012) and allowed us to better document the timing and context of adaptations and evolutionary histories (e.g., Campisano and Feibel 2007; DeMiguel et al. 2010; Levin et al. 2011; Eiler et al. 2014).

While these advances in techniques and methodologies have provided us with more data than ever to reconstruct paleobiology and paleoecology, they have also introduced new challenges. Most conspicuously, different types of data and analyses can provide conflicting evidence about the habits of extinct species and the habitats in which they lived. In some instances, this can simply reflect species or habitats that have no close modern analog. In other cases, such apparently conflicting data can be reconciled through closer inspection of methodological assumptions and limitations of particular techniques. Here, we briefly discuss two key factors that should be considered when reconstructing the paleoecology of a fossil site – the reliability of the fossil record and the scale at which ecological proxies operate –

D. A. Croft et al. (eds.), *Methods in Paleoecology: Reconstructing Cenozoic Terrestrial Environments and Ecological Communities*, Vertebrate Paleobiology and Paleoanthropology, https://doi.org/10.1007/978-3-319-94265-0_18

and provide examples of how knowledge of these factors can inform and refine paleoecological studies.

Contextual Considerations

Reliability of the Fossil Record

Many factors act upon a fossil assemblage during accumulation and fossilization that can result in compositional unfaithfulness compared to the original community. It is not difficult to imagine why this would be of particular concern to those interested in paleoecological analysis; accuracy and precision in paleoecological inference are directly related to the fidelity of the fossil assemblage compared to the original community. The fossil record, by its very nature, is incomplete in terms of both the species and skeletal elements that are represented. This, in turn, impacts the types of analyses that can be conducted and the inferences that can be made. Taphonomy – the study of the processes that impact the remains of past living organisms from death through their transition into the lithosphere (Efremov 1940) – strives to understand the processes and agents at work in the past that are responsible for the species and elements preserved in a fossil assemblage. Understanding the taphonomy of a site acts like a lens by bringing the fossil assemblage into clearer focus, but this requires knowledge of several factors including the manner in which the remains accumulated and the time interval during which they became part of the fossil record (Fernández-Jalvo and Andrews 2016).

Mode of accumulation: The agent(s) responsible for the accumulation of a fossil assemblage can profoundly affect both species composition and element representation. Predators significantly alter these aspects by preferentially selecting certain prey and processing their remains (Brain 1980; Blumenschine 1988; Lyman 1994; Fernández-Jalvo and Andrews 2016), often resulting in a sample that is a biased subset of the species present on the landscape. For example, hyaena-ravaged samples often lack articular surfaces (Lyman 1994), and prey preference can heavily skew the species composition of raptor-generated rodent accumulations (Reed and Denys 2011). At the opposite end of the spectrum, species associations can be artificially inflated when the agent of accumulation is a river that can move remains from one region (habitat) to another. An example of such a phenomenon is the so-called bone conglomerate (conglomerado osífero) at the base of the Ituzaingó Formation in northeast Argentina (Cione et al. 2000), which preserves a wide variety of late Miocene fossil remains derived from an unknown proportion of the drainage basin of the Paraná River. Mode of accumulation can be thought of as a type of geographic sampling that either restricts or expands the area of the landscape that is actually sampled. The ideal scale of such sampling largely depends on the geographic extent of the modern faunas to which a paleofauna is being compared (Croft 2013).

Duration of accumulation and fossilization: As for mode of accumulation, the span of time over which an assemblage accumulates and becomes integrated into the fossil record can also affect the number of species represented. The ideal situation is a "snapshot" of an ancient ecosystem – an assemblage that was preserved quickly in a catastrophic event (e.g., Cladera et al. 2004; Tucker et al. 2014) – but such cases do not constitute the majority of terrestrial mammal sites. More commonly, bones accumulate and are fossilized over an extended period, resulting in time-averaging, a phenomenon that can inflate species richness and generate false associations among species that were separated in time by up to hundreds of thousands of years (or even more). If environmental conditions were changing during this interval and accompanied by shifts in habitats and vegetation, the implications for paleoecological reconstruction are clear: the site will not closely resemble a single modern habitat but rather a mixture of those that existed over an extended period. Such temporal "smearing" is the primary rationale for limiting an ecological diversity analysis to a relatively narrow stratigraphic interval within a thick fossiliferous geological formation (e.g., Kay and Madden 1997; Kay et al. 2012). How does one determine the duration over which a fossil assemblage accumulated? On a formation-wide scale, geochronological and sedimentological data can provide absolute ages for particular stratigraphic intervals. The degree of weathering on bones can provide clues about how long bones were accumulating prior to burial (Behrensmeyer 1975), as can breakage patterns (Behrensmeyer and Hill 1980; Lyman 1994) and the extent of root and insect damage. The time interval that must be taken into account must be extended further if bones are transported post-burial (uncovered and redeposited) through fluvial action, carnivore or rodent activities, and/or bioturbation (e.g., Voorhies 1969; Blumenschine 1988; Fernández-Jalvo and Andrews 2016).

Understanding the taphonomy of a fossil site is seldom straightforward, but clues can be gathered from both the specimens themselves as well as the species they represent. Much work has been done to better identify and understand taphonomic traces on bones, including Voorhies (1969), Behrensmeyer (1975, 1980), Behrensmeyer and Hill (1980), Blumenschine (1988), Fernández-Jalvo and Andrews (2016); the last of these references is an excellent, recent

Table 18.1 Types of evidence relating to reconstructing paleoecology

Evidence Type		Temporal Scale	Spatial Scale	Paleoenvironmental Inferences
Paleosol	Trackways, trails, nests, burrowing	Minutes–years	10–100 m^2	Substrate type, climate
	Biomarkers	Lifetime of an individual	Locality	Vegetation, temperature, climate, water availability
	Soil/sediment	10^3–10^5 years	Locality	Vegetation, climate, water availability, geomorphology
Skeletal remains	Microwear	Days–weeks	Habitat of an individual	Vegetation structure
	Isotope	Weeks–months; lifetime of an individual	Habitat of an individual	Vegetation, water availability, climate
	Mesowear	Months to years	Habitat of an individual	Vegetation structure
	Microstructure	Lifetime of an individual	Habitat of an individual	Ecological pressures
	Gross (functional) Morphology	10^5–10^6 years	Habitat of an individual	Vegetation, substrate type, climate
Paleobotanical remains	Phytoliths	Months–years; 10^2–10^5 years	Locality-region	Vegetation, climate
	Pollen	Months–years; 10^5–10^6 years	Locality-region	Vegetation, climate
	Spores	Months–years; 10^5–10^6 years	Locality-region	Vegetation, climate
	Macrobotanicals	10^5–10^6 years	Locality	Vegetation, climate

monograph that documents these modifications in atlas form. Since these modifications result from taphonomic processes, they provide a direct way of inferring potential biases in a fossil assemblage.

Another approach to taphonomy that can be used in conjunction with direct methods are inferential ones that aim to gauge the completeness (or rather incompleteness) of a community based on its ecological attributes. Body size, as detailed by Hopkins (2018), is a critical biological constraint and one of the three key variables typically included in an ecological diversity analysis (along with diet and locomotor style; Kovarovic et al. 2018). Body size dictates metabolic and energetic limits, and deviations from the expected patterns of body size distribution (based on extant communities from different habitats) can help gauge taphonomic bias (Damuth 1982; Soligo 2001; Soligo and Andrews 2005; Fernández-Jalvo and Andrews 2016). Combining body size distributions with dietary data can permit specific taphonomic biases to be identified. For example, many Pliocene-Pleistocene African faunal communities lack small primary consumers, which is not characteristic of any modern African habitat, suggesting that these fossil assemblages are taphonomically biased against this ecotype (Soligo and Andrews 2005). Basing an ecological diversity analysis on the entire assemblage would almost certainly result in spurious paleoenvironmental and paleoecological interpretations.

All fossil assemblages are almost certainly incomplete to some degree simply due to the nature of the fossilization process. A key question, then, is how complete a fossil assemblage must be to yield reliable paleoecological information. Many studies have examined the effects of taphonomic bias on community structures, particularly as it relates to time-averaging and faunal mixing (e.g., Dreyer 1984; Foote 1992; Mares and Willig 1994; Robb 2002; Andrews 2006; Louys et al. 2009); however, much more research is necessary to better understand how taphonomic bias impacts fossil faunal community structure and ecological signal. It is not an exaggeration to say that understanding taphonomy is fundamental to understanding faunal communities and environments of the past and, by extension, evolutionary processes (Fernández-Jalvo and Andrews 2016).

Questions of Scale: What Are the Data Telling Us?

The types of evidence gathered to reconstruct paleoenvironment and paleoecology reflect a wide range of temporal and spatial scales, and each can bring a different type of information to bear on environmental reconstructions. While integrating a variety of analytical approaches is essential for arriving at a complete and nuanced picture of an ancient community, it can also highlight conflicts between different lines of evidence. Such apparent conflicts are often partly due to the different temporal and spatial scales at which such analyses operate. Here, we briefly discuss the type of evidence commonly used in paleoecological analyses and the time scales they reflect (see Table 18.1).

Trace fossils: As Hembree (2018) describes, trace fossils are the preserved remains of behavioral interactions between an organism and a substrate (e.g., burrows, tracks, trails, nests, borings). These traces preserve an instant in time (hours to days) and have important behavioral and environmental implications.

Skeletal elements: Skeletal elements (including teeth) provide biological information about the individual and the species to which they belong and can provide information about body size (Hopkins 2018), functional morphology (Dunn 2018; Evans and Pineda-Munoz 2018), diet (Evans and Pineda-Munoz 2018; Green and Croft 2018; Higgins 2018), and life history (Hogg 2018), all of which can provide primary data for paleoecological reconstructions (see applications in Barr 2018; Curran 2018; Kovarovic et al. 2018; Vermillion et al. 2018). The spatial scale over which skeletal elements operate depends on the geographic range of the individual or, more commonly, on the population, since many studies based on skeletal elements pool data from multiple individuals within an assemblage. On a temporal scale, even though skeletal data derive from an individual, they can represent adaptations that evolved over millions of years (e.g., gross tooth morphology, Evans and Pineda-Munoz 2018), use over an extended portion of an individual's lifetime (e.g., mesowear, Green and Croft 2018; enamel isotopic composition, Higgins 2018), or even to short-term use (e.g., microwear, Green and Croft 2018).

Paleobotanical remains: Plant fossils are direct evidence of vegetation and include pollen and spores (Mander and Punyasena 2018), phytoliths (Strömberg et al. 2018), and macrobotanical remains such as fossil leaves (Peppe et al. 2018). Similar to skeletal remains, paleobotanical remains can represent varying spatial and temporal scales ranging from site-specific (phytoliths, pollen, spores, macrobotanicals) to regional or an even wider geographic area (pollen, spores). In addition to providing evidence of the vegetation present at a site, paleobotanical remains can be used to reconstruct paleoclimate and paleoenvironment, particularly if data are available from different geographic regions, basins, and stratigraphic intervals (see Mander and Punyasena 2018; Strömberg et al. 2018; Peppe et al. 2018).

Paleosols: Ancient soils preserve information about the climate, ecosystems, and sedimentary systems in the past. Contained within paleosols are macroscopic, micromorphological, mineralogical, and geochemical indicators of paleoenvironment that reflect the vegetation, climate, sedimentary processes, and landscape stability of a site (Beverly et al. 2018). In addition to skeletal elements and paleobotanical remains, paleosols often also preserve organic biomarkers, another useful indicator of paleoenvironmental and paleoclimatic conditions (Berke 2018). Data from paleosols are generally locality-specific, although pollen and spores can be carried over long distances by wind or water (Mander and Punyasena 2018).

None of the different types of data described above exists in isolation, and most are found together at a fossil site. The different spatial and temporal scales they represent (Table 18.1) can highlight different aspects of an ecosystem and introduce complications for arriving at an integrated understanding of the paleoecology of a fossil site. Below, we present a pair of very different case studies to illustrate the challenges and possibilities of integrating multiple datasets when reconstructing a past ecosystem.

Case Studies

Laetoli, Tanzania

Laetoli, located in northeastern Tanzania on the margin of the Eyasi Plateau, is one of the most intensively sampled and best-studied early hominin sites (see papers in Leakey and Harris 1987; Harrison 2011a, b). Fossils of *Australopithecus afarensis* have been recovered from the Pliocene Upper Laetolil Beds (ULB, 3.85–3.6 Ma) (Harrison 2011c) along with a diverse assemblage of associated fauna and flora. Laetoli is famous for preserving sets of hominin trackways along with those of other fauna (Leakey and Harris 1987; Harrison 2011a, b). Detailed studies of taphonomy and paleoecological proxies have been conducted for the site (papers in Harrison 2011a, b), the results of which are summarized and synthesized here.

Taphonomy: Vertebrate remains primarily consist of isolated skeletal elements, although partial skeletons of mammals <5 kg are common (Su and Harrison 2008). They are in various stages of disarticulation and weathering. Pre-fossilization damage, such as weathering, breakage, root etching, carnivore damage, rodent gnawing, insect damage, is common and indicates that many specimens were exposed on the land surface for a period of time before burial (Su and Harrison 2008). Carnivores were active agents in the accumulation of the Laetoli assemblage as evidenced by the commonness of toothmarks on fossils, high taxonomic diversity of carnivores, and abundance of carnivore coprolites (Su and Harrison 2008). There is no evidence of fluvial transport of the bones. The sedimentary setting (mostly aeolian tuffs) and bone preservation suggest that vertebrate skeletal remains were naturally scattered on successive land surfaces prior to being buried by subaerially deposited volcanic ashes (Hay 1987; Su and Harrison 2008). Thus, one can be reasonably sure that the Laetoli fossil assemblage is a good representation of the original community, although taphonomic biases against animals in the

25–50 kg weight range and micromammals (<1 kg) appear to be present (Su and Harrison 2008).

Paleoecological proxies and indicators: The Laetoli faunal assemblage is dominated by taxa generally associated with more open habitats (e.g., alcelaphin bovids, leporids, pedetid rodents, phasianid and numidid birds) (Su 2011). Traces of termitaries, signs of extensive termite bioturbation, burrows and nests of solitary hymenoptera, and achatinid gastropods all suggest that soils were generally dry and well drained for much of the year (Darlington 2011; Tattersfield 2011; Harrison et al. in press). The aridity of Pliocene Laetoli is also indicated by the absence of aquatic and hydrophilic vertebrates (e.g., hippopotamids, crocodylids, testidudines), which confirms that large permanent sources of water were not available on the landscape (Su and Harrison 2015). Community structure analysis indicates a fauna most similar to those from open woodland, bushland, and grassland habitats (Su 2011). Initial studies of the pollen from ULB identified herbaceous plant families common in the area today, with a high proportion of Poaceae and *Acacia* pollen, suggesting a grassland-open woodland habitat (Bonnefille and Riollet 1987). However, subsequent re-sampling of the ULB sediments recovered no palynomorphs, suggesting either poor preservation of pollen at Laetoli or modern pollen contamination in the initial study (Rossouw and Scott 2011). Phytoliths indicate that grasses were common but not dominant in the ULB; furthermore, most unclassified phytoliths pertain to trees and shrubs (Rossouw and Scott 2011). These phytolith results are supported by enamel isotopic and mesowear studies indicating that mixed-feeders dominate the ungulate fauna (Kingston and Harrison 2007; Kaiser 2011; Kingston 2011). Ecomorphological analyses of bovid postcranial remains suggest that most specimens reflect heavy or open woodland habitats (Kovarovic and Andrews 2007, 2011; Bishop et al. 2011). The terrestrial gastropod community is closely associated with woodland and forest vegetation today (Tattersfield 2011), and the primates, which include arboreal and semi-terrestrial taxa, would have required dense woodland (Harrison 2011d). Plant macrofossils consist of wood, root casts, leaf impressions, fruits and seeds from taxa found in forest and dry open woodlands (Bamford 2011a, b). A sedge rhizome with basal culms is the only element recovered commonly associated with water-margin environments.

Synthesis: The paleoecological indicators and proxies from Laetoli appear to present disparate strands of conflicting evidence, with certain proxies signaling arid, open conditions and others suggesting more mesic and wooded conditions. While the large mammal community is dominated by taxa generally associated with dry, open habitats, enamel isotopic and mesowear analyses suggest that these taxa were not as specialized in their dietary preferences in the Pliocene as their extant relatives are today and that substantial C3 resources (i.e., trees and shrubs) must have been available on the paleolandscape. Ecomorphological studies of bovids indicate that most of the specimens sampled represent species that were forest or heavy-cover adapted. Postcranial elements of dik-dik (*Madoqua*, the most common bovid at Laetoli) make up a substantial proportion of the sample, a finding that has the potential to skew interpretations; although extant *Madoqua* species inhabit dense thickets and bushes, they do so in more open habitats. The terrestrial gastropod community appears to provide more accurate information about microhabitats within the Laetoli area than large mammals, as they vary over relatively small spatial scales in relation to fine-grained ecological differences. These gastropod communities suggest that dense woodland vegetation was scattered across the Pliocene Laetoli landscape, possibly along seasonal river courses (see below).

The faunal and floral evidence suggest that the vegetation of Pliocene Laetoli included a complex mosaic of woodland, bushland, shrubland, grassland with riverine woodland in a relatively arid and seasonal environment. Sedimentological studies indicate that water-worked tuffs are a relatively minor component of the sedimentary sequence. A small proportion of these tuffs were deposited in shallow bodies of standing water, likely ephemeral ponds that dried up during the dry season (Ditchfield and Harrison 2011; Harrison and Kweka 2011). Most of the water-worked tuffs are fluvially reworked and found in minor channel fill sequences at several localities, suggesting a well-developed and relatively sizable seasonal local drainage system that flowed in the wet season (Ditchfield and Harrison 2011). In present-day Laetoli, water run-off from the volcanic highlands flows below the surface year-round, and the water emerges as permanent springs along the margin of the Eyasi Plateau escarpment where the sediments interface with the underlying impervious Precambrian basement rocks (Harrison 2011d). Given that similar geomorphological and topographic features were already in place in the Pliocene, it is likely that small, perennial springs were present in the Laetoli area. The Pliocene hydrological system described here would have been able to support the complex mosaic of vegetation suggested by the faunal and floral evidence.

Except in rare cases of mass death events, fossil assemblages are almost always time-averaged. The Laetoli fossil assemblage samples a 200–300-ky interval. How do we know that the hominins in this assemblage were living in the same environment as the fauna with which they are associated? Serendipitously, trackways of mammals and birds, along with those of hominins, were preserved in thin layers of ash, each of which was probably produced by successive eruptions over the course of a few weeks (Hay 1987). Most of the footprint layers show little lateral change in thickness; this observation, combined with the excellent preservation of the footprints, suggests that the layers were essentially undisturbed after deposition and quickly buried (Hay 1987). The remarkable preservation of raindrop imprints on these ash layers suggests that they were laid down during the transition from the dry to rainy season (Hay 1987). The trackways preserve prints of many taxa represented by skeletal assemblage and their abundances are similar to those of the skeletal assemblage (Leakey 1987). For example, *Serengetilagus* (hare) and *Madoqua* (dik-dik) are the most common taxa in the skeletal assemblage, and their prints are the most numerous in the trackways (Leakey 1987). These trackways represent what is essentially a single point in time. In combination with detailed analyses of changes in taxonomic composition (fauna and phytoliths) and community structure over time, they tell us that the paleoenvironment during Pliocene Laetoli was remarkably stable throughout the 200–300-ky interval (Su and Harrison 2007).

Tinguiririca (Termas del Flaco), Central Chile

The fossil site known as Tinguiririca is located in the eastern end of the Tinguiririca River valley in central Chile near the tourist town of Termas del Flaco, roughly 150 km south-southeast of the Chilean capital of Santiago. The fossils derive from the Abanico Formation (Charrier et al. 1996; Wyss et al. 1993, 1994) and pertain to the early Oligocene (33–31 Ma) based on high-precision $^{40}Ar/^{39}Ar$ radiometric dates within and below the fossil-bearing horizon and biostratigraphic correlation with other dated faunas from southern Argentina (Flynn et al. 2003; Croft et al. 2008). Twenty-six species of mammals have been identified at the site (Croft 2016: Appendix 7), most of which (80%) belong to extinct orders endemic to South America.

Taphonomy: Fossil preservation in the volcaniclastic Abanico Formation (including Tinguiririca) contrasts with that of most fossil sites. The sediments represent volcanic mudflows (lahars), and preserved specimens appear to represent elements that were on the landscape when the lahar took place (bone fragments) as well as living animals that fell victim to the flow (complete skulls and articulated elements). Preservation is generally excellent, though removing the extremely hard sediment is both challenging and time-consuming; consequently, only a subset of recovered specimens have been cleaned and identified, mainly those that preserve teeth. No animals other than mammals have yet been identified at Tinguiririca, but this may partly be due to preferential preparation of craniodental specimens; it is possible that postcranial bones of birds or other vertebrates remain encased in sediment. Since lahars are geologically instantaneous events, the fauna is thought to represent a very short temporal interval, an inference supported by the tight clustering of dates directly associated with the fauna (Flynn et al. 2003). The specimens derive from a small geographic area (<10 km^2), but it is not known how far the bones may have been transported prior to burial. Therefore, the full extent of the area sampled cannot be determined. The fauna is biased against small mammals (<800 g), which represent roughly 25% of the fauna (Croft et al. 2008).

Paleoecological proxies and indicators: The unusual preservation conditions at Tinguiririca preclude many types of paleoecological analysis; no paleosols are present, no ichnofossils have been identified, and botanical remains of all types are also lacking. Thus, the mammal fauna has provided the only means of inferring the site's paleoenvironment. Several lines of evidence, including faunal hypsodonty, body size distribution (cenogram) analysis, and ecological diversity (community structure) analysis suggest that the mammals of Tinguiririca inhabited a dry, relatively open landscape with few trees and abundant low, herbaceous vegetation (Croft 2001; Flynn et al. 2003; Croft et al. 2008).

Synthesis: A peculiar aspect of the fossil record of South America that has long been recognized is that hypsodont mammals – those with high-crowned cheek teeth – appear 10–15 million years earlier than on other continents (e.g., Scott 1937; Patterson and Pascual 1968). This "precocious hypsodonty" has generally been thought to correlate with the early appearance of grasses and open habitats in South America (Webb 1978; Stebbins 1981), and Tinguiririca has figured prominently in these discussions because it represents the earliest interval in which hypsodont mammals predominate (Wyss et al. 1994; Flynn et al. 2003; Croft et al. 2008). The paleoecological analyses of Flynn et al. (2003) provided additional support for the early appearance of open habitats in South America, but considering that most (nearly 90%) of Tinguiririca species lack close living relatives and/or clear ecological analogs, one might rightly question whether its paleoenvironment can be accurately reconstructed based on the presumed paleoecological attributes of its mammals. Indeed, concern about a potential lack of correlation between hypsodonty and open-habitat feeding in notoungulates, the predominant herbivores at Tinguiririca, prompted Croft et al. (2008) to exclude browsing, mixed feeding, and grazing categories from their ecological diversity analysis and conservatively code all non-frugivorous herbivores as folivores. Like the study of Flynn et al. (2003), that of Croft et al. (2008) concluded that Tinguiririca most likely represented an open habitat, but it also highlighted the fact that the Tinguiririca mammal community had an ecological structure quite unlike any modern South American community. Such "non-analog" mammal associations have long been documented at Pleistocene sites in North America (see Semken et al. 2010 and references therein), and it seems likely that many more examples will be documented as paleoecological analyses are applied to older intervals and other regions of the world. However, documenting such cases becomes increasingly challenging in deep time; as the evolutionary distance between fossil and modern species increases, so does the uncertainty associated with paleobiological reconstructions. As a consequence, it can be difficult to distinguish truly non-analog communities from those that appear to be non-analog due to inaccurate paleobiological reconstructions.

Although additional lines of paleoenvironmental evidence are lacking at Tinguiririca, fossil mammals of equivalent age are known from more traditional sedimentary settings about 1,000–1,500 km to the southeast in Patagonia, Argentina. Paleoenvironmental interpretations for these sites based on ichnofossil and paleobotanical evidence have varied considerably, with ichnofossil data suggesting that relatively open habitats were present as early as the late middle Eocene (Bellosi et al. 2010; Sánchez et al. 2010) and pollen and phytolith analyses indicating that such habitats did not appear until well into the Miocene (Palazzesi and Barreda 2012; Strömberg et al. 2013). However, it is important to point out that these latter studies were based on the assumption that abundant grasses – which were not present in Patagonia during the Eocene or Oligocene – are necessary for the existence of open habitats. A recent study demonstrated this assumption to be false by using the ecomorphology of phytoliths to interpret vegetation structure rather than taxonomic composition (Dunn et al. 2015); this innovative new technique, known as Leaf Area Index (LAI) analysis, concluded that late Eocene and early Oligocene habitats in Patagonia were dry to arid and had an open vegetational structure most similar to modern deserts and shrublands (Dunn et al. 2015). Mean annual precipitation (MAP) estimates for this interval (270–770 mm) based on stable carbon isotope data from mammal teeth (Kohn et al. 2015) are concordant with this interpretation. In fact, that study's MAP estimate for the Tinguiririican temporal interval of Patagonia (400 $\pm$ 160 mm) is remarkably similar to that of Tinguiririca itself (600 $\pm$ 500 mm) based on an ecological diversity analysis of the mammals of Tinguiririca, Chile (Flynn et al. 2003).

Concluding Remarks

Reconstructing the paleoecology of a fossil site optimally draws on multiple lines of evidence, but care must be taken at all steps of the process in order to reach an accurate interpretation. Fundamentally, the data upon which any analysis is based must be sound. Understanding the taphonomy of a site is essential for understanding possible biases in data (both temporal and geographic) and a necessary first step toward dealing with such biases. Integrating the interpretations of multiple types of analyses requires knowledge of the varying temporal, geographic, and/or other scales that each samples. Apparently contradictory results can often be reconciled by carefully examining the assumptions and biases that underlie each analytical approach. Finally, *a priori* assumptions must be critically examined when reaching a final interpretation; precise and accurate data and analyses are of little use when constrained to fit within pre-conceived notions about a site or its flora or fauna. Cenozoic terrestrial ecosystems are filled with examples of non-analog plant and animal communities that can only be recognized as such when investigators are open to acknowledging their existence.

The ultimate goal of paleoecology is to aggregate detailed reconstructions of many paleontological sites over broad geographic and temporal scales in order to understand macroevolutionary events. Only by documenting variation in depositional context, taphonomy, paleobiology, and paleoecology of faunal communities through time and space can we fully answer questions about origination, extinction, and adaptation.

Acknowledgements We thank many colleagues for all the discussions over the years that have informed many of the ideas in this paper, particularly: P. Andrews, J. Flynn, T. Harrison, R. Kay, K. Kovarovic, R. Madden, K. Reed, H. Semken, B. Townsend, S. Vizcaíno, A. Wyss. We are grateful to all the contributors to this edited volume and the workshop on which this volume is based as well as our home institutions, the Cleveland Museum of Natural History and Case Western Reserve University. We thank K. Kovarovic and S. Simpson for commenting on earlier versions of the paper. The authors also gratefully acknowledge financial support from the National Science Foundation (DFS: BCS 0216683, BCS 1035897, BCS 1227964; DAC: DEB 0317014, EAR 0819817, EAR 0958733, EAR 1423058).

References

Andrews, P. (2006). Taphonomic effects of faunal impoverishment and faunal mixing. *Palaeogeography, Palaeoclimatology, Palaeoecology, 241,* 572–589.

Bamford, M. K. (2011a). Fossil leaves, fruits and seeds. In T. Harrison (Ed.), *Paleontology and geology of Laetoli: Human evolution in context. Volume 1: Geology, geochronology, paleoecology and paleoenvironment* (pp. 235–252). Dordrecht: Springer.

Bamford, M. K. (2011b). Fossil woods. In T. Harrison (Ed.), *Paleontology and geology of Laetoli: Human evolution in context. Volume 1: Geology, geochronology, paleoecology and paleoenvironment* (pp. 217–233). Dordrecht: Springer.

Barr, W. A. (2018). Ecomorphology. In D. A. Croft, D. F. Su & S. W. Simpson (Eds.), *Methods in paleoecology: Reconstructing Cenozoic terrestrial environments and ecological communities* (pp. 337–347). Cham: Springer.

Behrensmeyer, A. K. (1975). The taphonomy and paleoecology of Plio-Pleistocene vertebrate assemblages east of Lake Rudolf, Kenya. *Bulletin of the Museum of Comparative Zoology, 146,* 473–578.

Behrensmeyer, A. K., & Hill, A. P. (1980). *Fossils in the making*. Chicago: University of Chicago Press.

Bellosi, E. S., Laza, J. H., Victoria Sánchez, M., & Genise, J. F. (2010). Ichnofacies analysis of the Sarmiento Formation (middle Eocene – early Miocene) at Gran Barranca, central Patagonia. In R. H. Madden, A. A. Carlini, M. G. Vucetich & R. F. Kay (Eds.), *The paleontology of Gran Barranca. Evolution and environmental change through the middle Cenozoic of Patagonia* (pp. 306–316). Cambridge: Cambridge University Press.

Berke, M. A. (2018). Reconstructing terrestrial paleoenvironments using sedimentary organic biomarkers. In D. A. Croft, D. F. Su & S. W. Simpson (Eds.), *Methods in paleoecology: Reconstructing Cenozoic terrestrial environments and ecological communities* (pp. 121–149). Cham: Springer.

Beverly, J. E., Lukens, W. E., & Stinchcomb, G. E. (2018). Paleopedology as a tool for reconstructing paleoenvironments and paleoecology. In D. A. Croft, D. F. Su & S. W. Simpson (Eds.), *Methods in paleoecology: Reconstructing Cenozoic terrestrial environments and ecological communities* (pp. 151–182). Cham: Springer.

Biasatti, D., Wang, Y., Gao, F., Xu, Y., & Flynn, L. (2012). Paleoecologies and paleoclimates of late cenozoic mammals from Southwest China: evidence from stable carbon and oxygen isotopes. *Journal of Asian Earth Sciences, 44,* 48–61.

Bishop, L. C., Plummer, T. W., Hertel, F., & Kovarovic, K. (2011). Paleoenvironments of Laetoli, Tanzania as determined by antelope habitat preferences. In T. Harrison (Ed.), *Paleontology and geology of Laetoli: Human evolution in context. Volume 1: Geology, geochronology, paleoecology and paleoenvironment* (pp. 327–337). Dordrecht: Springer.

Blumenschine, R. J. (1988). An experimental model of the timing of hominid and carnivore influence on archaeological bone assemblages. *Journal of Archaeological Science, 15,* 483–502.

Bonnefille, R., & Riollet, G. (1987). Palynological spectra from the Upper Laetoli Beds. In M. D. Leakey & J. M. Harris (Eds.), *Laetoli: A Pliocene site in northern Tanzania* (pp. 52–61). Oxford: Oxford University Press.

Brain, C. K. (1980). Some criteria for the recognition of bone-collecting agencies in African caves. In A. K. Behrensmeyer & A. P. Hill (Eds.), *Fossils in the making* (pp. 107–130). Chicago: University of Chicago Press.

Campisano, C., & Feibel, G. S. (2007). Connecting local environmental sequences to global climate patterns: evidence from the hominin-bearing Hadar Formation, Ethiopia. *Journal of Human Evolution, 53,* 515–527.

Charrier, R., Wyss, A. R., Flynn, J. J., Swisher, C. C., Norell, M. A., Zapatta, F., et al. (1996). New evidence for late Mesozoic–early Cenozoic evolution of the Chilean Andes in the upper Tinguiririca Valley (35°S), central Chile. *Journal of South American Earth Sciences, 9,* 393–422.

Cione, A. L., Azpelicueta, M. M., Bond, M., Carlini, A. A., Casciotta, J. R., Cozzuol, M. A., et al. (2000). Miocene vertebrates from Entre Ríos Province, eastern Argentina. *El Neógeno de Argentina, INSUGEO, Serie Correlación Geológica, 14,* 191–237.

Cladera, G., Ruigómez, E., Ortiz-Jaureguizar, E., Bond, M., & López, G. (2004). Tafonomía de la Gran Hondonada (Formación Sarmiento, Edad-mamífero Mustersense, Eoceno Medio) Chubut, Argentina. *Ameghiniana, 41,* 315–330.

Croft, D. A. (2001). Cenozoic environmental change in South America as indicated by mammalian body size distributions (cenograms). *Diversity and Distributions, 7,* 271–287.

Croft, D. A. (2013). What constitutes a fossil mammal community in the early Miocene Santa Cruz Formation? *Journal of Vertebrate Paleontology, 33,* 401–409.

Croft, D. A. (2016). *Horned armadillos and rafting monkeys: The fascinating fossil mammals of South America*. Bloomington, Indiana: Indiana University Press.

Croft, D. A., Flynn, J. J., & Wyss, A. R. (2008). The Tinguiririca Fauna of Chile and the early stages of "modernization" of South American mammal faunas. *Arquivos do Museu Nacional, Rio de Janeiro, 66,* 191–211.

Curran, S. C. (2018). Three-dimensional geometric morphometrics in paleoecology. In D. A. Croft, D. F. Su & S. W. Simpson (Eds.), *Methods in paleoecology: Reconstructing Cenozoic terrestrial environments and ecological communities* (pp. 317–335). Cham: Springer.

Damuth, J. (1982). Analysis of the preservation of community structure in assemblages of fossil mammals. *Paleobiology, 8,* 434–446.

Darlington, J. P. (2011). Trace fossils interpreted in relation to the extant termite fauna at Laetoli, Tanzania. In T. Harrison (Ed.), *Paleontology and geology of Laetoli: Human evolution in context. Volume 1: Geology, geochronology, paleoecology and paleoenvironment* (pp. 555–565). Dordrecht: Springer.

DeMiguel, D., Azanza, B., & Morales, J. (2010). Trophic flexibility within the oldest Cervidae lineage to persist through the Miocene Climatic Optimum. *Palaeogeography, Palaeoclimatology, Palaeoecology, 289,* 81–92.

Ditchfield, P., & Harrison, T. (2011). Sedimentology, lithostratigraphy and depositional history of the Laetoli area. In T. Harrison (Ed.), *Paleontology and geology of Laetoli: Human evolution in context. Volume 1: Geology, geochronology, paleoecology and paleoenvironment* (pp. 47–76). Dordrecht: Springer.

Dreyer, S. (1984). The Theory and Use of Methods for the Study of Mammalian Paleoecology. Ph.D. Dissertation, University of London.

Dunn, R. E., Strömberg, C. A. E., Madden, R. H., Kohn, M. J., & Carlini, A. A. (2015). Linked canopy, climate, and faunal change in the Cenozoic of Patagonia. *Science, 347,* 258–261.

Dunn, R. H. (2018). Functional morphology of the postcranial skeleton. In D. A. Croft, D. F. Su & S. W. Simpson (Eds.), *Methods in paleoecology: Reconstructing Cenozoic terrestrial environments and ecological communities* (pp. 23–36). Cham: Springer.

Efremov, J. E. (1940). Taphonomy: a new branch of paleontology. *American Geologist, 74,* 81–93.

Eiler, J. M., Berquist, B., Bourg, I., Cartigny, P., Farquhar, J., Gagnon, A., et al. (2014). Frontiers of stable isotope geoscience. *Chemical Geology, 372,* 119–143.

Evans, A. R., & Pineda-Munoz, S. (2018). Inferring mammal dietary ecology from dental morphology. In D. A. Croft, D. F. Su & S. W. Simpson (Eds.), *Methods in paleoecology: Reconstructing Cenozoic terrestrial environments and ecological communities* (pp. 37–51). Cham: Springer.

Fernández-Jalvo, Y., & Andrews, P. (2016). *Atlas of taphonomic identifications*. Dordrecht: Springer.

Flynn, J. J., Wyss, A. R., Croft, D. A., & Charrier, R. (2003). The Tinguiririca Fauna, Chile: biochronology, paleoecology, biogeography, and a new earliest Oligocene South American Land Mammal "Age". *Palaeogeography, Palaeoclimatology, Palaeoecology, 195,* 229–259.

Foote, M. (1992). Rarefaction analysis of morphological and taxonomic diversity. *Paleobiology, 18,* 1–16.

Green, J. L., & Croft, D. A. (2018). Using dental mesowear and microwear for dietary inference: a review of current techniques and applications. In D. A. Croft, D. F. Su & S. W. Simpson (Eds.), *Methods in paleoecology: Reconstructing Cenozoic terrestrial environments and ecological communities* (pp. 53–73). Cham: Springer.

Goillot, C., Blondel, C., & Peigné, S. (2009). Relationships between dental microwear and diet in Carnivora (Mammalia)—Implications for the reconstruction of the diet of extinct taxa. *Palaeogeography, Palaeoclimatology, Palaeoecology, 271,* 13–23.

Harrison, T. (Ed). (2011a). *Paleontology and geology of Laetoli: Human evolution in context. Volume 1: Geology, geochronology, paleoecology and paleoenvironment*. Dordrecht: Springer.

Harrison, T. (Ed). (2011b). *Paleontology and geology of Laetoli: Human evolution in context. Volume 2: Fossil hominins and the associated fauna*. Dordrecht: Springer.

Harrison, T. (2011c). Hominins from the Upper Laetoli and Upper Ndolanya Beds, Laetoli. In T. Harrison (Ed.), *Paleontology and geology of Laetoli: Human evolution in context. Volume 2: Fossil hominins and the associated fauna* (pp. 141–188). Dordrecht: Springer.

Harrison, T. (2011d). Cercopithecids (Cercopithecidae, Primates). In T. Harrison (Ed.), *Paleontology and geology of Laetoli: Human evolution in context. Volume 2: Fossil hominins and the associated fauna* (pp. 83–139). Dordrecht: Springer.

Harrison, T., & Kweka, A. (2011). Paleontological localities on the Eyasi Plateau, including Laetoli. In T. Harrison (Ed.), *Paleontology and geology of Laetoli: Human evolution in context. Volume 1: Geology, geochronology, paleoecology and paleoenvironment* (pp. 17–45). Dordrecht: Springer.

Harrison, T., Kwekason, A., & Su, D. F. (in review). Paleoecology of the Upper Laetoli Beds, Laetoli, Tanzania. In S. Reynolds & R. Bobe (Eds.), *African paleoecology and human evolution*. Cambridge: Cambridge University Press.

Hay, R. L. (1987). Geology of the Laetoli area. In M. D. Leakey & J. M. Harris (Eds.), *Laetoli: A Pliocene site in northern Tanzania* (pp. 23–47). Oxford: Oxford University Press.

Hembree, D. (2018). The role of continental trace fossils in Cenozoic paleoenvironmental and paleoecological reconstructions. In D. A. Croft, D. F. Su & S. W. Simpson (Eds.), *Methods in paleoecology: Reconstructing Cenozoic terrestrial environments and ecological communities* (pp. 183–212). Cham: Springer.

Higgins, P. (2018). Isotope ecology from biominerals. In D. A. Croft, D. F. Su & S. W. Simpson (Eds.), *Methods in paleoecology: Reconstructing Cenozoic terrestrial environments and ecological communities* (pp. 99–120). Cham: Springer.

Hogg, R. (2018). Permanent record: the use of dental and bone microstructure to assess life history evolution and ecology. In D. A. Croft, D. F. Su & S. W. Simpson (Eds.), *Methods in paleoecology: Reconstructing Cenozoic terrestrial environments and ecological communities* (pp. 75–98). Cham: Springer.

Hopkins, S. S. B. (2018). Estimation of body size in fossil mammals. In D. A. Croft, D. F. Su & S. W. Simpson (Eds.), *Methods in paleoecology: Reconstructing Cenozoic terrestrial environments and ecological communities* (pp. 7–22). Cham: Springer.

Hopley, P. J., Marshall, J. D., Weedon, G. P., Latham, A. G., Herries, A. I., & Kuykendall, K. L. (2007). Orbital forcing and the spread of C_4 grasses in the late Neogene: stable isotope evidence from South African speleothems. *Journal of Human Evolution, 53,* 620–663.

Kaiser, T. M. (2011). Feeding ecology and niche partitioning of the Laetoli ungulate faunas. In T. Harrison (Ed.), *Paleontology and geology of Laetoli: Human evolution in context. Volume 1: Geology, geochronology, paleoecology and paleoenvironment* (pp. 321–354). Dordrecht: Springer.

Kay, R. F., & Madden, R. H. (1997). Mammals and rainfall: paleoecology of the middle Miocene at La Venta (Colombia, South America). *Journal of Human Evolution, 32,* 161–199.

Kay, R. F., Vizcaíno, S. F., & Bargo, M. S. (2012). A review of the paleoenvironment and paleoecology of the Miocene Santa Cruz Formation. In S. F. Vizcaíno, R. F. Kay & M. S. Bargo (Eds.), *Early Miocene paleobiology in Patagonia: High-latitude paleocommunities of the Santa Cruz Formation* (pp. 331–365). Cambridge: Cambridge University Press.

Kingston, J. D. (2011). Stable isotopic analyses of Laetoli fossil herbivores. In T. Harrison (Ed.), *Paleontology and geology of Laetoli: Human evolution in context. Volume 1: Geology, geochronology, paleoecology and paleoenvironment* (pp. 293–328). Dordrecht: Springer.

Kingston, J. D., & Harrison, T. (2007). Isotopic dietary reconstructions of Pliocene herbivores at Laetoli: implications for early hominin paleoecology. *Palaeogeography, Palaeoclimatology, Palaeoecology, 243,* 272–306.

Kohn, M. J., Strömberg, C. A. E., Madden, R. H., Dunn, R. E., Evans, S., Palacios, A., et al. (2015). Quasi-static Eocene-Oligocene climate in Patagonia promotes slow faunal evolution and mid-Cenozoic global cooling. *Palaeogeography, Palaeoclimatology, Palaeoecology, 435,* 24–37.

Kovarovic, K., & Andrews, P. (2007). Bovid postcranial ecomorphological survey of the Laetoli paleoenvironment. *Journal of Human Evolution, 52,* 663–680.

Kovarovic, K., & Andrews, P. (2011). Environmental change within the Laetoli fossiliferous sequence: vegetation catenas and bovid ecomorphology. In T. Harrison (Ed.), *Paleontology and geology of Laetoli: Human evolution in context. Volume 1: Geology, geochronology, paleoecology and paleoenvironment* (pp. 355–366). Dordrecht: Springer.

Kovarovic, K., Su, D. F., & Lintulaakso, K. (2018). Mammal community structure analysis. In D. A. Croft, D. F. Su & S. W. Simpson (Eds.), *Methods in paleoecology: Reconstructing Cenozoic terrestrial environments and ecological communities* (pp. 349–370). Cham: Springer.

Leakey, M. D. (1987). Animal prints and trails. In M. D. Leakey & J. M. Harris (Eds.), *Laetoli: A Pliocene site in northern Tanzania* (pp. 451–489). Oxford: University of Oxford Press.

Leakey, M. D., & Harris, J. (Eds.). (1987). *Laetoli: A Pliocene site in northern Tanzania*. Oxford: University of Oxford Press.

Levin, N. E., Brown, F. H., Behrensmeyer, A. K., Bobe, R., & Cerling, T. E. (2011). Paleosol carbonates from the Omo Group: isotopic records of local and regional environmental change in East Africa. *Palaeogeography, Palaeoclimatology, Palaeoecology, 307,* 75–89.

Louys, J., Travouillon, K. J., Bassarova, M., & Tong, H. (2009). The use of protected natural areas in palaeoecological analyses: assumptions, limitations and application. *Journal of Archaeological Science, 36,* 2274–2288.

Lyman, R. L. (1994). *Vertebrate taphonomy*. Cambridge: Cambridge University Press.

Mander, L., & Punyasena, S. W. (2018). Fossil pollen and spores in paleoecology. In D. A. Croft, D. F. Su & S. W. Simpson (Eds.), *Methods in paleoecology: Reconstructing Cenozoic terrestrial environments and ecological communities* (pp. 213–232). Cham: Springer.

Mares, M. A., & Willig, M. R. (1994). Inferring biome associations of Recent mammals from samples of temperate and tropical faunas: paleoecological considerations. *Historical Biology, 8,* 31–48.

Merceron, G., Costeur, L., Maridet, O., Ramdarshan, A., & Göhlich, U. B. (2012). Multi-proxy approach detects heterogeneous habitats for primates during the Miocene climatic optimum in Central Europe. *Journal of Human Evolution, 63,* 150–161.

Palazzesi, L., & Barreda, V. (2012). Fossil pollen records reveal a late rise of open-habitat ecosystems in Patagonia. *Nature Communications, 3,* 1294.

Patterson, B., & Pascual, R. (1968). The fossil mammal fauna of South America. *Quarterly Review of Biology, 43,* 409–451.

Peppe, D. J., Baumgartner, A., Flynn, A., & Blonder, B. (2018). Reconstructing paleoclimate and paleoecology using fossil leaves. In D. A. Croft, D. F. Su & S. W. Simpson (Eds.), *Methods in paleoecology: Reconstructing Cenozoic terrestrial environments and ecological communities* (pp. 287–315). Cham: Springer.

Reed, D., & Denys, C. (2011). The taphonomy and paleoenvironmental implications of the Laetoli micromammals. In T. Harrison (Ed.), *Paleontology and geology of Laetoli: Human evolution in context. Volume 1: Geology, geochronology, paleoecology and paleoenvironment* (pp. 265–278). Dordrecht: Springer.

Robb, C. (2002). Missing mammals: the effects of simulated fossil preservation biases on the paleoenvironmental reconstruction of hominid sites. *American Journal of Physical Anthropology, 34*, 132 suppl.

Rossouw, L., & Scott, L. (2011). Phytoliths and pollen, the microscopic plant remains in Pliocene volcanic sediments around Laetoli, Tanzania. In T. Harrison (Ed.), *Paleontology and geology of Laetoli: Human evolution in context. Volume 1: Geology, geochronology, paleoecology and paleoenvironment* (pp. 201–215). Dordrecht: Springer.

Sánchez, M. V., Laza, J. H., Bellosi, E. S., & Genise, J. F. (2010). Ichnostratigraphy of middle Cenozoic *Coprinisphaera* from central Patagonia: insights into the evolution of dung beetles, herbivores and grass-dominated habitats. *Palaeogeography, Palaeoclimatology, Palaeoecology, 297,* 633–648.

Scott, W. B. (1937). *A history of land mammals in the Western Hemisphere* (2nd ed.). New York: MacMillan Co.

Semken Jr. H. A., Graham, R. W., & Stafford Jr. T. W. (2010). AMS ^{14}C analysis of Late Pleistocene non-analog faunal components from 21 cave deposits in southeastern North America. *Quaternary International, 217,* 240–255.

Soligo, C. (2001). Adaptations and ecology of the earliest primates. Ph.D. Dissertation, University of Zurich.

Soligo, C., & Andrews, P. (2005). Taphonomic bias, taxonomic bias and historical non-equivalence of faunal structure in early hominin localities. *Journal of Human Evolution, 49,* 206–299.

Stebbins, G. L. (1981). Coevolution of grasses and herbivores. *Annals of the Missouri Botanical Garden*, 75–86.

Strömberg, C. A. E., Dunn, R. E., Madden, R. H., Kohn, M. J., & Carlini, A. A. (2013). Decoupling the spread of grasslands from the evolution of grazer-type herbivores in South America. *Nature Communications, 4,* 1478.

Strömberg, C. A. E., Dunn, R. E., Crifò, C., & Harris, E. B. (2018). Phytoliths in paleoecology: Analytical considerations, current use, and future directions. In D. A. Croft, D. F. Su & S. W. Simpson (Eds.), *Methods in paleoecology: Reconstructing Cenozoic terrestrial environments and ecological communities* (pp. 233–285). Cham: Springer.

Su, D. F. (2011). Large mammal evidence for the paleoenvironment of the Upper Laetoli and Upper Ndolanya Beds of Laetoli, Tanzania. In T. Harrison (Ed.), *Paleontology and geology of Laetoli: Human evolution in context. Volume 1: Geology, geochronology, paleoecology and paleoenvironment* (pp. 381–392). Dordrecht: Springer.

Su, D. F., & Harrison, T. (2007). The paleoecology of the Upper Laetoli Beds at Laetoli: a reconsideration of the large mammal evidence. In R. Bobe, Z. Alemseged & A. K. Behrensmeyer (Eds.), *Hominin environments in the East African Pliocene: An assessment of the faunal evidence* (pp. 279–313). Dordrecht: Springer.

Su, D. F., & Harrison, T. (2008). Ecological implications of the relative rarity of fossil hominins at Laetoli. *Journal of Human Evolution, 55,* 672–681.

Su, D. F., & Harrison, T. (2015). The paleoecology of the Upper Laetoli Beds, Laetoli, Tanzania: a review and synthesis. *Journal of African Earth Sciences, 101,* 405–419.

Tattersfield, I. (2011). Gastropoda. In T. Harrison (Ed.), *Paleontology and geology of Laetoli: Human evolution in context. Volume 2: Fossil hominins and the associated fauna* (pp. 567–587). Dordrecht: Springer.

Tucker, S., Otto, R., Joeckel, R., & Voorhies, M. (2014). The geology and paleontology of Ashfall Fossil Beds, a late Miocene (Clarendonian) mass-death assemblage, Antelope County and adjacent Knox County, Nebraska, USA. In J. T. Korus (Ed.), *Geologic field trips along the boundary between the Central Lowlands and Great Plains. GSA Field Guide 36* (pp. 1–22). Geological Society of America.

Vermillion, W. A., Polly, P. D., Head, J. J., Eronen, J. T., & Lawing, A. M. (2018). Ecometrics: A trait-based approach to paleoclimate and paleoenvironmental reconstruction. In D. A. Croft, D. F. Su & S. W. Simpson (Eds.), *Methods in paleoecology: Reconstructing Cenozoic terrestrial environments and ecological communities* (pp. 371–392). Cham: Springer.

Voorhies, M. R. (1969). Taphonomy and population dynamics of an early Pliocene vertebrate fauna, Knox County, Nebraska. *Rocky Mountain Geology, 8,* 1–69.

Webb, S. D. (1978). A history of savanna vertebrates in the New World. Part II: South America and the Great Interchange. *Annual Reviews of Ecology and Systematics, 9,* 393–426.

Wyss, A. R., Flynn, J. J., Norell, M. A., Swisher III, C. C., Charrier, R., Novacek, M. J., et al. (1993). South America's earliest rodent and recognition of a new interval of mammalian evolution. *Nature, 365,* 434–437.

Wyss, A. R., Flynn, J. J., Norell, M. A., Swisher III, C. C., Novacek, M. J., McKenna, M. C., et al. (1994). Paleogene mammals from the Andes of central Chile: a preliminary taxonomic, biostratigraphic, and geochronologic assessment. *American Museum Novitates, 3098,* 1–31.

Index

Note: Page numbers followed by *f* and *t* indicates figures and tables respectively

D. A. Croft et al. (eds.), *Methods in Paleoecology: Reconstructing Cenozoic Terrestrial Environments and Ecological Communities*, Vertebrate Paleobiology and Paleoanthropology, https://doi.org/10.1007/978-3-319-94265-0

MIX
Papier aus verantwortungsvollen Quellen
Paper from responsible sources
FSC® C105338

If you have any concerns about our products,
you can contact us on
ProductSafety@springernature.com

In case Publisher is established outside the EU,
the EU authorized representative is:
Springer Nature Customer Service Center GmbH
Europaplatz 3, 69115 Heidelberg, Germany

Printed by Libri Plureos GmbH
in Hamburg, Germany